Peptides

Peptides

Synthesis, Structures, and Applications

Edited by

Bernd Gutte

Biochemisches Institut der Universität Zürich
CH-8057 Zürich, Switzerland

Academic Press

San Diego New York Boston London Sydney Tokyo Toronto

Cover art: Conformation of cyclosporin A observed in the crystal.
See Chapter 5 by Marshall *et al.*

This book is printed on acid-free paper. ♾

Academic Press, Inc.
A Division of Harcourt Brace & Company
525 B Street, Suite 1900, San Diego, California 92101-4495

United Kingdom Edition published by
Academic Press Limited
24-28 Oval Road, London NW1 7DX

Library of Congress Cataloging-in-Publication Data

Peptides : synthesis, structures, and applications / edited by Bernd Gutte.
p. cm.
Includes bibliographical references and index.
ISBN 0-12-310920-5 (alk. paper)
1. Peptides. 2. Peptides--Synthesis. 3. Peptides--Structure.
I. Gutte, Bernd.
QP552.P4P487 1995
574.19'2456--dc20 95-15417
CIP

PRINTED IN THE UNITED STATES OF AMERICA
95 96 97 98 99 00 BB 9 8 7 6 5 4 3 2 1

Contents

6

Structure–Function Studies of Peptide Hormones: An Overview

Victor J. Hruby and Dinesh Patel

7

Neuropeptides: Peptide and Nonpeptide Analogs

Andrzej W. Lipkowski and Daniel B. Carr

8

Reversible Inhibitors of Serine Proteinases: *Naturally Occurring Miniproteins, Semisynthetic Variants, Recombinant Homologs, and Synthetic Peptides*

Herbert R. Wenzel and Harald Tschesche

12

Synthesis and Applications of Branched Peptides in Immunological Methods and Vaccines

James P. Tam

Contributors

Numbers in parentheses indicate the pages on which the authors' contributions begin.

Robert L. Baldwin (171), Department of Biochemistry, Stanford University School of Medicine, Stanford, California 94305

Denise D. Beusen (193), Center for Molecular Design, Washington University, St. Louis, Missouri 63130

Hans Rudolf Bosshard (419), Biochemisches Institut der Universität Zürich, CH-8057 Zürich, Switzerland

Daniel B. Carr (287), Departments of Anesthesia and Medicine, New England Medical Center and Tufts University School of Medicine, Boston, Massachusetts 02111

Bernd Gutte (363), Biochemisches Institut der Universität Zürich, CH-8057 Zürich, Switzerland

Richard A. Houghten (395), Torrey Pines Institute for Molecular Studies and Houghten Pharmaceuticals, Inc., San Diego, California 92121

Victor J. Hruby (247), Department of Chemistry, University of Arizona, Tucson, Arizona 85721

Yoshiaki Kiso (39), Department of Medicinal Chemistry, Kyoto Pharmaceutical University, Kyoto 607, Japan

Stephan Klauser (363), Biochemisches Institut der Universität Zürich, CH-8057 Zürich, Switzerland

Andrzej W. Lipkowski (287), Medical Research Centre, Polish Academy of Sciences, 00-784 Warsaw, Poland and Industrial Chemistry Research Institute, 01-793 Warsaw, Poland

Garland R. Marshall (193), Center for Molecular Design, Washington University, St. Louis, Missouri 63130

Bruce Merrifield (93), The Rockefeller University, New York, New York 10021

Gregory V. Nikiforovich (193), Center for Molecular Design, Washington University, St. Louis, Missouri 63130

Dinesh Patel (247), Department of Chemistry, University of Arizona, Tucson, Arizona 85721

J. Martin Scholtz (171), Department of Medical Biochemistry and Genetics, Texas A&M University, College Station, Texas 77843

James P. Tam (455), Vanderbilt University Medical Center, Department of Microbiology and Immunology, Nashville, Tennessee 37232

Harald Tschesche (321), Universität Bielefeld, Fakultät für Chemie/Biochemie I, D-33501 Bielefeld, Germany

Herbert R. Wenzel (321), Universität Bielefeld, Fakultät für Chemie/Biochemie I, D-33501 Bielefeld, Germany

Theodor Wieland (1), Max-Planck-Institut für Medizinische Forschung, D-69120 Heidelberg, Germany

Haruaki Yajima (39), Niigata College of Pharmacy, Niigata 950-21, Japan

Preface

Almost 100 years ago, Emil Fischer and Franz Hofmeister made the simultaneous discovery that the amino acids in peptides and proteins are linked by the amide bond. Like many other areas of research, the field of peptide chemistry has grown rapidly during the past 30 to 40 years. It would be impossible to present a complete account of the developments in a single volume.

Peptides: Synthesis, Structures, and Applications provides a wide range of information for the interested reader on modern peptide synthetic methods and structure–activity relationships of synthetic polypeptides, and a look into the future of peptide synthesis and application.

The first peptide of biological importance, the tripeptide glutathione, was synthesized in 1935. The continuous development of novel synthetic methods allowed the preparation of ever larger peptides, culminating in the landmark syntheses of oxytocin (1953), insulin (1963–1966), and several small proteins [for example, ribonuclease (1969, 1981), and HIV-1 protease (1988)].

Major driving forces behind these activities were the scarcity and difficulty in isolating natural peptide hormones and neuropeptides urgently needed for therapeutic purposes. To date, numerous analogs of these peptides, most of which are not longer than 30 to 40 residues, have been prepared in order to study their molecular interactions with the corresponding receptors, to modify their biological activity, and to increase their stability in the circulation. Larger peptides or small proteins have been synthesized using either stepwise solid-phase synthesis or fragment condensation. Two new, promising approaches to the latter have been recently described (Jackson *et al.*, 1994; Dawson *et al.*, 1994).

Another interesting group of peptides that may become therapeutically important are peptide inhibitors of proteases. These peptides, obtained by computer-aided design or by screening phage display systems and peptide libraries, may find application as inhibitors of renin or angiotensin-converting enzyme, HIV-1 protease, elastase, collagenase, to name a few. The development of advanced targeting methods for these and other potentially therapeutic polypeptides is in progress.

Peptides are also valuable tools in localizing epitopes of anti-protein antibodies, in raising antibodies of pre-determined specificity, in studies of the T- and B-cell response to MHC molecule–peptide complexes, and undoubtedly will become important as synthetic vaccines.

Peptide chemists who are not satisfied with the repertoire of naturally occurring proteins try to obtain artificial polypeptides with novel activities by rational design or by screening methods (Choo *et al.,* 1994). Polypeptide design was stimulated mainly by the availability of secondary structure prediction methods and by the finding that protein fragments smaller than 100 residues may assume or retain native-like structures and activities.

A book on peptides must cover conformational aspects because the biological activity of a peptide depends on its conformational properties, which in turn are determined by the amino acid sequence, the polarity of the medium, and interactions with ligands such as receptors, nucleic acids, and metal ions. Examples of experimental and computational techniques to study peptide conformational stability and dynamics are given in two chapters of this book.

It is noteworthy that some of the important tools and concepts of modern biomedical research have been developed by peptide chemists. The principle of solid-phase synthesis originally worked out for the efficient preparation of polypeptides has been successfully adapted to the synthesis of oligonucleotides, which are needed in large numbers for the polymerase chain reaction, for site-directed mutagenesis, and as antisense oligonucleotides and ribozymes. Protein engineering, currently performed at the level of DNA by molecular biologists, was initiated by peptide chemists who used chemically or enzymatically cleaved model proteins and synthetic miniproteins such as a miniribonuclease to study structure–activity relationships. Peptide chemists also were the first to prepare large mixtures of synthetic peptides (peptide libraries) for subsequent functional screening, an idea that was quickly adapted to oligonucleotides (RNA libraries). Finally, the work of peptide chemists forms the prerequisite for the development of peptide mimetics.

References

Choo, Y., Sanchez-Garcia, I., and Klug, A. (1994). *Nature* **372,** 642–645.

Dawson, P. E., Muir, T. W., Clark-Lewis, I., and Kent, S. B. H. (1994). *Science* **266,** 776–779.

Jackson, D. Y., Burnier, J., Quan, C., Stanley, M., Tom, J., and Wells, J. A. (1994). *Science* **266,** 243–247.

Bernd Gutte

1

The History of Peptide Chemistry

Theodor Wieland
Max Planck Institute for Medical Research
D-69120 Heidelberg, Germany

I. Introduction

In the mid 1930s, when I began research in organic chemistry, the class of peptides was almost unknown. Studies of alkaloids and carbohydrates dominated the field of natural substances. There was little room for peptides, as traceable amounts of free peptides in natural material were practically absent, except glutathione and carnosine, and the few existing artificial peptides (as substrates for protein-cleaving enzymes) were interesting only to specialists. The significance and wide role of peptides in all life processes have only become apparent since the 1950s, owing to the continuous development of increasingly sensitive

Peptides: Synthesis, Structures, and Applications

analytical, mainly chromatographic, methods. The still-increasing body of knowledge and compounds renders a brief account of the history of peptide chemistry, including the second half of the twentieth century difficult. It can be accomplished only by omitting important contributions. A brief history of peptide chemistry has been compiled by Wieland and Bodanszky (1991); for a review on the early synthesis of peptides, the collection by Fruton (1949) is instructive.

A. From Peptones to Peptides

By the mid-nineteenth century, egg white (albumen), milk casein, blood fibrin, gelatin, and cereal gluten were classified along with albumins as nitrogen-containing nutrients essential for life. In studies on the nature of these materials, it was also found that they were degraded by extracts from living organisms, such as pepsin from gastric juice and trypsin from pancreas. The products formed, called albumoses and peptones, were considered to represent the form in which the products of digestion were transported across the intestinal wall and used for the formation of proteins in the blood. Only in 1901 did Cohnheim show that the intestinal mucosa contained an enzyme, called erepsin, which cleaves peptones to amino acids. Amino acids had been isolated as components of proteins long before. Leucine was isolated from fermented wheat gluten and from casein by Proust in 1819 and tyrosine from an alkaline hydrolyzate of cow horn by J. Liebig in 1846. The first natural compound of this class, although recognized only 80 years later as an amino acid, was asparagine, which was crystallized from shoots of *Asparagus* by Vauquelin and Robiquet in 1806. For an historical review on discovery of amino acids, see Vickery (1972) and Vickery and Schmidt (1931); for a history of the resolution of the structure of proteins, see Fruton (1979).

This was the situation when Emil Fischer in 1899 embraced the field of protein chemistry that had been taboo for chemists because of the lack of adequate methods. Together with E. Fourneau he prepared the first free dipeptide, glycylglycine, by partial hydrolysis of the diketopiperazine, and at the fourteenth meeting of the German Natural Scientists and Physicians in Karlsbad, in 1902, he introduced the name "peptides."

B. The Structure of Proteins: Hofmeister–Fischer Theory

At the same meeting, on the same day, in a plenary lecture entitled "Über den Bau der Eiweißstoffe" ("On the structure of proteins") the physiologist and pharmacologist Franz Hofmeister (1850–1922) presented his views on the structure of proteins as long chains of α-amino acids linked to one another through amide bonds between carboxyl and amino groups (Hofmeister, 1902). Franz Hofmeister, then professor of physiological chemistry in Strassburg, among other topics had studied systematically the physical and chemical behavior of peptones and proteins (e.g., salting out, biuret reaction). The chemist Emil

Fischer (1852–1919), who in 1896 had succeeded A. W. von Hofmann on the famous chemistry chair in Berlin and had entered the protein field only a few years previously, in his lecture "Über die Hydrolyse der Proteinstoffe" reported on the isolation of amino acids and additional compounds from protein hydrolysates and proposed the names dipeptide and tripeptide for dimeric and trimeric compounds. In his autoreferat (Fischer, 1902) he also wrote: "Finally the speaker discussed the coupling of the amino acids in proteins. The idea that acid–amide-like groups play the principal role most readily comes to mind [*liegt am nächsten*] as Hofmeister also assumed in his general lecture this morning."

The structure of proteins as macromolecular polypeptides has been established and proved in several ways. The sequence of amino acids (primary structure) of innumerable peptides has been recognized by stepwise degradation or derived from corresponding DNA sequences, conformations of a great many of peptides have been determined (e.g., pleated sheet, α helix; Pauling *et al.*, 1951) by optical or nuclear magnetic resonance (NMR) methods, and the molecular architecture of hundreds of proteins has been revealed by X-ray crystallography after the pioneering work of hemoglobin by Perutz *et al.* (1960) and on myoglobin by Kendrew *et al.* (1960). The success of such experiments depends on the quality of the protein crystals. Since the mid-1980s attempts have been made to grow better crystals in the low gravity environment of space. A review on this costly, only partially successful project has appeared (Stoddard *et al.*, 1992).

In his Karlsbad lecture, Fischer also said that the idea of the amide-like connection had led him "more than $1\frac{1}{2}$ years ago to initiate experiments to effect the synthetic linkage of amino acids." That date, however, does not mark the origin of synthetic research with amino acids; this work began in 1882 and is connected with hippuric acid and the name of Theodor Curtius.

II. Early Peptide Syntheses

A. The Work of Theodor Curtius: The Azide Method

Theodor Curtius (1857–1928), the son of a dye manufacturer in Duisburg, took a doctorate in 1881 with Hermann Kolbe in Leipzig and on his suggestion reexamined the structure of hippuric acid, for which Dessaignes had proposed benzoylglycine as the formula. Curtius reacted glycine–silver with benzoyl chloride and obtained, besides hippuric acid, products with more than one glycine per benzoyl residue. He characterized one, the "β-Säure," as benzoylglycylglycine, and the second, "γ-Säure," he later recognized as benzoylhexaglycine. While continuing the glycine studies in the laboratory of Adolf Baeyer in Munich, he invented the esterification of amino acids in ethanol and hydrochloric acid and observed the easy formation of diketopiperazine and polymeric glycine esters on heating of the free ethyl ester. The glycine ester was found to yield a yellow oil on reaction with nitrous acid, diazoacetic ester, a compound that was destined to play a great role in Curtius' future work apart from peptides.

In 1887, in an unexpected reaction involving the heating of diazoacetic acid with dilute H_2SO_4, Curtius obtained hydrazine, H_2N-NH_2, and later showed (Curtius, 1902) that the novel base reacted with hippuric acid ethyl ester to give the hydrazide and that the latter was converted to the azide with nitrous acid. In reactivity, carboxylic acid azides proved to be comparable to acid chlorides. The transfer of the hippuryl residue from the azide to the amino group of amino acids (esters) or peptides was the first peptide coupling reaction and is still in use at present, mainly because azide-activated amino acids are resistant to racemization (Fig. 1). Under the influence of E. Fischer's increasingly exciting studies on protein structure, Curtius extended the method to the synthesis of (benzoylated) peptides containing alanine and aspartic acid, but he left the field after 1905. A biography of Curtius has been written (Freudenberg, 1963).

Unfortunately, there was no means at that time to remove the benzoyl residue from the nitrogen without destroying the amide (peptide) bonds generated during the peptide syntheses. Therefore, the method did not permit peptide chain elongation by coupling of the activated carboxyl component to the amino group of the growing chain. Once easily removable N-protecting groups were constructed, however, the azide coupling method was improved and modified by several authors. Mentioned here is the method of Honzl and Rudinger (1961), who improved the conversion of hydrazides to azides by replacing aqueous sodium nitrite with alkyl nitrite and HCl in organic solvents.

Joseph Rudinger was a remarkable peptide chemist. He was born in Jerusalem in 1924, grew up in Prague, and after studying chemistry in England returned to Prague to become Head of the Laboratory of Peptide Chemistry of the Czechoslovak Academy of Sciences. In 1958 he invited the handful of European investigators active in peptide research to participate in a symposium in Prague. In 1968 he and his family emigrated to Switzerland, where Rudinger was appointed professor in the Institute of Molecular Biology and Biophysics of the ETH (Eidgenossische Technische Hochschule) until his untimely death in 1975. The first European Peptide Symposium was followed by meetings held in various locations in Europe. Their excellence led to an unprecedented growth in research and understanding of peptide chemistry. The twenty-second symposium was held in September 1992 in Interlaken, Switzerland, with nearly a thousand participants. The example of the European meetings was followed in the United States and in Japan. Rudinger's many valuable contributions to the chemistry

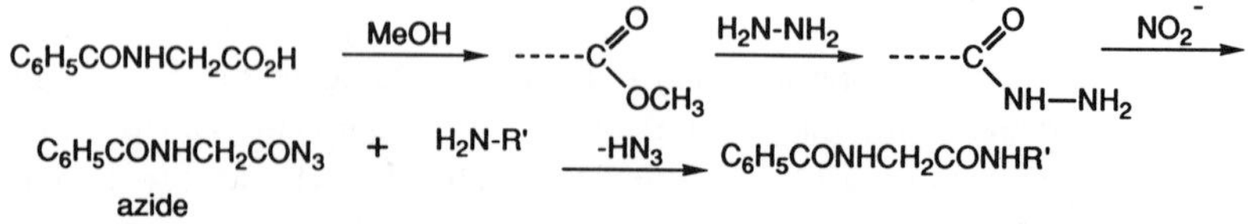

$$C_6H_5CONHCH_2CO_2H \xrightarrow{MeOH} \text{-----}C(=O)OCH_3 \xrightarrow{H_2N\text{-}NH_2} \text{-----}C(=O)NH{-}NH_2 \xrightarrow{NO_2^-}$$

$$\underset{\text{azide}}{C_6H_5CONHCH_2CON_3} + H_2N\text{-}R' \xrightarrow{-HN_3} C_6H_5CONHCH_2CONHR'$$

Figure 1 Formation of hippuric acid azide and application for peptide synthesis.

and bioactivity of peptides cannot be enumerated here. We shall discuss his work again in connection with peptide hormones.

B. Emil Fischer's Fundamental Efforts: Synthesis of an Octadecapeptide

In 1899 when Emil Fischer, triumphant after his mastery of purines and carbohydrates, entered the problematic field of protein chemistry, 14 of the protein-forming amino acids had been characterized, but no method for systematically connecting amino acids to one another existed. The azide method, which was just being developed by Curtius, was never applied by Fischer. His aim was *a priori* to handle peptides having a free amino group in order to elongate the chain by CO to NH coupling, as is customary with present-day peptide syntheses. Lacking a method for removing the benzoyl or acetyl residue under mild conditions (Curtius' dilemma), Fischer tried making use of the ethoxycarbonyl group, C_2H_5O-CO, assuming that it could be removed from the amino nitrogen by reaction with mild alkali or ammonia (Fig. 2). Unfortunately, these expectations were in vain. An undesired alkali-catalyzed rearrangement converts *N*-carbethoxy peptides to derivatives of urea (via hydantoin), as Friedrich Wessely later showed (Wessely and Komm, 1927).

For coupling, the protected amino acids were converted to the acid chlorides. *N*-Ethoxycarbonyl amino acid chlorides are discussed later in this chapter in connection with the work of H. Leuchs, an associate of Fischer. After a mild removal of the urethane protecting group failed, as mentioned above, Fischer attempted to omit a protecting group altogether. For this purpose he worked out the preparation of amino acid chlorides by reacting the finely powdered amino acids with PCl_5 in the unusual solvent acetyl chloride (Fischer, 1905):

$$H_2N-CH(R)-CO_2H + PCl_5 \rightarrow H_3\overset{+}{N}\text{-}CH(R)-COCl\ Cl^-$$

The amino acid chloride hydrochlorides so formed, which are extremely reactive substances, could be coupled with amino acid esters in nonaqueous solvents to give peptides which themselves had a free amino group and so, theoretically, would have been ideal components for the synthesis of longer

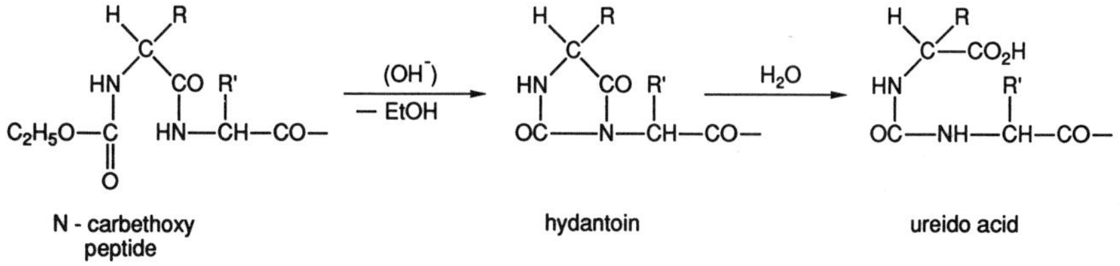

Figure 2 Attempt to remove the ethoxycarbonyl group from nitrogen and formation of hydantoin.

peptides by elongation at the amino end. In practice, however, self-condensation as predominant reaction rendered this approach illusory.

Fischer's most important contribution to peptide chemistry may be the introduction of α-halogen fatty acid chlorides, in which the halogen atom, after coupling to an amino component, can be converted to an amino group by treatment with ammonia to yield a new amino acyl terminus (Fischer and Otto, 1903) (Fig. 3). This strategy was adopted and extended by Fischer and numerous collaborators in the synthesis of about 100 simple peptides, mostly containing glycine and amino acids lacking side-chain functional groups. Difficulties also arose from the need for enantiomerically pure starting materials, particularly L-amino acids.

The resolution of DL-α-bromofatty acids as the alkaloid salts turned out to be troublesome. Otto Warburg (1883–1970), to whom this problem was offered, obtained 10% optically pure (−)-α-bromopropionic acid after 20 recrystallizations of the cinchonine salt from water. On aminolysis the product yielded optically pure D-alanine, the antipode of the desired natural L-alanine. In the aminolysis reaction, a so-called Walden inversion at the chiral center had occurred, which was a general finding with all optically active α-halogen fatty acid residues. In a simpler way, without inversion, enantiomeric α-bromofatty acids could be obtained by reaction of the corresponding α-amino acids with nitrosyl bromide, as in the above-mentioned example of (−)-α-bromopropionic acid from natural L-alanine. Because, however, the enantiomeric (+)-α- halogen fatty acids (D enantiomers) were required for natural L-amino acid residues in Fischer's peptides, he had to provide the corresponding D-α-amino acid for reaction with NOBr, which in turn resulted from previous resolution of the DL-α-amino acid mixture.

The complication of enantiomer resolution and the fact that amino acids with functional side chains could not be used in this method led Fischer to halt serious attempts to synthesize proteins after a decade of struggling. Nevertheless, he eventually did succeed (with Axhausen) in synthesizing an impressive octadecapeptide L-Leu-$(Gly)_3$-L-Leu-$(Gly)_3$-L-Leu-$(Gly)_8$-Gly (Fischer, 1907). In a famous lecture at the Deutsche Chemische Gesellschaft at Berlin in 1906, Fischer gave an impressive review of his and his students' efforts, which in the press aroused speculations on the imminent synthetic availability of living matter.

The large peptides may not have completely met today's criteria of purity. Half a century later, 34 peptides from Fischer's laboratory were made available by his son Hermann O. L. Fischer for tests of homogeneity. At Berkeley, Knight

$$BrCH(R)COCl + H_2N\text{-}CH(R')CO_2H \longrightarrow BrCH(R)CONHCH(R')CO_2H$$

$$\xrightarrow[\text{Inversion}]{+NH_3} H_2NCH(R)CONHCH(R')CO_2H + NH_4Br$$

Figure 3 Fischer's α-bromacyl method.

(1951) examined the samples by paper chromatography and, to everybody's surprise, found that all but 3 gave single spots. Fischer's great efforts were not in vain; in addition to a host of examples of the art of preparative chemistry, the protein studies, in the words of Kurt Hoesch (1921) in his Fischer biography, "also create definitive clarity about the final building elements of the carriers of life and the principles by which these are bound together."

C. The School of Emil Fischer: Abderhalden, Leuchs, and Bergmann

Several associates of Fischer pursued the peptide theme of their master's studies after his death in 1919 and advanced it further.

1. Emil Abderhalden

Emil Abderhalden (1877–1950), one of Fischer's most faithful and diligent associates (they published 30 joint papers), may be considered the "biochemical hand" in the laboratory. Born in 1877 and the son of a teacher in Oberuzwil near St. Gallen, Switzerland, he studied medicine in Basel and, after graduation in 1902, moved to Berlin to join Emil Fischer. He rapidly became well versed in chemistry, a science whose methodology together with that of physics he recognized as an inseparable part of an effective physiological research program. In Halle, where he was appointed Head of the Institute of Physiology in 1911, he continued Fischer's studies on substrates of proteolytic enzymes (of which he isolated and characterized new ones) and investigated the metabolism of amino acids in the framework of nutrition. On the chemical side, he continued experiments in peptide synthesis by Fischer's method with determination; however he was able to surpass the length of the famous octadecapeptide merely by a single L-leucine residue (Abderhalden and Fodor, 1916). In the late 1920s he resumed experiments with phenylisocyanate initiated by Max Bergmann in Fischer's laboratory and later in Dresden.

Bergmann had found that an N-terminal phenylureido amino acid residue of a peptide, on boiling with HCl, was transformed via ring closure to a phenylhydantoin with cleavage of the peptide bond. Abderhalden and Brockmann (1930) demonstrated that this cleavage occurred with methanol/HCl even at 60°C within 30 min and that the intact shortened peptide (glycylleucine) could, in turn, be subjected to the same operation. Application of the same reactions, but with phenylisothiocyanate, under milder conditions was the improvement introduced later by P. Edman.

Abderhalden, who for a long period also was president of the Academia Leopoldina in Halle, was transferred to the U.S. occupation zone after the second world war ended in 1945. Switzerland became his last refuge when he was appointed as an Honorary Professor of the University of Zurich, and he died in Zurich in 1950. For a biography of Abderhalden, see Hanson (1970).

2. *Hermann Leuchs*

Hermann Leuchs (1879–1945), a member of an old Franconian family, took his doctorate with E. Fischer in 1902 and beginning in 1904 worked with him as an assistant until Fischer's death in 1919. He became associate professor of Chemistry and in 1926 Associate Director of the Chemistry Institute of the Wilhelm von Humboldt University in Berlin.

In 1906, Leuchs observed that the *N*-ethoxycarbonyl amino acid chloride of glycine, on losing ethylchloride at 70°C, was converted to a compound, later called "Leuchs'scher Körper" (Leuchs' substance), which on heating in the presence of a little water was transformed to an insoluble material. The reaction, a polymerization, was formulated as starting from a supposed three-membered ring compound, which later was recognized as inner *N*-carbamoic acid anhydride (NCA). Leuchs extended the studies to additional amino acids like phenylalanine and leucine, but he did not realize the potential of his reaction as a method of directed peptide synthesis.

a. Peptide Synthesis Using N-Carboxy Anhydride The chemistry of peptide bond synthesis using NCAs is shown in Fig. 4. The free amino group of an amino acid or peptide attacks an amino acid oxazolidine-2,4-dione or *N*-carboxyanhydride (NCA), forming a peptide bond and releasing CO_2 from the terminal *N*-carboxy amino acid. This scheme is repeated until the desired sequence has been assembled.

It was the group of Friedrich Wessely (1897–1967) which systematically synthesized peptides, for example, phenylalanylglycine by the reaction of sodium glycine with phenylalanine *N*-carboxyanhydride (Sigmund and Wessely, 1926). Here the *N*-carboxydipeptide was fixed as the sodium salt and so was prevented from undergoing decarboxylation and further condensation. The trick of stopping the reaction after each step by salt formation with triethylamine led Bailey (1949) to produce a couple of peptides virtually without side reactions. Decarboxylation could be avoided completely (and a useful method elaborated) only when N-protected NCAs were introduced by Murray Goodman's group (Fuller *et al.*, 1990). By removal of the protecting group after each step, the peptide formed was made ready for the next condensation.

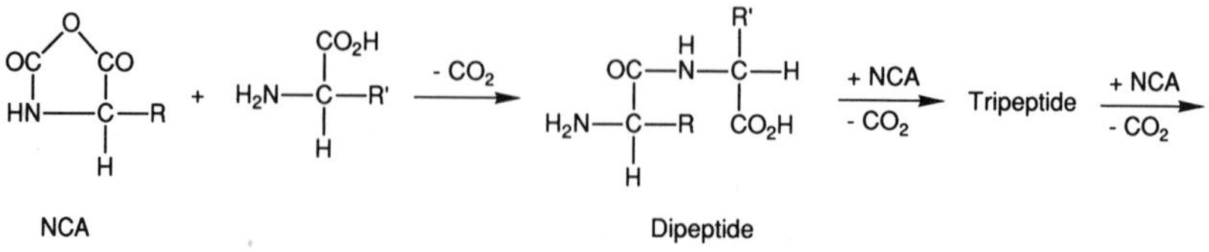

Figure 4 Peptide formation from an *N*-carboxyanhydride (NCA).

In the 1960s Hirschmann, Denkewalter, and co-workers published their studies on the total synthesis of an enzyme. They adapted the NCA method to develop a process carried out in aqueous solution, under precise pH control, which enabled rapid coupling in the cold. They thus produced about 20 peptide segments which when linked by the improved azide method yielded the S-protein (104 amino acids) and, by combination with the S-peptide (20 residues), a solution that clearly exhibited the enzymic activity of ribonuclease A (Hirschmann et al., 1969).

b. Polyamino Acids The most common method for the preparation of poly-α-amino acids is the repeated condensation of NCAs. Interest in such macromolecules as model substances for proteins already existed, to a certain extent, after Curtius' observation of solidification (oligomerization) of glycine ethylester, but it was revived by Woodward and Schramm (1947) who, on storage of a solution of NCAs of DL-phenylalanine and L-leucine in (moist) benzene, obtained films of a copolymer of a molecular weight of about 15,000. This was the time when the polyamides (nylon, perlon) started their triumphal way. These silklike fibers challenged peptide chemists to prepare analogous macromolecules from α-amino acids like natural silk, not as hydrophobic as the polyamides and yet with the mechanical strength of the latter. Of equal importance to these (never achieved) objectives was the concept of creating ideal models for the study of the dependence of the conformation of polypeptide chains on the amino acid composition and the nature of the solvent. Polymerization can be initiated by several agents, resulting in different mechanisms and leading to macromolecules of different sizes. With water at room temperature or with primary or secondary amines, polymers of several hundred components are formed. The highest degrees of polymerization (about 1000 to 3000 monomers) are obtained with strong bases in water-free solvents. Furthermore, polymerization can be initiated by salts like LiBr or by tertiary amines. The different chemical mechanisms of initiation of the polymerization of NCAs have been extensively discussed by Katchalski and Sela (1958).

The nature of the solvent has a strong influence on the conformation of peptides. Relationships between solvent polarity and proportion of α-helix, β-pleated sheet, and random chain can give information on the mechanism of folding of protein molecules. Oriented films of poly-α-L-glutamic acid and its esters were examined via X-ray analysis by Pauling *et al.* (1951). The reflections, related to a spacing of 1.5 Å, confirmed the existence of an α-helix predicted for fibrillar proteins. Noteworthy is the occurrence of a polyamino acid in nature: the capsule substance of certain bacteria like *Bacillus anthracis, B. licheniformis,* and *B. subtilis* with molecular weights of about 50,000 consists of exclusively D-glutamic acid residues which are linked only by γ-peptide bonds (Ivanovics and Bruckner, 1937). For a review of the biological properties of polyamino acids, see Sela and Katchalski (1959).

3. *Max Bergmann*

The most important contribution to the development of peptide chemistry after Emil Fischer's death came from his principal co-worker and trustee of his legacy, Max Bergmann (1886–1944). The son of a wholesaler merchant in Fürth (Bavaria), Bergmann started studies at the Polytechnic College in Munich in botany, but in 1907 he transferred to chemistry and to Berlin. After graduation in 1911 Bergmann was engaged by Fischer as a personal laboratory assistant to work in the field of sugars, tannins, amino acids, and peptides. After an appointment as Director at the Kaiser Wilhelm Institute (KWI) for Fiber Research in Berlin–Dahlem, he moved in 1922 to Dresden as Director of the newly founded KWI for Leather Research. There, he widened his studies of carbohydrates to the polymers cellulose and chitin, investigated *pro domo* the tannins, and also kept up on amino acids and peptides.

In 1926 Bergmann described a new principle of peptide synthesis, namely, opening of the azlactone ring by an amino acid in the presence of alkali. Azlactones can be easily prepared by the Erlenmeyer synthesis from (aromatic) aldehydes and *N*-acylglycine. From the azlactone of α-acetaminocinnamic acid and glutamic acid, Bergmann *et al.* (1926) obtained the dehydro peptide which was converted to the dipeptide by catalytic hydrogenation (Fig. 5).

The peptides so obtained were still blocked at the amino end by non-removable acyl residues (acetyl or benzoyl) and so suffered from the same failure as the first peptides of Curtius and Fischer. Although Bergmann found that treatment with dilute mineral acids at moderate temperature was able to preferentially cleave the acetyl residue from peptide amino groups this could not be adapted as a general satisfactory method. Moreover, because two chiral centers were generated by the hydrogenation step at the carbon, the method was not suitable for the synthesis of enantiomerically pure compounds.

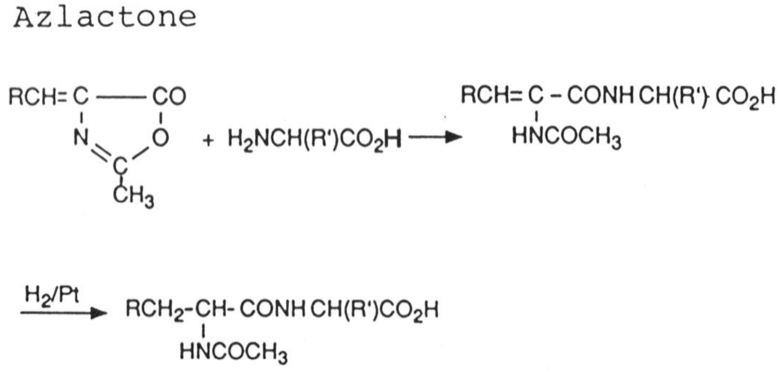

Azlactone

RCH= C —— CO
N O + $H_2NCH(R')CO_2H$ ⟶ RCH= C - CONH CH(R')-CO_2H
C $HNCOCH_3$
CH_3

$\xrightarrow{H_2/Pt}$ RCH_2-CH- CONH CH(R')CO_2H
$HNCOCH_3$

Figure 5 Peptide synthesis from azlactone.

A special aim of Bergmann was to find reagents with high specificity for the precipitation of individual amino acids. Such compounds had long been known for groups of amino acids, for instance, phosphotungstic acid for lysine, arginine, and histidine, or flavianic acid and picric acid. Bergmann enlarged the list of such precipitants with chromium complexes like "rhodanilic" acid for proline and later with several aromatic sulfonic acids.

a. Carbobenzoxy Method of Peptide Synthesis The search for a mildly removable amino-blocking group accompanied peptide chemistry through the early decades. Together with E. Fischer, Bergmann had already studied derivatization of amino acids and found that the tosyl residue can be removed from *N-p*-toluoylsulfonyl compounds at 70–100°C by reduction with hydroiodic acid (HI) in the presence of PH_4I. This reaction was applied in 1926 by R. Schoenheimer in the synthesis of a few small peptides, but it proved to be insufficiently mild for general application. Meanwhile, in the 1920s several publications appeared (from K. Rosenmund and from K. Freudenberg) dealing with cleavage of the *O*-benzyl group by catalytic hydrogenation over platinum metal catalysts. The method was adopted by Bergmann's laboratory in the study of carbohydrates. In 1931 Leonidas Zervas (1902–1980) used it in the synthesis of 1-benzoylglucose (Zervas, 1931), and he and Bergmann on this occasion had the idea of replacing ethyl in Fischer's unfavorable ethoxycarbonyl chloride by benzyl and thus creating the benzyloxycarbonyl protecting group, which at that time was called the carbobenzoxy (Cbz) group. This group is cleaved by catalytic hydrogenation under very mild conditions to yield toluene and the *N*-carbonic acid, which decomposes to CO_2 and the free amino acid (Bergmann and Zervas, 1932):

$$C_6H_5CH_2OCONHCH\,(R)\,CO- \xrightarrow{H_2/Pt} C_6H_5CH_3 + HO_2CNHCH\,(R)-CO- \xrightarrow{-CO_2} H_2N-CH\,(R)\,-CO-$$

The Cbz residue was introduced into amino acids by reaction with benzyloxycarbonyl chloride ($C_6H_5CH_2OCOCl$, derived from benzyl alcohol and phosgene).

The Cbz group, today termed the Z group, opened a new era in peptide synthesis. The new method was immediately put into use in Dresden for the preparation of peptides involving not only amino acids with protected terminal α-amino groups, but also lysine residues, whose ϵ-amino group was likewise blocked by the Z group. As one of several examples, *N*α,*N*ϵ-dibenzyloxycarbonyl-lysine methylester was coupled via hydrazide and azide to histidine methylester, and the dipeptide Lys-His was obtained after saponification and hydrogenation.

That the invention of the Z group did not immediately cause a flood of papers was merely due to the small number of laboratories, worldwide, working on peptides at that time. Also, the political situation greatly deteriorated. Under the Nazi regime, Max Bergmann and his family had to leave Germany in 1934.

He was warmly received by the Rockefeller Institute for Medical Research in New York City, where he successfully continued peptide and protein research.

III. A New Era in Peptide Chemistry

The invention of the carbobenzoxy group promoted a strong expansion of peptide chemistry. In the years following World War II, peptides became increasingly interesting, mainly in physiology and biochemistry, as a consequence of the development of methods not only for synthesis, but also for detection, purification, and structure analysis.

A. Again Max Bergmann, and the Rockefeller Institute

After his arrival in New York as a Member of the Chemical Laboratory of the Rockefeller Institute, Bergmann was able to continue peptide research in collaboration with Leonidas Zervas, who followed him from Dresden, and other coworkers, among whom Joseph S. Fruton and Heinz Fraenkel-Conrat were the first to join.

1. Analytical Work

As mentioned above, one of Bergmann's programs was to elaborate methods for the separation of amino acids and measuring their quantities in hydrolysates of proteins. Together with William H. Stein (1911–1980) and Stanford Moore (1913–1982), Bergmann found that certain amino acids could be precipitated, more or less selectively, by several aromatic sulfonic acids. The "solubility product method" for such precipitations was refined and applied for the analysis of the amino acid composition of several proteins (Bergmann and Stein, 1939). This method, however, was superseded after perfection of chromatography had been achieved by Spackman *et al.* (1958). Bergmann himself, regrettably, did not witness this breakthrough in analytical protein chemistry. He died in 1944 in New York at the age of 59 after a long illness. For a biography of Max Bergmann, see Helferich (1969).

a. Chromatographic Methods The history of chromatography started with the Russian botanist Michael S. Tswett (1872–1919), who in 1903 demonstrated the separation of leaf pigments by passing the solution in petroleum ether through a column of calcium carbonate (for a review, see Lederer, 1988). Trials involving application of the technique to amino acids and peptides, which are not soluble in lipophilic organic solvents, were not described until the early 1940s, when F. Turba in Prague and the author, independently, separated neutral from basic and from acidic amino acids by filtration in aqueous solution through ion-exchanging minerals, and Arne Tiselius in Uppsala used active carbon to retain the aromatic amino acids free from the other amino acids. In 1941, Martin and Synge reported a new principle of chromatography that was based on the different partition ratios of substances between two immiscible phases. One phase, an inert powder such as silica soaked with water, was filled into a tube, and a second liquid phase was run through the column. Here, partition between the phases replaced adsorption or ion exchange mechanisms. In 1944, using starch as an aqueous support medium and H_2O-saturated butanol as the developing solution, Synge presented a promising procedure. For a comprehensive review see Wieland (1949).

At the same time synthetic organic resins bearing cationic or anionic groups were being manufactured which, by means of their higher capacity, brought the breakthrough. It was Stein and Moore at the Rockefeller Institute who finally arrived at a perfect solution by using a sulfonated polystyrene resin in the Na^+ form and eluting the column by gradually increasing the pH from pH 3, simultaneously with the ionic strength of the buffer. With an automatic "analyzer" based on this procedure, the separation of all amino acids in a protein hydrolysate was achieved, culminating in the elucidation of the structure of an enzyme of 124 amino acids, namely, ribonuclease A from bovine pancreas (Smyth *et al.*, 1963). Stein and Moore received the 1972 Nobel Prize in Chemistry.

The successful development of amino acid analysis would not have been possible without the chromogenic reagent ninhydrin. Ninhydrin (triketohydrindene-hydrate; Fig. 6, **I**), which was first prepared by Ruhemann (1910), generates a dark blue coloration specifically with α-amino acids at a concentration of the reagent as low as 1 part in 15,000 parts of water. The color of "Ruhemann's purple" (Fig. 6, **II**) is due to an indigo-like reductive condensation of two molecules of ninhydrin with ammonia. An even more sensitive reagent was introduced 1973 by Udenfriend and co-workers (Stein *et al.*, 1973), namely fluorescamine (Fig. 6, **III**), which at room temperature quickly forms strongly fluorescent pyrrolinones with primary amines (Fig. 6, **IV**).

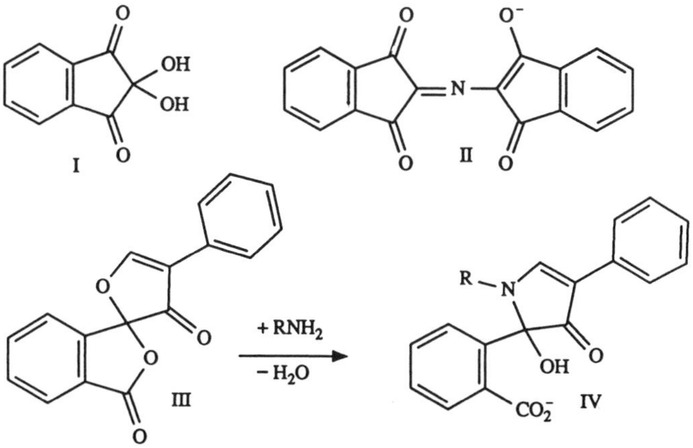

Figure 6 Formulas of ninhydrin (**I**), Ruhemann's purple (**II**), fluorescamine (**III**), and fluorescent product of fluorescamine (**IV**).

Partition chromatography achieved extreme success in the form of paper chromatography, invented by Consden *et al.* (1944). As a general approach, the new method has led to the discovery of a multitude of new substances in natural sources, not just amino acids and peptides.

Filter paper as a support for buffer electrolytes in electrophoretic separations of amino acids, peptides, and proteins, in a simple device, was suggested by Wieland and Fischer (1948). At the same time, Haugaard and Kroner (1948) had separated amino acid mixtures by descending chromatography on a paper sheet to which a perpendicular electric field was applied. To handle little more than microgram quantities of amino acids, Consden *et al.* (1946) suggested silica gel as a supporting medium in zone electrophoresis. For a short time starch layers or agar gels were in use, but all these materials were superseded by polyacrylamide. Today, no biological laboratory can do without polyacrylamide gel electrophoresis (PAGE) for the analysis of protein mixtures (after reaction with sodium dodecylsulfate, SDS) or mixtures of nucleotides. The optimal resolving power is achieved with modern capillary electrophoresis. For a review of early electrophoresis methods, see the chapter on application of zone electrophoresis by Wieland (1959).

Also in chromatography the better was the "enemy of the good." Paper chromatography has been replaced by thin-layer chromatography, first proposed by Stahl (1962), most frequently on glass plates covered with silica gel. Today, high-performance liquid chromatography (HPLC), a very rapid procedure, is gaining importance in peptide laboratories as well.

Gas chromatography, invented by James and Martin (1952) for separating volatile fatty acids, has also been extended to amino acid and peptide research. Because amino acids and peptides cannot be vaporized even at high temperature, they must be converted to appropriate derivatives. Ernst Bayer *et al.* (1957) showed that amino acid methyl esters can be subjected to gas chromatographic separation. Volatility was greatly enhanced by acylation of the polar amino groups, for example, as acetyl derivatives or by trifluoroacetylation. Trifluoroacetic acid (TFA) as a reagent in peptide chemistry had been introduced in 1952 by Friedrich Weygand (1911–1969) (Weygand and Csendes, 1952). Gas chromatography turned out to also be the method of choice for rapid separation of D- and L-amino acids. In 1965, Emanuel Gil-Av at the Weizmann Institute in Rehovot, Israel, showed that the selectivity of a gas chromatographic column can be extended to mobility differences of enantiomers by using chiral stationary phases, that is, by incorporation of chiral side chains (amino acids derivatives) into the polymeric polysiloxane matrix. For information on the capillary version of the method, see König (1987).

Gel chromatography, a relatively new development (see, e.g., Ackers, 1970), uses differences in the access of substances to pores or caves of a gel matrix to achieve separations. Smaller molecules, which more readily enter such matrices, will move more slowly than larger molecules. Among several gels are the well-known Sephadex series, products of Pharmacia (Uppsala, Sweden) derived from the initial work of Jerker Porath and Per Flodin.

As discussed, many of the modern separation methods for amino acids and peptides originated in Max Bergmann's laboratory at the Rockefeller Institute. However, a different separation technique was introduced in the same institute at the same time by Lyman C. Craig (1906–1974), in the laboratory of J. A. Jacobs. Starting in 1933, Craig worked on ergot alkaloids and developed a lasting interest in separation techniques. The well-known method of extracting substances from an aqueous phase into a non-water-miscible phase like chloroform, or vice versa, was systematically elaborated by Craig to a multiple extraction and reextraction procedure, called countercurrent distribution (Craig, 1949, 1967). With this method, performed in a specially constructed machine consisting of several hundred extraction units, isolation and purification of antibiotics, nucleic acids, as well as biologically active peptides became possible for amounts greater than those that could be handled by normal chromatographic techniques.

b. Determination of Amino Acid Sequences As a classic example of the determination of amino acid sequences, sequence analysis of the free amino acid constituents of the antibiotic gramicidin S by Consden *et al.* (1947) may be presented. On total hydrolysis the amino acids valine, ornithine, leucine, D-phenylalanine, and proline were formed (analyzed by paper chromatography). Partial hydrolysis (with HCl in acetic acid) yielded di- and tripeptides whose composition, after separation by paper chromatography, was determined by hydrolysis of samples before and after deamination of the N-terminal amino acid by nitrous acid (end group determination). Using overlapping fragments, the true sequence was obtained.

For end group determination of longer peptide chains, N-terminal amino acids had been reacted by heating with 2,4-dinitrochlorobenzene by Abderhalden and Blumberg as early as 1910, who found the *N*-2,4-dinitrophenyl (DNP)-amino acids withstand the conditions of acid peptide hydrolysis. Decisive progress, however, was not achieved until 1945, when Frederick Sanger introduced 2,4-dinitrofluorobenzene, which reacts much more readily than the chloro compound, and who applied paper chromatography, just invented, for separation and characterization of DNP-amino acids (Sanger, 1945). 5-Dimethylaminonaphthalene-1-sulfonyl chloride (DANSYL chloride), introduced by Gray and Heartley (1963), allows the determination of N-terminal amino acids in minute amounts by means of the fluorescence of DANSYL-amino acids on the chromatograms (Fig. 7). Of mostly historical interest are methods for the determination of carboxyl-terminal amino acids, such as the hydrazinolysis reaction proposed by Akabori (1956).

Longer peptide chains, prior to systematic degradation, are routinely split into fragments. For polypeptides containing methionine, the method of choice is cleavage by cyanogen bromide, as suggested by Gross and Witkop (1961). In this reaction a lactone ring is formed at the cost of an adjacent peptide bond.

Stepwise degradation techniques originated in 1950 when Edman showed that phenylisothiocyanate reacts with amino end groups more readily than does

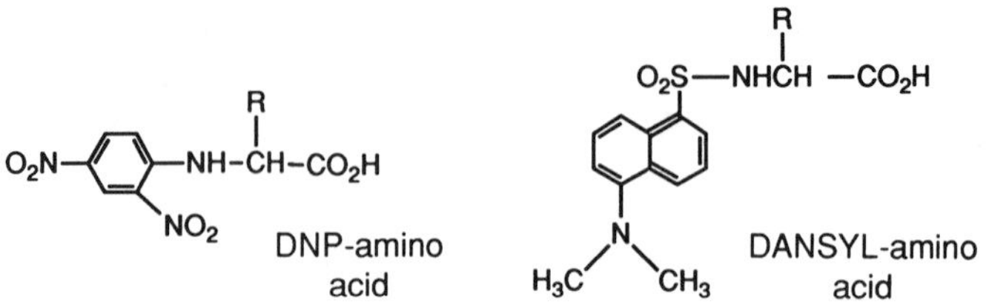

Figure 7 DNP-amino acid and DANSYL-amino acid derivatives.

Bergmann's isocyanate and that the thioureido compounds thus formed can be split by weak acids at room temperature to yield 2-anilinothiazolinones and shortened peptides, which are accessible to the same reaction cycle. In the automatic "sequenator," consequently developed by Edman and Begg (1967), the thiazolinones, on heating, are transformed to phenylthiohydantoins (Fig. 8) and characterized as such by chromatographic or mass spectrometric methods. The method has been improved and refined (gas-phase method) to allow nearly a hundred cycles with picomole quantities of polypeptides.

In more recent years, however, chemical approaches to peptide analysis have lost some of their significance. Nowadays, amino acid sequences can be easily obtained through speedy and precise sequencing of complementary DNAs.

2. *Synthesis of Peptides and Enzyme Research*

Synthetic work with peptides was resumed by Bergmann along the lines of the Fischer school in a systematic way with Joseph S. Fruton, who joined Bergmann in 1934, soon after his arrival at the Rockefeller Institute. Fruton utilized the potential of the new carbobenzoxy method to synthesize hitherto inaccessible peptides for testing as substrates for enzymatic hydrolysis. In the course of these studies he published, together with Bergmann, syntheses of simple substrates which were cleaved at defined peptide bonds by papain, chymotrypsin, trypsin, and pepsin, respectively (Bergmann and Fruton, 1941).

In parallel, the search for new peptidases continued. Emil L. Smith, later a famous authority on protein research in the United States, was also successfully engaged in this work. In Germany, similar studies were pursued at the Dresden Kaiser Wilhelm Institute for Leather Research by Bergmann's successor Wolfgang Grassmann (1898–1978). Although the institute was destroyed in 1945, it was reestablished by Grassmann, at first as a research laboratory in Regensburg, Bavaria, then as the Max Planck Institute for Protein and Leather Research in Regensburg, later in Munich, and finally as a department of peptide chemistry, directed by Erich Wünsch, incorporated into the large Max Planck Institute for Biochemistry in Martinsried, near Munich.

In the early years of enzyme research, stringent conclusions as to the exact catalytic sites of the respective protein- and peptide-splitting enzymes could not be drawn, but indications appeared of possible interactions of amino acid side chains of the substrate with those of the enzymes, providing affinity binding. More detailed information has accumulated from subsequent studies in various laboratories (e.g., see the review by Neurath, 1985). It is now apparent that the peptide bond hydrolyzed by a peptidase is determined rather specifically by the nature and relative position of the amino acid side chains involved. This also applies to the reverse reaction, namely bond formation by the respective proteases. In an overview on this aspect of peptide chemistry, Fruton (1982) discussed the specificity of numerous peptidases with respect to their condensing ability.

a. Enzymatic Peptide Syntheses At the beginning of the twentieth century, Sawyalow gave the name "plastein" to the insoluble material which appeared on the incubation of a soluble mixture of enzymatic digestion products

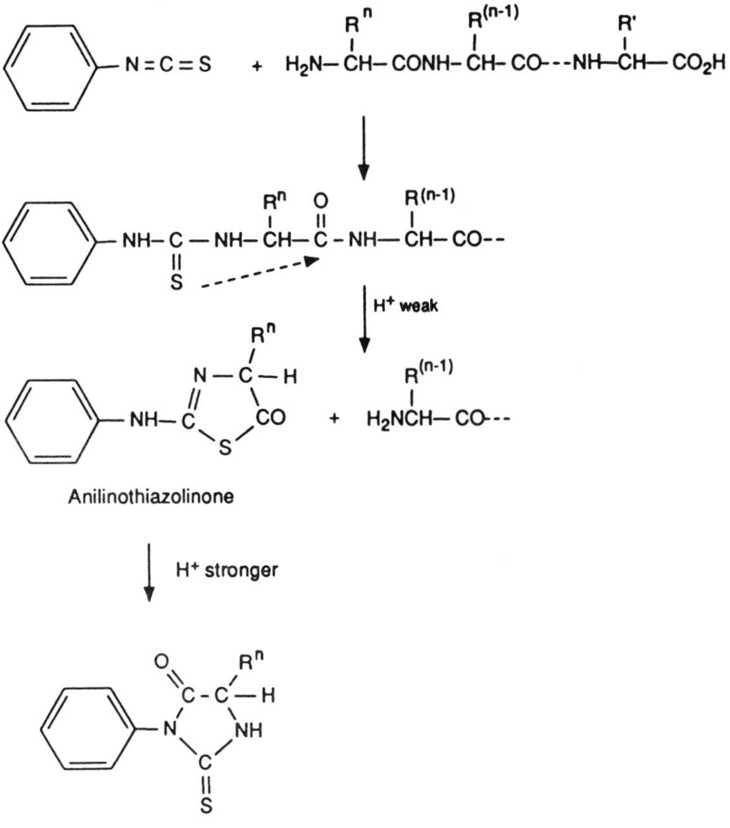

Figure 8 Peptide degradation after Edman.

of fibrin with rennin (a pepsin-like enzyme from calf stomach). This reaction, later also observed with enzymes other than rennin, was studied more intensively in the 1920s by Wasteneys and Borsook (1930), who showed that the products of peptic hydrolysis of egg albumin at pH 1.6 gradually formed a precipitate when the concentrated solution was incubated with pepsin at pH 4. The plastein reaction was not investigated further for more than 20 years, and it was completely elucidated only in the 1960s.

Studies of the enzymatic synthesis of peptide bonds were initiated with simple components by Heinz Fraenkel-Conrat, who was then working in the laboratory of Bergmann. The papain-catalyzed formation at pH 5 of sparingly soluble benzoylleucylleucine anilide from benzoylleucine and leucine anilide (Bergmann and Fraenkel-Conrat, 1938) was demonstrated:

$$C_6H_5COLeuOH + H-LeuNHC_6H_5 \underset{}{\overset{papain}{\rightleftarrows}} C_6H_5COLeuLeuNHC_6H_5 + H_2O$$

In these and similar studies, it appeared that proteinases established an equilibrium which is shifted to the side of synthesis by the (endergonic) precipitation of the slightly soluble product. Because proteinases have esterolytic properties as well, it is obvious that the reverse reaction, namely, peptide bond formation from amino acid esters, is also feasible. This type of transamidation was first observed by Brenner *et al.* (1950) with methionine isopropylester, which on incubation with chymotrypsin yielded dimethionine and trimethionine. As it turned out later in studies of the chemical mechanism of protein biosynthesis in ribosomes, ester formation (at a ribose hydroxyl group) is the mode of activation of amino acids in the living cell as well.

The "plastein reaction" and enzymatic peptide formation *in vitro* were reinvestigated by Virtanen *et al.* (1950) and systemically studied in the laboratory of the author during the 1960s (Determann *et al.*, 1963). It was found, for example, that synthetic H-Tyr-Ile-Leu-Gly-Glu-Leu-OH at pH 5 in the presence of pepsin was converted to an insoluble oligomer with an average amino acid number of 35. The minimal chain length of monomeric plastein-forming peptides was four amino acids, and the amino-terminal and the carboxyl-terminal residues must be hydrophobic. The plastein reaction is a true reversion of hydrolytic fission, however, not a transpeptidation. Proteinase-catalyzed peptide synthesis, owing to the total lack of racemization and minimal need for side chain protection, is still the subject of interest and investigation in several laboratories (for a review, see Jakubke *et al.*, 1983).

B. Development of Additional Protecting Groups

Protecting groups are required in syntheses where multifunctional molecules are to react only with one defined functional group while the others are made inert. The peptide chemist has to handle amino groups, carboxyl groups, and diverse reactive side chains, and so would welcome a choice of protecting groups removable by conditions as different as possible. An ample review on differential protection and selective deprotection in peptide synthesis has appeared (Fauchère and Schwyzer, 1981). In this chapter, only the major events in the development of this subject, and mainly N-protecting groups, are discussed.

The carbobenzoxy (Z) group, as mentioned above, can be easily removed by hydrogenolysis over Pd or Pt. Because such catalysts are poisoned by sulfur, which is present in cysteine-containing peptides, alternative methods of removal have been sought. An alternate reagent, sodium in liquid ammonia, which was introduced by du Vigneaud *et al.* (1930) for the reduction of cystine to cysteine, was later used for reductive removal of the Z group (Sifferd and du Vigneaud, 1935). A further reagent for the relatively selective, reductive removal of the Z group is the mixture of HI and PH_4I, already mentioned, which Harington and Mead (1935) successfully applied in their famous synthesis of glutathione (GSH).

1. Acidolytic Deprotection

Ben-Ishai and Berger (1952) introduced a solution of concentrated hydrobromic acid in glacial acetic acid as a reagent for relatively mild cleavage of the Z group. The acidolytic cleavability of the Z group gave rise in the following years to the construction of more or less easily removable urethane-type protecting groups. By the introduction of electron-withdrawing substituents into the benzene ring (NO_2, Cl) the rate of acidolysis could be reduced by factors of 10 to 1000, whereas electron-donating substituents like OCH_3 increased it more than 100-fold.

An important extension of the methods was the development of the *tert*-butyloxycarbonyl group (Boc) by Carpino (1957) and McKay and Albertson (1957). The Boc group surpasses the Z group in its rate of acidolytic fission by a factor of 10^3. Owing to its resistance to catalytic hydrogenation and the action of sodium in liquid ammonia and of strong alkali, it is an ideal partner for Z groups, which can be removed while conserving Boc protection. The rate of acidolytic fission (i.e., the stability of an intermediate carbonium ion) is increased by replacing a methyl group in the *tert*-butyl part with a phenyl group. A hybrid of Z and Boc groups, the "dimethylated" benzyloxycarbonyl (Dmz) group, is removed acidolytically from an amino acid 700 times faster than Boc (Sieber and Iselin, 1968). The 3,5-dimethoxy derivative, the Ddz group of Birr *et al.* (1972), which can also be split in neutral solution by photolysis, exhibits about equal sensitivity.

The earliest acid-labile N-protecting group, triphenylmethyl, which was suggested for peptide chemistry as early as in 1925 by Helferich *et al.* (1925), was too sensitive to acids to survive multistep syntheses completely. The powerful anhydrous liquid hydrogen fluoride reagent has been introduced by Sakakibara *et al.* (1967); peptide bonds are stable in the presence of the reagent, but the N-tosyl bond or the highly resistant dichloro-Z group will be quickly cleaved.

2. Deprotection under Nonacidic Conditions

Basic reagents for the removal of protecting groups have been applied since the early days of peptide chemistry. Removal of the phthalyl group by hydrazinolysis was described very early (Radenhausen, 1895), but phthalyl amino acids were introduced in peptide synthesis only about 50 years later (Kidd and King, 1948; Sheehan and Frank, 1949). Weygand and Csendes (1952) demonstrated that the trifluoroacetyl residue can be removed from nitrogen by treatment with dilute aqueous hydroxides or with piperidine. Removal of N-protecting groups by bases as a consequence of β-elimination is readily achieved under weakly basic conditions if the leaving group contains an "acidic" CH proton in a position β to the urethane oxygen atom. 2-(4-Toluenesulfonylmethyl)-ethoxycarbonyl was described by Kader and Stirling (1964), and several similarly acting groups, such as 9-fluorenylmethoxycarbonyl (Fmoc) (Fig. 9) proposed by Carpino and Han (1972), attained great popularity, particularly in connection with the solid-phase synthesis technique.

For the sake of completeness, a few less popular principles may be mentioned. At first, photolysis as a means of deblocking an amino group was suggested by Chamberlin (1966), using the UV-labile 3,5-dimethoxy-Z group. The Ddz-group proved to be more sensitive, as applied in the author's laboratory by Birr. For references as well as a description of parallel attempts at the Weizmann Institute, see Wieland and Birr (1967).

An elegant method for removing the allyloxycarbonyl residue under neutral conditions was demonstrated by Kunz (1987). He made use of the allyl transfer to nucleophiles by Pd(0) complexes, such as $[(C_6H_5)_3P]_4Pd$, for deblocking amino groups in very sensitive glycopeptides.

Silicon-containing organic compounds proved valuable in peptide chemistry and were developed in the laboratory of Birkofer and Müller (1968). Among other interesting observations and reactions, the trimethylsilyl group $(CH_3)_3Si$ was found to be easily removable by solvolysis.

C. New Methods for Forming the Peptide Bond

In the first half of the twentieth century, coupling of carboxyl to amino groups was achieved only via the chlorides (E. Fischer) and azides (Th. Curtius). Preparation of chlorides of multifunctional amino acids or peptides, however,

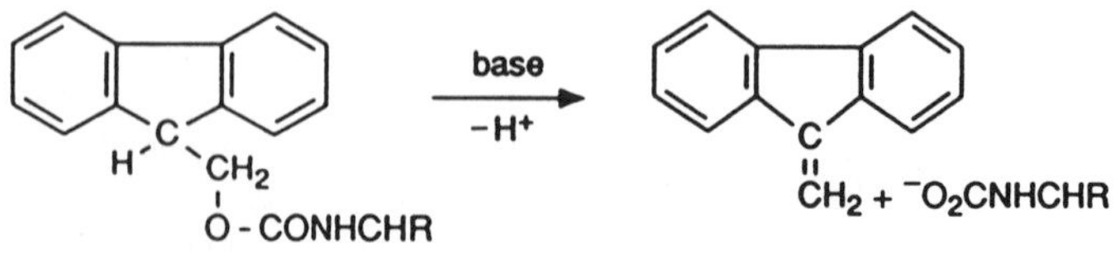

Figure 9 Removal of the Fmoc group by β-elimination.

often gave unsatisfactory results or even failed, and the preparation of azides required several reaction steps. Therefore, soon after 1945 attempts were initiated in several laboratories to simplify peptide coupling, attempts that were stimulated in no small part by the elucidation of the structures of interesting peptides such as insulin, oxytocin, and angiotensin within a few years of one another in the early 1950s. With respect to studies of the biological mechanism of acylation, peptide formation with anhydrides of amino acids and derivatives of phosphoric acid was verified by Chantrenne (1949) and Sheehan and Frank (1950), but such approaches did not come to practical use.

1. Thiol Esters

At the same time the structure of the "activated acetic acid" was recognized by Lynen and Reichert (1951) as *S*-acetyl coenzyme A. Therefore, the idea of aminoacyl thiols as energy-rich intermediates of biological peptide syntheses came to the mind of the author, who showed that N-protected as well as N-unprotected amino acid thiophenyl esters and other S-aminoacyl compounds were reactants suited for very mild peptide coupling (Wieland and Schäfer, 1951; Wieland *et al.*, 1951):

$$\text{Z-NH-CH(R)CO-SC}_6\text{H}_5\text{+ H}_2\text{N-CH(R}'\text{)CO=}\rightarrow$$
$$\text{Z-NH-CH(R) CONHCH(R}'\text{)CO-+ C}_6\text{H}_5\text{SH}$$

The coupling reaction of thioesters in aqueous solution at pH > 7 runs particularly well with cysteine: the activated aminoacyl residue by transacylation is caught by the thiol group of cysteine and immediately switches over to the neighboring amino group thus forming a peptide bond (Fig. 10; Wieland *et al.*, 1953; Wieland 1988). This principle has been used for ligation of 2 appropriate segments to yield a protein, an interleukin (Dawson *et al.*, 1994).

Indeed, the thiol ester activation of amino acids was later found to be the basis of the nonribosomal biosynthesis of gramicidin S and other microbial peptides by Gevers *et al.* (1969). These compounds were the prototypes of active esters discussed below. The N-protected amino acid thiophenylesters were prepared using the mixed anhydride method.

2. Anhydrides

Except for inner anhydrides, the NCAs, and the inner anhydrides of aminodicarboxylic acids (Bergmann and Zervas, 1932), anhydrides of amino acids or peptides had not been prepared and applied in peptide syntheses before 1950.

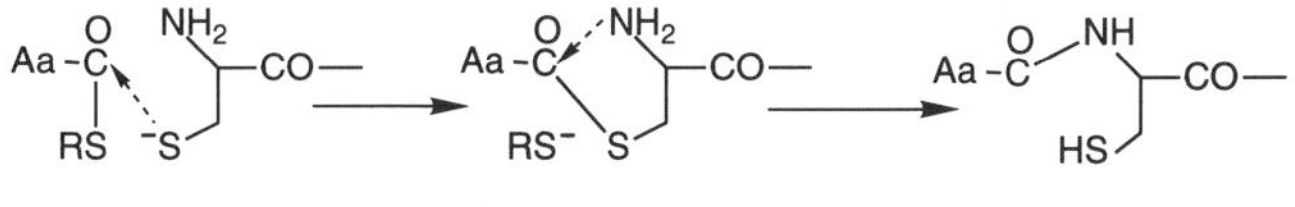

Figure 10 Easy peptide coupling of amino acid thioester to cysteine via S-aminoacyl-cysteine.

Mixed anhydrides from Z-glycine and acetic acid and from Z-alanine and benzoic acid as well as symmetrical anhydrides of the Z-amino acids have been obtained (Wieland *et al.*, 1950) and used for peptide syntheses (Wieland and Sehring, 1950). The mixed anhydrides were obtained from N-protected amino acids and acyl chlorides in an inert solvent in the presence of a tertiary base and were reacted without isolation at room temperature with the respective amino compounds (Fig. 11).

Based on different electron density, the aminoacyl moiety of the mixed anhydrides reacts preferentially with the amino group to be coupled. In the final modification, alkyloxycarbonyl chlorides (e.g., *i*-BuOCOCl) were reacted with the aminoacyl compounds to be "activated" (Wieland and Bernhard, 1951; Boissaonnas, 1951; Vaughan and Osato, 1951). The coupling of urethane-protected amino acids runs nearly without racemization. The mixed anhydride method has been successfully applied in thousands of peptide syntheses.

Symmetrical anhydrides of N-protected amino acids have the advantage that on coupling no foreign product and no second acylation product are formed. They have been obtained as mentioned above by disproportionation of mixed anhydrides, and later from the reaction of the sodium salts of Boc-amino acids with phosgene via a diacyl carbonate (Wieland *et al.*, 1973). Using thionyl chloride (SOCl,) instead of phosgene gives lower yields:

$$\underset{}{\text{RCO-O-}\underset{(\text{SO})}{\text{CO}}\text{-O-COR}} \rightarrow \text{RCOOCOR} + \underset{(\text{SO}_2)}{\text{CO}_2}$$

A convenient synthesis of symmetrical anhydrides has been found in the reaction of carbodiimides with protected amino acids more than 10 years earlier by Muramatsu and Hagitani (1959). Symmetric anhydrides are the acylating agents in the "activating reagent mode" of carbodiimide coupling.

Mixed anhydrides of blocked amino acids and sulfuric acid were introduced by Kenner and Stedman (1952). The reactive agents, applied in the form of a complex with dimethylformamide, could not compete, in spite of the excellent yields obtained, with the alkyl carbonate mixed anhydride method. At the same time diethylchlorophosphite, $(EtO)_2POCl$, was introduced by Anderson *et al.* (1951) followed by several diester chlorides of phophorous acid and finally tetraethylpyrophosphite, $(EtO)_2P{-}O{-}P(OEt)_2$ (Anderson *et al.*, 1952).

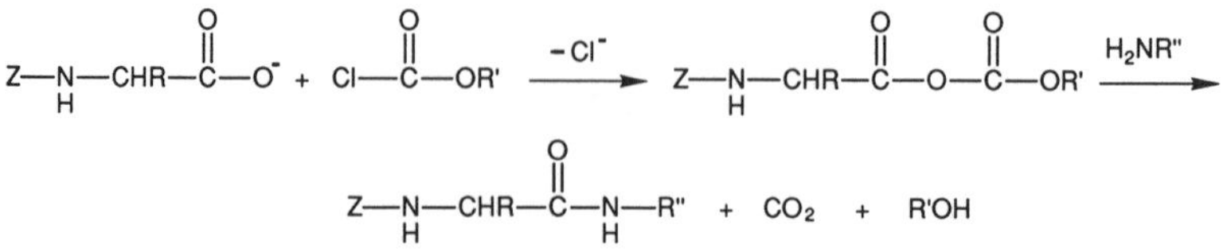

$$\text{Z—}\underset{\text{H}}{\text{N}}\text{—CHR—}\overset{\text{O}}{\overset{\|}{\text{C}}}\text{—O}^- + \text{Cl—}\overset{\text{O}}{\overset{\|}{\text{C}}}\text{—OR'} \xrightarrow{-\text{Cl}^-} \text{Z—}\underset{\text{H}}{\text{N}}\text{—CHR—}\overset{\text{O}}{\overset{\|}{\text{C}}}\text{—O—}\overset{\text{O}}{\overset{\|}{\text{C}}}\text{—OR'} \xrightarrow{\text{H}_2\text{NR''}}$$

$$\text{Z—}\underset{\text{H}}{\text{N}}\text{—CHR—}\overset{\text{O}}{\overset{\|}{\text{C}}}\text{—}\underset{\text{H}}{\text{N}}\text{—R''} + \text{CO}_2 + \text{R'OH}$$

Figure 11 Formation of a mixed anhydride and its use for peptide synthesis.

3. *Active Esters*

Aliphatic amino acid esters are acylating agents, although very slowly reacting. Their reactivity is increased by electron-attracting substituents in the ester component. In the early 1950s Schwyzer and co-workers examined systematically the acylating potency of protected amino acid methyl esters substituted with various electron-withdrawing groups with respect, for example, to their rate of aminolysis (Schwyzer *et al.*, 1955). Cyanomethyl esters, readily prepared through the reaction of acylamino acid salts with chloroacetonitrile, showed satisfactory properties in peptide syntheses:

$$\text{Z}-\text{NHCH(R)CO}_2^- + \text{ClCH}_2\text{CN} \rightarrow \text{Z}-\text{NHCH(R)COOCH}_2\text{CN}$$

The reactivity of cyanomethyl esters, however, was surpassed by phenyl esters, which Gordon *et al.* (1948) had studied with respect to their reactivity in ammonolysis. To further increase the electronic effect, nitrophenyl esters were prepared and examined by Bodanszky (1955). Their usefulness was demonstrated in stepwise syntheses of various peptides. The esters can be obtained as stable crystals from the protected amino acid and nitrophenol (*o*- or *p*-) by dicyclohexylcarbodiimide-mediated condensation.

The substituted aryl esters examined as potential acylating agents in the following years are too numerous to be described here. They are discussed in a comprehensive review article by Bodanszky (1979). Included therein are also the highly reactive *O*-acyl derivatives of hydroxylamine practically designated as "active esters," for example, the *O*-aminoacyl derivatives of *N*-hydroxyphthalimide of Nefkens and Tesser (1961) and of *N*-hydroxysuccinimide (Anderson and Zimmermann, 1964). Owing to their solubility in water, *N*-hydroxysuccinimide esters generally are well suited for the introduction of acyl residues into functional side chains of water-soluble polypeptides (proteins).

4. *Carbodiimides*

The dehydrating effect of carbodiimides (Zetzsche and Lindlar, 1938) was applied for peptide syntheses by Sheehan and Hess (1955) and Khorana (1955). Dicyclohexylcarbodiimide (DCC), the most popular reagent, can be added to a mixture of both components to be coupled, and therefore has been called a

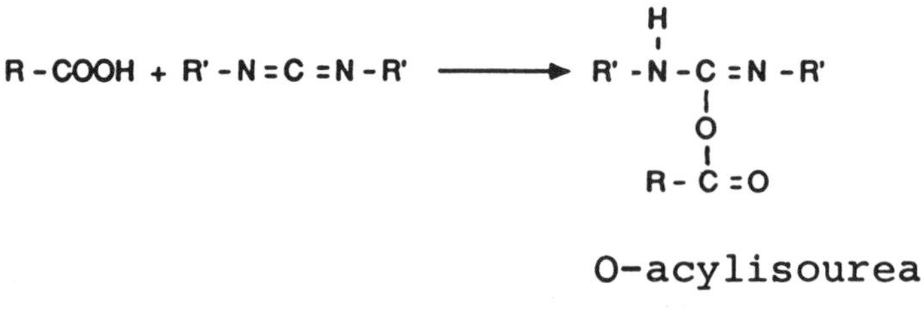

```
                                          H
                                          |
R-COOH + R'-N=C=N-R'  ------>       R'-N-C=N-R'
                                           |
                                           O
                                           |
                                         R-C=O

                                     O-acylisourea
```

Figure 12 Use of carbodiimide in peptide synthesis.

"coupling reagent." According to Khorana, the condensation reaction starts with the addition of the carboxyl component to the carbodiimide to yield the active *O*-acylisourea that transfers the acyl group to the amino component (Fig. 12).

In the absence of an amino compound the *o*-acylisourea reacts with a second carboxyl component, forming a symmetrical anhydride. As a side reaction rearrangement in the *O*-acylisourea can occur, yielding the *N*-acylurea derivative. This and a small extent of racemization can be suppressed by the addition of heterocyclic compounds, particularly 1-hydroxbenzortriazole (HOBt) (Fig. 13), as proposed by König and Geiger (1970). The function of this compound is to take over the strongly activated acyl residue from the *O*-acylisourea intermediate, thereby forming the less reactive, but sufficiently reactive, *O*-acyl--HOBt as the peptide-forming compound.

A distinctly accelerating and improving effect of imidazole on yields of peptide synthesis with active esters can be attributed to the potent acylation-promoting effect of *N*-acyl imidazoles (Wieland and Schneider, 1953). Staab (1957) introduced carbonyldiimidazole, which transfers one imidazole moiety to carboxylic acids to yield the active acylating species (Fig. 13).

5. *Miscellaneous Methods*

Further methods of peptide syntheses mentioned briefly, but not only of theoretical interest, have been described by different laboratories. Great expectations were roused by "Woodward's Reagent K," an isoxazolium salt (Woodward *et al.*, 1961). In the presence of bases it activates carboxylic acids via a ketene–imine (similar to carbodiimide). The reagent did not prove superior to existing methods, however, and so was not widely adopted among peptide chemists.

In "redox" methods of peptide synthesis, the strongly reducing (electron-donating) property of trivalent phosphorus in $(R)_3P$ compounds is the moving

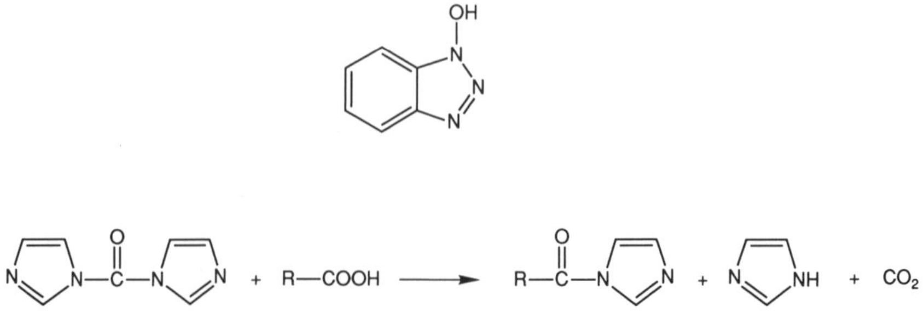

Figure 13 *N*-Hydroxybenzotriazole (top) and carbonylbisimidazole as a peptide-forming reagent via acylimidazolide (bottom).

force for peptide bond formation. Various electron-deficient species can serve as the electron acceptor A^+ (oxidant). A plausible mechanism could be the formation of a reactive acyloxy phosphonium intermediate, via the phosphonium adduct $[(R)_3PA]^+$), which through aminolysis would yield the peptide:

$$(R)_3\overset{+}{P}-O-CO-CHR'-NHZ + H_2N-R'' \xrightarrow{B^-} (R)_3PO + ZNHCHR'CONHR''$$

Oxidant A^+ can be the positive part of an arylsulfenyl compound like ArS−Cl, ArS−NHR, etc. (Mitin and Vlasov, 1968), one half of a disulfide (e.g., 2-mercaptopyridine or Py-S-S-Py) (Mukaiyama *et al.*, 1968), Br^+ from Cl_3CBr (Barstow and Hruby, 1971), or Cl^+ from CCl_4 (Wieland and Seeliger, 1971).

The four-center condensation (4CC) procedure, perhaps the most surprising approach to peptide bond formation, was invented by Ugi (1962). The reaction of an acylamino acid or peptide with an amine, an aldehyde, and an isonitrile leads to a peptide derivative in which a newly formed residue (−NH−CHR−CO−) appears whose side chain R is the side chain of the aldehyde in the admixture. For discussion of putative mechanism of the multicenter reaction and most of the procedures proposed for peptide synthesis from 1959 to 1962, see Wieland and Determann (1963), who summarize an important period in the development of peptide chemistry.

D. Solid-Phase Peptide Synthesis

The idea of linking an amino acid to an insoluble support, then condensing it with a second amino acid followed by a third one, and thus stepwise elongating the peptide chain was caught up simultaneously and independently but verified differently by two American laboratories (Letsinger and Kornet, 1963; Merrifield, 1963). Whereas the work of Merrifield gained wide acceptance, the Letsinger–Kornet approach had no followers. The reason for this difference lies in the choice of strategies adopted. In both methods a polystyrene resin was functionalized by chloromethylation (polybenzylchloride, Merrifield) followed by hydrolysis (polybenzylalcohol, Letsinger) (Fig. 14).

a) [P]$-C_6H_4-CH_2-O-CONH-CH(R)-CO_2H$

Poly - Z - amino acid is activated, coupled with next amino acid . . . etc.

b) [P]$-C_6H_4-CH_2-O-CO-CH(R)-NH_2$

Polybenzylester of amino acid, coupled with urethane-type protected amino acid . . . etc.

Figure 14 Solid-phase peptide synthesis: (a) Letsinger's approach and (b) Merrifield's approach. The symbol P within a box represents the polymer matrix.

Letsinger and Kornet treated the benzyl alcohol resin with phosgene and used the polybenzyloxycarbonyl chloride formed as N-protection and anchor of the N-terminal amino acid. The carboxyl group was then "activated" (with DDC) and reacted with the next amino acid, followed by again activating the resin-bound dipeptide and so forth. A shortcoming of this procedure is that the intermediates, from the second coupling step on, are subject to racemization because only the first amino acid, anchored to the resin by the racemization-preventing Z group, stays unchanged. Therefore, a complex mixture of diastereomeric peptides was to be expected. In contrast, the approach chosen by Merrifield was based on stepwise chain-lengthening starting with the C-terminal residue and adding activated single amino acids with urethane-type N-protection. Therefore, the peptides cleaved from the resin should be chirally homogeneous. For a detailed description, see Chapter 3 by Merrifield in this book.

IV. Classic Peptide Syntheses

The development of peptide synthesis has always depended on the invention of novel protecting groups and coupling methods, and it can best be seen illustrated by the length of the synthetic peptides, which has increased with time. In the following, a few examples of classic syntheses are discussed.

The synthesis of the tripeptide glutathione by Harington and Mead (1935) was a masterpiece in the early days of peptide chemistry. Nearly 20 years later du Vigneaud published syntheses of the hormones oxytocin and vasopressin, cyclic disulfide peptides of nine amino acids which had been isolated from the posterior lobe of the pituitary. After a further decade the methods had been developed to such an extent that Schwyzer and Sieber (1963) could report on the synthesis of corticotropin (ACTH), whose sequence of 39 amino acids had been published shortly before. At the same time the ambitious synthesis of insulin had been performed, at first by Helmut Zahn's group at the German Institute for Wool Research in Aachen, thereafter by Katsoyannis *et al.* in Pittsburgh, and by more than 20 scientists in China at the Chemistry Institute of Academia Sinica, Shanghai and at the Peking University; researchers started to aim at even larger polypeptides. On the route to the crystalline enzyme ribonuclease A (124 amino acids; Yajima and Fujii, 1981) were syntheses of the same enzyme in 1969 by Gutte and Merrifield on solid-phase supports and by Hirschmann *et al.* using *N*-carboxy anhydrides. The synthesis of proteins from natural amino acids today is usually considered a domain of biotechnology.

A. Glutathione

Glutathione (GSH) is the most widespread and probably best studied among the naturally occurring peptides. Observed as a reducing agent in yeast by Rey-Pailhade as early as 1888, it was isolated from yeast, liver, and muscles by F.

G. Hopkins (1921) and formulated as a γ-peptide, namely γ-L-glutamyl-L-cysteinylglycine (Hopkins, 1929). GSH has a multitude of functions within the cell.

The synthesis of GSH by Harington and Mead (1935) was the first event to draw the attention of organic chemists to peptides after the Emil Fischer era, and it profitted from the invention of the carbobenzoxy group a few years earlier although the Z group could not be removed by catalytic hydrogenation because of the sulfur present in the molecules. Instead, the hydroiodic acid–phosphonium iodide mixture was applied. Di-Z-L-cystine was converted with PCl_3 to the diacid chloride and the latter coupled with 2 equivalents of glycine ethylester. Treatment with HI/PH_4I removed the Z groups and reduced the disulfide, yielding cysteinylglycine ethylester that was acylated with Z-L-glutamic acid α-methylester γ-chloride. The tripeptide derivative, after saponification and removal of the Z group, afforded GSH (Fig. 15).

B. Oxytocin and Vasopressin

1. Oxytocin

Oxytocin is the uterus-contracting hormone produced in the posterior lobe of the pituitary gland. It was isolated by Livermore and du Vigneaud (1949) mainly by countercurrent distribution, and its structure was analyzed (du Vigneaud *et al.*, 1953a) using the fragmentation method and the new stepwise degradations of Edman (1950). Oxytocin is a cyclic disulfide peptide consisting of 9 amino acids (Fig. 16).

After the determination of the structures of oxytocin and vasopressin, it took less than 1 year to synthesize oxytocin (du Vigneaud *et al.*, 1953b, 1954a).

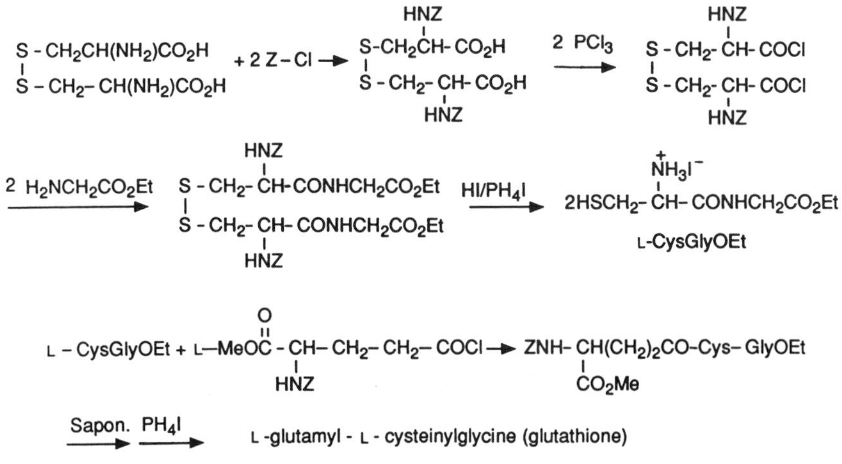

S - $CH_2CH(NH_2)CO_2H$ / S - CH_2- $CH(NH_2)CO_2H$ + 2 Z-Cl → S-CH_2CH(HNZ)-CO_2H / S-CH_2-CH(HNZ)-CO_2H —2 PCl_3→ S-CH_2-CH(HNZ)-COCl / S-CH_2-CH(HNZ)-COCl

—2 $H_2NCH_2CO_2Et$→ S-CH_2-CH(HNZ)-$CONHCH_2CO_2Et$ / S-CH_2-CH(HNZ)-$CONHCH_2CO_2Et$ —HI/PH_4I→ 2HSCH_2-CH($\overset{+}{N}H_3I^-$)-$CONHCH_2CO_2Et$ (L-CysGlyOEt)

L-CysGlyOEt + L–MeOC(=O)-CH(HNZ)-CH_2-CH_2-COCl → ZNH-CH(CO_2Me)$(CH_2)_2$CO-Cys-GlyOEt

—Sapon.→ —PH_4I→ L-glutamyl-L-cysteinylglycine (glutathione)

Figure 15 First synthesis of glutathione.

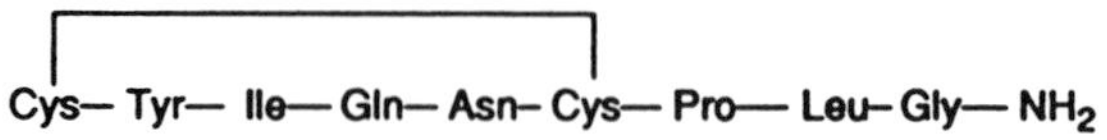

Cys— Tyr— Ile— Gln— Asn— Cys— Pro— Leu— Gly— NH_2

Figure 16 Structure of oxytocin.

At that time, the construction of a nonapeptide with an internal disulfide bond was a formidable task. Model studies using the natural hormone were very helpful. Thus, the disulfide bond was opened by reduction with sodium in liquid ammonia (a reaction invented by du Vigneaud), and oxytocin was regenerated by oxidation of the dithiol.

The synthesis of a nonapeptide (Fig. 17) from "difficult" amino acids as in oxytocin was breaking new ground at the time. Only the partially blocked N-terminal dipeptide *N*-Z-*S*-benzyl-L-cysteinyl-L-tyrosine (segment 1) had been prepared previously, by the pioneers Harington and Pitt Rivers (1944), whose procedure was applied. For assembling the C-terminal tetrapeptide (segment 3, Fig. 17), the new mixed anhydride method was of inestimable value. The central part of oxytocin (segment 2) was combined with segment 3 and the resulting heptapeptide was joined with segment 1 to yield the desired nonapeptide using the pyrophosphite method of Anderson *et al.* (1952). The final product (Fig. 17) was obtained in low yield but was identical with natural oxytocin (Fig. 16) after removal of the protecting groups by sodium in liquid ammonia, air oxidation, and purification by countercurrent distribution. The Nobel Prize 1955 in Chemistry was the well-deserved recognition of this milestone in the history of peptide chemistry.

The synthesis of oxytocin was soon improved. Bodanszky and du Vigneaud (1959) obtained oxytocin in 38% overall yield by stepwise chain elongation using protected amino acid nitrophenyl esters.

With respect to structure–activity relations, numerous analogs of the hormone have been synthesized. Of great interest was the work of Rudinger and Jost (1964), who synthesized an analog in which one of the sulfur atoms of the disulfide bridge was replaced by a methylene group. This carba analog had high potency in various biological assays, thus demonstrating that the disulfide grouping is an architectural component that does not participate directly in the biological effects.

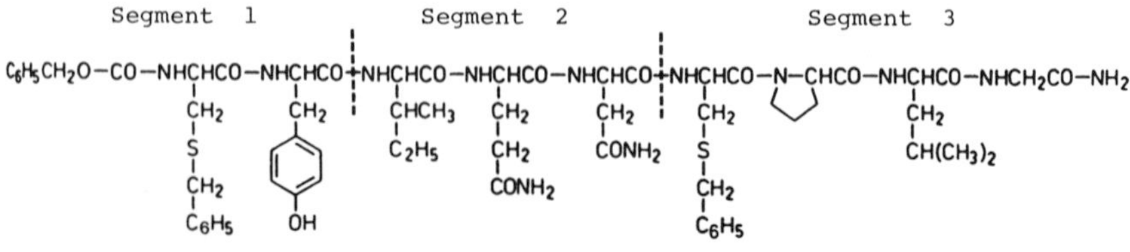

Figure 17 Molecular segments for du Vigneaud's synthesis of oxytocin.

2. *Vasopressin*

In the course of the isolation of oxytocin, the second principal hormone of the pituitary gland, vasopressin, a peptide with pressor and antidiuretic effects, was also obtained in pure form. The bovine hormone differs from oxytocin only at two positions; Ile-3 is replaced by Phe and Leu-8 by Arg. The synthesis of vasopressins, first by du Vigneaud *et al.* (1954b) and later by many other laboratories, opened the way to the large-scale manufacture of vasopressin and analogs for the control of diabetes insipidus in patients and to pharmacologically interesting inhibitors.

C. Insulin

Because of its great medical importance, insulin became the object of broad and continuous studies by natural scientists. Isolation, crystallization, structure elucidation, and synthesis of this pancreas hormone were major challenges. Here we restrict discussion to the chemical aspects of this work, which began with the determination of the primary structure of the bovine hormone by Frederick Sanger and associates in Cambridge, England.

1. *Structure*

Insulin consists of two peptide chains (A and B) cross-linked by two disulfide bridges. A third disulfide bridge is situated in the A chain. To separate the chains, Sanger oxidized insulin with performic acid, thereby converting each cystine to two cysteic acid residues, $HO_3S-CH_2-CH(NH_2)-CO_2H$. Since sequencing work had already started in 1950 (Sanger and Tuppy, 1951), when Edman degradation did not yet exist, partial hydrolysis, analysis of the fragments generated using Sanger's dinitrophenylation procedure, and skillful evaluation of the data were applied (Sanger and Thompson, 1953). After determination of the position of the disulfide bridges, the formula of bovine insulin (Fig. 18) could be presented (Ryle *et al.*, 1955). An account of this historical study can be found in the Nobel lecture of Sanger (1959).

Later, the structures of insulins from other animal species were elucidated by various laboratories. Human insulin, for example, differs from bovine insulin by three exchanges (A-8, Thr for Ala; A-10, Ile for Val; B-30, Thr for Ala) and from porcine insulin by only one exchange (B-30, Thr for Ala).

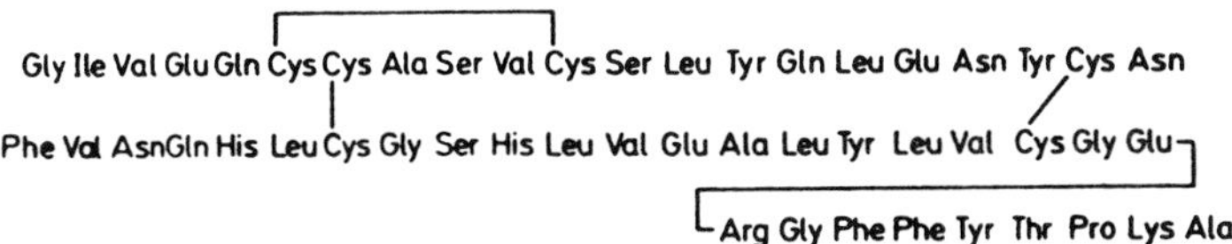

Figure 18 Formula of bovine insulin.

2. *Synthesis*

In the late 1950s the total synthesis of the two chains of insulin became a challenging goal. However, even more difficult seemed the formation of the three disulfide cross-links in the same arrangement as in the natural hormone, as random cooxidation of the reduced chains was expected to yield a large number of non-insulin oxidation products including polymers. Nevertheless, such experiments were conducted by Dixon and Wardlaw (1960) and Du *et al.* (1961), leading to insulin yields that were higher than expected. A total synthesis of insulin therefore seemed feasible. For a review on the history of the synthesis of insulin, see Lübke and Klostermeyer (1970).

Synthesis of the individual chains was started in three laboratories, in Aachen, Germany, in Pittsburgh, Pennsylvania, and in Peking and Shanghai, China. By mid 1963 the American group reported the synthesis of an A chain (Katsoyannis *et al.*, 1963) which was successfully combined a little later with a native B-chain by Dixon to yield semisynthetic insulin. The same year the group led by Helmut Zahn published the synthesis of both chains of sheep insulin and their oxidative combination to yield insulin-containing preparations (Meienhofer *et al.*, 1963). Shortly thereafter the total synthesis of insulin was also accomplished by Katsoyannis *et al.* (1964). In both laboratories the chains were built from almost identical segments using routine protection and coupling methods (DCC, mixed anhydride, azide; to avoid racemization). Cysteine was applied as the *S*-benzyl derivative. The synthetic chains were deprotected and reduced with sodium in liquid ammonia and then cooxidized by air.

Great admiration was called forth by the publication of the total synthesis of crystalline bovine insulin by a group of Chinese chemists in 1965 (Kung *et al.*, 1965). Of the collective group of scientists, Wang Yu, *primus inter pares,* of the Academia Sinica, Shanghai, and Hsing Chi-yi of Peking University are mentioned here as early fellow students of the author in Munich and Heidelberg before the second world war, which he met again 40 years later in China.

Further remarkable events in the insulin story were the application of the solid-phase technique for synthesis of the bovine hormone by Marglin and Merrifield (1966) and an elegant total synthesis of human insulin by a Swiss group in Basel (Sieber *et al.*, 1974). In the latter, two of the fragments contained a preformed disulfide bridge and two other fragments a (protected) cysteine, thus allowing the specific formation of the third disulfide bond of insulin.

The conversion of porcine insulin to human insulin with the help of an enzyme was patented as early as 1966 by Bodanszky and Fried (1966). Carboxypeptidase A, in the presence of excess threonine, catalyzes the production of equilibrium concentrations of human insulin (Thr B-30). Versions of this approach with other enzymes and threonine esters have been adapted to technical processes, but preparation of human insulin with the help of the protein synthesizing machinery of microorganisms is the preferred means of industrial production today (see Frank and Chance, 1983).

D. Depsipeptides and Cyclopeptides

1. Depsipeptides

Depsipeptides are compounds in which amino acids and hydroxy acids and thus peptide bonds and ester bonds alternate. Interest in this class of substances was aroused as early as 1948 when Plattner and Nager isolated two antibiotic compounds, called enniatins, from culture media of a *Fusarium* mold. Cyclic structures were assigned consisting of D-α-hydroxyvaleric acid and *N*-methyl-L-isoleucine (in enniatin A) or *N*-methyl-L-valine (in enniatin B, Fig. 19).

The formulas of the cyclic antibiotics were originally assumed to consist of four components (cyclotetradepsipeptides) but were revised at the Institute for Chemistry of Natural Products of the Soviet Academy of Sciences in Moscow by Shemyakin *et al.* (1963a), who, by comparison with synthetically prepared probes, found that enniatins are cyclohexadepsipeptides. Another antibiotic, valinomycin, isolated by Brockmann and Schmidt-Kastner (1955) from extracts of the mold *Streptomyces fulvissimus,* was also assigned a cyclodepsipeptide structure containing four residues of valine. Shemyakin *et al.* (1963b), by comparison with synthetic material, again disproved the original formula and identified the natural produce as a cyclic dodecadepsipeptide containing three L- and three D-valine, and three D-α-hydroxy-valeric acid and three L-lactic acid residues, as shown in Fig. 19.

The size of the 36-membered ring of valinomycin provides a specific cavity for K^+ ions, which are complexed 10^6 times stronger than Na^+. Valinomycin was the first compound recognized as an ionophore with antibiotic activity. The work on valinomycin has stimulated research on natural and artificial "complexones" such as cyclodextrins, crown ethers, and cryptands by chemists in many laboratories.

2. Cyclopeptides

Tyrothricin was the name give by René Dubos as early as 1939 to an antibiotic peptide mixture secreted by a strain of *Bacillus brevis* (Hotchkiss and Dubos, 1940). In the following years, gramicidins A, B, and C and tyrocidines were isolated from the mixture as homogeneous substances by countercurrent distribution (Craig, 1949). Craig's gramicidins containing ethanolamine at the C terminus later turned out to be linear peptides (Sarges and Witkop, 1965). In 1944 another strain of *B. brevis* yielded a crystalline antibiotic substance that was named by the Soviet scientists Gauze and Brazhnikova as gramicidin S (Gauze and Brazhnikova, 1944).

The nature and sequence of the amino acids constituting this cyclic peptide were determined by Consden *et al.* (1946) with the help of the newly invented technique of paper chromatography (Fig. 20). The occurrence of ornithine, an amino acid not present in proteins, was a little surprising. Gramicidin S turned out to be a cyclic dimer of the pentapeptide D-Phe-Pro-Val-Orn-Leu.

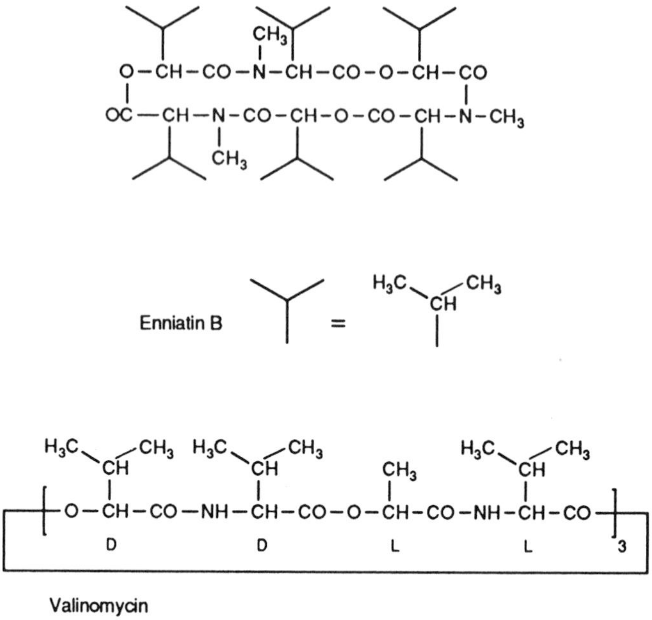

Enniatin B

Valinomycin

Figure 19 Structures of enniatin B and valinomycin.

The synthesis of gramicidin S was accomplished by Schwyzer and Sieber (1957). After assembly of the linear decapeptide H-Val-Orn(Tos)-Leu-D-Phe-Pro-Val-Orn(Tos)-Leu-D-Phe-Pro-OH, the amino end was temporarily protected by tritylation, the carboxyl end activated as a *p*-nitrophenyl ester, the trityl group removed by weak acid, and the molecule cyclized in highly dilute solution.

Studies on the biosynthesis of gramicidin S in *B. brevis* by several laboratories revealed a principle of microbial peptide synthesis different from that occurring in ribosomes of eukaryotic cells. Gevers *et al.* (1969) recognized that the activated form of the amino acids were thioesters linked with cysteine side chains of the synthesizing enzymes. This mode of activation was already known for acetic acid in *S*-acetyl coenzyme A from the work by Lynen and Reichert (1951) and had been anticipated for preparative peptide synthesis in the author's laboratory at about the same time. The nonribosomal mechanism of polypeptide synthesis also holds for the biosynthesis of the tyrocidins and many additional microbial peptides, and most likely for the peptides from *Amanita* mushrooms as well.

3. Amanita Peptides

The extremely poisonous mushrooms of the genus *Amanita* (*Amanita phalloides, A. virosa, A. bisporigera,* and others) contain the amatoxins (Wieland and

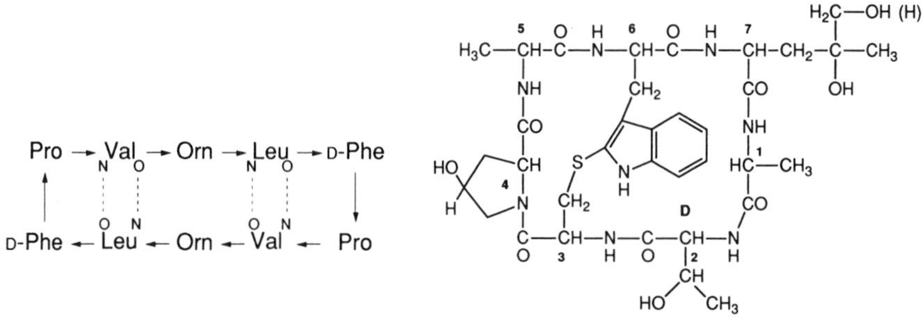

Gramicidin S

Phalloidin

Figure 20 Structures of gramicidin S and phalloidin. In phalloin, the primary hydroxyl group of residue 7 is replaced by hydrogen.

Faulstich, 1991), the phallotoxins (Wieland, 1987), and antamanide, a cyclic decapeptide (Wieland *et al.*, 1968) that competitively inhibits the transport of various substances into liver cells (see Wieland, 1986). Of these components, only the phallotoxins are briefly considered.

Phalloidin was the first cyclic peptide discovered. It was isolated from the poisonous green death cap fungus *Amanita phalloides* and crystallized by Lynen and U. Wieland (1938). Its structure was elucidated in the 1950s in the laboratory of the author in Frankfurt (Wieland, 1987). Phalloidin is a bicyclic peptide because the heptapeptide ring is cross-linked by an indolyl-2-thioether bridge from cystein to tryptophan (Fig. 20). As a consequence, the molecule is very rigid and has the same conformation in the crystal and in solution. It is a rapidly acting toxic compound that leads to destruction of the liver. At the molecular level it stabilizes actin in the polymerized (F-actin) state and thus inhibits actin activities that are essential for the life of the cell. Phalloin and the other phallotoxins interact with actin in the same way. As a minimal difference in the structure, phalloin has a hydrogen atom instead of the primary hydroxyl group in the side chain of residue 7 of phalloidin; therefore, phalloin was less difficult to synthesize.

To generate the thioether bridge a suitable cysteine-containing tetrapeptide was linked via its *S*-chloride to the indole ring of a tryptophan-containing tripeptide that carried the lactone of γ-hydroxyleucine at the C terminus. As shown in Fig. 21, the first cyclization led to the "secolactone" which, after the removal of the Boc group and opening of the lactone ring, provided the H_2N^- and ^-COOH groups needed for the second ring closure. However, only the mixed anhydride method (MA) produced the desired peptide bond in addition to reconverting part of the intermediate to the lactone. All other coupling methods formed the lactone exclusively (Munekata *et al.*, 1977).

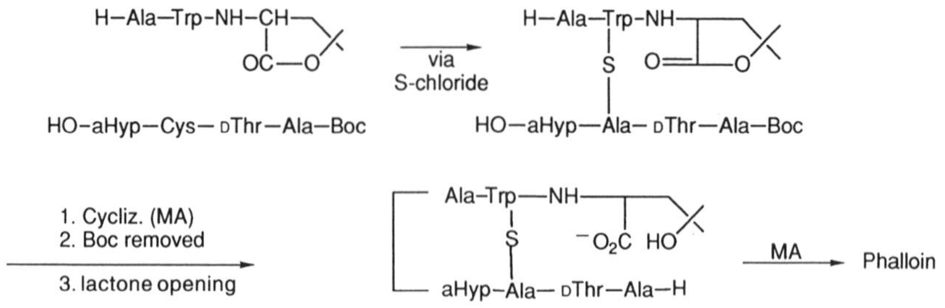

H–Ala–Trp-NH–CH– OC—O / HO-aHyp–Cys– DThr–Ala–Boc → via S-chloride → H–Ala–Trp-NH– S O= –O / HO–aHyp—Ala– DThr–Ala–Boc

1. Cycliz. (MA)
2. Boc removed
3. lactone opening
→ [Ala-Trp—NH– S $^{-}O_2C$ HO / aHyp–Ala– DThr–Ala–H] →MA Phalloin

Figure 21 Synthesis of phalloin.

V. Conclusion

The history of peptide chemistry stretches over a century, from compounds as simple as diglycine to enzymes with more than a hundred amino acids. On the basis of the individual character of each amino acid and enhanced by the combination of many amino acids to form larger molecules, peptides exert an extremely great variety of specific functions. They can act as chemical messengers, hormones in the strict sense, intra- or intercellular mediators, highly specific stimulators and inhibitors, and as biologically active peptides in the brain and nervous system. Numerous antibiotic compounds from bacteria, molds, and amphibian skin are peptidic in nature, and so are compounds toxic not only for pathogens but also for cells of higher organisms. Peptides actively involved in reactions of the immune system are attracting increasing attention from biomedical researchers. Peptide science is still in full development.

Several years ago, peptide chemists were concerned that recombinant protein technology would replace the art of classic manufacturing. Looking back over the 1980s and 1990s, it now appears that an increasing number of valuable polypeptides are produced by genetic methods but that innumerable syntheses of medium size peptides (mainly via solid-phase methods) are still performed in the chemical laboratory. In addition, the need for incorporation of unconventional building blocks into interesting products will require the skill and techniques involved in chemical peptide synthesis.

References

Abderhalden, E., and Brockmann, H. (1930). *Biochem. Z.* **225,** 386–425.
Abderhalden, E., and Fodor, A. (1916). *Ber. Dtsch. Chem. Ges.* **49,** 561–578.
Ackers, G. K. (1970). *Adv. Protein Chem.* **24,** 343–446.
Akabori, S. (1956). *Bull. Chem. Soc. Jpn.* **29,** 507–512.
Anderson, G. W., and Zimmermann, J. E. (1964). *J. Am. Chem. Soc.* **86,** 1839–1842.
Anderson, G. W., Welcher, A. D., and Young, R. W. (1951). *J. Am. Chem. Soc.* **73,** 501–502.
Anderson, G. W., Blodinger, J., and Welcher, A. D. (1952). *J. Am. Chem. Soc.* **74,** 5309–5312.
Bailey, J. L. (1949). *Nature (London)* **164,** 889.

Barstow, L. E., and Hruby, V. J. (1971). *J. Org. Chem.* **36,** 1305–1306.
Bayer, E., Reŭther, K. H., and Boon, F. (1957). *Angew. Chem.* **69,** 640.
Ben-Ishai, D., and Berger, A. (1952). *J. Org. Chem.* **17,** 1564–1570.
Bergmann, M., and Fraenkel-Conrat, H. (1938). *J. Biol. Chem.* **124,** 1–6.
Bergmann, M., and Fruton, J. S. (1941). *Adv. Enzymol.* **1,** 63–98.
Bergmann, M., and Stein W. H. (1939). *J. Biol. Chem.* **28,** 217–232.
Bergmann, M., and Zervas, L. (1932). *Ber. Dtsch. Chem. Ges.* **65,** 1192–1201.
Bergmann, M., Stern, F., and Witte, C. (1926). *Liebigs Ann. Chem.* **449,** 277–302.
Birkofer, L., and Müller, F. (1968). *In* "Peptides 1968" (E. Bricas, ed), pp. 151–155 (Proc. 9th Eur. Pept. Symp.). North-Holland Publ., Amsterdam.
Birr, C., Lochinger, W., Stahnke, G., and Lang, P. (1972). *Liebigs Ann. Chem.* **763,** 162–172.
Bodanszky, M. (1955). *Nature (London)* **175,** 685–686.
Bodanszky, M. (1979). *In* "The Peptides" (E. Gross and J. Meienhofer, eds.), Vol. 1, pp. 165–196. Academic Press, New York.
Bodanszky, M., and Fried, J. (1966). U.S. Patent 3,276,961 (to E. R. Squibb).
Bodanszky, M., and du Vigneaud, V. (1959). *Nature (London)* **183,** 1324–1325.
Boissonnas, R. A. (1951). *Helv. Chim. Acta* **34,** 874–879.
Brenner, M., Müller, R. H., and Pfister, R. W. (1950). *Helv. Chim. Acta* **33,** 568–591.
Brockmann, H., and Schmidt-Kastner, G. (1955). *Chem. Ber.* **88,** 57–61.
Carpino, L. A. (1957). *J. Am. Chem. Soc.* **79,** 98–101.
Carpino, L. A., and Han, G. (1972). *J. Org. Chem.* **37,** 3404–3409.
Chamberlin, J. W. (1966). *J. Org. Chem.* **31,** 1658–1660.
Chantrenne, H. (1949). *Nature (London)* **164,** 576–577.
Cohnheim, O. (1901). *Z. Physiol. Chem.* **33,** 451–465.
Consden, R., Gordon, A. H., and Martin, A. J. P. (1944). *Biochem. J.* **38,** 224–232.
Consden, R., Gordon, A. H., and Martin, A. J. P. (1946). *Biochem. J.* **40,** 33–41.
Consden, R., Gordon, A. H., Martin, A. J. P. and Synge, R. L. M. (1947). *Biochem. J.* **41,** 596–602.
Craig, L. C. (1949). *In* "Fortschritte der Chem. Forschung" (H. Mayer-Kaupp, ed.), Vol. 1, pp. 312–324. Springer-Verlag, Berlin, Heidelberg and Göttingen.
Craig, L. C. (1967). *In* "Methods in Enzymology" (C. H. W. Hirs, ed.) Vol. 11, pp. 870–895. Academic Press, New York.
Curtius, T. (1902). *Ber. Dtsch. Chem. Ges.* **35,** 3226–3228.
Dawson, P. E., Muir, T. W., Clark-Lewis, I., and Kent, S. B. H. (1994). *Science* **266,** 776–779.
Determann, H., Bonhard, K., Köhler, R., and Wieland, T. (1963). *Helv. Chim. Acta* **42,** 2489–2509, and references cited therein.
Dixon, G. H., and Wardlaw, A. C. (1960). *Nature (London)* **188,** 721.
Du, Y. C., Chang, Y. S., Lu, Z. X., and Tsou, C. L. (1961). *Scientia Sinica* **10,** 84.
Edman, P. (1950). *Acta Chem. Scand.* **4,** 283–293.
Edman, P., and Begg, C. (1967). *Eur. J. Biochem.* **1,** 80–91.
Fauchère, J. L., and Schwyzer, R. (1981). *In* "The Peptides" (E. Gross and J. Meienhofer, eds.), Vol. 3, pp. 203–253. Academic Press, New York.
Fischer, E. (1902). *Chem. Ztg.* **26,** 93–93.
Fischer, E. (1905). *Ber. Dtsch. Chem. Ges.* **39,** 530–610.
Fischer, E. (1907). *Ber. Dtsch. Chem. Ges.* **40,** 1754–1767.
Fischer, E., and Otto, E. (1903). *Ber. Dtsch. Chem. Ges.* **36,** 2106–2116.
Frank, B., and Chance, E. R. (1983). *Muench. Med. Wochenschr.* **125,** (Suppl. 1), 14–20.
Freudenberg, K. (1963). *Chem. Ber.* **96,** 1–25.
Fruton, J. S. (1949). *Adv. Protein Chem.* **5,** 1–82.
Fruton, J. S. (1979). *Ann. N. Y. Acad. Sci.* **325,** 1–18.
Fruton, J. S. (1982). *Adv. Enzymol.* **53,** 239–306.
Fuller, W. D., Cohen, M. P., Shabankareh, M., Blair, R. K., Goodman, M., and Naider, F. R. (1990). *J. Am. Chem. Soc.* **112,** 7414–7416.
Gauze, G. F., and Brazhnikova, M. G. (1944). *Ann. Rev. Sovjet. Med.* **2,** 134–138.

Gevers, W., Kleinkauf, H., and Lipmann, F. (1969). *Proc. Natl. Acad. Sci. U.S.A.* **63,** 1334–1342.
Gordon, M., Miller, J. G., and Day, A. R. (1948). *J. Am. Chem. Soc.* **70,** 1946–1953.
Gray, W., and Heartley, B. S. (1963). *Biochem. J.* **89,** 59.
Gross, E., and Witkop, B. (1961). *J. Am. Chem. Soc.* **83,** 1510–1511.
Gutte, B., and Merrifield, R. B. (1969). *J. Am. Chem. Soc.* **91,** 501–502.
Hanson, K. (1970). *Nova Acta Leopold.* **36,** 257–317.
Harington, C. R., and Mead, T. H. (1935). *Biochem. J.* **29,** 1602–1611.
Harington, C. R., and Pitt Rivers, R. V. (1944). *Biochem. J.* **38,** 417–428.
Haugaard, G., and Kroner, T. (1948). *J. Am. Chem. Soc.* **70,** 2135–2137.
Helferich, B. (1969). *Chem. Ber.* **102,** 1–21.
Helferich, B., Moog, L., and Jünger, A. (1925). *Ber. Dtsch. Chem. Ges.* **58,** 872–886.
Hirschmann, R., Nutt, R. F., Veber, D. F., Vitali, R. A., Varga, S. L., Jacob, T. A., Holly, F. W., and Denkewalter, R. G. (1969). *J. Am. Chem. Soc.* **91,** 507–508.
Hoesch, K. (1921). *Ber. Dtsch. Chem. Ges.* Sonderheft.
Hofmeister, F. (1902). *Naturwiss. Rundsch.* **17,** 529–545.
Honzl, J., and Rudinger, J. (1961). *Collect. Czech. Chem. Commun.* **26,** 2333–2344.
Hopkins, G. F. (1921). *Biochem. J.* **15,** 286–292.
Hopkins, G. F. (1929). *J. Biol. Chem.* **84,** 269–277.
Hotchkiss, R. D., and Dubos, R. J. (1940). *J. Biol. Chem.* **136,** 803–804.
Ivanovics, G., and Bruckner, V. (1937). *Z. Immunitaetsforsch.* **90,** 304–482.
Jakubke, H. D., Kuhl, P., and Könnecke, A. (1983). Angew. Chem. **97,** 79–140.
James, A. T., and Martin, A. J. P. (1952). *Biochem. J.* **50,** 679–690.
Kader, A. T., and Stirling, C. J. (1964). *J. Chem. Soc.* **1964,** 258–266.
Katchalski, E., and Sela, M. (1958). *Adv. Protein Chem.* **13,** 234–492.
Katsoyannis, P. G., Tometsko, A., and Fukuda, K. (1963). *J. Am. Chem. Soc.* **85,** 2836–2865.
Katsoyannis, P. G., Fukuda, K., Tometsko, A., Suzuki, M., and Tilak, M. (1964). *J. Am. Chem. Soc.* **86,** 930–932.
Kendrew, J. C., Dickerson, R. E., Strandberg, B. E., Hart, R. G., Davies, D. R., Phillips, D. C., and Shore, V. S. (1960). *Nature (London)* **185,** 422–427.
Kenner, G. W., and Stedman, R. J. (1952). *J. Chem. Soc.* **1952,** 2067–2076.
Khorana, H. G. (1955). *Chem. Ind.,* 1087–1088.
Kidd, D. A. A., and King, F. E. (1948). *Nature (London)* **162,** 776–776.
Knight, C. A. (1951). *J. Biol. Chem.* **190,** 753–756.
König, W. A. (1987). "The Practice of Enantiomer Separation by Capillary Gas Chromatography." Huthig Verlag, Heidelberg, Basel, and New York.
König, W., and Geiger, R. (1970). *Chem. Ber.* **103,** 788–798.
Kŭng, Y. T., Du, Y. C., Huang, W. T., Chen, C. C., Ke, L. T., Hu, S. C., Jiang, R. Q., Chu, S. Q., Niu, C. I., Hsu, J. Z., Chang, W. C., Chen, L. L., Li, H. S., Wang, Y., Loh, T. P., Chi, A. H., Li, C. H., Shi, P. T., Yieh, Y. H., Tang, K. L., and Hsing, C. Y. (1965). *Scientia Sinica* **14,** 1710–1716.
Kunz, H. (1987). *Angew. Chem.* **99,** 297–311.
Lederer, E. (1988). *New J. Chem.* **12,** 249–252.
Letsinger, R. L., and Kornet, M. J. (1963). *J. Am. Chem. Soc.* **85,** 3045–3046.
Leuchs, H. (1906). *Ber. Dtsch. Chem. Ges.* **39,** 857–861.
Livermore, A. H., and du Vigneaud, V. (1949). *J. Biol. Chem.* **180,** 365–373.
Lübke, K., and Klostermeyer, W. (1970). *Adv. Enzymol.* **33,** 445–525.
Lynen, F., and Reichert, E. (1951). *Angew. Chem.* **63,** 47–48.
Lynen, F., and Wieland, U. (1938). *Liebigs Ann. Chem.* **533,** 93–117.
McKay, F. C., and Albertson, N. F. (1957). *J. Am. Chem. Soc.* **79,** 4686–4690.
Marglin, A., and Merrifield, R. B. (1966). *J. Am. Chem. Soc.* **88,** 5051–5052.
Meienhofer, J., Schnabel, H., Bremer, H., Brinkhoff, O., Zabel, R., Sroka, W., Klostermeyer, W., Brandenburg, D., Okuda, T., and Zahn, H. (1963). *Z. Naturforsch.* **18b,** 1120–1121.
Merrifield, R. B. (1963). *J. Am. Chem. Soc.* **85,** 2149–2154.

Mitin, Yu. W., Vlasov, G. P., and Shelykh, G. I. (1967). *Izv. Akad. Nauk. Ser. Khim.* 1397–1398.
Mukaiyama, T., Ueki, N., Maruyama, H., and Matsueda, E. (1968). *J. Am. Chem. Soc.* **90,** 4490–4491.
Munekata, E., Faulstich, H., and Wieland, T. (1977). *Liebigs Ann. Chem.* **1977,** 1758–1765.
Muramatsu, I., and Hagitani, A. (1959). *J. Chem. Soc. Jpn. Pure Chem. Sect.* **80,** 1497–1497.
Nefkens, G. H. L., and Tesser, G. I. (1961). *J. Am. Chem. Soc.* **83,** 1263–1263.
Neurath, H. (1985). *Fed. Proc.* **44,** 2907–2913.
Pauling, L., Corey, R. B., and Branson, H. R. (1951). *Proc Natl. Acad. Sci. U.S.A.* **37,** 205–211.
Perutz, M. F., Rossmann, M. G., Cullis, A. F., Muirhead, H., Will, G., and North, A. C. T. (1960). *Nature (London)* **185,** 416–422.
Plattner, P. A., and Nager, U. (1948). *Helv. Chim. Acta* **31,** 665–671.
Radenhausen, R. (1895). *J. Prakt. Chem.* **52,** 4452–4459.
Rudinger, J., and Jost, K. (1964). *Experientia* **20,** 570–571.
Ruhemann, S. (1910). *J. Chem. Soc.* **99,** 2025–2031.
Ryle, A. P., Sanger, F., Smith, L. F., and Kitai, R. (1955). *Biochem. J.* **60,** 541–556.
Sakakibara, S., Shimonishi, Y., Kishida, Y., and Okada, M. (1967). *In* "Peptides 1966" (H. C. Beyerman, A. van de Linde, and A. Maassen van den Brink, eds.), pp. 44–49 (Proc. 8th Eur. Pept. Symp.). North-Holland Publ., Amsterdam.
Sanger, F. (1945). *Biochem. J.* **39,** 507–515.
Sanger, F. (1959). *Science* **129,** 1340–1344.
Sanger, F., and Tuppy, H. (1951). *Biochem. J.* **49,** 463–481.
Sanger, F., and Thompson, E. O. P. (1953). *Biochem. J.* **53,** 353–366.
Sarges, R., and Witkop, B. (1965). *Biochemistry* **4,** 2491–2494.
Schwyzer, R., and Sieber, P. (1957). *Helv. Chim. Acta* **40,** 624–639.
Schwyzer, R., and Sieber, P. (1963). *Nature (London)* **199,** 172–174.
Schwyzer, R., Feurer, M., and Iselin, B. (1955). *Helv. Chim. Acta* **38,** 83–98, and references cited therein.
Sela, M., and Katchalski, E. (1959). *Adv. Protein Chem.* **14,** 392–477.
Sheehan, J. C., and Frank, V. S. (1949). *J. Am. Chem. Soc.* **71,** 1856–1861.
Sheehan, J. C., and Frank, V. S. (1950). *J. Am. Chem. Soc.* **72,** 1312–1316.
Sheehan, J. C., and Hess, G. P. (1955). *J. Am. Chem. Ges.* **77,** 1067–1068.
Shemyakin, M. M., Ovchinnikov, Yu. A., Ivanov, V. T., and Kiryushkin, A. A. (1963a). *Tetrahedron* **19,** 581–591.
Shemyakin, M. M., Aldanova, N. A., Vinogradova, E. I., and Feigina, M. V. (1963b). *Tetrahedron Lett.* **1963,** 1921–1925.
Sieber, P., and Iselin, B. (1968). *Helv. Chim. Acta* **51,** 614–621.
Sieber, P., Kamber, B., Hartmann, A., Johl, A., Riniker, B., and Rittel, W. (1974). *Helv. Chim. Acta* **57,** 2617–2621.
Sifferd, R. H., and du Vigneaud, V. (1935). *J. Biol. Chem.* **108,** 753–761.
Sigmund, F., and Wessely, F. (1926). *Hoppe-Seyler's Z. Physiol. Chem.* **157,** 91–97.
Smyth, D. G., Stein, W. H., and Moore, S. (1963). *J. Biol. Chem.* **238,** 227–234.
Spackman, D. H., Stein, W. H., and Moore, S. (1958). *Anal. Chem.* **30,** 1190–1206.
Staab, H. A. (1957). *Liebigs Ann. Chem.* **609,** 75–83.
Stahl, E. (1962). "Dünnschichtchromatographie." Springer-Verlag, Berlin, New York, and Heidelberg.
Stein, S., Böhlen, P., Stone, J., Dairman, W., and Udenfriend, S. (1973). *Arch. Biochem. Biophys.* **155,** 203–212.
Stoddard, B. L., Strong, R. K., Arrot, A., and Farber, K. (1992). *Nature (London)* **360,** 293–294.
Ugi, I. (1962). *Angew. Chem.* **74,** 9–22.
Vaughan, J. R., Jr., and Osato, R. L. (1951). *J. Am. Chem. Soc.* **73,** 5553–5555.
Vickery, H. B. (1972). *Adv. Protein Chem.* **26,** 82–113.
Vickery, H. B., and Schmidt, C. L. A. (1931). *Chem. Rev.* **9,** 169–318.
du Vigneaud, V., Audrieth, L. F., and Loring, H. S. (1930). *J. Am. Chem. Soc.* **52,** 4500–4504.

du Vigneaud, V., Ressler, C., and Trippet, S. (1953a). *J. Biol. Chem.* **205,** 949–957.

du Vigneaud, V., Ressler, C., Swan, J. M., Roberts, C. W., Katsoyannis, P. G., and Gordon, S. (1953b). *J. Am. Chem. Soc.* **75,** 4879–4880.

du Vigneaud, V., Ressler, C., Swan, J. M., Roberts, C. W., and Katsoyannis, P. G. (1954a). *J. Am. Chem. Soc.* **76,** 3115–3121.

du Vigneaud, V., Gish, D. T., and Katsoyannis, P. G. (1954b). *J. Am. Chem. Soc.* **76,** 4751–4752.

Virtanen, A. I. (1951). *Makromol. Chem.* **6,** 94–103.

Wasteneys, H., and Borsook, H. (1930). *Physiol. Rev.* **10,** 110–145.

Wessely, F., and Komm, E. (1927). *Hoppe-Seyler's Z. Physiol. Chem.* **174,** 306–318.

Weygand, F., and Csendes, E. (1952). *Angew. Chem.* **64,** 136–136.

Wieland, T. (1949). *Fortschritte Chem. Forsch.* **1,** 212–291.

Wieland, T. (1959). *In* "Electrophoresis, Theory, Methods and Application" (M. Bier, ed.), pp. 493–530. Academic Press, New York.

Wieland, T. (1986). *In* "Peptides of Poisonous Amanita Mushrooms" (A. Rich, ed.), p. 58. Springer-Verlag, Berlin.

Wieland, T. (1987). *Naturwissenschaften* **74,** 367–373.

Wieland, T. (1988). *In* "The Roots of Modern Biochemistry" (H. Kleinkanf, H. von Dohren, and L. Jaenicke, eds.), p. 213. de Gruyter, Berlin.

Wieland, T., and Bernhard, H. (1951). *Liebigs Ann. Chem.* **572,** 190–194.

Wieland, T., and Birr, C. (1967). *In* "Peptides 1966" (H. C. Beyerman, A. van de Linde, and A. Maassen van den Brink, eds.), (Proc. 8th Eur. Pept. Symp.). North-Holland Publ., Amsterdam.

Wieland, T., and Bodanszky, M. (1991). "The World of Peptides: A Brief History of Peptide Chemistry." Springer-Verlag, Berlin.

Wieland, T., and Determann, H. (1963). *Angew. Chem.* **75,** 539–551.

Wieland, T., and Faulstich, H. (1991). *Experientia* **47,** 1186–1191.

Wieland, T., and Fischer, E. (1948). *Naturwissenschaften* **35,** 29–30.

Wieland, T., and Schäfer, W. (1951). *Angew. Chem.* **63,** 146–147.

Wieland, T., and Schneider, G. (1953). *Liebigs Ann. Chem.* **580,** 159–168.

Wieland, T., and Seeliger, A. (1971). *Chem. Ber.* **104,** 3992–3994.

Wieland, T., and Sehring, R. (1950). *Liebigs Ann. Chem.* **569,** 122–129.

Wieland, T., Kern W., and Shering, R. (1950). *Liebigs Ann. Chem.* **569,** 117–121.

Wieland, T., Schäfer, W., and Bokelmann, E. (1951). *Liebigs Ann. Chem.* **573,** 99–104.

Wieland, T., Bokelmann, E., Bauer, L., Lang, H. U., and Lau, H. (1953). *Liebigs Ann. Chem.* **583,** 129–149.

Wieland, T., Lüben, G., Ottenheym, H. C. J., Faesel, J., de Vries, J. X., Konz, W., Prox, A., and Schmid, J. (1968). *Angew. Chem. Int. Ed. Engl.* **7,** 204–208.

Wieland, T., Flor, F., and Birr, C. (1973). *Liebigs Ann. Chem.* **1973,** 1595–1600.

Woodward, R. B., and Schramm, C. H. (1947). *J. Am. Chem. Soc.* **69,** 1551–1552.

Woodward, R. B., Olofson, R. A., and Mayer, H. (1961). *J. Am. Chem. Soc.* **83,** 1010–1012.

Yajima, H., and Fujii, N. (1981). *In* "Chemical Synthesis and Sequencing of Peptides and Proteins" (Y. Liu, A. N. Schlechter, R. L. Henrikson, and P. G. Condliffe, eds.), pp. 21–39. Elsevier North-Holland Publ., New York.

Zervas, L. (1931). *Ber. Dtsch. Chem. Ges.* **64,** 2289–2296.

Zetzsche, F., and Lindlar, H. (1938). *Ber. Dtsch. Chem. Ges.* **71,** 2095–2102.

2

Amide Formation, Deprotection, and Disulfide Formation in Peptide Synthesis

Yoshiaki Kiso
Department of Medicinal Chemistry
Kyoto Pharmaceutical University
Yamashina-ku, Kyoto 607, Japan

Haruaki Yajima
Niigata College of Pharmacy
Niigata 950-21, Japan

I. Amide-Forming Reactions

The amide-forming reaction, or coupling reaction, is one of the most important steps in peptide synthesis. Because harsh chemical conditions can damage complex peptides, appropriate **activation** is required to perform the coupling under mild conditions. Theoretically, two types of reactions are available, namely, amino activation and carboxyl activation. The former methods, however, have rarely been applied to practical peptide synthesis because of the drastic reaction conditions and racemization. Carboxyl activation has been used in almost all the more recent peptide couplings. The general mechanism of carboxyl activation is illustrated in Scheme 1, where X represents any kind of activating group or atom.

Scheme 1 Peptide bond formation by carboxyl activation.

There are two possible synthetic strategies: **stepwise elongation** and **fragment condensation.** A synthetic scheme which employs either or both methods is elaborated according to the size and nature of the peptide to be synthesized. In stepwise elongation, each amino acid bearing an N^{α}-amino protecting group is activated and coupled to the carboxyl-terminal amino acid one by one. This method is usually applied to the synthesis of relatively small peptides or peptide fragments of a larger peptide. Fragment condensation is used to construct a larger peptide from two peptide fragments. The method, however, is often accompanied by racemization unless appropriate care is taken. Antonovics and Young (1967) studied in detail the coupling reactions which proceeded via azlactone (oxazolone) formation after carboxyl activation (Scheme 2). From this result, it is clear that the carboxyl activation on an amino acid residue other than glycine or proline produces a racemized peptide unless means to inhibit racemization are provided.

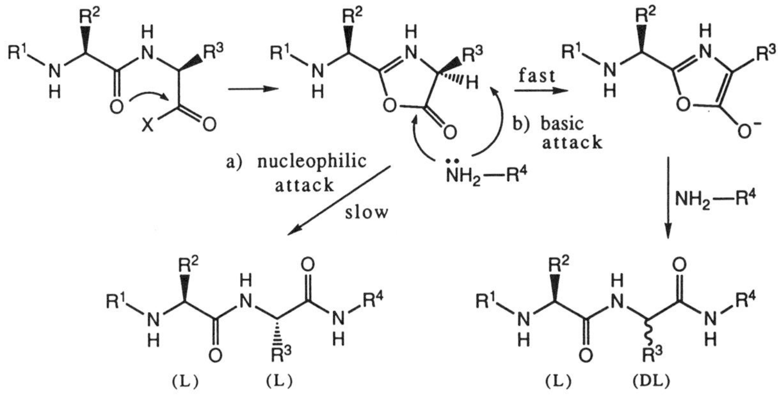

Scheme 2 Racemization mechanism during fragment condensation.

Stepwise elongation and fragment condensation methods have been developed to circumvent the inevitable racemization. The latter methods, including the useful azide method, require distinctive practical considerations.

A. Acid Chloride and Fluoride Methods

The acid chloride method was first introduced by E. Fischer (1903). Most acid chlorides are prepared by treating amino acids with PCl_5, PCl_3, $SOCl_2$ (Pizey, 1974; Matsuda *et al.*, 1985), or $(COCl)_2$. Any side-chain hydroxyl, thiol, as well as amino groups should be protected during coupling. The carboxyl group, however, is activated so strongly that some side reactions limit the application of this method.

Acid chloride methods have been used efficiently in coupling of amino acids bearing N^{α}-protecting groups such as 9-fluorenylmethoxycarbonyl (Fmoc; Carpino *et al.*, 1986) and 2-(triphenylphosphino)ethyloxycarbonyl (Peoc; Kunz and Bechtolsheimer, 1982) which are stable under acidic conditions and cleavable under alkaline conditions. The stability of the protecting groups to acidic conditions made it possible to survive conversion to an acid chloride.

Amino acid fluorides used for solution and solid-phase syntheses are obtained by treatment with cyanuric fluoride (Bertho *et al.*, 1991; Carpino *et al.*, 1991) under mild conditions.

B. Active Ester Methods

The characteristics of the active ester method are as follows: (i) Two pathways of aminolysis are possible in the mixed anhydride method, whereas only one is found in the active ester method. (ii) The active ester can be stored. (iii) The ester can be coupled with an amine component that has free carboxyl groups. (iv) The method is especially useful for the synthesis of glutamine- and/or asparagine-containing peptides (Bodanszky and du Vigneaud, 1959). The side product generated on carboxyl activation, a nitrile derivative, can be eliminated by purification.

Active esters are usually prepared by the dicyclohexylcarbodiimide (DCC) or mixed anhydride method from N-protected amino acids and alcohol. Transesterification methods using trifluoroacetate (Sakakibara and Inukai, 1965; Rzeszotarska and Vlasov, 1967) and trichloroacetate (Fujino and Hatanaka, 1968) have also been reported.

1. Phenyl Esters

Generally, phenols which have electron-withdrawing substituent in the ortho or para position can be used as active esters (Wieland and Jaenicke, 1956; Wieland *et al.*, 1962). In practice, however, *p*-nitrophenol (Bodanszky, 1955; Bodanszky *et al.*, 1957), 2,4,5-trichlorophenol (Pless and Boissonnas, 1963a,b,c; Guttmann *et al.*, 1965; Strumer *et al.*, 1965; Huguenin, 1964), pentachlorophenol (Kovacs and Ceprini, 1965; Kovacs *et al.*, 1967a; Kupryszewski and Formela, 1961), and pentafluorophenol (Kovacs *et al.*, 1967b) are widely used because of their easy availability.

After the coupling reaction with a phenol-type active ester, the liberated acidic phenol must be removed by an appropriate method such as washing with aqueous sodium carbonate or recrystallization. If the desired peptide product is an oil, the phenol cannot be removed completely. Water-soluble active esters, such as 4-hydroxyphenyldimethylsulfonium methylsulfate (Kouge *et al.*, 1986; 1987), overcome these difficulties not only by allowing easy removal of the by-products, but also by undergoing the reaction in aqueous solvents. It is noteworthy that an active ester acylates the phenolic hydroxyl group on tyrosine in the presence of a base (Agarwal *et al.*, 1969; Ramachandran and Li, 1963).

2. *N-Hydroxylamine Esters*

A series of *N*-hydroxylamine-type active esters represented by *N*-hydroxysuccinimide (HOSu) ester (Anderson *et al.*, 1963; 1964) have been described. These types of active esters are regarded as bifunctional active esters which rarely induce racemization of the α-carbon atom. In contrast to the above phenol esters, these esters eliminate *N*-hydroxylamine after coupling, which is easily soluble in water and thus removable by simple washing.

HOSu esters are prepared by coupling N-protected amino acids and HOSu by DCC. An undesired side reaction was reported to take place during the preparation (Gross and Bilk, 1968; Weygand *et al.*, 1968): the binding of DCC and HOSu produces β-alanine via Lossen rearrangement (Scheme 3). *N*-Hydroxy-5-norbornene-2,3-dicarboximide (HONB, **1** in Fig. 1) introduced by Fujino *et al.* (1974), rarely undergoes such side reactions because of its rigid structure.

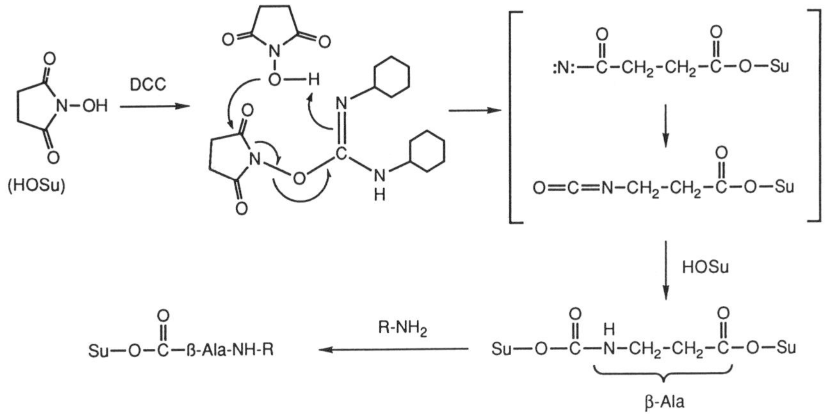

Scheme 3 Side reaction on reacting HOSu with DCC.

N-Hydroxybenzotriazole (HOBt, **2** in Fig. 1; König and Geiger, 1970a,b) esters are widely used in peptide synthesis and are prepared using DCC. The racemization-suppressing effect and coupling accelerating effect of HOBt are excellent. Many groups have reported the reaction mechanism involving HOBt ester (Horiki, 1977; König and Geiger, 1973). The effects of HOBt as an additive are described in a later section (D.1.). 3-Hydroxy-4-oxo-3,4-dihydro-1,2,3-benzotriazine (HOOBt, **3** in Fig. 1; Konig and Geiger, 1970c) is a reagent superior to HOBt with regard to racemization. However, HOOBt esters have the disadvantage of undergoing Lossen rearrangement.

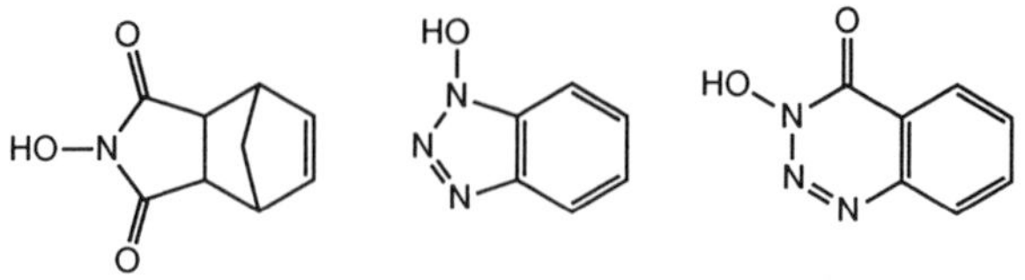

(1) (2) (3)

Figure 1 Structures of HONB (**1**), HOBt (**2**), and HOObt (**3**).

3. Bifunctional Active Esters

Various bifunctional active esters such as hydroxyquinoline (HOQ, **4** in Fig. 2; Jakubke, 1965; Jakubke and Voigt, 1966a,b), catechol (**5** in Fig. 2, Lloyd and Young, 1968, 1971), and hydroxypyridine (**6** in Fig. 2; Taschner *et al*., 1965a,b), have been developed as racemization-free reagents for peptide fragment coupling. They can, of course, be applied to coupling by stepwise elongation and are readily removable after the reaction, taking advantage of their bifunctionality.

C. Unsymmetrical Anhydride Methods

1. Mixed Anhydride Method

The mixed anhydride method of peptide bond formation involves aminolysis of an anhydride consisting of an N-protected amino acid and another acid (Albertson, 1962; Meienhofer, 1979). The carboxylic–carboxylic mixed anhydride method and carbonic–carboxylic mixed anhydride methods were developed. Isobutychloroformate (Vaughan and Osato, 1951) is commonly used for the preparation of carbonic–carboxylic mixed anhydrides (Scheme 4).

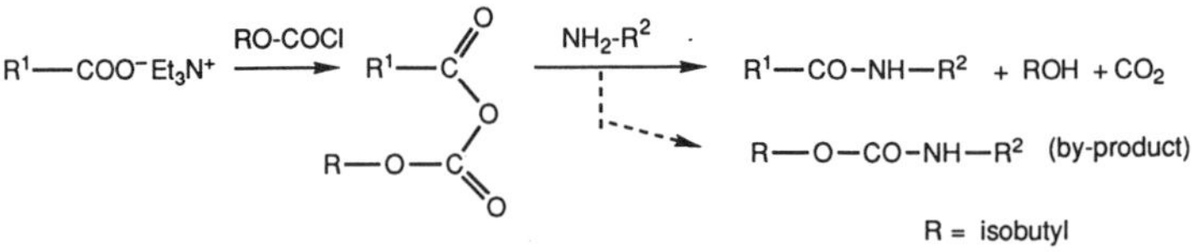

R^1—COO$^-$Et$_3$N$^+$ —(RO-COCl)→ R^1—C(=O)—O—C(=O)—O—R —(NH_2-R^2)→ R^1—CO-NH—R^2 + ROH + CO_2

⇢ R—O—CO-NH—R^2 (by-product)

R = isobutyl

Scheme 4 Mixed anhydride method using isobutylchloroformate.

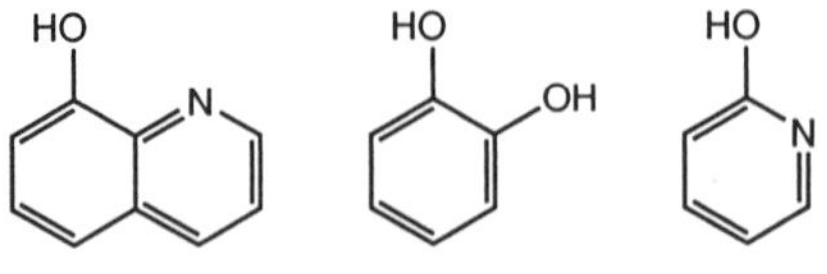

(4) (5) (6)

Figure 2 Structures of HOQ (**4**), catechol (**5**), and hydroxypyridine (**6**).

In mixed anhydride methods, the amide-forming reaction is complete within several minutes in an ice bath. The side-chain functional groups in the carboxyl component should be protected. The carboxyl groups in the amine component could be either in ester form or protected tentatively with a tertiary amine. In the latter case, the reaction can be performed in water-containing solvents, but the product should be carefully separated because of its acidity.

As a side reaction in mixed anhydride methods, urethane formation occurs in the amine component (Albertson, 1962; Wieland *et al.*, 1962; Wieland and Stimming, 1953), but it can be neglected under usual experimental conditions. Addition of HOBt to the reaction mixture before the amine component is added reduces urethane formation (Prasad *et al.*, 1985). Also, when Z-Gly (where Z represents the benzyloxycarbonyl group) is used as the carboxyl component, attention must be paid to the spontaneous formation of acylimide (Kopple and Renik, 1958; Schellenberg and Ullrich, 1959).

Mixed anhydrides are also obtained by the reaction of amino acids and dihydroquinoline derivatives. On reaction with a carboxyl component, 1-ethoxycarbonyl-2-ethoxy-1,2-dihydroquinoline (EEDQ; Belleau and Malek, 1968) or 1-isobutyloxycarbonyl-2-isobutyloxy-1,2-dihydroquinoline (IIDQ; Kiso and Yajima, 1972; Kiso *et al.*, 1973) is converted via a six-membered ring intermediate to quinoline and a carbonic–carboxylic mixed anhydride which immediately reacts with an amine component to give a peptide in good yield (Scheme 5). The by-products are carbon dioxide and quinoline as long as undesired urethane formation is negligible.

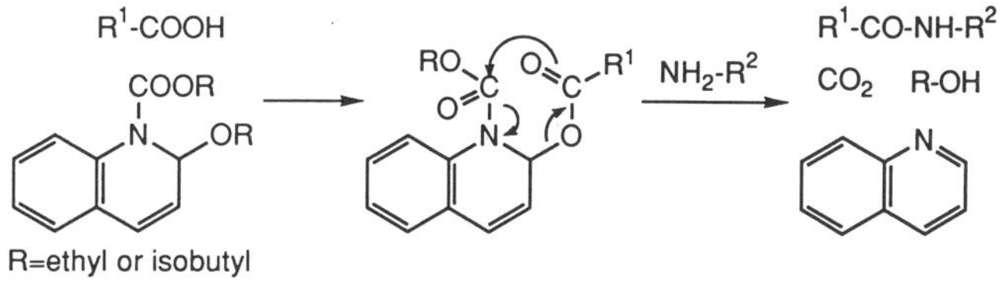

R^1-COOH

COOR

N OR

R=ethyl or isobutyl

RO C O C R^1 O N O

NH_2-R^2

R^1-CO-NH-R^2

CO_2 R-OH

N

Scheme 5 Mixed anhydride method using EEDQ or IIDQ.

2. *Phosphoric Mixed Anhydride Methods*

The mixed anhydrides of amino acid and phosphoric acid are also useful for formation of the peptide bond. Many kinds of phosphoric reagents, for example, $(PhO)_2P(O)Cl$ (Albertson, 1962; Meienhofer, 1979) and $Me_2P(S)Cl$ (Ueki *et al.*, 1979; 1988; Ueki and Inazu, 1982), have been developed. In this category, 3,3′-(chlorophosphoryl)bis(1,3-oxazolidin-2-one) (BOP-Cl, **7** in Fig. 3; Meseguer *et al.*, 1980; Cabre and Palomo, 1984; Omodei-Sale *et al.*, 1984) is particularly useful for coupling of imino acids (Van Der Auwera and Anteunis, 1987; Tung *et al.*, 1986) such as *N*-methyl amino acids (Tung and Rich, 1985) because of the strong reactivity and selectivity toward the amine component.

Norborn-5-ene-2,3-dicarboximido diphenyl phosphate (NDPP, **8** in Fig. 3) is a reagent for the so-called active ester-type mixed anhydride method (Kiso *et*

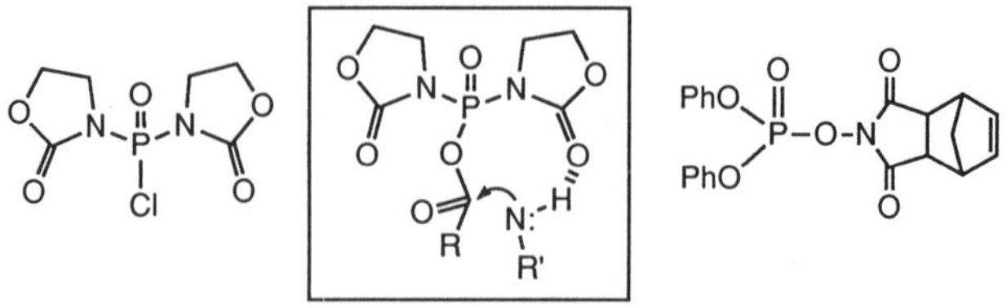

(7) (8)

Figure 3 Structures of BOP-Cl (**7**) and NDPP (**8**).

al., 1980a). In the reaction of NDPP and an amino acid, the HONB ester is formed via a carboxylic–phosphoric mixed anhydride intermediate (Fig. 3).

D. Carbodiimide Methods

N,N′-Dicyclohexylcarbodiimide (DCC) has been widely used (Rich and Singh, 1979; Williams and Ibrahim, 1981) since Sheehan and Hess (1955) first applied it to peptide synthesis. The reaction mechanism involving DCC is illustrated in Scheme 6 (Schüssler and Zahn, 1962; Detar *et al.*, 1966). Briefly, an activated intermediate, *O*-acylisourea, which is produced by reaction of a protected amino acid and DCC, undergoes aminolysis to form a peptide and dicyclohexylurea (DCU). The reaction of an N-protected amino acid and DCC without an amine component gives another activated species, a symmetrical anhydride, which also reacts with amine components. The DCU produced in this reaction is insoluble in almost all common organic solvents and is thus easily removed from the reaction mixture by filtration.

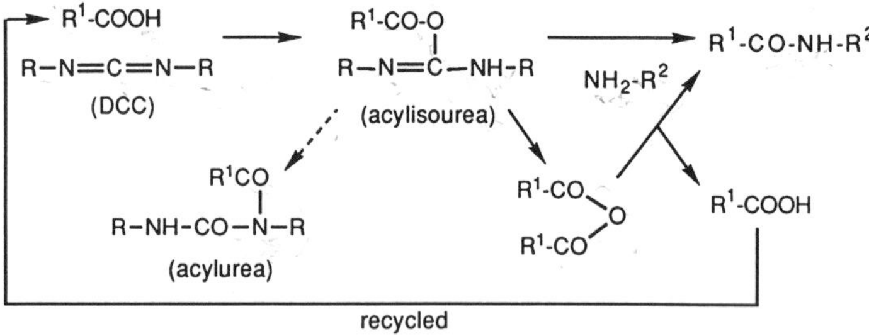

Scheme 6 Reaction mechanism of DCC.

The noticeable side reaction is the formation of an acylurea by intramolecular rearrangement of the acylisourea. The extent of the side reaction is enhanced when the proton concentration of N-protected amino acid is decreased by excess triethylamine (which is usually used for neutralization of chloride ions bound to the amine component). The side reaction is profoundly influenced by the choice of solvent as well as the base (Sheehan *et al.*, 1956). In the fragment condensation by DCC, this side reaction has to be considered seriously, but in the coupling of N-protected amino acids, it can be ignored because of its easy removal by recrystallization. Dehydration of the side-chain amide group of glutamine or asparagine also occurs during DCC activation.

To simplify the removal of DCU and unreacted carbodiimide, water-soluble carbodiimides such as 1-ethyl-3-(3′-dimethylaminopropyl)carbodiimide were developed; in such systems, both the reagent and the resulting urea derivative are soluble and easily removable by washing (Sheehan *et al.*, 1961). Diisopropylcarbodiimide (DIPCDI) is a useful reagent in solid-phase peptide synthesis (SPPS) because the diimide and the corresponding urea are soluble in organic solvents (Sheehan and Hess, 1955; Sheehan and Yang, 1958).

1. DCC-Additive Methods

Wünsch's group found that fragment condensation using DCC and HOSu under strictly controlled conditions preserves the chiral purity almost completely, with less than 1% racemization being reported (Weygand *et al.*, 1966; Wünsch and Drees, 1966). The method was applied to the synthesis of glucagon to obtain an active peptide (Wünsch, 1967; Wünsch and Wendlberger, 1968; Wünsch *et al.*, 1968). Strictly speaking, however, the synthesis was carried out with suppression of racemization, and an accurate racemization test is required to judge whether the product is absolutely optically pure.

The disadvantage of the DCC–HOSu method is the side reaction, Lossen rearrangement, which gives a β-alanine derivative by reaction with amine components (Gross and Bilk, 1968; Weygand *et al.*, 1968). To remedy this, König and Geiger (1970a,b,c) examined various analogous reagents and reported DCC–HOBt as the most suitable. This reagent suppresses not only racemization but also acylurea formation in fragment condensation. In these syntheses, use of water-soluble carbodiimide instead of DCC was recommended (Kimura *et al.*, 1981, 1988), as the urea or acylureas formed are more easily removed from the reaction products.

These methods are also useful for the coupling of amino acids with respect to increases in yield and suppression of side reactions. The suppression of racemization by this methodology has been reported by many researchers using various additives (Yajima *et al.*, 1973a,b; Fujino, 1976). However, the racemization test is carried out in a model system, and the results do not always agree with practical synthesis.

2. Fragment Condensation at Glycine or Proline Residues

Activation of glycine residues does not cause racemization because glycine has no chiral center, nor does activation of proline residues because proline has no imino hydrogen which would participate in azlactone formation. Therefore, whatever kind of coupling method is used, fragment condensation by activating the carboxyl group of glycine or proline never fails to preserve chiral purity of the peptide, and these tactics are commonly used in synthetic schemes for larger peptides. When DCC is used as a coupling reagent in fragment condensation, however, acylurea formation becomes a problem. The removal of acylurea is fairly difficult. The addition of racemization-suppressing reagents such as *N*-hydroxysuccinimide also suppresses acylurea formation.

More recently, coupling reagents such as benzotriazole-1-yloxytris(dimethylamino)phosphonium hexafluorophosphate (BOP) have been employed instead of DCC (Castro *et al.*, 1975, 1977). The strong reactivity of the coupling reagents is advantageous for fragment condensation, since the reaction rate is slower than that of amino acid coupling. In addition, 4-(dimethylamino)pyridine can be used as a catalyst because of the lack of the problem of racemization at Gly or Pro residues.

E. Azide Methods

The azide method, first developed by Curtius (1902, 1904), is currently used not only for the stepwise coupling of troublesome amino acids such as serine or histidine, but also for fragment condensation, where it shows real merit. It is the most suitable method for fragment condensation because little racemization occurs, and side-chain protection can be kept at a minimum (only the ϵ-amino group of lysine and the thiol group of cysteine have to be protected) because azide is less reactive with side-chain functional groups such as hydroxyl groups.

First, the hydrazide is prepared by treating an N-protected amino acid or peptide ester with excess hydrazine hydrate. The hydrazide is converted to azide by *tert*-butyl or isoamyl nitrite (Rudinger's variation; Honzl and Rudinger, 1961), and the azide reacts with the amine component without extraction with organic solvent. Because the azide is unstable at room temperature, the reaction with the amine component is carried out below 0°C for 40 hr after neutralization of hydrogen azide with a base such as triethylamine or *N*-methylmorpholine.

The resistance to racemization in this method is explained by the electron distribution of the N_3 group, which reduces the tendency of the R−CO−NH group to participate in azlactone formation (Scheme 7). However, using an isotope dilution assay, Kemp *et al.* (1970) detected racemization even in the azide method.

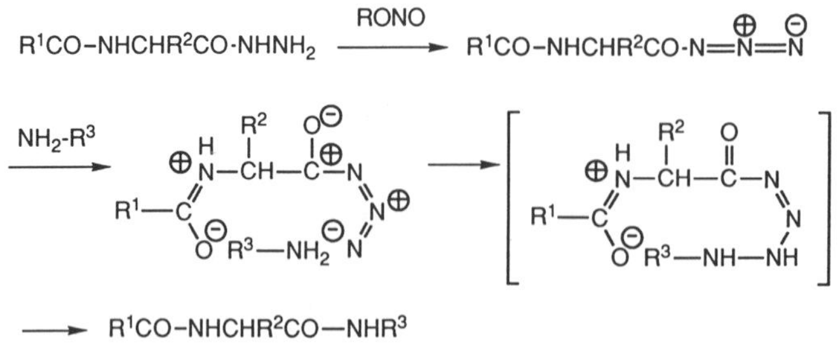

$$R^1CO\text{-}NHCHR^2CO\text{-}NHNH_2 \xrightarrow{RONO} R^1CO\text{-}NHCHR^2CO\text{-}N{=}\overset{\oplus}{N}{=}\overset{\ominus}{N}$$

$$\xrightarrow{NH_2\text{-}R^3} \text{[cyclic intermediate: } R^1, R^2, R^3\text{–}NH_2\text{]} \longrightarrow \left[\text{cyclic intermediate: } R^1, R^2, R^3\text{–NH–NH}\right]$$

$$\longrightarrow R^1CO\text{-}NHCHR^2CO\text{—}NHR^3$$

Scheme 7 Suppression of racemization of azide.

The side reactions that generally occur in the azide procedure are as follows: (i) Curtius rearrangement at higher temperature, producing a urea derivative or an amine (Hofmann *et al.*, 1960) (Scheme 8), and (ii) hydrolysis of azide, giving an amide and nitric oxide (Honzl and Rudinger, 1961). In addition, side reactions which involve particular amino acids have also been reported (Gregory *et al.*, 1968). In conclusion, the azide procedure is an excellent coupling method; however, some side reactions have been reported, and the procedure is not simple.

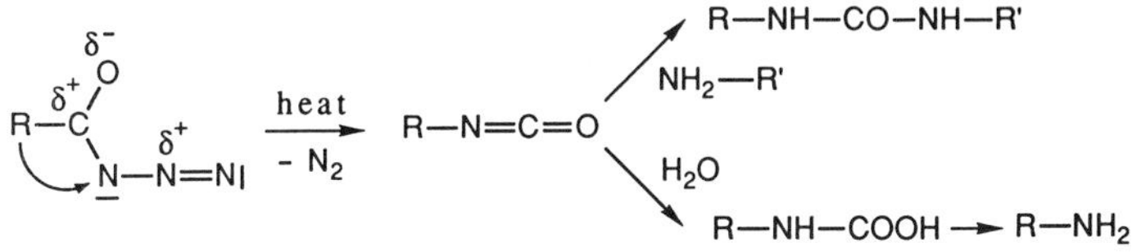

$$\text{R-C}^{\delta+}(\text{O}^{\delta-})\text{-N-N}^{\delta+}\text{=N} \xrightarrow[\ -\ N_2]{\text{heat}} \text{R-N=C=O} \begin{cases} \xrightarrow{NH_2\text{-R'}} \text{R-NH-CO-NH-R'} \\ \xrightarrow{H_2O} \text{R-NH-COOH} \rightarrow \text{R-NH}_2 \end{cases}$$

Scheme 8 Curtius rearrangement of azide.

1. Protected Hydrazide Methods

Peptide esters containing Arg(NO_2), Asp(O-*t*-Bu), Asp(OBzl), or Glu(OBzl) (where *t*-Bu and Bzl represent *tert*-butyl and benzyl, respectively) cannot be converted to the corresponding hydrazide by hydrazinolysis. In such cases, a protected hydrazide is introduced at the C terminus of a peptide or carboxyl group of an amino acid. After elongation of the peptide chain, the substituent on the hydrazide is removed, and the resulting hydrazide is converted to an azide.

Various protected hydrazides have been reported. Naturally, the substituent chosen has to have a selectivity of removal different from that of the α-amino protecting group. The usual protecting groups for hydrazine are Z (Hofmann *et al.*, 1950, 1952), *tert*-butoxycarbonyl (Boc; Boissonnas *et al.*, 1960; Schwyzer *et al.*, 1960), 2,2,2-trichloroethyloxycarbonyl (Troc; Yajima and Kiso, 1971; Watanabe *et al.*, 1974), and 2,2,2-trichloro-*tert*-butoxycarbonyl (Tcboc; Shimokura *et al.*, 1985).

2. Azide Method Using Diphenylphosphorylazide

Azides have long been obtained exclusively via hydrazides since their first preparation by Curtius. Shioiri *et al.* (1972), however, devised an interesting alternative to the hydrazide-mediated azide method, and reported that

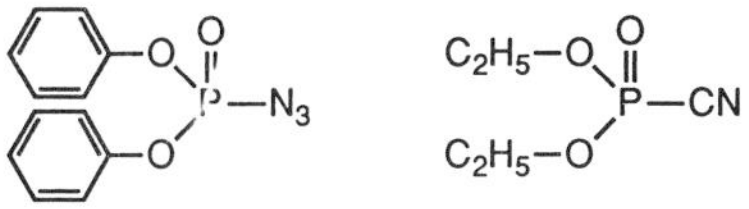

$$(C_6H_5O)_2P(=O)\text{-}N_3 \qquad (C_2H_5O)_2P(=O)\text{-CN}$$

(9) **(10)**

Figure 4 Structures of DPPA (**9**) and DEPC (**10**).

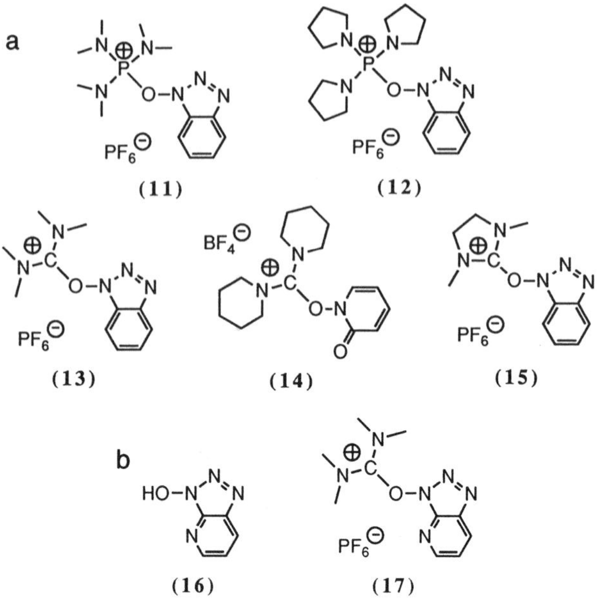

Figure 5 (a) Structures of BOP (**11**), PyBop (**12**), HBTU (**13**), TOPPipU (**14**), and BOI (**15**). (b) Structures of HOAt (**16**) and HATU (**17**).

diphenylphosphorylazide (DPPA, **9** in Fig. 4) directly converts a free carboxyl group to an azide (Shioiri and Yamada, 1974). The coupling reaction is considered to proceed via an azide rather than an acyloxyphosphoric mixed anhydride, which is also a likely intermediate. A similar reagent, diethylphosphorocyanidate (DEPC, **10** in Fig. 4) was also reported (Yamada *et al.*, 1973; Shioiri *et al.*, 1976a,b). The advantages of the reaction using DPPA or DEPC are no formation of insoluble by-products and less racemization in fragment condensation.

F. Phosphonium and Uronium Salt Type Coupling Reagents

Very high performance of coupling reagents has been found to be necessary, especially in the field of SPPS. Benzotriazole-1-yloyxtris(dimethylamino) phosphonium hexafluorophosphate (BOP, **11** in Fig. 5a), reported by Castro *et al.* (1975, 1977), is an excellent coupling reagent offering high reactivity and easy handling. The reaction between BOP and an N-protected amino acid proceeds via an acyloxyphosphonium as an intermediate to produce the corresponding benzotriazole ester and easily removable by-products such as hexamethylphosphoric triamide (HMPA) and salts (Scheme 9a). Under basic conditions, no side reaction accompanies the main reaction, in contrast to the use of DCC, and the reaction rate is generally very fast. For example, in the case of coupling of Boc-Ile with Val-OBzl, the reaction is completed within a few minutes. BOP is used in both solution and solid-phase syntheses.

However, HMPA, the starting material in the synthesis and a by-product in the reaction of BOP, has been reported to have respiratory toxicity. It is therefore recommended to replace BOP by benzotriazole-1-yloxytrispyrrolidinophosphonium hexafluorophosphate (PyBOP, **12** in Fig. 5a), which has similar reactivity to BOP and forms no carcinogenic by-products (Martinez *et al.*, 1988; Coste *et al.*, 1990a).

2-(1*H*-Benzotriazole-1-yl)-oxy-1,1,3,3-tetramethyluronium hexafluorophosphate (HBTU, **13** in Fig. 5a; Knorr *et al.*, 1989), 2-[2-oxo-1(2*H*)-pyridyl)-1,1,3,3-bispentamethyleneuronium tetrafluoroborate (TOPPipU, **14** in Fig. 5a; Henklein *et al.*, 1991), and 2-(benzotriazole-1-yl)-oxy-1,3-dimethylimidazolidinium hexafluorophosphate (BOI, **15** in Fig. 5a; Kiso *et al.*, 1992b) have similar reactivities to BOP. Phosphonium is substituted by uronium in these reagents. Carpino (1993) reported that 1-hydroxy-7-azabenzotriazole (HOAt, **16** in Fig. 5b), an additive, and 2-(1*H*-7-azabenzotriazole-1-yl)-oxy-1,1,3,3-tetramethyluronium hexafluorophosphate (HATU, **17** in Fig. 5b), a uronium-type coupling reagent containing the HOAt moiety, enhanced the coupling reaction rate and reduced the loss of chiral integrity. The nitrogen atom at the 7 position of HOAt has a neighboring group effect to accelerate the coupling reaction.

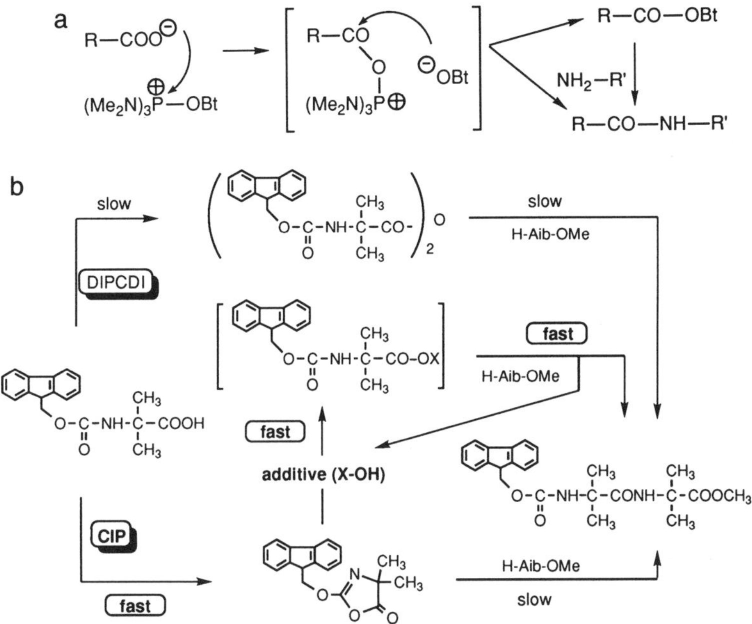

Scheme 9 (a) Reaction mechanism of BOP. (b) Reaction mechanism of CIP.

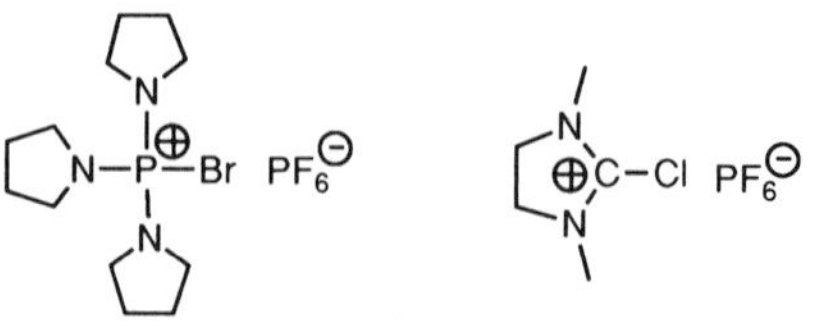

(18) (19)

Figure 6 Structures of PyBroP (**18**) and CIP (**19**).

Bromotrispyrrolidinophosphonium hexafluorophosphate (PyBroP, **18** in Fig. 6: Coste *et al.*, 1990b) and 2-chloro-1,3-dimethylimidazolidinium hexafluorophosphate (CIP, **19** in Fig. 6; Akaji *et al.*, 1992c, 1994) have been developed as coupling reagents for *N*-methylamino acids or α-dialkylamino acids and as esterification reagents (Scheme 9b).

G. Other Amide-Forming Reactions

Many reactions that form amide bonds are known besides those described above. In this section, those which are interesting from the viewpoint of organic chemistry or synthesis of large peptides are described.

Kemp *et al.* reported a series of studies on peptide synthesis referred to as the "prior thiol capture method" (Kemp and Galakatos, 1986; Kemp *et al.*, 1986; McBride and Kemp, 1987; Kemp and Fotouhi, 1987). This method consists of intermolecular disulfide formation and subsequent intramolecular acyl rearrangement (Scheme 10).

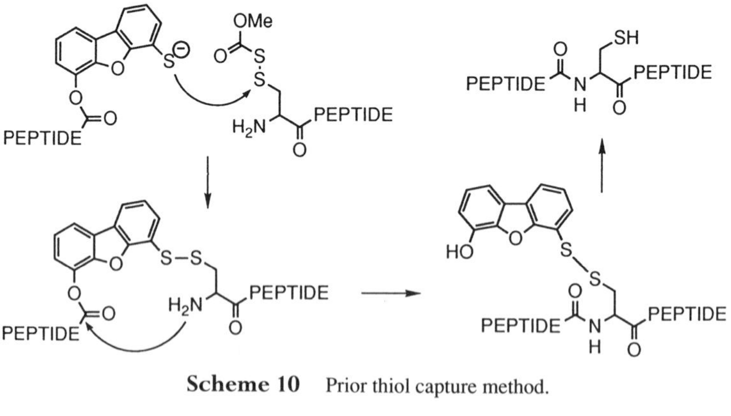

Scheme 10 Prior thiol capture method.

Blake (1981) proposed a new method for the assembly of peptide segments in aqueous medium by the use of the thiocarboxyl function. A partially protected C-terminal Gly thiocarboxyl segment is coupled with an amine component by the aid of silver ion or 2,2′-dipyridyl disulfide as an activator. In this reaction, the ε-amino group of lysine residues is protected by the citraconyl group, but other functional side-chain groups are not protected. Thiocarboxyl peptides have been synthesized by the solid-phase method starting from Boc-Gly-S-resin. This method was successfully applied to the syntheses of human β-lipotropin (Scheme 11; Blake and Li, 1983), *S*-carbamoylmethyl bovine apocytochrome *c* (Blake, 1986) and α-inhibin-92 (Yamashiro and Li, 1986). In addition, Hojo and Aimoto (1991) employed an *S*-alkyl thioester of partially protected peptide segments for the synthesis of the DNA-binding domain of c-Myb protein(142–193)-amide.

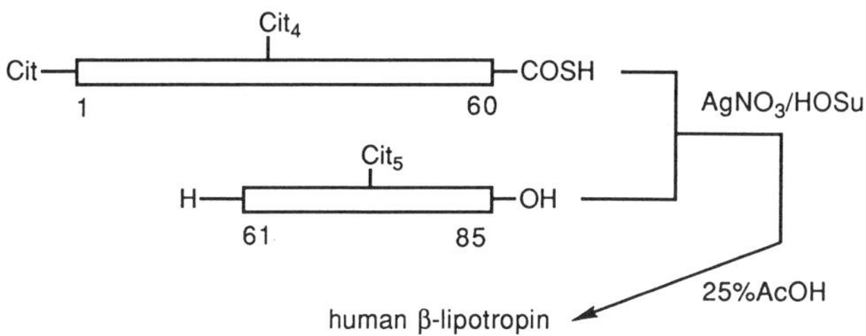

Scheme 11 Thiocarboxylic acid method.

To obtain a small protein, Schnölzer and Kent (1992) reported an interesting method using substitution of a peptide bond and ligation at that site (Scheme 12). A thioester was employed instead of the peptide bond between Gly-51 and Gly-52 of human immunodeficiency virus type 1 (HIV-1) protease. This region has been reported to bear no relation on activity. The peptide carrying C-terminal thiocarboxyl glycine and the N-bromoacetylated peptide were prepared by a solid-phase technique. Ligation of the two segments in aqueous media gave an HIV-1 protease analog which had activity similar to that of the native enzyme.

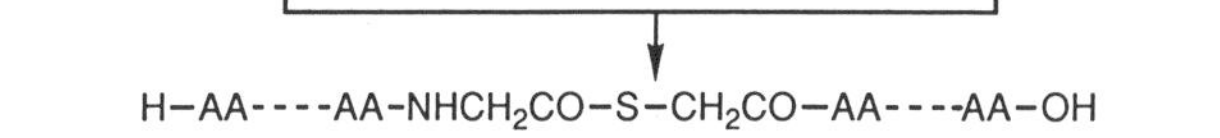

Gly-S-CH(Ph)-C$_6$H$_4$-OCH$_2$CONH—resin

1) SPPS
2) HF

H–AA- - - -AA-NHCH$_2$CO-SH

AA–OCH$_2$–Pam—resin

1) SPPS
2) HF

Br–CH$_2$CO–AA- - - -AA–OH

H–AA- - - -AA-NHCH$_2$CO–S–CH$_2$CO–AA- - - -AA–OH

Scheme 12 Chemical ligation.

II. Deprotection

Protection and deprotection steps are important in peptide synthesis. Deprotecting steps can be briefly categorized into two types: (1) deprotection of α-amino groups to give peptides for further extension and (2) final deprotection to give free peptides. The results of peptide syntheses are critically influenced by the combination of these two types of deprotection along with the selection of protecting groups for α-amino and side-chain functional groups in a so-called strategy. Most frequently, strong acids such as HF or trifluoromethanesulfonic acid (TFMSA) have been used for the final deprotection in combination with selective deprotection of temporary α-amino protecting groups such as Boc using trifluoroacetic acid (TFA). Alkaline conditions have also been used for selective deprotection of temporary α-amino protecting Fmoc groups and acidic reagents such as TFA for the final deprotection. However, the synthesis of higher peptides with various functional groups requires new and different types of protection and deprotection. In this section, the deprotection methods are described in relation to their strategies.

A. Deprotection of α-Amino Protecting Groups

Selectivity is highly important in deprotection of α-amino protecting groups. Cleavage of the side-chain protecting groups is one of the most serious problems during peptide synthesis. Such side reactions are particularly troublesome in SPPS, which is carried out without purification of the reaction intermediates. Cleavage of the protected peptides from the resin is also a serious problem in solid-phase synthesis. In other words, the side-chain protecting groups should be kept intact up to the final deprotection stage, and selectivity of the α-amino deprotecting method is crucial for peptide chain elongation. When a peptide contains sensitive amino acids such as methionine or cysteine, it becomes even harder to cleave the α-amino protecting groups selectively.

1. Mild Acids

a. Trifluoroacetic Acid Trifluoroacetic acid is one of the most commonly used reagents for cleavage of α-protecting groups such as Boc and Z(OMe). Good solubility of peptides in TFA, fewer side reactions during deprotection, and mildness make it useful not only for removal of temporary protecting groups but also for the final deprotection step. The reagent in the form of 20–60% TFA in CH_2Cl_2 is often used for removal of α-Boc in solid-phase synthesis (Gutte and Merrifield, 1971). The side reactions encountered during the deprotection are (i) partial elimination of side-chain protecting groups, (ii) pyroglutamyl formation from peptides containing glutamine at the N terminus (Takashima *et al.*, 1968), and (iii) *tert*-butylation of the indole ring in tryptophan-containing peptides (Alakhov *et al.*, 1970).

b. Hydrochloric Acid/Dioxane The HCl/dioxane reagent is prepared by introducing hydrogen chloride gas into dioxane, and it is often used for removal of acid-sensitive temporary protecting groups. In the form of 4 *N* HCl/dioxane, it is used for cleavage of Boc groups (Guttmann, 1961; Stewart and Woolley, 1965). Pyroglutamyl formation from N-terminal glutamine-containing peptides is reduced greatly (Beyerman *et al.*, 1973).

c. Dilute Methanesulfonic Acid in Dioxane–Dichloromethane This diluted methanesulfonic acid (MSA) system, which uses 0.5 *M* MSA in 1:9 (v/v) dioxane–CH_2Cl_2 (Kiso *et al.*, 1992b), is primarily used in SPPS. The advantages are as follows: (i) elimination of side-chain protecting groups is reduced compared to the conventional 45% TFA/CH_2Cl_2 method, and (ii) pyroglutamyl formation from glutamine-containing peptides is similarly decreased relative to the use of 4 *N* HCl/dioxane. Using the MSA deprotection system, Kiso *et al.* (1990a) developed an efficient method for SPPS consisting of *in situ* neutralization and the rapid coupling reaction using BOP or BOI reagent activation (Kiso *et al.*, 1990a) (Fig. 7). Porcine brain natriuretic peptide (pBNP) was synthesized successfully using this method (Kiso *et al.*, 1992b).

2. Alkaline Conditions

a. Piperidine in Dimethylformamide The combination of 20–55% piperidine in dimethylformamide (DMF) is used mainly for removal of Fmoc groups in SPPS (Atherton *et al.*, 1981; Heimer *et al.*, 1981). Diketopiperazine formation is the most serious side reaction occurring during peptide chain elongation at the second residue (Roth and Mazanek, 1974; Bray *et al.*, 1990), and it is especially serious in peptides containing proline at the C terminus.

b. Tetra-n-butylammonium Fluoride Trihydrate in Dimethylformamide Dilute (0.02 *M*) tetra-*n*-butylammonium fluoride trihydrate (TBAF) in DMF is

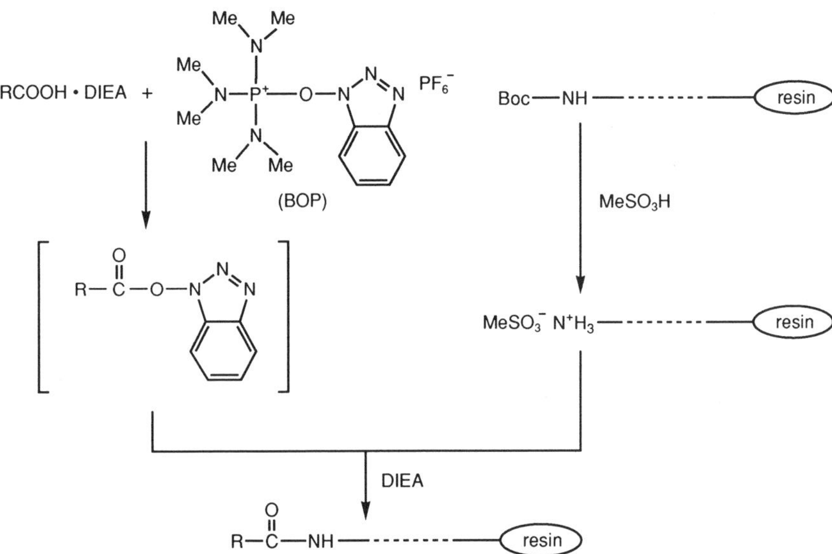

Figure 7 Coupling pathway for efficient SPPS employing *in situ* neutralization.

used for removal of Fmoc groups (Ueki and Amemiya, 1987) as exemplified in the synthesis of Leu-Ala-Gly-Val in continuous-flow system.

c. 1,8-Diazabicyclo[5.4.0]undec-7-ene in Dimethylformamide Another reagent employed for cleavage of Fmoc groups is 2% 1,8-diazabicyclo[5.4.0]undec-7-ene in DMF (Wade *et al.*, 1991). Peptides with about 20 residues have been synthesized in continuous-flow systems using this reagent.

B. Final Deprotection

The reagent for final deprotection should cleave all the protecting groups as completely as possible without side reactions. Particularly in the synthesis of higher peptides using various side-chain protecting groups, selection of appropriate protecting groups in combination with an unfailing final deprotecting reagent is extremely important.

1. *tert-Butoxycarbonyl/Benzyl*

a. Anhydrous Hydrogen Fluoride Anhydrous HF cleaves Boc or *t*-Bu groups within several minutes and Bzl-based, N^G-NO_2 and N^G-Tos groups in 30–60 min (Sakakibara *et al.*, 1967; Mazur, 1968). Owing to its ability to remove all the protecting groups, including that of cysteine, HF is widely utilized both in solid-phase and in solution synthesis. The reported side reactions are binding of anisole to glutamic acid (Sano and Kawanishi, 1975; Feinberg and Merrifield, 1975), formation of 3-benzyltyrosine (Sakakibara *et al.*, 1967: Erickson and Merrifield, 1973), and alkylation of tryptophan indole (Masui *et al.*, 1980). These side reactions can be greatly reduced, however, by adjusting the reaction conditions and by adding scavengers.

Tam *et al.* (1983a,b) introduced a two-step HF treatment referred to as the low–high HF method. 3-Benzylation of Tyr residues and succinimide formation from Asp(OBzl) residues are notably suppressed in comparison with the orthodox HF method, and reduction from Met(O) to Met and deformylation of Trp(CHO) are effectively accomplished by the low–high HF method.

b. Trifluoromethanesulfonic Acid in Trifluoroacetic Acid The 1 *M* TFMSA–thioanisole/TFA system completely cleaves almost all protecting groups currently used in peptide synthesis within 1 hr (Yajima *et al.*, 1974). Many biologically active peptides have so far been synthesized by this method, demonstrating its usefulness. The total synthesis of bovine ribonuclease A (Yajima and Fujii, 1980, 1981; Fujii and Yajima, 1981) can be regarded as a comprehensive study of this method. This facile method, which needs no special equipment, is widely applied in both solution and solid-phase syntheses.

c. Trimethylsilyl Trifluoromethanesulfonate in Trifluoroacetic Acid The 1 *M* trimethylsilyl trifluoromethanesulfonate (TMSOTf)–thioanisole/TFA system almost completely cleaves various protecting groups currently used in peptide synthesis (Fujii *et al.*, 1987b,c) in a short time. High recovery of bulky isoleucines from Boc-Ile-phenylacetamidomethyl (PAM) resin after 1 *M* TMSOTf–thioanisole/TFA treatment indicates that this is a promising final deprotecting method in solid-phase synthesis. This system has been employed in solution-phase syntheses of glucose-dependent insulinotropic polypeptide (GIP; Yajima *et al.*, 1985b, 1986), neuromedin U-25 (Fujii *et al.*, 1987a), sauvagine (Nomizu *et al.*, 1988), and human pancreatic polypeptide (Sugiyama *et al.*, 1987) as well as in the solid-phase synthesis of magainin (Matsuzaki *et al.*, 1989).

d. Tetrafluoroboric Acid in Trifluoroacetic Acid The system 1 *M* HBF_4 in TFA is a weaker acid than HBr in TFA. Kiso *et al.* (1989b) found the usefulness of this mild reagent in the presence of thioanisole (1 *M* HBF_4–thioanisole in TFA at 4°C) as a final deprotecting reagent. Together with α-amino protecting groups such as Boc or 4-methoxybenzyloxycarbonyl [Z(OMe)], the Z or 2-chlorobenzyloxycarbonyl (Cl-Z) group at the side chain of Lys and the Bzl groups at Ser, Thr, Asp, and Glu were cleaved quantitatively within 60 min. Complete removal of the Cl_2-Bzl group from Tyr, the benzyloxymethyl (Bom) group from His, and the diphenylphosphinothioyl (Ppt) group from Trp was also achieved within 60 min. However, a 60-min treatment at an elevated temperature (25°C) was necessary for complete removal of the cyclohexyl (cHex) group from Asp and the mesitylene-2-sulfonyl (Mts) group from Arg. HBF_4 in TFA also cleaves the amino acid amide from 4-methylbenzhydrylamine resin more effectively than TFMSA in TFA. Lamprey gonadotropin-releasing hormone (a 10-residue peptide amide) was synthesized using 1 *M* HBF_4–thioanisole in TFA by both solution-phase and solid-phase methods (Kiso *et al.*, 1989b). Dynorphin A was also synthesized using this deprotecting reagent in the solid phase (Kiso *et al.*, 1991a).

2. 9-Fluorenylmethoxycarbonyl/tert-Butyl

a. Trifluoroacetic Acid As mentioned above, TFA is commonly used for cleavage of α-amino protecting groups such as Boc, whereas it is used as a final deprotecting reagent in the Fmoc/*t*-Bu strategy. The most notable advantage of TFA in Fmoc/*t*-Bu strategy is the reduced incidence of side reactions such as formation of 3-benzyltyrosine and succinimide formation on aspartate residues compared to strong acids such as HF and TFMSA. Addition of ethanedithiol can suppress the *tert*-butylation of indole rings to a considerable extent (Wünsch *et al.*, 1977).

With regard to protecting groups for arginine, it has been reported (Atherton *et al.*, 1983, 1985; Akaji *et al.*, 1990) that TFA cannot completely cleave the widely used 4-methoxy-2,3,6-trimethylbenzenesulfonyl (Mtr) group (Fujino *et al.*, 1980, 1981). 2,2,5,7,8-Pentamethylchroman-6-sulfonyl (Pmc; Ramage and Green, 1988) has been developed as an N^G-protecting group efficiently cleavable by TFA. In addition, other reagents mentioned below can be used more efficiently for synthesis of longer Arg-containing peptides.

b. Trimethylsilyl Bromide in Trifluoroacetic Acid The 1 *M* trimethylsilyl bromide (TMSBr)–thioanisole/TFA system cleaves Bzl and Z groups nearly completely, and 2,6-dichlorobenzyloxycarbonyl (Cl_2Bzl) and Mts groups up to 90% (Fujii *et al.*, 1987d). It also efficiently (80%) reduces Met(O) to Met and suppresses succinimide formation of Asp nearly perfectly.

c. Tetrafluoroboric Acid in Trifluoroacetic Acid In Fmoc-based peptide synthesis, 1 *M* HBF_4–thioanisole in TFA was found to be useful (Akaji *et al.*, 1990). Complete removal of *t*-Bu groups from the side chains of Ser, Thr, Tyr, Asp, and Glu, removal of the Boc group from Lys, removal of the *tert*-butoxymethyl (Bm) group from His, the removal of 4,4-dimethoxylbenzhydryl (Mbh) groups from Asn and Gln, and the cleavage of Mtr groups from Arg could be achieved within 30 min. In the presence of ethanedithiol, Trp was quantitatively regenerated from Boc-Trp after a 120-min treatment. Of the various S-protecting groups of Cys, 4-methylbenzyl (MeBzl) and *t*-Bu were cleaved quantitatively within 60 min, but 4-methoylbenzyl (MeOBzl) was cleaved incompletely. *S*-Linked acetamidomethyl (Acm) groups remained intact after 120 min of treatment even in the presence of thioanisole. Bulky amino acids including Ile could be cleaved from *p*-benzyloxybenzyl alcohol resin (Wang resin) by this reagent. Amino acid amides were cleaved from *p*-methylbenzhydrylamine (MBHA) resin within 60 min except for Val-NH_2 and Phe-NH_2, for which longer treatment (120 min) was necessary for complete cleavage. However, Val-NH_2 was quantitatively cleaved from acid-labile 4-(2′,4′-dimethoxyphenylaminomethyl)phenoxy-type (DMAMP) resin by treatment with this reagent for 30 min at 4°C. Human glucagon, oxytocin, human brain natriuretic peptide (BNP), and eel atrial natriuretic peptide (ANP) were synthesized in the solid phase using this deprotecting reagent (Kiso *et al.*, 1991a; Akaji *et al.*, 1992b).

3. Other Strategies

a. Fluoride Ion Deprotection Strategy In this method, the acid-sensitive Boc group is used for α-amino protection, and tetra-*n*-butylammonium fluoride trihydrate (TBAF)-sensitive protecting groups are employed for side chain functional groups. A TBAF-sensitive resin is also used in this orthogonal strategy (Kiso *et al.*, 1988a) (Fig. 8). Bradykinin potentiating peptide 5a (BPP5a) and α-neo endorphin have been synthesized in high yield (Kiso *et al.*, 1990a).

b. Reductive Acidolysis Strategy Kiso *et al.* (1979, 1980b,c) reported that scavengers such as thioanisole promote deprotection under acidic conditions, and that the deprotection reaction proceeds by the cooperative action of a hard electrophile and a soft nucleophile, the "push–pull" mechanism, well agreeing with Pearson's hard soft acids bases principle. In the course of exploring new deprotecting procedures based on this concept, Kiso *et al.* (1987a) found that chlorinated silanes–TFA scavengers had strong reductivity along with acidolytic activities.

A series of safety-catch type protecting groups based on the 4-methylsulfinylbenzyl group have been developed; 4-methylsulfinylbenzyloxycarbonyl

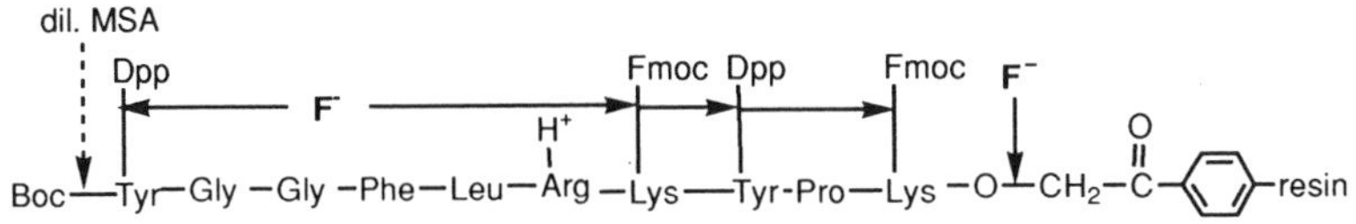

Figure 8 Two-dimensional orthogonal protection scheme for peptide synthesis using a fluoride ion final deprotection strategy.

(Msz; Kiso *et al.*, 1989a) for amino groups, and 4-methylsulfinylbenzyl (Msob; Samanen and Brandeis, 1985, 1988; Kiso *et al.*, 1987b, 1988b, 1989a) for carboxyl and hydroxyl groups. These protecting groups are stable under acidic conditions but can be smoothly removed in a one-pot reaction involving reductive acidolysis using tetrachlorosilane ($SiCl_4$)–TFA scavengers (Kiso and Kimura, 1991). In the deprotecting reaction, the Msz group is first reduced to the TFA-labile 4-methylthiobenzyloxycarbonyl (Mtz) group and then cleaved by acidolysis; in other words, the Msz group is deprotected by the reduction of the sulfoxide moiety and subsequent acidolysis (reductive acidolysis). Deprotection of the Msob group proceeds in a similar manner (Scheme 13).

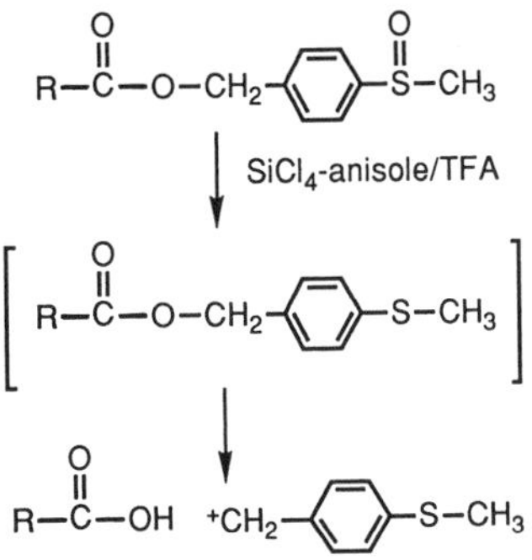

Scheme 13 Mechanism of reductive acidolysis.

Using the acid-stable Msz group as the semipermanent side-chain protecting group in combination with an acid-labile temporary Nα-protecting group such as Boc, a novel two-dimensional (2D) protection methodology in peptide synthesis has been developed by Kiso *et al.* (1989a). The method is based on a final deprotection strategy using reductive acidolysis. Scyliorhinin I was synthesized in solution using this system (Kiso *et al.*, 1989a) (Scheme 14).

Furthermore, Kiso *et al.* (1992c) have also developed a new 2D protection strategy in SPPS employing the safety-catch linkage to the resin, which is an orthogonal system based on an acid-labile temporary and reductive acidolysis-labile semipermanent protection scheme. Two types of resins carrying ester and substituted benzhydrylamide linkages are prepared and applied to practical peptide synthesis. To introduce a safety-catch ester linkage to aminomethylated-polystyrene resin, a new handle reagent 4-(2,5-dimethyl-4-methylsulfinylphenyl)-4-

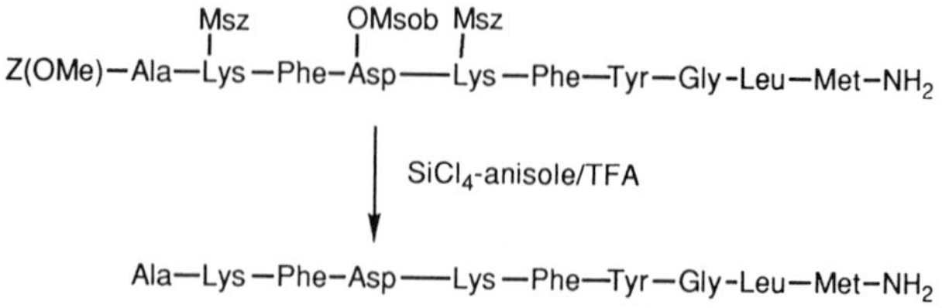

Z(OMe)—Ala—Lys(Msz)—Phe—Asp(OMsob)—Lys(Msz)—Phe—Tyr—Gly—Leu—Met—NH_2

↓ $SiCl_4$-anisole/TFA

Ala—Lys—Phe—Asp—Lys—Phe—Tyr—Gly—Leu—Met—NH_2

Scheme 14 Synthesis of scyliorhinin I by reductive acidolysis.

hydroxybutanoic (DSB) acid, was developed. γ-Endorphin (17-amino acid peptide) was synthesized using the DSB-resin (Kiso *et al.*, 1992a,c, 1994) (Scheme 15).

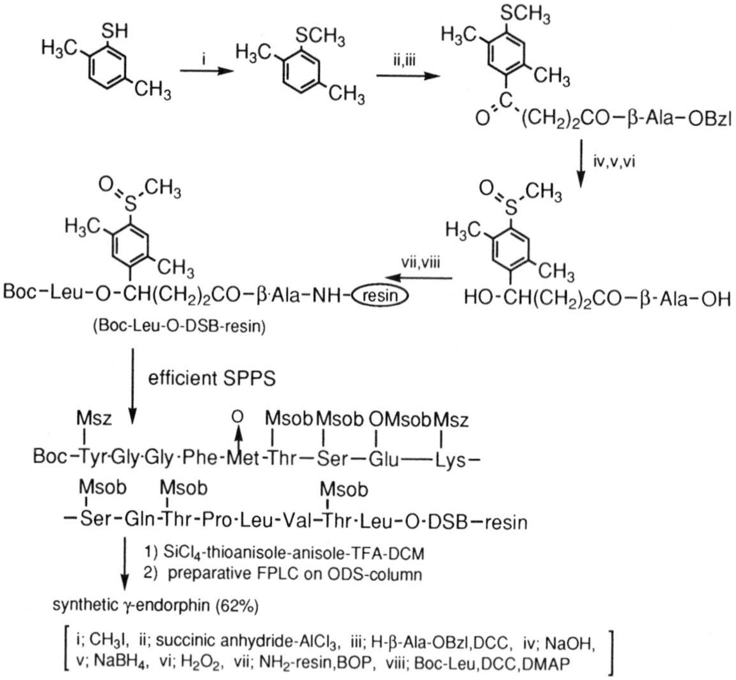

Boc-Leu-O-CH$(CH_2)_2$CO-β-Ala-NH-resin (Boc-Leu-O-DSB-resin)

↓ efficient SPPS

Boc-Tyr(Msz)-Gly-Gly-Phe-Met(O)-Thr(Msob)-Ser(Msob)-Glu(OMsob)-Lys(Msz)-Ser(Msob)-Gln-Thr(Msob)-Pro-Leu-Val-Thr(Msob)-Leu-O-DSB-resin

↓ 1) $SiCl_4$-thioanisole-anisole-TFA-DCM 2) preparative FPLC on ODS-column

synthetic γ-endorphin (62%)

[i; CH_3I, ii; succinic anhydride-$AlCl_3$, iii; H-β-Ala-OBzl,DCC, iv; NaOH, v; $NaBH_4$, vi; H_2O_2, vii; NH_2-resin,BOP, viii; Boc-Leu,DCC,DMAP]

Scheme 15 Preparation of DSB-resin and synthesis of γ -endorphin.

A safety-catch type of resin suitable for peptide amide synthesis by incorporation of an alkylsulfinyl group to the conventional benzhydrylamine type resin was also designed by Kiso *et al.* (1992c). A new handle reagent, *N*-Boc-(2 or 4)-(4′-methoxyphenylaminomethyl)-5-methoxyphenylsulfinylhexanoic acid, was prepared in seven steps starting from 3-methoxythiophenol. Coupling of the acid with β-Ala-NH-resin gave *N*-Boc-dimethoxyalkylsulfinylbenzhydrylamine (DSA) resin. Using DSA-resin, buccalin (Gly-Met-Asp-Ser-Leu-Ala-Phe-Ser-Gly-Gly-Leu-NH_2) was successfully synthesized (Kiso *et al.*, 1992c). Another safety-catch acid labile (SCAL) handle was prepared for the synthesis of peptide amides by the Boc/Fmoc strategy and was successfully applied to the synthesis of a portion of human calcitonin (Patek and Lebl, 1990,1991, 1992; Patek, 1993).

In addition, using a TBAF-sensitive resin, γ-endorphin (Kiso *et al.*, 1991b) and thymic humoral factor γ_2 (THFγ2) (Kiso *et al.*, 1993) were prepared in high yield by a stepwise deprotection method using reductive acidolysis followed by fluoride ion.

III. Disulfide Bond Formation

Disulfide bond formation is an important step that critically influences the synthesis of a cystine-containing peptide. Particularly in the syntheses of peptides and proteins containing two or more disulfides, the final disulfide forming reactions often occur in poor yields. This is attributed to the formation of side products such as polymers or isomers with disulfide locations different from those of the desired product. These side reactions are unavoidable whenever peptide chains are synthesized either by chemical means or by genetic engineering. For example, when a protein is synthesized by recombinant gene technology in *Escherichia coli,* the product forms granules in the bacteria, and production of the desired molecule with the correct spatial structure and the right disulfide combination is the essential step. In this section, aspects of the folding and disulfide formation of proteins are briefly introduced, and disulfide forming reactions in the chemical synthesis of cystine-containing peptides, including (1) air oxidation, (2) iodine oxidation, and (3) stepwise methods, are described.

The 20 kinds of amino acids contained in proteins determine the spatial structure and the minimum free energy by hydrogen-bonding and hydrophobic interactions between amino acid residues or between an amino acid residue and solvent. Among these amino acids, cysteines greatly contribute to the stabilization of the conformation by forming covalent disulfide bonds. Anfinsen (1973) showed using ribonuclease that the tertiary structure of proteins is predetermined basically by the primary structure. After reduction of the four disulfide bonds in ribonuclease in the presence of denaturant, removal of the denaturant and air oxidation regenerated predominantly the original protein with the correct disulfide bonds. Thus, they showed that refolding of the reduced and denaturated protein produced mainly the natural form among the 105 possible isomers with different disulfide combinations.

This result indicates that the appropriate disulfide bonds can be spontaneously formed from linear peptides with the correct amino acid sequence. In the laboratory, disulfide bonds are mostly formed by such natural refolding processes, although the reduction–reoxidation reaction requires a relatively long time. In biological systems, disulfide formation involves several accelerating factors, such as metal ions, disulfide exchanging enzymes, and low molecular weight SH or S−S compounds. These factors are applicable to chemical syntheses, and air oxidation in the presence of oxidized–reduced glutathione is often used in practical syntheses.

Disulfide formation processes have been intensively examined in some proteins, and the intermediates have been identified or inferred. For instance, Creighton (1980) conducted intensive research on bovine pancreatic trypsin inhibitor (BPTI) and described the formation process of three disulfides and the conformation of several intermediates (Creighton and Goldenberg, 1984). The natural type BPTI was obtained by air oxidation of reduced BPTI in the presence of oxidized–reduced glutathione. The intermediates produced during the reaction were trapped and identified by iodoacetic acid, and the disulfide bond formation process shown in Scheme 16 was deduced. Thus, among the six sulfhydryl groups in BPTI, formation of the first two disulfides occurs between residues 30 and 51 and between positions 5 and 30. The second group of bonds occurs between 14 and 38, between 5 and 14, and between 5 and 38. The natural disulfide between residues 5 and 55 has not yet been formed at this step and is not generated from the second three intermediates with two disulfides. The compound $[N]_{SH}^{SH}$ (Scheme 16) with the nativelike conformation and two disulfide bonds (30–51 and 5–55) is formed only after the second intermediates undergo intramolecular rearrangement. The formation of $[N]_{SH}^{SH}$ is the rate-limiting step in the whole oxidation reaction. Thus, disulfide formation in BPTI follows a predetermined process; that is, intermediates with nonnative type disulfide orientations refold to form native type disulfide combinations. Furthermore, States *et al.* (1987) examined the intermediates by ^{1}H nuclear magnetic resonance (NMR) and discussed the relationship between disulfide formation and conformation.

The studies mentioned above concern the disulfide formation process *in vitro,* whereas recent experiments (Taniyama *et al.*, 1990) on expression and excretion of human lysozyme in yeast by genetic recombination imply that a similar process involving S−S exchange reactions also occurs *in vivo.* Human lysozyme consists of 130 amino acid residues and four disulfide bridges. Expression and excretion experiments involving a series of mutant proteins in which cysteines were converted to alanines showed the disulfide forming process of lysozyme via intermediates with nonnatural disulfide combinations.

A. Reduction–Reoxidation of Disulfide Bond-Containing Peptides and Proteins

As mentioned above, the amino acid sequence is supposed to dictate formation of the correct disulfide combination. However, not all proteins are oxidized as smoothly as ribonuclease. For example, oxidation of insulin or epidermal growth factor gives only a low yield, because these peptides are synthesized by proteolytic cleavage after the disulfide bonds are formed in the linear precursor protein. In the case of insulin, the disulfide forming reactions in proinsulin proceed more efficiently than in a mixture of the A and B chains after separate synthesis (Dixon and Wardlaw, 1960; Crea *et al.*, 1978; Santere *et al.*, 1981; Blohm *et al.*, 1988).

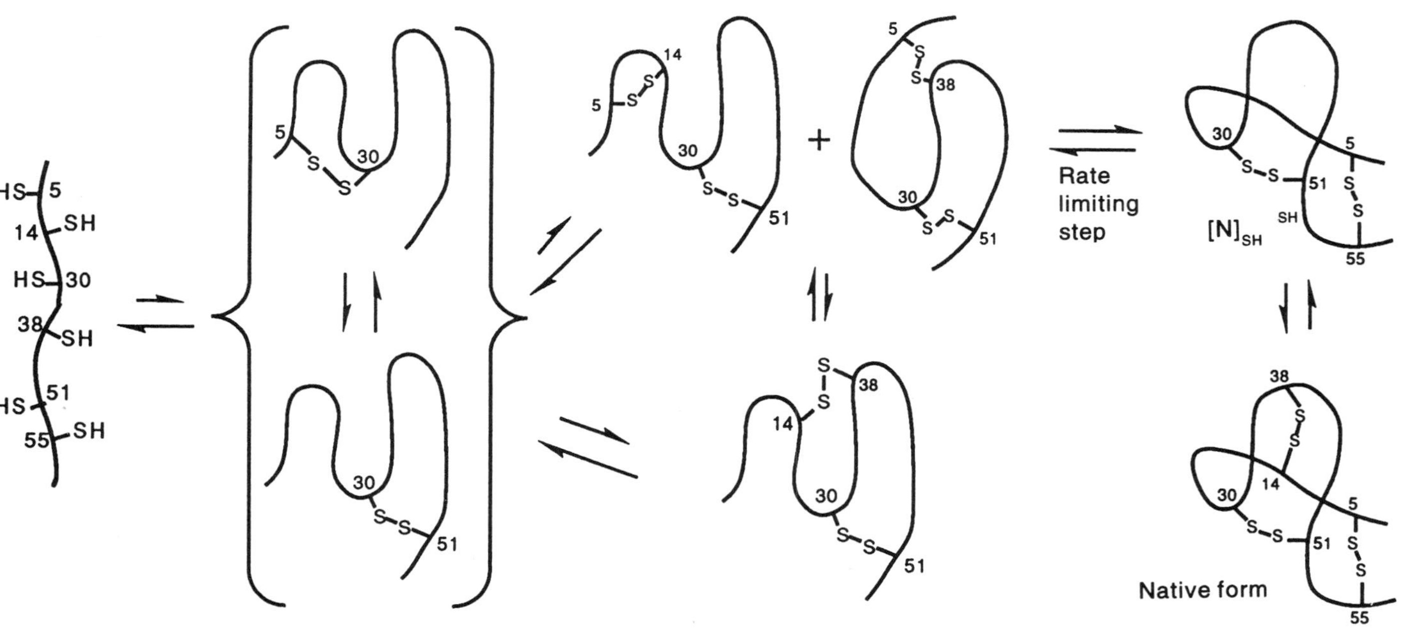

Scheme 16 Disulfide bond forming process in BPTI.

The formation of polymers and isomers to a certain extent during the oxidation reaction is inevitable and reduces the yield. In chemical synthesis, the yield is highly influenced by the purity of the protected peptide chain and the extent of side reactions during the final deprotection step. Therefore, for peptides or proteins with more than two disulfide bonds, it is very difficult to establish definite reaction conditions under which the air oxidation always proceeds smoothly, and it is necessary to find the best oxidation conditions for each compound. Generally, various factors concerning the reaction such as the extent of polymer and isomer formation, the solvent, pH, temperature, and the choice and concentration of additives such as glutathione are examined by reduction–reoxidation experiments.

1. Examination of Oxidation Conditions

Since Anfinsen (1973) performed the first reduction–reoxidation experiment, several peptides and proteins such as snake venom toxins (Menz *et al.*, 1980; Galat *et al.*, 1981; Gacond *et al.*, 1984), lysozyme (Acharya and Taniuchi, 1976; Perraudin *et al.*, 1983), and BPTI (Creighton and Goldenberg, 1984; Creighton, 1980) have been examined. This section describes the reduction–reoxidation experiments using scorpion toxin AaHII, in terms of improving the yield in chemical synthesis (Sabatier *et al.*, 1987).

AaHII, a peptide amide isolated from scorpion venom, consists of 64 amino acid residues and four disulfide bonds (Fig. 9). Normal air oxidation after reduction of AaHII gave mainly associated compounds along with very small amounts of the active peptide. To avoid the undesirable reactions, the reduced peptide was subjected to disulfide formation while it was dialyzed in buffer solution to solubilize the desired product, and the pH of the dialysis buffer and the influence of additives on the reaction were examined. The progress of the oxidation reaction was monitored by HPLC, and the product yield was calculated taking the toxicity of the soluble product as an indicator. As a result, the best yield of the active peptide was obtained when dialysis and air oxidation were carried out in the presence of dilute denaturant and oxidized–reduced glutathione at pH 8.0. The formation of associated compounds is often observed in the syntheses of other disulfide bond-containing peptides and proteins such as transforming growth factor (TGF) and epidermal growth factor (EGF), which are described later.

2. Formation of Isomers

Besides polymer formation, the formation of isomers which have the same amino acid sequence but have disulfide positions different from those of the native peptide is one of the most serious causes of reduced yield. In some cases, the isomers have been identified and the positions of disulfide bridges determined. Such identification of isomers is exemplified by the synthesis of insulin-like growth factor I (IGF-I), composed of 70 amino acid residues and three disulfide bridges, produced by recombinant methods in *E. coli*. Air oxidation of the reduced IGF-I synthesized in *E. coli* gives an isomer which elutes just before IGF-I, the main product, on HPLC (Saito *et al.*, 1987; Meng *et al.*, 1988). Comparison of peptide maps of the two products strongly implied that the isomer has

Figure 9 Structure of scorpion toxin AaHII.

two different disulfide positions relative to the native form. The disulfide positions were identified by chemical synthesis of the disulfide-containing peptide fragments with all possible disulfide combinations (Iwai *et al.*, 1989).

The synthesis of TGF-α, a peptide with 50 amino acids and three disulfide linkages, provides another example of the identification of isomers produced during chemical synthesis. Several groups synthesized TGF-α as described later. Among them, the group at Monsanto isolated and identified an isomer with two different disulfide bridges which was eluted after the main peak on HPLC (Tou *et al.*, 1990).

This section has briefly mentioned disulfide formation and the formation of polymers and isomers. In the next section, several important disulfide forming methods used in chemical synthesis of peptides and proteins are described along with some examples.

B. Disulfide Formation by Air Oxidation

Air oxidation is the most commonly used disulfide forming reaction. In particular, peptides and proteins with more than three disulfide bonds are synthesized by air oxidation in almost all cases. Depending on the thiol protecting groups used in peptide chain elongation, the reduced peptide for oxidation can be obtained either by one- or by two-step final deprotection.

1. Disulfide Formation by One-Step Deprotection

Scheme 17 shows the common synthetic scheme for disulfide-containing peptides using this method. At first, the protected peptide chain is constructed using solid-phase or solution methods. The MeBzl (Erickson and Merrifield, 1973) or MeOBzl groups (Akabori *et al.*, 1964) are employed as the thiol protecting group on cysteine, and benzyl alcohol-type protecting groups are used for other side-chain functional groups. After all the protecting groups, including those of cysteines, are removed by HF or TFMSA, reduction yields the peptide containing free SH groups. In almost all cases, the reduced peptide is subjected to air oxidation without purification to give the desired disulfide-containing peptide. An additive such as glutathione is included in the reaction mixture to accelerate the reaction and to improve the reaction yield. Generally, in this method the purity of the protected peptide critically influences the final yield, and only solution methods so far have given successful results in the synthesis of cystine-containing proteins of approximately 100 residues.

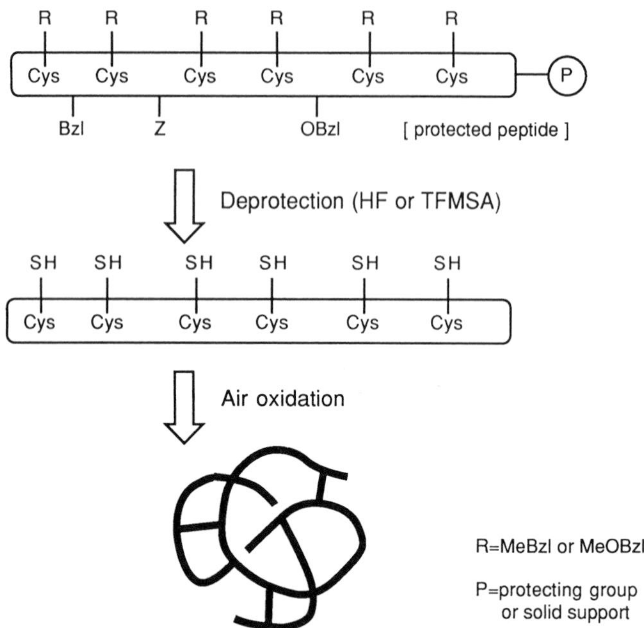

Scheme 17 Disulfide bond formation by one-step deprotection.

One of the examples of disulfide formation by high-dilution air oxidation in the presence of glutathione is the total synthesis of ribonuclease A by Yajima and Fujii (1981). Oxidized–reduced glutathione is considered to accelerate the disulfide forming reaction by S−S exchange and is often used as an additive for disulfide formation involving higher peptides and proteins (Szajewski and Whitesides, 1980). In particular, Chavez and Scheraga (1980) and Ahmed *et al.* (1975) studied ribonuclease A intensively and found that the addition of glutathione increased the reaction yield by approximately 20% in comparison to the reaction without glutathione.

Although the above-mentioned high-dilution air oxidation in the presence of glutathione is a generally applicable method, glutathione interferes with proper disulfide formation in some cases. Yajima *et al.* (1984, 1985a) reported that the correct disulfide bonds are not formed by air oxidation in the presence of glutathione in the synthesis of mouse EGF, which has 53 amino acid residues and three disulfides. Reduced EGF has a high tendency to undergo association and polymer formation during air oxidation. Air oxidation of reduced EGF without glutathione does give the objective peptide, although the yield is low (4.1%). However, in the presence of glutathione, air oxidation gives exclusively glutathione–EGF adducts without any formation of the desired peptide. Heath and Merrifield (1986) also observed such glutathione adduct formation and association in solid-phase synthesis of mouse EGF. Addition of a denaturant to the air oxidation reaction mixture is also often used.

Human, rat, and bovine TGF-α have 50 amino acid residues and three intramolecular disulfide bonds, and they have been synthesized on a solid phase separately by the groups of Tam, Kent, and Tou (Monsanto). All three groups added a denaturant such as urea or guanidine during the air oxidation reaction. Kent and co-workers (Woo *et al.*, 1989) used 1 *M* guanidine hydrochloride in the synthesis of human TGF-α. They reported that air oxidation without guanidine gave a mixture which showed a complicated elution pattern on HPLC. In the syntheses of human and rat TGF-α by Tam's group (Tam *et al.*, 1986; Tam, 1987) and of bovine TGF-α by the Monsanto group (Tou *et al.*, 1990), the urea concentration of the reaction mixture in which the reduced TGFα was dissolved was gradually reduced by dialysis, and finally air oxidation was performed in the presence of 1–2 *M* urea.

In the air oxidation of TGF-α, oxidized–reduced glutathione is also added to the reaction mixture to facilitate disulfide formation by S−S exchange reaction. The Monsanto group reported that the overall yield without glutathione in the air oxidation step was 57% of that with glutathione. In these syntheses overall yields remained between 4.5 and 10%, whereas Tam *et al.* (1986) described a slightly higher yield. Further attempts included synthesis of a peptide corresponding to positions 26 to 81 of Shope fibromavirus growth factor (SFGF) using the same method (Lin *et al.*, 1988) as used for the synthesis of human and rat TGF-α. However, most of the reduced peptides were converted to the insoluble associated form. Although SFGF(26–80) is of similar size and has similar disulfide positions as does TGF-α, the reduced SFGF(26–80) peptide precipitates in 5 *M* urea. Tam *et al.* (1986) finally obtained the desired product by air oxidation in 1.5 *M* guanidine hydrochloride solution (peptide concentration of 0.1 mg/ml) but in only 3% overall yield.

As usual air oxidation reactions require a considerably long time, ranging from 1 to several days, oxidants such as $K_3Fe(CN)_6$ are often used to facilitate disulfide formation. This method is often employed in the synthesis of peptides with a single disulfide bond (Minamitake *et al.*, 1990; Spear *et al.*, 1989; Kazmierski *et al.*, 1988).

Le-Nguyen and Rivier (1986) used carboethoxy sulfenyl chloride (Sce-Cl) as an additive to accelerate the oxidation reaction, effectively utilizing the disulfide exchange reaction (Scheme 18). When this reagent was applied to the synthesis of turnip trypsin inhibitor containing 28 amino acid residues and three intramolecular disulfide bonds, the oxidation reaction was completed within 10 min, but the product yield was lowered by approximately 25% compared to usual air oxidation (50 hr) (Le-Nguyen *et al.*, 1989). Air oxidation has been employed in syntheses of many peptides with 20 to 50 residues and two or three disulfide bonds, such as conotoxins (Cruz *et al.*, 1989) and echistatin (Garsky *et al.*, 1989). The methods of syntheses of relatively small peptides with a single disulfide bond are extensively described in a review by Hiskey (1981).

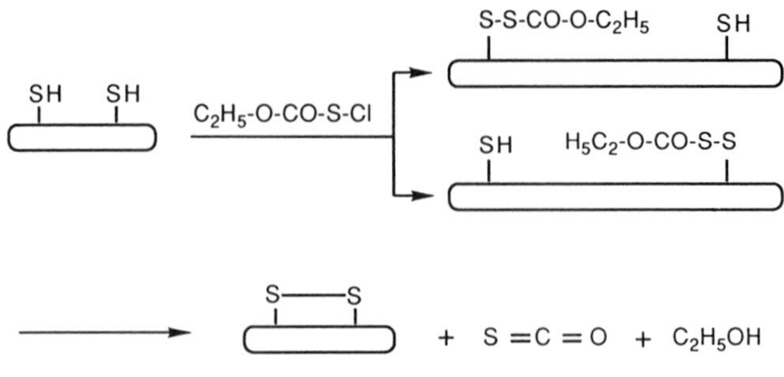

Scheme 18 Disulfide bond formation using Sce-Cl.

2. *Disulfide Formation by Two-Step Deprotection*

Scheme 19 shows a synthetic scheme for disulfide formation by two-step deprotection. A protected peptide chain is synthesized by the solid-phase or solution method based on either the Boc strategy (using benzylalcohol-type protecting groups) or the Fmoc strategy (*tert*-butylalcohol-type protecting groups). The Acm group (Veber *et al.*, 1968) is used as the thiol protecting group of cysteine.

At the first deprotection step, protecting groups on side-chain functional groups are cleaved by acidolysis. This deprotection gives the thiol-protected peptide because the Acm group is resistant to acid but is cleavable by $Hg(OAc)_2$ or I_2 (Ac represents acetyl). The trimethylacetamidomethyl (Tacm) group (Kiso *et al.*, 1990b), which has chemical properties similar to those of Acm, is also used for this type of synthesis.

Subsequently, after the Acm group is cleaved from the thiol-protected peptide, air oxidation forms the required disulfide bridges. This method has the advantage that the thiol-protected peptide can be purified before air oxidation. The syntheses of disulfide-containing higher peptides or proteins carry the possibility that the proper disulfide bridges will not be formed when purification is insufficient. In fact, Kenner's group (Galpin *et al.*, 1981) reported that the active peptide could not be obtained because of the insufficient removal of eight Acm groups by $Hg(OAc)_2$ in the synthesis of a lysozyme analog consisting of 129 amino acids and containing four disulfides. It is presumed that the purification of protected peptides before the final deprotection was insufficient. The progress of purification methods using HPLC has made it possible to synthesize small proteins up to 123 residues long.

Removal of the Acm or Tacm groups used in this method is conducted by mercury salt such as $Hg(OAc)_2$ (Veber *et al.*, 1972) or $Hg(OCOCF_3)_2$ (Fujino and Nishimura, 1976; Nishimura *et al.*, 1978) or silver salt. $Hg(OAc)_2$ is most commonly used. The peptide mercuric salt obtained by deprotection is treated with a thiol compound such as mercaptoethanol to give a reduced peptide with free thiol groups, which is then subjected to air oxidation. Generally, mercuric compounds are highly toxic and require careful handling. Therefore, silver salts have been developed to remove the Acm group. The counteranion of the silver cation greatly influences cleavage of the thiol protecting group, and silver trifluoromethanesulfonate (AgOTf; Fujii *et al.*, 1989) or silver tetrafluoroborate ($AgBF_4$; Yoshida *et al.*, 1990a) have been used in practical syntheses.

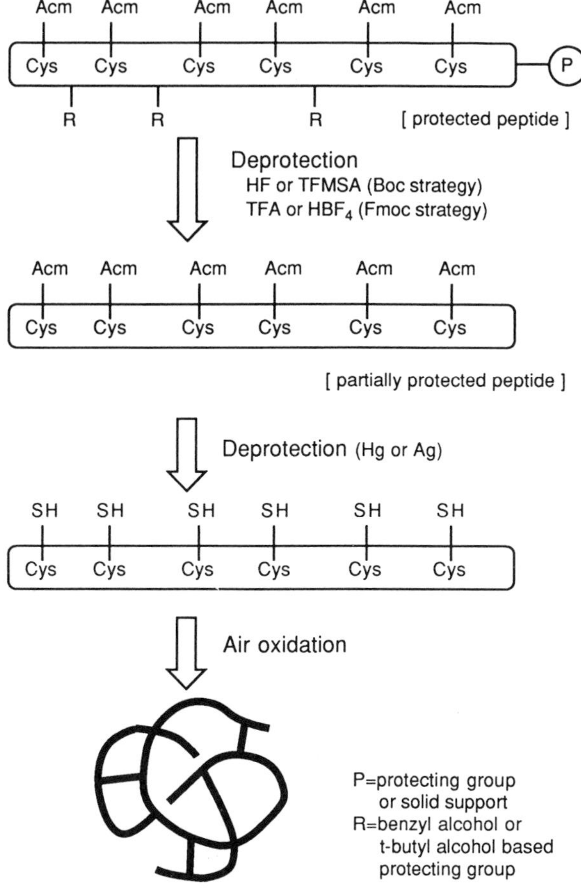

Scheme 19 Disulfide bond formation by two-step deprotection.

In contrast to mercuric compounds, AgOAc cannot cleave Acm. The Acm group is cleavable by iodine as mentioned below, although modification of tryptophan and oxidation of methionine also occur. The side reactions are largely avoided in the synthesis of Trp- or Met-containing peptides using two-step deprotection by $AgBF_4$ (Yoshida *et al.*, 1990c). AgOTf has been applied to the synthesis of chicken calcitonin (Fujii *et al.*, 1989), endothelin (Nomizu *et al.*, 1989), and tachyplesin (Akaji *et al.*, 1989), among others, and $AgBF_4$ has been used in the synthesis of somatostatin (Yoshida *et al.*, 1990c), porcine brain natriuretic peptide (pBNP-32; Yoshida *et al.*, 1990b), and others.

C. Disulfide Formation by Iodine Oxidation

The iodine oxidation method is based on the iodine treatment of Cys(Acm) or Cys(Trt) derivatives (where Trt is trityl) to give cystine via sulfenyl iodide (Scheme 20) (Kamber *et al.*, 1980). The general synthetic scheme using iodine

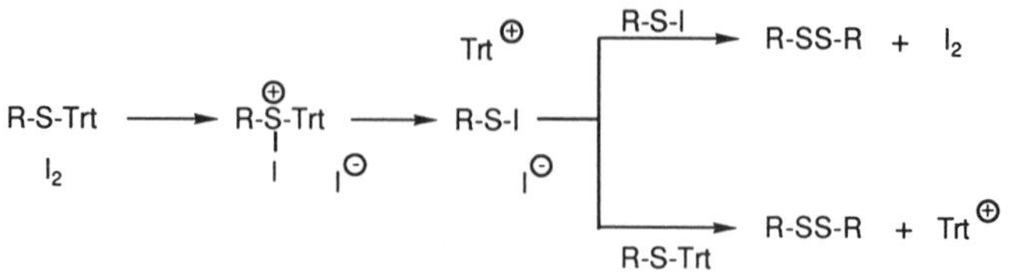

R-S-Trt $\xrightarrow{I_2}$ R-$\overset{\oplus}{S}$(I)-Trt $I^{\ominus}$ ⟶ R-S-I $I^{\ominus}$ + $Trt^{\oplus}$

R-S-I $\xrightarrow{\text{R-S-I}}$ R-SS-R + I_2

R-S-I $\xrightarrow{\text{R-S-Trt}}$ R-SS-R + $Trt^{\oplus}$

Scheme 20 Mechanism of the iodine oxidation method.

oxidation is illustrated in Scheme 21. As in the case of the two-step deprotection mentioned above, the peptide chain is constructed with Acm for thiol protection and then subjected to acidolysis to produce the thiol-protected peptide. Subsequent iodine treatment simultaneously facilitates cleavage of Acm groups and disulfide formation. The advantages of this method are the purification before SS formation and the simultaneous thiol deprotection and SS formation.

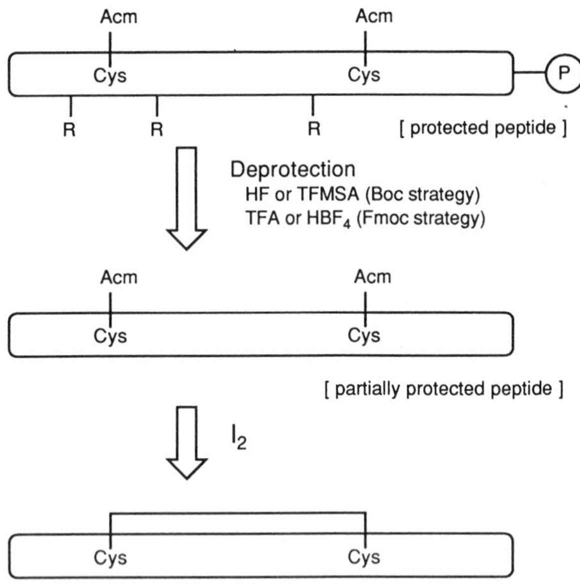

Scheme 21 Disulfide bond formation by iodine oxidation.

However, disulfide formation by iodine is not a regioselective reaction, so it is not easy to build up more than two disulfides in one step. In fact, in the synthesis of human TGF-α with three disulfides by Scanlon *et al.* (1987), using iodine treatment of Acm removal and disulfide formation, gave a mixture of many products. The desired peptide was obtained by reoxidation after reduction, but the overall yield remained as low as 0.1%. Because iodine treatment was reported (Yoshida *et al.*, 1990c) to be accompanied by oxidation of methionine and modification of tryptophan (mentioned later), the method shown in Scheme 21 has been mainly applied to the synthesis of peptides containing one cystine and lacking tryptophan and methionine. However, in the case of small proteins such as HIV protease (99 amino acids), it is difficult to apply the method, owing to very slow cleavage of Acm by iodine (Nutt *et al.*, 1988).

Natriuretic peptides (Lyle *et al.*, 1987), BNP (Kiso *et al.*, 1990c), and some other peptides have been synthesized by the iodine method. More recently, iodine oxidation has been used instead in the stepwise disulfide formation mentioned in the following section, whereas mercuric salts and silver salts are exclusively employed for the deprotection of Acm groups.

D. Stepwise Disulfide Formation

The methods described so far construct disulfide bonds in one step, whereas the stepwise method builds one disulfide bridge at each step. Theoretically, disulfide bonds can be formed at any intended positions by this method. Therefore, stepwise disulfide formation has been applied to the syntheses of newly isolated cystine-containing peptides and isomers having different disulfide positions, and thus has been used for the determination of disulfide positions in the native form (Iwai *et al.*, 1989).

In this method, each disulfide is bridged regioselectively. The second and subsequent disulfides must be formed under conditions which do not cleave the already connected disulfide bonds. Thus, alkaline conditions, use of thiol reagents, or long reaction times with the possibility of SS exchange should be avoided. Because of these limitations, established procedures to construct more than three disulfides have not yet been reported, although several papers have used the method to build two disulfide linkages using a combination of two types of protecting groups cleavable by different reagents. Three types of combinations of pairs of protecting groups are discussed in this section.

1. Combination of 4-Methylbenzyl and Acetamidomethyl Groups

The common MeBzl–Acm combination is used in the stepwise disulfide forming method using air oxidation and iodine oxidation. The general synthetic scheme is shown in Scheme 22. The protected peptide chain is built up using MeBzl and Acm groups for the protection of thiols corresponding to the two disulfide positions, and benzyl alcohol-type groups for the protection of other side-chain functional groups. HF treatment cleaves MeBzl and benzylalcohol type groups to give the partially protected peptide with two free SH groups, and subsequent air oxidation forms the first disulfide bond. $K_3Fe(CN)_6$ is added to accelerate the reaction at this stage. Tryptophan is often protected by the formyl group to avoid modification of the indole ring during iodine treatment to form the second disulfide. The formyl group is cleavable by brief treatment with NaOH after formation of two disulfide bonds. This method was applied to the synthesis of endothelin (Kumagaye *et al.*, 1988), conotoxins and isomers with different disulfide positions (Nishiuchi and Sakakibara, 1982, 1983; Almquist *et al.*, 1989), mast cell degranulating peptide (Buku *et al.*, 1989), and others.

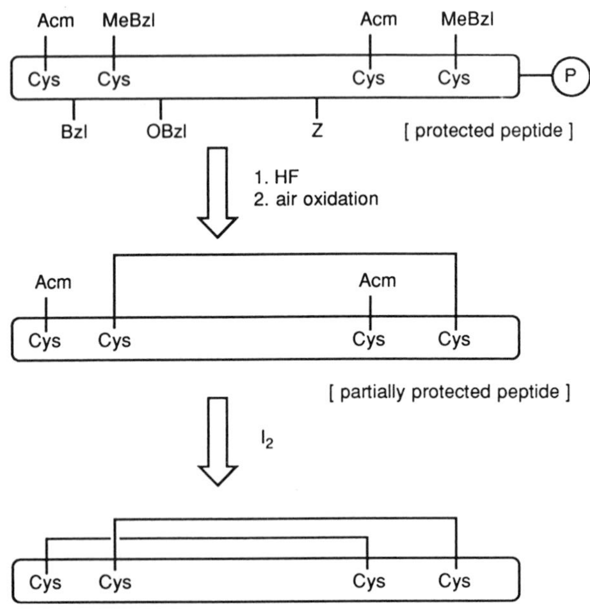

Scheme 22 Stepwise disulfide formation from MeBzl- and Acm-protected cysteines.

2. *Combination of Trityl and Acetamidomethyl Groups*

As mentioned in the section on iodine oxidation (Section III,C), the reaction rate of oxidative cleavage by iodine is greatly influenced by the reaction solvent (Scanlon *et al.*, 1987). Immer *et al.* (1988, 1989) utilized these properties to perform stepwise disulfide formation by selectively forming the first disulfide by cleavage of trityl group by iodine, leaving the Acm group intact. Endothelin and sarafotoxin S6b were synthesized by this method. The synthetic scheme for endothelin is illustrated in Scheme 23. The first disulfide is formed between two Cys(Trt) residues by treatment with iodine in methylene chloride–trifluororethanol (TFE) for 5 min to cleave the Trt group, and then the second disulfide is formed between Cys(Trt) and Cys(Acm) by treating with iodine in methanol for 5 min. The yields of cyclizations are 70% for the first disulfide and 65% for the second.

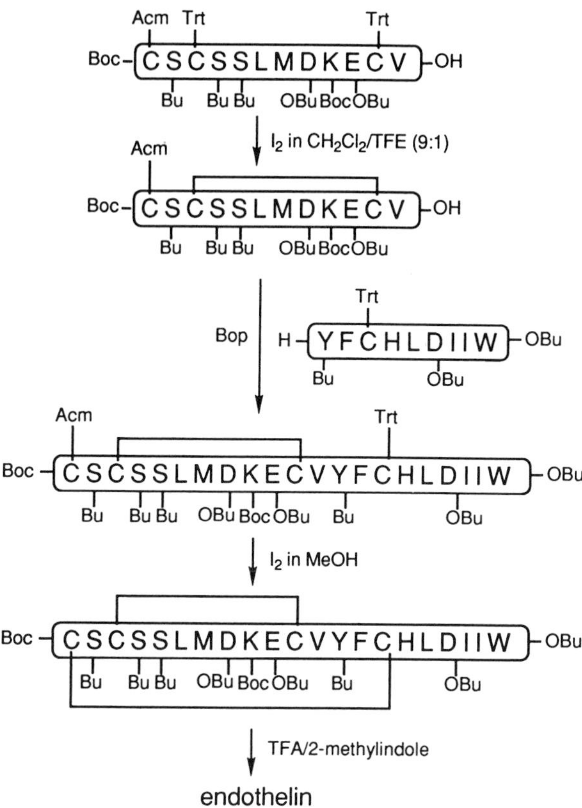

Scheme 23 Synthesis of endothelin using Trt- and Acm-protected cysteines.

3. Combination of 3-Nitro-2-pyridosulfenyl and Acetamidomethyl Groups

3-Nitro-2-pyridosulfenyl (Npys), a thiol protecting group developed by Matsueda and Walter (1980), is stable under acidic conditions and is cleavable by tri-*n*-butylphosphine in the presence of water (Matsueda *et al.*, 1981). It has another important characteristic in that it reacts readily with free sulfhydryl groups to form disulfide bonds (Matsueda *et al.*, 1982: Bernatowicz *et al.*, 1986) (Scheme 24). This property is applicable to regioselective intermolecular disulfide formation between Cys(Npys) and Cys with a free sulfhydryl group (Ruiz-Gayo *et al.*, 1988).

Sakakibara's group (Chino *et al.*, 1986) applied this stepwise method to the solution-phase synthesis of β-human atrial natriuretic peptide (β-hANP) with an antiparallel dimeric structure. They formed the first disulfide between Cys(Npys) and Cys, and the second by iodine oxidation. The free cysteine for the first step is generated by HF treatment of Cys(MeBzl). Albericio *et al.*

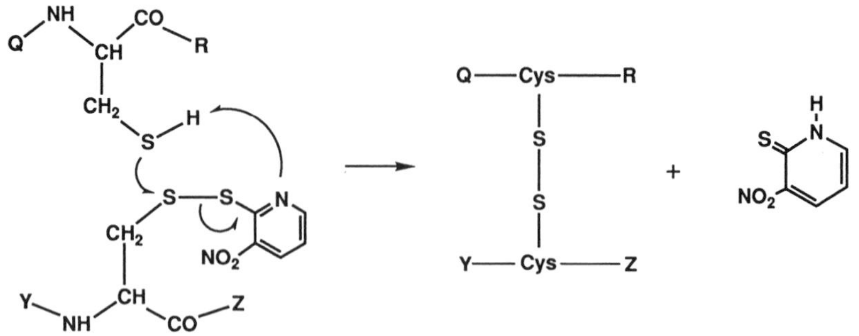

Scheme 24 Mechanism of disulfide formation using the Npys group.

(1989) intensively examined the stability of the Npys group under the conditions used in solid-phase synthesis.

E. Disulfide Formation by Silyl Chloride–Sulfoxide System

Akaji *et al.* (1991, 1992a,b, 1993a,b) demonstrated that methyltrichlorosilane in the presence of diphenylsulfoxide can cleave various thiol protecting groups of cysteine to form cystine directly (Scheme 25). This reaction is rapid and completed within 10 to 30 min in TFA. Most of the nucleophilic amino acids can be recovered unchanged after treatment with chlorosilane.

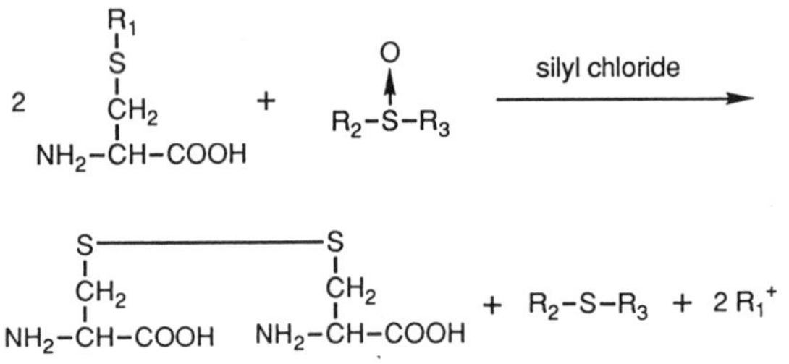

Scheme 25 Disulfide bond formation using silyl chloride–sulfoxide system.

1. Disulfide Bond Formation Using the Silyl Chloride–Sulfoxide System

As summarized in Table I, all three protecting groups of the amidomethyl type, namely, Acm (Veber *et al.*, 1972), Tacm (Kiso *et al.*, 1990b), and benzamidomethyl (Bam; Chakravarty and Olsen, 1978), were cleaved to form cystine as a sole product within 30 min at 4°C. Among the benzyl type protecting groups, MeOBzl (Akabori *et al.*, 1964) was cleaved quantitatively to form cystine within 10 min at 4°C; however, MeBzl (Erickson and Merrifield, 1973) was cleaved incompletely, with 26% of the parent compound remaining after the 30 min treatment, and no cleavage of 4-nitrobenzyl (NO_2Bzl; Bachi and Ross-Petersen, 1972) was observed even after the 60 min treatment. Similarly, *t*-Bu

Table I Cleavage of Various Thiol Protecting Groups of Cysteine and Formation of Cystine by Methyltrichlorosilane–Diphenylsulfoxide at 4°C[a]

	10 min		30 min		60 min	
Protected Cysteine	Starting compound (%)	Cystine (%)	Starting compound (%)	Cystine (%)	Starting compound (%)	Cystine (%)
Cys(Acm)	15	79	0	93		
Cys(Tacm)	9	82	0	88		
Cys(Bam)	4	98	0	100		
Cys(*t*-Bu)	0	99				
Cys(*i*-Pr)	94	0	93	0	100	0
Cys(MeOBzl)	0	95				
Cys(MeBzl)	51	49	26	76	16	83
Cys(NO_2Bzl)	97	0	96	0	95	0

[a]Acm, Acetamidomethyl; Tacm, trimethylacetamidomethyl; Bam, benzamidomethyl; MeOBzl, 4-methoxybenzyl; MeBzl, 4-methylbenzyl; NO_2Bzl, 4-nitrobenzyl.

(Callahan *et al.*, 1963) was cleaved quantitatively within 10 min, whereas the more acid-stable isopropyl group (*i*-Pr; Gawron and Lieb, 1952) remained intact even after 60 min of treatment with silyl chloride–sulfoxide. Thus, the acid stability of the S-protecting group appears to influence the rate of cystine formation by silyl chloride–sulfoxide. Among the benzyl and alkylthioether type S-protecting groups, only those stable to HF were quantitatively recovered unchanged. Other less acid-stable S-protecting groups were cleaved to form cystine in more than 80% yield.

Met, His, and Tyr, as well as cystine, were recovered quantitatively even after 60 min of treatment with methyltrichlorosilane–diphenylsulfoxide at 25°C (Table II). No Met(O) was formed. Quantitative recovery of Trp(CHO) (Previero *et al.*, 1967) under the same reaction conditions was also confirmed.

2. *Possible Mechanism*

A proposed pathway for the synthesis of cystine peptides using the silyl chloride–sulfoxide system is shown in Scheme 26. The first interaction should be the formation of a sulfonium cation (**21**) derived from diphenylsulfoxide and an oxygenophilic silyl compound. The formation of sulfonium ions of this type has been shown and utilized for the reduction of sulfoxides using the trimethylsilylchloride/thiol system (Numata *et al.*, 1979). Subsequent electrophilic attack of **21** on the sulfur atom of S-protected cysteine (**20**) will lead to the formation of intermediate **22.** The nature of the silyl chloride employed should be one of the main factors which influence the electrophilicity of **21.** The assumed intermediate sulfenyl compound (**22**) may then function as the electrophile and react with another S-protected cysteine to provide disulfide **23** and diphenylsulfide. This final step would be analogous to the reaction of a sulfenyl iodide with an S-protected cysteine in the iodine oxidation reaction studied by Kamber *et al.* (1980). In addition, the mechanism suggests that the ability of the S-protecting group to form a stabilized moiety on cleavage of the bond between the sulfur atom and itself, like the iminium ion from the *S*-Acm group (Kamber *et al.*, 1980), may influence the rate of disulfide bond formation in the silyl chloride–sulfoxide system. We consider the proposed mechanism to provide an explanation for our observation that the reaction rate of cystine formation largely depends on the nature of the silyl chloride and the S-protecting group.

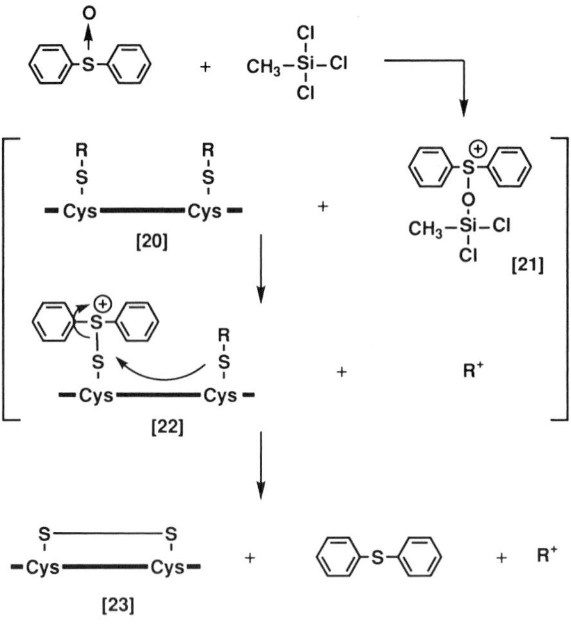

Scheme 26 Possible mechanism of disulfide formation by silyl chloride–sulfoxide system.

3. Synthesis of Human Brain Natriuretic Peptide (hBNP)

Human BNP consists of 32 amino acid residues, including two Met and a His residue, and it contains one disulfide bond in the molecule (Sudoh *et al.*, 1989) (Fig. 10). The synthesis of hBNP appears to be a suitable test for the silyl chloride–sulfoxide method in practical synthesis of a cystine-containing peptide, as we have observed the formation of a large amount of the Met(O) derivative during synthesis of the structurally homologous porcine BNP using the iodine oxidation method (Yoshida *et al.*, 1990c).

The S-protected hBNP in TFA was treated with methyltrichlorosilane–diphenylsulfoxide at 25°C for 10 min to form the disulfide bond. The

Table II Recoveries of Various Nucleophilic Amino Acids After Treatment of Boc-Cys(Acm)-OH with Methyltrichlorosilane–Diphenylsulfoxide

Amino acid	30 min	60 min
Cys(Acm)	0	0
Cystine	1.05	1.02
Met	1.02	1.02
His	1.00	1.01
Tyr	1.02	1.02
Trp(CHO)	0.92	0.95
Gly[a]	1.00	1.00

[a] Internal standard.

crude product gave a single main peak on HPLC, showing that no significant modification occurred on the peptide chain. After purification, the homogeneous peptide had physicochemical properties identical to those of an authentic sample prepared by air oxidation (Yoshida *et al.*, 1990c).

4. Directed Formation of Double Intrachain Disulfides

Using silyl chloride, we have established a scheme for the directed formation of double intrachain disulfide bonds and synthesized conotoxin M1 (McIntosh *et al.*, 1982), a peptide neurotoxin consisting of 14 amino acid residues and containing two disulfide bonds per molecule (Scheme 27). The S-protected conotoxin MI (**24**) was treated with iodine at 25°C for 15 min to construct the first disulfide bond between the cysteine residues protected by Acm groups, leaving the two *t*-Bu groups intact. The product (**25**) in TFA was then treated

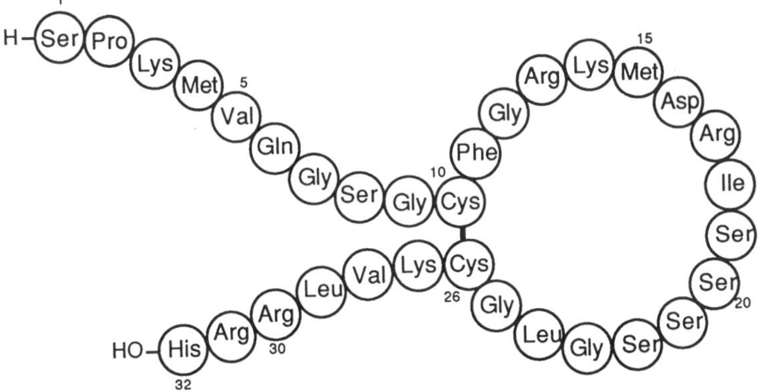

Figure 10 Amino acid sequence and disulfide pairing of hBNP.

with CH_3SiCl_3 in the presence of PhS(O)Ph at 25°C for 10 min to construct the second disulfide bond between the cysteine residues with the *t*-Bu groups. The crude product (**26**) gave a single main peak on HPLC, showing that no significant disulfide exchange occurred during the silyl chloride treatment.

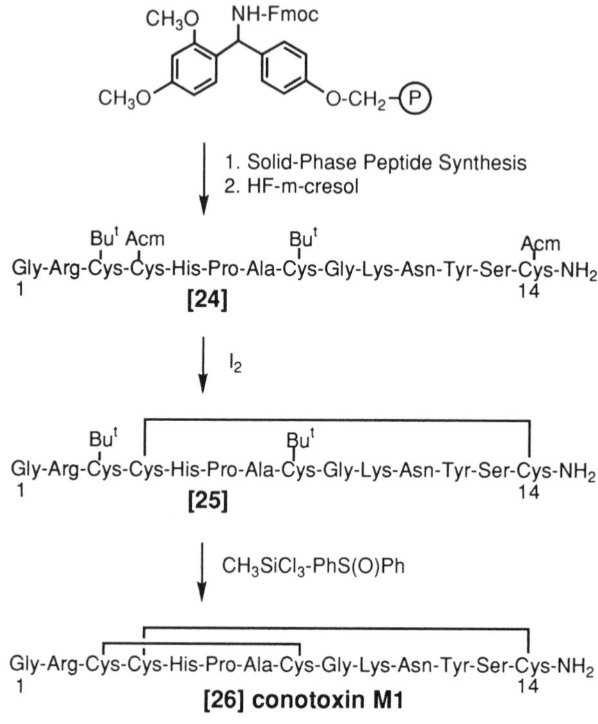

Scheme 27 Synthesis of conotoxin M1.

5. Directed Formation of Double Interchain Disulfides

Next, the silyl chloride method was applied to the stepwise formation of two interchain disulfide bonds. The Acm and Trt groups (Zervas and Photaki, 1962) were employed as the orthogonal thiol protecting groups of cysteine. The Trt group can be cleaved by a weak acid in the presence of scavengers (Pearson *et al.*, 1989; König and Kernebeck, 1979; Hiskey *et al.*, 1975; Photaki *et al.*, 1970), keeping the Acm group intact. By subsequent treatment with 2,2′-dithiodipyridine (Grassetti and Murray, 1967), the liberated thiol group can be converted to the corresponding *S*-2-pyridinesulfenyl cysteine [Cys(*S*-Py)] derivative. Thus, to achieve the directed formation of double interchain disulfides, the silyl chloride method was combined with disulfide formation by thiolysis of the *S*-2-pyridinesulfenyl cysteine. According to this scheme, we have synthesized β-hANP (Kangawa *et al.*, 1985), an antiparallel dimer of α-hANP (Scheme 28).

The [Cys(Acm)^7, Cys^{23}]-α-hANP (**27**) was allowed to react with 2,2′-dithiodipyridine to give [Cys(Acm)^7, Cys(*S*-Py)^{23}]-α-hANP (**28**), which was

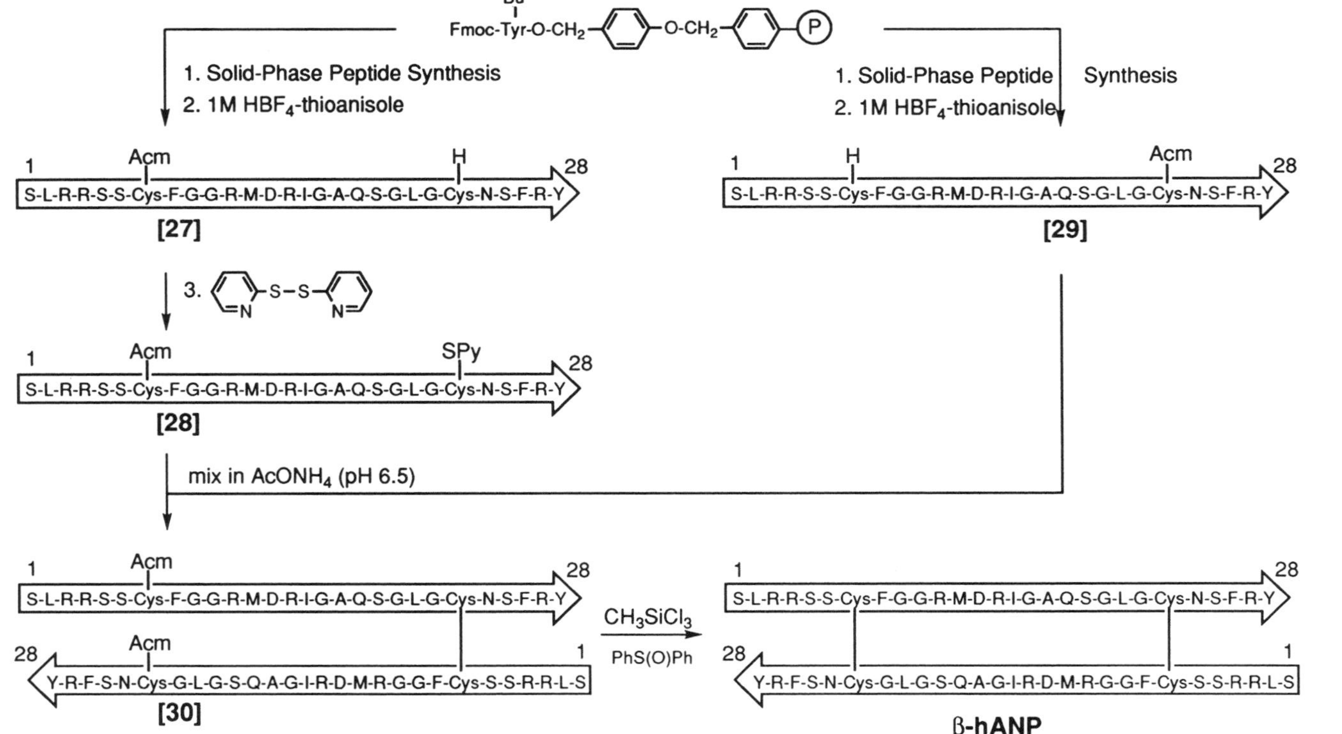

Scheme 28 Synthesis of β-hANP by regioselective disulfide formation.

then mixed with [Cys^7,$Cys(Acm)^{23}$]-α-hANP (**29**) at pH 6.5 for 30 min to construct the first disulfide bridge. After purification, the product (**30**) in TFA was treated with CH_3SiCl_3–PhS(O)Ph at 25°C for 10 min to form the second disulfide bond between the cysteine residues carrying the Acm groups. The crude product had a single main peak on HPLC, and no production of α-hANP was detected, indicating that no disulfide exchange had occurred during the silyl chloride treatment.

6. *Regioselective Formation of the Three Disulfide Bonds of Human Insulin*

By a combination of the above two schemes for directed double disulfide formation, we established a regioselective scheme to form the three disulfide bonds of human insulin, in which the disulfide bonds are constructed stepwise by successive reactions using thiolysis, iodine, and silyl chloride (Scheme 29). The three orthogonal thiol protecting groups of cysteine, Trt, Acm, and *t*-Bu groups are employed. As confirmed in both double disulfide formations, the Trt group is selectively cleavable with an acid in the presence of Acm and *t*-Bu groups, and the Acm group can be oxidatively cleaved with iodine to form cystine, leaving the *t*-Bu group.

According to this novel scheme, cleavage of the Trt group with a weak acid to liberate the cysteine side chain in position A20 and subsequent thiolysis of the *S*-2-pyridinesulfenyl cysteine in position B19 direct the formation of the first interchain disulfide bond (A20–B19). The second interchain disulfide bond (A7–B7) is formed between the cysteine residues carrying the Acm groups by treatment of the resulting [di-Acm, di-*t*-Bu, monodisulfide]human insulin (**31**) with iodine. The starting peptide (**31**) disappeared completely after the 60-min treatment to give a di-disulfide insulin (**32**). Finally, the product (**32**) was treated with silyl chloride to construct the intrachain disulfide bond (A6–A11) between the cysteine residues with the *t*-Bu groups. After the 15-min reaction, the starting peptide (**32**) disappeared completely to give a crude product which was easily purified by preparative HPLC. No significant formation of isomers or polymers was detected during the silyl chloride treatment. Synthetic human insulin gave a sharp single peak on an analytical HPLC column and had the same retention time as that of an authentic sample prepared by a semisynthetic method (Inouye *et al.*, 1979).

We consider that the scheme using silyl chloride makes it more feasible and efficient to construct three disulfide bonds regioselectively in peptides or proteins. The present regioselective method will be useful for the preparation of unnatural isomers with disulfide bonds at different positions and, hence, thermodynamical stability different from that of the corresponding natural peptide.

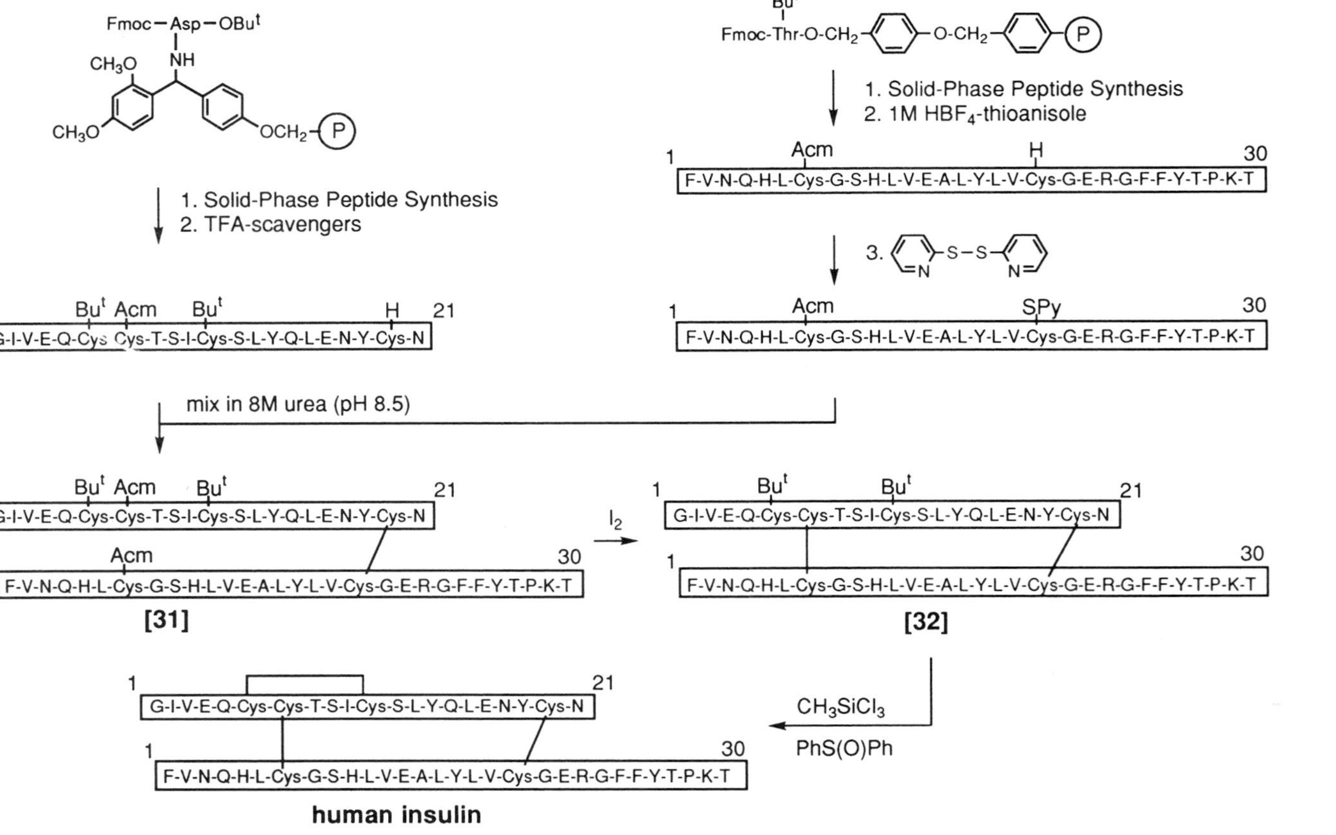

Scheme 29 Total synthesis of human insulin by regioselective disulfide formation.

Abbreviations

Acm	Acetamidomethyl
ANP	Atrial natriuretic peptide
Bam	Benzamidomethyl
Bm	*tert*-butoxymethyl
BNP	Brain natriuretic peptide
Boc	*tert*-butoxycarbonyl
BOI	2-(Benzotriazole-1-yl)-oxy-1,3-dimethylimidazolidinium hexafluorophosphate
Bom	Benzyloxymethyl
BOP	Benzotriazole-1-yloxytris(dimethylamino)phosphonium hexafluorophosphate
BOP-Cl	3,3′-(Chlorophosphoryl)bis(1,3-oxazolidin-2-one)
BPP5a	Bradykinin potentiating peptide 5a
BPTI	Bovine pancreatic trypsin inhibitor
t-Bu	*tert*-Butyl
Bzl	Benzyl
cHex	Cyclohexyl
CIP	2-Chloro-1,3-dimethylimidazolidinium hexafluorophosphate
Cl-Z	2-Chlorobenzyloxycarbonyl
Cl_2Bzl	2,6-Dichlorobenzyloxycarbonyl
DBU	1,8-Diazabicyclo[5.4.0]undec-7-ene
DCC	Dicyclohexylcarbodiimide
DCU	Dicyclohexylurea
DEPC	Diethylophosphorocyanidate
DIPCDI	Diisopropylcarbodiimide
DMAMP	4-(2′,4′-dimethoxyphenylaminomethyl)phenoxy
DMF	Dimethylformamide
DPPA	Diphenylphosphorylazide
DSA	Dimethoxyalkylsulfinylbenzhydrylamine
DSB	4-(2,5-Dimethyl-4-methylsulfinylphenyl)-4-hydroxybutanoic acid
EEDQ	1-Ethoxycarbonyl-2-ethoxy-1,2-dihydroquinoline
EGF	Epidermal growth factor
Fmoc	9-Fluorenylmethoxycarbonyl
GIP	Glucose-dependent insulinotropic polypeptide
hANP	Human atrial natriuretic peptide
HATU	2-(1*H*-7-azabenzotriazole-1-yl)-oxy-1,1,3,3-tetramethyl-uronium hexafluorophosphate
hBNP	Human brain natriuretic peptide
HBTU	2-(1*H*-Benzotriazole-1-yl)-oxy-1,1,3,3-tetramethyluronium hexafluorophosphate
HIV	Human immunodeficiency virus
HMPA	Hexamethylphosphoric triamide

HOAt	1-Hydroxy-7-azabenzotriazole
HOBt	*N*-Hydroxybenzotriazole
HONB	*N*-Hydroxy-5-norbornene-2,3-dicarboximide
HOOBt	3-Hydroxy-4-oxo-3,4-dihydro-1,2,3-benzotriadine
HOSu	*N*-Hydroxysuccinimide
HOQ	Hydroxyquinoline
HPLC	High-performance liquid chromatography
IGF	Insulin-like growth factor
IIDQ	1-Isobutyloxycarbonyl-2-isobutyloxy-1,2-dihydroquinoline
Mbh	4,4-Dimethoxybenzhydryl
MBHA	*p*-Methylbenzhydrylamine
MeBzl	4-Methylbenzyl
MeOBzl	4-Methoxybenzyl
MSA	Methanesulfonic acid
Msob	4-Methylsulfinylbenzyl
Msz	4-Methylsulfinylbenzyloxycarbonyl
Mtr	4-Methoxy-2,3,6-trimethylbenzenesulfonyl
Mts	Mesitylene-2-sulfonyl
Mtz	4-Methylthiobenzyloxycarbonyl
NDPP	Norborn-5-ene-2,3-dicarboximido diphenyl phosphate
NO_2Bzl	4-Nitrobenzyl
Npys	3-Nitro-2-pyridylsulfenyl
PAM	Phenylacetamidomethyl
pBNP	Porcine brain natriuretic peptide
Peoc	2-(Triphenylphosphino)ethyloxycarbonyl
Pmc	2,2,5,7,8-pentamethylchroman-6-sulfonyl
Ppt	Diphenylphosphinothioyl
i-Pr	Isopropyl
PyBOP	Benzotriazole-1-yloxytrispyrrolidinophosphonium hexafluorophosphate
PyBroP	Bromotrispyrrolidinophosphonium hexafluorophosphate
SCAL	Safety-catch acid labile
Sce-Cl	Carboethoxy sulfenyl chloride
SFGF	Shope fibromavirus growth factor
SPPS	Solid-phase peptide synthesis
Tacm	Trimethylacetamidomethyl
TBAF	Tetra-*n*-butylammonium fluoride trihydrate
Tcboc	2,2,2-Trichloro-*tert*-butoxycarbonyl
TFA	Trifluoroacetic acid
TFMSA	Trifluoromethanesulfonic acid
TGF	Transforming growth factor
TMSOTf	Trimethylsilyl trifluoromethanesulfonate
TMSBr	Trimethylsilyl bromide
THFγ2	Thymic humoral factor γ_2

TOPPipU	2-[2-Oxo-1(2*H*)-pyridyl]-1,1,3,3-bispentamethyleneuronium tetrafluoroborate
Tos	*p*-Toluenesulfonyl
Troc	2,2,2-Trichloroethyloxycarbonyl
Trt	Trityl
Z	Benzyloxycarbonyl
Z(OMe)	4-Methoxybenzyloxycarbonyl

Acknowledgments

We are grateful to Dr. Kenichi Akaji, Mr. Yoichi Fujiwara, and Mr. Tooru Kimura for valuable assistance in preparation of the manuscript.

References

Acharya, A. S., and Taniuchi, H. (1976). *J. Biol. Chem.* **251,** 6934–6946.

Agarwal, K. L., Kenner, G. W., and Sheppard, R. C. (1969) *J. Chem. Soc. C,* 2213–2217.

Ahmed, A. K., Shaffer, S. W. and Wetlaufer, D. B. (1975). *J. Biol. Chem.* **250,** 8477–8482.

Akabori, S., Sakakibara, S., Shimonishi, Y., and Nobuhara, Y. (1964). *Bull. Chem. Soc. Jpn.* **37,** 433–434.

Akaji, K., Fujii, N., Tokunaga, F., Miyata, T., Iwanaga, S., and Yajima, H. (1989). *Chem. Pharm. Bull.* **37,** 2661–2664.

Akaji, K., Yoshida, M., Tatsumi, T., Kimura, T., Fujiwara, Y., and Kiso, Y. (1990). *J. Chem. Soc., Chem. Commun.,* 288–290.

Akaji, K., Tatsumi, T., Yoshida, M., Kimura, T., Fujiwara, Y., and Kiso, Y. (1991). *J. Chem. Soc., Chem. Commun.,* 167–168.

Akaji, K., Fujino, K., Tatsumi, T., and Kiso, Y. (1992a). *Tetrahedron Lett.* **33,** 1073–1076.

Akaji, K., Tatsumi, T., Yoshida, M., Kimura, T., Fujiwara, Y., and Kiso, Y. (1992b). *J. Am. Chem. Soc.* **114,** 4137–4143.

Akaji, K., Kuriyama, N., Kimura, T., Fujiwara, Y., and Kiso, Y. (1992c). *Tetrahedron Lett.* **33,** 3177–3180.

Akaji, K., Nakagawa, Y., Fujiwara, Y., Fujino, K., and Kiso, Y. (1993a). *Chem. Pharm. Bull.* **47,** 1244–1248.

Akaji, K., Fujino, K., Tatsumi, T., and Kiso, Y. (1993b). *J. Am. Chem. Soc.* **115,** 11384-11392.

Akaji, K., Kuriyama, N., and Kiso, Y. (1994). *Tetrahedron Lett.* **35,** 3315–3318.

Alakhov, Y. B., Kiryushkin, A. A., Lipkin, V. M., and Milne, G. W. (1970). *J. Chem. Soc., Chem. Commun.*,406–407.

Albericio, F., Andreu, D., Giralt, E., Navalpotro, C., Pedroso, E., Ponsati, B., and Ruiz-Gayo, M. (1989). *Int. J. Pept. Protein Res.* **34,** 124–128.

Albertson, N. F. (1962). *In* "Organic Reactions," Vol. 12, pp. 157–355. Wiley, New York.

Almquist, R., Kadambi, S. R., Yasuda, D. M., Weitl, F. L., Polgar, W. E., and Toll, L. R. (1989) *Int. J. Pept. Protein Res.* **34,** 455–462.

Anderson, G. W., Zimmerman, J. E., and Callahan, F. (1963). *J. Am. Chem. Soc.* **85,** 3039.

Anderson, G. W., Zimmerman, J. E., and Callahan, F. (1964). *J. Am. Chem. Soc.* **86,** 1839–1842.

Anfinsen, C. B. (1973). *Science* **181,** 223–230, and references cited therein.

Antonovics, I., and Young, G. T. (1967). *J. Chem. Soc.* 595–601.

Atherton, E., Logan, C. J., and Sheppard, R. C. (1981). *J. Chem. Soc., Perkin Trans. 1,* 538–546.

Atherton, E., Sheppard, R. C., and Wade, J. D. (1983). *J. Chem. Soc., Chem. Commun.,* 1060–1062.

Atherton, E., Sheppard, R. C., and Wade, J. D. (1985). *J. Chem. Soc., Perkin Trans. 1,* 2065–2073.

Bachi, M. D., and Ross-Petersen, K. J. (1972). *J. Org. Chem.* **37,** 3550–3551.

Belleau, B., and Malek, G. (1968). *J. Am. Chem. Soc.* **90,** 1651–1652.

Bernatowicz, M. S., Matsueda, R., and Matsueda, G. R. (1986). *Int. J. Pept. Protein Res.* **28,** 107–112.

Bertho, J.-N., Loffet, A., Pinel, C., Reuther, F., and Sennyey, G. (1991). *Tetrahedron Lett.* **32,** 1303–1306.

Beyerman, H. C., Lie, T. S., and van Veldhuizen, C. J. (1973). *In* "Peptides 1971" (H. Nesvadba, ed.), pp. 162–164. North-Holland Publ., Amsterdam.

Blake, J. (1981). *Int. J. Pept. Protein Res.* **17,** 273–274.

Blake, J. (1986). *Int. J. Pept. Protein Res.* **27,** 191–200.

Blake, J., and Li, C.H. (1983). *Proc. Natl. Acad. Sci. U.S.A.* **80,** 1556–1559.

Blohm, D., Bollschweiler, C., and Hillen, H. (1988). *Angew. Chem., Int. Ed. Engl.* **27,** 207–225.

Bodanszky, M. (1955). *Nature (London)* **175,** 685–686.

Bodanszky, M., and du Vigneaud, V. (1959). *J. Am. Chem. Soc.* **81,** 5688–5691.

Bodanszky, M., Szelke, M., Tomorkiny, E., and Weisz, E. (1957). *Acta Chim. Acad. Sci. Hung.* **11,** 179–184.

Boissonnas, R. A., Guttmann, S., and Jaquenoud, P. A. (1960). *Helv. Chim. Acta* **43,** 1349–1358.

Bray, A. M., Maeji, N. J., and Geysen, H. M. (1990). *Tetrahedron Lett.* **31,** 5811–5814.

Buku, A., Blandina, P., Birr, C., and Gazis, D. (1989). *Int. J. Pept. Protein Res.* **33,** 86–93.

Cabre, J., and Palomo, A. L. (1984). *Synthesis,* 413–417.

Callahan, F. M., Anderson, G. W., Paul, R., and Zimmerman, J. E. (1963). *J. Am. Chem. Soc.* **85,** 201–207.

Carpino, L. A. (1993). *J. Am. Chem. Soc.* **115,** 4397–4398.

Carpino, L. A., Cohen, B. J., Stephens, Jr., K. E., Sadat-Aalaee, S. Y., Tien, J.-H., and Langridge, D. C. (1986). *J. Org. Chem.* **51,** 3732–3734.

Carpino, L. A., Monsour, E. M. E., and Sadat-Aalaee, D. (1991). *J. Org. Chem.* **56,** 2611–2614.

Castro, B., Dormoy, J. R., Evin, G., and Selve, C. (1975). *Tetrahedron Lett.,* 1219–1222.

Castro, B., Dormoy, J. R., Evin, G., and Selve, C. (1977). *J. Chem. Res. (Suppl.)* 182.

Chakravarty, P. K., and Olsen, P. K. (1978). *J. Org. Chem.* **43,** 1270–1271.

Chavez, L. G., Jr., and Scheraga, H. A. (1980). *Biochemistry* **19,** 996–1004.

Chino, N., Kumagaye, K., Noda, Y., Watanabe, T., Kimura, T., and Sakakibara, S. (1986). *Biochem. Biophys. Res. Commun.* **141,** 665–672.

Coste, J., Le-Nguyen, D., and Castro, B. (1990a). *Tetrahedron Lett.* **31,** 205–208.

Coste, J., Dufour, M.-N., Pantaloni, A., and Castro, B. (1990b). *Tetrahedron Lett.* **31,** 669–672.

Crea, R., Kraszewski, A., Hirose, T., and Itakuta, K. (1978). *Proc. Natl. Acad. Sci. U.S.A.* **75,** 5765–5769.

Creighton, T. E. (1980). *J. Mol. Biol.* **144,** 521–550.

Creighton, T. E., and Goldenberg, D. P. (1984). *J. Mol. Biol.* **179,** 497–526.

Cruz, L. J., Kupryszewski, G., LeCheminant, G. W., Gray, W. R., Olivera, B. M., and Rivier, J. (1989). *Biochemistry* **28,** 3437–3442.

Curtius, T. (1902). *Ber.* **35,** 3226–3228.

Curtius, T. (1904). *J. Prakt. Chem.* **70,** 57–128.

Detar, D. F., Silverstein, R., and Roger, F. F., Jr. (1966). *J. Am. Chem. Soc.* **88,** 1024–1030.

Dixon, G. H., and Wardlaw, A. C. (1960). *Nature (London)* **188,** 721–724.

Erickson, B. W., and Merrifield, R. B. (1973). *J. Am. Chem. Soc.* **95,** 3750–3756.

Feinberg, R. S., and Merrifield, R. B. (1975). *J. Am. Chem. Soc.* **97,** 3485–3496.

Fischer, E. (1903). *Chem. Ber.* **36,** 2094–2106.

Fujii, N., and Yajima, H. (1981). *J. Chem. Soc., Perkin Trans. 1,* 789–841.

Fujii, N., Ikemura, O., Funakoshi, S., Matsuo, H., Segawa, T., Nakata, Y., Inoue A., and Yajima, H. (1987a). *Chem. Pharm. Bull.* **35,** 1076–1084.

Fujii, N., Otaka, A., Ikemura, O., Akaji, K., Funakoshi, S., Hayashi, Y., Kuroda, Y., and Yajima, H. (1987b). *J. Chem. Soc., Chem. Commun.,* 274–275.

Fujii, N., Otaka, A., Ikemura, O., Hatano, M., Okamachi, A., Funakoshi, S., Sakurai, M., Shioiri, T.,

and Yajima, H. (1987c). *Chem. Pharm. Bull.* **35,** 3447–3452.
Fujii, N., Otaka, A., Sugiyama, N., Hatano, M., and Yajima, H. (1987d). *Chem. Pharm. Bull.* **35,** 3880–3883.
Fujii, N., Otaka, A., Watanabe, T., Okamachi, A., Tamamura, H., Yajima, H., Inagaki, Y., Nomizu, M., and Asano, K. (1989). *J. Chem. Soc., Chem. Commun.*, 283–284.
Fujino, M. (1976). *PRF (Protein Research Foundation Peptide Institute Ho)* **2,** 21.
Fujino, M., and Hatanaka, C. (1968). *Chem. Pharm. Bull.* **16,** 929–932.
Fujino, M., and Nishimura, O. (1976). *J. Chem. Soc., Chem. Commun.*, 998.
Fujino, M., Kobayashi, S., Obayashi, M., Fukuda, T., Shinagawa, S., and Nishimura, O. (1974). *Chem. Pharm. Bull.* **22,** 1857–1863.
Fujino, M., Nishimura, O., Wakimasu, M., and Kitada, C. (1980). *J. Chem. Soc., Chem. Commun.*, 668–669.
Fujino, M., Wakimasu, M., and Kitada, C. (1981). *Chem. Pharm. Bull.* **29,** 2825–2831.
Gacond, J. J., Bargetzi, J. P., and Juillerat, M. A. (1984). *FEBS Lett.* **178,** 114–118.
Galat, A., Degelaen, J. P., Yang, C. C., and Blout, E. R. (1981). *Biochemistry* **20,** 7415–7423.
Galpin, I. J., Hancock, F. E., Handa, B. K., Jackson, A. G., Kenner, G. W., McDowell, P., Noble, P., and Ramage, R. (1981). *Tetrahedron* **37,** 3043–3050.
Garsky, V. M., Lumma, P. K., Freidinger, R. M., Pitzenberger, S. M., Randall, W. C., Veber, D. F., Gould, R. J., and Friedman, P. A. (1989). *Proc. Natl. Acad. Sci. U.S.A.* **86,** 4022–4026.
Gawron, O., and Lieb, J. A. (1952). *J. Am. Chem. Soc.* **74,** 834.
Grassetti, D. R., and Murray, J. F., Jr. (1967). *Arch. Biochem. Biophys.* **119,** 41–49.
Gregory, J., Laird, A. H., Morley, J. S., and Smith J. M. (1968). *J. Chem. Soc. C*, 522–531.
Gross, H., and Bilk, L. (1968). *Tetrahedron* **24,** 6935–6939.
Gutte, B., and Merrifield, R. B. (1971). *J. Biol. Chem.* **246,** 1922–1941.
Guttmann, S. (1961). *Helv. Chim. Acta* **44,** 721–744.
Guttmann, S., Pless, J., and Boissonnas, R. A. (1965). *Acta Chim. Acad. Sci. Hung.* **44,** 141–142.
Heath, W. F., and Merrifield, R. B. (1986). *Proc. Natl. Acad. Sci. U.S.A.* **83,** 6367–6371.
Heimer, E. P., Chang, C. D., Lambros, T., and Meienhofer, J. (1981). *Int. J. Pept. Protein Res.* **18,** 237–241.
Henklein, P., Beyerman, M., Bienert, M., and Knorr, R. (1991). *In* "Peptides 1990" (E. Giralt and D. Andreu, eds.), p. 67–68. ESCOM, Leiden.
Hiskey, R. G. (1981). *In* "The Peptides" (E. Gross and J. Meienhofer, eds.), Vol. 3, pp. 137–167. Academic Press, Orlando, Florida.
Hiskey, R. G., Li, C., and Vunnam, R. R. (1975). *J. Org. Chem.* **40,** 3697–3703.
Hofmann, K., Magee, M. Z., and Lindenmann, A. (1950). *J. Am. Chem. Soc.* **72,** 2814–2815.
Hofmann, K., Lindenmann, A., Magee, M. Z., and Khan, N. H. (1952). *J. Am. Chem. Soc.* **74,** 470–476.
Hofmann, K., Thompson, T. A., Yajima, H., Schwartz, E. T., and Inouye, H. (1960). *J. Am. Chem. Soc.* **82,** 3715–3721.
Hojo, H., and Aimoto, S. (1991). *Bull Chem. Soc. Jpn.* **64,** 111–117.
Honzl, J., and Rudinger, J. (1961). *Collect. Czech. Chem. Commun.* **26,** 2333–2344.
Horiki, K. (1977). *Tetrahedron Lett.*, 1897–1900.
Huguenin, R. L. (1964). *Helv. Chim. Acta* **47,** 1934–1941.
Immer, H., Eberle, I., Fischer, W., and Moser, E. (1988). *In* "Peptides 1988" (G. Jung, and E. Bayer, eds.), pp. 94–96 (Proc. 20th Eur. Pept. Symp.) de Gruyter, Berlin.
Immer, H., Eberle, I., Moser, E., Bernath, E., and Pipkorn, R. (1989). *In* "Peptides," (J. E. Rivier and G. R. Marshall, eds.), pp. 1054–1056 (Proc. 11th Am. Pept. Symp.). ESCOM, Leiden.
Inouye, K., Watanabe, K., Morihara, K., Tochino, Y., Kanaya, T., Emura, J., and Sakakibara, S. (1979). *J. Am. Chem. Soc.* **101,** 751–752.
Iwai, M., Kobayashi, M., Tamura, K., Ishii, Y., Yamada, H., and Niwa, M. (1989). *J. Biochem.* **106,** 949–951.
Jakubke, H. D. (1965). *Z. Naturforsch.* **20B,** 273–274.
Jakubke, H. D., and Voigt, A. (1966a). *Chem. Ber.* **99,** 2419–2429.

Jakubke, H. D., and Voigt, A. (1966b). *Chem. Ber.* **99,** 2944–2954.

Kamber, B., Hartmann, A., Eisler, K., Riniker, B., Rink, H., Sieber, P., and Rittel, W. (1980). *Helv. Chim. Acta* **63,** 899–915.

Kangawa, K., Fukuda, A., and Matsuo, H. (1985). *Nature (London)* **313,** 397–400.

Kazmierski, W., Wire, W. S., Lui, G. K., Knapp, R. J., Shook, J. E., Burks, T. F., Yamamura, H. I., and Hruby, V. (1988). *J. Med. Chem.* **31,** 2170–2177.

Kemp, D. S., and Fotouhi, N. (1987). *Tetrahedron Lett.* **28,** 4637–4640.

Kemp, D. S., and Galakatos, N. G. (1986). *J. Org. Chem.* **51,** 1821–1829.

Kemp, D. S., Wand, S. W., Busby, G., and Hugel, G. (1970). *J. Am. Chem. Soc.* **92,** 1043–1055.

Kemp, D. S., Galakatos, N. G., Bowen, B., and Tan, K. (1986). *J. Org. Chem.* **51,** 1829–1838.

Kimura, T., Takai, M., Masui, Y., Morikawa, T., and Sakakibara, S. (1981). *Biopolymers* **20,** 1823–1832.

Kimura, T., Takai, Yoshizawa, K., and Sakakibara, S. (1988). *Biochem. Biophys. Res. Commun.* **151,** 1285–1292.

Kiso, Y., and Kimura, T. (1991). *J. Synth. Org. Chem. Jpn.,* 1046–1047.

Kiso, Y., and Yajima, H. (1972). *J. Chem. Soc., Chem. Commun.,* 942–943.

Kiso, Y., Kai, Y., and Yajima, H. (1973). *Chem. Pharm. Bull.* **21,** 2507–2510.

Kiso, Y., Nakamura, S., Ito, K., Ukawa, K., Kitagawa, K., Akita, T., and Moritoki, H. (1979). *J. Chem. Soc., Chem. Commun.,* 971–972.

Kiso, Y., Miyazaki, T. Y., Satomi, M., Hiraiwa, H., and Akita, T. (1980a). *J. Chem. Soc., Chem. Commun.,* 1029–1030.

Kiso, Y., Ukawa, K., and Akita, T. (1980b). *J. Chem. Soc., Chem. Commun.,* 101–102.

Kiso, Y., Ukawa, K., Nakamura, S., Ito, K., and Akita, T. (1980c). *Chem. Pharm. Bull.* **28,** 673–676.

Kiso, Y., Shimokura, M., Hosoi, S., Fujisaki, T., Fujiwara, Y., and Yoshida, M. (1987a). *J. Protein. Chem.* **6,** 147–162.

Kiso, Y., Shimokura, M., Kimura, T., Mimoto, T., and Fujisaki, T. (1987b). *In* "Peptide Chemistry 1986" (T. Miyazawa, ed.), pp. 211–216. Protein Research Foundation, Osaka, Japan.

Kiso, Y., Kimura, T., Fujiwara, Y., Shimokura, M., and Nishitani, A. (1988a). *Chem. Pharm. Bull.* **36,** 5024–5027.

Kiso, Y., Yoshida, M., Kimura, T., Shimokura, M., and Fujisaki, T. (1988b). *In* "Peptides: Chemistry and Biology" (G. R. Marshall, ed.), pp. 229–231. ESCOM, Leiden.

Kiso, Y., Kimura, T., Yoshida, M., Shimokura, M., Akaji, K., and Mimoto T. (1989a). *J. Chem. Soc., Chem. Commun.,* 1511–1513.

Kiso, Y., Yoshida, M., Tatsumi, T., Kimura, T., Fujiwara, Y., and Akaji, K. (1989b). *Chem. Pharm. Bull.* **37,** 3432–3434.

Kiso, Y., Kimura, T., Fujiwara, Y., Sakikawa, H., and Akaji, K. (1990a). *Chem. Pharm. Bull.* **38,** 270–272.

Kiso, Y., Yoshida, M., Fujiwara, Y., Kimura, T., Shimokura, M., and Akaji, K. (1990b). *Chem. Pharm. Bull.* **38,** 673–675.

Kiso, Y., Yoshida, M., Kimura, T., Fujiwara, Y., Shimokura, M., and Akaji, K. (1990c). *Chem. Pharm. Bull.* **38,** 1192–1199.

Kiso, Y., Akaji, K., Yoshida, M., Tatsumi, T., Fujiwara, Y., and Kimura, T. (1991a). *In* "Peptides 1990" (E. Giralt and D. Andreu, eds.), pp. 92–93. ESCOM, Leiden.

Kiso, Y., Tanaka, S., Kimura, T., Itoh, H., and Akaji, K. (1991b). *Chem. Pharm. Bull.* **39,** 3097–3099.

Kiso, Y., Kimura, T., Itoh, H., Tanaka, S., and Akaji, K. (1992a). *In* "Peptides," (J. A. Smith and J. E. Rivier, eds.), pp. 533–534 (Proc. 12th Am. Pept. Symp.). ESCOM, Leiden.

Kiso, Y., Fujiwara, Y., Kimura, T., Nishitani, A., and Akaji, K. (1992b). *Int. J. Pept. Protein Res.* **40,** 308–314.

Kiso, Y., Fukui, T., Tanaka, S., Kimura, T., and Akaji, K. (1992c). *In* "Peptide Chemistry 1991" (A. Suzuki, ed.), pp. 187–192. Protein Research Foundation, Osaka, Japan.

Kiso, Y., Itoh, H., Tanaka, S., Kimura, T., and Akaji, K. (1993). *Tetrahedron Lett.* **34,** 7599–7602.

Kiso, Y., Fukui, T., Tanaka, S., Kimura, T., and Akaji, K. (1994). *Tetrahedron Lett.* **35,** 3571–3574.

Knorr, R., Trzecial, A., Bannwarth, W., and Gillesen, D. (1989). *Tetrahedron Lett.* **30,** 1927–1930.

König, W., and Geiger, R. (1970a). *Chem. Ber.* **103,** 788–798.
König, W., and Geiger, R. (1970b). *Chem. Ber.* **103,** 2024–2034.
König, W., and Geiger, R. (1970c). *Chem. Ber.* **103,** 2034–2040.
König, W., and Geiger, R. (1973). *Chem. Ber.* **106,** 3626–3635.
Konig, W., and Kernebeck, K. (1979). *Liebigs Ann. Chem.,* 227–247.
Kopple, K. D., and Renik, R. J. (1958). *J. Org. Chem.* **23,** 1565–1567.
Kouge, K., Koizumi, T., and Okai, H. (1986). *Chem. Express* **1,** 499–502.
Kouge, K., Koizumi, T., Okai, H., and Kato, T. (1987). *Bull Chem. Soc. Jpn.* **60,** 2409–2418.
Kovacs, J., and Ceprini, M. Q. (1965). *Chem. Ind. (London),* 2100.
Kovacs, J., Ceprini, M. Q., Dupraz, C. A., and Schmit, G. N. (1967a). *J. Org. Chem.* **32,** 3696–3698.
Kovacs, J., Kisfaludy, L., and Ceprini, M. Q. (1967b). *J. Am. Chem. Soc.* **89,** 183–184.
Kumagaye, S., Kuroda, H., Nakajima, K., Watanabe, T., Kimura, T., Masaki, T., and Sakakibara, S. (1988). *Int. J. Pept. Protein Res.* **32,** 519–526.
Kunz, H., and Bechtolsheimer, H.-H. (1982). *Liebigs Ann. Chem.,* 2068–2078.
Kupryszewski, G., and Formela, M. (1961). *Rocz. Chem.* **35,** 1533–1536.
Le-Nguyen, D., and Rivier, J. (1986). *Int. J. Pept. Protein Res.* **27,** 285–292.
Le-Nguyen, D., Nalis, D., and Castro, B. (1989). *Int. J. Pept. Protein Res.* **34,** 492–497.
Lin, Y. Z., Caporaso, G., Chang, P. Y., Ke, X.-H., and Tam, J. P. (1988). *Biochemistry* **27,** 5640–5645.
Lloyd, K., and Young, G. T. (1968). *J. Chem. Soc., Chem. Commun.,* 1400–1401.
Lloyd, K., and Young, G. T. (1971). *J. Chem. Soc., C.,* 2890–2896.
Lyle, T. A., Brady, S. F., Ciccarone, T. M., Colton, C. D., Paleveda, W. J., Veber, D. F., and Nutt, R. F. (1987). *J. Org. Chem.* **52,** 3752–3759.
McBride, N. J., and Kemp, D. S. (1987). *Tetrahedron Lett.* **28,** 3435–3438.
McIntosh, M., Cruz, L. J., Hunkapiller, M. W., Gray, W. R., and Olivera, B. M. (1982). *Arch. Biochem. Biophys.* **218,** 329–334.
Martinez, J., Bali, J.-P., Rodriguez, M., Castro, B., Laur, J., and Lignon, M.-F. (1988). *J. Med. Chem.* **28,** 1874–1879.
Masui, Y., Chino, N., and Sakakibara, S. (1980). *Bull Chem. Soc. Jpn.* **53,** 464–468.
Matsuda, F., Itoh, S., Hattori, N., Yanagiya, M., and Matsumoto, T. (1985). *Tetrahedron* **41,** 3625–3631.
Matsueda, R., and Walter, R. (1980). *Int. J. Pept. Protein Res.* **16,** 392–401.
Matsueda, R., Kimura, T., Kaiser, T. T., and Matsueda, G. R. (1981). *Chem. Lett.,* 737–740.
Matsueda, R., Higashida, S., Ridge, R. J., and Matsueda, G. R. (1982). *Chem. Lett.,* 921–924.
Matsuzaki, K., Harada, M., Handa, T., Funakoshi, S., Fujii, N., Yajima, H., and Miyajima, K. (1989). *Biochim. Biophys. Acta* **981,** 130–134.
Mazur, R. H. (1968). *Experientia* **24,** 661.
Meienhofer, J. (1979). *In* "Peptides: Analysis, Synthesis, Biology" (E. Gross and J. Meienhofer, eds.), Vol. 1, pp. 263–314. Academic Press, New York.
Meng, H., Burleigh, B. D., and Kelly, G. M. (1988). *J. Chromatogr.* **443,** 183–192.
Menz, A., Bouet, F., Guschlbauer, W., and Fromageot, P. (1980). *Biochemistry* **19,** 4166–4172.
Meseguer, J. D., Liizarbe, J. F., Palomo, A. L., and Bilbao, A. Z. (1980). *Synthesis,* 547–551.
Minamitake, Y., Furuya, M., Kitajima, Y., Takehisa, M., and Tanaka, S. (1990). *Chem. Pharm. Bull.* **38,** 1920–1926.
Nishimura, O., Kitada, C., and Fujino, M. (1978). *Chem. Pharm. Bull.* **26,** 1576–1585.
Nishiuchi, Y., and Sakakibara, S. (1982). *FEBS Lett.* **148,** 260–262.
Nishiuchi, Y., and Sakakibara, S. (1983). *In* "Peptide Chemistry 1983" (E. Munekata, ed.), pp. 191–196. Protein Research Foundation Osaka, Japan.
Nomizu, M., Akaji, K., Fukata, J., Imura, H., Inoue, A., Nakata, Y., Segawa, T., Fujii, N., and Yajima, H. (1988). *Chem. Pharm. Bull.* **36,** 122–133.
Nomizu, M., Inagaki, Y., Iwamatsu, A., Kashiwabara, T., Ohta, H., Morita, A., Nishikori, K., Otaka, A., Fujii, N., and Yajima, H. (1989). *In* "Peptides" (J. E. Rivier, G. R. Marshall, eds.), pp. 276–278 (Proc. 11th Am. Pept. Symp.). ESCOM, Leiden.
Numata, T., Togo, H., and Oae, S. (1979). *Chem. Lett.,* 329–332.
Nutt, R. F., Brady, S. F., Darke, P. L., Ciccarone, T. M., Colton, C. D., Nutt, E. M., Rodkey, J. A.,

Bennett, C. D., Waxman, L. H., Sigal, I. S., Anderson, P. S., and Veber, D. F. (1988). *Proc. Natl. Acad. Sci. U.S.A.* **85,** 7129–7133.
Omodei-Sale, A., Sindona, G., Sola, D., and Uccella, N. (1984). *J. Chem. Res. (Suppl.),* 50–51.
Patek, M. (1993). *Int. J. Pept. Protein Res.* **42,** 97–117.
Patek, M., and Lebl, M. (1990). *Tetrahedron Lett.* **31,** 5209–5212.
Patek, M., and Lebl, M. (1991). *Tetrahedron Lett.* **32,** 3891–3894.
Patek, M., and Lebl, M. (1992). *Collect. Czech. Chem. Commun.* **57,** 508–524.
Pearson, D. A., Blanchette, M., Baker, M. L., and Guindon, C. A. (1989). *Tetrahedron Lett.* **30,** 2739–2742.
Perraudin, J. P., Torchia, T. E., and Wetlaufer, D. B. (1983). *J. Biol. Chem.* **258,** 11834–11839.
Photaki, I., Taylor-Papadimitriou, J., Sakarellos, C., Mazarakis, P., and Zervas, L. (1970). *J. Chem. Soc. (C),* 2683–2687.
Pizey, S. S. (1974). *In* "Synthetic Reagents," Vol. 1, Chap. 4, p. 335. Wiley, New York.
Pless, J., and Boissonnas, R. A. (1963a). *Helv. Chim. Acta* **46,** 1609–1625.
Pless, J., and Boissonnas, R. A. (1963b). *Helv. Chim. Acta* **46,** 1625–1636.
Pless, J., and Boissonnas, R. A. (1963c). *Helv. Chim. Acta* **46,** 1637–1669.
Prasad, K. U., Iqbal, M. A., and Urry, D. W. (1985). *Int. J. Pept. Protein Res.* **25,** 408–413.
Previero, A., Antonia, M., Previero, C., and Cavadore, J. C. (1967). *Biochim. Biophys. Acta* **147,** 453–461.
Ramachandran, J., and Li, C. H. (1963). *J. Org. Chem.* **28,** 173–177.
Ramage, R., and Green, J. (1988). *Tetrahedron Lett.* **28,** 2287–2290.
Rich, D. H., and Singh, J. (1979). *In* "The Peptides: Analysis, Synthesis, Biology," Vol. 1, pp. 241–261. Academic Press, New York.
Roth, M., and Mazanek, J. (1974). *Liebigs Ann. Chem.,* 439–459.
Ruiz-Gayo, M., Albericio, F., Pons, M., Royo, M., Pedroso, E., and Giralt, E. (1988). *Tetrahedron Lett.* **29,** 3845–3848.
Rzeszotarska, B., and Vlasov, G. P. (1967). *Bull. Acad. Pol. Sci. Ser. Sci. Chim.* **15,** 143–147.
Sabatier, J.-M., Darbon, H., Fourquet, P., Rochat, H., and Rietschoten, J. V. (1987). *Int. J. Pept. Protein Res.* **30,** 125–134.
Saito, Y., Yamada, H., Niwa, M., and Ueda, I. (1987). *J. Biochem. (Tokyo)* **101,** 123–134.
Sakakibara, S., and Inukai, N. (1965). *Bull Chem. Soc. Jpn.* **38,** 1979–1984.
Sakakibara, S., Shimonishi, Y., Kishida, Y., Okada, M., and Sugihara, H. (1967). *Bull Chem. Soc. Jpn.* **40,** 2164–2167.
Samanen, J. M., and Brandeis, E. (1985). *In* "Peptides: Structure and Function" (C. Deber and K. Kopple, eds.), pp. 225–228. Pierce Chemical, Rockford, Illinois.
Samanen, J. M., and Brandeis, E. (1988). *J. Org. Chem.* **53,** 561–569.
Sano, S., and Kawanishi, S. (1975). *J. Am. Chem. Soc.* **97,** 3480–3484.
Santere, R. F., Cook, R. A., Crisel, R. M. D., Sharp, J. D., Schmidt, R. J., Williams, D. C., and Wilson, C. P. (1981). *Proc. Natl. Acad. Sci. U.S.A.* **78,** 4339–4343.
Scanlon, D. B., Eefting, M. A., Lloyd, D. J., Burgess, A. W., and Simpson, R. J. (1987). *J. Chem. Soc., Chem. Commun.,* 516–518.
Schellenberg, P., and Ullrich, J. (1959). *Chem. Ber.* **92,** 1276–1287.
Schnolzer, M., and Kent, S. B. H. (1992). *Science* **256,** 221–225.
Schussler, H., and Zahn, H. (1962). *Chem. Ber.* **95,** 1076–1080.
Schwyzer, R., Wegman, E. S., and Dietrich, H. (1960). *Chimia* **14,** 366.
Sheehan, J. C., and Hess, G. P. (1955). *J. Am. Chem. Soc.* **77,** 1067–1068.
Sheehan, J. C., and Yang, D. D. (1958). *J. Am. Chem. Soc.* **80,** 1154–1158.
Sheehan, J. C., Goodman, M., and Hess, G. P. (1956). *J. Am. Chem. Soc.* **78,** 1367–1369.
Sheehan, J. C., Cruickshank, P. A., and Boshart, G. L. (1961). *J. Org. Chem.* **26,** 2525–2528.
Shimokura, M., Hosoi, S., Okamoto, K., Fujiwara, Y., Yoshida, M., and Kiso, Y. (1985). *Peptide Chemistry 1984,* 235–240.
Shioiri, T., and Yamada, S. (1974). *Chem. Pharm. Bull.* **22,** 855–858.
Shioiri, T., Ninomiya, K., and Yamada, S. (1972). *J. Am. Chem. Soc.* **94,** 6203–6205.
Shioiri, T., Yokoyama, Y., Kasai, Y., and Yamada, S. (1976a). *Tetrahedron* **32,** 2211–2217.

Shioiri, T., Yokoyama, Y., Kasai, Y., and Yamada, S. (1976b). *Tetrahedron* **32,** 2854.
Spear, K. L., Brown, M. S., Olins, G. M., and Patton, D. R. (1989). *J. Med. Chem.* **32,** 1094–1098.
States, D. J., Creighton, T. E., Dobson, C. M., and Karplus, M. (1987). *J. Mol. Biol.* **195,** 731–739.
Stewart, J. M., and Woolley, D. W. (1965). *Nature (London)* **206,** 619–620.
Strumer, E., Huguenin, R. L., Boissonnas, R. A., and Berde, B. (1965). *Experientia* **21,** 583–585.
Sudoh, T., Maekawa, K., Kojima, M., Minamino, N., Kangawa, K., and Matsuo, H. (1989). *Biochem. Biophys. Res. Commun.* **159,** 1427–1434.
Sugiyama, N., Fujii, N., Funakoshi, S., Funakoshi, A., Miyasaka, K., Aono, M., Moriga, M., Inoue, K., Kogire, M., Sumi, S., Doi, R., Tobe, T., and Yajima, H. (1987). *Chem. Pharm. Bull.* **35,** 3585–3596.
Szajewski, R. P., and Whitesides, G. M. (1980). *J. Am. Chem. Soc.* **102,** 2011–2026.
Takashima, H., du Vigeaud, V., and Merrifield, R. B. (1968). *J. Am. Chem. Soc.* **80,** 1323–1325.
Tam, J. P. (1987). *Int. J. Pept. Protein Res.* **29,** 421–431.
Tam, J. P., Heath, W. F., and Merrifield, R. B. (1983a). *J. Am. Chem. Soc.* **105,** 6442–6455.
Tam, J. P., Heath, W. F., and Merrifield, R. B. (1983b). *Int. J. Pept. Protein Res.* **21,** 57–65.
Tam, J. P., Sheikh, M. A., Solomon, D. D., and Ossowski, L. (1986). *Proc. Natl. Acad. Sci. U.S.A.* **83,** 8082–8086.
Taniyama, Y., Yamamoto, Y., Kuroki, R., and Kikuchi, M. (1990). *J. Biol. Chem.* **265,** 7570–7575.
Taschner, E., Rzeszotarska, B., and Lubiewska, L. (1965a). *Ann. Chem.* **690,** 177–181.
Taschner, E., Rzeszotarska, B., and Lubiewska, L. (1965b). *Angew. Chem.* **77,** 619.
Tou, J., McGrath, M. F., Zupec, M. E., Byatt, J. C., Violand, B. N., Kaempfe, L. A., and Vineyard, B. D. (1990). *Biochem. Biophys. Res. Commun.* **167,** 484–491.
Tung, R. D., and Rich, D. H. (1985). *J. Am. Chem. Soc.* **107,** 4342–4343.
Tung, R. D., Dhaon, M. K., and Rich, D. H. (1986). *J. Org. Chem.* **51,** 3350–3354.
Ueki, M., and Amemiya, A. (1987). *Tetrahedron Lett.* **28,** 6617–6620.
Ueki, M., and Inazu, T. (1982). *Chem. Lett.*, 45–48.
Ueki, M., Inazu, T., and Ikeda, S. (1979). *Bull Chem. Soc. Jpn.* **52,** 2424–2427.
Ueki, M., Saito, T., Sasaya, J., Ikeda, S., and Oyamada, H. (1988). *Bull Chem. Soc. Jpn.* **61,** 3653–3657.
Van Der Auwera, C., and Anteunis, M. J. O. (1987). *Int. J. Pept. Protein Res.* **29,** 574–588.
Vaughan, J. R., Jr., J. R., and Osato, R. L. (1951). *J. Am. Chem. Soc.* **73,** 5553–5555.
Veber, F. F., Milkowski, J. D., Denkewalter, R. G., and Hirschmann, R. (1968). *Tetrahedron Lett.*, 3057–3058.
Veber, D. F., Milkowski, J. D., Varga, S. L., Denkewalter, R. G., and Hirschmann, R. (1972). *J. Am. Chem. Soc.* **94,** 5456–5461.
Wade, J. D., Bedford, J., Sheppard, R. C., and Tregear, G. W. (1991). *Pept. Res.* **4,** 194–199.
Watanabe, H., Kubota, M., Yajima, H., Tanaka, A., Nakamura, M., and Kawabata, T. (1974). *Chem. Pharm. Bull.* **22,** 1889–1894.
Weygand, F., Hoffmann, D., and Wünsch, E. (1966). *Z. Naturforsch.* **21B,** 426–428.
Weygand, F., Steglich, W., and Chytil, N. (1968). *Z. Naturforsch.* **23B,** 1391–1392.
Wieland, T., and Jaenicke, F. (1956). *Angew. Chem.* **599,** 125–130.
Wieland, T., and Stimming, D. (1953). *Angew. Chem.* **579,** 97–106.
Wieland, T., Heinke, B., Vogeler, K., and Morimoto, H. (1962). *Angew. Chem.* **655,** 189–194.
Williams, A., and Ibrahim, I. T. (1981). *Chem. Rev.* **81,** 589–636.
Woo, D. D. L., Clark-Lewis, I., Chait, B. T., and Kent, S. B. H. (1989). *Protein Eng.* **3,** 29–37.
Wünsch, E. (1967). *Z. Naturforsch.* **22B,** 1269–1276.
Wünsch, E., and Drees, F. (1966). *Chem. Ber.* **99,** 110–120.
Wünsch, E., and Wendlberger, G. (1968). *Chem. Ber.* **101,** 3659–3663.
Wünsch, E., Jaeger, E., and Schart, R. (1968). *Chem. Ber.* **101,** 3664–3670.
Wünsch, E., Jaeger, E., Kisfaludy, L., and Low, M. (1977). *Angew. Chem.* **89,** 330–331.
Yajima, H., and Fujii, N. (1980). *J. Chem. Soc., Chem. Commun.*, 115–116.
Yajima, H., and Fujii, N. (1981). *J. Am. Chem. Soc.* **103,** 5867–5871.
Yajima, H., and Kiso, Y. (1971). *Chem. Pharm. Bull.* **19,** 420–423.

Yajima, H., Kurobe, M., and Koyama, K. (1973a). *Chem. Pharm. Bull.* **21,** 1612.

Yajima, H., Kurobe, M., and Koyama, K. (1973b). *Chem. Pharm. Bull.* **21,** 2566–2567.

Yajima, H., Fujii, N., Ogawa H., and Kawatani, H. (1974). *J. Chem. Soc., Chem. Commun.*, 107–108.

Yajima, H., Akaji, K., Fujii, N., Hayashi, K., Mizuta, K., Aono, M., and Moriga, M. (1984). *J. Chem. Soc., Chem. Commun.*, 1103.

Yajima, H., Akaji, K., Fujii, N., Hayashi, K., Mizuta, K., Aono, M., and Moriga, M. (1985a). *Chem. Pharm. Bull.* **33,** 184–201.

Yajima, H., Fujii, N., Akaji, K., Sakurai, M., Nomizu, M., Mizuta, K., Aono, M., Moriga, M., Inoue, K., Hosotani, R., and Tobe, T. (1985b). *Chem. Pharm. Bull.* **33,** 3578–3581.

Yajima, H., Fujii, N., Akaji, K., Sakurai, M., Nomizu, M., Mizuta, K., Aono, M., Moriga, M., Inoue, K., Hosotani, R., and Tobe, T. (1986). *Chem. Pharm. Bull.* **34,** 2397–2410.

Yamada, S., Kasai, Y., and Shioiri, T. (1973). *Tetrahedron Lett.*, 1595–1598.

Yamashiro, D., and Li, C. H. (1986). *Int. J. Pept. Protein Res.* **31,** 322–334.

Yoshida, M., Akaji, K., Tatsumi, T., Iinuma, S., Fujiwara, Y., Kimura, T., and Kiso, Y. (1990a). *Chem. Pharm. Bull.* **38,** 273–275.

Yoshida, M., Akaji, K., Tatsumi, T., Fujiwara, Y., Kimura, T., and Kiso, Y. (1990b). *In* "Peptide Chemistry 1989" (N. Yanaihara, ed.), pp. 33–38. Protein Research Foundation Osaka, Japan.

Yoshida, M., Tatsumi, T., Fujiwara, Y., Iinuma, S., Kimura, T., Akaji, K., and Kiso, Y. (1990c). *Chem. Pharm. Bull.* **38,** 1551–1557.

Zervas, L., and Photaki, I. (1962). *J. Am. Chem. Soc.* **83,** 3887–3897.

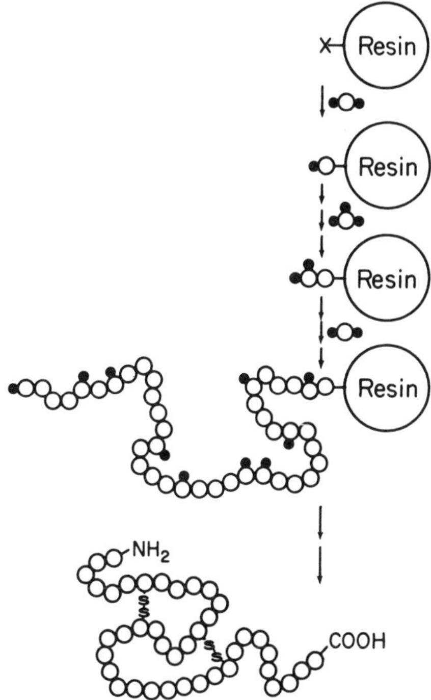

Figure 1 General idea of solid-phase synthesis. Open circles represent amino acids and filled circles are protecting groups.

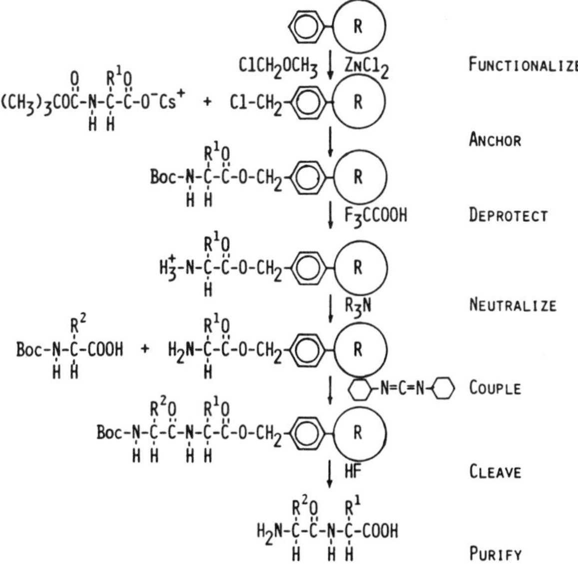

Figure 2 A synthetic scheme for solid-phase peptide synthesis.

Table I Polymeric Supports

Entry	Support[a]	Type	Load (mmol/g)	Ref.
1	Copoly(S–2% DVB)	Gel, 50μm beads	0.005–1	Merrifield (1963)
2	Copoly S–HOCH$_2$S–DVB)	Popcorn	0.1–1	Letsinger *et al.* (1964)
3	Copoly(S–15% DVB) (Polyhipe)	Macroporous, bead Rigid	0.3	Merrifield (1963) Small and Sherrington (1989)
4	Poly S–Kel F	Rigid, particulate	0.3	Tregear (1972)
5	Poly S–glass	Rigid, pellicular	0.03	Bayer *et al.* (1970)
6	PEO-grafted S–DVB (Tenta Gel)	Gel, beads	0.1	Bayer and Rapp (1986) Rapp *et al.* (1990)
7	POE coupled to S–DVB	Gel, beads	0.1–0.4	Hellermann *et al.* (1983 a, b)
8	Poly S–graft polyethylene	Sheets	0.35	Berg *et al.* (1991)
9	Polyamide (Pepsyn)	Gel, beads	01.–1	Sheppard (1973) Arshady *et al.* (1981)
10	Polyamide in kieselguhr (Pepsyn K)	Rigid, 350μm beads	0.05–0.25	Atherton *et al.* (1981)
11	Poly(HO–C$_6$H$_4$CH$_2$CH$_2$-acrylamide) (Core Q)	Gel, high load	1–6	Epton *et al.* (1987)
12	Controlled pore glass	Rigid, beads	0.07	Parr and Grohmann (1972) Büttner *et al.* (1988)
13	Cellulose	Powder	0.13	Merrifield (unpublished, 1960). See Merrifield (1993)
		Paper disk	0.01–0.09	Frank and Döring (1988)
14	Cotton	Sheet, thread	0.1	Lebl and Eichler (1989)
15	LH-Sephadex	Gel	0.3	Vlasov and Bilibin (1969)

[a] Copoly (S–2% DVB refers to copoly(styrene–2% divinylbenzene); PEO, polyethylene oxide; POE, polyoxyethylene.

substituted silica supports, and carbohydrates have also been examined. There is still room, however, for much new polymer chemistry to be applied to this field.

A. Copoly(styrene–divinylbenzene)

The earliest and by far the most widely used supports have been based on free radical-initiated suspension copolymers of styrene and *m*-divinylbenzene (S-DVB). Preparations containing from 0.2 to 16% cross-linker have been studied, but those containing 1% DVB have the best overall properties. They are physically stable and completely insoluble in all usual solvents. They form approximately 50-μm beaded gels, which become (reversibly) highly solvated and swollen in organic solvents of intermediate polarity such as dichloromethane, toluene, and dimethylformamide (DMF), but do not swell in water, methanol, and hexane. The physical and chemical properties of these resins have been reviewed in detail (Merrifield, 1984).

The styrene resins can be readily derivatized by nearly all methods of aromatic chemistry to allow modifications in their properties and to provide functional groups for the attachment of amino acids for peptide synthesis. The peptide chains are not restricted to the surface of the beads, but are uniformly distributed throughout the resin matrix. The swollen S-DVB beads are rapidly permeated by dissolved reagents, and the coupling reactions are very fast. They follow good second-order reactions rates for at least 90% completion and are generally 99% complete within 10 to 100 sec, depending on the amino acid and the method of activation. Mass transfer rates showed that diffusion is at least 10 times faster than coupling and is not rate limiting (Hetnarski and Merrifield, 1988; Pickup *et al.*, 1986). Kinetics show that most sites within the beads are rapidly accessible but do not provide good evidence for coupling efficiency. This is best done by quantitating the amount of deletion peptides produced from incomplete coupling during the synthesis. Under good conditions, less than 0.05% of deletion peptides were formed, and no effect of chain length between 4 and 60 residues could be detected in model systems (Sarin *et al.*, 1984). It must be pointed out that certain peptides contain "difficult sequences" and under usual conditions will give poorer results. The problem, which resides in the peptide not the polymeric support, and methods to correct it are discussed later (Section VI).

Further study of the physical properties of the lightly cross-linked styrene–divinylbenzene beads led to several useful conclusions. When a peptide of about 6000 molecular weight was synthesized at a loading that gave a peptide–resin containing 80% peptide by weight, the relative volume of the dry beads increased from 1 to 5.04, indicating that the volume of a bead could be greatly enlarged by the addition of peptide chains to it. The mean volume of the swollen beads increased from 6.2 to 11.7 ml in CH_2Cl_2 and from 3.3 to 26.3 ml in DMF. There were no damaged or ruptured beads. The data showed that the volume of a swollen resin bead containing a large load of peptide could be much larger than the volume of the fully swollen unsubstituted bead in a given solvent. There was no evidence that the resin matrix was becoming filled with peptide or that the synthesis was becoming limited by an overcrowding or steric effect. The maximum swelling of the 1% cross-linked polymer beads was estimated to be 200 ml/g when the polymer chains are fully extended, which is an order of magnitude larger than the volume observed even when the ratio of peptide to resin was 4:1. Therefore, the capacity of the resin is very large. Both polymer and peptide solvation contribute to the swelling, and even in a solvent that is relatively poor for one of the components the system does not collapse. These observations have been described and discussed (Sarin *et al.*, 1980).

Alternatives to the highly swollen, lightly cross-linked polystyrene gels have been examined. Letsinger and Kornet (1963), used popcorn copolymers of styrene and divinylbenzene. These contained as little as 0.01% cross-linking but were a low-density material that did not swell in solvents such as benzene, pyridine, and DMF. In contrast, very highly cross-linked S-DVB copolymers having large pores and rigid nonswelling structures were prepared by Kunin *et al.*

(1962) and Millar *et al.* (1963). These macroreticular, or macroporous, resins have very large surfaces that are several hundred times that of a smooth sphere of the same diameter. They are effective supports for synthesis of small peptides but have been less effective for long peptides.

To circumvent the diffusion problems present in highly cross-linked S-DVB polymer beads, pellicular resins were examined (Horvath *et al.*, 1967; Bayer *et al.*, 1970; Scott *et al.*, 1971). Such resins contain a shell of polystyrene formed around a core of glass beads. These preparations were not very stable, however, and have found little use in peptide synthesis.

B. Polyacrylamides

In an effort to overcome problems sometimes encountered in solid-phase synthesis on polystyrene supports, Sheppard and Atherton introduced a new polymer support based on polyacrylamide (Sheppard, 1973; Atherton *et al.*, 1975). It was specifically selected to have much less hydrophobic character than polystyrene and to more closely resemble polypeptides in polarity (see, however, discussion by Stewart and Klis, 1990).

Cross-linked polydimethylacrylamide resin was prepared by persulfate-initiated emulsion copolymerization of dimethylacrylamide and *N,N*′-bisacryloylethylenediamine in water and cellulose acetate butyrate together with Boc-β-alanyl-*N*-acryloylhexamethylenediamine in dichloroethane. The largely beaded polymer swells about 10-fold in DMF and acetic acid, and also in water, but much less in CH_2Cl_2. Anhydride or active ester couplings were carried out in polar solvents. Cleavage was in HF. The composition and conditions for synthesis of those resins, together with continuous modifications and improvements, were described in a large series of papers and summarized by Arshady *et al.* (1981). The synthesis was changed to a suspension polymerization method, as is used for polystyrene–divinylbenzene, β-alanine was introduced as an internal reference, the substitution was substantially increased, and the resin was produced in a good beaded form.

The resins and related gel supports were used in the discontinuous mode but were not satisfactory for continuous flow systems because of excessive compression. However, this deficiency was eventually overcome (Atherton *et al.*, 1981) by polymerizing the dimethylacrylamide/ethylene bisacrylamide/acryloylsarcosine within the pores of a rigid, inert kieselguhr particle. The particles, called Pepsyn K, were packed into a column, derivatized with ethylenediamine, then with 9-fluorenylmethyloxycarbonyl (Fmoc)-Nle, followed by trichlorophenyl hydroxymethylbenzoate. Finally, Fmoc-amino acid anhydride in *N*-methylpyrrolidinone (NMP) plus dimethylaminopyridine (DMAP) in DMF was coupled (Fig. 3). There was essentially no back pressure in the continuous flow mode.

The Pepsyn K support is being widely used. With heavy peptide loads some difficulties have been encountered, presumably caused by swelling of the peptide–resin within the kieselguhr matrix. In 1992 Mendre *et al.* patented a new polyamide resin called Expansin and have applied it to the continuous flow synthesis of several peptides. Sherrington (1990) prepared a new resin, Polyhipe, in which a polydimethylacrylamide gel is adsorbed into a rigid macroporous polystyrene matrix for use in a flow mode, and a macroporous polymethacrylate has been described for continuous flow (Andersson and Lindqvist, 1993).

Kanda *et al.* (1991) have prepared polyamide supports by free radical-initiated emulsion copolymerization of *N-N*-dimethylacrylamide, *N,N'*-bisacrylyl-1,3-diaminopropane, and 2-(methylsulfonyl)ethyloxycarbonyl allyl amide in CCl_4/hexane/water. The resin was deprotected and a peptide was assembled by standard methods. It remained attached to the resin following deprotection in HF. The peptide–resin was an effective immunogen in rabbits. Chitin was used in a similar way to support the synthesis of peptides that could be used without cleavage as antigens.

C. Polyethylene Glycol

A soluble polyethylene glycol (PEG) with an average molecular weight of 20,000 was introduced as a support for peptide synthesis by Bayer *et al.* (1970). The synthesis was conducted entirely in the liquid phase and provided certain benefits. Polyethyleneglycol has been adapted to solid-phase supports in two principal ways. Ethylene oxide can be polymerized onto a derivatized, preformed copoly(styrene–divinylbenzene) resin, or a performed PEG polymer can be attached to a derivatized S-DVB resin. Unfortunately, the terminology used to describe these new supports has not been rigorous or clear in several instances. Both types are often referred to as graft copolymers without making a distinction between their modes of formation.

Hellermann *et al.* (1983a,b) prepared PEG-grafted polystyrenes from polyethylene glycol + $ClCH_2$copoly(S-DVB) + NaH. Various linkers were inserted

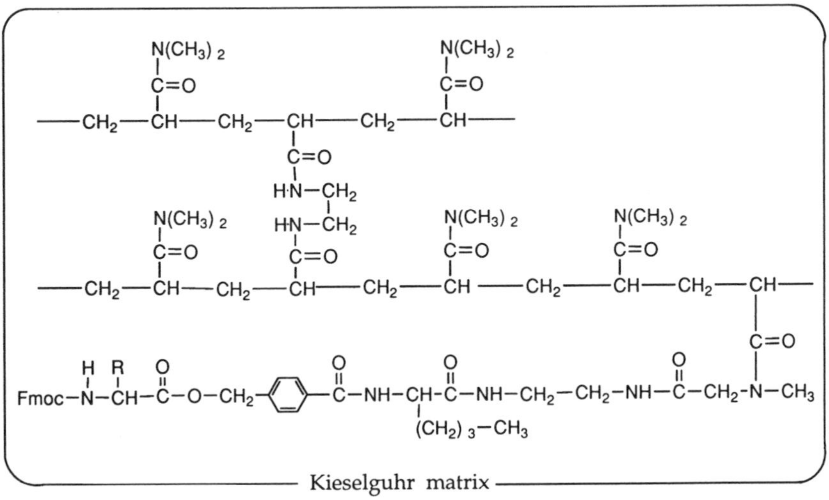

Kieselguhr matrix

Figure 3 Pepsyn K support with protected amino acid.

between the S-DVB and PEG or between PEG and the peptide to provide flexibility in synthetic design. Polyethylene glycol of only 400 molecular weight was adequate to give the desired swelling properties.

Zalipsky *et al.* (1985) and Barany *et al.* (1993) also made PEG–polystyrene graft supports. They attached several carboxyl group-containing PEG polymers of molecular weight 600–4000 onto amine-functionalized S-DVB by *N,N′*-dicyclohexylcarbodiimide (DCC)/*N*[1]-hydroxybenztriazole (HOBt) or benzotriazolyloxy tris(dimethylamino)phosphonium hexafluorophosphate (BOP)/-HOBt/*N,N*-diisopropylethylamine (DIEA) activation. Supports with 20–70% PEG were produced, with substitutions of 0.1 to 0.4 mmol/g (Fig. 4). A variety of chemistries were used to obtain suitable functionalities. These supports were considered to have nearly ideal physical properties. The investigators also believe that the improvements obtained were due to solvation and environmental effects rather than to a spacer arm effect.

Bayer and Rapp (1986) derivatized chloromethyl(styrene–1% divinylbenzene) with tetraethylene glycol in NaOH and then polymerized ethylene oxide in dioxane by anion initiation onto the resin beads. A polyether chain of molecular

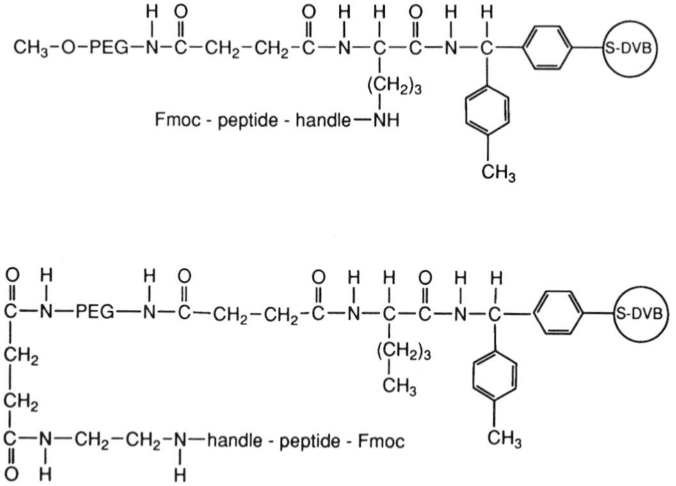

Figure 4 Polyethylene glycol-grafted copoly(styrene–divinylbenzene) resin supports.

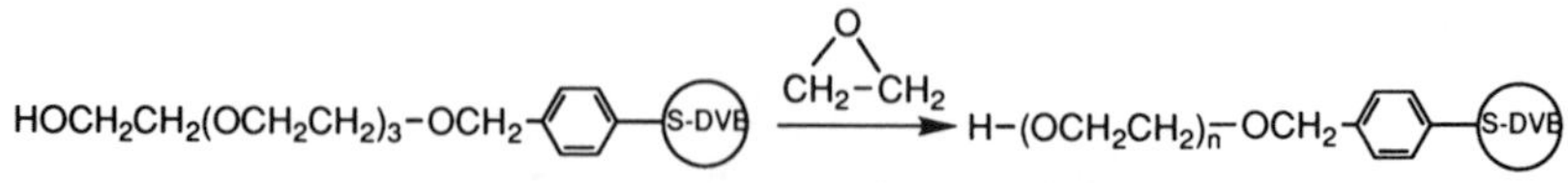

Figure 5 Tentagel; prepared by graft polymerization of ethylene oxide onto tetraethylene glycol–copoly(styrene–divinylbenzene) beads.

weight 6900 was obtained with an OH substitution of 0.13 mequiv/g. Rapp *et al.* (1990) have called this support Tentagel (Fig. 5). It swells approximately 3-fold in several organic solvents and somewhat less in water or methanol, but very little in ethanol or ether. Like S-DVB beads it expands with growing peptide chains. It maintains good flow rates in columns.

A flow-stable copolymer of bisacrylamido polyethylene glycol, monoacrylamido polyethylene glycol and *N,N*-dimethylacrylamide has also been prepared by radical polymerization (Meldal, 1992). The beads swelled to 8 ml/g in DMF and contained 0.1 mmol amino groups/ml. They were derivatized with the Rink linker by the 2-(benzotriazol-1-yl)-1,1,3,3-tetramethyluronium tetrafluoroborate (TBTU) procedure (see Section VE). The acyl carrier(65–74) peptide was satisfactorily synthesized using 3-hydroxy-4-oxo-3,4-dihydrobenzotriazine (Dhbt) ester coupling (Section VD), and the reactions were reported to be faster than with kieselguhr-supported polyamide resin.

Other graft copolymers have also been examined. For example, Tregear (1972) polymerized styrene onto small fluorocarbon particles with initiation by cobalt-60 radiation. The core did not swell in organic solvents, but the pendant linear polystyrene was well solvated and served as a useful support for peptide synthesis. The material was interesting because its high density allowed physical separation from peptide–resins made on copoly(styrene–divinylbenzene) (Van Rietschoten *et al.*, 1975). The efficiency of syntheses of some model peptides, however, was not always as good as those on the S-DVB or polyamide supports (Kent and Merrifield, 1978).

A similar styrene graft polymer has been made on polyethylene films and on nonwoven felt by irradiation using ^{60}Co (Berg *et al.*, 1989, 1991). The molecular weight of the linear polystyrene was approximately 10^6. Aminomethylation and a 4-hydroxymethylphenylacetamidomethyl (Pam) handle were used to synthesize a segment of human parathyroid hormone (hPTH) (residues 70–84) at 1 mmol/g substitution in an overall yield of 85%. This support was useful for simultaneous multiple syntheses. A new membrane support composed of an ethylene–vinyl alcohol copolymer (EVAL) has been introduced by Anwer and Spatola (1992) which is suitable for Fmoc-based solid-phase peptide synthesis. The relative merits of EVAL membrane, RAPP membrane, $ClCH_2$-(S-DVB) beads, and Pam-resin beads have been evaluated using phase-transfer catalysis and catalytic transfer hydrogenation cleavage methods (Wen *et al.*, 1994). By this methodology the rates were faster for the membranes than for the beads.

D. Other Supports

In 1970, Bayer *et al.* introduced silica as a suitable support for peptide synthesis. They prepared the mono ester of 1,4-(dihydroxymethyl)phenyl-silica, which is stable in anhydrous solvents. The first amino acid was esterified with carbonyldiimidazole, giving substitutions of 0.006 to 0.06 mmol/g. The synthesis was carried out in a packed-column flow system. Controlled-pore glass has also been recommended (Büttner *et al.*, 1988).

A number of other polymers have been prepared and tested, but none gained popularity. They include a phenol–formaldehyde polymer, a polymer made by heating resorcinol dimethyl ether and formaldehyde (Wissmann *et al.*, 1969), and an amorphous polymer prepared from phenol, *s*-trioxane, and a catalytic amount of toluenesulfonic acid (Inukai *et al.*, 1968). Flanigan and Marshall (1970) polymerized 4-(methylthio)phenol and formaldehyde by refluxing with a catalytic amount of NaOH and also heated 4-mercaptophenol and KOH with $ClCH_2$(S-DVB). A copolymer of isobutylene and maleic anhydride was examined (Wildi and Johnson, 1968).

Carbohydrates have been appealing as supports for solid-phase peptide synthesis since inception of the process in 1959, and were in fact the first supports studied. The initial attempts to use cellulose powder and regenerated cellulose were considered to be only marginally successful because of low loading (~0.1 mmol/g) and chemical reactivity of the cellulose. However, Sephadex LH-20 was later found to be a usable support (reviewed by Merrifield, 1993; see also Vlasov and Bilibin, 1969). Cellulose paper has proved to be an effective carrier. Frank and Döring (1988) and Frank *et al.* (1991) esterified Fmoc-Gly to sheets of Whatman cellulose paper in the presence of HOBt and *N*-methylimidazole in DMF and used it for the rapid, multiple synthesis of a series of peptide antigens. The antigens were tested without removal from the paper.

Cotton, a very pure form of cellulose, has been developed for solid-phase synthesis in the form of sheets (Eichler *et al.*, 1990, 1991), threads (Lebl and Eichler, 1989), or beads called Perloza (Englebretsen and Harding, 1992). The cellulose sheets and threads are insoluble and do not swell in water or organic solvents. The amino acid can be attached directly as an ester to the hydroxyls of the cellulose and, after synthesis of a peptide, the bond can be cleaved by saponification or aminolysis. Alternatively, an acid-labile handle can be attached to the amino acyl-substituted carrier, followed by the C-terminal amino acid of the peptide. Substitution levels were above 1 to 3 $\mu mol/cm^2$, or 0.1 to 0.3 mmol/g. Small peptides of good quality have been prepared using multiple synthesis techniques or by a continuous belt method. The Fmoc/*t*-Bu strategy was preferred.

Perloza beaded cellulose is swollen in solvents such as water, dioxane, DMF, dimethyl sulfoxide (DMSO), or tetrahydrofuran (THF) to the extent of about 10 ml/g. It is available in beads ranging from 80 to 500 μm in diameter with porosities of 100 to 500 kDa cutoff. The beads were cyanoethylated up to 4.2 mmol/g with acrylonitrile in NaOH, then reduced with diborane to give aminopropylcellulose. Conventional methods were then used to synthesize several peptides successfully.

Finally, it can be mentioned that bovine serum albumin (BSA) has been examined as a support for solid-phase synthesis (Hansen *et al.*, 1992, 1993). All 100 residues of Glu and Asp were converted to the diethyl amide by *N,N'*-diisopropylcarbodiimide (DIC) activation. Hydroxymethylphenylacetic acid Dhbt ester was coupled to the protein amino groups, and the first Fmoc-amino acid was esterified as the anhydride. After assembly of the peptide by standard methods, it was cleaved in trifluoroacetic acid (TFA). Thyroglobulin was also used as a protein support. The first amino acid was coupled directly to the protein carrier and extended to a peptide–protein conjugate containing 35 peptide chains per protein molecule, which could be used as an antigen.

III. Handles, Linkers, and Spacers

The number of derivatives introduced to modify the physical or chemical properties of polymeric supports has been quite large. This is possible because almost any chemical reaction that can be conducted in homogeneous solution can be carried out on the polymeric materials, especially those derived from polystyrene. It has been possible to prepare supports with altered polarity and solvation properties by introducing functional groups or, for example, by grafting one polymer to another. It has also been possible to fine-tune the resin derivatives so that the anchoring bond holding the peptide chain to the support can have any desired sensitivity to cleavage conditions. The modifications are based on physical-organic principles, and the rates of reaction can be easily and rather accurately predicted. The resin derivatives can be designed for cleavage by strong, medium, or weak acids, by bases and other nucleophiles, and by photolysis, hydrogenolysis, or other catalytic reagents. They can also be designed to yield free carboxyl groups, esters, amides, or hydrazides. This kind of modification began early in the development of solid-phase peptide synthesis but has been greatly accelerated in recent years. Some linkers have been introduced, not for the purpose of changing cleavage conditions, but to serve as spacers, with the idea that the peptide could be assembled better if it were farther removed from the resin support. However, there is little or no evidence that this is required or beneficial. The design of linkers is closely associated with protecting group strategies, which will be covered in Section IV.

A. Acid Cleavage Leading to Peptide Acids

Some of the linkers or handles designed to yield peptide acids under a wide range of acidic conditions are shown in Table II. Those linkers used with strong acid cleavage, that is, with HBr, HF, and trifluoromethanesulfonic acid (TFMSA), began with simple benzyl esters (entry 1, in Table II) prepared from chloromethyl-copoly(styrene–divinylbenzene) or ring-nitrated or ring-brominated derivatives (Merrifield, 1963). These were suitable for use with N^{α}-*tert*-butyloxycarbonyl (Boc) temporary protection and side-chain benzyl-based protecting groups.

The following brief section is a summary of the newer methods for preparing linkers designed to yield peptides with free acids after strong, moderate, or weak acid cleavage (Table II), or peptides with C-terminal amides after strong, medium, or weak acid cleavage (Table III), or handles for nonacidic cleavage methods (Table IV). Introduction of the 4-hydroxymethylphenylacetamidomethyl (Pam) linker (entry 2 in Table II) greatly improved anchoring-bond stability and

Table II Acid-Labile Linkers for Peptide Acids

Entry	Structure[a]	Ester	Cleavage reagent	Ref.
1	$R{-}C(=O){-}OCH_2{-}C_6H_4{-}(R)$	Benzyl	HBr, HF, TFMSA	Merrifield (1963)
2	$R{-}C(=O){-}OCH_2{-}C_6H_4{-}CH_2C(=O){-}NHCH_2{-}C_6H_4{-}(R)$ PAM	Hydroxymethyl-phenyl acetamidomethyl	HF	Mitchell *et al.* (1978)
3	$R{-}C(=O){-}OCH_2{-}C_6H_4{-}OCH_2{-}C_6H_4{-}(R)$	Alkoxybenzyl	TFA	Wang (1973)
4	$R{-}C(=O){-}OCH_2{-}C_6H_4{-}OCH_2{-}C(=O){-}NH{-}$Poly Amide HMPA/PAB	Acetoxybenzyl	TFA	Sheppard and Williams (1982)
5	$R{-}C(=O){-}OCH_2{-}C_6H_4{-}OCH_2CH_2C(=O^-){-}NHCH_2{-}(R)$ PAB	Hydroxymethyl-phenoxypropionyl	TFA	Albericio and Barany (1985)
6	$R{-}C(=O){-}OCH_2{-}C_6H_3(CH_3O){-}O{-}CH_2C(=O)NH{-}$Poly Amide	Methoxyacetoxybenzyl	Dilute TFA	Sheppard and Williams (1982)

(continues)

(continued)

7	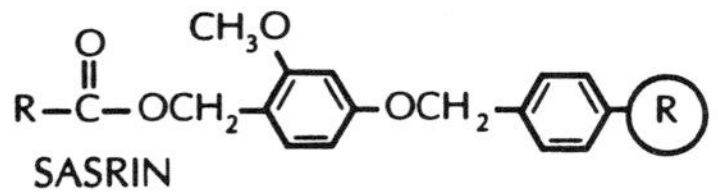 SASRIN	Methoxyalkoxybenzyl	Dilute TFA	Mergler *et al.* (1988)
8	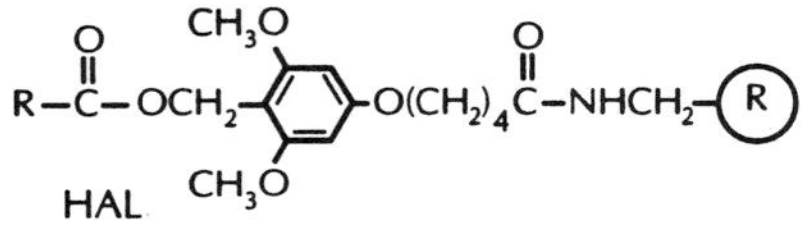 HAL	Hydroxymethyl-dimethoxyphenoxyvaleryl	Dilute TFA	Albericio and Barany (1991)
9	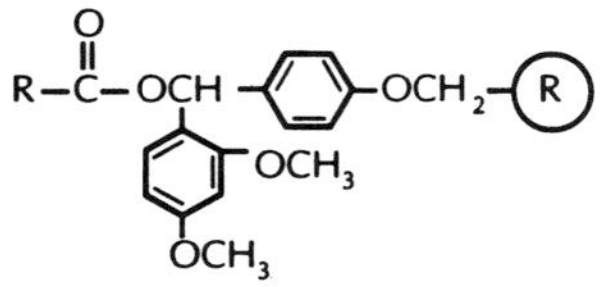	Alkoxy-dimethoxybenzhydryl	Dilute TFA	Rink (1987)
10	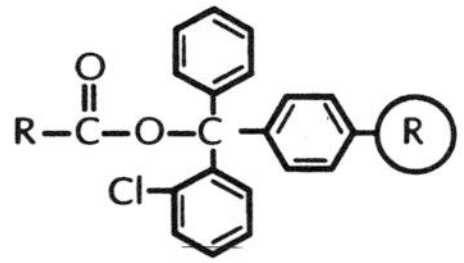	Chlorotrityl	Dilute TFA	Barlos *et al.* (1989)

[a] (R) = S-DVB resin

Table III Acid-Labile Linkers for Amides

Entry	Structure[a]	Type	Cleavage Reagent	Ref.
1	R–C(=O)–NH–CH(C$_6$H$_5$)–C$_6$H$_4$–Ⓡ BHA	Benzhydrylamine	HF	Pietta and Marshall (1970)
2	R–C(=O)–NH–CH(C$_6$H$_4$–CH$_3$)–C$_6$H$_4$–Ⓡ MBHA	Methylbenzhydrylamine	HF	Matsueda and Stewart (1981)
3	R–C(=O)–NH–CH(C$_6$H$_4$–OCH$_3$)–C$_6$H$_4$–OCH$_2$C(=O)–OCH$_2$–Ⓡ MeOBHA	Methoxybenzhydrylamine	TFA	Stüber *et al.* (1989)
4	R–C(=O)–NH–CH(C$_6$H$_3$(OCH$_3$)$_2$)–C$_6$H$_3$(OCH$_3$)–NHC(=O)CH$_2$CH$_2$C(=O)NHCH$_2$–Ⓡ SAMBHA	Trimethoxybenzhydrylamine	TFA	Penke and Nyerges (1989)

(continues)

(continued)

5	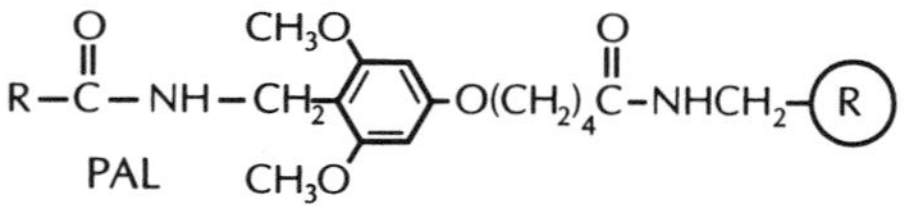 $R{-}C({=}O){-}NH{-}CH_2{-}$[2,6-di($CH_3O$)phenyl]$-O(CH_2)_4C({=}O){-}NHCH_2{-}$Ⓡ PAL	Aminomethyl-dimethoxyphenoxyvaleryl	TFA	Albericio and Barany (1987)
6	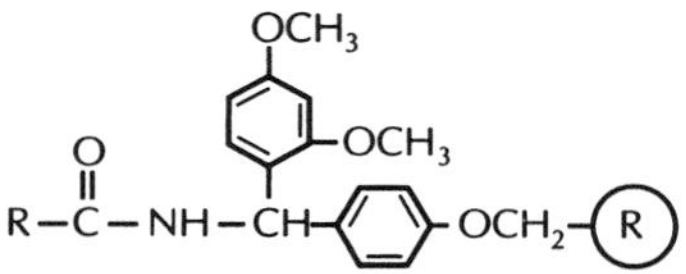 $R{-}C({=}O){-}NH{-}CH$(2,4-di(OCH_3)phenyl)$-$[phenyl]$-OCH_2{-}$Ⓡ	Dimethoxybenzhydryl	Dilute TFA	Rink (1987)
7	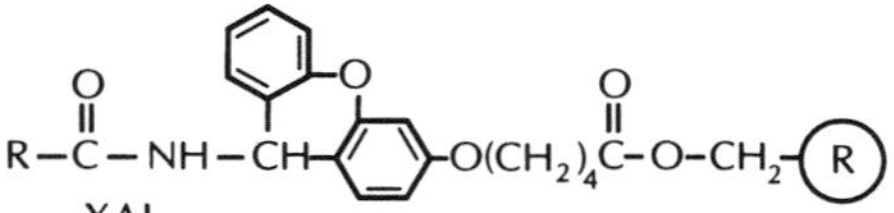 $R{-}C({=}O){-}NH{-}CH$(xanthyl)$-O(CH_2)_4C({=}O){-}O{-}CH_2{-}$Ⓡ XAL	Aminoxanthyl-oxyvaleryl	Dilute TFA	Sieber (1987)

[a] Ⓡ = Copoly (styrene - 1% divinylbenzyne) resin

Table IV Linkers Removable by Nonacidic Reagents

Entry	Structure	Type	Cleavage reagent	Ref.
1	$R-C(=O)-OCH_2-C_6H_3(NO_2)-$Ⓡ	Nitrobenzyl	*hν*	Rich and Gurwara (1975)
2	$R-C(=O)-OCH(CH_3)-C(=O)-C_6H_4-$Ⓡ	Phenacyl	*hν*	Wang (1976)
3	$R-C(=O)-OCH_2CH_2-C_6H_3(NO_2)-C(=O)-NH(CH_2)_n-$Glass	Nitrophenylethyl	Piperidine	Eritja *et al.* (1991)
4	$R-C(=O)-O-CH_2-$(fluorenyl C–H)$-C(=O)-NHCH_2-$Ⓡ	Fluorenylmethyl	Piperidine	Mutter and Bellof (1984)
5	$R-C(=O)-O-CH_2-CH{=}CH-C(=O)-NHCH_2-C_6H_4-$Ⓡ HYCRAM	Allyl ester	Pd(0)	Kunz and Dombo (1988)

selectivity for extended stepwise synthesis procedures (Mitchell *et al.*, 1976a, 1978; Tam *et al.*, 1979a). Sparrow (1976) used the same 4-acetamido modification to place a spacer between the peptide and the resin. The Pam-resin was prepared from an aminomethyl-resin (Mitchell *et al.*, 1976b) synthesized by a greatly improved Friedel–Crafts reaction of hydroxymethylphthalimide or chloromethylphthalimide with the polystyrene resin followed by hydrazinolyis. A modified procedure has been recommended for the preparation of Pam-resins (Tóth and Penke, 1991). The Pam-resin is a fully satisfactory support for those wishing to use the Boc/benzyl protection strategy and final cleavage with HF.

The design of anchoring bonds that could be cleaved by TFA or other acids of intermediate strength, and thus avoid strong acids, was begun by Wang and Merrifield (1971), who prepared a *tert*-alkoxycarbonylhydrazide-resin and a *tert*-alkyl alcohol-resin, which could be used in conjunction with the very acid-labile N^{α}-2-biphenylisopropyloxycarbonyl (Bpoc)-amino acids to produce protected fragments for further coupling. Subsequently hydroxymethyl-4-alkoxystyrene resin was introduced (entry 3 in Table II) (Wang, 1973). This resin, which is cleaved by TFA, has been finding renewed use, especially in combination with N^{α}-Fmoc-amino acids.

Some years later, Sheppard and Williams (1982) also prepared a TFA-labile hydroxymethyl-4-alkoxy derivative, but attached to the polyamide resin introduced earlier in their laboratory (entry 4 in Table II). Albericio and Barany (1985) prepared the corresponding phenoxypropionylaminomethyl-polystyrene handle, phenoxypropionylaminomethyl (PAB) (entry 5 in Table II). Fmoc-amino acids were coupled to 2,4,5-trichlorophenyl 3-(4′-hydroxymethylphenoxy)propionate by use of *N*, *N*-dimethylformamide dineopentyl acetal, and then coupled to an aminomethyl-polystyrene resin. Substitution of the propionyl group for the acetyl gave a linker that was 2–3 times more acid labile.

It was a straightforward matter to further modify the linkers with electron-donating groups to become even more acid labile. Sheppard and Williams (1982) added one methoxyl group to their acetoxybenzyl resin, giving a linker that can be cleaved by 1% TFA; Mergler *et al.* (1988) added one more methoxyl to the Wang resin giving the *s*uper *a*cid-*s*ensitive SASRIN *resin* (2-methoxy-4-alkoxy benzyl alcohol) (entry 7 in Table II), removable by 0.5–1% TFA/CH_2Cl_2; and Albericio and Barany (1991) added two methoxyls to their hydroxymethyl-phenoxylvaleryl-resin giving 5-(4-hydroxymethyl-3,5-dimethoxyphenoxyvaleryl) (HAL) resin (entry 8 in Table II), also removed by 1% TFA in CH_2Cl_2. The substitution was further modified (Rink, 1987) with a benzhydryl derivative, 4-(2′,4′-dimethoxyhydroxymethyl)phenoxymethyl-polystyrene (entry 9 in Table II). The Fmoc-aminoacyl-linker-resin was prepared as follows: 2,4-Dimethoxy-4′-hydroxybenzophenone cesium salt was coupled to $ClCH_2$-polystyrene. The ketone was reduced with $LiBH_4$. Fmoc-Gly was esterified to the benzhydryl with DCC in dichloroethane, DMAP, and *N*-methylmorpholine. A 12-residue Fmoc peptide containing side-chain *tert*-butyl groups was assembled stepwise by symmetrical anhydride (SA) couplings in DMAP. Final cleavage was in acetic acid/CH_2Cl_2 (1:9; 90 min, room temperature). The *t*-Bu groups on Lys and Tyr were stable. A 3-min cleavage in 0.2% TFA in CH_2Cl_2 was also effective.

Fréchet and Haque (1975) derivatized copoly(S–1% DVB) by Friedel–Crafts reaction with benzoyl chloride, treatment with a phenyl Grignard, followed by treatment with acetylchloride to yield a trityl chloride-resin. Barlos *et al.* (1989) added a 2-chloro substituent and obtained a very acid-sensitive linker for peptide synthesis (entry 10 in Table II). It can even be removed in 15 min by acetic acid/trifluoroethanol (TFE)/CH_2Cl_2 (1:2:7). A modified trityl linker has also been used by Bayer.

A further fine-tuning was studied by Flörsheimer and Riniker (1991). For their purpose, linker 6 (Table II) was somewhat too stable and not completely selective with respect to Tyr(*t*-Bu) and Lys(Boc), whereas SASRIN resin and the 2-chlorotrityl linker were so acid-labile they required a tertiary amine during coupling and were not compatible with benzotriazolyloxy tris(dimethylamino)phosphonium hexafluorophosphate (BOP) or TBTU. The intermediate level of acid lability was achieved by condensing 4-(4′-hydroxymethyl-3′-methoxyphenoxy)butyric acid (HMPB) to benzhydrylamine-polystyrene resin. The Fmoc-peptide-HMPB-benzhydrylamine (BHA)-resin (**I**) was cleaved by four 2-min treatments with 1% TFA in CH_2Cl_2. Side chain *tert*-butyl and trityl (Trt) groups were stable, and BOP or TBTU with 2 equiv DIEA in DMF could be used.

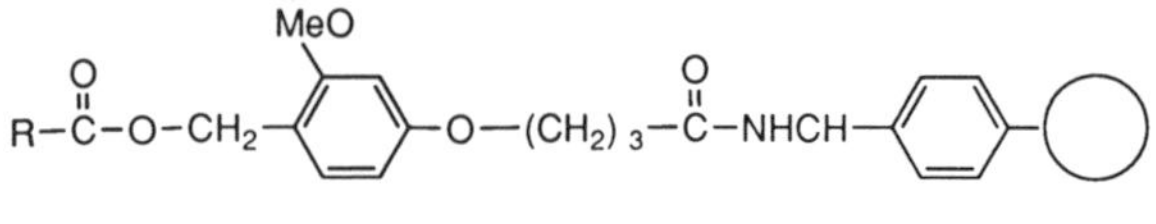

MeO

R–C(=O)–O–CH_2–C₆H₃–O–$(CH_2)_3$–C(=O)–NHCH–C₆H₄–(resin)

(I)

All of the acid labile linkers require N^{α}-protecting groups that are removed by reagents other than acid (e.g., Fmoc). For some of them the peptide containing *tert*-butyl side chains can be removed from the resin in a fully protected form for further segment condensations.

B. Acid Cleavage Leading to Peptide Amides

In a similar way, linkers with graded acidolytic rates have been prepared that give rise to peptide amides (Table III). The first was a benzhydrylamine-substituted polystyrene (entry 1 in Table III) that was cleavable in HF (Pietta and Marshall, 1970). A more acid-labile 4-methylbenzhydryl derivative MBHA (4-methoxybenzhydrylamine; entry 2 in Table III) was eventually prepared (Matsueda and Stewart, 1981). It still requires HF for cleavage but soon became the linker of choice. Again, proper substitution with methoxyl groups gave benzhydryl amines with predictably more acid sensitivity. For example, MeOBHA (entry 3 in Table III) (Stüber *et al.*, 1989) and SAMBHA (entry 4 in Table III) (Penke and Hyerges, 1989) are cleavable in 90% TFA. Funakoshi *et al.*

(1988) used 3-(α-Fmoc-peptidyl-amino-4-methoxybenzyl)-4-methoxyphenylpropionylaminomethyl-resin (**II**). The linker was also attached to Ile-Pam-resin instead of aminomethyl-resin. These were cleaved in 1 hr at 25°C with TFA containing 1 *M* thioanisole (cleavage in 1:1 TFA/CH_2Cl_2 was ineffective). A trimethoxybenzhydryl derivative attached through an alkyl carboxamide linkage to polystyrene (Breipohl *et al.*, 1989) is cleavable in 60% or less TFA containing a small amount of water. The methylphenoxy derivative PAL [5-(4-aminomethyl-3,5-dimethoxyphenoxy)valeryl-resin; entry 5 in Table III] (Albericio and Barany, 1987; Albericio *et al.*, 1990) is cleavable in 50% TFA/CH_2Cl_2.

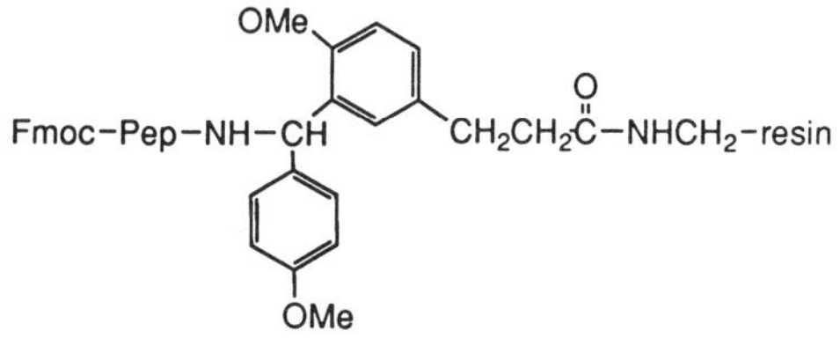

OMe

Fmoc–Pep–NH–CH ... $CH_2CH_2\overset{O}{\overset{\|}{C}}$–$NHCH_2$–resin

OMe

(II)

For very acid-sensitive linkers the Rink benzhydryl-resin was reacted with Fmoc-Pro-NH_2 or Fmoc-NH_2 to give the substituted benzhydrylamine derivatives (entry 6 in Table III) (Rink, 1987). An alkoxyxanthylamine [5-(9-aminoxanthene-2-oxy)valeryl (XAL)] attachment was also cleavable in very dilute acid (entry 7 in Table III) (Sieber, 1987). It was removed in TFA/dichloroethane (2:98). Barlos *et al.* (1991) have adapted their chlorotrityl-resin to the synthesis of peptide amides by anchoring the side-chain β-carboxyl of an Asp residue and adding an amino acid amide at the C terminus. Gastrin(1–17) amide was quantitatively cleaved in high purity with acetic acid/TFE/CH_2Cl_2 (1:1:8), in 30 min. Van Vliet *et al.* (1993) have used new trityl-resin derivatives to prepare peptide amides, hydrazides, and ω-aminoalkylamides following cleavage with 0.5% TFA/CH_2Cl_2/4% methanol.

Chao *et al.* (1993) have sought to eliminate the dangers of stable carbocations produced from many of the linkers during TFA cleavage, as judged from the observations that scavengers are not always effective in trapping them. The silyl-linkage agent 4-{1-[*N*-(9-fluorenylmethyloxycarbonyl)-amino]-2-(trimethylsilyl)ethyl}phenoxy acetic acid (SAL) was designed to undergo deblocking by a β-elimination mechanism, which would neutralize the transient carbocation and form a stable styrene derivative. The linker (**III**) was obtained in reasonable yield after a total of nine steps.

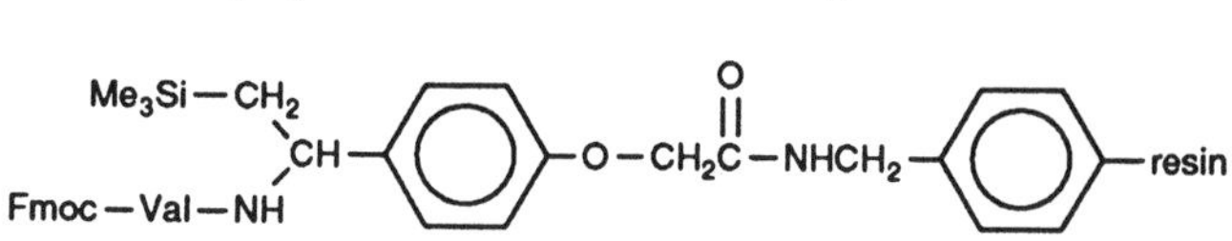

Me_3Si–CH_2

Fmoc–Val–NH ... CH–⟨ ⟩–O–$CH_2\overset{O}{\overset{\|}{C}}$–$NHCH_2$–⟨ ⟩–resin

(III)

The Fmoc-Val-NH_2 was cleaved in 97% yield with TFA–5% phenol in 15 min. The mechanism involves an attack of trifluoroacetate on silicon to give peptide amide plus $CF_3COOSi(Me_3)CH_2{}^+CHPhOCH_2CONHCH_2$-resin, which rapidly eliminates $CF_3COOSiMe_3$, with formation of the styrene-bound by-product $CH_2{=}CHPhOCH_2CONHCH_2$-resin. C-Terminal Trp peptides can be extensively alkylated by the carbocation intermediates, but yields of 90% could be obtained when the cleavage was with TFA/ethane dithiol (EDT)/phenol/thioanisole (90:5:3:2). The PAL and AM linkers gave only 20–30% yields under these conditions because of more extensive Trp alkylation.

C. Cleavage by Nonacidic Reagents

Among the nonacidic reagents that can be designed for the release of peptides from a resin support, fluoride ion has been particularly interesting. This reagent has been discussed in some detail by Mullen and Barany (1988). They prepared a bifunctional linker, 2,4,5-trichlorophenyl *N*-(3 or 4)-{[4-(hydroxymethyl)phenoxy]-*tert*-butylphenylsilyl}phenylpentanedioate monoamide (Pbs) (**IV**). It can be coupled to an aminomethyl-resin and the C-terminal amino acid attached as a benzyl ester. The latter is removable by fluoridolysis of the oxygen–silicon bond with 1 equiv tetrabutylammonium fluoride. Ramage *et al.* (1987), Kiso *et al.* (1988b), and Ueki *et al.* (1988) have also used F^- cleavages in various ways.

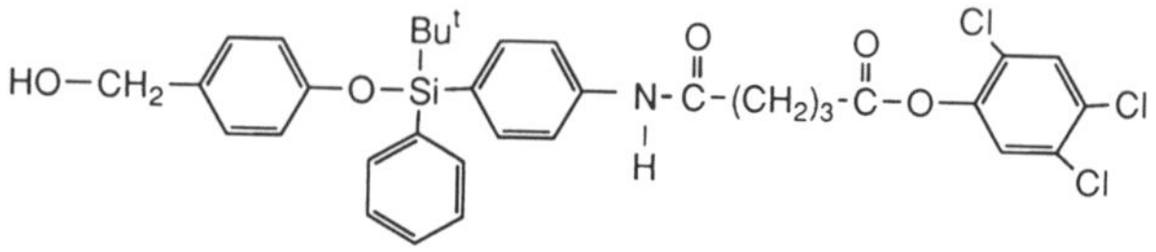

(IV)

An important new nonacidic cleavage method utilizes a linker containing an allyl group between the peptide and the copoly(S-DVB) resin as shown in entry 5 of Table IV. The support has been called HYCRAM-resin (hydroxycrotonylaminomethy-resin) (Birr *et al.*, 1991). The noble metal-catalyzed activation of the allylic ester at neutral pH results in the transfer of the allylic moiety onto an acceptor such as morpholine, *N,N'*dimethylbarbituric acid, or dimedone, with the release of the fully protected peptide acid. The catalyst was tetrakis(triphenylphosphino)palladium in DMSO/DMF.

Osborn and Robinson (1993) have reported a new resin linkage that allows a peptide fully protected with *t*-Bu or Fmoc groups to be cleaved under neutral conditions. The linker is a 2-azidomethyl-4-hydroxy-6,*N*-dimethylbenzamide moiety which is coupled to polystyrene resin as a phenol ether and to the peptide by an *N*-hydroxymethyl group (Scheme 1). The aryl azidomethyl group, after reduction to an aminomethyl group with *n*-Bu_3P, undergoes an intramolecular reaction to release the protected peptide.

An important new linker based on a kinetically inert cobalt(III) amine complex has been developed by Isied and co-workers (Mensi and Isied, 1987;

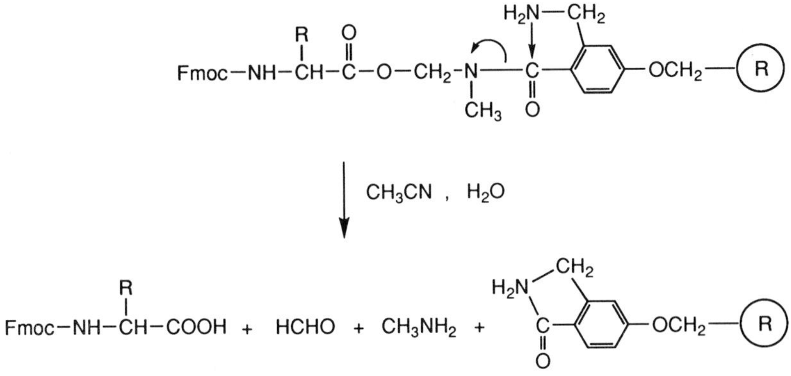

Fmoc−NH−CH(R)−C(=O)−O−CH_2−N(CH_3)−C(=O)−C_6H_4(CH_2−NH_2)−OCH_2−(R)

↓ CH_3CN, H_2O

Fmoc−NH−CH(R)−COOH + HCHO + CH_3NH_2 + (lactam: H_2N−CH_2/C=O)−C_6H_3−OCH_2−(R)

Scheme 1

trans [$(NH_3)_4Co(SO_3)OH$] + NH_2CH_2−C_6H_4−COOH —pH 9.7→ [trans $(NH_3)_4$−Co(SO_3)(NH_2CH_2−C_6H_4−CO_2^-)]

—1. TFMSA, -12°C; 2. pH 1, 60°C→ cis [$(NH_3)_4$−Co(OH_2)(NH_2CH_2−C_6H_4−COOH)] $(TFMSA)_3$ —Boc−AA[1]-OBt, DIEA / DMF→

cis [$(NH_3)_4$Co(O−AA[1]−Boc)(NH_2CH_2−C_6H_4−CO_2H)] $(TFMSA)_2$ —DIC, HOBt; CH_3CN, 0°C→ cis [$(NH_3)_4$−Co(O−AA[1]−Boc)(NH_2CH_2−C_6H_4−COOBt)] $(TFMSA)_2$

—Nle−$NHCH_2$−(R)→ Boc−AA[1]−O−Co(cis $(NH_3)_4$)−NH_2CH_2−C_6H_4−C(=O)−Nle−$NHCH_2$−(R) $^{2+}$

↓ DMF, DTT / DIEA

Boc−AA[1] + Co(II)−DTT complex + NH_2CH_2−C_6H_4−C(=O)−Nle−$NHCH_2$−(R)

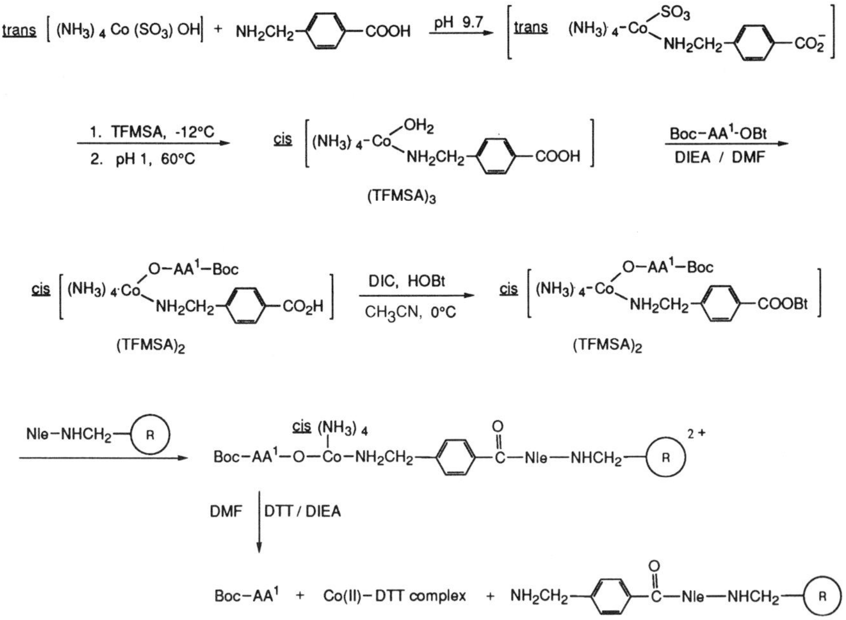

Scheme 2

Arbo and Isied, 1993). The comp!ex is quite stable to the peptide synthesis conditions for both Boc/Bzl and Fmoc chemistry, with chain losses of 0.1% or less per cycle after treatment with 50% TFA, 5% DIEA, or 20% piperidine. The protected peptide can be released by mild reduction to cobalt(II) with thiols such as 0.5 *M* dithiothreitol (DTT), 0.3 *M* DIEA in DMF. The synthetic steps leading to anchoring of the first amino acid and its cleavage are outlined in Scheme 2. A similar scheme was developed using bisethylenediamine in place of tetraammonium in the cobalt(III) complex. The technique has been successfully applied to the synthesis of peptides such as Leu-enkephalin in overall purified yields of approximately 68%. Peptide chemists, however, have been slow to adopt this organometallic complex chemistry.

IV. Protecting Group Strategy

The successful synthesis of peptides requires an overall protecting group strategy. For solid-phase synthesis at least two levels of protection are required. The peptide chain must be held to the polymeric support by a cleavable bond that will be stable to the repetitive N^{α}-deprotection steps. Side-chain protecting groups must also be stable to the N^{α}-deprotection, and for certain purposes they must be stable to the C^{α}-cleavage reagent. The selectivity was traditionally achieved by quantitative kinetic differences in rates of removal. For example, the N^{α}-Boc group could be removed by 4 *N* HCl in dioxane or 50% TFA in CH_2Cl_2, which would leave the benzyl-derived side-chain and anchoring groups intact. The latter were then removed by strong acids such as HBr, HF, or TFMSA. The Boc group can also be removed in 6 *M* aqueous HCl (Naharisson *et al.*, 1992) or 10 *N* H_2SO_4 in dioxane (Houghten *et al.*, 1986).

Much has been done to fine-tune this system to obtain larger differences in rates. For example, introduction of halogens onto the benzyl ring as in *O*-2,6-dichlorobenzyl-tyrosine (Erickson and Merrifield, 1973a) or N^{ε}-2-ClZ-lysine (Erickson and Merrifield, 1973b) made them sufficiently stable to HCl or TFA to avoid losses during repetitive Boc removal. Similarly, cyclohexyl (Tam *et al.*, 1979b) or cyclopentyl esters (Blake, 1979) of aspartic acid or glutamic acid were more acid stable and in addition were less subject to other side reactions. Alkylations by benzyl carbonium ions are avoided because the corresponding cyclohexyl ion rearranges rapidly to an unreactive tertiary methylcyclopentyl ion.

Better acid selectivity was also possible by making the N^{α}-amino-protecting group more acid labile. Thus, the biphenylisopropyloxycarbonyl group (**V**) on nitrogen was about 1000 times more susceptible to HCl or TFA than Boc, and it could be completely removed by 1% TFA/CH_2Cl_2 in 30 min without measurable loss of side-chain groups (Sieber and Iselin, 1968; Wang and Merrifield, 1969). The *p*-nitrophenylsulfenyl group (**VI**) (Zervas *et al.*, 1963; Najjar and Merrifield, 1966) was similar, being removed by 1% TFA in acetic acid. Other acid-labile N^{α}-protecting groups include Ddz (**VII**) (Birr, 1972), Azoc (**VIII**) (Kyi and Schwyzer, 1976), and Adpoc (Kalbacher and Voelter, 1978).

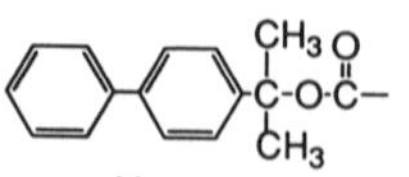

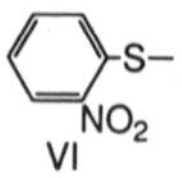

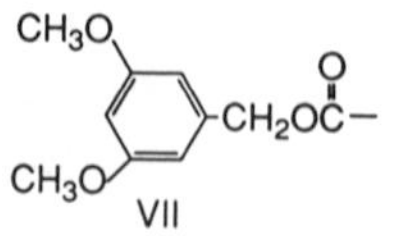

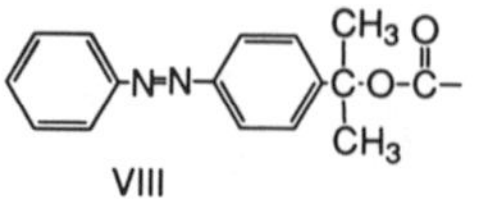

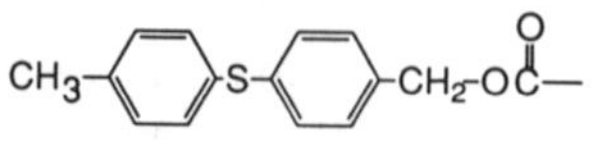

The 4-methylsulfinylbenzyloxycarbonyl (Msz) group (**IX**) (Kiso *et al.*, 1989a) is stable to both acidic and basic conditions. Because of a safety-catch mechanism, the group can be readily removed by reductive acidolysis with 10 equiv $SiCl_4$ plus 10 equiv anisole in TFA. Thus, the sulfoxide of Msz is first reduced to the sulfide, which is then labile to acidolysis by TFA.

A silane-based reagent using 1 *M* chlorotrimethylsilane and 3 *M* phenol in CH_2Cl_2 has been introduced. It is much more selective for *tert*-butyl-versus benzyl-protecting groups (Kaiser *et al.*, 1993) (Fig. 6).

A. Alternative α-Amino-Protecting Groups

Acid-sensitive N$^{\alpha}$-protecting groups have gradually been supplemented with groups cleavable by bases, nucleophiles, or other reagents. The 9-fluorenylmethyloxycarbonyl (Fmoc) derivative of Carpino (Carpino and Han, 1972) has become an extremely important and widely used N$^{\alpha}$-protecting group for solid-phase synthesis (Wang, 1975; Chang *et al.*, 1980; Atherton *et al.*, 1981; for a review, see Fields and Noble, 1990). It is removable by β-elimination with mild bases such as piperidine or morpholine, yielding the free NH_2-peptide-resin without the need for a neutralization step. The by-product, dibenzofulvene, is then trapped by excess secondary amine. The Fmoc group can also be removed in DMF containing 2% of the strong base, nonnucleophilic amidine, 1,8-diazabicyclo[5.4.0]undec-7-ene (DBU) (Wade *et al.*, 1991). DBU was particularly useful in continuous-flow applications. DBU had also been useful in solid-phase applications involving cleavage of protected peptides from 2-[4′-(hydroxymethyl)phenylacetoxy]propionyl-resin (Whitney *et al.*, 1984).

2,2-Bis(4′-nitrophenyl)ethan-1-ol has been proposed as another base-labile urethane protecting group (Bnpeoc) by Ramage *et al.* (1991b). It is also cleaved by piperidine, DBU, or DBU/acetic acid. The Bnpeoc-amino acids are more soluble in CH_2Cl_2 than some of the Fmoc derivatives.

One of the most important new classes of protecting groups is that based on allyl derivatives, which are rapidly cleaved under nearly neutral conditions by palladium(0)-catalyzed allyl transfer. These include allyl esters (Kunz and Waldmann, 1984) and allyloxycarbonyl groups (Kunz and Unverzagt, 1984). Even the sensitive *O*-glycosyl-serine bond remains intact. Kunz and März (1988) have further modified the allyl protection principle by introducing the *p*-nitrocinnamyloxycarbonyl (Noc) group (**X**) for N^{α}-amine protection. The Noc-amino acids have increased acid stability, improved crystallizability, and a

$$(CH_3)_3CO\overset{O}{\overset{\|}{C}}NH\overset{R}{\overset{|}{C}H}\overset{O}{\overset{\|}{C}}OCH_2\text{–}Ⓡ \xrightarrow[CH_2Cl_2]{(CH_3)_3SiCl\ +\ C_6H_5OH}$$

$$(CH_3)_2C = CH_2 + CO_2 + (CH_3)_3SiOC_6H_5 + HCl\cdot NH_2\overset{R}{\overset{|}{C}H}\overset{O}{\overset{\|}{C}}OCH_2\text{–}Ⓡ$$

Figure 6 Cholortrimethylsilane-phenol cleavage of *tert*-butyl protecting groups.

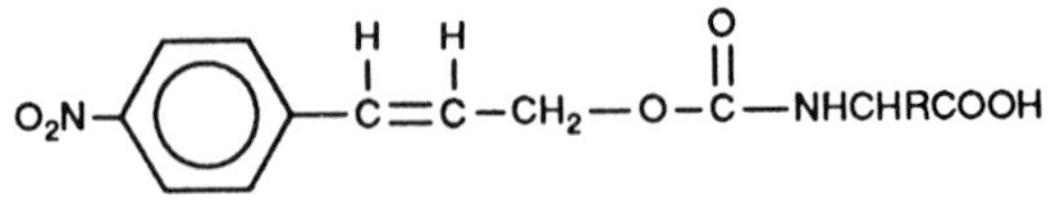

$O_2N-C_6H_4-CH{=}CH-CH_2-O-CO-NHCHRCOOH$

(X)

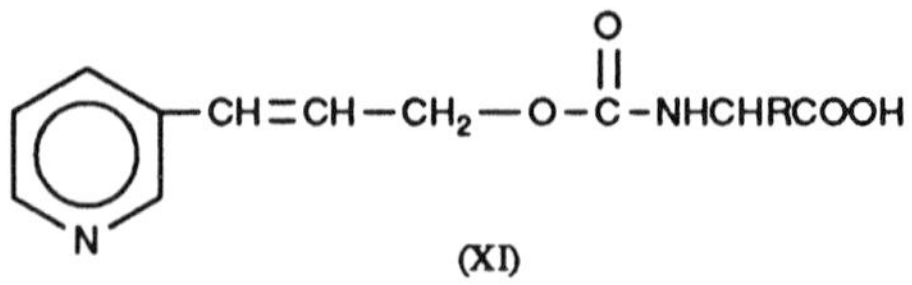

$C_5H_4N-CH{=}CH-CH_2-O-CO-NHCHRCOOH$

(XI)

S–S ring: CH_2 / CH_2, N–CHRCOOH

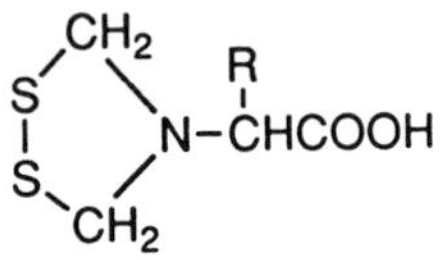

(XII)

strong UV absorption, but they retain the ease of cleavage by Pd(0) using $Pd(PPh_3)_4$ (10 mol%) in THF containing 0.78 equiv *N,N′*-dimethylbarbituric acid and 1 equiv triphenylphosphine. The Noc group is also stable to the cleavage of allyl esters by $(Ph_3P)_3RhCl$, an important feature.

A second allyl derivative for amino protection was developed by Kunz (von dem Bruck and Kunz, 1990). The Paloc group, 3-(3′-pyridyl)allyloxycarbonyl (**XI**), combines the advantages of stability under acid and basic reaction conditions, cleavability under neutral conditions, and solubility in water or organic solvents. The amino acid derivatives are made from the nitrophenylcarbonate, which is prepared from 3-(3′-pyridyl)allyl alcohol and bis(4-nitrophenyl)carbonate.

The Dts or dithiasuccinoyl group (**XII**) of Barany (Barany and Merrifield, 1977) is also an effective alternative to acid-labile N^{α}-protecting groups. It is very stable to acid but is cleanly removed by thiols.

B. Side-Chain-Protecting Groups

Protecting groups for solid-phase synthesis have been reviewed repeatedly (e.g., Barany and Merrifield, 1979a; Fields *et al.*, 1992). Therefore, only a few of the most important or recently described groups are mentioned here.

1. Arginine

Simple guanidine protonation or the classic nitro-, dicarbobenzoxy-, and tosyl-guanidine protecting groups have served well for synthesis of arginine peptides; however, each has limitations, and much has been done to improve the protection of arginine, especially in conjunction with N^{α}-Fmoc systems where more acid-labile groups are desirable. The work of Yajima *et al.* (1978) and Fujino *et al.* (1980) led to the proposal of N^g-4-methoxy-2,6-dimethylbenzenesulfonyl (Mds) and later 4-methoxy-2,3,6-trimethyl (Mtr) (Fujino *et al.*, 1981) together with other variants. Atherton *et al.* (1983) confirmed that Fmoc-Arg-(Mtr) had the best properties and was deprotected by TFA/thioanisole in 1 hr, although in some difficult sequences 6 hr was required.

The problem was examined in detail by the Ramage laboratory with attention to electronic and steric requirements. N^g-2,2,5,7,8-Pentamethylchroman-6-sulfonyl-L-arginine (Pmc-Arg) (**XIII**) was selected as the preferred derivative. It was deprotected with 50% TFA/CH_2Cl_2 in 1 hr and was useful in combination with N^α-Z (Ramage and Green, 1987) or N^α-Fmoc groups (Green *et al.*, 1988) groups. It was successfully applied to the solid-phase synthesis of ubiquitin(48–76) containing three arginine residues (Ramage *et al.*, 1991a), and is now the most frequently used arginine protecting group. Fischer *et al.* (1992) find that TFA does not always completely deprotect peptides containing multiple Arg(Pmc) groups and strongly recommend hard-acid deprotection with trimethylsilyl bromide. The five-member analog of Pmc, 2,2,4,6,7-pentamethyldihydrobenzofuran-5-sulfonyl (Pbf) was shown to be slightly more acid labile than Pmc and may represent a slightly better protecting group (Shroff *et al.*, 1994).

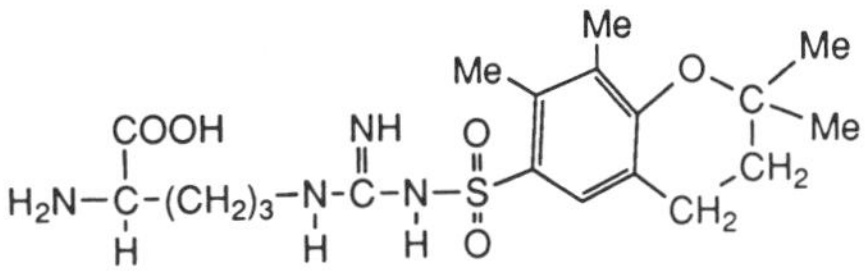

(XIII)

A 9-anthracenesulfonamido (Ans) group for guanidino protection has been explored by Arzeno and Kemp (1988). The group is stable to solid-phase synthetic procedures but is removed slowly by TFA. It is better removed by various reductive procedures, including aluminum amalgam and photoinduced ruthenium-catalyzed reduction with 1-benzyl-1,4-dihydronicotin-amide with a tungsten lamp at 25°C for 20 hr. A solid-phase synthesis of bradykinin, containing two arginine residues, was achieved in this way.

2. *Histidine*

Unlike other N^α-urethane-protected amino acids, histidine is prone to serious racemization during the activation and coupling reactions (Windridge and Jorgensen, 1971). The pros (π) imidazole nitrogen is properly situated and sufficiently basic to abstract intramolecularly the proton from the α carbon when the carboxyl is activated, leading to loss of chirality. The problem could be largely or completely avoided in two ways: (1) N^{im} electron withdrawing groups such as tosyl or 2,4-dinitrophenyl sufficiently reduced the basicity of the ring (Lin *et al.*, 1972) to give only 0.3 to 0.7% of the D isomer of a decapeptide made by DCC activation; (2) specific substitution of the π nitrogen with a benzyloxymethyl, N^{α}-Boc-N^{π}-His (Bom), or *tert*-butoxymethyl, N^{α}-Fmoc-N^{π}-His (Bum), gave racemization-free couplings (Brown *et al.*, 1982; Colombo *et al.*, 1984). Because formaldehyde is liberated on deprotection with HF or TFA, special precautions are necessary to avoid methylation of some side chains or thiazolidine formation from N-terminal Cys residues. Formaldehyde scavengers such as Cys or resorcinol are effective (Mitchell *et al.*, 1990; Kumagaye *et al.*, 1991). The addition of HOBt to a DCC coupling of Boc-His(Bzl)-OH with H-Glu(OBzl)-resin gave only 0.3% of the D-His isomer (Windridge and Jorgensen, 1971).

3. *Aspartic and Glutamic Acids*

Boc-dibasic amino acids are usually protected as β- and γ-benzyl esters, but cyclohexyl or cyclopentyl esters are finding greater use in solid-phase synthesis because of their increased acid stability and decreased tendency for alkylation side reactions. Okada and Iguchi (1988) introduced β-1- and β-2-adamantyl (Ada) aspartates. The β-1 derivative was labile to TFA, but the β-2 derivative was stable to TFA; both were unaffected by 55% piperidine. Both derivatives, as well as the cyclohexyl esters, suppressed aspartimide formation under acidic and basic conditions. The Ada group was cleaved by methanesulfonoic acid in a few minutes.

For convergent solid-phase synthesis of large peptides, solubility becomes a problem with many of the protected peptide segment intermediates, owing in part to hydrophobic amino acid side chains and to most of the common protecting groups. To overcome this problem the picolyl esters of Geoffrey Young have been applied to the side-chain protection of Asp and Glu and also serine and threonine (Rizo *et al.*, 1988). The 3-picolyl ester was used for Asp and Glu and the 4-picolyl ether for serine and threonine. The esters were made from 3-pyridylmethanol and DCC. They are stable to 30% TFA, 5% DIEA in CH_2Cl_2, high or low HF, and photolysis. Quantitative deprotection was achieved with 10% palladium on carbon in 80% acetic acid or galvanostatic electrolysis in 25 m*M* H_2SO_4 at 40 mA for 1 hr. The picolyl ethers were similarly stable and were also cleaved in an analogous way. The pyridine ring provided improved aqueous solubility in the protected peptides and allowed them to be purified by ion-exchange chromatography.

4. *Asparagine and Glutamine*

Asparagine and glutamine can be incorporated into peptide chains successfully with active ester coupling (Mojsov *et al.*, 1980). However, nitrile and amidine formation is substantial with DCC activation, and Asn dehydration is observed even in the presence of HOBt during BOP acylations (Gausepohl *et al.*, 1989). The nitrile dehydration product is readily reversed by treatment with HF followed by aqueous workup (Mojsov *et al.*, 1980). However the other by-product, aspartamidinoglycine, is not reconverted to the amide and remains as a contaminant. Treatment of the nitrile with TFA did not cause rehydration. Coupling with DCC plus HOBt gave only 1% of nitrile and only a trace of amidine. Amino-terminal Gln gives rise to a pyroglutamyl residue when it is deprotected and exposed to weak acids such as HOBt or the Boc-amino acid of the next coupling step (DiMarchi *et al.*, 1982). The coupling to Gln with a nonacidic species such as an active ester or preformed symmetrical anhydride was therefore recommended.

The alternative for introducing Asn and Gln into peptide chains is a suitable side-chain protecting group. The 9-xanthenyl (Xan) (Shimonishi *et al.*, 1962), 2,4-dimethoxybenzyl (Dmob) (Pietta *et al.*, 1976), and trityl (Trt) groups (Sieber and Riniker, 1991) have all been shown to minimize or eliminate the dehydration and cyclization reactions. Alkylation of Trp and other side reactions were not observed with the Trt group. The Trt group also minimizes hydrogen-bonded β-sheet structures. The 4-methyltrityl (Mtt) group has been recommended to replace Trt because of 2- to 3-fold enhanced acid lability (Sax *et al.*, 1992). The Asn(Mtt) was stable to 90% acetic acid or HOBt. It was introduced onto Z-Asn or Z-Glu via 4-methyltriphenylcarbinol. The products were then hydrogenated and reacylated with Fmoc-OSu.

5. *Tyrosine*

For N^{α}-Boc schemes, the phenolic hydroxyl of tyrosine has usually been protected as the benzyl ether or, better, as the 2,6 dichlorobenzyl (Erickson and Merrifield, 1973a) or cyclohexyl ethers (Engelhard and Merrifield, 1978), which increase acid stability and minimize ring alkylation during the HF deprotection step. 2-Bromobenzyloxycarbonyl-Tyr is an effective alternative to the ether derivatives (Yamashiro and Li, 1973). For N^{α}-Fmoc chemistry the acid labile *t*-Bu ether is commonly used.

Bpoc-*O*-β-ethyltrimethylsilyl-tyrosine (Tmse) (**XIV**), representing a novel class of silyl-based protecting groups compatible with Bpoc or *t*-Bu, has been developed for the phenolic group of tyrosine. The Bpoc group is removed selectively in 0.5% TFA/CH_2Cl_2, and the Tmse was quantitatively removed in 5 min in neat TFA. The products are ethylene and trimethylsilyltrifluoroacetate. The reaction is very clean because no alkylating species is generated.

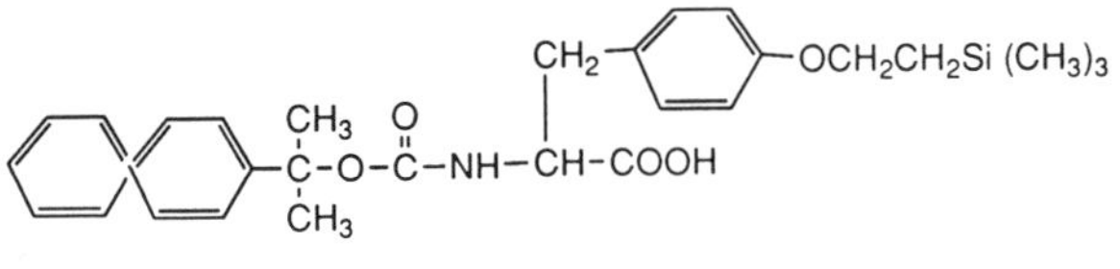

CH_2–C_6H_4–$OCH_2CH_2Si\,(CH_3)_3$

C_6H_5–C_6H_4–$C(CH_3)_2$–O–C(=O)–NH–CH–COOH

(XIV)

6. *Tryptophan*

The free Boc or Fmoc derivative of tryptophan is frequently used without indole protection, but some prefer to protect with the formyl group to avoid ring *tert*-butylation. The N^{in}-formyl group is readily removed by piperidine in DMF before HF cleavage. It is also removed under low HF conditions (25% HF/65% dimethylsulfide/10% *p*-cresol) (Tam *et al.*, 1983).

Kiso *et al.* (1988a) have used N^{in}-diphenylphosphenothiotryptophan to protect against side reactions and have shown that the Ppt group can be removed by 0.25 *M* methanesulfonic acid–thioanisole–TFA and also by tetra-*n*-butylammonium fluoride in DMF. More recently, Boc-Trp(Aloc)-OH and Fmoc-Trp(Aloc)-OH were prepared (Trzeciak *et al.*, 1993). The Aloc (allyloxycarbonyl) group was stable to acid. Selective removal of the Aloc group was achieved with $Pd^0(PPh_3)_4$ or 50% piperidine in water. Use with Fmoc synthesis required the stepwise cleavage of Fmoc with 2% DBU in DMF, where Aloc was stable, instead of piperidine.

7. *Cysteine*

Protection, anchoring, deprotection, disulfide bond formation, and side reactions of cysteine have been discussed in some detail by Barany and Merrifield (1979a). The following protecting groups, and method of removal, are compatible with N^{α}-Boc synthetic schemes: 4-methylbenzyl (Meb), HF or thallium (III) tristrifluoroacetate [$Tl(Tfa)_3$]; acetamidomethyl (Acm), mercuric acetate, silver tetrafluoroborate, or I_2; *tert*-butylsulfenyl (S-*t*-Bu), thiols or phosphines; 3-nitro-2-pyridinesulfenyl (npys), thiol or base; fluorenylmethyl thio ether (Fm), thiol or base. For use with Fmoc chemistry, Acm, S-*t*-Bu, Trt, and 2,4,6-trimethoxybenzyl (Tmob) are suitable. Studies have established that Cys(Npys) is stable to HOBt and is compatible with Boc/benzyl solid-phase synthesis (Matsueda *et al.*, 1992). The *S*-2,4,6-trimethoxybenzyl-cysteine derivative has been further examined (Munson *et al.*, 1992). Fmoc-Cys(Tmob) was synthesized from the alcohol by acid catalysis. Cleavage from model peptides was achieved in CH_2Cl_2 containing 30% TFA, 5% phenol, 5% thioanisole, and 5% water, or 6% TFA, 0.5% Et_3SiH in CH_2Cl_2, to give the free thiol. Oxidative deblocking with I_2 or $Tl(Tfa)_3$ gave the intramolecular disulfide.

Royo *et al.* (1992) introduced a new cysteine-protecting group, *S*-(2,4-dinitrophenyl)ethyl (Dnpe), to give Boc-Cys(Dnpe) (**XV**). The group is stable to acid (50% TFA/CH_2Cl_2, 90% HF/*p*-cresol or TFMSA/TFA/*p*-cresol, 1:10:3) and can be removed by β-elimination with base (1:1 piperidine–DMF, 30 min), but it was stable to 5% DIEA/CH_2Cl_2. The Dnpe group is also stable to oxidative conditions with $Tl(Tfa)_3$ or I_2 in 80% acetic acid. These are properties similar to those of the fluorenylmethyl (Fm) group (Bodanszky and Bednarek, 1982), but the derivatives are less hindered, more soluble, and give better yields of protected peptides after HF cleavage from resins.

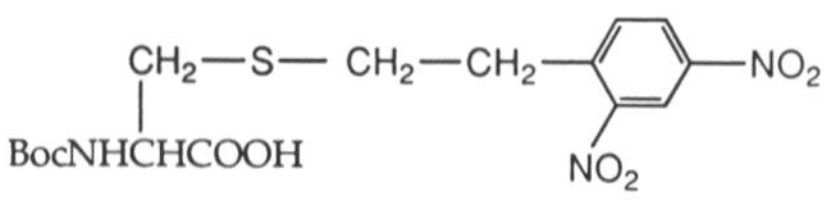

CH_2—S— CH_2—CH_2—(2,4-dinitrophenyl: NO_2, NO_2)
BocNHCHCOOH

(XV)

C. Orthogonal Schemes of Protection

Following the development of new protecting groups, alternatives to kinetic methods to achieve selectivity during synthesis became possible via

Table V Solid-Phase Orthogonal Schemes

Entry		Dimensions[a]	Ref.
1	$CH_3-C(CH_3)_2-O-C(=O)-NH-CH(CH_2-O-CH_2-C_6H_5)-C(=O)-OCH_2-C_6H_4-$(R); TFA; HF; HF	1 + *k*	Merrifield (1963)
2	$C_6H_5-C_6H_4-C(CH_3)_2-O-C(=O)-NH-CH(CH_2-O-CH_2-C_6H_5)-C(=O)-O-C(CH_3)_2-CH_2CH_2-C_6H_{10}-$(R); 1% TFA; HF; 50% TFA	1 + *k*	Wang and Merrifield (1971)
3	$CH_3-C(CH_3)_2-O-C(=O)-NH-CH(CH_2-O-CH_2-C_6H_5)-C(=O)-OCH_2-C(=O)-C_6H_4-$(R); TFA; HF; hν	2 + *k*	Wang (1976)

(continues)

Table V Solid-Phase Orthogonal Schemes *(continued)*

Entry		Dimensions[a]	Ref.
4	RSH → S–C(=O), S–C(=O) ring N–CH(CH$_2$–O–Bu^t ← TFA)–C(=O)–O–CH$_2$–Ⓡ (↑ HF)	$2 + k$	Barany and Merrifield (1977)
5	Fmoc – NH –CH(CH$_2$–O–Bu^t ← TFA)–C(=O)–OCH$_2$–C$_6$H$_4$–OCH$_2$–C$_6$H$_4$–Ⓡ (Piperidine ↑ at Fmoc–NH; TFA ↑ at C–OCH$_2$)	$2 + k$	Chang and Meienhofer (1978) Atherton *et al.* (1978)
6	Fmoc – NH– CH(CH$_2$–O–Bu^t ← TFA)–C(=O)– O – CH(C$_6$H$_3$(OCH$_3$)$_2$)–C$_6$H$_4$–OCH$_2$–Ⓡ (Piperidine ↑ at Fmoc–NH; 1%TFA ↑ at O–CH)	$2 + k$	Rink (1987)

(continues)

(continued)

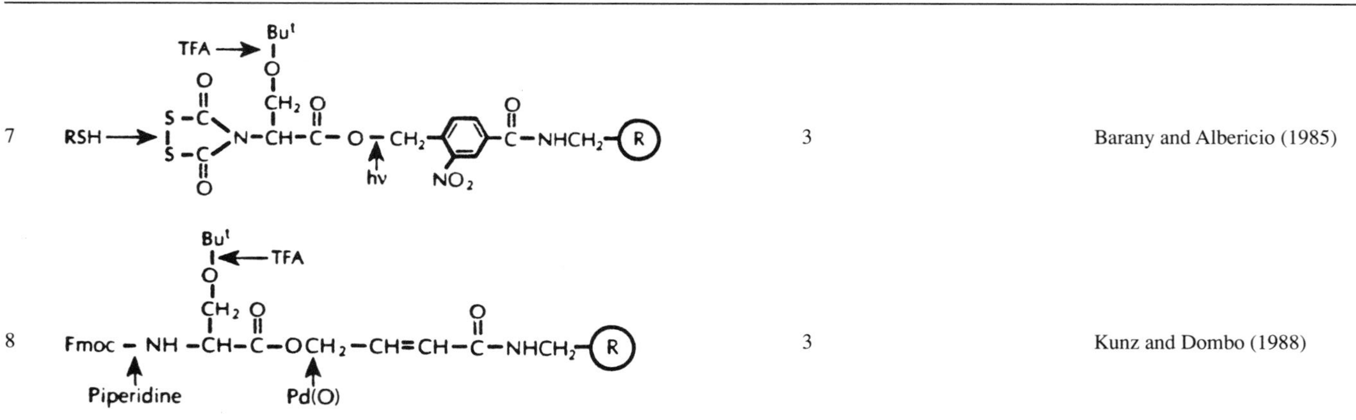

7		3	Barany and Albericio (1985)
8		3	Kunz and Dombo (1988)

[a] k means that selectivity by a single type of reagent for two or more groups is achieved by kinetic differences.

orthogonal schemes of protection in which each type of protecting group is removed by a unique kind of chemical reagent. An orthogonal system has been defined (Barany and Merrifield, 1977) as a set of completely independent classes of protecting groups in which the members of each class can be selectively removed in any order and in the presence of members of all other classes. The concept of orthogonality has also been discussed by Kemp *et al.* (1975), Merrifield *et al.* (1977), Fauchére and Schwyzer (1981), and Bodanszky and Bodanszky (1984) and is now being widely utilized. It is possible to develop two, three, or more degrees of orthogonality (Table V). In general, these will be based on protecting groups that are very selectively removed by acid, base, thiols, cyanide, fluoride ion and other nucleophiles, silanes, photolysis, hydrogenolysis, reduction, oxidation, and Pd(0) catalysis.

Many combinations of reagents have resulted in two-dimensional orthogonality, with two of the three groups being differentiated by kinetic (*k*) differences with the same class of reagent. The most frequently used system for solid-phase synthesis is the Fmoc/*t*-Bu strategy, in which the N^{α}-Fmoc group is removed by piperidine or other base in the presence of acid-labile protecting groups. The Fmoc group is very stable to acid (TFA/HF/refluxing HCl), and most acid-labile groups are resistant to piperidine. Of the other two-dimensional schemes that have been reported, the following are representative of the variety of chemistry that has been used: Boc/Bzl/ONb-R (H^+, H^+, *hν*) (Rich and Gurwara, 1975); Fmoc/*t*-Bu/$OCH_2C_6H_4OCH_2$-R (base, H^+, H^+) (Chang *et al.*, 1980; Atherton *et al.*, 1981); Dts/t-Bu/OBzl (RSH, H^+, H^+) (Barany and Merrifield, 1977); ONPS/*t*-Bu/$OCH_2C_6H_4OCH_2$-R (RSH, H^+, H^+) (Fries *et al.*, 1979); 2-sulfo-9-fluorenylmethyloxycarbonyl (Sulfmoc)/Bzl/OBzl (base, H^+, H^+) (Merrifield and Bach, 1978); NPS/*t*-Bu/$OCH_2C_6H_4OCH_2$-R [$(C_6H_5)_3P$, H^+, H^+] (Wang *et al.*, 1982); Boc/allyl-R [H^+, Pd(0)] (Kunz and Dombo, 1988); Boc/Bzl (H^+, catalytic H_2) (Schlatter *et al.*, 1977); Boc/Fmoc/phenacyl (H^+, F^-, F^-) (Kiso *et al.*, 1988a,b).

Only a few demonstrations of three-dimensional orthogonal protection and removal during solid-phase synthesis have been reported, although many other schemes could be designed using only those protecting groups already available. The first of these appears to be by Barany and Albericio (1985). They combined the N^{α}-dithiasuccinoyl (Dts) group, which is removed by thiols but stable to acid or photolysis, with side-chain *tert*-butyl groups that are removed by acid, but stable to thiolysis or photolysis, with an *ortho*-nitrobenzyl (ONb) ester that is removed by light but stable to thiols or acid (entry 7 in Table V). Thus, a Dts-amino acid was esterified to *tert*-butyl 4-(hydroxymethyl)-3-nitrobenzoate, treated with TFA, and coupled to aminomethyl-copoly(styrene–1% divinylbenzene)-resin. The test peptide Dts-Tyr(*t*-Bu)-Gly-Gly-Phe-Leu-ONb-resin was then assembled stepwise by deprotection with β-mercaptoethanol-*N*,*N*-diisopropylethylamine (0.5/0.5 *M*) in CH_2Cl_2 and coupling of Dts-amino acids with DCC in CH_2Cl_2. The completed, fully protected peptide-resin was then treated in all possible orders with the following three orthogonal reagents; (1) thiolytic removal of the Dts group, (2) acidolytic cleavage of the *tert*-butyl ether, and

(3) photolytic cleavage at 350nm of the ONb ester. One deprotected and two partially deprotected peptide-resins as well as three partially protected free peptides and one fully deprotected peptide were obtained pure and in good yield. This demonstrated the feasibility and benefits of a mild three-dimensional orthogonal protection scheme.

An alternative three-dimensional orthogonal scheme (entry 8 in Table V) combines N^{α}-Fmoc, C^{α}-allyl ester, and side-chain *tert*-butyl groups that are selectively removed by base, Pd(0), and acid, respectively (Kunz and Dombo, 1988). The allyl group has the advantages that it is rapidly and quantitatively removed and avoids alkylation side reactions because it does not give rise to reactive carbocations.

Allyl-based orthogonal schemes have also been developed by Albericio *et al.* (1993) and Kates *et al.* (1993). The allyl group was used in combination with Fmoc and *t*-Bu groups and led to the synthesis of side-chain lactams, head-to-tail cyclic peptides, branched peptides, and glyco- and sulfopeptides. For cyclic lactams the starting resin was (a) Fmoc-Glu(OPAC-PEG-PS)-OAl, (b) Fmoc-Asp(OPAC-PEG-PS)-OAl, (c) Fmoc-Asp(PAL-PEG-PS)-OAl. After stepwise synthesis using piperidine deprotection of Fmoc, the allyl ester was deprotected with 1 to 2 equiv $Pd(PPh_3)_4$ in DMSO/THF/0.5 *N* HCl/morpholine (2:2:1:0.1). Cyclization was with BOP/HOBt/*N*-methylmorpholine (NMM). Final deprotection and cleavage was with TFA/thioanisol/2-mercaptoethanol/anisol. To automate the process, the allyl group was removed in the solvent $CHCl_3$/acetic acid/NMM (20:1:0.5), because $Pd(PPh_3)_4$ is readily soluble. Trzeciak *et al.* (1993) and Kates *et al.* (1993) have used related approaches.

An interesting orthogonal scheme was devised for the synthesis of endothelin-converting enzyme substrate (Handa and Keech, 1992). A side-chain Glu was protected as the allyl ester while other side-chain groups were protected with *t*-Bu or Trt groups. The peptide was anchored to a Pepsyn KA resin and N^{α}-Fmoc was the temporary N^{α}-protecting group. The allyl group was removed with Pd(0) and reacted with 5-(2-aminoethylamino)-1-naphthalenesulfonic acid to serve as a quencher. The peptide was cleaved and further deprotected with TFA.

Baeza and Undén (1992) used Boc-Lys(Aloc)-OH side-chain protection in an orthogonal scheme to allow the attachment of biotin. It was used in conjunction with Boc-, Dnp-, and MBHA-resin groups. Selective orthogonal release of peptides from resins following the synthesis of peptide libraries has also been described (Salmon, *et al.*, 1994).

D. Multidetachable Resins

All of the resin derivatives discussed so far have only a single cleavable bond between the peptide and the resin support. It is possible, however, to devise derivatives with more than one cleavable bond. Tam prepared several such derivatives and called them multidetachable resins (Tam *et al.*, 1980). The design provided one labile benzyl ester bond between the peptide and the linker that was cleavable by HF, CN^-, or OH^-, and one phenacyl ester bond between the linker and the resin that was cleavable by RSH, *hν*, CN^-, or OH^- (Fig. 7). It was then possible, after a single synthesis, to prepare either a free peptide, a protected peptide, or a protected peptide containing a C-terminal linker that could be purified and reattached to another support for further synthesis.

Hellermann *et al.* (1983a,b) devised a related scheme (Fig. 8) in which the central linker was polyoxyethylene. This provided a way to prepare unprotected peptides with a C-terminal polyoxyethylene (POE) ester that were very useful for studying the physical properties of the peptide.

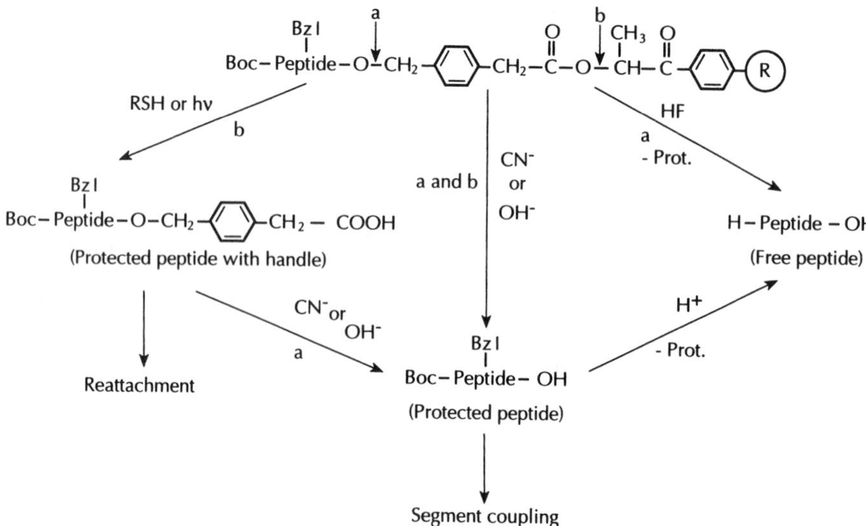

Figure 7 Multidetachable resin support on styrene–divinylbenzene.

Bu^t ←TFA O NO$_2$ O O

Fmoc-NH-Peptide-C-O-CH$_2$-⟨⟩-C-NH-PEG-NHC-O-CH$_2$-⟨⟩-OCH$_2$-(R)

Piperidine hυ or OH⁻ TFA

1. Free peptide
2. Protected peptide
3. Peptide-handle

Figure 8 Multidetachable PEG-S-DVB resin support (Hellerman *et al.*, 1983a, b).

V. The Coupling Reaction

The third major parameter in solid-phase peptide synthesis is the peptide-forming reaction. Ideally, it must be rapid and complete, even with hindered components, and must be free of racemization or other side reactions. Thus, there should be no deletion, insertion, or modification peptides. The coupling reaction requires the activation of the incoming amino acid and subsequent condensation with the resin-bound amino acid or peptide. In practice it is always the carboxyl component that is activated, and in nearly all cases it is the soluble component, either a protected amino acid or peptide. It then reacts with the support-bound amino component to form the new peptide bond. The carboxyl group may be preactivated in a stable, crystalline form, or it may be activated *in situ* with a suitable reagent.

A. Carbodiimides

The first fully successful activating reagent for solid-phase synthesis (Merrifield, 1963) was dicyclohexylcarbodiimide (Sheehan and Hess, 1955), and it is still frequently used. It is thought to proceed via the *O*-acylisourea, which can (1) react directly with the amine component, (2) react with another mole of the carboxyl component to give the symmetrical anhydride intermediate, (3) rearrange to the oxazolone, (4) rearrange to the inactive *N*-acylurea, or (5) in the presence of an added acidic alcohol, such as N^1-hydroxybenzotriazole, react to give an intermediate active ester. These are competing reactions, which depend on the exact reaction conditions.

The carbodiimide coupling reaction was studied in great detail by DeTar *et al.* (1966), with the conclusion that it proceeds by a mixed mechanism. Hudson (1988) designed a competitive coupling experiment in which the relative extent of incorporation of preformed Fmoc-Phe anhydride was compared with the competing incorporation of Fmoc-Tyr(Bzl) when activated in various ways. Complete preactivation to the symmetrical anhydride by diisopropylcarbodiimide in CH_2Cl_2/DMF (1:1) required at least 10 min, but preactivation for only 2 min produced a better relative coupling rate (>1 for Tyr versus Phe). The data were interpreted to mean that there was an active intermediate other than the anhydride, and it was concluded that the *O*-acylisourea (OAIU) was formed rapidly and reacted with the amine nucleophile faster than with another mole of carboxylate. Thus, the reaction proceeded directly from the OAIU without going through the anhydride. This agrees with the previously expressed belief that the anhydride does not couple more rapidly than the *O*-acylisourea (Merrifield, 1969).

The existence of a true coupling reagent has been described by Bodanszky (1992) as a "myth," because in nearly all cases the agent simply serves to convert the carboxyl component to an activated species, for example, an ester or anhydride, which then is subject to nucleophilic attack by the amine. By his definition a coupling reagent is one that "forms a ternary complex with the acid and amine, and in a concerted reaction forms the peptide bond directly by a quasi-intramolecular activation and concomitant coupling." Bodanszky feels that such a reaction has not been described. It is probable, however, that the very first successful activating reagent for solid-phase peptide synthesis, dicyclohexylcarbodiimide, can actually proceed in that way and qualify as a true coupling reagent. When the salt, or ion pair, between the carboxyl component and resin-bound amine is formed in a nonpolar organic solvent, it is stable to washing and couples rapidly in high yield on addition of the diimide.

The effect was discovered by Klostermeyer and demonstrated by Esko and Karlsson (1970). This occurs with the amino acid-resin suspended in either a small or large volume and follows pseudo-first-order kinetics (R. B. Merrifield, unpublished), suggesting that the activated carboxyl component is not released into free solution and then, after dilution with bulk solvent, returns for reaction with the amine, but, rather, it participates in a "concerted quasi-intramolecular activation and concomitant peptide formation induced by the coupling reagent."

The diimide method for solid-phase synthesis was soon supplemented by a large variety of other activating reagents, some of which are shown in Fig. 9.

B. Active Esters

Bodanszky and Sheehan (1964) quickly demonstrated that *p*-nitrophenyl esters would couple under modified solid-phase conditions. This reagent is frequently used to couple Asn and Gln to avoid nitrile formation (Mojsov *et al.*, 1980). For general synthesis the ONp esters were soon replaced by more reactive esters prepared from *N*-hydroxysuccinimide (Weygand and Ragnarsson, 1966), N^1-hydroxybenzotriazole (König and Geiger, 1972), 3-hydroxy-4-oxo-3,4-dihydrobenzotriazine (Dhbt) (König and Geiger, 1970), 5-chlor-8-hydroxyquinoline (Jakubke and Baumert, 1974), pentafluorophenol (Kisfaludy *et al.*, 1967; Kisfaludy and Schön, 1983), and many others.

The Fmoc-amino acid pentafluorophenyl esters were adopted by Atherton *et al.* (1988a) for solid-phase synthesis in DMF on a polydimethylacrylamide support with good results. Coupling rates were accelerated by HOBt. The Fmoc-amino acid Dhbt esters were also prepared by Atherton *et al.* (1988b) and used for automated syntheses. These coupling rates were also accelerated by HOBt. Ionization of the liberated alcohol by the resin-bound amine gave a yellow color, which was useful for monitoring the coupling reaction. When all the amine has coupled, the solution becomes colorless.

C. Anhydrides

Following the carbodiimide and active ester period, amino acid anhydrides began to be explored for solid-phase synthesis. The mixed anhydrides were effective reagents, but they were never extensively used because of the concern over "wrong

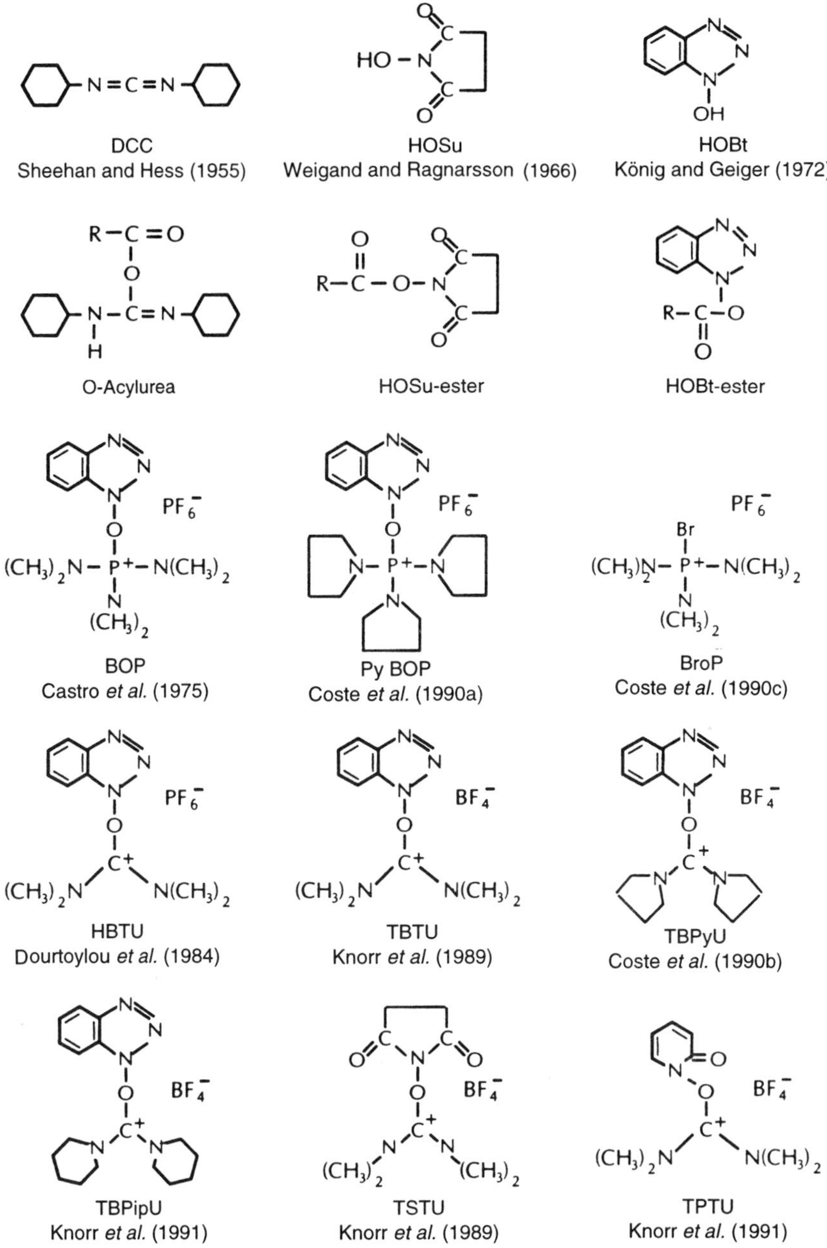

Figure 9 Peptide coupling reagents.

way" addition, a concern that has not been well documented experimentally. However, when the symmetrical anhydrides were introduced (Wieland *et al.*, 1973; Yamashiro and Li, 1974) they were recognized to be excellent reagents. They can be prepared, in crystalline form, from the amino acid salt and phosgene, or from 2 equiv of Boc-amino acid and 1 equiv of DCC in CH_2Cl_2 and used without isolation.

Because of the existence of a class of "difficult sequences" there has been a widespread effort to design more efficient activating reagents. For example, Tung and Rich (1985) introduced a new reagent for synthesis of peptides containing *N*-alkylamino acids, such as cyclosporin A. Thus, bis(2-oxo-3-oxazolidinyl)phosphinyl chloride (BopCl) (**XVI**) was found to be very reactive, to give only low levels of racemization, and to be convenient to use. The activated Boc-amino acid is a carboxylic-phosphinic anhydride, which undergoes aminolysis directly; however, with highly hindered systems, it may also disproportionate to the less reactive symmetrical anhydride.

O O O C C O N—P—N O Cl

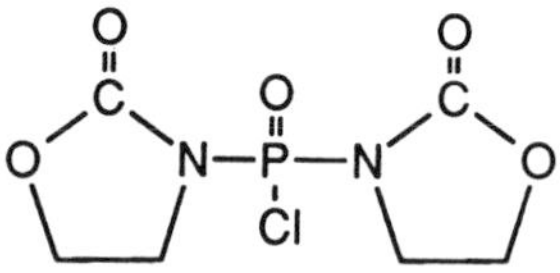

(XVI)

D. Phosphonium Salts

Castro and co-workers made major advances in the development of new, highly active reagents. The BOP reagent [benzotriazolyloxytris(dimethylamino)phosphonium hexafluorophosphate] (Fig. 9) (Castro *et al.*, 1975) was especially important. It promotes rapid and highly efficient couplings and soon became the reagent of choice in many laboratories. It is prepared from tris(dimethylamino)phosphine and phosgene (Castro and Dormoy, 1972) followed by reaction with HOBt and KPF_6 (Scheme 3). This phosphonium derivative was soon modified by replacing the hexafluorophosphate counterion with tetrafluoroborate.

The BOP reagent was used extensively for solution synthesis and for solid-phase synthesis using the Boc/TFA strategy (e.g., Seyer *et al.*, 1990; Bouhnik *et al.*, 1987) and with the Fmoc/morpholine strategy (Felix *et al.*, 1988). BOP reacts only with carboxylate salts and is stable in the presence of carboxylic acids. Therefore, in the coupling of a Boc-amino acid, a C-terminal protected peptide TFA salt and BOP, 3 equiv of a tertiary base are required. In segment coupling some racemization is detected (Le-Nguyen *et al.*, 1981). Because preneutralization of the TFA salt is not required, diketopiperazine formation (and probably pyroglutamate formation) can be minimized. This is possible because TFA is not

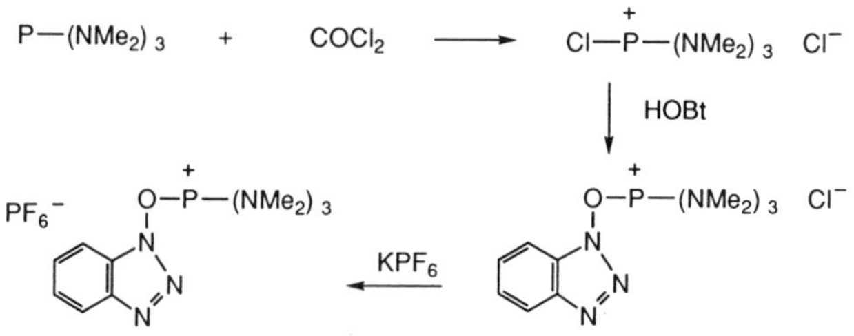

$$P{-}(NMe_2)_3 + COCl_2 \longrightarrow Cl{-}\overset{+}{P}{-}(NMe_2)_3\ Cl^-$$

$$\xrightarrow{HOBt} BtO{-}\overset{+}{P}{-}(NMe_2)_3\ Cl^- \xrightarrow{KPF_6} BtO{-}\overset{+}{P}{-}(NMe_2)_3\ PF_6^-$$

Scheme 3

activated by BOP (Coste *et al.*, 1990a). The mechanism was originally thought to proceed directly through the HOBt ester of the incoming Boc-amino acid. It was later postulated (Coste *et al.*, 1990a) that the coupling proceeds directly with the acylphosphonium salt via a cyclic complex (Scheme 4).

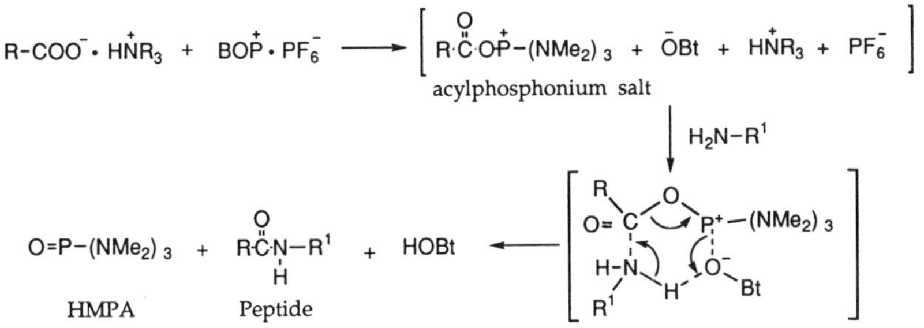

$$R{-}COO^- \cdot H\overset{+}{N}R_3 + BO\overset{+}{P} \cdot PF_6^- \longrightarrow \left[R{\cdot}\overset{O}{\overset{\|}{C}}{\cdot}O\overset{+}{P}{-}(NMe_2)_3 + \bar{O}Bt + H\overset{+}{N}R_3 + PF_6^- \right]$$

acylphosphonium salt

$$\xrightarrow{H_2N{-}R^1}$$

$$O{=}P{-}(NMe_2)_3 + R{\cdot}\overset{O}{\overset{\|}{C}}{\cdot}\underset{H}{N}{-}R^1 + HOBt$$

HMPA Peptide

Scheme 4

Hexamethylphosphoramide (HMPA) is carcinogenic and was therefore replaced in various ways. Substitutions of the six methyl groups by ethyl, piperidino, or morpholino groups all gave slow coupling, but pyrrolidino groups on phosphorus gave a reagent, benzotriazolyloxytrispyrrolidinophosphonium hexafluorophosphate (PyBOP) (Fig. 9), that was at least as active as BOP and was more effective for hindered *N*-Me-Val couplings. A segment of acyl carrier protein, ACP(65–74), was made in excellent yields. It was known that HOBt accelerates active ester- or DCC-mediated coupling of normal α-amino acids and of *N*-methylamino acids, but it is actually inhibitory for BOP coupling, suggesting that the OBt component was responsible for the poor results. Consequently, Coste *et al.* (1990c) began to examine phosphonium derivatives lacking HOBt and concentrated on bromo and chloro analogs. They prepared bromotris(dimethylamino)phosphonium hexafluorophosphate (BroP), bromotrispyrrolidinophosphonium hexafluorophosphate (PyBroP), and chlorotrispyrrolidinophosphonium hexafluorophosphate (PyCloP).

The BroP reagent was equivalent to BOP in unhindered syntheses but vastly better (70 versus 5%) for coupling Z-MeVal to MeValOMe. The new bromo derivative is interesting because the activated amino acid intermediate is not the ester, but the transient acylphosphonium salt, Z-NHCHRC(=0)0P$^+$[N$(CH_3)_2]_3$ PF_6^- after loss of the small leaving group, Br^-. Aminolysis by the amine component gives the peptide. If $^-$OBt were present it would compete favorably with amine and give the HOBt ester, which for steric reasons reacts much more slowly. This reagent was particularly good for coupling *N*-methyl amino acids, whereas all couplings containing HOBt were poor (Coste *et al.*, 1990c). It was interesting to note that DCC was as good as BroP in this study.

To avoid HMPA, the pyrrolidine derivatives PyBroP and PyCloP were prepared and found to be effective reagents. They were much better than PyBOP or DCC/HOBt for hindered couplings. The results of coupling two α-aminoisobutyric acid (AIB) residues were different. In this case PyBOP was recommended over PyBroP (Coste *et al.*, 1991). The bromo derivative was thought to go through a symmetric anhydride because the reaction was accelerated by DMAP. It was observed by Coste *et al.* (1993), however, that yields could be decreased when coupling Boc-protected amino acids to *N*-methyl or α-methyl amino acid methyl esters, because of the conversion of part of the activated Boc-amino acid to the *N*-carboxyanhydride via an *N*-methyloxazolonium intermediate. It seems likely that the NCA would introduce multiple residues of the activated amino acid, but this side reaction should be minimized under solid-phase conditions.

E. Uronium Salts

Following the development of the phosphonium salts it was found that uronium salts were also effective activating reagents (Fig. 9). Dourtoglou *et al.* (1978, 1984) reported the preparation of 2-(benzotriazol-1-yl)-1,1,3,3-tetramethyluronium hexafluorophosphate (HBTU), and Knorr *et al.* (1989) reported the corresponding TBTU, containing the tetrafluoroborate counterion, and also the uronium salt containing *N*-hydroxysuccinimide as the hydroxyl component [succinoyltetramethyluronium tetrafluoroborate (TSTU)] (Fig. 10). Knorr *et al.* (1991) later produced the bispiperidinyluronium derivative, 2-(benzotriazol-1-yl)-bispiperidinyluronium (TBPipU), and the pyridone-containing *N*-(2-pyridonyl)-1,1,3,3-tetramethyluronium tetrafluoroborate (TPTU) derivative. Coste *et al.* (1990b) reported the bispyrrolidine uronium derivative, benzotriazolyl-

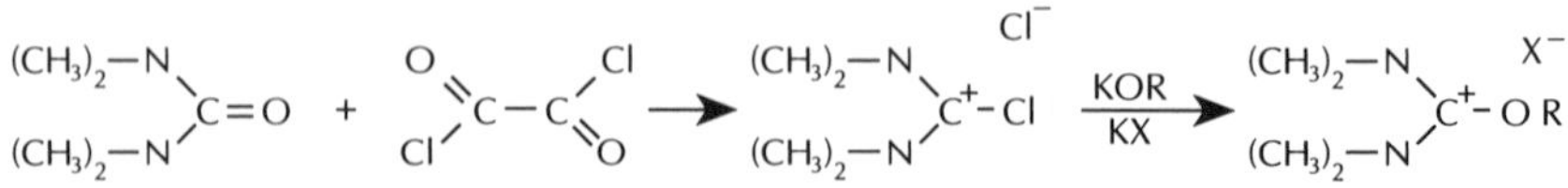

$$\begin{matrix}(CH_3)_2-N \\ (CH_3)_2-N\end{matrix}\!\!>C{=}O \;+\; ClC(=O)-C(=O)Cl \longrightarrow \begin{matrix}(CH_3)_2-N \\ (CH_3)_2-N\end{matrix}\!\!>C^+{-}Cl\;\; Cl^- \xrightarrow[KX]{KOR} \begin{matrix}(CH_3)_2-N \\ (CH_3)_2-N\end{matrix}\!\!>C^+{-}OR\;\; X^-$$

R= –Bt, –DHBt, OPy, –Su, etc.

X= PF_6^-, BF_4^-

Figure 10 Synthesis of uronium salts.

bispyrrolidinouronium (TBPyU) (Fig. 9). These compounds bear a resemblance to the reactive, but unstable, *O*-acylisourea intermediate in the carbodiimide activations, as well as to the Bates reagent, $(Me_2N)_3$-P^+-O-P^+-$(NMe_2)_3$, developed in the laboratory of George Kenner (Bates *et al.*, 1975).

Most of these new reagents proceed by way of the active ester, which is formed very rapidly *in situ* and quickly yields the peptide bond with the amino component. They apparently do not form symmetrical anhydrides under the synthetic conditions. Knorr made the following recommendations: (1) TBTU and HBTU are best for solid-phase synthesis; (2) TPTU is best for segment coupling; (3) TDBTU is best for suppression of racemization.

Carpino (1993) has introduced the aza analog of HOBt, 1-hydroxy-7-azabenzotriazole (HOAt) (Fig. 11), which as an additive can accelerate DCC coupling reactions. The compound was also incorporated into the new phosphonium and uronium salts, giving, for example, *O*-(7-azabenzotriazol-1-yl)-1,1,3,3-tetramethyluronium hexafluorophosphate (HATU) or its BF_4^- analog *O*-(7-azabenzotriazol-1-yl)-1,1,3,3-tetramethyl tetrafluoroborate (TATU) (Carpino *et al.*, 1994). The increased reactivity of these aza reagents is thought to result from the neighboring group effect of the nitrogen at position 7, which can assist in the aminolysis by hydrogen bonding.

F. *N*-Carboxyanhydrides

Shortly after the introduction and development of the phosphonium and uronium coupling reagents, attention was focused again on *N*-carboxyanhydrides (NCAs). These amino acid derivatives were first studied by Leuchs (1906) for the preparation of polymers. Conditions were later found for coupling single amino acid residues (Denkewalter *et al.*, 1966), but they were not adapted to solid-phase methods. Early attempts to prepare protected NCAs using phosgene on ethoxycarbonyl-glycine were not general. In 1990 Fuller *et al.* reported a general synthesis of crystalline urethane-protected *N*-carboxyanhydrides (UNCAs) by treatment of NCAs in dry THF at 2°C with Fmoc chloride and slow addition of *N*-methylmorpholine. This class of activated amino acids can

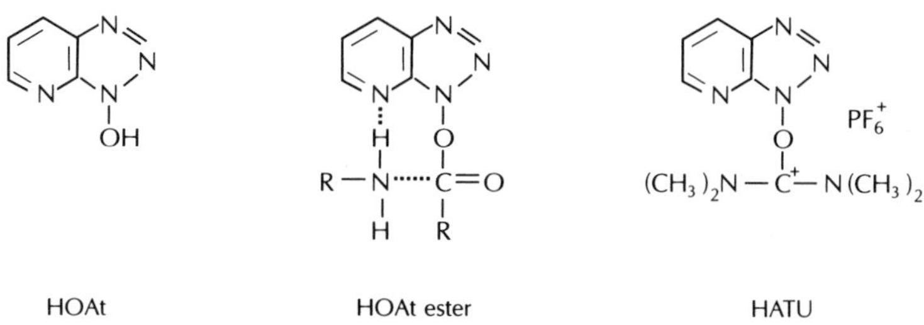

Figure 11 1-Hydroxy-7-azabenzotriazole.

be used directly for forming peptide bonds. They couple rapidly, even with hindered peptides, and the only by-product is CO_2. Synthesis of ACP(65–74) was prepared on 4-alkoxy-2′,4′-dimethoxy benzhydrol–2% polystyrene resin in a flow reactor in good yield and purity. The UNCAs were also effective for esterifying the Wang and Rink hydroxy resins with both Fmoc- and Boc-amino acids (Fuller *et al.*, 1993a). UNCAs are very soluble in a range of organic solvents and are quite stable, even at 50° in DMF for 6 hr, provided water or other nucleophiles are excluded (Fuller *et al.*, 1993b).

G. Acid Halides

Acid halides were among the earliest known activated amino acid derivatives; however, they are complicated by conversion to *N*-carboxyanhydrides, and they also became associated with problems of racemization during solution synthesis and were generally abandoned. They were never seriously used for solid-phase synthesis. More recently Fmoc amino acid chlorides (but not Boc forms) have been prepared with TMSCl/Fmoc-Cl (Bolin *et al.*, 1989; Carpino *et al.*, 1986). Perlow *et al.* (1992) found that Fmoc amino acid chlorides were very useful for solid-phase synthesis of peptides containing several secondary amino acids. For very hindered coupling, yields of 30% were raised to 76% by addition of AgCN, where the activated derivative was an oxazolone. No racemization was detected in these reactions. In a hindered coupling of Fmoc-pipicolic acid to Pro-resin, BOP activation gave 60% diketopiperazine (DKP), whereas the more reactive Fmoc-pipicolic acid chloride gave a high yield of product with little DKP. The cyclic hexapeptide cyclo(D-Phe-Ile-D-Pip-Pip-D-NMePhe-Pro) was prepared from the acid chlorides in 70% overall yield and 99% purity.

It was soon found that Fmoc, Boc, and Z groups are stable to acid fluoride preparation with cyanuric fluoride (Carpino *et al.*, 1990, 1991). They are more stable than the chlorides to oxygen nucleophiles but are equally active toward amines. During solid-phase synthesis in the presence of DIEA they react rapidly and safely, even in the presence of *tert*-butyl derivatives, again with no racemization. In the coupling of Fmoc-AIB residues the acid fluorides were remarkably good compared with symmetrical anhydrides and UNCAs (Wenschuh *et al.*, 1994).

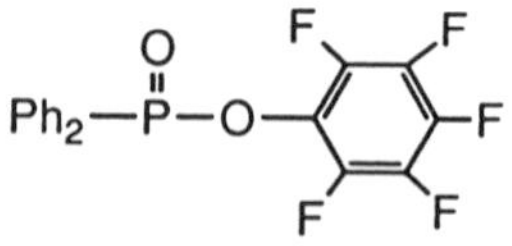

(XVII)

H. Other Activation Methods

A variety of other activating reagents have been introduced. For example, Chen and Xu (1991) described pentafluorophenyldiphenylphosphinate (FDPP)

(**XVII**), which was prepared from the phosphinate chloride. It required no separate preactivation step and was an effective condensing reagent for the solid-phase synthesis of Leu-enkephalin in 98.5% yield. There was no racemization in the Anderson or Young tests. Ueda and Mori (1992) have proposed diphenyl (2,3-dihydro-2-thioxo-3-benzoxazolyl)phosphonate as a condensing agent, and Gruszecki *et al.* (1988) proposed Boc-NH-CH(R)-CON(C_6H_5)CO-C_6H_3(2,6-Cl_2) as a mild acylating agent.

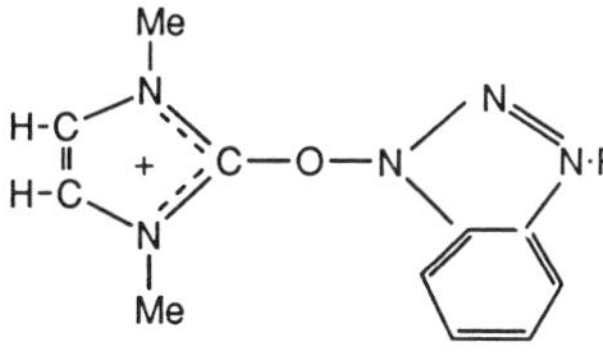

(XVIII)

Kiso *et al.* (1992) have reported a procedure for efficient solid-phase synthesis that includes an activation step with 2-benzotriazol-1-yloxy-1,3-dimethylimidazolium hexafluorophosphate (BOI) (**XVIII**). Reaction with a Boc-amino acid in the presence of DIEA proceeds through the acyluronium to the HOBt ester and finally to coupling with the amine methylsulfonate component. Synthesis of natriuretic peptide (28 residues) was achieved in 69% yield at the protected peptide step and a final 10% yield of pure product.

I. Relative Rate and Extent of Coupling Reactions

Many studies on the relative rates and coupling efficiencies have been conducted. They are in general agreement, but with a number of differences.

Table VI Relative Coupling Rates of General Activating Reagents[a]

Activation	Relative rate	Activation	Relative rate
BOP + HOBt	1.6	DICD + Pfp	0.9
DBTO[b]	1.8	BopCl	0.8
OAIU	1.4	EEDQ	0.34
BOP[c]	1.3	DICD + tetrazole	0.27
DICD + HOBt	1.3	Bates + HOBt	0.24
BOP[d]	1.0	Bates	0.05
Presymmetric anhydride	1.0	Woodward K	0.05

[a]Based on Hudson (1988).
[b]Dibenzotriazolyl oxalate.
[c]Requires a 10-min activation.
[d]Requires 1 equiv. and a 2-min activation.

On the basis of the competitive coupling system (Hudson, 1988) the rates of several activation methods were compared relative to the preformed symmetrical anhydride (Table VI). In these experiments there was a 10-min preactivation period followed by a 10-min coupling period. In the test system, dibenzotriazolyl oxalate (DBTO) was fastest, but it gave poor results in the ACP(65–74) synthesis. Hudson recommended BOP plus HOBt in DMF because of its high relative rate and good ACP synthesis.

Hudson (1990) compared the efficiency of 26 active esters of Fmoc-Val for solid-phase synthesis. This study also used a competition protocol in which equal amounts of the test esters of Fmoc-Val and Fmoc-Ile-OPfp in DMF were added to H-Gln(Tmob)-Ala-Ala-Ile-Asp(OtBu)-Ile-Asn-Gly-PAC-resin. The Val/Ile ratio gave a measure of relative coupling rates. Of the esters tested, 2,4-dinitrophenyl was fastest, being about 20 times as active as the pentafluorophenyl ester (OPfp). 3,4-Dihydro-3-hydroxy-4-oxo-1,2,3-benzotriazine ester was also faster, whereas all other esters were significantly slower, including *p*-nitrophenol, which was only about 10% as active as the pentafluorophenyl ester.

Beyermann *et al.* (1991) found little to choose between DCC/HOBt/DIEA, BOP, TBTU, or PyBOP in the coupling of Fmoc-MeLeu to MeLeu-BHA resin. Spencer *et al.* (1992) observed that coupling of 5 equiv Boc-Phe with 1 equiv (*N*-Me)Aib-Phe-OBzl for 24 hr was best when activated with PyBroP or NCA (~50%), slower with HBTU (35%) and BOP (23%), whereas OPfp, F, or pivaloyl mixed anhydride were very poor. The TBTU/HOBt combination was found to be equivalent to DCC/HOBt for synthesis of a 29-residue peptide on a Pam-resin in DMF. Overall yields of 98 and 96% were obtained (Reid and Simpson, 1992).

Wenschuh *et al.* (1994) have compared the extent of reaction after 15 min of the coupling of four successive α-aminoisobutyric residues to Lys-Leu-Met-Glu-Ile-Ile-MBHA-S-DVB by four activation methods in DMF. They were Fmoc-AIB-F/DIEA, Fmoc-AIB-NCA, (Fmoc-AIB)$_2$O/DIEA, and Fmoc-AIB/PyBroP/DIEA/DMAP. In each case the first AIB coupled well, and with Fmoc-AIB-NCA the second, third, and fourth residues also coupled nearly quantitatively. However, when the other three activated AIB derivatives were used, the yields of the last three couplings were very low. This was attributed to a smaller leaving group of the NCA. These results relate to a sterically hindered incoming amino acid and not necessarily to an amino component in a nonreactive β-sheet conformation. They found some by-product resulting from wrong way opening of the hindered Fmoc-AIB-NCA, which was not observed for Boc-AIB-NCA.

Coupling rates have been compared for seven different phosphonium and uronium activating reagents (3 equiv) for the reaction of Fmoc-Ile (3 equiv) with Gly-1,4,4′-trimethoxybenzhydrylamine-Pepsyn K in DIEA/DMF (Knorr *et al.*, 1989; 1991). At several time intervals aliquots were capped with Fmoc-Ala. After cleavage from the resin in 97% TFA/water the Fmoc-Ile-Gly-NH$_2$ and Fmoc Ala-Gly NH$_2$ were separated and quantitated by high-performance liquid chromatography (HPLC). The ratio was a direct measure of the coupling yield. Small rate differences were observed, with TBTU being about 10% faster than TPTU or BOP, which were a few percent faster than TBPipU, N-(2-pyridonyl)bispiperidenyluronium tetrafluoroborate (TOPPipU), TBPyU, or PyBOP. The symmetrical anhydride was somewhat slower than the others. After 30 min all gave nearly quantitative yields. Choices between the reagents may depend more on solubility than rates.

Ehrlich *et al.* (1994) studied the cyclization of gonadotropin-releasing hormone (GnRH) decapeptide to give rings of 3 to 10 residues. The new aza uronium activating reagent gave complete cyclization of the decapeptide within 2 min, whereas TBTU, TOPPipU, and DPPA gave 75, 30, and 25% after 100 min.

Fields *et al.* (1990) have compared coupling rates resulting from several acylating reagents for the reaction of Fmoc-Val with Val-HMP resin. In DMF, acylations utilizing HBTU or PyBOP were rapid (10–20 min), whereas DIC/HOBt was much slower (~2 hr), but none reached 100% under their conditions.

VI. Cause and Correction of Incomplete Coupling Reactions

From the outset, it was clear the new solid-phase peptide synthesis method (Merrifield, 1963) could only succeed if the coupling and deprotection steps were near quantitative, and most of the reactions were actually shown to proceed in high yields. However, there have been peptide sequences that were uniformly difficult to synthesize and where coupling yields were not quantitative under the usual conditions. The first such observation was with angiotensinylbradykinin (Merrifield, 1967), which proceeded well up to the eleventh amino acid but gave low yields at residues 12 to 17. The problem was thought to be due to chain termination by acylation and to poor swelling of the resin. It was overcome by changing the deprotection reagent from 1 *N* HCl in acetic acid to 4 *N* HCl in dioxane and by replacing 2% cross-linked polystyrene with 1%, both of which permit much greater swelling of the resin. But this was not a general solution to the problems with other peptides.

Hancock *et al.* (1973) soon found a very difficult sequence in residues 63–74 of the acyl carrier protein. Changing solvent, additives, and coupling reagents and introducing swell–shrink washing cycles were helpful but could only partially correct the problem, which was then attributed to poor solvation of the peptide chains. Sheppard (1972) attributed such difficulties to an incompatibility of the peptide chain and the resin support and was able to achieve nearly 100% reactions on a cross-linked polydimethylacrylamide support, but not on polystyrene. However, we were able to achieve excellent results on copoly(styrene–1% divinylbenzene) using our standard procedures for Boc/benzyl chemistry and symmetrical anhydride coupling in DMF (Kent and Merrifield, 1981).

It is now generally acknowledged that the problems are not with interactions between the peptide and the resin support, but with the conformation of the peptide itself. The evidence is consistent with the formation of intermolecular β sheets through hydrogen bonding, although for longer sequences intramolecular sheets and bends may also play a role (Deber *et al.*, 1989). The intermolecular hydrogen bonds are equivalent to introducing additional cross-linking in the region of the peptide, which results in reduced solvation of the peptide and therefore poor access to the incoming activated amino acid, leading to slow and incomplete coupling. Because the hydrogen bonding is partially reversible, it results eventually in a series of deletion peptides. In solution synthesis this kind of β-sheet formation also results in peptide aggregation, poor solvation, insolubility, and slow and incomplete coupling.

The modifications of resin supports, linkers, protecting groups, and coupling reagents have all contributed to overall improvements in solid-phase synthesis. However, the major remaining challenge lies in the development of ways to overcome the problem of difficult sequences. Many attempts have been made to understand and correct this phenomenon, and significant improvements have resulted.

Nuclear magnetic resonance (NMR) studies of the acyl carrier decapeptide (Live and Kent, 1983), gonadotropin-releasing hormone fragments (Milton *et al.*, 1986), and a growth hormone releasing factor (Deber *et al.*, 1989) have shown decreases in swelling in CH_2Cl_2 and DMF at the points of incomplete coupling. The change in swelling is reversible, however, when the amine is protonated. These results indicate that the problem is not due to increased covalent cross-linking, but are consistent with the idea that effective noncovalent cross-linking results from intermolecular hydrogen bonding between amide bonds of the peptide chains. Such bonds are increased in β structures and minimized in helical or random coil structures. Those peptides that coupled without difficulty gave sharp signals in NMR and rotational rates comparable to the same peptide when free in solution. The aromatic rings of the styrene were also highly mobile.

Mutter and colleagues have studied the relation between primary sequence of peptides, preferred conformation, and physicochemical properties such as aggregation, solubility, and reactivity. This was enabled by liquid-phase synthesis of peptides on linear, soluble polyethylene glycol (PEG). A series of oligo- and cooligopeptides containing 9 to 20 residues was prepared (El Rahman *et al.*, 1980). The solubility of the peptides was strongly enhanced by the C-terminal PEG, allowing conformational studies in a variety of solvents. Oligomers with tendencies to form α-helical or unordered structures showed no strong changes in solubility or coupling kinetics, whereas the onset of a β structure was accompanied by a drastic decrease in solubility and reactivity of the terminal amino groups. Insertion of a structure-disrupting proline gave a marked increase in solubility. It was noted that PEG exerted no influence on the conformation of bound peptide chains (Toniolo *et al.*, 1979). The general conclusion was that, for hydrophobic homopolymers, a transition from random coil to β structure occurs at about 5 residues and at about 9 residues another transition to α helix occurs. This was confirmed by gel-permeation chromatography (Schmitt and Mutter, 1983). The PEG was effective in solubilizing single chains, but not able to solvate fully aggregated structures. A protonated $H_2^+Ala_n$-PEG was more β at $n = 5$ or 6 but helical at $n = 7$ or 8, whereas the Boc peptide was β even at $n = 8$. The β structure was enhanced by high concentration, temperature, and ionic strength, suggesting intermolecular, hydrophobic interactions between peptide chains. The results of conformational studies of poly(oxyethylene)-bound peptides are well reviewed by Pillai and Mutter (1981).

Extensive work from the laboratory of Narita on the conformation of free peptides in solution documents and supports this general point of view derived from the PEG-supported peptides. Fourier transform infrared (FTIR) spectroscopy showed that protonation of a 23-residue peptide broadened the amide A band (3500–3100 cm^{-1}) owing to more non-hydrogen-bonded NH. The amide I band (1800–1600 cm^{-1}) showed a shift of the C–O stretching from 1650 cm^{-1} in the neutral sample to 1670 cm^{-1} in the protonated form, again indicating more β-sheet in the neutral sample. Narita *et al.* (1984a) derived a number of general conclusions: (1) the problem in fragment coupling is low solubility leading to low coupling; (2) solubility depends on competition between interpeptide interactions and peptide–solvent interaction; (3) the interactions are ionic, hydrophobic, and hydrogen bonding; (4) hydrophobic interactions depend on polarity of solvents; (5) helical or coil structures have high solubility in organic solvents owing to fewer intermolecular hydrogen bonds; (6) β sheets are less soluble; and (7) insertion in a central position of a tertiary peptide bond, for example, Pro or N^{α}-benzyl, and also C^{α}-alkyl groups, greatly improves solubility by restriction of dihedral angles.

The results from solution synthesis and solid-phase synthesis arise from the same phenomenon of segmented insolubility, leading to poor coupling reactions. However, the improved solvation of the pendant peptide chains by the polystyrene markedly reduces the degree of aggregation during the supported synthesis and effectively improves the synthesis of long-chain peptides (Sarin *et al.*, 1980).

Based on the assumption that normal solid-phase couplings occur with the peptide in a random coil conformation and that difficult couplings are caused by intermolecular hydrogen bonding due to β-sheet formation, Narita *et al.* (1984c) devised a method to predict which sequences are likely to be soluble and useful for fragment condensations in solution. It is based on Chou and Fasman (1978) parameters and depends on the coil parameter $<P_c>$. It predicts that peptides with a low potential (<0.9) for random coil conformation are likely to be insoluble and difficult to synthesize. If $<P_c> >1$, the peptide will be soluble and easy to synthesize. Peptides with α-helical structure will be soluble for any $<P_c>$, and the presence of a tertiary peptide bond will improve solubility. Milton *et al.* (1990) have discussed the general problem of difficult sequences and have extended the Narita method to a solid-phase predictive method using the expression $^{1}/_{P_c} = aP_{\alpha} + bP_{\beta}$. Correlation of their refined P_c* values with a large number of previously determined coupling reactions was good. Fortunately, most peptides are predicted to couple well.

To measure deletion and insertion peptides with high precision, the model peptide Ala_{10}Val-Pam-S-DVB resin was selected (Merrifield *et al.*, 1988) so that synthetic conditions could be compared and evaluated. The analytical method was fission fragment mass spectrometry (MS). Because the deletion or insertion of an Ala residue at *any* position would give the same Ala_9Val or $Ala_{11}Val$, there was an amplification of one order of magnitude in the detection of the modified peptide, and the wide MS dynamic range allowed the detection of trace amounts in the presence of very large amounts of the target peptide. The analysis could therefore be extended to multiple modifications, which occur in much lower levels. For the all-L peptide the average level of single deletions was 3.5% per step. Ninhydrin monitoring of individual steps showed that the major decrease in yield was at the couplings of Ala^5 and Ala^6, whereas couplings of residues 1 to 4 and 8 to 10 were essentially quantitative. This fits well with the idea that the short sequences do not provide enough hydrogen bonding to produce stable structures, but at Ala^5, Ala^6, and Ala^7 appreciable amounts of β-conformations are produced, which form stable intermolecular sheets with nearby peptide chains. These structures are reversible at each synthetic cycle, and by residues 8, 9, and 10 α-helical structures predominate, without formation of cross-links. The swelling and solvation return to normal and quantitative couplings are again achieved.

The introduction of D-Ala at positions 2, 5, and 8 completely avoided the β-sheet structures, and the resulting coil structures gave highly quantitative reactions averaging 99.97% per step! Extension to [D-$Ala^{3,6,9,12,15,18}$]Ala_{20}Val-resin gave equally good results. The minimum number of D residues to disrupt β-sheet structure not yet been established, but may be much smaller than was chosen here. The system allowed the testing of many solvents, coupling reagents, temperatures, additives, and chaotropic agents, but none were found that completely eliminated the coupling problem in this very sensitive and demanding all-L test peptide.

It is useful to be able to predict which sequences may be difficult to synthesize and to have sensitive ways to detect incomplete reactions, but it is more important to find ways to overcome the fundamental problem. Several methods leading to improvements have been reported.

In solid-phase synthesis the peptide aggregation is increased by peptide concentration, that is, high peptide substitution, whereas reactions in difficult sequences can be driven nearly to completion at very low substitution (Kent and Merrifield, 1981). Melting out of structures at higher temperature, up to 90°C, also is beneficial (Lloyd *et al.*, 1990; Rabinovich and Rivier, 1994). Yu *et al.* (1992) found that microwave radiation (72 W, 2 to 6 min, 55°C) increased the rate and extent of coupling for ACP(65–74) in DMF. Ultrasonic fields during coupling have also been reported to increase the rate and extent of synthesis (Vágner *et al.*, 1991). Solvent effects have been studied repeatedly, with the general finding that more polar solvents, especially those which are good proton donors or acceptors, such as DMF, NMM, DMSO, hexafluoroisopropanol, and hexamethylphosphoramide, improve solvation and swelling of the peptide–resin and in the case of difficult sequences can increase the rate and extent of coupling.

Some of the newer activation methods using phosphonium or uronium salts, acid fluorides, or urethane-protected carboxyanhydrides have been reported to improve coupling, but again none of the procedures has fully solved the problem with the most difficult sequences. Neither have multiple couplings and capping experiments. It was shown many years ago that the pseudo-first-order rate of coupling of protected amino acids is slower for β-branched than for unhindered amino acids (Erickson and Merrifield, 1976) but this alone is not the primary cause of poor and incomplete couplings.

Beginning in 1967, there have been efforts to disrupt peptide structures that were thought to be interfering with coupling reactions by means other than solvent changes. For example, Westall and Robinson (1970) found that 1.5 *M* urea in DMF raised the yield of coupling with active esters of Gln to Ser-Pro-Phe-Gly-Lys-resin from 70 to 98%.

Notable success was achieved by Stewart and Klis (1990) when they studied the use of chaotropic salt additives during or preceeding the coupling step. Their test peptide was RNase(1–13)-S-1% DVB which gave poor couplings when alanines 5, 4, and 3 were added (the seventh, eighth, and ninth synthetic steps). $NaClO_4$, LiBr, or KSCN in DMF (0.4 *M* each) greatly increased the coupling rate by DCC, HOBt esters, or symmetrical anhydrides and reduced unreacted amino groups to low levels within 1 to 6 hr. In examples with side-chain functional groups it was better to prewash with the salt and not include it in the coupling. Therefore, for this test peptide these chaotropic salts essentially solved the problem. With the test peptide Ala_{10}Val-resin (Merrifield *et al.*, 1988; R. B. Merrifield and B. Cunningham, unpublished data, 1989) had obtained only marginal improvements using LiCl, LiBr, KSCN, and $NaClO_4$, suggesting that the peptide had much more stable intermolecular β-sheet structures than RNase (1–13).

Independently, Seebach studied the solubilization of peptides in organic solvents containing inorganic salts (Seebach *et al.*, 1989). They found remarkable increases (100-fold) in the solubility of Boc-Ala-Gly-Gly-Gly-OH and other peptides in THF when 0.4 *M* LiCl or other salts were added. Hendrix *et al.* (1990) also found that 2 *M* LiBr in THF was a very effective solvent for dissolving peptides such as LMVGGVVIA. The solvent disrupted the peptide antiparallel β-sheets, as indicated by FTIR bands at 1630 cm^{-1}, and increased resin swelling.

Thaler *et al.* (1991) applied this solubilizing effect to the solid-phase synthesis of Ala_{12}Phe-S-DVB. They also found low coupling and only small improvements by addition of salts. All isomers from Ala_5 to Ala_{12} were seen in HPLC and MS. The coupling of Fmoc-Ala to Ala_5Phe-resin was studied in detail using OPfp esters or symmetrical anhydride activation in DMF, DMF/CH_2Cl_2/THF, NMP, or DMPU with and without LiCl, LiBr, $LiClO_4$, or KSCN. When S-1% DVB was the resin, the active ester efficiency remained at 35 to 45% with little salt effect. With anhydride coupling in DMF/CH_2Cl_2, LiBr raised the conversion from 66 to 76%. Under similar conditions on PEO-PS resin yields went from 89 to 97%, and on a polyacrylamide resin LiCl raised the yield from 78 to 94%. As pointed out by Thaler *et al.* (1991), the better yields for the more polar resins may be due to the resin loading, which was 0.097 mmol/g for PEO-PS, 0.088 mmol/g for the polyacrylamide, but 0.54 mmol/g for the S-DVB! Substitution levels are known to have a large effect on the synthesis of difficult sequences on S-DVB (Kent and Merrifield, 1981).

This was an important study, and significant coupling improvements were obtained by modifying solvents and supplementing the coupling mixture with chaotropic salts. The comment concerning the improved coupling (Seebach, 1992) that the "magic number" of 95% had been achieved is, of course, not quite correct because the number we are all seeking is closer to 99.95%

Barany *et al.* (1992) have confirmed the polyethylene oxide-S-DVB effect. In the absence of salts the synthesis of Ala_{10}Val-PEO-S-DVB was reported to be very good relative to poly S-DVB. Oliveira *et al.* (1992) have also confirmed the fact that KSCN in DMF disrupts peptide–resin structure, increases swelling and improves coupling efficiency.

As amide bond substitution or modification is known to influence the conformation of a peptide chain, it is a potentially fruitful approach to the design of systems that will not form aggregated peptides strongly stabilized by hydrogen bonds. Weygand *et al.* (1968) used 2,4-dimethoxybenzyl for this purpose. It is also known, for example, that insertion of a Pro residue, as well as an N^{α}-2,4-dimethoxybenzyl group, in the middle of a peptide chain will strongly disrupt β conformations (Narita *et al.*, 1984b). C^{α} substitution, for example, α-aminoisobutyric acid (AIB) results in a similar effect (Narita *et al.*, 1985). Even intermittent substitution by D-amino acid residues can disrupt otherwise serious aggregation of all-L peptides (Merrifield *et al.*, 1988).

Eckert and Seidel (1986) substituted the N^{α}-amine of amino acids with a ferrocenemethyl (Fem) group, by catalytic reduction of a Schiff base, and could condense Boc-amino acids with it by DCC in yields up to 90% to give **XIX**. The Fem group can be removed by TFA at the same rate as Boc. In addition, 2,2,2-Cl_3-Boc-N(-Fem)CHR-COOH + H-Leu-O-*t*-Bu coupled in high yield.

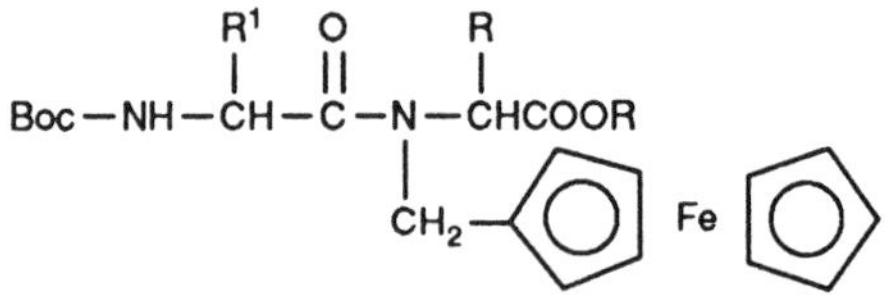

```
               R¹   O      R
               |    ||     |
Boc—NH—CH—C—N—CHCOOR
                       |
                      CH₂—⟨◯⟩  Fe  ⟨◯⟩
```

(XIX)

Through a Mannich reaction, Bartl *et al.* (1992) replaced the proton of the urethane nitrogen with CH_3OCH_2-, $C_2H_5CH_2OCH_2$-, or $C_6H_5SCH_2$-(Ptm) groups. They prepared Ala_{13} on a cellulose support with Fmoc(Ptm)-Ala-OH in 85 to 100% yield at each step. The yields of an Fmoc-Ala-OH control synthesis dropped steadily from 100% at step 3 to 45% at step 12. Unfortunately this very promising procedure did not improve yields in the synthesis of ACP(65–74) or gp41(81–103) Bartl *et al.*, 1993). This was partly due to the instability of Ptm to 20% piperidine.

Sheppard and co-workers have developed an elegant solution to the difficult sequence problem. They used the reversible N^{α} substitution approach studied by Weygand, Narita, and Bartl and also found that acylation of terminal secondary amino groups resulted in serious steric difficulties (Bedford *et al.*, 1992). They used Ala_nVal on Pepsyn polyamide gel resin or the more rigid Pepsyn K support, with N^{α}-Fmoc protection. The second N^{α}-protecting groups of the Fmoc amino acids were methyl- and methoxy-substituted benzyl and furfuryl compounds, which are readily cleaved by 95% TFA. The 2,4,6-trimethoxybenzyl(Tmb) and 5-methoxythienylmethyl groups were most labile. These Fmoc-imino acids coupled readily to resin-bound peptides, and the Fmoc was readily removed. The addition of the next residue, however, was massively inhibited, and satisfactory syntheses by this route were not obtained.

An ortho effect, which increased nucleophilicity of the imine, was recognized and Bedford *et al.* (1992) therefore prepared N^{α}-Fmoc-N^{α}-(2-hydroxy-4-methoxyl)benzyl-L-alanine and coupled it to a peptide–resin. After removal of Fmoc, Fma-Ala coupled completely in 1 hr as shown in Scheme 5. With this method ACP(65–74) was synthesized with less than 1% of the Val^{65} deletion peptide. They also prepared in good yield Ala_{17}Val-resin with the 2-OH,4-MeO-Bzl protection only at Ala^{4} and Ala^{9} (Johnson *et al.*, 1993). These are impressive results. For peptides in general some limitations and precautions were pointed out, but with careful selection of sites to be substituted this appears to be a good solution to this problem.

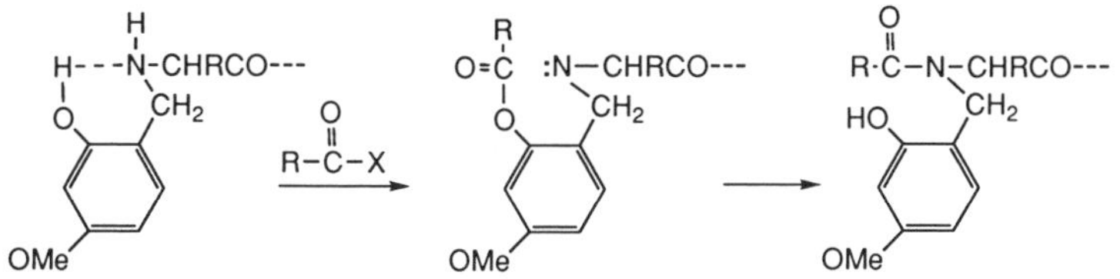

Scheme 5

With the N^{α} substitution effect, the chaotropic salt effect, the PEG effect, and the strong polar solvent effects, we appear to be close to the elimination of the last major impediment to solid-phase synthesis.

VII. The Cleavage Reaction

Methods to cleave peptide chains from resin supports have already been discussed in Sections III and IV on linkers and protecting group schemes and need only be summarized here.

The HF method of Sakakibara and Shimonishi (1965) was adapted to solid-phase synthesis by Lenard and Robinson (1967) and has been used for many years. The method of choice now is the low/high HF procedure of Tam (Tam *et al.* 1983). Very strong acids like HF, with acidity function H_0 −11, are well known to catalyze various side reactions under certain conditions. It was recognized that under normal conditions for cleavage of benzyl-derived protecting groups (90%

HF plus 10% anisole) the reaction proceeds by an S_N1 ($A_{AL}1$) mechanism. The resulting benzylcarbonium ions or acylium ions are then powerful alkylating and acylating species that can produce by-products by reaction with nucleophilic amino acids or scavengers. It was reasoned that mixtures of anhydrous HF with a suitable weak base should reduce the acidity function of the reagent, but leave enough of the base unprotonated for it to be effective as a nucleophile. The nucleophile could then assist in an acid-catalyzed deprotection reaction that would proceed by an S_N2 mechanism. In this way the benzyl group or acyl group would be transferred to the base and the free benzylcarbonium or acylium ions would not be formed. The trapped cation would be a much weaker alkylating reagent, and the deleterious side reactions would be avoided. An ideal weak base but powerful nucleophile was found to be dimethyl sulfide (DMS) (pK_a of ~7). A 3:1 (v/v) mixture of DMS and HF, containing 10% *p*-cresol, lowered the acidity to −5.2. This level of acidity was still strong enough to protonate the groups to be cleaved (esters, amides, ethers, and urethanes), but a large part of the DMS remained unprotonated and nucleophilic. This low-HF reagent removed most of the protecting groups and avoided most of the known side reactions associated with strong acids.

The reagent also reduced methionine sulfoxide to the sulfide, so the sulfoxide can be used to avoid accumulation of the *tert*-butylsulfonium by-product (Tam *et al.*, 1982). In addition, N^{in}-formyltryptophan is cleanly deprotected if 5% *p*-thiocresol is included in the reagent (Heath *et al.*, 1982). Peptides containing Arg(Tos), Cys(MeBzl), or those attached to benzhydrylamine-resins require a second treatment, with high-HF (90% HF, 5% cresol, and 5% thiocresol). However, most of the harmful carbonium ion sources will already have been removed in the first step, and far fewer side reactions occur.

Yajima and colleagues have pioneered in the application of trifluoromethanesulfonic acid (TFMSA) in TFA as a deprotecting agent for peptide synthesis (Yajima *et al.*, 1974). More recently, this hard acid method has been further refined by the addition of thioanisole as a weak soft base to function in a push–pull S_N2 mechanism (Irie *et al.*, 1976; Kiso *et al.*, 1979; Yajima *et al.*, 1982; Nomizu *et al.*, 1991). However, if the acidity is too high a partial S_N1 reaction can occur. Because thioanisole has been found to be an alkylating agent under conditions of moderate acidity it may not be the best weak base. Kiso *et al.* (1989b) have also found that 1 *M* tetrafluoroboric acid plus 1 *M* thioanisole in TFA is an effective cleavage reagent.

In an extension of the HF methodology, Tam *et al.* (1986) showed that mixtures of TFMSA, TFA, and Me_2S can be used for solid-phase deprotection and cleavage under S_N2 conditions. A three-component diagram was generated (Fig. 12), showing which compositions will give an S_N2 reaction mechanism and which will fall in the S_N1 area. A recommended composition (v/v) was 10% CF_3SO_3H, 60% CF_3COOH, and 30% CH_3SCH_3 (4 hr, 0°C), but many other compositions will fall in the desired S_N2 area. The compositions recommended by Irie *et al.* (1976) and by Bergot *et al.* (1987) appear to be just on the border between the two mechanisms.

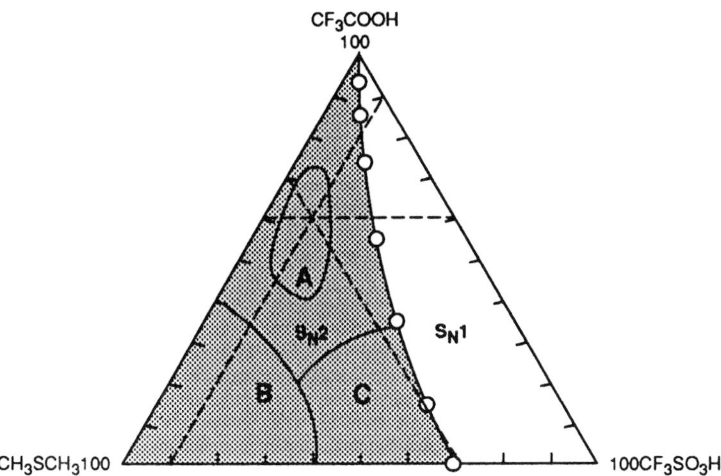

Figure 12 Three-component diagram for cleavage of peptides with mixtures of CF_3SO_3H, CF_3COOH, and CH_3-S-CH_3. Each apex of the triangle represents 100 volume percent of the appropriate reagent. CF_3SO_3H is on scale at base of triangle.

R—C(=O)—O—CH₂ … H … —CH₂C(=O)NHCH₂—(R)

(XX)

The use of weaker acids for final cleavage of protected or free peptides has already been discussed. Linkers to the resin have been designed to allow cleavage by 100% TFA down to as little as 1% TFA, or even acetic acid. These schemes depend on temporary N^{α} protection that is removed by reagents other than acid (e.g., Fmoc, Dts, and Paloc). Some details and precautions have been well discussed by Fields *et al.* (1992).

Nonacidolytic cleavages have been used for many years but less frequently than acidic methods. These have included benzyl ester ammoniolysis (Manning, 1968) to give the amide and treatment with 2-dimethylaminoethanol in DMF and subsequent hydrolysis of the DMAE ester to give the free acid (Barton and Lemieux, 1971; Barton *et al.*, 1973). Piperidine treatment of esters of 4-(2-hydroxyethyl)-3-nitrobenzoyl aminomethyl resin (Liu *et al.*, 1990) and 9-(hydroxymethyl)-2-fluorene acetylaminomethyl resin (**XX**) has been used (Albericio *et al.*, 1991).

Peptides were cleaved by catalytic hydrogenolysis with $Pd(OAc)_2$ reduced *in situ* (Schlatter *et al.*, 1977) or by phase-transfer hydrogenolysis (Anwer and Spatola, 1980). Finally, the elegant, selective, and mild method of Kunz and

Dombo (1988) for cleavage of allyl esters using Pd catalysis by tetrakis(triphenylphosphino)palladium is becoming widely utilized. The allyl group is transferred to a suitable basic receptor such as morpholine, *N,N*-dimethylbarbituric acid, or dimedone.

VIII. Solid-Phase Synthesis of Peptides and Proteins

From the beginning, large peptides and small proteins were synthesized by the fragment condensation method in homogeneous solution. The work was difficult and time-consuming, but many outstanding syntheses of pure peptides were achieved (see Schröder and Lübke, 1966). This fragment (or segment) approach continues today and in the hands of experts produces excellent results. The early problems, especially insolubility of intermediates and partial racemization at the carboxyl residues of the activated segment, have been largely overcome by proper choices of splicing points, improved solvent mixtures, and new activation methods (see Sakakibara, 1994). These syntheses have included most of the known peptide hormones such as gastrin, glucagon, and insulin, and the enzymes RNase A, RNase T1, lysozyme, and many others. Specific examples are covered in other chapters of this volume.

The synthesis of proteins by the stepwise solid-phase method began in the late 1960s and led to the synthesis of ribonuclease A, staphylococcal nuclease, cytochrome *c*, and lysozyme, and continues to be the method of choice, but applications of solid-phase segment condensations in convergent syntheses are appearing more frequently. Automation has been valuable for the synthesis of the larger molecules.

A. Examples of Synthetic Large Peptides and Small Proteins

Table VII contains a partial list of large peptides and small proteins that have been prepared by stepwise solid-phase methods. A cutoff at about 50 residues was arbitrarily selected. Most have been prepared by the discontinuous batch procedure on styrene–divinylbenzene supports with Boc/benzyl protection schemes and HF cleavage, although we are beginning to see more use of polyamide, polyethylene glycol, and other supports, together with Fmoc/*t*-Bu protection schemes and weak acid cleavage. Some syntheses have been by continuous free-flow procedures. Early preparations used mainly DCC, active ester, and symmetrical anhydride couplings, but the many new activation methods are now playing important roles.

Some of the syntheses listed in Table VII were excellent, resulting in reasonable yields of highly purified products of correct composition, structure, and activity. And at least one protein was obtained in crystalline form. Others, especially some of the older syntheses, were less satisfactory. It is sometimes said that the upper limit for a satisfactory solid-phase synthesis is only about 20 residues, whereas others will acknowledge a limit of approximately 50 residues. However, there is abundant evidence that proteins in the range of 100 to 150 residues can be prepared by the stepwise solid-phase approach.

Table VII. Large Peptides and Small Proteins Synthesized by Stepwise Solid-Phase Methods

Number of residues	Name	Ref.
48	Bombyxin	Maruyama *et al.* (1992)
51	Insulin	Marglin and Merrifield (1966); Okuda and Zahn (1969)
55	Ferredoxin	Bayer *et al.* (1968), Smith *et al.* (1991)
58	Trypsin inhibitor	Noda *et al.* (1971)
64	Prepro-cecropin A-Gly	Boman *et al.* (1989)
66	Cytochrome *c* segment	DiBello and Gozzini (1993)
70	Insulin-like growth factor	Bagley *et al.* (1990)
72	Interleukin 8 (monocyte)	Clarke-Lewis *et al.* (1991)
74	Acyl carrier protein	Hancock *et al.* (1971)
74	β-Chorionic gonadotropin	Wu *et al.* (1989)
76	Ubiquitin	Briand *et al.* (1990)
77	Interleukin 8 (endothelial)	Sueiras-Diaz and Horton (1992)
79	P9 cell regulation protein	Valembois *et al.* (1992)
84	Parathyroid hormone	Fairwell *et al.* (1983)
86	Tat protein	Mascagni (1990)
88	Cardiodilatin	Nokihara and Semba (1988)
92	Prepro-gonadotropin-releasing hormone	Milton *et al.* (1992)
97	Gro ES chaperonin	Lim *et al.* (1990)
99	HIV-1 protease	Nutt *et al.* (1988), Schneider and Kent (1988),
99	HIV-2 protease	Copeland and Oroszlan (1988)
99	SIV protease	Tomasselli *et al.* (1992)
104	Cytochrome *c*	Sano and Kurihara (1969)
104	Ribonuclease T	Waki *et al.* (1974)
104	Core gag p24, C terminus	Mascagni *et al.* (1990)
109	Prothymosin α	Barlos *et al.* (1991)
115	McPC603 V_H(1–115) segment	Martin *et al.* (1993)
124	Ribonuclease A	Gutte and Merrifield (1969, 1971)
126	Leukemia virus protease	Bláha *et al.* (1992)
126	Cardiodilatin	Rapp and Nokihara (1991)
127	Colony-stimulating factor	Clark-Lewis *et al.* (1988)
129	Lysozyme	Barstow *et al.* (1972)
134	HIV-1 nucleoprotein p24 segment	Chong *et al.* (1993)
140	Interleukin 3	Clark-Lewis *et al.* (1986)
149	Staphylococcal nuclease	Ontjes and Anfinsen (1969)
166	Interferon-α_1	Merrifield (1983)
188	Growth hormone analog	Li and Yamashiro (1970)

B. Peptide Ligation

Many peptide chemists feel that to synthesize small proteins in greater yield and purity, and especially to make even larger proteins, it will be necessary to develop better segment condensation methods and to combine solid-phase and solution techniques. Therefore, objectives have focused on the development of new ligation methods for combining large peptide segments. Considerable progress has been achieved. The four main approaches to ligation are as follows.

(1) Strong chemical activation of the carboxyl group of protected segments and coupling at, necessarily, low concentration with the protected amino component in organic solvents: This is the classic solution-phase fragment coupling method except the peptide segments are made by solid-phase methods. The couplings are by second-order intermolecular reactions and are relatively slow, and yields tend to decrease with increasing peptide size. To help overcome these problems, an excess of one of the valuable peptide components is usually employed. Low solubility and a tendency for racemization are additional problems, but the ability to purify intermediates is an important advantage.

(2) Enzyme-catalyzed coupling of α-carboxyl and α-amine components of unprotected peptides in aqueous or aqueous-organic solvents: The principle, which was demonstrated in the laboratory of Bergmann in the 1940s, has developed into a sophisticated and effective method for moderate-sized peptides, although the synthesis of small proteins still poses severe problems.

(3) Coupling of protected or free peptides by activation of one or both components with functional groups that selectively react with one another: Blake (1981) introduced peptide thioacids that, after activation with silver ion, would react in aqueous solution with the α-amine of a peptide blocked only at its N^{ε}-lysine or cysteine sulfhydryl groups (Fig. 13). This is also a second-order reaction, but the high selectivity avoids competing side reactions. Apocytochrome *c* (104 residues) was synthesized from four segments in this way (Blake, 1981). A variation of this idea was developed by Schnölzer and Kent (1992) in which the C^{α}-thioacid of a peptide was coupled with a bromoacetyl peptide (Fig. 14). Because the reaction is so selective, no side-chain protection is required, and the ligation is carried out in high yield in aqueous solution at concentrations of 4 m*M*. The product, however, carries a thioester bond at the site of ligation rather than the normal amide bond. For HIV-1 protease (99 residues), this was not a problem, and fully active enzyme was obtained in good yield and purity. In the latter two applications the peptides were all prepared by solid-phase synthesis, although they need not be limited to that method.

(4) Ligation via cyclic transition state intermediates: The couplings by methods 1, 2, and 3 involve second-order intermolecular reactions of two peptide components, which are necessarily present in low concentrations and, unless highly and selectively activated, have a low potential for effective reaction. New methods are being designed to overcome this defect. Kemp and students have developed, over a number of years, techniques leading to intramolecular amide

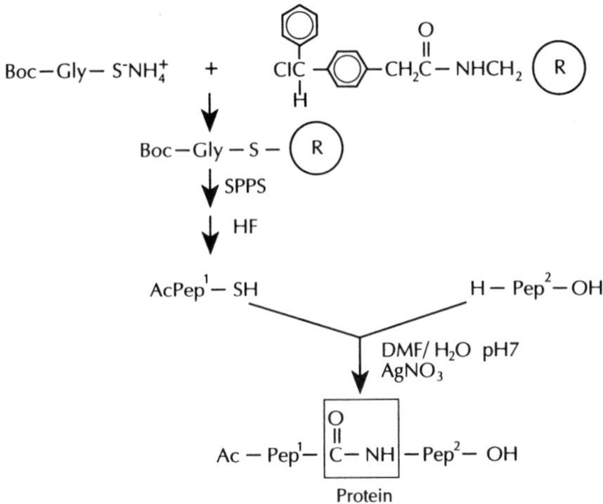

Figure 13 Silver ion-activated coupling between a thiocarboxyl peptide and the α-amine of another peptide segment.

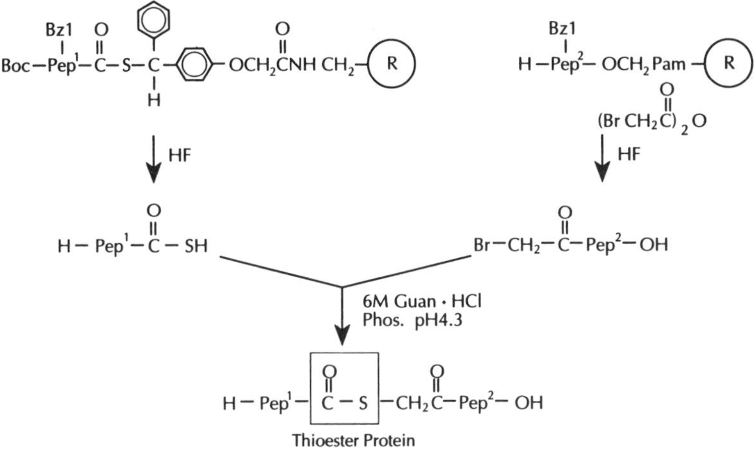

Figure 14 Peptide ligation by thiolester formation.

bond formation which occurs at very high effective concentrations. Based on kinetic data, the amide bond-forming reaction occurs at concentrations of 10 *M*. The scheme shown in Fig. 15 (Fotouhi *et al.*, 1989) involves the solid-phase synthesis of a protected peptide on a template containing a properly spaced thiol attached to a resin support. After removal of the support another, unprotected peptide containing an *N*-terminal cysteine derivative undergoes a "thiol capture" reaction, thus placing both peptides on one template. That brings the carboxyl

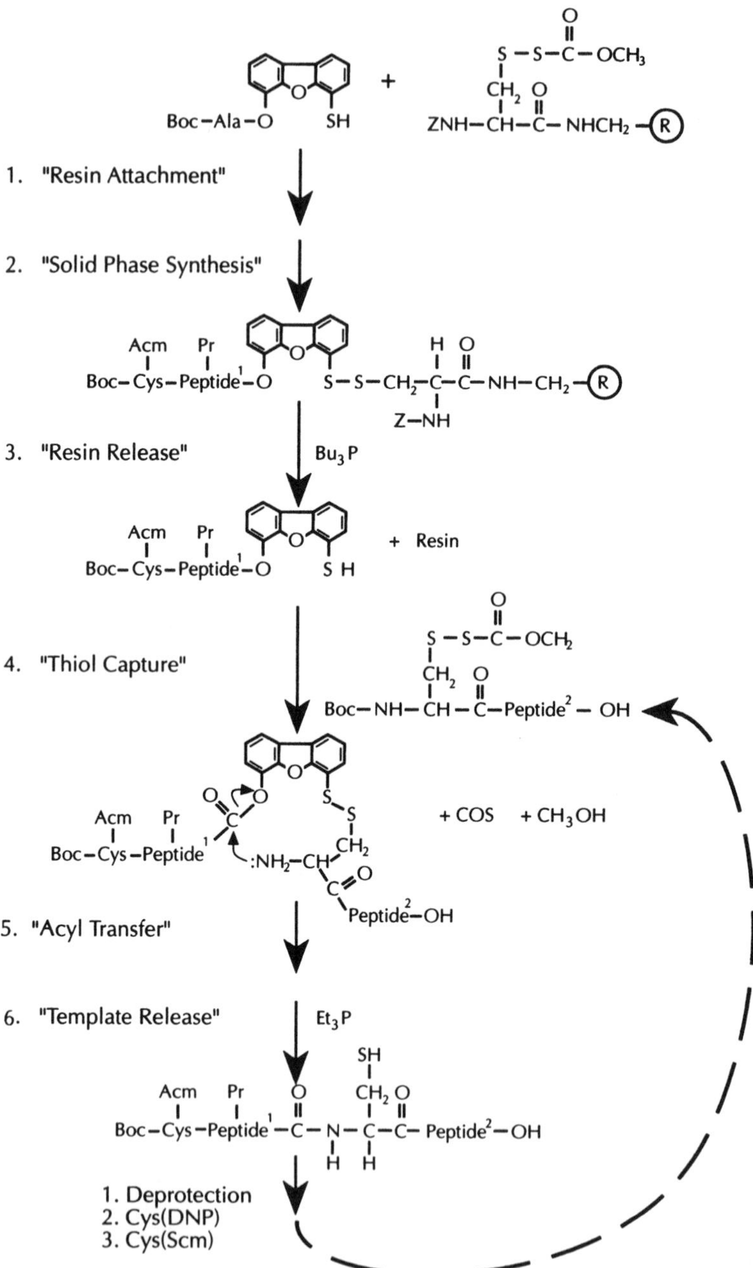

Figure 15 Segment synthesis by thiol capture.

and amino components into close proximity for a rapid monomolecular reaction via an O to N acyl shift. This elegant method has been effectively applied to the synthesis of a 29-residue segment of bovine pancreatic trypsin inhibitor.

Tam (Liu and Tam, 1994) has also developed an intramolecular ligation strategy (Fig. 16). The carboxyl component is prepared synthetically (or enzymatically from a natural peptide) in the form of a glycolaldehyde ester of a completely *un*protected peptide. The amine component is an N-terminal serine-, threonine-, or cysteine-peptide. The latter is "captured" by the aldehyde in aqueous solution of pH 4, to produce an oxazolidine or a thiazolidine derivative (Fig. 16). The ring nitrogen is then in close proximity to the C-terminal carbonyl of the other peptide segment and forms a 3,3,0 fused ring system in the transition state of an O to N acyl transfer reaction. This reaction occurs at high effective concentration, without further activation, in aqueous solution at pH 6 to 9. Subsequent release of aldehyde from the *N*-acylthiazolidine gives the peptide ligation product, peptide^1-Cys-peptide^2, in very high yield. This method is somewhat more general than the thiol capture method, requires fewer steps, is applicable to completely unprotected peptides, operates in aqueous solution, is racemization free, and proceeds in good yield. It has been evaluated by the synthesis of a 50-residue analog of transforming growth factor containing one disulfide bond.

It is reasonable to expect that these new ligation methods will lead to efficient syntheses of considerably longer proteins than have been possible in the past.

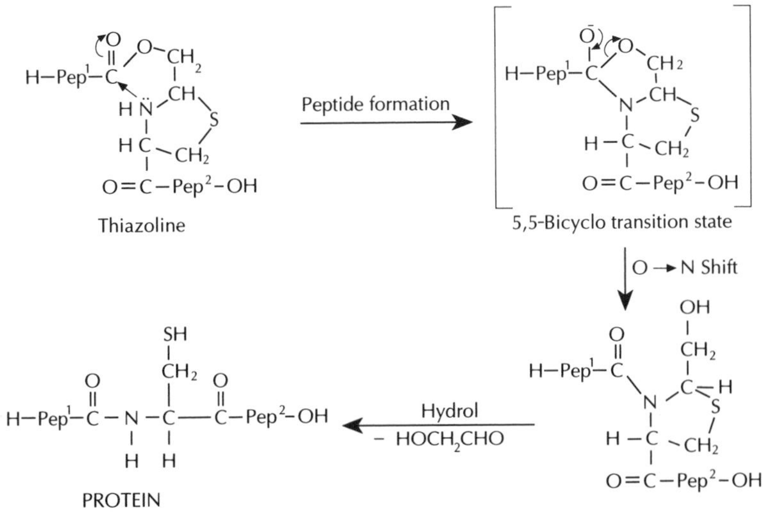

Thiazoline

Peptide formation

5,5-Bicyclo transition state

O → N Shift

Hydrol
– $HOCH_2CHO$

PROTEIN

Figure 16 Segment synthesis by aldehyde capture.

C. Multiple Simultaneous Peptide Synthesis and Construction of Peptide Libraries by Combinatorial Methods

These techniques have simplified our ability to synthesize peptides by orders of magnitude, and they represent major advances in the application of peptide synthesis to other fields of research. These important and exciting developments must be left for other chapters of this volume.

D. Design and Synthesis of Protein–Nucleic Acid Hybrids

Protein–nucleic acid hybrids (PNAs) are a new class of compounds in which individual nucleobases are linked to an achiral peptide backbone. They were designed to compete specifically for binding to one strand of a DNA duplex (Nielsen *et al.*, 1991; Egholm *et al.*, 1993). It was estimated, using computer graphics, that the optimal number of bonds between the bases of a DNA duplex is six and the number of bonds between the backbone and the base is two to three. To simulate such a structure, N^1-carboxymethylthymine was coupled to N^3 of Boc-aminoethylglycine to give a thymine-containing amino acid derivative that could then be extended by solid-phase synthesis to an oligomer in which the sugar phosphates of a polynucleotide were replaced by branched-chain amino acids (Fig.17) (Berg *et al.*, 1993). Binding to double-stranded A and T DNA chains was shown to take place by strand displacement, in which the PNA is bound to the Watson–Crick complementary adenine-containing strand. Hoogsteen base pairing probably can also take place. In this example the thymine-containing strand of the duplex was excluded in a single-strand conformation. The binding was 2:1 PNA–DNA and was much strong than DNA–DNA binding, probably by the elimination of repulsive phosphate anions. The change in melting temperature (ΔTm) was about 10°C/unit. The specificity for A-T and G-C should allow the selection of any oligonucleotide sequence by the appropriate synthetic PNA structure. Such compounds would provide tools for developing gene-targeted drugs and other sequence-specific gene modulators.

E. *De Novo* Synthesis of Peptides and Proteins

One of the most important contributions of peptide synthesis, and especially solid-phase synthesis, has been in the area of *de novo* design and synthesis of peptides and proteins. The objectives of this emerging field have been the design of peptides, not based on any known natural sequence, that will (1) give a predicted tertiary structure and/or (2) have a predicted physical, chemical, or biological activity. The first attempt that I know of was by Merrifield and Woolley (1958), who made homopolymers of histidine and copolymers with other amino acids by the *N*-carboxyanhydride method initiated with triethylamine, diethyl glutamate, or diethyl serylglutamate and looked for catalytic activity. The hope was that some species within the mixture would have activity. The polymers

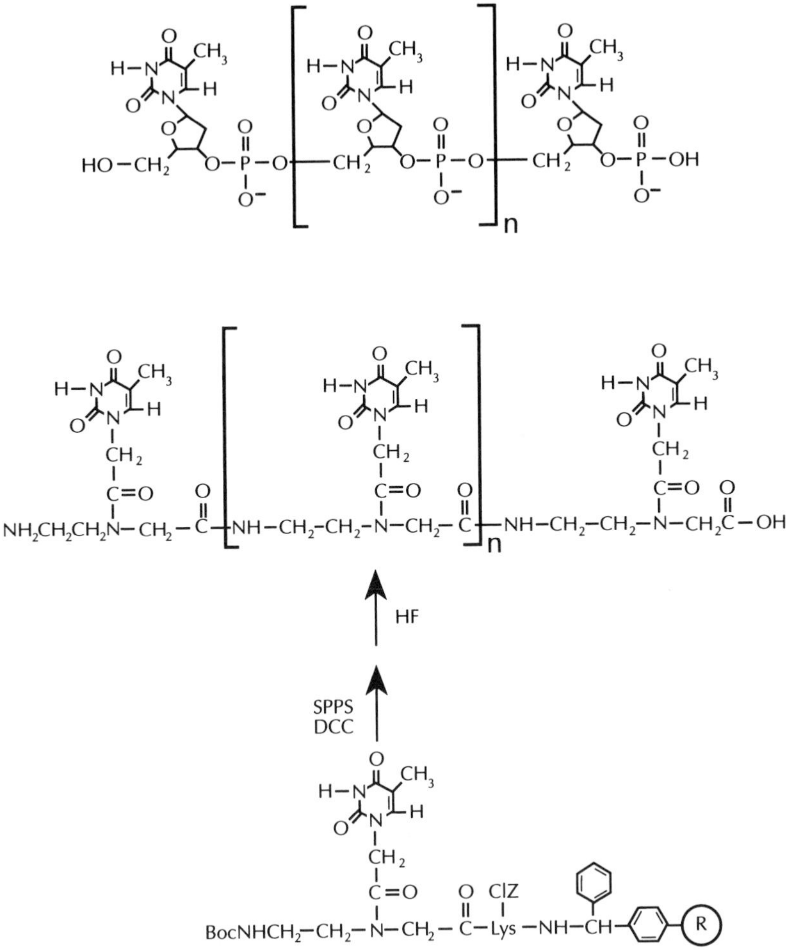

Figure 17 Synthesis of a protein-nucleic acid hybrid.

were in fact catalysts for the hydrolysis of ethyl nitrophenyl carbonate but were not effective toward more stable esters. Our plans to define a precise structure by iterative methods was therefore abandoned in favor of working on the development of solid-phase peptide synthesis, which I thought might be useful for the stepwise synthesis of oligomers of defined structure.

The pioneering work in the field can be traced to Bernd Gutte (see Richardson and Richardson, 1989, for an excellent review). The first protein Gutte designed was based on RNase A. He removed five segments of the protein that occurred as loops projecting from the globular core and joined the exposed ends with short connecting segments (Gutte, 1975). A 70-residue and a 63-residue protein (only half the size of RNase A) showed up to 15% of RNase A activity. A truly *de novo* peptide of 34 residues was then designed which was expected to form a ββα secondary structure (Gutte *et al.*, 1979). This peptide and the 68-residue S–S dimer did have a definite tertiary structure, which could be denatured, and it bound GAA and other oligonucleotides. A 24-mer, MTFIRPNVGAMSNFYHYPNIIITF, was then designed to form a four-stranded β sheet that would bind DDT (Moser *et al.*, 1983). It bound 10^3 better than controls.

Hodges *et al.* (1972) learned that tropomyosin is a coiled coil of two α-helical chains with a 7-residue repeat that places two large hydrophobic residues at positions 1 and 4 where they provide a packing surface. To study such α-helical coiled coils, Hodges *et al.* (1981) synthesized several models, such as $(LEALEGK)_5$. They were made by synthesis of the first heptapeptide on a S-DVB resin and coupling of successive protected segments to give a 43-residue peptide and its 86-residue S–S dimer. Subsequently Lau *et al.* (1984) synthesized by stepwise solid-phase synthesis a series of peptides, Ac(Lys-Leu-Glu-Ala-Leu-Glu-Gly)$_n$-Lys-NH_2, with n = 1–5. The 29- and 36-residue peptides were large enough to form the two-stranded α-helical coiled coil (parallel and in register) structures found in tropomyosin, and therefore confirmed the predicted secondary and tertiary structures. This approach allowed the study of both intra- and interchain interactions of amino acid side chains and provided new insights into the rules for the stabilization of tropomyosin and other coiled-coil proteins.

In 1981 B. W. Erickson began a collaboration with J. S. Richardson and D. C. Richardson to design and synthesize a β-sheet protein that was predicted to form an antiparallel β sandwich of two identical four-stranded sheets. It was called betabellin because of its expected bell-shaped structure. The two linear chains were synthesized simultaneously on a template containing a resin-bound branched linker (Unson *et al.*, 1984). The criteria for the design included secondary-structure predictions, location of hydrophilic and hydrophobic residues, location of turn residues, residue pairing on adjacent strands, preferred side-chain conformation, and strand twist. There was no homology with any known protein. From the outset the goal of this design was to add a catalytic site at one end of the molecule and thus produce an enzyme *de novo*. Because of synthetic problems, low solubility, a deficiency of charged residues, and unexpected folding, the molecule was redesigned and resynthesized many times. Most of the peptides showed a high content of β structure by circular dichroism (CD) and converted to a random coil CD spectrum under denaturing conditions. The sequences of betabellin 1 and betabellin 9, containing 12 changes were as follows:

1 S T V T A R Q P N V T Y S I S P N J A T V R L P N ϕ T L S I G
9 H T L T A S I P D L T Y S I D P N T A T C K V P D ϕ T L S I G B
(where ϕ is iodophenylalanine and B is β-alanine)

Based on the amphipathic helical structures derived from lipoproteins by Segrest *et al.* (1974), Kaiser and Kézdy (1983) synthesized several peptide hormone analogs containing little homology with the native molecule and showed that they can form amphipathic helix secondary structures at membrane surfaces.

About 1985 Mutter began an extensive series of studies on the *de novo* design of proteins that would assume various predetermined secondary and tertiary structures. He also adopted the template approach, which assisted in the assembly and folding of the synthetic products. The main synthetic tool was stepwise solid-phase synthesis, but fragment syntheses and solution syntheses were used when convenient. An early design (Mutter, 1985) made use of a linear heptapeptide–resin template and the stepwise synthesis of 15-mers on each of the four lysine ε-amino groups. The technique was called template assembled peptide synthesis (TASP) and gradually evolved into a sophisticated design. For example, Mutter *et al.* (1991) have prepared a helix to sheet switch peptide in which the transition is pH controlled. A four-α-helical bundle, designed to resemble the major histocompatibility complex (MHC) receptor, was assembled on a restricted circular decapeptide template (Tuchscherer *et al.*, 1992). After deprotection and cleavage the product was carefully purified to homogeneity and found to react with antibodies to the MHC, thus verifying the initial design. Mutter and Vuilleumier (1989) have written an excellent review on the structure and parameters for folding of proteins and on the role that chemically synthesized peptides can play in the understanding of protein folding.

The branched handle or template approach has allowed Tam (1988) to synthesize MAPs (multiple antigen peptides) containing four or eight peptide T-cell and B-cell epitopes held together at one end by a branched oligolysine. These are highly immunogenic and have led to the development of several synthetic vaccines, for example, against malaria, foot-and-mouth disease virus, and hepatitis.

Urry (1988) succeeded in making many models based on the helical $(VPGVG)_n$ theme of elastin, with tailored temperature transitions or pH transitions. This work is a clear example of the *de novo* design of proteins.

During this same time frame DeGrado also contributed significantly to the protein folding problem. His model protein was a designed four-helix bundle motif found in some natural proteins. The first phase was to design peptides of 12 to 16 residues that form α-helical secondary structures and would self-assemble to produce α-helical tetramers. In the second phase, a single hairpin loop was inserted between two identical helices to give a peptide that dimerized in aqueous solution and produced an α helical tetramer (Ho and DeGrado, 1987). In the third phase of the work, a linear protein was synthesized containing four 16-mers connected by three trimer linkers. The monomer segment was GELEELLKKLKELLKG, and the linker was PRR. In the helical tetramer the Leu residues were in the interior and the polar Glu and Lys projected to the aqueous exterior. These peptides were first synthesized by automated solid-phase methods, and later the linear tetramer was made by recombinant DNA technology from oligonucleotides prepared by solid-phase methods (Regan and DeGrado, 1988). This *de novo* design was very successful.

The tetramer was further engineered to produce a zinc-binding protein. Metal-coordinating histidine residues were substituted at positions 7 and 11 of one segment and at residue 7 of another segment (Handel and DeGrado, 1990). The long-term intent is to engineer a catalytic site and eventually produce an enzyme *de novo*.

Hahn *et al.* (1990) have reported an example of the *de novo* design and chemical synthesis of a protein possessing a predetermined enzymatic activity. They began with a resin-bound branched α,ε-dilysylornithine selectively blocked so that each of four peptide chains (15 to 22 residues) could be assembled independently by solid-phase methods. The peptides were designed to be helices, with considerable stabilizing hydrophobic interactions toward the C terminus and with looser packing at the N terminus to allow a binding pocket. At each of the amino ends was placed a residue (Ser, His, or Asp) derived from the active center triad of chymotrypsin, and Glu was at the end of the fourth chain. The cleaved and purified four-stranded protein bundle was homogeneous and highly helical. It was found to catalyze specifically the hydrolysis of the chymotrypsin substrates, Ac-Tyr-OEt, Z-Tyr-ONp, and benzoyl-Tyr-OEt, but not the trypsin substrate benzoylarginine ethyl ester. The enzyme was saturable, pH dependent, and denatured at elevated temperature. The catalysis was 10^5 greater than spontaneous hydrolysis, but, of course, this model protein was not nearly as active as chymotrypsin itself.

Pessi *et al.* (1993) have used the 19–83 region of an immunoglobulin variable domain heavy chain as a template for internal packing. Their 61-residue synthetic protein has two loops, corresponding to the hypervariable regions H1 and H2, and a β framework with a novel fold consisting of two three-stranded β sheets packed face to face. Because of very low solubility (20 μ*M*), the synthesis of the model gave less than 4% overall yield. Introduction of a solubilizing motif at either end raised the solubility by a factor of 50 or more (Pessi *et al.*, 1994). Three histidine residues in the two loops formed a zinc-binding site (K_d of 5×10^{-6} *M*).

The ultimate refinement of *de novo* design of proteins is still far away, but developments in the field have been very encouraging. It is satisfying to see that solid-phase peptide synthesis has played such an important role in these advances.

Abbreviations

Acm	Acetamidomethyl
ACP	Acyl carrier protein
Ada	Adamantyl
Adpoc	1-(1-Adamantyl)-1-methylethoxycarbonyl
AIB	α-Aminoisobutyric acid
Aloc	Allyloxycarbonyl
AM	2-[4-Fmoc-aminomethyl (2,4-dimethoxyphenyl)phenoxy]acetic acid
Ans	9-Anthracenesulfonamido
Azoc	2-(Phenylazophenylyl)propyl-2-oxycarbonyl
BHA	Benzhydrylamine
Bnpeoc	2,2-Bis(4′-nitrophenyl)ethan-1-oxycarbonyl
Boc	*tert*-Butyloxycarbonyl

BOI	2-Benzotriazolyloxy-1,3-dimethylimidazolium hexafluorophosphate
Bom	Benzoxymethyl
BOP	Benzotriazolyloxytris(dimethylamino)phosphonium hexafluorophosphate
BopCl	Bis(2-oxo-3-oxazolidinyl)phosphinyl chloride
Bpoc	2-Biphenylisopropyloxycarbonyl
BroP	Bromotris(dimethylamino)phosphonium hexafluorophosphate
BSA	Bovine serum albumin
Bum	*tert*-Butoxymethyl
Bzl	Benzyl
DBU	1,8-Diazabicyclo[5.4.0]undec-7-ene
DBTO	Dibenzotriazolyl oxalate
DCC	*N,N′*-Dicyclohexylcarbodiimide
Ddz	2-(3,5-Dimethoxyphenyl)propyl-2-oxycarbonyl
Dhbt	3-hydroxy-4-oxo-3,4-dihydrobenzotriazine
DIC, DICD	*N,N′*-Diisopropylcarbodiimide
DIEA	*N,N*-Diisopropylethylamine
DKP	Diketopiperazine
DMAP	Dimethylaminopyridine
DMF	*N,N*-Dimethylformamide
Dmob	2,4-Dimethoxybenzyl
DMPU	3,4,5,6-Tetrahydro-1,3-dimethylpyrimidine-2(1*H*)-one
DMSO	Dimethyl sulfoxide
Dnpe	(2,4-Dinitrophenyl)ethyl
DPPA	Diphenyl phosphorylazid
Dts	Dithiasuccinoyl
DTT	Dithiothreitol
EEDQ	*N*-Ethyloxycarbonyl-2-ethyloxy-1,2-dihydroquinoline
EVAL	Ethylene–vinyl alcohol copolymer membrane support
FDPP	Pentafluorophenyldiphenylphosphinate
Fem	Ferrocenemethyl
Fm	Fluorenylmethyl
Fmoc	9-Fluorenylmethyloxycarbonyl
HAL-resin	5-(4-Hydroxymethyl-3,5-dimethoxyphenoxy)valeryl-resin
HATU	0-(7-Azabenzotriazol-1-yl)-1,1,3,3-tetramethyluronium hexafluorophosphate
HBTU	2-(Benzotriazol-1-yl)-1,1,3,3-tetramethyluronium hexafluorophosphate
HMPA	Hexamethylphosphoramide
HMPB	4-(4′-Hydroxymethyl-3′-methoxyphenoxy)butyric acid

HOAt	1-Hydroxy-7-azabenzotriazole
HOBt	*N*[1]-Hydroxybenztriazole
hPTH	Human parathyroid hormone
HYCRAM-resin	Hydroxycrotonylaminomethyl-resin
MAP	Multiple antigen peptide
MBHA	4-Methoxybenzhydrylamine
Mds	4-Methoxy-2,6-dimethylbenzenesulfonyl
Meb	4-Methylbenzyl
Msz	4-Methylsulfinylbenzyloxycarbonyl
Mtr	4-Methoxy-2,3,6-trimethyl
Mtt	4-Methyltrityl
NCA	*N*-Carboxyanhydride
NMM	*N*-Methylmorpholine
NMP	*N*-Methylpyrrolidinone
Noc	*p*-Nitrocinnamyloxycarbonyl
NPS	4-Nitrophenylsulfenyl
Npys	3-Nitro-2-pyridinesulfenyl
OAIU	*O*-Acylisourea
ODhbt	1-Oxo-2-hydroxydihydrobenzotriazine
ONb	*O*-Nitrobenzyl
PAB-resin	Phenoxypropionylaminomethyl-resin
PAC	Phenacyl
PAL-resin	5-(4-Aminomethyl-3,5-dimethoxyphenoxy)valeryl-resin
Paloc	3-(3′-Pyridyl)allyloxycarbonyl
Pam-resin	4-Hydroxymethylphenylaceteamidomethyl-resin
Pbf	2,2,4,6,7-Pentamethyldihydrobenzofuran-5-sulfonyl
Pbs	2,4,5-trichlorophenyl *N*-(3 or 4)-{[4-(hydroxymethyl)phenoxy]-*tert*-butylphenylsilyl}-phenylpentanedioate monoamide
PEG	Polyethylene glycol
PEO	Polyethylene oxide
Pfp	Pentafluorophenyl
Pmc	2,2,5,7,8-Pentamethylchroman-6-sulfonyl
PNA	Protein–nucleic acid hybrid
POE	Polyoxyethylene
Ptm	Phenylthiomethyl
PyBOP	Benzotriazolyloxytrispyrrolidinophosphonium hexafluorophosphate
PyBroP	Bromotrispyrrolidinophosphonium hexafluorophosphate
PyClop	Chlorotrispyrrolidinophosphonium hexafluorophosphate
RAPP	Rapp polymer (Tentagel)
SA	Symmetrical anhydride

S-DVB	Copoly(styrene–divinylbenzene)
SAL-resin	4-{1-[*N*-(9-Fluorenylmethyloxycarbonyl)-amino]-2-(trimethyl)ethyl}phenoxyacetyl-resin
SASRIN resin	Super acid sensitive resin; 2-methoxy-4-alkoxy benzyl alcohol
Sulfmoc	2-Sulfo-9-fluorenylmethyloxycarbonyl
TASP	Template assembled peptide synthesis
TATU	*O*-(7-Azabenzotriazol-1-yl)-1,1,3,3-tetramethyl tetrafluoroborate
TBPipU	2-(Benzotriazol-1-yl)-bispiperidinyluronium
TBPyU	Benzotriazolyl-bispyrrolidinouronium
TBTU	2-(Benzotriazol-1-yl)-1,1,3,3-tetramethyluronium tetrafluoroborate
TFA	Trifluoroacetic acid
Tfa	Trifluoroacetyl
TFE	Trifluoroethanol
TFMSA	Trifluoromethanesulfonic acid
THF	Tetrahydrofuran
Tmob	2,4,6-Trimethoxybenzyl
TMSCl	Trimethylsilyl chloride
Tmse	β-Ethyltrimethylsilyl
TPCK	Tosylphenylalanine chloromethylketone
TPTU	*N*-(2-Pyridonyl)-1,1,3,3-tetramethyluronium tetrafluoroborate
Trt	Triphenylmethyl; trityl
TSTU	Succinoyl tetramethyluronium tetrafluoroborate
UNCA	Urethane-protected *N*-carboxyanhydride
XAL-resin	5-(9-Aminoxanthene-2-oxy)valeryl-resin
Xan	Xanthyl
Z	Benzyloxycarbonyl

References

Albericio, F., and Barany, G. (1985). *Int. J. Pept. Protein Res.* **26,** 92–97.

Albericio, F., and Barany, G. (1987). *Int. J. Pept. Protein Res.* **30,** 206–216.

Albericio, F., and Barany, G. (1991). *Tetrahedron Lett.* **32,** 1015–1018.

Albericio, F., Kneib-Cordonier, N., Brancalana, S., Masada, R., Hudson, D., and Barany, G. (1990). *J. Org. Chem.* **55,** 3730–3743.

Albericio, F., Giralt, E., and Eritja, R. (1991). *Tetrahedron Lett.* **32,** 1515–1518.

Albericio, F., Barany, G., Fields, G. B., Hudson, D., Kates, S. A., Lyttle, M. H., and Sole, N. A. (1993). *In* "Peptides 1992" (C. H. Schneider and A. N. Eberle, eds.), pp. 191–193. ESCOM, Leiden.

Andersson, L., and Lindqvist, M. (1993). *In* "Peptides 1992" (C. H. Schneider and A. M. Eberle, eds.), pp. 265–266. ESCOM, Leiden.

Anwer, M. K., and Spatola, A. F. (1980). *Synthesis,* 929–932.

Anwer, M. K., and Spatola, A. F. (1992). *Tetrahedron Lett.* **33,** 3121–3124.

Arbo, B. E., and Isied, S. S. (1993). *Int. J. Pept. Protein Res.* **42,** 138–154.

Arshady, R., Atherton, E., Cleve, D. L. J., and Sheppard, R. C. (1981). *J. Chem. Soc., Perkin Trans. 1*, 529–537.

Arzeno, H. B., and Kemp, D. S. (1988). *Synthesis*, 32–36.

Atherton, E., and Sheppard, R. C. (1989). "Solid Phase Peptide Synthesis: A Practical Approach." Oxford Univ. Press, London and New York.

Atherton, E., Clive, D. L. J., and Sheppard, R. C. (1975). *J. Am. Chem. Soc.* **97,** 6584–6585.

Atherton, E., Fox, H., Harkiss, D., Logan, C. J., Sheppard, R. C., and Williams, B. J. (1978). *J. Chem. Soc., Chem. Commun.*, 537–539.

Atherton, E., Hübscher, W., Sheppard, R. C., and Wooley, W. (1981). *Hoppe Seyler's Z. Physiol. Chem.* **362,** 833–839.

Atherton, E., Sheppard, R. C., and Wade, J. D. (1983). *J. Chem. Soc., Chem. Commun.*, 1060–1062.

Atherton, E., Cameron, L. R., and Sheppard, R. C. (1988a). *Tetrahedron* **44,** 843–857.

Atherton, E., Holder, J. L., Meldal, M., Sheppard, R. C., and Valerio, R. M. (1988b). *J. Chem. Soc., Perkin Trans. 1*, 2887–2894.

Baeza, C. R., and Undén, A. (1992). *Int. J. Pept. Protein Res.* **39,** 195–200.

Bagley, C. J., Otteson, K. M., May, B. L., McCurdy, S. N., Pierce, L., Ballard, F. S., and Wallace, J. C. (1990). *Int. J. Pept. Protein Res.* **36,** 356–361.

Barany, G., and Albericio, F. (1985). *J. Am. Chem. Soc.* **107,** 4936–4942.

Barany, G., and Merrifield, R. B. (1977). *J. Am. Chem. Soc.* **99,** 7363–7365.

Barany, G., and Merrifield, R. B. (1979a). *Anal. Biochem.* **95,** 160–170.

Barany, G., and Merrifield, R. B. (1979b). *In* "Peptides" (E. Gross and J. Meienhofer, eds.), Vol. 2, pp. 1–284. Academic Press, New York.

Barany, G., Kneib-Cordonier, N., and Mullen, D. B. (1987). *Int. J. Pept. Protein Res.* **30,** 705–739.

Barany, G., Albericio, F., Biancalana, S., Bontems, S. L., Chang, J. L., Eritja, R., Ferrer, M., Fields, C. G., Fields, G. B., Lyttle, M. H., Solé, N. A., Tian, Z., Van Abel, R. J., Wright, P. B., Zalipsky, S., and Hudson, D. (1992). *In* "Peptides" (J. A. Smith and J. E. Rivier, eds.), pp. 603–604. ESCOM, Leiden.

Barany, G., Albericio, F., Solé, N. A., Griffin, G. W., Kates, S. A., and Hudson, D. (1993). *In* "Peptides 1992" (C. H. Schneider and A. N. Eberle, eds.), pp. 267–268. ESCOM, Leiden.

Barlos, K., Gatos, D., Kallitsis, J., Papaphotiu, G., Sotiriu, P., Wenging, Y., and Schäfer, W. (1989). *Tetrahedron Lett.* **30,** 3943–3946.

Barlos, K., Gatos, D., Kapolos, S., Poulos, C., Schäfer, W., and Yao, W. (1991). *Int. J. Pept. Protein Res.* **38,** 555–561.

Barlos, K., Gatos, D., and Schäfer, W. (1991a). *Angew. Chem., Int. Ed. Engl.* **30,** 590–593.

Barstow, L. E., Cornelius, D. A., Hruby, V. J., Shimoda, T., Rupley, J. A., Sharp, J. S., Robinson, A. B., and Kamen, M. D. (1972). *In* "Chemistry and Biology of Peptides" (J. Meienhofer, ed.), pp. 231–233. Ann Arbor Science Publ., Ann Arbor, Michigan.

Bartl, R., Klöppel, K.-D., and Frank, R. (1992). *In* "Peptides" (J. A. Smith and J. E. Rivier, eds.), pp. 505–506. ESCOM, Leiden.

Bartl, R., Klöppel, K.-D., and Frank, R. (1993). *In* "Peptides 1992" (C. H. Schneider and A. N. Eberle, eds.), pp. 277–278. ESCOM, Leiden.

Barton, M. A., and Lemieux, R. U. (1971). *Proc. Can. Fed. Biol. Sci.* **13,** 14.

Barton, M. A., Lemieux, R. U., and Savoie, J. Y. (1973). *J. Am. Chem. Soc.* **95,** 4501–4506.

Bates, A. J., Galpin, I. J., Hallett, A., Hudson, D., Kenner, G. W., Ramage, R., and Sheppard, R. C. (1975). *Helv. Chim. Acta* **58,** 688–696.

Bayer, E., and Rapp, W. (1986). *In* "Chemistry of Peptides and Proteins" (W. Voelter, E. Bayer, Y. A. Ovchinikov, and V. T. Ivanov, eds.), Vol. 3, pp. 3–7. de Gruyter, New York.

Bayer, E., Jung, G., and Hagemaier, H. (1968). *Tetrahedron* **24,** 4853–4860.

Bayer, E., Jung, G., Halasz, I., and Sebastian, I. (1970). *Tetrahedron Lett.* **5,** 4503–4505.

Bayer, E., Clausen, N., Goldammer, C., Henkel, B., Rapp, W., and Zhang, L. (1994). *In* "13th American Peptide Symposium, Edmonton, Alberta, Canada" (R. S. Hodges, ed.), in press.

Bedford, J., Hyde, C., Johnson, T., Jun, W., Owen, D., Quibell, M., and Sheppard, R. C. (1992). *Int. J. Pept. Protein Res.* **40,** 300–307 (1992).

Berg, R. H., Almdal, K., Pedersen, W. B., Holm, A., Tam, J. P., and Merrifield, R. B. (1989). *J. Am. Chem. Soc.* **111,** 8024–8026.

Berg, R. H., Almdal, K., Pedersen, W. B., Holm, A., Tam, J. P., and Merrifield, R. B. (1991). *In* "Peptides 1990" (E. Giralt and D. Andreu, eds.), pp. 149–150. ESCOM, Leiden.

Berg, R. H., Nielsen, P. E., Buchardt, O., and Egholm, M. (1993). *In* "Peptides 1992" (C. H. Schneider and A. M. Eberle, eds.), pp. 152–153. ESCOM, Leiden.

Bergot, B. J., Noble, R. F., and Geiser, T. (1987). *In* "Peptides 1986" (D. Theodoropoulos, ed.), pp. 97–101. de Gruyter, Berlin.

Beyermann, M., Henklein, P., Klose, A., Sohr, R., and Bienert, M. (1991). *In* "Peptides 1990" (E. Giralt and D. Andreu, eds.), pp. 59–61. ESCOM, Leiden.

Birr, C. (1972). *Justus Liebigs Ann. Chem.* **763,** 162–172.

Birr, C. (1978). "Aspects of the Merrifield Peptide Synthesis." Springer-Verlag, Berlin.

Birr, C. Nguyen-Trong, H., Becker, G., Muller, T., Schramm, M., Kunz, H., Dombo, B., and Kosch, W. (1991). *In* "Peptides 1990" (E. Giralt and D. Andreu, eds.), pp. 127–128. ESCOM, Leiden.

Bláha, I., Tözsér, J., Kim, Y., Copeland, T. D., and Oroszlan, S. (1992). *FEBS Lett.* **309,** 389–393.

Blake, J. (1979). *Int. J. Pept. Protein Res.* **13,** 418–425.

Blake, J. (1981). *Int. J. Pept. Protein Res.* **17,** 273–274.

Bodanszky, M. (1992). *Pept. Res.* **5,** 134–139.

Bodanszky, M., and Bednarek, M. A. (1982). *Int. J. Pept. Protein Res.* **20,** 434–437.

Bodanszky, M., and Bodanszky, A. (1984). "The Practice of Peptide Synthesis," See pages 202–205. Springer-Verlag, Berlin.

Bodanszky, M., and Sheehan, J. T. (1964). *Chem. Ind. (London),* 1423–1424.

Bolin, D. R., Sytwu, I.-I., Humiec, F., and Meienhofer, J. (1989). *Int. J. Pept. Protein Res.* **33,** 353–359.

Boman, H. G., Boman, I. A., Andreu, D., Li, Z.-Q., Merrifield, R. B., Schlenstedt, G., and Zimmermann, R. (1989). *J. Biol. Chem.* **264,** 5852–5860.

Bontems, R. J., Hegyes, P., Bontems, S. L., Albericio, F., and Barany, G. (1992). *In* "Peptides" (J. A. Smith and J. E. Rivier, eds.), pp. 601–602. ESCOM, Leiden.

Bouhnik, J., Galen, F.-X., Menard, J., Corvol, P., Seyer, R., Fehrentz, J.-A., Nguyen, D. L., Fulcrand, P., and Castro, B. (1987). *J. Biol. Chem.* **262,** 2913–2918.

Breipohl, G., Knolle, J., and Stüber, W. (1989). *Int. J. Pept. Protein Res.* **34,** 262–267.

Briand, J. P., Van Dorsselaer, A., and Muller, S. (1990). *In* "Peptides: Structure, Biology" (J. E. Rivier and G. R. Marshall, eds.), pp. 1010–1011. ESCOM, Leiden.

Brown, T., Jones, J. H., and Richards, J. D. (1982). *J. Chem. Soc., Perkin Trans. 1,* 1553–1561.

Büttner, K., Zahn, H., and Fischer, W. H. (1988). *In* "Peptides" (G. R. Marshall, ed.), pp. 210–211. ESCOM, Leiden.

Carpino, L. A., and Han, G. Y. (1972). *J. Org. Chem.* **37,** 3404–3409.

Carpino, L. A., Cohen, B. J., Stephens, K. E., Jr., Sadat-Aalaee, S. Y., Tien, J.-H., and Langridge, D. C. (1986). *J. Org. Chem.* **51,** 3732–3734.

Carpino, L. A., Sadat-Aalaee, D., Chao, H. G., and De Selms, R. H. (1990). *J. Am. Chem. Soc.* **112,** 9651–9652.

Carpino, L. A., El-Faham, A., Truran, G., Triolo, S. A., Shroff, H., Griffin, G. W., Minor, C. A., Kates, S. A., and Albericio, F. (1994). *In* "Peptides" (R. S. Hodges and J. A. Smith, eds.), pp. 124–126. ESCOM, Leiden.

Carpino, L. A., Mansour, E. M. E., and Sadat-Aelee, D. (1991). *J. Org. Chem.* **56,** 2611–2614.

Castro, B., and Dormoy, J. R. (1972). *Tetrahedron Lett.* **47,** 4747–4750.

Castro, B., Dormoy, J. R., Evin, G., and Selve, C. (1975). *Tetrahedron Lett.* **14,** 1219–1222.

Chang, C.-D., and Meienhofer, J. (1978). *Int. J. Pept. Protein Res.* **11,** 246–249.

Chang, C.-D., Waki, M., Akman, J. M., Meienhofer, J., Lundell, E. O., and Haug, J. D. (1980). *Int. J. Pept. Protein Res.* **15,** 59–66.

Chao, H.-G., Bernatowicz, M. S., and Matsueda, G. R. (1993). *J. Org. Chem.* **58,** 2640–2644.

Chen, S., and Xu, J. (1991). *Tetrahedron Lett.* **32,** 6711–6714.

Chou, P. Y., and Fasman, G. D. (1978). *Adv. Enzymol.* **2,** 45–148.

Chong, P., Sia, C., Tam, E., Kandil, A., and Klein, M. (1993). *Int. J. Pept. Protein Res.* **41,** 21–27.

Clark-Lewis, I., Aebersold, R., Ziltener, H., Schrader, J. W., Hood, L. E., and Kent, S. B. H. (1986). *Science* **231,** 134–139.

Clark-Lewis, I., Lopez, A. F., To, L. B., Vadas, M. A., Schrader, J. W., Hood, L. E., and Kent, S. B. H. (1988). *J. Immunol.* **141,** 881–889.

Clark-Lewis, I., Moser, B., Walz, A., Baggiolini, M., Scott, G. J., and Aebersold, R. (1991). *Biochemistry* **30,** 3128–3135.

Colombo, R., Colombo, F., and Jones, J. H. (1984). *J. Chem. Soc., Chem. Commun.*, 292–293.

Copeland, T. D., and Oroszlan, S. (1988). *Gene Anal. Tech.* **5,** 109–115.

Coste, J., Dufour, M.-N., Le-Nguyen, D., and Castro, B. (1990a). *In* "Peptides" (J. E. Rivier and G. R. Marshall, eds.), pp. 885–888. ESCOM, Leiden.

Coste, J., Le-Nguyen, D., and Castro, B. (1990b). *Tetrahedron Lett.* **31,** 205–208.

Coste, J., Dufour, M.-N., Pantaloni, A., and Castro, B. (1990c). *Tetrahedron Lett.* **31,** 669–672.

Coste, J., Frérot, E., Dufour, M.-N., Pantaloni, A., and Jouin, P. (1991) *In* "Peptides 1990" (E. Giralt and D. Andreu, eds.), pp. 76–77. ESCOM, Leiden.

Coste, J., Frérot, E., Jouin, P., and Castro, B. (1993). *In* "Peptides 1992" (C. H. Schneider and A. N. Eberle, eds.), pp. 245–246. ESCOM, Leiden.

Deber, C. M., Lutek, M. K., Heimer, E. P., and Felix, A. M. (1989). *Pept. Res.* **2,** 184–188.

Denkewalter, R. B., Schwam, H., Strachan, R. G., Beesley, T. E., Veber, D. F., Schoenwaldt, E. F., Barkemyer, H., Paleveda, W. J., Jr., Jacob, T. A., and Hirschmann, R. (1966). *J. Am. Chem. Soc.* **88,** 3163–3164.

DeTar, D. F., Silverstein, R., and Rogers, F. F., Jr. (1966). *J. Am. Chem. Soc.* **88,** 1024–1030.

Di Bello, C., and Gozzini, L. (1993). *Int. J. Pept. Protein Res.* **41,** 34–42.

DiMarchi, R. D., Tam, J. P., Kent, S. B. H., and Merrifield, R. B. (1982). *Int. J. Pept. Protein Res.* **19,** 88–93.

Dourtoglou, V., Zielger, J.-C., and Gross, B. (1978). *Tetrahedron Lett.* **2,** 1269–1272.

Dourtoglou, V., Gross, B., Lambropoulou, V., and Zioudrou, C. (1984). *Synthesis,* 572–574.

Eckert, H., and Seidel, C. (1986). *Angew. Chem., Int. Ed. Engl.* **25,** 159–160.

Egholm, M., Buchardt, O., Berg, R. H., and Nielsen, P. E. (1993). *Proc. Natl. Acad. Sci. U.S.A.* **90,** 1667–1670.

Ehrlich, A., Rothemund, S., Beyermann, M., Carpino, L. A., and Bienert, M. (1994). *In* "Peptides" (R. S. Hodges and J. A. Smith, eds.), pp. 95–96. ESCOM, Leiden.

Eichler, J., Bienert, M., Sepetov, N. F., Štolba, P., Krchňák, V., Smékal, O., Gut, V., and Lebl, M. (1990). *In* "Solid Phase Synthesis" (R. Epton, ed.), pp. 337–343. SPCC, Birmingham, U. K.

Eichler, J. M., Bienert, M., Stierandova, A., and Lebl, M. (1991). *Pept. Res.* **4,** 296–307.

El Rahman, S. A., Anzinger, H., and Mutter, M. (1980). *Biopolymers* **19,** 173–187.

Engelhard, M., and Merrifield, R. B. (1978). *J. Am. Chem. Soc.* **100,** 3559–3563.

Englebretsen, D. R., and Harding, D. R. K. (1992). *Int. J. Pept. Protein Res.* **40,** 487–496.

Epton, R., Wellings, D. A., and Williams, A. (1987). *React. Polym.* **6,** 143–157 (1987).

Erickson, B. W., and Merrifield, R. B. (1973a). *J. Am. Chem. Soc.* **95,** 3750–3756.

Erickson, B. W., and Merrifield, R. B. (1973b). *J. Am. Chem. Soc.* **95,** 3757–3763.

Erickson, B. W., and Merrifield, R. B. (1976). *In* "The Proteins" (H. Neurath and R. L. Hill, eds.), 3rd Ed., Vol. 2, pp. 255–527. Academic Press, New York.

Eritja, R., Robles, J., Fernandez-Forner, D., Albericio, F., Giralt, E., and Pedroso, E. (1991). *Tetrahedron Lett.* **32,** 1511–1514.

Esko, K., and Karlsson, S. (1970). *Acta Chem. Scand.* **24,** 1415–1422.

Fairwell, T., Hospattankar, A. V., Ronan, R., Brewer, H. B., Jr., Chang, J. K., Shimizu, M., Zitzner, L., and Arnaud, C. D. (1983). *Biochemistry* **22,** 2691–2697.

Fauchére, J. L., and Schwyzer, R. (1981). *In* "The Peptides" (E. Gross and J. Meienhofer, eds.), Vol. 3, pp. 203–252. Academic Press, New York.

Felix, A. M., Wang, C. T., Heimer, E. P., and Fournier, A. (1988). *Int. J. Pept. Protein Res.* **31,** 231–238.

Fields, G. B., and Noble, R. L. (1990). *Int. J. Pept. Protein Res.* **35,** 161–214.

Fields, G. B., Otteson, K. M., Fields, C. G., and Noble, R. L. (1990). *In* "Solid Phase Synthesis" (R. Epton, ed.), pp. 241–260. SPCC, Birmingham, U.K.

Fields, G. B., Tian, Z., and Barany, G. (1992). *In* "Synthetic Peptides, A Users Guide" (G. A. Grant, ed.), pp. 77–183. Freeman, New York.

Fischer, P. M., Retson, K. V., Tyler, M. I., and Howden, M. E. H. (1992). *Int. J. Pept. Protein Res.* **40,** 19–24.

Flanigan, E., and Marshall, G. R. (1970). *Tetrahedron Lett.* **3,** 2403–2406.

Flörsheimer, A., and Riniker, B. (1991). *In* "Peptides 1990" (E. Giralt and D. Andreu, eds.), pp. 131–133. ESCOM, Leiden.

Fotouhi, N., Galakatos, N. G., and Kemp, D. S. (1989). *J. Org. Chem.* **54,** 2803–2817.

Frank, R., and Döring, R. (1988). *Tetrahedron Lett.* **44,** 6031–6040.

Frank, R., Güler, S., Krause, S., and Lindenmaier, W. (1991). *In* "Peptides 1990" (E. Giralt and D. Andreu, eds.), pp. 151–152. ESCOM, Leiden.

Fréchet, J. M. J., and Haque, K. E. (1975). *Tetrahedron Lett.* **35,** 3055–3056.

Fries, J. L., Coy, D. H., Huang, W. Y., and Meyers, C. A. (1979). *In* "Peptides: Structure and Biological Function" (E. Gross and J. Meienhofer, eds.), pp. 499–502. Pierce Chemical Co., Rockford, Illinois.

Fujino, M., Nishimura, O., Wakimasu, M., and Kitada, C. (1980). *J. Chem. Soc., Chem. Commun.,* 668–669.

Fujino, M., Wakimasu, M., and Kitada, E. (1981). *Chem. Pharm. Bull. Jpn.* **29,** 2825–2831.

Fuller, W. D., Cohen, M. P., Shabankarek, M., Blain, R. K., Goodman, M., and Naider, F. R. (1990). *J. Am. Chem. Soc.* **112,** 7414–7416 (1990).

Fuller, W. D., Krotzer, N. J., Naider, F. R., Xue, C.-B., and Goodman, M. (1993a). *In* "Peptides 1992" (C. H. Schneider and A. N. Eberle, eds.), pp. 229–230. ESCOM, Leiden.

Fuller, W. D., Krotzer, N. J., Swain, P. A., Anderson, B. L., Comer, D., and Goodman, M. (1993b). *In* "Peptides 1992" (C. H. Schneider and A. N. Eberle, eds.), pp. 231–232. ESCOM, Leiden.

Funakoshi, S., Murayama, E., Guo, L., Fujii, N., and Yajima, H. (1988). *J. Chem. Soc., Chem. Commun.,* 382–384.

Gausepohl, H., Kraft, M., and Frank, R. W. (1989). *Int. J. Pept. Protein Res.* **34,** 287–294.

Green, J., Ogunjobi, O. M., Ramage, R., and Stewart, A. S. J. (1988). *Tetrahedron Lett.* **29,** 4341–4344.

Gruszecki, W., Gruszecka, M., and Bradaczek, H. (1988). *Liebigs Ann. Chem.,* 331–336.

Gutte, B. (1975). *J. Biol. Chem.* **250,** 889–904.

Gutte, B., and Merrifield, R. B. (1969). *J. Am. Chem. Soc.* **91,** 501–502.

Gutte, B., and Merrifield, R. B. (1971). *J. Biol. Chem.* **246,** 1922–1944.

Gutte, B., Däumigen, M., and Wittschieber, E. (1979). *Nature (London)* **281,** 650–655.

Hahn, K. W., Klis, W. A., and Stewart, J. M. (1990). *Science* **248,** 1544–1547.

Hancock, W. S., Prescott, D. J., Nulty, W. L., Weintraub, J., Vagelos, P. R., and Marshall, G. R. (1971). *J. Am. Chem. Soc.* **93,** 1799–1800.

Hancock, W. S., Prescott, D. J., Vagelos, P. R., and Marshall, G. R. (1973). *J. Org. Chem.* **38,** 774–781.

Handa, B. K., and Keech, E. (1992). *Int. J. Pept. Protein Res.* **40,** 66–71 (1992).

Handel, T., and DeGrado, W. F. (1990). *J. Am. Chem. Soc.* **112,** 6710–6711.

Hansen, P. R., Holm, A., and Houen, G. (1992). *In* "Peptides" (J. A. Smith and J. E. Rivier, eds.), pp. 637–638. ESCOM, Leiden.

Hansen, P. R., Holm, A., and Houen, G. (1993). *Int. J. Pept. Protein Res.* **41,** 237–245.

Heath, W. F., Tam, J. P., and Merrifield, R. B. (1982). *J. Chem. Soc., Chem. Commun.,* 896–897.

Hellermann, H., Lucas, H. N., Maul, J., Pillai, V. N. R., and Mutter, M. (1983a). *Int. J. Pept. Protein Res.* **18,** 237–241.

Hellermann, H., Lucas, H.-W., Maul, J., Pillai, V. N. R., and Mutter, M. (1983b). *Makromol. Chem.* **184,** 2603–2617.

Hendrix, J. C., Halverson, K. J., Jarrett, J. T., and Lansbury, P. T., Jr. (1990). *J. Org. Chem.* **55,** 4517–4518.

Hetnarski, B., and Merrifield, R. B. (1988). *In* "Peptides, Chemistry and Biology" (G. R. Marshall, ed.), pp. 220–222. ESCOM, Leiden.

Ho, S. P., and DeGrado, W. F. (1987). *J. Am. Chem. Soc.* **109,** 6751–6758.

Hodges, R. S., Sodek, J., Smillie, L. B., and Jurasek, L. (1972). *Cold Spring Harbor Symp. Quant. Biol.* **37,** 299–310.

Hodges, R. S., Saund, A. K., Chong, P. C. S., St. Pierre, S. A., and Reid, R. D. (1981). *J. Biol. Chem.* **256,** 1214–1224.

Horvath, C. G., Preiss, B. A., and Lipsky, S. R. (1967). *Anal. Chem.* **39,** 1422–1428.

Houghten, R. A., Beckman, A., and Ostresh, J. M. (1986). *Int. J. Pept. Protein Res.* **27,** 653–658.

Hudson, D. (1988). *J. Org. Chem.* **53,** 617–624.

Hudson, D. (1990). *Pept. Res.* **3,** 51–55.

Inukai, N., Nakano, K., and Murakami, M. (1968). *Bull Chem. Soc. Jpn.* **41,** 182–186.

Irie, H., Fujii, N., Ogawa, H., Yajima, H., Fujino, M., and Shinagawa, S. (1976). *J. Chem. Soc., Chem. Commun.,* 922–923.

Jakubke, Von H.-D., and Baumert, A. (1974). *J. Prakt. Chem. Bd.* **316,** 241–248.

Johnson, T., Quibell, M., Owen, D., and Sheppard, R. C. (1993). *J. Chem. Soc., Chem. Commun.,* 369–372.

Kaiser, E. T., and Kézdy, F. J. (1983). *Proc. Natl. Acad. Sci. U.S.A.* **80,** 1137–1143.

Kaiser, E., Picart, F., Kubiak, T., Tam, J. P., and Merrifield, R. B. (1993). *J. Org. Chem.* **58,** 5167–5175.

Kalbacher, H., and Voelter, W. (1978). *Angew. Chem., Int. Ed. Engl.* **17,** 944–945.

Kanda, P., Kennedy, R. C., and Sparrow, J. T. (1991). *Int. J. Pept. Protein Res.* **38,** 385–391.

Kates, S. A., Solé, N. A., Johnson, C. R., Hudson, D., Barany, G., and Albericio, F. (1993). *Tetrahedron Lett.* **34,** 1549–1552.

Kemp, D. S., Roberts, D., Hoyng, C., Crallan, J., Vellaccio, F., and Reczek, J. (1975). *In* "Peptides: Chemistry, Structure and Biology" (R. Walter and J. Meienhofer, eds.), pp. 295–305. Ann Arbor Science Publ., Ann Arbor, Michigan.

Kent, S. B. H. (1988). *Annu. Rev. Biochem.* **57,** 957–989.

Kent, S. B. H., and Merrifield, R. B. (1978). *Isr. J. Chem.* **17,** 243–247.

Kent, S. B. H., and Merrifield, R. B. (1981). *In* "Peptides 1980" (K. Brunfeldt, ed.), pp. 328–333. Scriptor, Copenhagen.

Kisfaludy, L., and Schon, I. (1983). *Synthesis,* 325–327.

Kisfaludy, L., Ceprini, M. Q., Rakoczy, B., and Kovacs, J. (1967). *In* "Peptides" (H. C. Beyerman, A. van der Linde, and W. M. van der Brink, eds.), pp. 25–27. North-Holland Publ., Amsterdam.

Kiso, Y., Ukawa, K., Nakamura, S., Ito, K., and Akita, T. (1980). *Chem. Pharm. Bull. Jpn.* **28,** 673–676.

Kiso, Y., Kimura, T., Shimokura, M., and Narukami, T. (1988a). *J. Chem. Soc., Chem. Commun.,* 287–289.

Kiso, Y., Kimura, T., Fujiwara, Y., Shimokura, M., and Nishitani, A. (1988b). *Chem. Pharm. Bull.* **36,** 5024–5027.

Kiso, Y., Kimura, T., Yoshida, M., Shimokura, M., Akaji, K., and Mimoto, T. (1989a). *J. Chem. Soc., Chem. Commun.,* 1511–1513.

Kiso, Y., Yoshida, M., Tatsumi, T., Kimura, T., Fujiwara, Y., and Akaji, K. (1989b). *Chem. Pharm. Bull.* **37,** 3432–3434.

Kiso, Y., Fujiwara, Y., Kimura, T., Nishitani, A., and Akaji, K. (1992). *Int. J. Pept. Protein Res.* **40,** 308–314.

Knorr, R., Trzeciak, A., Bannwarth, W., and Gillessen, D. (1989). *Tetrahedron Lett.* **30,** 1927–1930.

Knorr, R., Trzeciak, A., Bannwarth, W., and Gillessen, D. (1991). *In* "Peptides 1990" (E. Giralt and D. Andreu, eds.), pp. 62–64. ESCOM, Leiden.

König, W., and Geiger, R. (1970). *Chem. Ber.* **103,** 3024–2040.

König, W., and Geiger, R. (1972). *Chem. Ber.* **105,** 2872–2882.

Kumagaye, K. Y., Inui, T., Nakajima, K., Kimura, T., and Sakakibara, S. (1991). *Pept. Res.* **4,** 84–87.

Kunin, R., Meitzner, E., and Bortnick, N. (1962). *J. Am. Chem. Soc.* **84,** 305–306.

Kunz, H., and Dombo, B. (1988). *Angew. Chem., Int. Ed. Engl.* **27,** 711–713.

Kunz, H., and März, J. (1988). *Angew. Chem., Int. Ed. Engl.* **27,** 1375–1377.

Kunz, H., and Unverzagt, C. (1984). *Angew. Chem., Int. Ed. Engl.* **23,** 436–437.

Kunz, H., and Waldmann, H. (1984). *Angew. Chem., Int. Ed. Engl.* **33,** 71–72.

Kyi, A. T., and Schwyzer, R. (1976). *Helv. Chim. Acta* **59,** 1642–1646.
Lau, S. Y. M., Taneja, A. K., and Hodges, R. S. (1984). *J. Biol. Chem.* **259,** 13253–13261.
Lebl, M., and Eichler, J. (1989). *J. Pept. Res.* **2,** 232.
Lenard, J., and Robinson, A. B. (1967). *J. Am. Chem. Soc.* **89,** 181–183.
Le-Nguyen, D., Dormoy, J.-R., Castro, B., and Prevot, D. (1981). *Tetrahedron* **37,** 4229–4238.
Letsinger, R. L., and Kornet, M. J. (1963). *J. Am. Chem. Soc.* **85,** 3045–3046.
Letsinger, R. L., Kornet, M. J., Makadevan, V., and Jerina, D. M. (1964). *J. Am. Chem. Soc.* **86,** 5163–5165.
Leuchs, H. (1906). *Chem. Ber.* **39,** 857.
Li, C. H., and Yamashiro, D. (1970). *J. Am. Chem. Soc.* **93,** 7608–7609.
Lim, M. S. L., Ball, H. L., Tonolo, M., Coates, A. R. M., and Mascagni, P. (1991). *In* "Peptides 1990" (E. Giralt and D. Andreu, eds.), pp. 176–178. ESCOM, Leiden.
Lin, M. C., Gutte B., Caldi, D. G., Merrifield, R. B., and Moore, S. (1972). *J. Biol. Chem.* **247,** 4763–4767.
Liu, C.-F., and Tam, J. P. (1994). *Proc. Natl. Acad. Sci. U.S.A.* **91,** 6584–6588.
Liu, Y.-Z., Ding, S.-H., Chu, J.-Y., and Felix, A. M. (1990). *Int. J. Pept. Protein Res.* **35,** 95–98.
Live, D., and Kent, S. B. H. (1983). *In* "Peptides: Structure and Function (V. Hruby and D. Rich, eds.), pp. 65–68. Pierce Chemical Co., Rockford, Illinois.
Lloyd, D. H., Petrie, G. M., Noble, R. L., and Tam, J. P. (1990). *In* "Peptides" (J. Rivier and G. R. Marshall, eds.), pp. 909–910. ESCOM, Leiden.
Manning, M. (1968). *J. Am. Chem. Soc.* **90,** 1348–1349.
Marglin, A., and Merrifield, R. B. (1966). *J. Am. Chem. Soc.* **88,** 5051–5052.
Martin, L. M., Rotondi, K. S., and Merrifield, R. B. (1994). *In* "Peptides" (R. S. Hodges and J. A. Smith, eds.), pp. 745–747. ESCOM, Leiden.
Maruyama, K., Nagata, K., Tanaka, M., Nagasawa, H., Isogai, A., Ishizaki, H., and Suzuki, A. (1992). *J. Protein Chem.* **11,** 11, 1–12.
Mascagni, P., Dwo, S., Coates, A. R. M., and Gibbons, W. A. (1990). *Tetrahedron Lett.* **31,** 4637–4640.
Matsueda, G. R., and Stewart, J. M. (1981). *Peptides* **2,** 45–50.
Matsueda, R., Higashida, S., Albericio, F., and Andreu, D. (1992). *Pept. Res.* **5,** 262–264.
Meienhofer, J., and Chang, C.-D. (1979). *In* "Peptides 1978" (I. Z. Siemion and G. Kupryszewski, eds.), pp. 573–575. Warsaw Univ. Press, Warsaw, Poland.
Meienhofer, J., Waki, M., Heimer, E. P., Lambros, T. J., Makofske, R., and Chang, C.-D. (1979). *Int. J. Pept. Res.* **13,** 35–42.
Meldal, M. (1992). *Tetrahedron Lett.* **33,** 3077–3080.
Mendre, C., Sarrade, V., and Calas, B. (1992). *Int. J. Pept. Protein Res.* **39,** 278–284.
Mensi, N., and Isied, S. S. (1987). *J. Am. Chem. Soc.* **109,** 7882–7884.
Mergler, M., Tanner, R., Gosteli, J., and Grogg, P. (1988). *Tetrahedron Lett.* **29,** 4005–4008.
Merrifield, R. B. (1963). *J. Am. Chem. Soc.* **85,** 2149–2154.
Merrifield, R. B. (1965). *Endeavor* **24,** 3–7.
Merrifield, R. B. (1967). *In* "Recent Progress in Hormone Research" (G. Pincus, ed.), Vol. 23, pp. 451–482. Academic Press, New York.
Merrifield, R. B. (1969). *Adv. Enzym.* **32,** 221–296.
Merrifield, R. B. (1983). *In* "Peptides, Structure and Function" (V. J. Hruby and D. H. Rich, eds.), pp. 33–44. Pierce Chemical Co., Rockford, Illinois.
Merrifield, R. B. (1984). *Bri. Polym. J.* **16,** 173–178.
Merrifield, R. B. (1993). "Life during a Golden Age of Peptide Chemistry" (J. I. Seeman, ed.), pp. 1–297. American Chemical Society, Washington, D. C.
Merrifield, R. B., and Bach, A. E. (1978). *J. Org. Chem.* **43,** 4808–4816.
Merrifield, R. B., and Woolley, D. W. (1958). *Fed. Proc.* **17,** 275.
Merrifield, R. B., Barany, G., Cosand, W. L., Engelhard, M., and Mojsov, S. (1977). *In* "Peptides" (M. Goodman and J. Meienhofer, eds.), pp. 488–502. Wiley, New York.
Merrifield, R. B., Singer, J., and Chait, B. T. (1988). *Anal. Biochem.* **174,** 399–414.

Millar, J. R., Smith, D. G., Marr, W. E., and Kressman, T. R. E. (1963). *J. Chem. Soc.* **1,** 218–224.

Milton, R. C., Wormald, P. J., Brandt, W., and Millar, R. P. (1986). *J. Biol. Chem.* **261,** 16990–16997.

Milton, R. C. de L., Milton, S. C. F., and Adams, P. A. (1990). *J. Am. Chem. Soc.* **112,** 6039–6046.

Milton, S. C. F., Brandt, W. F., Schnölzer, M., and Milton R. C. de L. (1992). *Biochemistry* **31,** 8799–8809.

Mitchell, A. R., Erickson, B. W., Ryabtsev, M. N., Hodges, R. S., and Merrifield, R. B. (1976a). *J. Am. Chem. Soc.* **98,** 7357–7362.

Mitchell, A. R., Kent, S. B. H., Erickson, B. W., and Merrifield, R. B. (1976b). *Tetrahedron Lett.* **42,** 3795–3798.

Mitchell, A. R., Kent, S. B. H., Engelhard, M., and Merrifield, R. B. (1978). *J. Org. Chem.* **43,** 2845–2852.

Mitchell, M. A., Runge, T. A., Mathews, W. R., Ichhpurani, A. K., Harn, N. K., Dobrowolski, P. J., and Eckenrode, F. M. (1990). *Int. J. Pept. Protein Res.* **36,** 350–355.

Mojsov, S., Mitchell, A. R., and Merrifield, R. B. (1980). *J. Org. Chem.* **45,** 555–560.

Moser, R., Thomas, R. M., and Gutte, B. (1983). *FEBS Lett.* **157,** 247–251.

Mullen, D. G., and Barany, G. (1988). *J. Org. Chem.* **53,** 5240–5248.

Munson, M. C., García-Echeverría, C., Albericio, F., and Barany, G. (1992). *J. Org. Chem.* **57,** 3013–3018.

Mutter, M. (1985). *Angew. Chem., Int. Ed. Engl.* **24,** 639–653.

Mutter, M., and Bellof, D. (1984). *Helv. Chim. Acta* **67,** 2009–2016.

Mutter, M., and Vuilleumier, S. (1989). *Angew. Chem., Int. Ed. Engl.* **28,** 535–676.

Mutter, M., Gassmann, R., Buttkus, U., and Altmann, K.-H. (1991). *Angew. Chem.* **30,** 1514–1516.

Naharisson, H., Sarrade, V., Follet, M., and Calas, B. (1992). *Pept. Res.* **5,** 293–299.

Najjar, V. A., and Merrifield, R. B. (1966). *Biochemistry* **5,** 3765–3770.

Narita, M., Fukunaga, T., Wakabayashi, A., Ishikawa, K., and Nakano, H. (1984a). *Int. J. Pept. Protein Res.* **23,** 306–314.

Narita, M., Ishikawa, K., Nakano, H., and Isokawa, S. (1984b). *Int. J. Pept. Protein Res.* **24,** 14–24.

Narita, M., Ishikawa, K., Chen, J.-Y., and Kim, Y. (1984c). *Int. J. Pept. Protein Res.* **24,** 580–587.

Narita, M., Doi, M., Sugasawa, H., and Ishikawa, K. (1985). *Bull Chem. Soc. Jpn.* **58,** 1473–1479.

Neugebauer, W., Brzezinski, R., and Willick, G. (1994). *In* "Peptides" (R. S. Hodges and J. A. Smith, eds.), ESCOM, Leiden.

Nielsen, P. E., Egholm, M., Berg, R. H., and Buchardt, O. (1991). *Science* **254,** 1497–1500.

Noda, K., Terada, S., Mitsuyasu, N., Waki, M., Kato, T., and Izumiya, N. (1971). *Naturwissenschaften* **58,** 147–148.

Nokihara, K., and Semba, T. (1988). *J. Am. Chem. Soc.* **110,** 7847–7854.

Nomizu, M., Inagaki, Y., Yamashita, T., Ohkubo, A., Otaka, A., Fujii, N., Roller, P. P., and Yajima, H. (1991). *Int. J. Pept. Protein Res.* **37,** 145–152.

Nutt, R. F., Brady, S. F., Darke, P. L., Ciccarone, T. M., Colton, C. D., Nutt, E. M., Rodkey, J. A., Bennett, C. D., Waxman, L. H., Sigal, I. S., Anderson, P. S., and Veber, D. F. (1988). *Proc. Natl. Acad. Sci. U.S.A.* **85,** 7129–7133.

Okada, Y., and Iguchi, S. (1988). *J. Chem. Soc., Perkin Trans. 1,* 2129–2136.

Okuda, T., and Zahn, H. (1969). *Makromol. Chem.* **121,** 87–101.

Oliveira, E., Marchetto, R., Jubilut, G. N., Paiva, A. C. M., and Nakaie, C. R. (1992). *In* "Peptides" (J. A. Smith and J. E. Rivier, eds.), pp. 569–570. ESCOM, Leiden.

Ontjes, D. A., and Anfinsen, C. B. (1969). *Proc. Natl. Acad. Sci. U.S.A.* **64,** 428–435.

Osborn, N. J., and Robinson, J. A. (1993). *Tetrahedron* **49,** 2873–2884.

Parr, W., and Grohmann, K. (1972). *Angew. Chem., Int. Ed. Engl.* **11,** 314–315.

Penke, B., and Nyerges, L. (1989). *In* "Peptides 1988" (G. Jung and E. Bayer, eds.), pp. 142–144. de Gruyter, Berlin.

Penke, B., and Nyerges, L. (1991). *In* "Peptides 1990" (E. Giralt and D. Andreu, eds.), pp. 158–159. ESCOM, Leiden.

Perlow, D. S., Erb, J. M., Gould, N. P., Tung, R. D., Freidinger, R. M., Williams, P. D., and Veber, D. F. (1992). *J. Org. Chem.* **57,** 4394–4400.

Pessi, A., Bianchi, E., Crameri, A., Tramontano, A., and Sollazzo, M. (1993). *In* "Peptides 1992" (C. H. Schneider and A. N. Eberle, eds.), pp. 89–90. ESCOM, Leiden.

Pessi, A., Bianchi, E., Venturini, S., Barbato, G., Tramontano, A., and Sollazzo, M. (1994). *In* "13th American Peptide Symposium, Edmonton, Alberta, Canada" (R. S. Hodges, ed.), in press.

Pickup, S., Blum, F. D., Ford, W. T., and Periyasamy, M. (1986). *J. Am. Chem. Soc.* **108,** 3987–3990.

Pietta, P. G., and Marshall, G. R. (1970). *J. Chem. Soc. B.,* 650.

Pietta, P. G., Biondi, P. A., and Brenna, O. (1976). *J. Org. Chem.* **41,** 703–704.

Pillai, V. N. R., and Mutter, M. (1981). *Acc. Chem. Res.* **14,** 122–130.

Rabinovich, A. K., and Rivier, J. E. (1994). *In* "13th American Peptide Symposium, Edmonton, Alberta, Canada" (R. S. Hodges, ed.), in press.

Rahman *et al.* (1980).

Ramage, R., and Green, J. (1987). *Tetrahedron Lett.* **28,** 2287–2290.

Ramage, R., Barron, C. A., Bielecki, S., and Thomas, D. W. (1987). *Tetrahedron Lett.* **28,** 4105–4108.

Ramage, R., Green, J., and Blake, A. J. (1991a). *Tetrahedron* **47,** 6353–6370.

Ramage, R., Blake, A. J., Florence, M. R., Gray, T., Raphy, G., and Roach, P. L. (1991b). *Tetrahedron* **47,** 8001–8024.

Rapp, W., and Nokihara, K. (1991). *In* "Peptides 1990" E. Giralt and D. Andreu, eds.), pp. 194–195. ESCOM, Leiden.

Rapp, W., Zhang, L., and Bayer, E. (1990). *In* "Solid Phase Synthesis" (R. Epton, ed.), pp. 205–210. SPCC, Birmingham, U. K.

Regan, L., and DeGrado, W. F. (1988). *Science* **241,** 976–978.

Reid, G. E., and Simpson, R. J. (1992). *Anal. Biochem.* **200,** 301–309.

Rich, D. H., and Gurwara, S. K. (1975). *J. Am. Chem. Soc.* **97,** 1575–1579.

Richardson, J. S., and Richardson, D. C. (1989). *Trends Biochem. Sci.* **14,** 304–309.

Rink, H. (1987). *Tetrahedron Lett.* **28,** 3787–3790.

Rizo, J., Albericio, F., Romero, G., García-Echeverría, C., Claret, J., Muller, C., Giralt, E., and Pedroso, E. (1988). *J. Org. Chem.* **53,** 5386–5389.

Royo, M., García-Echeverría, C., Giralt, E., Eritja, R., and Albericio, F. (1992). *Tetrahedron Lett.* **33,** 2391–2394.

Sakakibara, S. (1994). *In* "Ringberg Conference, Kreuth am Tegernsee, Germany, 1993" (E. Wünsch, ed.), in press.

Sakakibara, S., and Shimonishi, Y. (1965). *Bull Chem. Soc. Jpn.* **38,** 1412–1413.

Salmon, S. E., Lam, K. S., Lebl, M., Kandola, A., Khattri, P., Healy, S., Wade, S., Patek, M., Kocis, P., Krchnak, V., Thorpe, D., and Felder, S. (1994). *In* "Peptides" (R. S. Hodges and J. A. Smith, eds.), pp. 1001–1002. ESCOM, Leiden.

Sano, S., and Kurihara, M. (1969). *Hoppe-Seyler's Z. Physiol. Chem.* **350,** 1183–1187.

Sarin, V. K., Kent, S. B. H., and Merrifield, R. B. (1980). *J. Am. Chem. Soc.* **103,** 5463–5470.

Sarin, V. K., Kent, S. B. H., Mitchell, A. R., and Merrifield, R. B. (1984). *J. Am. Chem. Soc.* **106,** 7845–7850.

Sax, B., Dick, F., Tanner, R., and Gosteli, J. (1992). *Pept. Res.* **5,** 245–247.

Schlatter, J. M., Mazur, R. H., and Goodmonson, O. (1977). *Tetrahedron Lett.* **4,** 2851–2852.

Schmitt, J., and Mutter, M. (1983). *Biopolymers* **22,** 1849–1852.

Schneider, J., and Kent, S. B. H. (1988). *Cell (Cambridge, Mass.)* **54,** 363–368.

Schnölzer, M., and Kent, S. B. H. (1992). *Science* **256,** 221–225.

Schröder, E., and Lübke, K. (1966). "The Peptides," Vol. 2, Academic Press, New York.

Scott, R. P. W., Chan, K. K., Kucera, P., and Zolty, S. (1971). *J. Chromatogr. Sci.* **9,** 577–591.

Seebach, D. (1992). *Aldrichimica Acta* **25,** 59–66.

Seebach, D., Thaler, A., and Beck, A. K. (1989). *Helv. Chim. Acta* **72,** 857–867.

Segrest, J. P., Jackson, R. L., Morrisett, J. D., and Gotto, A. M., Jr. (1974). *FEBS Lett.* **38,** 247–253.

Seyer, R., Aumelas, A., Caraty, A., Rivaille, P., and Castro, B. (1990). *Int. J. Pept. Protein Res.* **35,** 465–472.

Sheehan, J. C., and Hess, G. P. (1955). *J. Am. Chem. Soc.* **77,** 1067–1068.

Sheppard, R. C. (1973). *In* "Peptides 1971" (H. Nesvadba, ed.), pp. 111–125. North-Holland Publ., Amsterdam.

Sheppard, R. C., and Williams, B. J. (1982). *Int. J. Pept. Protein Res.* **20,** 451–454.

Sherrington, D. C. (1990). *In* "Solid Phase Synthesis" (R. Epton, ed.), pp. 71–86. SPCC, Birmingham, U. K.

Shimonishi, Y., Sakakibara, S., and Akabori, S. (1962). *Bull Chem. Soc. Jpn.* **35,** 1966–1970.

Shroff, H. N., Carpino, L. A., Wenschuh, H., Mansour, E.-S., Triolo, S. A., Griffin, G. W., and Albericio, F. (1994). *In* "Peptides" (R. S. Hodges and J. A. Smith, eds.), pp. 121–123. ESCOM, Leiden.

Sieber, P. (1987). *Tetrahedron Lett.* **28,** 2107–2110.

Sieber, P., and Iselin, B. (1968). *Helv. Chim. Acta* **51,** 614–622.

Sieber, P., and Riniker B. (1991). *Tetrahedron Lett.*, 739–742.

Small, P. W., and Sherrington, D. C. (1989). *J. Chem Soc. Chem. Commun*, 1589–1591.

Smith, E. T., Tomich, J. M., Iwamoto, T., Richards, J. H., Mao, Y., and Feinberg, B. A. (1991). *Biochemistry* **30,** 11669–11676.

Sparrow, J. T. (1976). *J. Org. Chem.* **41,** 1350–1353.

Spencer, J. R., Antonenko, V. V., Delaet, N. G. J., and Goodman, M. (1992). *Int. J. Pept. Protein Res.* **40,** 282–293.

Stewart, J. M., and Klis, W. A. (1990). *In* "Solid Phase Synthesis" (R. Epton, ed.), pp. 1–9. SPCC, Birmingham, U. K.

Stewart, J. M., and Young, J. D. (1969). "Solid Phase Peptide Synthesis." Freeman, San Francisco.

Stewart, J. M., and Young, J. D. (1984). "Solid Phase Peptide Synthesis," 2nd Ed. Pierce Chemical Co., Rockford, Illinois.

Stüber, W., Knolle, J., and Breipohl, G. (1989). *Int. J. Pept. Protein Res.* **34,** 215–221.

Sueiras-Diaz, J., and Horton, J. (1992). *Tetrahedron Lett.* **33,** 2721–2724.

Tam, J. P. (1988). *Proc. Natl. Acad. Sci. U.S.A.* **85,** 5409–5413.

Tam, J. P., Kent, S. B. H., Wong, T.-W., and Merrifield, R. B. (1979a). *Science* **12,** 955–957.

Tam, J. P., Wong, T.-W., Riemen, M. W., Tjoeng, F. S., and Merrifield, R. B. (1979b). *Tetrahedron Lett.* **42,** 4033–4036.

Tam, J. P., Tjoeng, F. S., and Merrifield, R. B. (1980). *J. Am. Chem. Soc.* **102,** 6117–6127.

Tam, J. P., Heath, W. F., and Merrifield, R. B. (1982). *Tetrahedron Lett.* **23,** 2939–2942.

Tam, J. P., Heath, W. F., and Merrifield, R. B. (1983). *J. Am. Chem. Soc.* **105,** 6442–6455.

Tam, J. P., Heath, W. F., and Merrifield, R. B. (1986). *J. Am. Chem. Soc.* **108,** 5242–5251.

Thaler, A., Seebach, D., and Cardinaux, F. (1991). *Helv. Chim. Acta* **74,** 628–643.

Tomasselli, A. G., Bannow, C. A., Deibel, M. R., Jr., Hui, J. O., Zurcher-Neely, H. A., Reardon, I. M., Smith, C. W., and Heinrikson, R. L. (1992). *J. Biol. Chem.* **267,** 10232–10237.

Toniolo, C., Bonora, G. M., Salardi, S., and Mutter, M. (1979). *Marcromolecules* **12,** 620–625.

Tóth, G. K., and Penke, B. (1991). *In* "Peptides 1990" (E. Giralt and D. Andreu, eds.), pp. 125–126. ESCOM, Leiden.

Tregear, G. W. (1972). *In* "Chemistry and Biology of Peptides" (J. Meienhofer, ed.), pp. 175–178. Ann Arbor Sci. Publ., Ann Arbor, Michigan.

Trzeciak, A., Vorherr, T., and Bannwarth, W. (1993). *In* "Peptides 1992" (C. H. Schneider and A. N. Eberle, eds.), pp. 342–344. ESCOM, Leiden.

Tuchscherer, G., Servis, C., Corradin, G., Blum, U., Rivier, J., and Mutter, M. (1992). *Protein Sci.* **1,** 1377–1386.

Tung, R. D., and Rich, D. H. (1985). *J. Am. Chem. Soc.* **107,** 4342–4343.

Ueda, M., and Mori, H. (1992). *Bull Chem. Soc. Jpn.* **65,** 1636–1641.

Ueki, M., Kai, K., Amemiya, M., Horino, H., and Oyamada, H. (1988). *J. Chem. Soc., Chem. Commun.*, 414–415.

Unson, C. G., Erickson, B. W., Richardson, D. C., and Richardson, J. S. (1984). *Fed. Proc.* **43,** 1937.

Urry, D. W. (1988). *J. Prakt. Chem.* **7,** 1–34.

Vágner, J., Kocna, P., and Krhňák, V. (1991). *Pept. Res.* **4,** 284–288.

Valembois, C., Mendre, C., Cavadore, J. C., and Calas, B. (1992). *Tetrahedron Lett.* **33,** 4005–4008.

Van Rietschoten, J., Tregear, G. W., Leeman, S., Powell, D., Niall, H., and Potts, J. T. (1975). *In* "Peptides 1974" (Y. Wolman, ed.), pp. 113–116. Wiley, New York, and Universities Press, Jerusalem.

van Vliet, A., Smulders, R. H. P. H., Rietman, B. H., Klink, A.-M. E., Rijkers, D. T. S., Eggen, I. F., van de Werken, G., and Tesser, G. I. (1993). *In* "Peptides 1992" (C. H. Schneider and A. N. Eberle, eds.), pp. 279–280. ESCOM, Leiden.

van Vliet, A., Rietman, B. H., Karkdijk, S. C. F., Adams, P. I. H. M., and Tesser, G. I. (1994). *In* "Peptides" (R. S. Hodges and J. A. Smith, eds.), pp. 151–152, ESCOM, Leiden.

Vlasov, G. P., and Bilibin, A. Y. (1969). *Izv. Akad. Nauk SSSR, Ser. Khim.*, 1400.

von dem Bruch, K., and Kunz, H. (1990). *Angew. Chem., Ed. Engl.* **29,** 1457–1459.

Wade, J. D., Bedford, J., Sheppard, R. C., and Tregear, G. W. (1991). *Pept. Res.* **4,** 194–199.

Waki, M., Mitsuyasu, N., Terada, S., Matsuura, S., Kato, T., and Izumiya, N. (1974). *Biochem. Biophys. Res. Commun.* **61,** 526.

Wang, S. S. (1973). *J. Am. Chem. Soc.* **95,** 1327–1333.

Wang, S. S. (1975). *J. Org. Chem.* **40,** 1235–1239.

Wang, S. S. (1976). *J. Org. Chem.* **41,** 3258–3261.

Wang, S. S., and Merrifield, R. B. (1969). *J. Am. Chem. Soc.* **91,** 6488–6491.

Wang, S. S., and Merrifield, R. B. (1971). *In* "Peptides 1969" (E. Scoffone, ed.), pp. 74–83. North-Holland Publ., Amsterdam.

Wang, S. S., Matsueda, R., and Matsueda, G. R. (1982). *In* "Peptide Chemistry 1981" (T. Shioiri, ed.), pp. 37–40. Protein Research Foundation, Osaka, Japan.

Wen, J. J., and Spatola, A. F. (1994). *In* "Peptides" (R. S. Hodges and J. A. Smith, eds.), pp. 153–155. ESCOM, Leiden.

Wenschuh, H., Beyermann, M., Krause, E., Carpino, L. A., and Bienert, M. (1994). *In* "Peptides" (R. S. Hodges and J. A. Smith, eds.), pp. 130–132. ESCOM, Leiden.

Westall, F. C., and Robinson, A. B. (1970). *J. Org. Chem.* **35,** 2842–2844.

Weygand, F., and Ragnarsson, U. (1966). *Z. Naturforsch. B.* **21,** 1141–1144.

Weygand, F., Steglich, W., and Bjarnason, J. (1968). *Chem. Ber.* **101,** 3642–3648.

Whitney, D. B., Tam, J. P., and Merrifield, R. B. (1984). *Tetrahedron* **40,** 4237–4244.

Wieland, T., Flor, F., and Birr, C. (1973). *Justus Liebigs Ann. Chem.*, 1595–1600.

Wildi, D. S., and Johnson, J. H. (1968). *155th Natl. Meet. Am. Chem. Soc.*, No. A008.

Windridge, G. C., and Jorgensen, E. C. (1971). *J. Am. Chem. Soc.* **93,** 6318–6319.

Wissmann, H., Siedel, W., and Geiger, R. (1969). *Chem. Abstr.* **72,** 55891.

Wu, C.-R., Stevens, V. C., Tregear, G. W., and Wade, J. D. (1989). *J. Chem. Soc., Perkin Trans. 1*, 81–87.

Yajima, H., Fujii, N., Ogawa, H., and Kawatani, H. (1974). *J. Chem. Soc., Chem. Commun.*, 107–108.

Yajima, H., Takeyama, M., Kanaki, J., and Mitani, K. (1978). *J. Chem. Soc., Chem. Commun.*, 482–483.

Yajima, H., Minamitake, Y., Funakoshi, S., Katajima, I., Fujii, N., Segawa, T., Nakata, Y., Yasuhara, T., and Natajima, T. (1982). *Chem. Pharm. Bull.* **30,** 344–348.

Yamashiro, D., and Li, C. H. (1973). *J. Org. Chem.* **38,** 591–592.

Yamashiro, D., and Li, C. H. (1974). *Proc. Natl. Acad. Sci. U.S.A.* **71,** 4945–4949.

Yu, H.-M., Chen, S. T., and Wang, K.-T. (1992). *J. Org. Chem.* **57,** 4781–4784.

Zalipsky, S., Albericio, F., and Barany, G. (1985). *In* "Peptides Structure and Function" (C. M. Deber, V. J. Hruby, and K. D. Kopple, eds.), pp. 257–260. Pierce Chemical Co., Rockford, Illinois.

Zervas, L., Borovas, D., and Gazis, E. (1963). *J. Am. Chem. Soc.* **85,** 3660–3666.

4

α-Helix Formation by Peptides in Water

J. Martin Scholtz
Department of Medical Biochemistry and Genetics
Texas A&M University
College Station, Texas 77843-1114

Robert L. Baldwin
Department of Biochemistry
Stanford University School of Medicine
Stanford, California 94305-5307

I. Introduction

The systematic study of helix formation by peptides of defined length and sequence began in the early 1980s (for review, see Scholtz and Baldwin, 1992). Most of the advances in the field are a result of the increased availability of solid-phase peptide synthesis methods. The initial studies in the field were directed toward an understanding of helix formation in water by the 13-residue C-peptide from the N terminus of ribonuclease A (Brown and Klee, 1971). The C-peptide, and other model systems derived from proteins, provided excellent systems for investigating the qualitative roles of side-chain interactions as well

Peptides: Synthesis, Structures, and Applications

as the effects of specific amino acid substitutions on helix formation. The C-peptide system, however, proved too complex as a host peptide with which to study all the specific elements of helix formation in peptides.

The major advance in the peptide helix field came with the design of short peptide sequences that show good α-helix formation in water, first by using ion-pair interactions to stabilize the helix (Marqusee and Baldwin, 1987) and later by making use of the unexpectedly strong helix-forming tendency of alanine (Marqusee *et al.*, 1989). These studies brought into question the prediction from the original host–guest studies (Wójcik *et al.*, 1990) that short peptides (<20 residues) should not be able to form α helices in water. With the availability of several *de novo* designed peptide systems, including the alanine-based peptides as well as peptides designed with overlapping salt bridges similar to those found in troponin C (Lyu *et al.*, 1989), there has been a renewed interest in measuring the intrinsic helix-forming tendencies, or helix propensities, of all of the amino acids, and also in analyzing the side-chain interactions that can help to stabilize α helices in aqueous solution.

There are several uses of peptide helix studies. Peptide fragments from proteins often exhibit specific binding to receptors, and, if the fragment comes from a helical segment of the protein, then it is of interest to enhance helix formation by the isolated peptide. Factors that stabilize proteins often can be studied advantageously in peptide helices, where a given side-chain interaction can be isolated from other interactions present in the protein. Peptide helices in water typically have marginal stability, so that a small change in free energy resulting from making a specific side-chain interaction gives an easily measurable change in helix content. Currently there is wide interest in examining the factors that control helix propensities, and in relating the helix propensities found in peptide studies to results found by directed mutagenesis of proteins. Even the N-cap interactions found in studies of protein helices can be observed and analyzed in peptide helices. A main focus of this chapter is on using peptide helices to analyze factors that determine protein stability: helix propensities and specific interactions between side chains.

II. Methods of Studying Helix Formation

To determine the amount of helical structure in a peptide we must use a spectroscopic technique that provides the necessary resolution and discrimination. Unfortunately, many optical techniques such as ultraviolet and fluorescence spectroscopy, which are popular techniques for the study of other biopolymers, are not suitable methods for monitoring helix formation in most cases. This is because the transition from helix to coil does not produce a change in the optical properties measured by either fluorescence or ultraviolet spectroscopy. Fortunately, there is a convenient optical technique that can be used to monitor the formation of helical structure, namely, circular dichroism spectroscopy (CD).

The CD spectrum of a peptide is very sensitive to the secondary structure. A complete description of the CD behavior of a peptide is beyond the scope of this chapter; interested readers are referred to a number of excellent review articles (Johnson, 1990; Woody, 1985) for a complete description. The book by Cantor and Schimmel (1980) provides a general introduction. Figure 1 shows CD spectra of a 50-residue peptide recorded at several different temperatures where the amount of helical structure changes with temperature. At low temperatures, the spectra show the characteristic features of a helical conformation, namely, minima at 222 and 208 nm and a maximum at approximately 192 nm. At higher temperatures, these strong signals are diminished, and an isodichroic point for the transition is observed at about 202 nm. The amount of helical structure in a peptide is generally determined from the magnitude of the CD signal at 222 nm. Although there is still some controversy regarding the values of $[\theta]_{222}$ for 0 and 100% helix as a function of both chain length and temperature, it is nonetheless possible to use the observed CD signal at 222 nm to measure the amount of helical structure present in the peptide (see Scholtz *et al.*, 1991a).

A typical thermally induced helix to coil transition, as monitored by CD at 222 nm, is shown in Fig. 2. The transition is very broad, and the helix and coil baselines are not well determined, even for this very long peptide. Circular

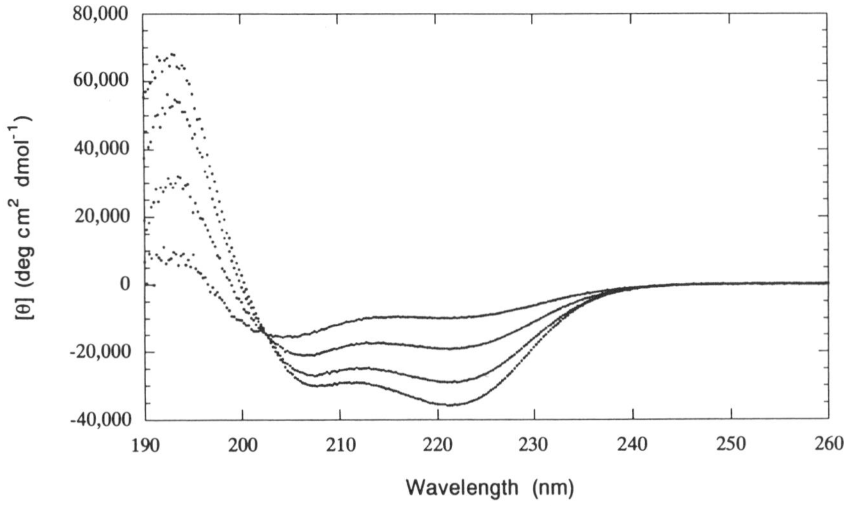

Figure 1 Helix unfolding monitored by CD spectra for a 50-residue peptide. Spectra of AcY-(AEAAKA)$_8$-F(NH$_2$) were recorded at various temperatures from 0°C (bottom curve at 222 nm) to 60°C (top curve at 222 nm) every 20°C. The peptide concentration was 8.5 μ*M* in 1 m*M* potassium phosphate and 100 m*M* potassium fluoride at pH 7.00. The spectra were recorded on an Aviv 60DS spectropolarimeter in a 1.0-cm path length cuvette. Data are from Scholtz *et al.* (1991b).

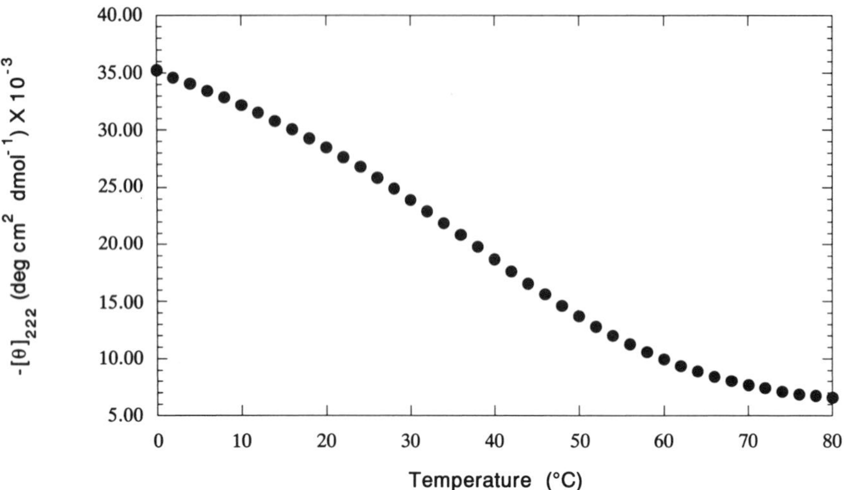

Figure 2 Thermal unfolding curve of a 50-residue helical peptide [AcY-$(AEAAKA)_8$-$F(NH_2)$)], monitored by the change in ellipticity at 222 nm. The conditions are the same as described in Fig. 1. Data are from Scholtz *et al.* (1991a,b).

dichroism spectroscopy can also be used to explore other properties of the transition, including the reversibility and the concentration dependence. An analysis similar to that shown in Fig. 2 can be used to illustrate both aspects of structure formation in this peptide. The transition is reversible and independent of the total peptide concentration over the entire temperature range (data not shown). The latter fact is used as evidence that structure formation is caused only by intramolecular interactions and not by intermolecular associations. Further evidence on the molecularity of structure formation can be achieved by molecular weight determinations using ultracentrifugation techniques (Padmanabhan *et al.*, 1990).

Circular dichroism spectroscopy is a very powerful technique for monitoring structure formation in α-helical peptides. The chief advantages of CD over the other techniques discussed below are its ease and simplicity, the need for only small amounts of sample (2 ml of 0.2 m*M* total amide bond concentration is sufficient), and its excellent sensitivity. There are, however, a few disadvantages to CD. First of all, it is a macroscopic technique; that is, it gives only a measure of the average amount of structure formed by the sample (see below). Second, the calibration of the technique, in terms of $[\theta]_{222}$ values for 0 and 100% helix, must be determined empirically. Also, certain aromatic residues and disulfide bonds may contribute to the CD signal at 222 nm, making the quantification of the amount of helical structure problematic in some cases (Chakrabartty *et al.*, 1993b).

In addition to CD, there are several other techniques that can be used to analyze helical structure in a peptide. Chief among these is nuclear magnetic resonance (NMR) spectroscopy. NMR is a very powerful technique that can look at individual magnetic nuclei (^{1}H, ^{13}C, ^{15}N, etc.) and, through the use of a number of different techniques (Wüthrich, 1986), determine distances and bond angles which can lead to detailed structural information. NMR has been utilized to determine the structure of a variety of helical peptides (Bradley *et al.*, 1990; Liff *et al.*, 1991; Osterhout *et al.*, 1989). The direct determination of the structure of a peptide using NMR can be very labor intensive, but it does provide the only high-resolution structural tool that is available, as crystallographic methods have failed because of the inability to grow good crystals for analysis.

Nuclear magnetic resonance spectroscopy has been used, in combination with hydrogen–deuterium exchange techniques, to probe the amide hydrogen bonding pattern in helical peptides. This technique can be used to infer the structure of a peptide through the kinetics of exchange of the backbone amide protons with D_2O in the bulk solvent. If the resonances for all the amide protons in the NMR spectrum have been assigned, it is possible to quantify the amount of structure present at the level of each individual amide. It is possible to use NMR to quantify the "bulk" hydrogen exchange rate by integrating the area under the total amide region as a function of exchange-out time. The latter technique has been employed by Rohl *et al.* (1992) and is illustrated in Fig. 3, where the total occupancy of the amide proton region is shown as a function of exchange-out time after dissolving each peptide in D_2O. The peptides used in this study range in length from 6 (fastest exchanging peptide) to 51 residues (slowest exchanging peptide).

In addition to CD and NMR, there are a few other techniques that are being used to study α-helix formation in peptides. Miick *et al.* (1992) have used Fourier transform infrared (FTIR) spectroscopy to try to distinguish α-helical structure from 3_{10} helix and have also employed electron spin resonance (ESR) to look at structure formation and dynamics of a peptide modified with an appropriate spin label (Miick *et al.*, 1993). These techniques are complementary to the traditional CD and NMR methods and, in the case of ESR, can give a picture of the dynamics of helical structure in solution.

III. Models for Helix–Coil Transition

In the 1950s and early 1960s several models for the helix to coil transition were developed. These models are based on a statistical mechanical treatment of the transition that allows for a description of the transition in terms of populations of molecules. This situation is different from most reactions in biophysics in that the reaction is not two-state when the entire peptide chain is considered. The transition is not from a completely helical peptide to the random coil but, rather, from a population of helical molecules, with strongly frayed ends, to the random coil conformation (see Fig. 4). Zimm and Bragg (1959) and Lifson and Roig (1961) gave related theoretical analyses of the transition, based on different criteria for whether a residue is in the helix or random

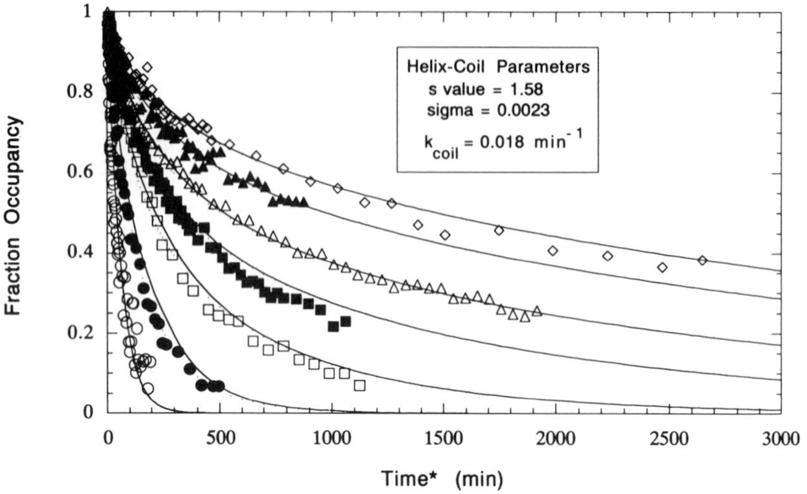

Figure 3 Effect of chain length on the hydrogen exchange kinetics of a series of peptides of the form Ac-(AAKAA)$_m$-Y(NH_2). The total length of each peptide is (○) 6, (●) 16, (□) 21, (■) 26, (Δ) 31, (▲) 36, and (◊) 50. For experimental details and the generation of the theoretical curves, see Rohl *et al.* (1992). Reprinted with permission from Rohl *et al.*, 1992. Copyright 1992 American Chemical Society.

coil state. A good introduction is given by Cantor and Schimmel (1980), and a complete discussion of the theories can be found in the book by Poland and Scheraga (1970). A recent article compares the Zimm-Bragg and Lifson–Roig models (Qian and Schellman, 1992).

The basic forms of the two models are similar. Helix formation is described as a function of three parameters: an initiation parameter, a propagation parameter, and the chain length. Most workers have used the Zimm–Bragg formalism, in which the nucleation parameter is σ,the propagation parameter is s, and the chain length is n. The nucleation parameter, σ, is treated as an entropic penalty that must be overcome in ordering four residues before a hydrogen bond can be formed that stabilizes the helical conformation. Helix propagation, as measured by the s value, is simply an equilibrium constant for the addition of a residue onto an existing stretch of helical residues. The intrinsic helix forming tendency, or helix propensity, of a residue is identified with its s value; residues with s values greater than 1 are good helix formers, whereas residues with s values less than 1 prefer a nonhelical conformation. Because s can be treated as an equilibrium constant, the enthalpy of helix formation is the variation of s with temperature described by the van't Hoff relationship:

$$s = s_0 \times \exp\left[\frac{\Delta H^{\circ}}{} \left(\frac{1}{} - \frac{1}{}\right)\right] \tag{1}$$

In this expression ΔH° is assumed to be independent of temperature. There are advantages to using the Lifson–Roig formalism for peptides that

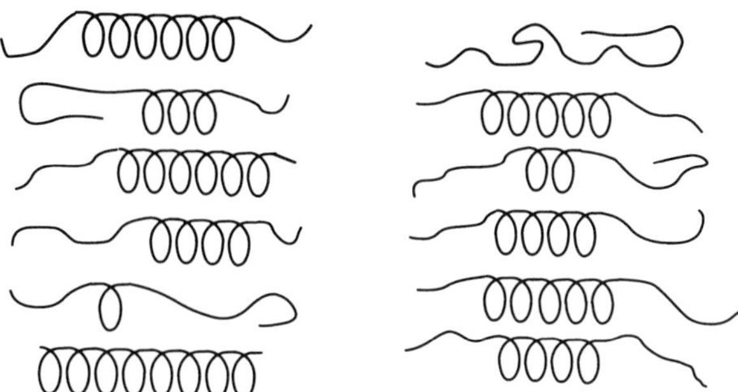

Figure 4 Diagram of frayed ends in partly helical peptide molecules. When the overall helix content of a peptide is measured to be 50% by circular dichroism, the system contains a distribution of molecules like those shown here.

contain two or more different amino acid residues (Qian and Schellman, 1992). Relations for computing the Zimm–Bragg parameters from measured values of the Lifson–Roig parameters have been given by Qian and Schellman (1992).

With the advent of *de novo* designed short peptides that show substantial helix formation in water (Marqusee and Baldwin, 1987; Marqusee *et al.*, 1989), a renewed interest in helix formation as a model for protein structure has emerged. To provide a working framework for the characterization of the stability of these short α-helical peptides, the classic theories for the helix–coil transition must be tested. It was unclear initially if the theories, in their original forms, would adequately describe the transitions of short peptides in water. To test the theories, two key approaches were taken: a series of peptides of identical composition but different chain lengths were used to check the validity of the helix–coil models, and a single 50-residue peptide was subjected to differential scanning calorimetry (DSC) in order to determine the enthalpy change (ΔH°) for helix formation and to compare the calorimetric, or true, enthalpy change with that determined using the helix-coil models for the transition.

The DSC analysis (Scholtz *et al.*, 1991b) of the 50-residue AEK peptide (see Figs. 1 and 2) revealed the expected broad transition. The breadth of the thermal transition precluded estimating ΔC_p and, therefore, the data were analyzed assuming $\Delta C_p = 0$. With this assumption, it was possible to determine ΔH_{cal}, the true enthalpy change for helix formation, and ΔH_{vH}, the enthalpy change based on a two-state (all-or-none) treatment of the transition. These two values were quite different, as expected: $\Delta H_{cal} \approx -50$–$65$ kcal/mol and $\Delta H_{vH} \approx -11$ kcal/mol, indicating that the helix to coil transition is not two-state at the level of the entire peptide chain. The large enthalpy change for helix formation (about −1 kcal/mol per residue) favors helix formation and explains why all *de novo* designed helical peptides show thermal unfolding.

The investigation of the length dependence of helix formation (Scholtz *et al.*, 1991a) for the AEK peptides provided a test of helix–coil transition theories and demonstrated their applicability to helix formation in short peptides. In addition, a direct measure of σ, the helix nucleation parameter, was obtained ($\sigma = 0.003$). This value of σ was later confirmed using a second set of peptides and a different probe for helix formation (Rohl *et al.*, 1992) as seen in Fig. 3. The key proof of the applicability of the theory was obtained by using helix–coil theory to determine ΔH^o, –0.8 to –0.9 kcal/mol per residue. The agreement between the true calorimetric ΔH^o_{cal} and the ΔH^o determined using helix–coil theory indicates that the theory provides a good model for the effect of temperature on the helix to coil transition.

Because the basic models used in helix–coil theory are unable to take specific side-chain interactions into account, several modifications of the theories have been developed in order to achieve more realistic accounting of all the interactions that are known to affect helix formation. For example, the energetics of electrostatic interactions between charged side chains can be assessed, and the interaction of a charged residue with the helix macrodipole can be demonstrated (see below). These models for the helix-coil transition provide the framework for a more complete description of helix formation by peptides in water.

IV. Helix Propensities

The first systematic study of α-helix formation aimed at measuring the intrinsic helix forming tendencies of the amino acids was initiated by the Scheraga group in the early 1970s. These host-guest studies utilized each of the 20 natural amino acids as guests in a random copolymer with hydroxybutyl- or hydroxypropy-L-glutamine as the host residue. By assuming a truly random distribution of guest residues in the copolymers, the Scheraga group was able to obtain estimates of the intrinsic helix forming tendencies of all 20 amino acids (Wójcik *et al.*, 1990). The surprising result was that all the *s* values, with the exception of those for Gly and Pro, were very similar and near unity. The implication was that no peptide less than 20 residues could be expected to form a helix in water without the use of stabilizing side-chain interactions (Shoemaker *et al.*, 1985). Evidence has called these *s* values into question (see review by Chakrabartty and Baldwin, 1993), and a suggestion for the possible discrepancy has been put forward (Marqusee *et al.*, 1989), based on the indication that the host residue forms a helix in water that is stabilized by hydrophobic interactions among its side chains.

The first *de novo* designed peptide was described by Marqusee and Baldwin (1987). It contains three pairs of Glu and Lys residues in an otherwise Ala-based peptide (AEK host). The Glu and Lys residues form ion pairs and provide water solubility to the otherwise insoluble Ala-based peptide. Later it was found that single Lys residues in an otherwise all-Ala-based peptide (AK host) could provide the necessary water solubility (Marqusee *et al.*, 1989), and the alanine residues themselves are sufficient to provide strong helix formation.

The Kallenbach group has taken a different approach to the design of the host peptide (Lyu *et al.*, 1989). Instead of using the high helix forming tendency of Ala to provide helical structure, a set of overlapping ion pairs between Glu and Lys were used (E_4K_4 host). The Stellwagen group has used a modified version of the original AEK host peptide described by Marqusee and Baldwin (1987) in determining *s*-values (Park *et al.*, 1993a,b). A water-soluble Ala-based peptide has been described that does not contain any charged residue (Scholtz *et al.*, 1991c). This Ala-Gln host has been used to determine the *s* values for some of the charged amino acids. A summary of the host peptides, along with their sequences, is shown in Table I.

Helix propensities are a determinant, and often the major determinant, of stability in peptide helices. This conclusion can be seen directly by examining the effect of substituting a single amino acid in a reference peptide (Table II). The helix content of a 17-residue reference peptide drops from 70 to 25% when a single glycine residue is substituted for alanine, and there are also large changes in helix content when other amino acids are substituted. Consequently, the intrinsic helix forming tendency, or helix propensity, of even a single amino acid in a helix can have a large effect on helix stability. The probability of helix formation by a sequence of amino acids is proportional to the product of the *s* values of the amino acids in the sequences, excluding the end residues. This means that when the *s* values of two amino acids differ by a factor of 2, the probability of helix formation by a 12-residue sequence of each amino acid differs by a factor of 2^{10}, or approximately 1000.

In studies of peptide helices, the helix propensity of amino acid X is determined by substituting X for another amino acid, usually alanine, in a reference peptide. When a single substitution A → X is analyzed in a peptide that contains chiefly alanine, the helix content of the reference peptide can be represented adequately by the homopolymer approximation, by assigning single average values of *s* and σ to the different amino acids in the reference peptide. When the homopolymer approximation is used, the *s* value of X can be com-

Table I Reference Peptides for Studies of Helix Propensities

Host name	Peptide sequence	Host design[a]	*s* value[a]
AEK	Ac-Y-EAAAK-EAXAK-EAAAK-A(NH_2)	1	5
AK	Ac-Y-KAAXA-KAAXA-KAAXA-K(NH_2)	2	6
E_4K_4	Ac-YS-EEEE-KKKK-XXX-EEEE-KKKK(NH_2)	3	7
AQ	Ac-AAQAA-AAQAA-AAQAA-Y(NH_2)	4	8

[a]Key to references: 1, Marqusee and Baldwin (1987); 2, Marqusee *et al.* (1989); 3, Lyu *et al.* (1989); 4, Scholtz *et al.* (1991c); 5, Park *et al.* (1993a,b); 6, Chakrabartty *et al.* (1991), Padmanabhan *et al.* (1990) and Chakrabartty *et al.* (1994); 7, Lyu *et al.* (1990); 8, Armstrong and Baldwin (1993), Huyghues-Despointes *et al.* (1993), and Scholtz *et al.* (1993).

Table II Change in Helix Content Produced by Substitution of a Single Amino Acid

Amino Acid[a]	$-[\theta]_{222}$[b]
Ala	24,300
Leu	22,100
Met	21,100
Gln	20,400
Ile	18,000
Ser	16,000
Asn	13,800
Thr	13,700
Val	13,700
Gly	8,700

[a]The 17-residue reference peptide is AcY-$(EAAAK)_3$-$A(NH_2)$ and the data are from Stellwagen *et al.* (1992). The substitution Ala → X is made at residue 9, the central residue.
[b]The mean residue ellipticity is given in units of deg cm^2 $dmol^{-1}$. The conditions are 10 m*M* KCl, pH 7.0, 0°C. The most helical peptide (Ala) shows approximately 70% helix, and the least helical (Gly) shows approximately 25% helix.

puted straightforwardly from the change in helix content caused by the A → X substitution.

The position of the A → X substitution has a major effect on the change in helix content if X is a strong helix breaking residue. This point is illustrated in Fig. 5 for X being glycine. A series of 17-residue peptides was synthesized in which the position of a single A → G substitution was varied across the length of the peptide.

The reference peptide (AK in Table I) was chosen for three reasons: (1) its helix content is affected only slightly (in 1 *M* NaCl) by side-chain interactions, (2) it has a high helix content as well as good solubility in water, and (3) the neighboring residues at each site of the A → G substitution are chiefly alanine, which interacts minimally with other amino acid side chains. The measurements of helix content are made in 1 *M* NaCl to damp the electrostatic interactions of the 4 Lys^+ residues (needed for water solubility) with one another and with the helix dipole.

Figure 5 shows that the effect of the A → G substitution is largest when it is made at the center of the peptide, and the effect is relatively small close to either end. This position effect may be considered as a consequence of fraying

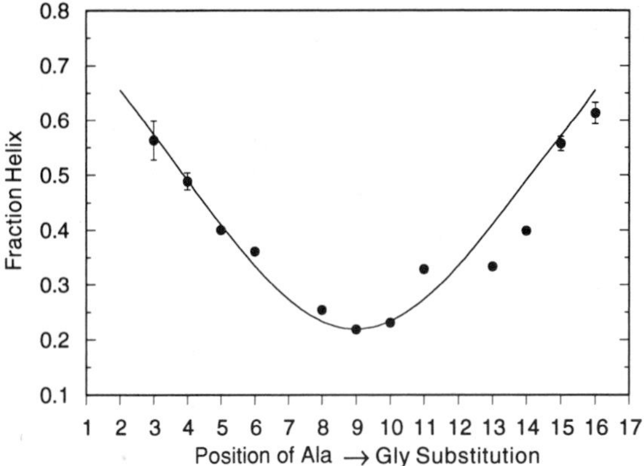

Figure 5 Helix content versus position of a single glycine residue in the reference peptide AcY-$(KAAAA)_3$-$K(NH_2)$. Reprinted with permission from *Nature* (Chakrabartty *et al.*, 1991). Copyright 1991 Macmillan Magazines Limited.

at the helix ends: the A → G substitution causes a small reduction in the helix content near the helix ends because the fraction helix, per residue, is small near the ends. Figure 5 illustrates why the two-state approximately (helix–coil) cannot be used to determine helix propensities. If it were used, the value found for the helix propensity would depend strongly on the position of the substitution. The Lifson–Roig theory, on the other hand, gives the same value of the helix propensity, almost within error, regardless of where the substitution is made: this behavior is indicated by the solid line in Fig. 5.

Helix propensities (s values) measured with different reference peptides of defined sequence show agreement as regards rank order, but the numerical values do not yet agree satisfactorily. This situation is illustrated in Fig. 6 by plotting s values measured with peptides AK and AQ versus the ones found with E_4K_4 (Fig. 6A) or EAK (Fig. 6B). These reference peptides differ notably in their content of potential ion-pair interactions: peptide E_4K_4 has eight pairs of oppositely charged residues, EAK has three pairs, while AK and AQ have none. The s values found with these different references peptides are linearly related to one another, as shown in Fig. 6. The line in Fig. 6B has an intercept, close to 0,0 and a slope near 1, indicating that the s values found with peptide EAK are close to those found with AK and AQ. The s values found for ionizing amino acids, particularly the ones with longer side chains (Arg, Lys, Glu) do not show good agreement, however, between EAK and AK and AQ.

There are some possible explanations for the numerical differences between these values of s. The free energy term assigned to the ion-pair interactions in E_4K_4 affects the s values obtained with this peptide, including the intercept of the line shown in Fig. 6A (Chakrabartty and Baldwin, 1993). Ion-pair and other charge interactions in peptide EAK may affect the s values of ionizing amino acids found with this peptide, because side-chain interactions

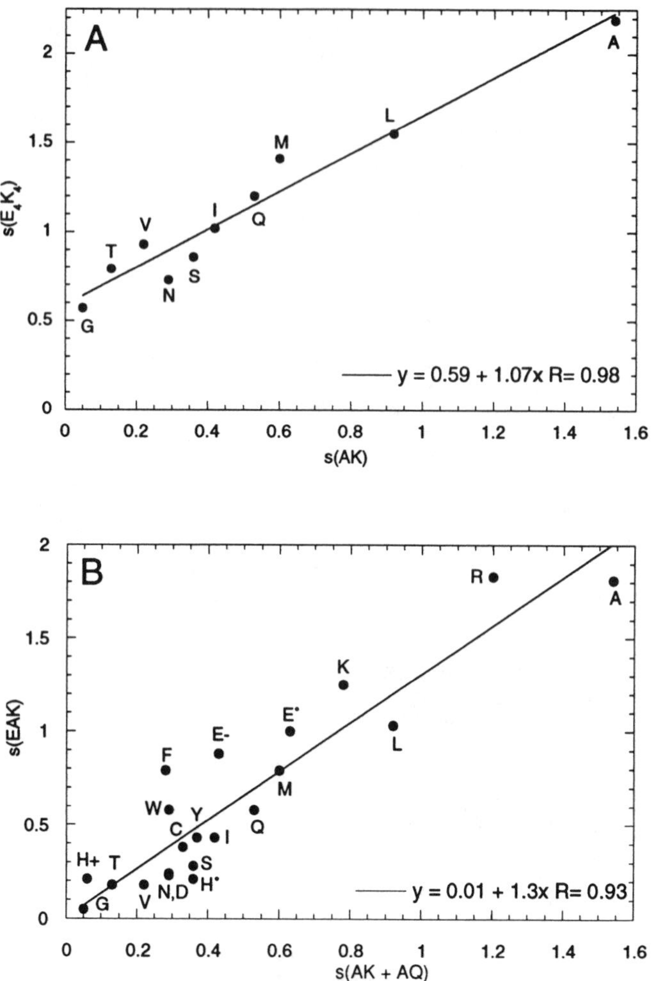

Figure 6 Comparison of the *s* values measured with peptide AK versus E_4K_4(A) and (AK + AQ) versus EAK (B). From Chakrabartty *et al.* (1994). Copyright 1994 Cambridge University Press. Reprinted with the permission of Cambridge University Press.

contribute to the apparent helix propensity. A different problem appears with the aromatic amino acids: their side chains contribute to the CD spectrum at 222 nm and thus affect the measurement of helix content; this side-chain contribution is itself affected, in a major way, by the conformation (Chakrabartty *et al.*, 1993b). Experiments with an amphiphilic helix (Zhou *et al.*, 1993) indicate that neighboring residue, or context-dependent, effects may also affect the measured *s* values, because different changes in helix content were found when the same substitution (A $\rightarrow$ X) was made on either the hydrophobic or the hydrophilic face of the helix.

In agreement with the proposal by Blout (1962), based on studies of polyamino acids in nonpolar solvents, modern studies of peptide helices in water indicate that nonpolar β-branched and aromatic residues are helix-breaking relative to other nonpolar amino acids. This effect was discovered in 1990 by several workers, all of whom attributed it to restriction of side-chain rotamers in the helical conformation, resulting in a loss of side-chain conformational entropy. This conclusion was tested further by Padmanabhan and Baldwin (1991) and by Lyu *et al.* (1991), who tested straight-chain amino acids not found in proteins; they proved to be relatively good helix-formers. Creamer and Rose (1992) used Monte Carlo simulations to compute the entropy loss from restricted side-chain conformations in a helix and obtained a good correlation with the helix propensities of naturally occurring amino acids.

The correlation between helix propensity and side-chain entropy loss provides a simple explanation for the unusually high helix propensity of alanine, which does not suffer any loss in side-chain entropy on forming a helix. Calorimetric experiments, discussed above, show that helix formation in water is enthalpy-driven, and the enthalpy change is primarily associated with forming the helix backbone, not with specific properties of the side chain. Again, this picture fits the fact that alanine has the highest propensity although it has the smallest side chain, excluding glycine. The very low helix propensity of glycine is attributed chiefly to its highly flexible backbone when glycine is in peptide linkage. Interaction of the C^{β} atom with the helix backbone is though to be helix stabilizing (see Gō *et al.*, 1971), and this effect may also contribute to the large difference in helix propensity between alanine and glycine.

The question of whether the hydrophobic interaction contributes substantially to helix propensities for isolated helices in water has not yet been settled. The fact that alanine has a substantially higher helix propensity than other nonpolar amino acids suggests that any helix-stabilizing effect of the hydrophobic interaction does not extend much beyond C^{β}. Richards and Richmond (1978) investigated the question by computing for the various amino acids the amount of nonpolar surface buried on helix formation. They found that the effect does not vary greatly among the different amino acids. The question has been considered further by Blaber *et al.* (1993, 1994), who compared helix propensities measured in protein helices with peptide helix studies. They concluded that the hydrophobic interaction can be an important factor in determining helix propensities.

V. Side-Chain Interactions

The amount of helical structure in a peptide is governed not only by the intrinsic helix forming tendencies of the residues in the peptide, but also by specific and nonspecific interactions between side-chain residues and between the side chains and the helix backbone. Many of these side-chain interactions were observed in the early work on the S- and C-peptides from RNase A, and a partial description of their properties was given. For example, the pH dependence of helix formation in C-peptide indicated that the protonation states of Glu-2 and His-12 were important determinants of the observed helicity (for review, see Scholtz and Baldwin, 1992). Later, substitution experiments were used to determine the precise roles of these residues in helix formation in the C-peptide system (Shoemaker *et al.*, 1985). Although the C- and S-peptide systems were excellent vehicles for determining the qualitative role of side-chain interactions in helix formation, they were found to be too complicated to use as models for quantifying the energetics of side-chain interactions.

Using helical peptides of *de novo* design that contain Ala and three pairs of Glu and Lys residues, Marqusee and Baldwin (1987) studied the effects of spacing, (i, i + 3) and (i, i + 4), and orientation, Glu-Lys versus Lys-Glu, as a function of both pH and NaCl concentration. The results indicated that (i, i + 4) ion pairs were helix stabilizing for both orientations of the charged residues, but that (i, i + 3) ion pairs were considerably less effective. Furthermore, the persistence of substantial helix formation at low pH suggested a $Glu^0 \cdots Lys^+$ singly charged hydrogen bond interaction.

Merutka and Stellwagen (1991) later used a related alanine-based peptide to study the differences in side-chain interactions for all combinations of Glu and Asp with Lys, Orn, and Arg in multiple ion-pair peptides. They found helix stabilization by the (i, i + 4) arrangement of all these ion pairs; the differences in helix content between the pairs were attributed chiefly to the different helix propensities (s values) of the charged residues. Gans *et al.* (1991) investigated the energetics of ion-pair interactions in their peptide containing multiple Glu and Lys residues by comparing the helicities of two peptides with identical compositions but different spacings of the Glu and Lys residues:

Suc-Y-S-E-E-E-E-K-K-K-K-E-E-E-E-K-K-K-K(NH_2) E_4K_4
Suc-Y-S-E-E-K-K-E-E-K-K-E-E-K-K-E-E-K-K(NH_2) E_2K_2

Using a modification of a helix–coil transition model, they were able to relate the differences in helix content between E_2K_2 and E_4K_4 to the ability of the latter to form helix-stabilizing ion pairs. They calculated the energetics of ion-pair formation to be 0.5 kcal/mol per ion pair. The helical structure of E_4K_4 results entirely from ion-pair formation as E_2K_2 shows no measurable helix formation in water at neutral pH.

The role of electrostatic interactions between charged side chains in stabilizing helical peptides has also been studied by several other investigators (Armstrong and Baldwin, 1993; Huyghues-Despointes *et al.*, 1993; Scholtz *et al.*, 1993). The use of a neutral peptide comprising primarily Ala and Gln as a host peptide (Scholtz *et al.*, 1991c) allowed the investigation of the effects of a charged residue on the stability of an otherwise neutral peptide and gave a quantitative estimate of the energetics of the electrostatic interaction of a charged residue with the helix macrodipole. To quantify the effects of side-chain interactions, several modifications of standard helix–coil theories have been developed, since the classic theories for helix formation in peptides only consider the helix propensities (s values) and do not take any side-chain interactions into account. Both these modifications are described in an article by Scholtz *et al.* (1993). By comparing a series of peptides with a single charged residue substituted at various positions along the chain, it is possible to evaluate the charge–helix macrodipole interaction and to determine the intrinsic helix forming tendency of the charged residue (Armstrong and Baldwin, 1993; Huyghues-Despointes *et al.*, 1993; Scholtz *et al.*, 1993). Figure 7 illustrates this procedure for Asp (Huyghues-Despointes *et al.*, 1993) in the Ala-Gln reference peptide (Scholtz *et al.*, 1991c).

At low pH, where Asp is uncharged, the symmetric curve of fraction helix as a function of the position of the single Asp replacement (solid circles and line in Fig. 7) can be described by the unmodified Lifson–Roig equation in the same fashion as that described for the glycine substitutions in Fig. 5. Substituting a helix-breaking residue (Asp, $s = 0.29$) destabilizes the peptide helix the most at the center of the chain, whereas a substitution at the ends of the chain has little or no effect on the helix content of the peptide. This is caused by fraying of the ends of the helix (see Fig. 4). On ionization of the Asp at higher pH, the charged side chain interacts with partial charges in the helix backbone, and this charge–helix dipole interaction affects helix stability in an asymmetric manner. The altered profile of peptide helix content versus position of the substitution is

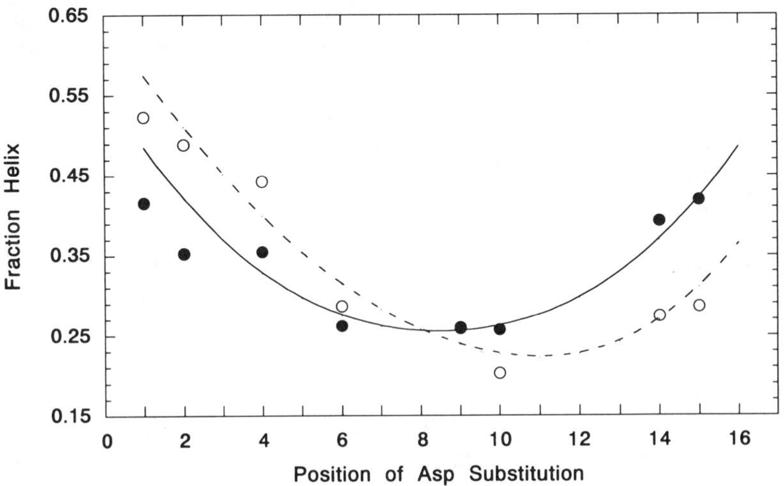

Figure 7 Helix content versus position of a single aspartate residue and the effect of ionization of the aspartate. The neutral host peptide is Ac-$(AAQAA)_3$-$Y(NH_2)$ and the pH is 2.5 (●) or 7.0 (○). The lines through the data are generated using standard helix–coil theory modified to include the interaction of charged side chain with the helix macrodipole. For details, see Huyghues-Despointes *et al.* (1993) and Scholtz *et al.* (1993).

shown by the dashed line and open circles (Fig.7). When the Asp is present near the N terminus of the helix, an increase in the helical content of the peptide results from the favorable interaction of the charged side chain with the positive pole of the helix macrodipole. The opposite effect is observed when Asp is substituted near the C terminus. The dashed line in Fig. 7 is the best fit of the Asp^- data to a model for the helix–coil transition that includes the charge–helix macrodipole interaction (Scholtz *et al.*, 1993). It should be noted that the Asp^0 and Asp^- curves cross in the center of the peptide, where the charge–helix macrodipole interaction is near zero; this implies that the *s* values for the charged and uncharged forms of Asp are nearly the same (Huyghues-Despointes *et al.*, 1993).

The same type of experiment has been performed for His (Armstrong and Baldwin, 1993) and for Glu (Scholtz *et al.*, 1993). In these cases the *s* values of the charged and uncharged forms of the residues are not identical; however, the interaction of the charged residue with the helix macrodipole can still be evaluated.

To assess the energetics of the interaction between charged side chains, a complete accounting of the charge–helix dipole interaction must be made, and the helix propensities (*s* values) of each charged residue must be determined. Scholtz *et al.* (1993) provided a general method for determining the energetics of side-chain interactions. The model is based on the hierarchical "nesting" approach outlined by Robert (1990); related theories for determining the energetics of ion-pair formation have been described by Gans *et al.* (1991) and by Vásquez and Scheraga (1988).

The approach used by Scholtz *et al.* (1993) is to place a single pair of charged residues in the neutral Ala-Gln host peptide. The effects of the spacing between charged groups and the orientation of the ion pair with respect to the helix macrodipole were investigated. After a complete accounting for the charge–helix macrodipole interaction, as well as determining the *s* values of the charged guest residues, it was possible to calculate the energetics of the interaction between the side chains. Some very interesting results were obtained for the Glu-Lys pair of charged side chains. When the spacing between the charged side chains is $(i, i + 3)$ or $(i, i + 4)$, which is appropriate for the α-helix, a substantial side-chain interaction is observed at neutral pH ($Glu^-\cdots Lys^+$), as well as at low pH ($Glu^0\cdots Lys^+$), for both orientations of the Glu-Lys residues in the helical peptide. In contrast, when the spacing between the residues is $(i, i + 1)$ or $(i, i + 2)$, there is no interaction between the side chains at low pH, and there is a small, but measurable, coil-stabilizing interaction at neutral pH. This latter interaction can be screened by added salt, whereas the helix-stabilizing interactions are still observed at 2.5 *M* NaCl. This result, together with other data, suggests that singly charged hydrogen bonds play important roles in stabilizing these α-helical peptides, in addition to the ion-pair interactions. More work is required to test these ideas and to quantify the energetics of other side-chain interactions in simple helical peptides.

In addition to electrostatic interactions and hydrogen bonds between side chains, and the interaction of a charged side chain with the helix macrodipole, there are a few unique interactions that have been studied in isolated peptides. These include an interaction between Phe and His^+ (Armstrong *et al.*, 1993), first identified in C-peptide from RNase A (Shoemaker *et al.*, 1985, 1990), and a hydrophobic interaction between Tyr and Leu or Val (Padmanabhan and Baldwin, 1994). These initial studies represent the types of interactions that are possible. More work is needed to catalog and evaluate all the other interactions that are possible; however, it appears that a viable experimental system is in place to perform these studies.

VI. Capping Interactions

The N-cap and C-cap residues that have been analyzed in protein helices by Richardson and Richardson (1988) are formally outside the helix by the Lifson–Roig definition, which requires that both backbone angles (ϕ, ψ) of a residue be helical. In the Lifson–Roig theory, the two helical amino acid residues at either end of a helical segment are formally responsible for initiating the helix: each one is assigned a nucleation parameter v. The v^2 term plays the same role as the Zimm–Bragg parameter σ, although the value of v^2 is somewhat different from σ (see Qian and Schellman, 1992). In both the Lifson–Roig and Zimm–Bragg models for helix formation, the N-cap and C-cap residues are treated as being in the random coil state, and so these models predict that the choice of N-cap and C-cap residues has little or no effect on peptide helix stability. Recent experiments show otherwise, however (Bruch *et al.*, 1991; Lyu *et al.*, 1993; Forood *et al.*, 1993; Chakrabartty *et al.*, 1993a). The choice of the N-cap residue has a major effect on the stability of a peptide helix.

This development was foreseen by Presta and Rose (1988) who proposed that side chain–main chain hydrogen bonds at either end of a helix are helix stabilizing and may be involved in initiating helix formation during protein folding. Studies of peptide helices indicate that other factors are also involved: various nonpolar amino acids have different N-cap propensities. For example, Gly has a substantially higher N-cap propensity than Ala, although at interior helix positions, the helix propensity of Gly is much smaller than that of Ala (see Serrano *et al.*, 1992a,b). Thus, both the Zimm–Bragg and Lifson–Roig models of helix formation need to be extended to include the N-cap effect. This has been done for the Lifson–Roig model by assigning new parameters, n and c, to the two coil residues at either end of a helical segment (Doig *et al.*, 1994). These residues, which in protein helices are the N-cap and C-cap residues, may be regarded as playing a supplemental role in initiating the helix.

The choice of C-cap residue has little effect on the stability of a peptide helix (Chakrabartty *et al.*, 1993a). In protein helices, only Gly has been identified as playing a special role at the C-cap position (Schellman, 1980; Richardson and Richardson, 1988) and a special effect of Gly at the C-cap position has not yet been observed in peptide helix studies.

VII. Protein Helices

A. General Considerations

Amino acid frequencies in protein helices provide important clues about the role of various amino acids in stabilizing protein helices. This has been found for the charge–helix dipole interaction, N-cap and C-cap interactions, and the capping box. It has been widely expected that amino acid frequencies in the central regions of protein helices should be closely correlated with peptide helix propensities. At present, the correlation is not as close as expected, and further study is needed. Experiments by Zhong and Johnson (1992) show that equivocal sequences, predicted by amino acid frequencies to be α helical but found in β strands in the X-ray structures of proteins, can form either helices or strands, depending on the solvent.

Tertiary interactions in protein helices are much stronger than the interactions that stabilize an isolated helix. Measurements of peptide NH exchange rates show that the S-peptide helix is about 1000 times more stable in RNase S than in isolation (Kuwajima and Baldwin, 1983). Moreover, typical protein helices have one solvent-exposed face and one buried face. Burial of both polar and nonpolar groups has major energetic consequences in protein folding.

There are, however, common factors that affect the stability of peptide and protein helices. A side-chain interaction that stabilizes a peptide helix will also contribute to protein stability. For example, the S-peptide helix of RNase S can be stabilized to a varying extent through the charge–helix dipole interaction by changing the charge on the N-terminal residue. When the S-peptide derivatives continaung various charges on the N-terminal residue are incorporated into RNase S, the thermal stability (T_m) of the protein changes in the same manner as the stability of the isolated S-peptide helix (Mitchinson and Baldwin, 1986). Second, the helix propensity of a solvent-exposed residue in a protein helix is expected to affect the stability of the protein in the same manner that it affects an isolated helix. O'Neil and DeGrado (1990) examined the effects of substituting all 20 amino acids at a solvent-exposed position in a dimeric coiled-coil helix, which may be considered as a very small protein. They found essentially the same rank order of helix propensities that has been found in peptide helix studies.

B. Helix Propensities

Comparison between the helix propensities found in peptide and protein helices reveals a basic problem, whose solution is not understood at this time. If the helix propensity is s and $\Delta G^{\circ} = -RT \ln s$, then the comparison is limited to values of $\Delta\Delta G^{\circ}$, the difference in ΔG° between a reference amino acid (alanine) and X, because substitution experiments with proteins do not yield values of s itself. The basic problem that appears when this comparison is made is that values of $\Delta\Delta G^{\circ}$ are roughly twice as large in studies of some peptide helices as they are in protein helices.

Studies of two different proteins, T4 lysozyme (Blaber *et al.*, 1993, 1994) and barnase (Serrano *et al.*, 1992a), and also of a coiled-coil dimer helix (O'Neil and DeGrado, 1990) yield similar values of $\Delta\Delta G^{\circ}$ in magnitude. The agreement regarding the rank order of helix propensities found in these studies is only fair, probably because it is difficult to find a site in a protein helix that is solvent exposed and does not allow interactions with neighboring residues (see Blaber *et al.*, 1993, 1994). These substitution experiments agree, both among themselves and with short peptide studies, that alanine is the most helix-stabilizing residue and glycine and proline are the most helix-breaking residues.

C. Side-Chain Interactions

In general, the results from substitution experiments involving side-chain interactions in protein helices agree closely with peptide helix studies. The best-studied interaction is probably the charge–helix dipole interaction, which has been found to stabilize RNase S (Mitchinson and Baldwin, 1986), barnase (Sali *et al.*, 1988; Sancho *et al.*, 1992), and T4 lysozyme (Nicholson *et al.*, 1988, 1991). An important feature of this interaction is that it can be used reliably to increase the stability of a protein through introduction of a designed charge–helix dipole interaction (Nicholson *et al.*, 1991). The status of ion-pair or salt-bridge interactions in proteins is curious. There are well-known examples of functionally and energetically important salt bridges in proteins: the salt bridges that stabilize the deoxy conformation of hemoglobin (Perutz and TenEyck, 1971) and the internal salt bridge formed when trypsin activates chymotrypsinogen (Fersht, 1972). There is also the striking salt bridge formed by His-31 and Asp-70 in T4 lysozyme, which contributes 3–5 kcal/mol to the stability of T4 lysozyme (Anderson *et al.*, 1990). These energetically strong salt bridges probably all involve buried, or partly buried, groups. On the other hand, attempts to increase the stability of T4 lysozyme by engineering solvent-exposed ion-pair interactions have met with failure (Dao-pin *et al.*, 1991).

Studies of side-chain interactions in proteins have the dual advantages that the two-state model can be used to derive the ΔG° of the interaction and the X-ray structure of the interacting groups can be obtained and analyzed. Studies of peptide helices, on the other hand, have other advantages: first, the side-chain interaction can be studied in isolation, apart from other interactions present in the whole protein, and, second, a side-chain interaction often has a large effect on the helix content of a peptide and therefore is easily measurable.

D. N-Cap and C-Cap Interactions

The probable existence of N-cap and C-cap interactions was inferred from studies of amino acid frequencies at these positions by Schellman (1980) and Richardson and Richardson (1988) (see also Dasgupta and Bell, 1993). This expectation was confirmed by substitution experiments with proteins (Serrano and Fersht, 1989; Serrano *et al.*, 1992a,b; Bell *et al.*, 1992), and these experiments confirmed the hypothesis by Presta and Rose (1988) of the energetic importance of side chain–main chain hydrogen bonds involving the peptide NH and CO groups that are not hydrogen bonded at either end of an α helix. These experiments show very clearly the difference between internal versus N-cap and C-cap positions in the relative helix-stabilizing effectiveness of alanine versus glycine (Serrano *et al.*, 1992a,b). They also show the necessity of taking into account the change in buried nonpolar surface area when a substitution is made in a protein helix (Serrano *et al.*, 1992a). On the other hand, these experiments demonstrate that structural features specific to the particular site of substitution are important in determining the results of a substitution experiment made in a protein; in practically every case, the amino acid found in the wild-type protein is the most stabilizing residue at the N-cap position (Bell *et al.*, 1992; Dasgupta and Bell, 1993).

A striking result is that the helix-stabilizing effectiveness of a N-cap residue (or N-cap propensity) in peptide helix studies (Chakrabartty *et al.*, 1993a) is closely correlated with the amino acid frequencies at the N-cap position in protein helices, as tabulated by Richardson and Richardson (1988) (see Fig. 8). This means that it is profitable to study in peptide helices problems such as the origin of the different N-cap propensities of various nonpolar amino acids.

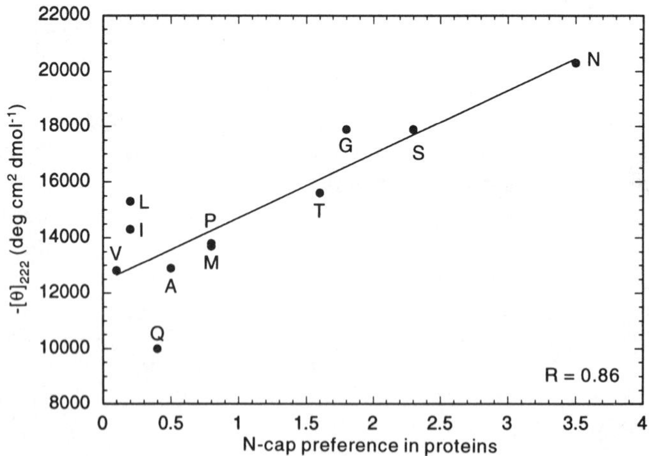

Figure 8 N-cap propensity measured in the reference peptide XAKAA-(AAKAA)$_2$-GY(NH$_2$), where X is the N-cap residue (with an uncharged α-NH$_2$ group) plotted against N-cap preference in proteins. Data for N-cap propensities are from Chakrabartty *et al.* (1993a), and data for N-cap preferences are from Richardson and Richardson (1988).

References

Anderson, D. E., Becktel, W. J., and Dahlquist, F. W. (1990). *Biochemistry* **29,** 2403–2408.

Armstrong, K. M., and Baldwin, R. L. (1993). *Proc. Natl. Acad. Sci. U.S.A.* **90,** 11337–11340.

Armstrong, K. M., Fairman, R., and Baldwin, R. L. (1993). *J. Mol. Biol.* **230,** 284–291.

Bell, J. A., Becktel, W. J., Sauer, U., Baase, W. A., and Matthews, B. W. (1992). *Biochemistry* **31,** 3590–3596.

Blaber, M., Zhang, X.-J., and Matthews, B. W. (1993). *Science* **260,** 1637–1640.

Blaber, M., Zhang, X.-J., Lindstrom, J. D., Pepiot, S. D., Baase, W. A., and Matthews, B. W. (1994). *J. Mol. Biol.* **235,** 600–624.

Blout, E. R. (1962). *In* "Polyamino Acids, Peptides and Proteins" (M. A. Stahmann, ed.), pp. 275–279. Univ. of Wisconsin Press, Madison.

Bradley, E. K., Thomason, J. F., Cohen, F. E., Kosen, P. A., and Kuntz, I. D. (1990). *J. Mol. Biol.* **215,** 607–22.

Brown, J. E., and Klee, W. A. (1971). *Biochemistry* **10,** 470–476.

Bruch, M. D., and Dhingra, M. M., and Gierasch, L. M. (1991). *Proteins* **10,** 130–139.

Cantor, C. R., and Schimmel, P. R. (1980). "Biophysical Chemistry. Part II: Techniques for the Study of Biological Structure and Function." Freeman, New York.

Chakrabartty, A., and Baldwin, R. L. (1993). "Protein Folding: *In Vivo* and *In Vitro.*" pp. 166–177. American Chemical Society, Washington, D. C.

Chakrabartty, A., Schellman, J. A., and Baldwin, R. L. (1991). *Nature (London)* **351,** 586–588.

Chakrabartty, A., Doig, A. J., and Baldwin, R. L. (1993a). *Proc. Natl. Acad. Sci. U.S.A.* **90,** 11332–11336.

Chakrabartty, A., Kortemme, T., Padmanabhan, S., and Baldwin, R. L. (1993b). *Biochemistry* **32,** 5560–5565.

Chakrabartty, A., Kortemme, T., Baldwin, R. L. (1994). *Protein Sci.* **3,** 843–852.

Creamer, T. P., and Rose, G. D. (1992). *Proc. Natl. Acad. Sci. U.S.A.* **89,** 5937–5941.

Dao-pin, S., Sauer, U., Nicholson, H., and Matthews, B. W. (1991). *Biochemistry* **30,** 7142–7153.

Dasgupta, S., and Bell, J. A. (1993). *Int. J. Pept. Protein Res.* **41,** 499–511.

Doig, A. J., Chakrabartty, A., Klinger, T. M., and Baldwin, R. L. (1994). *Biochemistry* **33,** 3396–3403.

Fersht, A. R. (1972). *J. Mol. Biol.* **64,** 497–509.

Forood, B., Feliciano, E. J., and Nambiar, K. P. (1993). *Proc. Natl. Acad. Sci. U.S.A.* **90,** 838–842.

Gans, P. J., Lyu, P. C., Manning, M. C., Woody, R. W., and Kallenbach, N. R. (1991). *Biopolymers* **31,** 1605–1614.

Go, M., Go, N., and Scheraga, H. A. (1971). *J. Chem. Phys.* **54,** 4489–4503.

Huyghues-Despointes, B. M. P., Scholtz, J. M., and Baldwin, R. L. (1993). *Protein Sci.* **2,** 1604–1611.

Johnson, W. C. (1990). *Proteins: Struct. Funct. Genet.* **7,** 205–214.

Kuwajima, K., and Baldwin, R. L. (1983). *J. Mol. Biol.* **169,** 299–323.

Liff, M. I., Lyu, P. C., and Kallenbach, N. R. (1991). *J. Am. Chem. Soc.* **113,** 1014–1019.

Lifson, S., and Roig, A. (1961). *J. Chem. Phys.* **34,** 1963–1974.

Lyu, P. C., Marky, L. A., and Kallenbach, N. R. (1989). *J. Am. Chem. Soc.* **111,** 2733–2744.

Lyu, P. C., Liff, M. I., Marky, L. A., and Kallenbach, N. R. (1990). *Science* **250,** 669–673.

Lyu, P. C., Sherman, J. C., Chen, A., and Kallenbach, N. R. (1991). *Proc. Natl. Acad. Sci. U.S.A.* **88,** 5317–5320.

Lyu, P. C., Wemmer, D. E., Zhou, H. X., Pinker, R. J., and Kallenbach, N. R. (1993). *Biochemistry* **32,** 421–425.

Marqusee, S., and Baldwin, R. L. (1987). *Proc. Natl. Acad. Sci. U.S.A.* **84,** 8898–8902.

Marqusee, S., Robbins, V. H., and Baldwin, R. L. (1989). *Proc. Natl. Acad. Sci. U.S.A.* **86,** 5286–5290.

Merutka, G., and Stellwagen, E. (1991). *Biochemistry* **30,** 1591–1594.

Miick, S. M., Martinez, G. V., Fiori, W. R., Todd, A. P., and Millhauser, G. L. (1992). *Nature (London)* **359,** 653–655.

Miick, S. M., Casteel, K. M., and Millhauser, G. L. (1993). *Biochemistry* **32,** 8014–8021.

Mitchinson, C., and Baldwin, R. L. (1986). *Proteins* **1,** 23–33.

Nicholson, H., Becktel, W. J., and Matthews, B. W. (1988). *Nature (London)* **336,** 651–656.

Nicholson, H., Anderson, D. E., Dao-pin, S., and Matthews, B. W. (1991). *Biochemistry* **30,** 9816–9828.

O'Neil, K. T., and DeGrado, W. F. (1990). *Science* **250,** 646–651.

Osterhout, J. J., Baldwin, R. L., York, E. J., Stewart, J. M., Dyson, H. J., and Wright, P. E. (1989). *Biochemistry* **28,** 7059–7064.

Padmanabham, S., and Baldwin, R. L. (1991). *J. Mol. Biol.* **219,** 135–137.

Padmanabhan, S., and Baldwin, R. L. (1994). *J. Mol. Biol.* **241,** 706–713.

Padmanabhan, S., Marqusee, S., Ridgeway, T., Laue, T. M., and Baldwin, R. L. (1990). *Nature (London)* **344,** 268–270.

Park, S.-H., Shalongo, W., and Stellwagen, E. (1993a). *Biochemistry* **32,** 7048–7053.

Park, S.-H., Shalongo, W., and Stellwagen, E. (1993b). *Biochemistry* **32,** 12901–12905.

Perutz, M. F., and TenEyck, L. F. (1971). *Cold Spring Harbor Symp. Quant. Biol.* **36,** 295–310.

Poland, D., and Scheraga, H. A. (1970). "Theory of Helix–Coil Transitions in Biopolymers." Academic Press, New York.

Presta, L. G., and Rose, G. D. (1988). *Science* **240,** 1632–1641.

Qian, H., and Schellman, J. A. (1992). *J. Phys. Chem.* **96,** 3987–3994.

Richards, F. M., and Richmond, T. (1978). *Ciba Found. Symp.* **60,** 23–45.

Richardson, J. S., and Richardson, D. C. (1988). *Science* **240,** 1648–1652.

Robert, C. H. (1990). *Biopolymers* **30,** 335–347.

Rohl, C. A., Scholtz, J. M., York, E. J., Stewart, J. M., and Baldwin, R. L. (1992). *Biochemistry* **31,** 1263–1269.

Sali, D., Bycroft, M., and Fersht, A. R. (1988). *Nature (London)* **335,** 740–743.

Sancho, J., Serrano, L., and Fersht, A. R. (1992). *Biochemistry* **31,** 2253–2258.

Schellman, C. (1980). *In* "Protein Folding" (R. Jaenicke, ed.), pp. 53–61. Elsevier North-Holland, New York.

Scholtz, J. M., and Baldwin, R. L. (1992). *Annu. Rev. Biophys. Biomol. Struct.* **21,** 95–118.

Scholtz, J. M., Qian, H., York, E. J., Stewart, J. M., and Baldwin, R. L. (1991a). *Biopolymers* **31,** 1463–1470.

Scholtz, J. M., Marqusee, S., Baldwin, R. L., York, E. J., Stewart, J. M., Santoro, M., and Bolen, D. W. (1991b). *Proc. Natl. Acad. Sci. U.S.A.* **88,** 2854–2858.

Scholtz, J. M., York, E. J., Stewart, J. M., and Baldwin, R. L. (1991c). *J. Am. Chem. Soc.* **113,** 5102–5104.

Scholtz, J. M., Qian, H., Robbins, V. H., and Baldwin, R. L. (1993). *Biochemistry* **32,** 9668–9676.

Serrano, L., and Fersht, A. R. (1989). *Nature (London)* **342,** 296–299.

Serrano, L., Neira, J.-L., Sancho, J., and Fersht, A. R. (1992a). *Nature (London)* **356,** 453–455.

Serrano, L., Sancho, J., Hirshberg, M., and Fersht, A. R. (1992b). *J. Mol. Biol.* **227,** 544–559.

Shoemaker, K. R., Kim, P. S., Brems, D. N., Marqusee, S., York, E. J., Chaiken, I. M., Stewart, J. M., and Baldwin, R. L. (1985). *Proc. Natl. Acad. Sci. U.S.A.* **82,** 2349–2353.

Shoemaker, K. R., Fairman, R., Schultz, D. A., Robertson, A. D., York, E. J., Stewart, J. M., and Baldwin, R. L. (1990). *Biopolymers* **29,** 1–11.

Stellwagen, E., Park, S.-H., Shalongo, W., and Jain, A. (1992). *Biopolymers* **32,** 1193–1200.

Vásquez, M., and Scheraga, H. A. (1988). *Biopolymers* **27,** 41–58.

Wójcik, J., Altman, K. H., and Scheraga, H. A. (1990). *Biopolymers* **30,** 121–134.

Woody, R. W. (1985). "The Peptides," Vol.7, pp. 15–114. Academic Press, Orlando, Florida.

Wüthrich, K. (1986). "NMR of Proteins and Nucleic Acids." Wiley, New York.

Zhong, L., and Johnson, W. C., Jr. (1992). *Proc. Natl. Acad. Sci. U.S.A.* **89,** 4462–4465.

Zhou, N. E., Kay, C. M., Sykes, B. D., and Hodges, R. S. (1993). *Biochemistry* **32,** 6190–6197.

Zimm, B. H., and Bragg, J. K. (1959). *J. Chem. Phys.* **31,** 526–535.

5

Peptide Conformation: Stability and Dynamics

Garland R. Marshall
Denise D. Beusen
Gregory V. Nikiforovich
Center for Molecular Design
Washington University
St. Louis, Missouri 63130

I. Introduction

The importance of peptides in regulating biological processes and their seemingly infinite conformational possibilities define both the impetus and the impediment to understanding their mechanism of action. In the spectrum of biological molecules, peptides are unique in having both a large number of degrees of freedom and a broad spectrum of functional groups. This is in contrast to steroids, which have limited conformational freedom and limited functionality; and to fatty acids, which have a great deal of conformational mobility

Peptides: Synthesis, Structures, and Applications

and limited functionality. As mediators in precisely regulated, complex biological systems, peptides have a combination of flexibility and functionality which gives them the countervailing properties of adaptability and specificity. The innate properties of peptides that allow them to be multipotent with regard to encoded three-dimensional message complement the expression of receptor subtypes having specific biological functions.

It is generally accepted that the biological activity of a peptide is coupled to its conformation. In other words, to trigger a specific biological activity, a peptide must adopt a conformation that aligns essential functional groups in a required spatial orientation. This is true regardless of the activity, whether it involves signal transduction at a protein receptor or aggregation to form an ion channel. Because of the linkage between conformation and activity, the field of conformationally directed peptide design has arisen in order to understand, elicit, or inhibit a given biological activity. There are two fundamental elements of this process: identifying the relevant conformation and developing means to stabilize that conformation. In the discussion that follows, the term "conformation" will refer to any three-dimensional structure of a peptide molecule, whereas the term "conformer" will describe a conformation representative of a local energetic minimum (IUPAC, 1979). Consequently, conformations within the same local minimum will spontaneously transit to their associated conformer, whereas the transition from one conformer to another requires significant time and energy to overcome the barrier separating them.

Any evaluation of peptide conformational stability and dynamics requires a description of the statistical distribution between equilibrium states (or conformers) of a peptide as well as the energetic barriers separating them. The former task is typically emphasized, since the overwhelming majority of experimental data on peptide conformations relates to thermodynamic equilibria rather than dynamics. Experimental observations of receptor–ligand complexes have reinforced the bias that peptides bind as single conformations. The role of dynamics (i.e., conformer interconversion) in the bioactivity of a peptide is uncertain, in part because experimentally this is a much harder question to answer and consequently the evidence is sparse. We know that hinged proteins, for example, have a region linking domains which must remain flexible to allow binding and subsequent release of substrates (Williams, 1989). It is not known if the ligand must be flexible, but it seems likely that ligand release may be affected by its innate flexibility. In simpler cases, such as the peptide ion channel alamethicin, segmental motion has been postulated to play a crucial role in bioactivity (Fox and Richards, 1982).

Identification of the bioactive conformation of a peptide is trivial in those few cases where direct experimental observation is possible. In other cases, the simplifying assumption is often made that the active conformation can be approximated by one of its low-energy conformers in solution. In principle, computational and experimental tools can then be used to identify these conformers. Numerous studies indicate that linear peptides in solution are a mixture of many conformations of several conformers which are rapidly interconverting. The small size of many peptides (<20 amino acids) is inadequate to generate the numerous long-range interactions which tend to stabilize proteins in a dominant conformer. It is generally assumed that the potential energy surface of a linear peptide is relatively featureless in most solvent environments, with barriers into and out of minima being relatively small. When a receptor binds one conformation, it in essence perturbs the potential surface, stabilizing one conformer by creating high barriers to its transition to other conformers. The intent in designed conformationally stable peptides is to perturb the potential surface of the free peptide to mimic that of the bioactive state. In the case of linear peptides, the binding site is preorganized to select a conformer; in the case of constrained peptides, we attempt to preorganize the ligand to mimic the bioactive conformer which should enhance affinity by decreasing entropy loss on binding.

In this chapter, we focus on computational and experimental tools used to elucidate the conformational stability and dynamics of peptides, and we discuss specific cases in which the conformational stability of peptides has been studied and/or rationally perturbed.

II. Theoretical Studies

A. Energetics of Peptides

1. Molecular Mechanics

Although quantum mechanics has a definite role to play, peptide systems are generally too large for thorough investigation. Nevertheless, studies employing an *ab initio* approach to map the potential energy surface of *N*-acetyl-*N′*-methylamides of glycine and alanine exist (Bohm, 1993; Perczel *et al.*, 1991a; Schafer *et al.*, 1993). Even with modern computational facilities, extension of rigorous quantum mechanical treatment to peptides of reasonable size seems improbable, although semiempirical methods such as AM1 (Dewar *et al.*, 1985) have been applied to octapeptides (Bindal and Marshall, unpublished). For this reason, a peptide molecule is usually treated classically as a collection of atoms whose interactions can be described by Newtonian mechanics, a process called molecular mechanics (Burkert and Allinger, 1982). Because the mass of the nuclei is much greater than the mass of the electrons, one can separate the Schrodinger equation into a product of two functions: one for electrons and one for nuclei (the Born–Oppenheimer approximation). For the purposes of molecular mechanics, which was initially developed to interpret spectroscopic data, the electronic function is ignored; that is, the charge distribution is assumed to remain constant during changes in the position of the nuclei. Because molecular mechanics is based on classical physics, it cannot provide information about the electronic properties of molecules, which are generally assumed fixed during the parameterization of the force field with experimental data.

A few words about the basics of molecular mechanics (Bowen and Allinger, 1991; Burkert and Allinger, 1982) may provide the elements of understanding for what follows. This is not meant to be comprehensive, but rather a simple overview to remind the reader of a few crucial points. The basic assumption underlying molecular mechanics is that classical physical concepts can be used to represent the forces between atoms. In other words, one can approximate the potential energy surface by the summation of a set of equations representing pairwise and multibody interactions. These equations represent forces between atoms involved in bonded and nonbonded interactions. Nonbonded forces between atoms are based on an attractive interaction which has a firm theoretical basis and varies as the inverse of the sixth power of the distance between the atoms. It is balanced by a repulsion between the electronic clouds as the atoms come close and this interaction has been represented empirically by a variety of functional forms: exponential, twelfth power, or ninth power of the distance between the atoms. The coefficients in the resulting expression, $[C_{12}(i,j)/r_{ij}^{12}] - [C_6(i,j)/r_{ij}^{6}]$ are parameterized for atom types, usually by element, so that the minimum of the combined functions corresponds to the sum of the experimental van der Waals radii for the two atoms.

The interaction of bonded atoms is represented by a harmonic potential obeying Hooke's law ($\frac{1}{2}K_b[b - b_0]^2$), with a "spring constant" determining the energy of deformation from experimental bond lengths. Atoms directly bonded to the same atom (one–three interactions) are eliminated from the van der Waals list and have a special energetic term ($\frac{1}{2}K_\theta[\theta - \theta_0]^2$) that penalizes deviation from an ideal bond angle (θ_0). Atoms having one–four interaction define a torsional relationship which is usually parameterized based on the types of the four connected atoms defining the torsion angle. The torsional potential depends on the dihedral angle $\{K_\phi[1 + \cos(\phi - \delta)]\}$ and is used to account for orbital delocalization and to compensate for other deficiencies in the force field. A harmonic term ($\frac{1}{2}K\xi[\xi - \xi_0]^2$) is often introduced for dihedral angles ξ which are relatively fixed, such as those in aromatic rings. Coulomb's law ($q_iq_j/4\pi\varepsilon_0\varepsilon_r r_{ij}$) is the simplest approach to the contribution of electrostatics to the potential. Equation (1) gives

$$\begin{aligned} V = {} & \Sigma\, \tfrac{1}{2}K_b[b - b_0]2 + \Sigma\, \tfrac{1}{2}K_\theta[\theta - \theta_0]^2 + \Sigma\, \tfrac{1}{2}K_\xi[\xi - \xi_0]^2 \\ & + K_\phi[1 + \cos(\phi - \delta)] + \Sigma[C_{12}(i,j)/r_{ij}^{12} - C_6(i,j)/r_{ij}^{6}] \\ & + \Sigma(q_iq_j/4\pi\varepsilon_0\varepsilon_r r_{ij}), \end{aligned} \qquad (1)$$

a set of equations comprising a generalized, molecular mechanics force field.

A central issue is the number of different atom types which are used in a particular force field. There is always a compromise between increasing the number to allow for the inclusion of more environmental effects and determining the increased number of parameters needed to adequately represent a new atom type. In general, the more subtypes of atoms (e.g., how many different kinds of nitrogens), the less likely that the parameters for a particular application will be available in the force field. The extreme, of course, would be a special atom type for each kind of atomic environment in which the parameters were chosen so that the calculated properties of each molecule would simply reproduce the experimental observations. One major assumption, therefore, is that the force constants (parameters) and equilibrium values of the equations are functions of a limited number of atom types and can be transferred from one molecular environment to another. This assumption holds reasonably well where one may be primarily interested in geometric issues, but is not so valid in molecular spectroscopy. This has led to the introduction of additional equations, the so-called cross-terms, which allow additional parameters to account for correlations between bond lengths and bond angles ($K_{b\theta}[b - b_0][\theta - \theta_0]$), dihedral angles and bond angles, etc.

Development of appropriate parameters from either theoretical (quantum mechanics) and/or experimental data (Bowen and Allinger, 1991; Palmo *et al.*, 1993) can require significant effort. For example, the amide bond is normally represented by one set of parameters whether the configuration is cis or trans. Recent experimental data are quite compelling that the electronic state is different between the two configurations and different parameter sets should be used for accurate results (Jorgensen and Gao, 1988; Mirkin and Krimm, 1991). Only AMBER/OPLS currently distinguishes between these two conformational states (Jorgensen and Gao, 1988). Simplified force fields in which the torsional parameters depend only on the atoms at the end of a bond have been developed to give approximate geometries for further refinement by quantum mechanics (Clark *et al.*, 1989). Certainly, the limited parameterization of simplified force fields would not allow accurate prediction of spectra, which is more reflective of the dynamic behavior of the molecule. Accurate estimates of energy may require accurate representation of the dynamics of molecules and justify derivation of the larger number of parameters. The new version (Lii and Allinger, 1991) of the Allinger force field, MM3, has the objective of reproducing spectral data more accurately than MM2. Much of chemistry remains to be incorporated into appropriate force fields. Only recently have adequate modifications been made to the force fields developed for organic molecules to include some metals (Allured *et al.*, 1991; Aqvist and Warshel, 1990; Hancock, 1990; Vedani and Huhta, 1990).

Because different force fields may use different mathematical representations of the forces between atoms and because the details of their parameterization will in general differ also, it is unwise to utilize parameters derived for one force field to replace missing parameters in another. One often hears of a "balanced" parameter set that reproduces well the phenomena under consideration, but which is often inadequate for other applications. A comparison by Burkert and Allinger (1982) shows the different van der Waals (VDW) potentials used in several of the popular force fields, and the situation has not improved in the intervening years (Dudek and Ponder, 1995). Because of other differences in parameters and functional forms of the equations used in the rest of the individual force fields, these quite different approaches to the VDW potential give excellent results when used in the correct combination. Indiscriminant combination of one part of a force field with another derived independently would lead to considerable divergence in the calculated results from experimental observation.

The largest difference between force fields arises in the method by which the hydrogen bond is included. Because atoms involved in a hydrogen bond are often closer than the sum of their VDW radii, they must be handled in a special manner. Several force fields have special functional forms with angular dependence that not only have special VDW parameters to ensure that the proper close approach of the atoms involved is calculated correctly, but also reproduce the correct angular distribution observed for hydrogen bonds. Hagler *et al.* (1974) used an amide hydrogen with a zero VDW radius for hydrogen bonding and a slightly greater nitrogen radius to give a correct amide hydrogen bond distance. The charges on the atoms involved (including the amide hydrogen) are adjusted to give an appropriate balance of VDW repulsion and dipole attraction. Clearly, the method for handling the electrostatic interaction is an integral part of each force field and cannot be modified independently.

2. Electrostatics

The most difficult aspect of molecular mechanics is electrostatics (Davis and McCammon, 1990; Dykstra, 1993; Harvey, 1989; van Gunsteren and Berendsen, 1990). In most force fields, the electronic distribution surrounding each atom is treated as a monopole with a simple Coulombic term for the interaction. The effect of the surrounding medium is generally treated with a continuum model by use of a dielectric constant. To evaluate atom–atom interactions using Coulomb's law, the concept of net atomic charge is invoked. This amounts to representing charge as a point, a monopole, and is an artificial construct. Nevertheless, this is the common method. Improvements in calculating an appropriate set of point charges to reproduce accurately the molecular electrostatic potential derived by accurate quantum calculations have been reported (Bayly *et al.*, 1993).

In an effort to increase the quality of electrostatic representations, dipole and higher multipole moments have been used. There are advantages in these more accurate representations with a relatively small computational increase due to the reductions in distances over which the higher moments have to be summed, but they do require additional effort in the derivation of the parameters for the higher moments themselves. A good example is the distributed multipole model of electrostatics (Stone and Alderton, 1985) derived for peptides (Faerman and Price, 1990; Price *et al.*, 1991, 1992). A review by Williams (1991) discusses the problems of deriving a distributed multipole expansion of charge representation that accurately reproduces the molecular electrostatic potential derived from quantum calculations. Comparisons were made between atomic multipoles, bond dipole, and restricted bond dipole models. Williams finds that a model for the electrostatic potential based on bond dipoles supplemented with monopoles (for ions) and atomic dipoles (for lone pairs) is most useful. Dipole–dipole energy converges much faster than monopole–monopole energy. Molecular charge at any desired position in a molecule is not a physically measurable quantity; one can only calculate a delocalized electron probability distribution from quantum theory. Clearly, the more complex the representation, the more accurately one can approximate the quantum mechanical results, and the more realistic should be the resulting calculations.

One complexity of electrostatics is the long distances over which interactions occur. Electrostatic interactions range from those operating only at very short distances which are nonspecific (dispersive interactions, r^{-6} dependence) to those operating at very long distances with a high degree of specificity (charge–charge interactions, r^{-1} dependence). Appropriate means of truncating the long-range forces to maintain the accuracy of simulations are necessary (Guenot and Kollman, 1993; Loncharich and Brooks, 1989; Tasaki *et al.*, 1993), and progress in better approximations has been reported (Shimada *et al.*, 1993). The difficulties with cutoff schemes were demonstrated (Schreiber and Steinhauser, 1992a,b) by significant variations in the behavior of a 17-residue helical peptide simulated with explicit waters as various electrostatic schemes were employed (Smith and Pettit, 1991) and by studies of a pentapeptide in aqueous ionic solution (Marlow *et al.*, 1993). In both cases, the Ewald approximation in which periodicity is assumed allows summation over much longer distances and gave superior results (Schreiber and Steinhauser, 1992a,b; Smith and Pettit, 1991).

3. The Dielectric Problem and Solvation

If one can realistically calculate the Boltzmann distribution of conformations for a peptide, then any measurable property of that peptide can be calculated by averaging over all states of the system. This requires knowledge of free energy of the system, not just the potential energy. There are several means by which one can calculate free energy in conformational analyses, all of them, attempting to approximate the entire conformational space available to a given peptide. However, the conformational states available to a flexible peptide in solution depend on complex interactions with the particular solvent. In other words, the only proper way to estimate the free energy of a peptide is to explicitly include solvent in the calculation.

The use of Coulomb's law as a simplified treatment of solvation is clearly of concern. The dielectric constant is a scaling factor related to the polarizability of the medium between the charges. At the molecular level, the dielectric is neither homogeneous nor continuous, nor even well-defined, and thus violates the basic assumption of Coulomb's law. Although the use of a low, uniform dielectric is more correct in dynamic simulations where all solute and solvent atoms are explicitly included, a variety of comparisons of experimental data with the results of calculation using a simplified solvent model have led to the realization that much better approaches are needed. Initial efforts (Whitlow and Teeter, 1986) led to the proposal of a variable dielectric (ε of $1/r$ or $1/4r$). Approaches that model the inhomogeneity of the dielectric at the interface between the solute and solvent using the Poisson–Boltzman equation have shown considerable promise (Gilson *et al.*,1987; Nicholls and Honig, 1991). An alternative approach using the mirror charge approximation has been described by Schaefer and Froemmel (1990). Excellent reviews (Davis and McCammon, 1990; Dykstra, 1993; Harvey, 1989; van Gunsteren and Berendsen, 1990) of the electrostatic problem have appeared, to which the reader is referred.

Much effort has been given to simple continuum models of solvation to explain the origin of solvent effects on conformational equilibria and reaction rates. The current status of such efforts, as well as simulations to rationalize solvation effects, has been reviewed by Richards *et al.* (1987). There are two general approaches to the continuum models. The first is reaction field theory (Bell, Kirkwood, Onsager), which follows the classical treatment of Debye–Huckel. The solvent is considered in terms of charge distribution, polarizability, and dielectric constant. The solvation energy is determined simply by considering the solute as a point dipole which interacts with the induced charge distribution in the solvent (Onsager reaction field). An extension by Sinangolou in the 1960s partitioned solvation energy into cavity formation, solvent–solute interaction, and the free volume of the solute. The logical extension of this approach is scaled-particle theory (Pierotti, 1976), where the free energy of formation of a hard-sphere cavity of diameter σ_2 in a hard-sphere solvent of diameter σ_1 and number density ρ is scaled to the exact solution for small cavity sizes. Alternatively, the virtual charge approach used a system of effective and virtual charges interacting in the gas phase. The Hamiltonian of the system is modified to include an imaginary particle, a "solvation" with an opposite charge for each of the solute atoms and solved by a self-consistent field (SCF) procedure. These continuum models have met with limited success (trends and relative effects of solvation can be predicted), but highly specific molecular interaction, such as those involving hydrogen bonding groups, cannot be accommodated.

In the equation for calculating the affinity of a drug for a receptor, the ligand is solvated either by the receptor or by the solvent. This competition means that accurate determination of the free energy of solvation is important in understanding differences in affinities. Solvation free energy (G_{sol}) can be approximated by three terms [Eq. (2)]: G_{cav}, the formation of a cavity in the solvent to hold the solute; and G_{vdw} and G_{pol}, the interaction between solute and solvent divided between van der Waals and electrostatic forces,

$$G_{sol} = G_{cav} + G_{vdw} + G_{pol}. \tag{2}$$

There are four theoretical approaches to the problem: scaled particle theory (Pierotti, 1976), virtual charge method (Schaefer and Froemmel, 1990), boundary element method (Zauhar and Morgan, 1988), and the Poisson–Boltzmann equation (Nicholls and Honig, 1991). Only the latter is in common use with peptides and proteins. Generalization of Debye–Huckel theory leads directly to the Poisson–Boltzmann equation that describes the electrostatic potential of a field of charges with dielectric discontinuities. This equation has been solved analytically for spherical and elliptical cavities, but it must be solved by finite difference methods on a grid for more complicated systems. One exciting advance in this area is the development of an approximate equation for the reaction field acting on a macromolecular solute due to the surrounding water and ions (Sharp, 1991). By combining these equations with conventional molecular dynamics, solvation free energies were obtained similar to those with explicit solvent molecules at little computational cost over vacuum simulations. This implies that a more correct solution to the electrostatics problem might minimize the solvation problem. Other approaches to evaluations of G_{sol} have appeared in the literature. Still *et al.* (1990) estimate $G_{cav} + G_{vdw}$ by the solvent-accessible surface area times 7.2 cal/mol-Å^2. The G_{pol} value is estimated from the generalized Born equation. Effective solvation terms have been added (Schiffer *et al.*, 1993; Stouten *et al.*, 1993) to molecular mechanics force fields in order to improve molecular dynamics simulations without the cost of modeling explicit solvent. Zauhar (1991) has combined the polarization–charge technique with molecular mechanics to effectively minimize a tripeptide in solvent.

4. *Polarizability*

One final refinement may be necessary in some situations, namely the inclusion of electric polarizability, for example, by inclusion of induced dipoles, or distributed polarizability (Stone, 1985), in the electrostatic representation of the model. The effects of this refinement in modeling crystal structures of polymorphs of ice have been examined (Kuwajima and Warshel, 1990). Inclusion of a bond dipole model with polarizability in molecular dynamics simulations has given excellent agreement in predicting physical properties of crystalline polymorphs of polymers (Sorensen *et al.*, 1988). The inclusion of implicit nonadditive polarization energies results in improved accuracy (Caldwell *et al.*, 1990). Dang *et al.* (1991) described nonadditive many-body potential models to calculate ion solvation in polarizable water with good agreement with experimental observation. It was necessary to include a three-body potential (ion–water–water) in the molecular dynamics simulation of the ionic solution in order to obtain quantitative agreement with solvation enthalpies and coordination numbers. A novel approach based on the concept of charge equilibration has been suggested (Rappe and Goddard, 1991) which allows the inclusion of polarizabilities in molecular dynamics calculations. At the semiempirical level of quantum theory, Cramer and Truhlar (Caldwell *et al.*, 1990; Cramer, 1991; Cramer and Truhlar, 1991, 1992a,b) have added solvation and solvent effects on polarizability to AM1 with impressive agreement between experimental and calculated solvation energies. Rauhut *et al.* (1993) have also introduced an arbitrary shaped cavity model using standard AM1 theory.

5. *The "Hydrophobic" Effect*

Water has been the nemesis of solvation modeling because of its rather unique thermodynamic properties (reviewed by Franks, 1975, and Stillinger, 1980). The biochemical literature discusses at length "hydrophobic effects" (Pratt, 1985). This effect is not hydrophobic at all, as the enthalpic interaction of nonpolar solutes with water is favorable. This, however, is counterbalanced by an unfavorable entropic interaction which is interpreted as due to an induced structuring of the water by the nonpolar solute. Water interacts less well with the nonpolar solute than it does with itself due to the lack of hydrogen bonding groups on the solute. This creates an interface similar to the air–water interface with a resulting surface tension due to the organization of the hydrogen-bonded patterns available. This is the so-called iceberg formation around nonpolar solutes in water. Studies by both molecular dynamics (Ohmine and Tanaka, 1993; Postma *et al.*, 1982; Rao and Singh, 1989) and Monte Carlo simulations (Jorgensen *et al.*, 1985) support this interpretation (Stillinger, 1980), although there is still considerable controversy in the interpretation of experimental data (Muller, 1992).

6. *The Potential Surface*

The set of equations which describe the sum of interactions between atoms in an ensemble is an analytical representation of the Born–Oppenhemier surface which describes the energy of the molecule as a function of the atomic positions. Many important properties of the molecule can be derived by evaluation of this function and its derivatives. For example, setting the value of the first derivative to zero and solving for the coordinates of the atoms leads to minima, maxima, and saddle points. Evaluation of the sign of the second derivative can determine which of the above have been found. It is a straightforward procedure to calculate the vibrational frequencies from the force constants by evaluation of the eigenvalues of the secular determinant (the mass-weighted matrix). Gradient methods for the location of energy minima and transition states are an essential part of any molecular modeling package. It is important to remember, however, that minimization is an iterative method of geometrical optimization which is dependent on the starting geometry unless the potential surface contains only one minimum (a condition not found for any system of sufficient complexity to be of real interest).

The ability to locate both minima and transition points enables one to determine the minimum energy reaction path between any two minima. In the case of flexible molecules, these minima could correspond to conformers, and the reaction path would correspond to the most likely reaction coordinate. One could estimate the rate of transition by determining the height of the transition states (the activation energy) between the minima. Czerminski and Elbers (1990) have developed a new protocol for the location of minima and transition states and applied it to the determination of reaction paths for the conformational transition of a tetrapeptide (Choi and Elber, 1991). Huston and Marshall

(1994) have used this approach to map the reaction coordinates of the α- to 3_{10}-helical transition in model peptides.

Thus, any force field developed for the description of peptide systems is only an approximation of real intramolecular interactions. This approximation can be calibrated to reproduce vibrational spectra, relative energetics of conformational minima, and rates of conformational transitions. Despite the limitations which curtail exact quantitative applications, molecular mechanics can provide three-dimensional insight as the geometric relations between molecules are adequately represented. Electrical field potentials can be calculated and compared to give a qualitative basis for rationalizing differences in activity. Molecular modeling and its graphical representation allow the chemist to explore the three-dimensional aspects of molecular recognition and to generate hypotheses which lead to design and synthesis of new ligands. The more accurate the representation of the potential surface of the molecular system under investigation, the more likely that the modeling studies will provide qualitatively correct solutions. Once a proper set of parameters for the system under study has been selected, the force field provides the fundamental basis for simulating the system either through Newtonian mechanics or molecular dynamics, or through a stochastic generation of the relevant partition function, by Monte Carlo simulations.

B. Conformational Analysis

Whereas interaction with a receptor will certainly perturb the conformational energy surface of a flexible peptide, high affinity would suggest that the peptide binds in a conformation which is not exceptionally different from one of its low-energy conformers (i.e., conformers whose energy is not dramatically higher than that of the global minimum-energy conformer). By determining the set of low-energy conformers, one attempts to generate a description of the equilibrium states of a peptide and to avoid the limitations inherent in static representations of the conformational possibilities of a peptide. Mapping the energy surface of the peptide in isolation to determine the low-energy conformers will, at the very least, provide a set of candidate conformations for consideration, or as starting points for further analyses.

The problem of finding the global minimum on a complicated potential surface is common to many disciplines, and lacks a general solution. In general, the difficulties with most molecular mechanics minimization methods are similar to those seen with optimization procedures. If one is in the area of the global minimum, then one is likely to converge to that solution. Otherwise, one will be trapped in some local minimum. The ability of a minimization procedure to locate the closest local minimum depends on the starting conformation.

Several strategies have been developed to map the potential surface and locate minima. Either stochastic methods, such as Monte Carlo, or molecular dynamics can be used to explore the potential energy surface and can be combined with simulated annealing (Moskowitz *et al.*, 1988; van Schaik *et al.*, 1992) to help overcome energy barriers between minima. Systematic, or grid, search samples conformations in a regular fashion, at least in the parameter space (usually torsional space) which is incremented. In a sense, the same approach is represented by various build-up procedures, whose aim is to explore the conformational states of the entire peptide molecule by combining results of conformational samplings for fragments of a peptide. Below we describe briefly such methods of conformational analysis. For an excellent overview of the different approaches, the reader is referred to the surveys by Leach (1993) and by Burt and Greer (1988).

Systematic search methods are algorithmic, so that all sterically allowed conformations are generated at the selected torsional grid points. Systematic search methods, therefore, do not have problems in sampling and are path independent, but they are combinatorial in complexity. Only in small systems such as cycloalkane rings have the potential energy hypersurfaces been mapped (Kolossvary and Guida, 1993). In the absence of detailed three-dimensional information regarding the receptor, no objective criterion function, such as relative free energy, can be used to identify the receptor-bound conformation. Systematic search is, therefore, appropriate for investigating potential bioactive conformations as it can identify the set of low-energy conformations which are likely candidates for receptor interaction. Adequate sampling of the potential surface to assure that the complete set of local minima is found is problematic due to the phenomenon known as grid tyranny. This relates to the compromise one must make between adequate sampling and the combinatorial explosion which results from decreasing the torsion angle scan increment. Using a 10º grid for a seven-rotatable bond problem the entire number of conformations explored is 7^{36}, or 2.65×10^{30}. Because the energetics of the system are very sensitive to interatomic distances, a conformation generated at 10º increments may be sterically disallowed, but very close to a minimum. Relaxation of the structure by allowing a search of torsional angles in 1º increments might find the missing minimum, but would require exploration of 7^{360}, or 1.72×10^{304}, conformations. Even using rather sophisticated computational algorithms to avoid this problem (Dammkoehler *et al.*, 1989), only a few short peptides have been investigated in this manner, although strategies have evolved to make this tool usable for larger systems (Beusen *et al.*, 1993).

Build-up procedures are based on the previous systematic search of small peptide fragments (i.e., *N*-methyl-*N*′-acetylamides of amino acid residues, Lewis *et al.*, 1973), revealing all of their local energetic minima. The number of starting backbone conformations for a given residue varies from 5 to 10 depending on its nature (see Section IV). To examine an octapeptide, one would start by performing an energy minimization of all possible combinations of the starting backbone conformations for each residue in the C-terminal pentapeptide (residues 4–8). The number of possible combinations would be in the range of 5^5 to 10^5. The backbone conformers selected for further consideration are those which lie within a selected threshold (ΔE) above the minimum energy. At each subsequent step of the build-up procedure, a residue is added (thereby yielding the hexapeptide 3–8, then heptapeptide 2–8, and finally the entire molecule). With the addition of each residue, all of its starting backbone conformations are considered. At each step the spatial arrangement of side chains can be optimized prior to energy minimization by several cycles of a stepwise grid search of side-chain dihedral angles (see Nikiforovich, *et al.*, 1991, for more details). A build-up procedure can start from a central peptide fragment equally well; it can also involve overlapping peptide fragments (Galaktionov *et al.*, 1976). Owing to the computational feasibility of this procedure, it has been applied to many peptides: opioid peptides (Nikiforovich *et al.*, 1991), angiotensin (Nikiforovich and Marshall, 1993c), cholecystokinin (CCK)-related peptides (Nikiforovich and Hruby, 1993), α-melanotropins (Nikiforovich *et al.*, 1992), gonadotropin-releasing hormone (GnRH) agonists and antagonists (Nikiforovich and Marshall, 1993a), and others. The main weak point of any build-up procedure is the need for an *a priori* estimate of ΔE at each step of the procedure. A value of ΔE that is too high will result in an unmanageable number of conformers, whereas a lower value may result in omission of significant low-energy conformers from the final result.

Molecular dynamics (Allen and Tildesley, 1989; McCammon and Harvey, 1987; van Gunsteren and Berendsen, 1990) is a deterministic process which solves Newton's equations of motion for each atom and increments the position and velocity of each atom using a small time increment. In this paradigm, atoms are essentially a collection of billiard balls with classical mechanics determining their positions and velocities at any moment in time. As the position of one atom changes with respect to the others, the forces which it experiences also change. The forces on any particular atom can be calculated using the appropriate force field. The time step chosen is smaller than the period of fastest local motion to ensure that the position of surrounding atoms does not change significantly per incremental move. The time increment is typically on the order of 10^{-15} sec, which reflects the need to adequately represent atomic vibrations of similar time scale. For simulations of molecules in solvent, sufficient solvent molecules must be included to adequately represent all classes of solvent–solute interactions. This requires several hundred solvent molecules for even small solutes, and as a result simulations of more than several nanoseconds are rare. Because of the short time steps of molecular dynamics, events requiring longer times, such as diffusion, are difficult to simulate at the molecular level of detail. In this case, Brownian dynamics are used, and the particles (consisting of many atoms) move under the Langevin equations which govern diffusion rather than Newton's equations (Madura *et al.*, 1994; Pastor *et al.*, 1988). Electrostatic forces are derived from the relative positions of the charged particles in the simulation by solution of the Poisson–Boltzmann equation which describes dielectric behavior in a nonhomogeneous system.

The Monte Carlo method (Allen and Tildesley, 1989) is based on statistical mechanics and generates sufficient different configurations of a system by computer simulation to allow the desired structural, statistical, and thermodynamic properties to be calculated as a weighted average of these properties over these configurations. Monte Carlo simulations are successfully performed by sampling only a limited set of the energetically feasible conformations, say, 10^6 out of 10^{100} theoretical possibilities. One could sample all states, calculate the energy of each, and then Boltzmann weight its contribution to the average. Instead, Monte Carlo operates by importance sampling and looking only at energetically feasible answers among all possibilities.

The term Monte Carlo comes from the random selection of the parameter (e.g., coordinate or torsion angle) which determines the next configuration. The energy of the new state is compared with the old state. If it is the same, or lower, the new configuration is kept and becomes the basis for calculation of the next configuration. At this point, we can see that the procedure functions as a crude minimizer. If the successive configuration has a higher energy than the previous configuration, then it is either kept or discarded depending on the energetic difference (ΔE) and a random number, x, chosen between 0 and 1. If $\exp(-\Delta E/kT) \geq x$, then the new configuration is accepted. In other words, there is a unitary probability of accepting a move which results in an energy decrease, and an exponential probability based on the Boltzmann factor of accepting a move with a higher energy. This procedure generates trajectories which sample configurations in accord with the canonical Boltzmann distribution, and average properties can be calculated by simply averaging the properties associated with each configuration.

One aspect shared by Monte Carlo methods and molecular dynamics is the ability to cross barriers. In the case of Monte Carlo, barrier crossing occurs by random change and acceptance of higher energy states as a function of temperature. Because it is difficult to simulate systems with explicit solvent for long enough to allow conformational transitions, there is always a concern that sampling of the potential surface is insufficient. One approach to this problem is to do multiple runs from different starting configurations of the system. One can examine convergence of the ensemble averaged properties of each run to determine if adequate sampling has occurred. Obviously, the ability of the system to make transitions over activation energy barriers depends on the temperature. To increase the efficiency of sampling of conformational space, one can elevate the temperature for a short period and then resume the simulation at the desired temperature, allowing time for reequilibration. This technique, called simulated annealing (Moskowitz *et al.*, 1988; Nilges *et al.*, 1988; van Schaik *et al.*, 1992), is useful for overcoming activation energy barriers between local minima.

Both molecular dynamics and Monte Carlo techniques have been extensively used for conformational analysis of peptides. Molecular dynamics has been used to identify likely conformers of peptides in unrestrained analyses (O'Connor *et al.*, 1992; Vega *et al.*, 1992); to assess the impact of steric or covalent constraints on the conformational properties of peptides [e.g., analogs of opioid peptides (Smith and Pettit, 1991; Wilkes and Schiller, 1992) and GnRH (Struthers *et al.*, 1990)]; and to identify conformations consistent with experimental restraints (Kessler *et al.*, 1988b, 1990b). Long, unrestrained simulations have also been used to evaluate the turn- and helix-forming properties of linear sequences (Daggett and Levitt, 1992; Hermans, 1993; Jacchieri and Richards, 1993; Soman *et al.*, 1991; Tobias and Brooks, 1991; Tobias *et al.*, 1990, 1991). The Monte Carlo approach has been used in similar protocols. For instance, 10,000 conformations were generated for enkephalin and angiotensin using the Monte Carlo technique (Olatunij and Premilat, 1987), and about 10^6 conformations for CCK-8 (Kreissler *et al.*, 1989). In the last case, only five types of low-energy backbone conformers were revealed for CCK-8, which presumably do not consist of the complete set of low-energy minima for this linear octapeptide. On the other hand, the Monte Carlo sample of 10^6 conformations is certainly not sufficient for an accurate representation of the entire conformational space of a peptide system like CCK-8, which involves about 30 degrees of freedom even with the assumption of rigid-valence geometry. Therefore, a special procedure has been developed for generating Monte Carlo samples only in the vicinity of low-energy minima determined previously by a build-up procedure (Betins and Nikiforovich, 1984). It appears that generation of 30,000–60,000 conformations for each low-energy region allows the *a priori* estimation of the statistical weights of the minima for enkephalin (17 degrees of freedom) with accuracy sufficient to reproduce the available nuclear magnetic resonance (NMR) and fluorescence data (Betins and Nikiforovich, 1984). The same procedure performed for angiotensin (Nikiforovich *et al.*, 1987) and dermenkephalin (Nikiforovich *et al.*, 1993) has been helpful in estimating the statistical weights for conformations of these peptides in solution.

Non-Boltzmann sampling is an approach to sampling high-energy states that are infrequently reached in simulations (Beveridge and DiCapua, 1989). Often we are interested in sampling only those conformations relevant to a particular chemical question, for example, the energetics associated with a conformational transition from conformer A to conformer B. By definition, transition states are higher energy and would not be frequently sampled in an unconstrained simulation. Often a series of individual simulations will be conducted with constraints to focus each on a discrete region of the reaction coordinate of interest. This procedure is called umbrella sampling, and the effects of the different constraining potentials for each individual simulation can be removed to generate a potential of mean force (pmf) for the reaction coordinate. Occasionally, there are sufficient crystal structure data that examination of the variation in the data will indicate the transition path (reaction coordinate) between two states of interest. A more general approach (Choi and Elber, 1991; Czerminski and Elber, 1990) is based on the force field of the system itself and attempts to map a low-energy pathway, a minimum potential energy pathway (MPEP), between the two states by a variety of mathematical techniques. The resultant MPEP on the potential energy surface is only one of many possible, and because the role of entropic stabilization has not been taken into account, this is not necessarily the path with the lowest free energy of activation. It is, however, a logical starting point for simulations, for example, umbrella sampling, to determine an upper bound (as there may be another path of lower free energy) on the free energy associated with the transformation represented by the reaction coordinate.

Distance geometry (Blaney and Dixon, 1994; Crippen and Havel, 1988; Havel and Wuthrich, 1984; Havel, 1991; Kuntz *et al.*, 1989) is yet another means of generating conformations of a molecule. There are three stages in this calculation: construction of bounds matrices and bounds smoothing; embedding; and optimization. In the first phase, matrices are created which enumerate the minimum and maximum distance between all pairs of nuclei in the molecule. Entries in the matrices originate in experimental observations [e.g., nuclear Overhauser effect (NOE)-derived distances] or are holonomic constraints derived from the covalent structure of the molecule (van der Waals radii, bond lengths, 1–3 distances, and 1–4 distances). For most molecules, only a small fraction of the required interatomic distances are available from these sources. Additional constraints are implicit in the geometric relationships (e.g., triangle inequalities) between the distances, and in the smoothing process entries in the bounds matrices are adjusted to satisfy these relationships. The result is a general tightening of distance limits. A trial distance matrix is created by randomly selecting internuclear distances within the smoothed upper and lower limits. This trial distance matrix is used in the embedding process by which Cartesian coordinates are calculated from the matrix of trial distances. The structure produced from the embedding process is only an approximate fit to the distance restraints. The optimization of x, y, and z coordinates in the final stage is with respect to the input distance constraints and not with respect to a molecular mechanics force field.

In principle, one could search for all possible conformations of a molecule by distance geometry. The problem is that the number of possible states can be enormous, and the sampling properties of the algorithm in the absence of sufficient constraints may be biased (Metzler *et al.*, 1989). Both the distribution of distances selected for the trial matrix and the order in which the bounds are smoothed impact the sampling properties of distance geometry (Havel, 1990; Oshiro *et al.*, 1991). Distance geometry has been used in the unconstrained conformational analysis of cyclic peptides (Peishoff *et al.*, 1990). By requiring that pharmacophores in different molecules be superimposed, distance geometry has been used to identify the common spatial orientation of functional groups for receptor binding (Sheridan *et al.*, 1986). This is analagous to the systematic search approach discussed earlier, in which one considers all possible conformations rather than just low-energy conformers. At present, the most common use of distance geometry is in the analysis of NMR data (Wagner *et al.*, 1992), largely because it is an efficient, path-independent means of generating initial structures for which a large number of experimental constraints are available.

III. Experimental Analysis

As an alternative to identifying possible bioactive conformations by computing the potential energy surface of a peptide, one can experimentally evaluate its conformational properties. The experimental data of most relevance to the conformation of a peptide is clearly derived from structural analyses in the environment of interest. For a peptide that functions as a receptor ligand, that would be the ligand–receptor complex; for an ion channel, the lipid bilayer is appropriate. The heterogeneity of these environments is difficult to reproduce in any symmetric solution or solid-state environment. Because the conformation of a peptide reflects a balance of intra- and intermolecular forces, there is no guarantee that observations made outside the receptor environment bear a one-to-one correspondence with those that could be made on the ligand–receptor complex. Unfortunately, structural information on complexes is rarely available, and one is forced to use any available experimental information on the conformation of the isolated peptide. In these cases, the strategy is to search by experimental means for any evidence of a stable conformation and to enhance its stability through the incorporation of steric and/or covalent constraints. If these constraints can be shown to stabilize a particular conformation and correlate with enhanced binding affinity and/or activity, confidence in the relevance of the conformer to activity may increase. This is especially true if the preferred conformer persists in different environments and can be corroborated by several structural methods. It is worth noting, however, that even seemingly well-constrained peptides can exhibit environmentally dependent conformations (Marshall *et al.*, 1990b).

Stability and dynamics can be viewed as extremes in a continuum defining the conformational properties of a molecule. Whether a conformation is to be considered stable depends on the time scale of interest. Experimental techniques which probe the conformational state of a peptide are distinguishable not only in the types of structural information they reveal, but also in the time scale of their measurement. Structural studies of linear peptides often report that they are undergoing "conformational averaging". This inclusive term is used to describe the case in which several discrete, stable conformational states are simultaneously present as well as the case in which interconversion between states is occurring during the time of observation. The experimental approaches described below differ in ability to discriminate these two possibilities, to estimate relative populations of stable conformers, and to provide information on the rates of interconversion between conformers.

In the early stages of conformationally directed peptide design, X-ray crystallography was used to identify low-energy conformations of peptides which were assumed to be relevant to the receptor-bound conformation (Horn and De Ranter, 1984). At the time, it was the only experimental technique that could provide enough detailed, quantitative information to define the three-dimensional structure of a peptide. It is now recognized that the forces required for torsional deformations are small and the crystalline packing environment may dictate a conformation which has no relevance to the receptor-bound state. In addition, other experimental methods such as NMR (see below) have matured and are now capable of providing this information in environments that are more relevant. Even with these limitations, the X-ray structure of a free peptide is still a viable tool for determining precise equilibrium bond lengths and bond angles, which are essential for modeling and are unlikely to be available for the unusual amino acids or surrogates that are generated in the design process (Talluri *et al.*, 1987). Multiple X-ray analyses of a peptide can elucidate a finite number of conformational classes accessible to a peptide (Rao, 1992). X-Ray structures of ligand–receptor complexes reveal functional groups on both entities essential for binding as well as the bound conformation of the ligand and are of unquestioned utility in the design of high-affinity ligands (Appelt, 1993). The requirement for a crystalline sample is still a barrier to X-ray analysis, even for peptides in the absence of receptor.

The nature of X-ray analysis is that information on the dynamics of a molecule is largely limited to thermal vibrations, as larger motions are excluded by the requirement of a crystalline lattice. Regions whose conformation is identical throughout the crystal lattice will produce a unique structure, whether that be one or more conformations in the unit cell. Conformational averaging is observed as reduced electron density due to partial occupancy of the unit cell, or a complete absence of density which reflects disorder—a lack of periodicity in the solid state. In the extreme, widespread disorder makes crystallization impossible.

Solution NMR spectroscopy is similar to X-ray analyses in providing large amounts of detailed, atom-based structural information on peptides. Most of the structural information derived from NMR studies is based on the NOE (Noggle and Schirmer, 1971), which detects interproton distances of 5 Å or less, and coupling constants (typically $^3J_{HH}$), which are a function of torsion angle (Bystrov, 1976). This information can be used in a qualitative sense to identify regions of secondary structure because turns, sheets, and helices all present different patterns of NOEs and coupling constants (Wuthrich, 1986). To determine atomic coordinates, the measured NOEs and $^3J_{HH}$ values are converted to distance and torsion angle restraints that are implemented in any of the conformational searching tools discussed earlier.

Solution NMR data coupled with computational methods (primarily distance geometry (have been used to solve the solution structures of proteins up to 18 kDa in size (Wagner, 1993b). Most studies have involved uncomplexed protein, but in some cases, complexes have been solved. The difficulties in sample preparation arising from solubility and/or aggregation problems at the sample concentrations required (0.1–2 m*M*) are not always appreciated. Three-

and four-dimensional experiments, which require isotopic labels (Bax and Grezsiek, 1993), have made NOE spectroscopy (NOESY) evaluation of proteins feasible by addressing the signal overlap problem. Transferred NOE (Clore and Gronenborn, 1982) and isotope-directed NOE experiments (Fesik, 1991) can provide information on the bound conformation of a ligand without solving the structure of the entire complex. In transferred NOE experiments, the ligand released from the receptor is observed. In essence, the released ligand retains a "memory" of its conformation in bound state. This experiment is most useful when exchange between bound and free forms of the ligand is 1000 or more times faster than T_1 for the free ligand. As a result, transfer NOE experiments are usually feasible only for low-affinity ligands. Isotope-directed NOE experiments (Fesik *et al.*, 1988) detect only those protons attached to isotopically labeled nuclei. By labeling only the ligand, one can edit the spectrum of the complex so only ligand protons are observed. Because one is observing the complex and not relying on exchange processes, the affinity of the ligand can be higher than in the case of the transferred NOE.

When analysis of the receptor–ligand complex by any of the methods listed above is not feasible. The ligand alone is often analyzed for any evidence of conformational preference. The behavior of the ^{1}H–^{1}H nuclear Overhauser effect as a function of molecule size is well known (Noggle and Schirmer, 1971): for small molecules in the extreme narrowing limit the NOEs are positive; for proteins in spin diffusion limit the NOEs are negative; and at the transition point between the two extremes, NOEs are zero. This is precisely the motional regime which often applies to peptides 3–20 residues in length. To sidestep this difficulty, ROESY experiments (Bothner-By *et al.*, 1984) can be performed. These are NOE experiments done in the rotating frame, and under the conditions of this experiment, all cross-relaxation peaks are positive over the entire range of molecular size. Another alternative is the use of cryoprotective solvents [e.g., dimethyl sulfoxide (DMSO) in water (Fesik and Olejniczak, 1987)] that have high viscosities at the temperatures typically used for NOESY experiments. The correlation time of the molecule is increased, effectively shifting the zero crossover point of the NOE to lower molecular weights.

Extracting distances and torsion angles from NMR data to use as restraints in conformational searches generally rests on a significant assumption: the molecule is present in solution as a single conformer. Most linear peptides display conformational averaging in solution (Dyson and Wright, 1991), and, in fact, cyclic peptides often do as well (Kopple *et al.*, 1993). Several short-range interproton distances (e.g., sequential αH–NH) are less than the detectable limit of 5 Å over many conformational states. The long time scale of the NMR measurement (10^{-3}–10^{-5} sec) allows contributions to the same chemical shift from conformations exchanging over a wide rate window. Consequently, the observed NOEs are population-weighted averages, and the use of distance and torsion angle restraints extracted from the data can give rise to a "virtual conformation"—a structure which satisfies the experimental restraints but in the absence of restraints would be a high-energy conformation (Jardetzky, 1980). Conformational averaging can be diagnosed by several features: (1) the failure of any energetically reasonable conformation to accommodate all distance and torsion angle restraints (this can be the case globally, representing complete disordering of the structure, or in local "hot spots" in a candidate structure, i.e., regions in which restraints are poorly satisfied, due to averaging); (2) coupling constant and NOE values that are not consistent with one another (these average differently, with $^3J_{HH}$ being a simple population-weighted average, whereas the NOE has an additional r^{-6} weighting); and (3) the absence of coordinated, characteristic information over a contiguous block of amino acids necessary to define a secondary structure in that region. This includes medium-range NOEs, coupling constants, and hydrogen bonds (Wuthrich, 1986) as well as chemical shift information (Wishart *et al.*, 1991).

Several methods have evolved for dealing with conformational averaging in the NMR spectrum which differ in whether any of the experimental data is used to select contributing conformers and how the populations are calculated. One strategy which uses the NMR data to aid in selecting candidate conformers is to implement restraints over just that part of the molecule which appears to be well-defined (Kessler *et al.*, 1988a) and allow the conformational search to generate possible conformations of the averaged portion. Another approach assumes that some population of structures must be present which reflect the observed NOE, because the r^{-6} weighting skews the observed NOE to short distances. From the data, subsets of restraints are randomly selected and satisfied to generate a set of candidate conformations, each of which explains a part of the data (Bruschweiler *et al.*, 1991). For conformational searching tools which incorporate a time step such as molecular dynamics, one could implement time-averaged restraints (Torda *et al.*, 1990) over parts of the molecule which are undergoing averaging. This relaxes the requirement that a conformation satisfy a given set of restraints at any instant and allows a broader range of conformational possibilities which, when averaged, could explain the data.

Other approaches do not incorporate the primary experimental data in the generation of candidate conformers. Instead, energetically reasonable candidates are used in a set of simultaneous equations to solve for the relative populations of each. In the best known example (Pachler, 1964), the three staggered states of an amino acid side chain are the candidate structures and the observed $J\alpha\beta'$, $J\alpha\beta''$, and $J\beta'\beta''$ are used to solve for relative populations of trans, gauche$^+$, and gauch$^-$ rotamers. Another option is to do an unconstrained conformational search, cluster the results, and solve for the coefficients which define the relative contribution of each cluster to the NMR results (Landis and Allured, 1991). Alternatively, instead of determining populations by fitting to the data, one could calculate the Boltzmann-weighted (Cumming and Carver, 1987) contribution from each cluster. In cases where the number of degrees of torsional freedom are too great (and as a result the number of candidates greater than the number of observables), one can also impose the requirement that the starting set of candidate conformations all lie within a given energy of the absolute minimum energy conformation (Nikiforovich *et al.*, 1988). Any energetic criterion for filtering candidate conformations or calculating relative populations carries with it the inherent limitations of the force field and the solvation model employed. Yet another option would be to select candidates based on the probabilities of the preferred conformational states of each amino acid as determined from statistical analyses of protein crystal structures (Sherman and Johnson, 1993). All of these approaches focus on identifying discrete conformational candidates that contribute to the observed spectrum. An entirely different approach uses the experimental data to solve for the continuous probability function about a given torsion angle (Dzakula *et al.*, 1992). In theory, this approach avoids the issue of inadequate sampling of conformational space implicit in the use of discrete candidates for conformational averaging.

The net effect of conformational averaging for linear peptides is to dilute signals which might identify conformational preferences for all or part of a molecule. In other words, the multiplicity of folded forms appears to be equivalent to an unstructured "random coil." In these cases, one option is to examine the data to search for subtle effects that indicate folded conformations. Examples would be weak, nonsequential NOEs such as the α_i–NH_{i+2} NOE typical of β turns (Wuthrich, 1986), or deviation of chemical shifts from canonical random values. Another option is to vary the solution environment, until one finds a solvent in which a single dominant conformation is observed. Cyclolinopeptide (Di Blasio *et al.*, 1989) and cyclosporin A (Kessler *et al.*, 1990b) are peptides which, in an appropriate solvent environment, adopt predominantly a single conformation. As noted earlier in discussing X-ray analyses, this approach may identify a low-energy conformation in a given environment, but its relevance to the bioactive conformation is uncertain.

The measurable relaxation properties of nuclei (T_1, T_2, and ^{13}C or ^{15}N NOEs) can aid in the interpretation of ^1H–^1H NOEs, coupling constants, and other structural data by indicating regions of a molecule undergoing internal motion and consequently subject to conformational averaging (Wagner, 1993a). Slow conformational exchange can be observed in NOE experiments, whereas broadened line shapes may arise from conformational exchange that is comparable to the NMR time scale (Ernst *et al.*, 1988; Jardetzky and Roberts, 1981). In the case of cyclosporin A (Dellwo and Wand, 1989) and alamethicin (Kelsh *et al.*, 1992), ^{13}C T_1 studies have been used to estimate rates and amplitudes of localized motions, to calculate the overall correlation time of the molecule, and to identify conformationally well-defined regions of the molecule. Characterization of the dynamics of peptides in some cases is sufficient to allow comparison of experimentally observed conformational interconversion with that predicted by molecular dynamics simulations, revealing both successes and limitations in current molecular mechanics force fields (Brunne *et al.*, 1993; Schmidt *et al.*, 1993). Other spectroscopic techniques such as circular dichroism (Merutka *et al.*, 1993) and time-resolved fluorescence (Palmer *et al.*, 1993) complement and corroborate assessments of conformational dynamics determined by NMR.

Solid-state NMR can provide diverse structural and dynamic information, particularly for samples which are difficult to crystallize for X-ray analysis or whose size precludes study by solution NMR. In solution, the random tumbling of molecules of molecular weight below 20 kDa averages out chemical shift anistropy and dipole–dipole interactions. As a result, lines are narrow, and detailed structural information from NOEs and scalar couplings can be extracted. In contrast, samples for solid-state NMR are relatively immobilized, and the orientation dependence of NMR parameters such as chemical shift and quadrupolar coupling is retained. Because many different orientations are sampled, solid-state NMR spectra exhibit broad lines. Internal motion reduces the width of the powder pattern, and analysis of line width and line shape, coupled with spin–lattice relaxation determinations (primarily using ^{2}H) can suggest the frequency and amplitude of internal dynamics (Sparks *et al.*, 1989; Torchia, 1984). The primary requirements for solid-state NMR are rare spins (^{2}H, ^{13}C, etc.) and, owing to the low sensitivity of these nuclei, large amounts of sample. The limited resolution in static, randomly oriented samples precludes the extraction of detailed structural information by solid-state NMR. To obtain geometric parameters necessary for structure generation, two approaches have been taken (Smith and Peersen, 1992). In one case, the sample is oriented within the magnetic field, and the orientation dependence of chemical shift tensors and dipolar interactions is used to extract structural information. In the second, chemical shift anisotropy is averaged by spinning the sample at the magic angle (MAS) to generate narrow lines, and specialized pulse sequences are used to recover the rotationally averaged, distance-dependent dipolar couplings.

Both the ^{15}N chemical shift tensor and the ^{15}N–^{1}H directly bonded dipolar coupling depend on the orientation of the peptide plane in the magnetic field (Opella *et al.*, 1987). A combination of these two pieces of information for a block of contiguous amino acids can define their conformation, and this approach has been used to establish that magainin 2, an antibiotic which disrupts the electrochemical gradient across cell membranes, adopts an α-helical conformation in oriented bilayer samples and lies parallel to the membrane surface (Bechinger *et al.*, 1993). Analysis of membrane-bound bacteriophage Pf1 coat protein reveals it to have two helical regions: one which is hydrophobic and spans the bilayer and a second which is shorter, amphipathic, and lies in the plane of the bilayer (Shon *et al.*, 1991). This approach has been used to solve the structure of gramicidin A in the lipid bilayer (see Section V,A)

Rotational echo double resonance (REDOR) NMR (Pan *et al.*, 1990) is a magic-angle spinning experiment in which the echoes that normally form during each rotor cycle are prevented from reforming by rotor-synchronized π pulses applied to another nucleus. The difference in intensity between spectra obtained with and without the perturbing π pulses is related to the dipolar coupling constant, which in turn has an r^{-3} distance dependence. To date, REDOR has been used to measure distances between a variety of heteronuclear pairs, including ^{13}C and ^{15}N, ^{13}C and ^{19}F, ^{13}C and ^{31}P, and ^{15}N and ^{31}P. The precision of measurement is much greater than NOE measurements (±0.3 Å), in part because there is no requirement for a motional model to extract the distance information. The maximum distance that can be measured depends on the gyromagnetic ratio, γ, of the nuclei involved. REDOR studies of helical channel-forming peptides (Holl *et al.*, 1992; Marshall *et al.*, 1990a) have measured distances of 8 Å. The conformation and relative orientation of glyphosate and shikimate 3-phosphate bound to 5-enol-pyruvyl-shikimate-3-phosphate synthase have been determined (Beusen *et al.*, 1994; Christensen and Schaefer, 1993) using several labeled forms of the ligands and enzyme. REDOR has been used to measure ^{13}C–^{15}N distances in the type II β turn of crystalline melanostatin (Garbow and McWherter, 1993).

Rotational resonance (RR) solid-state NMR (Peersen and Smith, 1993; Raleigh *et al.*, 1989) is another MAS experiment that extracts distance information by measuring homonuclear dipolar couplings. Magnetization transfer between two resonances is enhanced by setting the rotor speed (or an integral multiple) equal to the chemical shift difference. One of the resonances is inverted, and the transfer (measured as the difference in signal intensity between the inverted and noninverted resonances) is monitored as a function of time. Simulations of this curve for a variety of distances/dipolar couplings are done to determine the best match to the experimental data. The precision and accuracy of the technique is of the order of 0.5 Å, and ^{13}C–^{13}C distances up to 6.8 Å have been measured in a helical alamethicin fragment (Peersen *et al.*, 1992). Rotational resonance has also been used to measure ^{13}C–^{13}C distances in studying the conformation of retinal bound to bacteriorhodopsin (Creuzet *et al.*, 1991; Thompson *et al.*, 1992) and the conformation of a peptide derived from the Alzheimer β-protein (Spencer *et al.*, 1991). The measurement of a ^{31}P–^{31}P distance revealed the mechanism of inhibition of D-alanyl-D-alanine ligase by a slow-binding inhibitor (McDermott *et al.*, 1990).

Electron spin resonance (ESR) employs stable, free radical reporter groups whose resonance spectra are sensitive to the environment and can reflect mobility and exposure of the spin-labeled site (Millhauser, 1992). The dipolar interaction between two free radicals causes an increase in relaxation rate, which is manifested as an increase in ESR line width. This perturbation has an inverse r^6 distance dependence, and, like the NOE experiment, quantitation of the distance requires an estimate of correlation time. The large gyromagnetic ratio of the electron (658 times that of ^1H) means that the technique has high sensitivity, which can be translated into a decreased sample requirement and the ability to measure distances greater than the 5 Å observable by ^1H–^1H NOE. The time scale of the ESR measurement itself is fast, in some cases of the order of 10^{-9} sec.

Typically, ESR reporting groups are nitroxides, although paramagnetic metal ions are also used. To avoid disproportionation during synthesis, nitroxides are usually introduced at the *N* terminus or through a disulfide bridge to a cysteine substituted in the sequence, although this limitation has been overcome (Marchetto *et al.*, 1993). Cysteine substitution mutants were used in time-resolved studies of the binding of colicin E1 to membranes (Shin *et al.*, 1993), which demonstrated rapid adsorption of the protein to the bilayer and slow insertion of the peptide into the membrane interior. A periodicity in the solvent exposure of several site-specifically labeled colicin E1 derivatives (probed by a second paramagnetic species, CrOx) correlated with the mobility of the labels determined by ESR (Todd *et al.*, 1989) and was interpreted as arising from an amphipathic helix. ESR measurements have been used to establish the presence of 3_{10} helices in solution for short peptides (Miick *et al.*, 1992) by incorporating pairs of nitroxide labels at varying distances from one another in the sequence. The relative mobility of residues in secondary structures, such as helices, has been probed by analyzing location-dependent differences in line shape (Miick *et al.*, 1993). ESR studies have been used to characterize the aggregation state and orientation of peptides interacting with membrane bilayers (Archer *et al.*, 1991; Wille *et al.*, 1989). Paramagnetic species can also be used to perturb NMR line widths, allowing distance measurements of 20 Å or more (de Jong *et al.*, 1988; Esposito *et al.*, 1992; Mildvan, 1989). Distances determined in this fashion have been used to dock a spin-labeled steroid into the crystal structure of ketosteroid isomerase (Kuliopulos *et al.*, 1987).

Circular dichroism (CD) (Johnson, 1988, 1990; Yang *et al.*, 1986) is useful for identifying the secondary structure content of peptides in solution environments, even under conditions which are not suitable for solution NMR analysis. An additional advantage is the small sample requirement, a trait held in common with other optical methods. In contrast to NMR, the time scale of the CD measurement is rapid and should in theory yield the population of particular secondary structural types directly. Initially, spectra were evaluated as summed contributions from basis spectra which correlated with secondary structure type. These basis spectra were developed from peptides with pure secondary structure (Greenfield and Fasman, 1969) or by deconvolution of the complex spectra for proteins of known secondary structure content (Hennessey and Johnson, 1981). The dependence of the basis spectra on length of secondary structure and the presence of nonpeptide chromophores limits the accuracy of analyses based on a linear combination of basis spectra.

The accuracy of CD structural determinations has improved with extension to shorter wavelengths (<200 mm) and by fitting spectra to a combination of spectra from proteins whose secondary structure is known (Provencher and Glockner, 1981). Although the original consideration of helix, sheet, and random coil states has been extended to include turns (Brahms and Brahms, 1980), the latter class is subject to the greatest errors in determination. The difficulty in generating models which are purely a single turn type has limited experimental verification of predicted spectra (Woody, 1985). Basis spectra for turns generated by deconvolution of spectra from a family of peptides having mixed turn conformations have been used to quantitate turn populations of peptides (Perczel *et al.*, 1991b); however, exact determination of turn type still depends on corroborative information from another source such as NMR (Perczel *et al.*, 1993; Smith and Pease, 1980). A CD spectral analysis can readily determine helix screw sense (Pieroni *et al.*, 1993). Quantitative determination of secondary structure content by CD continues to be an active research area, driven by the questions of appropriate standards for calibration, the type of secondary structure descriptors (Johnson, 1988), and computational methods for correlating secondary structure with experimental spectra (Bohm *et al.*, 1992; Perczel *et al.*, 1992; Sreerama and Woody, 1993).

Fourier transform infrared spectroscopy (FTIR) has developed significantly since the early 1980s. The rapid time scale ($<10^{-12}$ sec) of the measurement is complementary to that of ESR and NMR, and allows direct determination of conformational populations. The primary lines of interest are the N–H stretch, which gives information on hydrogen bonding, and the position of the amide I band, which arises mainly from backbone C=O stretching and has been correlated with secondary structure. In the past, the broad lines and inability to extract site-specific information have limited the impact of FTIR to qualitative structural determinations, but resolution enhancement by derivative techniques and spectral deconvolution have now made it possible to measure line positions within 0.1 cm^{-1} (Arrondo *et al.*, 1992; Krimm and Bandekar, 1986). Correlations to helix and sheet structures (Bandekar, 1992; Byler and Susi, 1986) have been extended to include turns (Mantsch *et al.*, 1993). Isotopic labeling allows site-specific perturbation of the FTIR spectrum (Tadesse *et al.*, 1991), and as the correlation between local structure and amide I line position improves (Prestrelski *et al.*, 1992), it should be possible to extract detailed information on local conformation from FTIR spectra (Surewicz *et al.*, 1993). As in the case of circular dichroism, FTIR has proved particularly useful for examining peptides that tend to aggregate (Gasset *et al.*, 1992; Lansbury, 1992; Wei *et al.*, 1991) or that are active in membranous environments (Braiman and Rothschild, 1988; Jackson *et al.*, 1992; Zhang *et al.*, 1992).

Fluorescence spectroscopy provides information on the local environment (Blatt and Sawyer, 1985) and site-specific distance information (Stryer, 1978) through the use of appropriate reporter groups such as tyrosine and tryptophan. Environmental effects are seen in steady-state experiments and include shifts in excitation and emission frequencies as well as perturbation of fluorescent intensity. These approaches have been used to characterize association of peptides with lipid vesicles (Chung *et al.*, 1992; Pouny and Shai, 1992) and to probe ligand–receptor interaction at the binding site (Chapman *et al.*, 1992). Energy transfer between fluorescent groups, which is based on dipole–dipole interactions and has an r^{-6} dependence, enables distance determinations of 50 Å or more (Stryer, 1978). CD and fluorescence energy transfer have been used to develop a conformation–activity correlation for β-casomorphin analogs (Epps *et al.*, 1991). Time-resolved energy transfer experiments reveal the distribution of distances between reporting groups (Grinvald *et al.*, 1972; Haas *et al.*, 1975) and enable characterization of molecules present as a mixture of conformational states (Amir and Haas, 1987). Complementary information on local dynamics can be obtained both from time-resolved energy transfer experiments (Beechem and Haas, 1989) and from frequency-domain experiments (Lakowicz *et al.*, 1986). The time scales measurable by fluorescence spectrometers ($<10^{-9}$ sec) and those approachable by computational methods have converged, allowing a direct (and favorable) comparison of simulated and experimentally observable values (Axelsen *et al.*, 1991).

IV. Goal-Directed Modifications of Conformational Flexibility

The first systematic investigation of the conformational flexibility of the peptide backbone was that of Ramachandran (Ramachandran and Sasisekharan, 1968). A plot of the two torsional variables, ϕ(C′–N–C$^\alpha$–C′) and ψ(N–C$^\alpha$–C′–N), indicating energetically allowed combinations of the two backbone torsional angles adjacent to the α carbon, has become known as a Ramachandran, or ϕ, ψ, plot. On the basis of sterically allowed areas of the Ramachandran plot, amino acids fall into three basic categories (Fig. 1): glycine, which is the most flexible and has most regions of the Ramachandran plot available; proline, which is the only naturally occurring, conformationally restricted amino acid residue in proteins with ϕ constrained to approximately -60° owing to an intraresidue cycle linking Cδ and the α-amino nitrogen; and all other residues, whose Ramachandran plots are nearly identical to that of alanine, which has only 50% of the torsional space available to glycine.

A closer examination of the Ramachandran plot reveals that it can be roughly divided into four quadrants: the upper left, corresponding to ($\phi < 0°$, $\psi > 0°$); the lower left ($\phi < 0°$, $\psi < 0°$); the upper right ($\phi > 0°$, $\psi > 0°$); and the lower right ($\phi > 0°$, $\psi < 0°$). The upper left quadrant contains (ϕ, ψ) points corresponding to a β structure, and the lower left and upper right quadrants contain points corresponding to right- and left-handed α helices, respectively. At this rough "quadrant" approximation, all quadrants except the lower right can be considered as sterically allowed for L-amino acid residues. On the other hand, all quadrants except the upper left are allowed for a D-amino acid residue (i.e., the Ramachandran plots for the L- and D-amino acid residues are symmetric with respect to rotation by 180° around an axis at the center of the plot). Steric contacts involving the Cβ atom are the reason for differences in the Ramachandran plots of L- and D-amino acid residues. All four quadrants are allowed for a Gly residue, and only two of them, namely, those at the left side of the plot, are allowed for Pro.

Simple modification of the naturally occurring amino acids can dramatically change the accessible region of the Ramachandran plot. For example, the replacement of the α proton by a methyl group to yield α,α-dialkyl amino acids

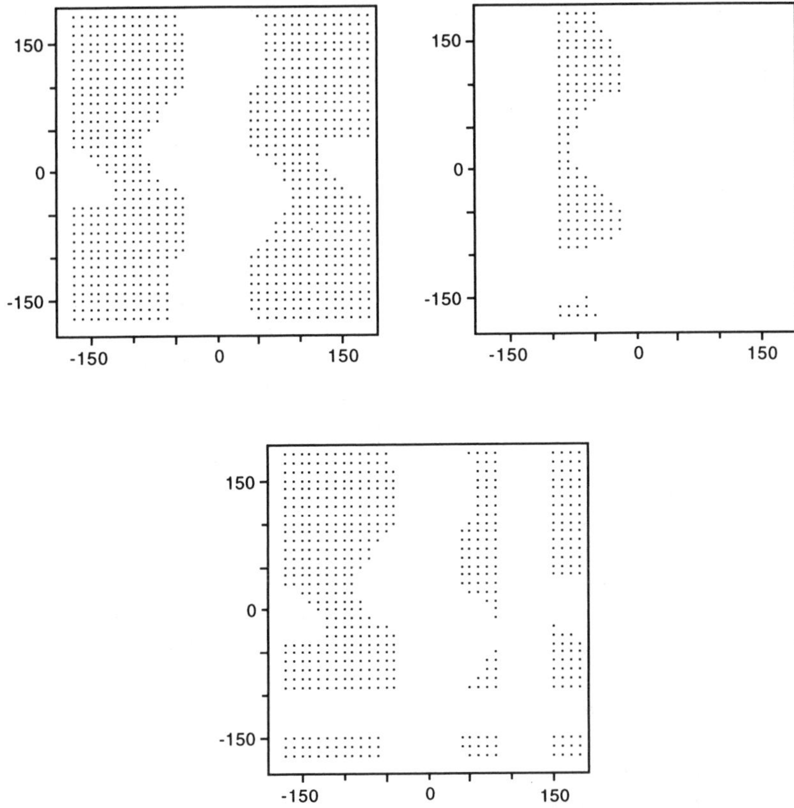

Figure 1 Ramachandran (ϕ,ψ) plots for glycine (top left), proline (top right), and alanine (bottom). Dots represent sterically allowed conformations found in a systematic conformational search using SYBYL.

results in a Ramachandran plot limited primarily, although not exclusively, to the left- and right-handed helical regions. A simple way of looking at the restriction imposed is to consider that α-methylalanine combines the steric effects of both L-alanine and D-alanine and, thereby, reduces the sterically allowed area of the Ramachandran plot to approximately 1% (Marshall and Bosshard, 1972). This modification can be useful for imposing a turn on a peptide, and incorporation of several of these residues leads to helical peptides. Other classes of amino acids which are useful in probing the conformational requirements of receptor recognition are *N*-methyl-amino acids, and α,β-dehydroamino acids in which the bond between the α and β carbons is unsaturated. In the latter case, the delocalized system introduces considerable rigidity in the peptide backbone, inducing turns and 3_{10} helices which have been documented both theoretically and experimentally (Chauhan *et al.*, 1987; Ciajolo *et al.*, 1990; Palmer *et al.*, 1992; Pietrzynski *et al.*, 1992).

Differences in the Ramachandran plots of the various types of amino acid residues outlined here can be useful for purposeful alterations of conformational flexibility of peptides in the process of peptide drug design. Normally in a design process, the investigator proposes a possible model for a "biologically active conformation", that is, the conformation in which functional groups are oriented in space to enable the specific biological response. Some elements of this model can be directly deduced using the Ramachandran plot to evaluate the structure–activity relationships in a series of analogs for a given peptide. The effect of a single amino acid substitution on the biological activity of an analog can be evaluated in terms of the backbone conformations accessible to that substituted residue. For instance, when a relatively flexible residue such as Gly is replaced by another one having a more restricted Ramachandran plot (e.g., Pro), and the resulting analog is still active, it is likely that the biologically active backbone conformer at the substituted position is characterized by allowed quadrants of the Ramachandran plot which are held in common (in this case, the left-hand side of the plot). Moreover, vice versa, when the resulting analog is inactive, it may indicate no overlap in sterically allowed quadrants for the analog and the parent peptide. This approach can be easily extended to substitutions of L- or D-amino acids for Gly, substitution of D-amino acid residues for L, or for any substitutions involving proline. (There is an assumption, namely, that the changes in biological activity of the analog can be assigned to conformational changes in the backbone only. It should be obvious that such an assumption has only limited application.)

Design considerations based on the Ramachandran plot have been applied, for instance, in studies of possible bioactive conformers of GnRH (LHRH, luliberin) and enkephalin. In both cases, substitution of a Gly residue (Gly^2 in enkephalin and Gly^6 in GnRH) by D-Ala resulted in an active analog, whereas replacement by L-Ala produced an inactive analog [see, e.g., Hansen and Morgan (1984) and Monahan *et al.* (1973), respectively]. These findings were interpreted as indicating that the backbone conformations of Gly in the corresponding bioactive conformers were not extended ones (such as those β-structure-like conformations in the upper left quadrant). Rather, it was suggested that the Gly residues in question can serve as "conformational hinges" leading to folding of the backbones. Evidence for such folding was obtained by spectroscopic measurements and by synthesis of active cyclic analogs of both GnRH (Rivier *et al.*, 1988) and enkephalin (Hansen and Morgan, 1984). In the case of bradykinin, substitution of Gly^4 for either isomer of alanine resulted in significant loss of activity, implying a possible turn in close proximity to the receptor (Kaczmarek *et al.*, 1994).

The presence of Pro (and any other *N*-methyl-amino acid residue) in a peptide chain may cause an additional effect in the Ramachandran plot for the preceding residue (MacArthur and Thornton, 1991). In this case, the quadrant containing right-handed α-helical conformers appear to be sterically forbidden for an L-amino acid residue, and the same is true for the quadrant containing left-handed α-helical conformers for a D-amino acid residue. This feature can be used to investigate the backbone conformation of the X residue in an X-Pro (or X-NMeAA) fragment). Examples in which this consideration has been employed in order to suggest the biologically active backbone conformers include tetragastrin, substance P, bradykinin, and angiotensin (for a review, see Nikiforovich, 1986). *N*-Methyl-amino acid residues can complicate the analysis by facilitating trans/cis isomerization of the peptide bond in the X-NMe-residue fragment, owing to the small energetic barrier to interconversion.

The considerations discussed above can also be applied in the design of peptide conformations involving backbone structures consisting of two or more neighboring residues. A propensity for secondary structures such as α helices, β sheets, and extended structures is evident for each amino acid type, and an analysis similar to that outlined above can be extended to *n* residues. Particularly instructive is the use of the Ramachandran plot in examining structures which reverse the peptide chain, now commonly known as β turns (Rose *et al.*, 1985). In looking at the definitions of β turns, it is evident that the $i + 1$ residue of β-II′ turns and the $i + 2$ residue of β-I′, β-II, and β-V turns are found mainly in the lower right quadrant of the Ramachandran map, a region more compatible with Gly or D-amino acid residues (Hruby and Nikiforovich, 1991). In contrast, the destabilization of β turns can be achieved by substituting Pro for the $i + 2$ residue in β-I and β-III turns (Hruby and Nikiforovich, 1991). In other words, it is possible to influence the conformation of a four-residue fragment by perturbing the conformational preference of one residue only. These considerations played major roles in the design of conformationally constrained, potent analogs of such peptides as somatostatin (Veber, 1992) and α-melanotropin (Sugg *et al.*, 1986).

Whereas the accessible areas outlined in the Ramachandran plot indicate the probable torsional values available to an individual residue, incorporation of that residue into a longer peptide allows for additional interactions which can modify the potential surface for the residue. The balance between short-range and long-range effects (with respect to sequence) has led to the development of covalently constrained structures whose intent is to make the short-range effects dominant, and in fact to propagate their effect throughout a sequence. The prevalence of the turn motif in molecular recognition has led to the development of numerous cyclic and bicyclic dipeptide analogs (Ball and Alewood, 1990; Ball *et al.*, 1993; Holzemann, 1991a,b) which involve covalent linkage of two or more residues and which help stabilize the peptide chain in a reverse turn. Examples include the dipeptide lactam of Freidinger *et al.* (1982), the bicyclic dipeptide BTD (Nagai *et al.*, 1993), and spirobicyclic systems based on proline (Genin and Johnson, 1992; Hinds *et al.*, 1988, 1991; Ward *et al.*, 1990). Another approach in designing turn mimetics is to replace the hydrogen bonding groups which stabilize the turn by covalent bonds (Chen *et al.*, 1992; Gardner *et al.*, 1993; Kahn *et al.*, 1991; Nakanishi *et al.*, 1992; Arrhenius and Satterthwait, 1990). A similar approach is seen in the γ turn mimetics of Callahan *et al.*

(1993). Benzodiazepines have also been used as turn mimetics (Ku *et al.*, 1993; Ripka *et al.*, 1993a,b). Other strategies involve the design of molecules which serve as nucleation sites for the hydrogen bonding patterns of secondary structures (Kemp, 1992) and may not be amino acids at all.

A paradigm for the applications of conformational energy calculations to molecular modeling in peptides has evolved (for review see, e.g., Marshall *et al.*, 1994) from many of the considerations discussed above. This paradigm is based on the assumption that the receptor-bound (biologically active) conformation for highly potent peptide analogs resembles one of the low-energy conformers. By systematically sampling the variety of low-energy conformers for each of a set of active analogs and comparing them for three-dimensional commonality, one can deduce a common low-energy conformation accessible to all the analogs that, hypothetically, resembles the bioactive conformation.

The exact approach to a particular question varies depending on the information available on a given peptide and its analogs, the calculation procedures involved, etc. The following steps, however, are useful in obtaining a reasonable starting model of the biologically active conformation employing the approach we have developed:

1. Decide on the specificity and selectivity of the peptide–receptor interaction to be modeled. In many cases, the modes of interaction of the same peptide with different specific receptors may be different, and consequently the models of the biologically active conformers should be different as well. Considerable evidence indicates that agonists and antagonists have different binding modes and, therefore, should be modeled independently [see, e.g., approaches to determine agonist and antagonist receptor-bound conformers for oxytocin/vasopressin (Hruby *et al.*, 1990) and LHRH (Nikiforovich and Marshall, 1993a,b)].

2. Determine (or hypothesize) which groups in the peptide are important for receptor recognition and activation. This set of groups can be refined by subsequent comparison of the spatial arrangements of these functional groups in various analogs.

3. Perform energy calculations for several active analogs as well as for the peptide itself and determine sets of low-energy conformers for each. Calculations for conformationally different analogs (e.g., those with substitutions of D-, *N*-Me-, dehydro-, or α-Me-amino acid residues, or those with cyclic constraints) are preferable at this step for the reasons mentioned above. Generally, a more conformationally diverse set will yield a more reliable conclusion as to a possible model of the receptor-bound-conformers. Needless to say, it is essential to consider sets of low-energy conformers for each compound, not just a single lowest-energy conformer.

4. Compare spatial arrangements of the functionally important groups for every low-energy conformer of the peptide and its active analogs. Those sets of

conformers with similar spatial arrangements of the functionally important groups common to all active compounds are candidate models for the biologically active conformation. Sometimes, it is possible to include nonactive analogs in the comparison as well, to eliminate common conformations from consideration. However, it should be remembered that the reasons for lack of activity may not be directly connected to conformational properties of the analog. For instance, any "conformational" modification of the Gly^3 residue in enkephalins, such as substitutions for L-, D-, αMe-, or *N*-Me-Ala, led to inactive analogs (Hansen and Morgan, 1984); at the same time, a large class of analogs with this residue deleted show high potency in binding and biological testing both at μ- and δ-opioid receptors (Schiller, 1991).

5. Use the derived model(s) to design new conformationally constrained analogs that test its predictive value. Energy calculations of the new analogs are then performed independently, and the model is refined against the biological results. At this step, it is very important to obtain the entire set of low-energy conformers for the newly designed analog, not just to check their compatibility with the proposed model of the receptor-bound conformer.

This approach was successfully employed to deduce a model of the receptor-bound conformation for δ-opioid peptides (Nikiforovich *et al.*, 1991). Three δ-selective analogs of enkephalin, namely, DPDPE (Tyr-*cyclo*[D-Pen-Gly-Phe-D-Pen]), JOM-13 (Tyr-*cyclo*[D-Cys-Phe-D-Pen]), and dermenkephalin (Tyr-D-Met-Phe-His-Leu-Met-Asp-NH_2, DRE) were selected for molecular modeling. Previous work in the field suggested a common spatial arrangement of the N-terminal α-amino group and the side chains of the Tyr^1 and $Phe^{4/3}$ residues in the receptor-bound conformation of δ-opioid peptides. Energy calculations for all three peptides were performed by the build-up procedure with the use of the ECEPP potential field. A comparison of the spatial arrangement of functionally important groups found a specific type of low-energy conformer similar to all δ-selective peptides in question, and allowed construction of a starting model for the δ-receptor-bound conformation. The most characteristic features of the model was the placement of the Phe side chain in space corresponding to a χ_1 rotamer that is gauche(−) for peptides containing Phe^4 and trans for peptides with Phe^3 (Nikiforovich *et al.*, 1991). The model was refined by considering new "hybrid" analogs of opioid peptides with pronounced shifts in affinities toward μ- and δ-opioid receptors (Sagan *et al.*, 1989). Energy calculations and geometrical comparison of these analogs together with previous findings led to the suggestion that there is a unique, common spatial arrangement of the α-amino group and Tyr and Phe aromatic moieties in the DRE δ-receptor-bound conformer. It corresponds to the gauche(−) rotamer of the Tyr and to the trans rotamer of the Phe side chains (Nikiforovich and Hruby, 1990).

The proposed model was confirmed by synthesis and biological testing of conformationally constrained analogs of deltorphin I, a δ-selective linear peptide (Tyr-D-Ala-Phe-Asp-Val-Val-Gly-NH_2, DT). Energy calculations performed for DT and subsequent comparison of its low-energy conformers to the proposed model showed the pronounced similarity of this model to one of the low-energy conformers of DT. The side chains of the D-Ala^2 and Val^5 residues in this particular conformer of DT are fairly close to one another, which suggested the possibility of linking positions 2 and 5 in DT by a disulfide bridge while preserving the chirality of the corresponding C^α atoms. Energy calculations on the [D-Cys^2,Cys^5]DT analog confirmed the compatibility of this cyclization with the model. Subsequently, several cylic analogs of DT were synthesized and tested for receptor binding and biological potency (Misicka *et al.*, 1992a,b). The results showed that [D-Cys^2,Cys^5]DT was as active as linear DT at δ-receptors, but much more active at μ-receptors (Misicka *et al.*, 1992b). Thus, the δ-selectivity of the [D-Cys^2,Cys^5]DT analog was diminished, but it was restored in the similar [D-Cys^2,Pen^5]DT cyclic analog (Misicka *et al.*, 1992a). Another confirmation of the proposed model came from the synthesis of conformationally restricted analogs of the μ-selective dermorphin peptide (Tyr-D-Ala-Phe-Gly-Tyr-Pro-Ser-NH_2, DRM) (Tourwe *et al.*, 1992a,b). Tourwe *et al.* (1992b) have fixed the Phe^3 side chain of DRM into the trans rotamer (which is predicted by the model for δ-selective opioid peptides) by using 4-amino-tetrahydro-2-benzazepin-3-one as a mimic for the Phe-Gly fragment. This subtle modification alone shifted the selectivity profile for this DRM analog from μ- to δ-selectivity. At the same time, [Tic^3]DRM, where the Phe side chain was fixed into the gauche(+) rotamer (see above), was inactive on both receptor types (Tourwe *et al.*, 1992a).

The examples above were obtained using the ECEPP force field, which has been calibrated specifically for peptide molecules and shown to be in good agreement with high-resolution X-ray data on proteins. The results of molecular modeling with different force fields and calculation techniques (such as build-up procedures, Monte Carlo, or molecular dynamics simulations) can differ in their estimates of energy values for low-energy conformers. However, the set of sterically accessible geometrical forms of the peptide backbone found by different calculations are in much better agreement. A survey of publications on DPDPE over the period 1988–1992 revealed that, despite the large differences in relative energies for the lowest energy conformers, different authors found very similar allowed geometrical forms of the DPDPE backbone (in terms of a common spatial arrangement of C^α and C^β atoms) by different protocols using a variety of force fields. Generally, the sets of low-energy geometrical forms for peptides are much more limited than the sets of low-energy conformers. At the same time, the backbone conformers for different analogs do not necessary have to be identical, or even very similar, to ensure the similar spatial arrangement of functionally important groups which is the dominant recognition pattern.

V. Examples of Experimental and Computational Techniques Applied to the Study of Peptide Conformational Stability and Dynamics

A. Gramicidin A

Gramicidin A (HCO-L-Val-Gly-L-Ala-D-Leu-L-Ala-D-Val-L-Val-D-Val-L-Trp-D-Leu-L-Trp-D-Leu-L-Trp-D-Leu-L-Trp-$CONHCH_2CH_2OH$) is a hydrophobic, linear peptide of 15 residues arranged in alternating chirality (Sarges and Witkop, 1964, 1965). Produced by *Bacillus brevis* (Hotchkiss and Dubois, 1940), it functions as a channel for monovalent alkali metal and small organic cations in lipid membranes (Hladky and Haydon, 1972; Mueller and Rudin, 1967). The channel consists of two molecules of gramicidin A which diffuse together to span the bilayer (Koeppe and Andersen, 1990). Numerous computational, chemical, and biophysical techniques have been used to study gramicidin A, and several recent review articles cover the resulting literature (Busath, 1993; Wallace, 1990, 1992; Woolley and Wallace, 1992).

Initial computational analysis of gramicidin A using maximal intramolecular hydrogen bonding and pore ion selectivity as restraints led to the prediction of a dimer of single-stranded, left-handed π(L,D) helices as the active conformation (Fig. 2A,B; Urry, 1971). A head-to-head orientation (Fig. 2A) was supported by the activity of an analog in which two gramicidin molecules were covalently linked together (Urry *et al.*, 1971). Of the theoretically possible forms, the $\beta^{6.3}$ dimer best fits expectations of length (25–30 Å) and pore size (4 Å) needed for channel activity and ion selectivity. To explain spectroscopic observations ^{1}H-NMR, infrared, CD) of solution conformations which did not fit this model, a family of double-stranded helices was proposed (Fig. 2C,D;

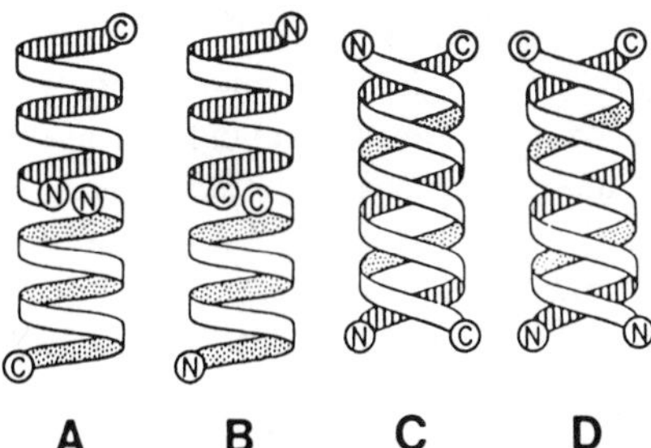

Figure 2 Schematic models of gramicidin A dimers: (A) head-to-head single-stranded helical dimer; (B) tail-to-tail single-stranded helical dimer; (C) antiparallel double-stranded helix; (D) parallel double-stranded helix. (Reproduced from Wallace, B. A. 1986, *Biophysical Journal*, **49**, 296, by permission of the *Biophysical* Society.)

Veatch *et al.*, 1974). These aggregates were right- or left-handed, had the two strands oriented in parallel or antiparallel fashion, and could vary in hydrogen bonding register. Both the single- and double-stranded forms have helically wound β-sheet structures in which the peptide dipoles alternate in orientation, resulting in no net dipole. Qualitatively, the two classes of structural models are similar in having side chains oriented to the lipid with the polar carbonyl groups lining the central pore, presumably acting to solvate the ion. Molecular dynamics simulations and molecular mechanics have been used (Partenskii and Jordan, 1992; Roux and Karplus, 1993; Busath, 1993) to estimate the relative energies of possible conformations; to investigate the behavior of channel water and/or the role of amide libration in solvating the ion; to characterize the free energy profile of the ion in the channel; as well as to address numerous other questions.

Structural studies of gramicidin A have demonstrated the considerable environmental dependence of its conformation, as well as its conformational heterogeneity in a given environment (Wallace, 1990). Technical difficulties in solution NMR or X-ray analyses (Wallace, 1992) of membrane-active peptides in the bilayer have led investigators to study gramicidin A in organic solvents. Two of the species observed in dioxane are a left-handed, antiparallel $\beta^{5.6}$ double helix and a right-handed parallel $\beta^{5.6}$ double helix (Arseniev *et al.*, 1984; Pascal and Cross, 1992). Gramicidin A in 1:1 methanol–chloroform is conformationally heterogeneous, but on addition of Cs^+ converges to a single predominant conformation, a right-handed antiparallel $\beta^{7.2}$ double helix with two ions bound (Arseniev *et al.*, 1985b). Solution NMR studies in 1:1 DMSO–acetone indicate the absence of any defined conformation (Roux *et al.*, 1990). In ethanol, four species are seen by two-dimensional NMR: two left-handed parallel $\beta^{5.6}$ double helices, differing only in hydrogen bond register; a left-handed antiparallel $\beta^{5.6}$ double helix; and a right-handed parallel $\beta^{5.6}$ double helix (Bystrov and Arseniev, 1988). In benzene–ethanol, an antiparallel $\beta^{5.6}$ double helix is observed (Pascal and Cross, 1993). X-Ray analyses of gramicidin A crystallized from ethanol (Langs, 1988, 1989) and from methanol–CsCl (Wallace and Ravikumar, 1988) reveal double-stranded helical forms.

Although no crystal structure of gramicidin A in the bilayer is yet available, spectroscopic methods have been used to study its conformation in lipid environments. Synthetic enrichment of specific carbonyl sites with ^{13}C made it possible to examine the perturbation in chemical shift induced by Na^+ and Tl^+ when gramicidin was incorporated into lipids (Urry *et al.*, 1982). These experiments revealed the ion binding site to be located at Trp^{11} and Trp^{13}, and similar studies were used to argue that the channel structure was the predicted left-handed, head-to-head $\beta^{6.3}$ dimer (Urry *et al.*, 1983). The fact that paramagnetic species perturb only the ^{19}F-NMR resonances of a C-terminal fluorinated gramicidin A analog in phospholipid vesicles, and not those of an N-fluorinated analog, can only be explained by N-terminal to N-terminal dimers (Weinstein *et al.*, 1985). An N-terminal to N-terminal dimer of right-handed, single-stranded $\beta^{6.3}$ helices was found by NOESY evaluation of gramicidin A in sodium dodecyl sulfate micelles (Arseniev *et al.*, 1985a). Solid-state NMR studies of gramicidin in oriented bilayers have resulted in a high-resolution structure with a right-handed, single-stranded head-to-head β-helical structure with 6 to 7 residues per turn (Ketchem *et al.*, 1993).

The conformational interconversions of gramicidin continue to be an active area of research. Several studies suggest that the conformation of gramicidin inserted into the membrane is that found in solution (Bano *et al.*, 1991; Bouchard and Auger, 1993; Killian *et al.*, 1988; LoGrasso *et al.*, 1988; Tournois *et al.*, 1987; Zhang *et al.*, 1992), whether that be a solvent that supports the double-stranded helix or the helical monomer. On heating, the membrane-bound, antiparallel, double-stranded helix converts to a helical dimer (Lepre *et al.*, 1993). Subsequent studies have suggested that the putative double-stranded helical forms initially inserted are unlikely to have any conductance capabilities (Hillis and Boghosian, 1993).

B. Cyclosporin A

Cyclosporin A (cyclo[MeBmt^1-Abu^2-Sar^3-MeLeu^4-Val^5-MeLeu^6-Ala^7-D-Ala^8-MeLeu^9-MeLeu^{10}-MeVal^{11}], CsA) is a fungal metabolite discovered in 1976 and subsequently found to have immunosuppressive activity (Borel, 1989). It was rapidly adopted for use in organ tranplantation and the treatment of autoimmune diseases (Faulds *et al.*, 1993). For several years, the target molecule for CsA was unidentified, although CsA was known to block lymphokine mRNA induction, thereby inhibiting T-cell activation (Elliott *et al.*, 1984; Kronke *et al.*, 1984). In the absence of any structural information about the receptor, efforts to improve the activity and therapeutic ratio of CsA focused on the conformation of CsA itself (Wenger, 1986). In polar organic solvents, CsA gives rise to a complex NMR spectrum thought to be due to a mixture of conformers (Kessler *et al.*, 1990a; Vine and Bowers, 1987) which may arise from cis–trans isomerization about the many *N*-methyl amide bonds. Its limited solubility precluded solution NMR studies of CsA in water. The crystal structure of CsA is similar to the solution structure in chloroform (Kessler *et al.*, 1990b; Loosli *et al.*, 1985), with a twisted β sheet at residues 1–7 and 11, and a type II′ β turn at Sar^3 and MeLeu^4 (Fig. 3A). In these structures, only one (MeLeu^9-MeLeu^{10}) of the seven *N*-methyl amide bonds is in a cis conformation. The fact that essential residues present a contiguous face in the CsA crystal, and the similarity of the crystal structure to that in chloroform, led to some acceptance of this conformation as the bound form. Incorporation of conformational constraints to stabilize the β-II′ turn resulted in analogs which were less active than the parent molecule (Lee *et al.*, 1990).

In the mid 1980s, a 17-kDa protein with peptidyl-prolyl cis/trans isomerase activity was found to specifically bind CsA ($<10^{-8}$ *M*) and to be inhibited by it (Fischer *et al.*, 1989; Takahashi *et al.*, 1989). Cyclophilin (CyP) (Handschumacher *et al.*, 1984; Harding and Handschumacher, 1988) was subsequently

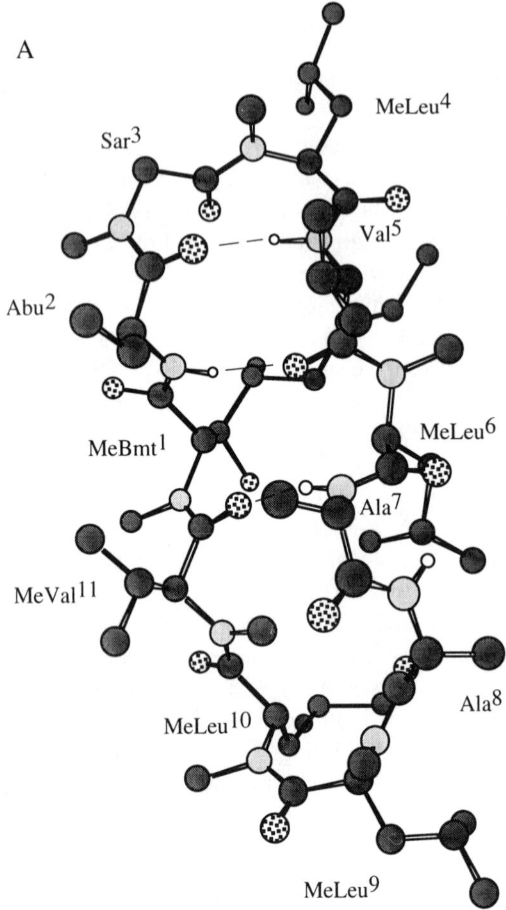

Figure 3 Conformation of cyclosporin A observed (A) in the crystal and (B) when bound to cyclophilin. Light gray spheres are nitrogen; dotted spheres are oxygen. Only amide hydrogens are shown, and hydrogen bonds are indicated by dashed lines.

recognized to be part of a larger family of proteins called the immunophilins (Schreiber, 1991). These proteins (CyP, FKBP) bind immunosuppressive ligands (CsA, FK506), and the resulting complex inhibits the phosphatase activity of calceneurin (Schreiber, 1992). The result is to block a cascade of events that normally results in transcription of the interleukin 2 (IL-2) gene and IL-2-stimulated activation of T cells (Walsh *et al.*, 1992).

Isotope-edited NMR experiments using cyclophilin and labeled CsA revealed that the MeLeu^9-MeLeu^{10} amide bond is trans (Fesik *et al.*, 1990) when bound, and that all other amide bonds are also trans (Fesik *et al.*, 1991a; Weber *et al.*, 1991). The identity of solvent-exposed residues in a CsA–

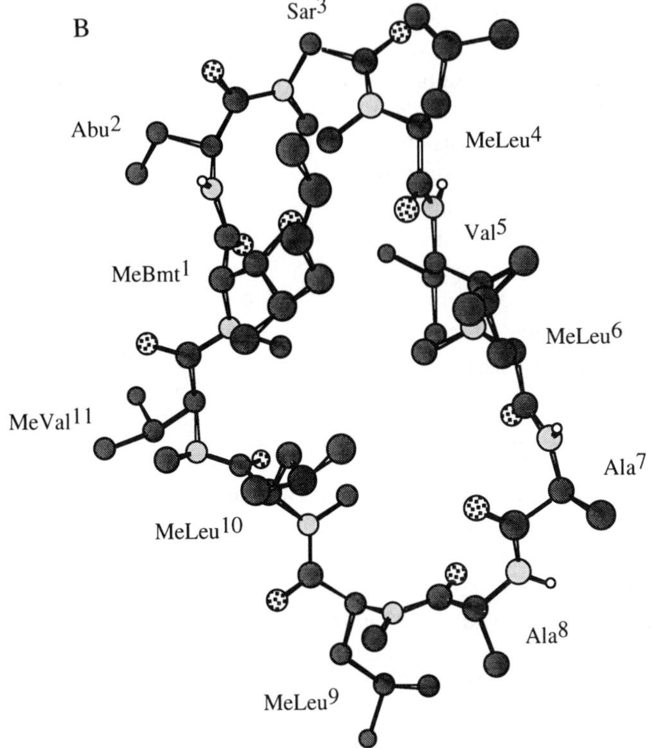

Figure 3 (*continued*)

cyclophilin complex was revealed by using a nitroxide radical to perturb proton T_1 values in the ^{13}C-labeled CsA (Fesik *et al.*, 1991b). The complete structure of the protein–ligand complex was subsequently determined by both NMR (Theriault *et al.*, 1993) and X-ray studies (Pflugl *et al.*, 1993). In the bound conformation, CsA has no elements of regular secondary structure, and no intramolecular hydrogen bonds (Fig. 3B). In the crystal structure of CsA, the backbone moieties point toward the center of the molecule and are essentially sequestered by the lipophilic side chains. In contrast, the bound form of CsA is "opened" to allow polar sites on the peptide backbone to interact with residues in the protein binding site. Notably, the conformation of CsA bound to an antibody determined by X-ray crystallography is similar to that seen in its complex with cyclophilin (Altschuh *et al.*, 1992) and led to the suggestion that the bound conformation of CsA preexists in aqueous solution, rather than being induced by the protein. Kinetic studies are consistent with this observation, in that CsA is a slow-binding inhibitor of the peptidylprolyl isomerase action, and the kinetics are sensitive to the solvent system in which CsA is dissolved (Kofron *et al.*,

1992). A similar suggestion has been made for an FK506 analog (Petros *et al.*, 1993). The addition of LiCl to a solution of CsA in tetrahydrofuran (THF) induces a change from the conformation seen in chloroform to one similar to the cyclophilin-bound state (Kock *et al.*, 1992). Although one molecular dynamics study (Lautz *et al.*, 1990) was not able to simulate a change in conformation with change in solvent, a more recent study in water and carbon tetrachloride supports the observation that CsA undergoes solvent-dependent conformational changes (Tayar *et al.*, 1993). The bound conformation of CsA has been used to design a constrained, tricyclic analog of CsA which exhibits 3-fold increased affinity for cyclophilin and when bound to cyclophilin has a higher affinity for calcineurin (Alberg and Schreiber, 1993). Unexpectedly, the analog shows multiple conformers in solution under conditions where CsA exists as a single conformer.

C. Somatostatin

Somatostatin (**I** in Table I) is a cyclic tetradecapeptide hormone widely distributed throughout the nervous system, the gastrointestinal tract, and the pancreas. It inhibits the release of a variety of other hormones, including gastrin, glucagon, insulin, and somatotropin (growth hormone) (Vale *et al.*, 1977). Early synthetic analogs established that residues outside the cycle were nonessential (Rivier *et al.*, 1975a) and that the side chain of Trp^8 was important in activity (Vale *et al.*, 1975). The D-Trp^8 analog was found to be 6–8 times more active than the natural sequence (Rivier *et al.*, 1975b). Circular dichroism studies of somatostatin in water suggested an ordered secondary structure, and a model of its bioactive conformation based on these analyses invoked an antiparallel sheet with a turn between residues 6 and 11 (Holladay and Puett, 1976). A refined model for the bioactive conformation included a β-II′ turn at Phe^7-Thr^{10}, and predicted proximity of the β carbons of residues 6 and 11 (Veber *et al.*, 1978).

To test the model for the bound structure of somatostatin, a disulfide bridge linking residues 6 and 11 and a D-Trp^8 were incorporated into a carbacyclic compound (**II** in Table I) which had previously been used to show that the 3–14 disulfide was not essential for activity (Veber *et al.*, 1976). The bicyclic compound **III** (Table I) was essentially equipotent to somatostatin itself (Veber *et al.*, 1978), although its reduced form had greatly decreased activity, suggesting that hydrophobic interaction between the aromatic rings of residues 6 and 11 was important in stabilizing the active conformer of somatostatin (**I**). The predicted β-II′ turn would make the D-Trp^8 and Lys^9 side chains proximal, with an upfield shift in resonance frequency for the Lys^9 side-chain protons due to the nearby aromatic ring. This was subsequently reported for the γ-methylene protons of Lys^9 in the 6,11-bicyclic-D-Trp^8 analog **III** (Arison *et al.*, 1978), and although somatostatin is a mixture of conformers by solution NMR, there is evidence for turn structure at residues 7–12 and shielding of the Lys^9 γ-CH_2 (Arison *et al.*, 1981; Buffington *et al.*, 1980; Hallenga *et al.*, 1980).

Table I Somatostatin and Analogs[a]

I	H-Ala^1-Gly^2-Cys^3-Lys^4-Asn^5-Phe^6-Phe^7-Trp^8-Lys^9-Thr^{10}-Phe^{11}-Thr^{12}-Ser^{13}-Cys^{14}-OH
II	Cyclo[Aha-Lys^4-Asn^5-Phe^6-Phe^7-Trp^8-Lys^9-Thr^{10}-Phe^{11}-Thr^{12}-Ser^{13}]
III	Bicyclo[Aha-Lys^4-Asn^5-Cys^6-Phe^7-Trp^8-Lys^9-Thr^{10}-Cys^{11}-Thr^{12}-Ser^{13}]
IV	Bicyclo[Aha-Cys^6-Phe^7-D-Trp^8-Lys^9-Thr^{10}-Cys^{11}]
V	Cyclo[Pro^6-Phe^7-D-Trp^8-Lys^9-Thr^{10}-Phe^{11}]
VI	Cyclo[NMeAla^6-Tyr^7-D-Trp^8-Lys^9-Val^{10}-Phe^{11}]
VII	D-Phe^5-Cys^6-Phe^7-D-Trp^8-Lys-Thr^9-Cys^{10}-Thr(OH)^{11}

[a] All Cys residues are present at their oxidized, intramolecularly cyclized form. Numbering of residues in each case corresponds to that of somatostatin (**I**). Aha = ω-aminoheptanoic acid

Eventually, compound **III** was truncated to yield **IV,** a cyclic hexapeptide nearly as potent as somatostatin (Veber *et al.*, 1979) but retaining only four of the original amino acids. Further design studies sought simple bridging groups to replace the Cys-Aha-Cys linker which would preserve a β-II′ turn at residues 7–10. This effort resulted in a cyclic hexapeptide, cyclo[Pro-Phe^7-D-Trp^8-Lys^9-Thr^{10}-Phe^{11}] (**V** in Table I), with potency greater than that of somatostatin both *in vivo* and *in vitro* (Veber *et al.*, 1981). Solution NMR studies in DMSO revealed two β turns in this cyclic hexapeptide: a type II′ β turn at residues 7–10, and a type VI β turn with a *cis*-amide at the Pro-Phe bridge (Kessler *et al.*, 1983; Veber, 1982). Subsequent optimization of this sequence produced a somatostatin analog, cyclo[NMeAla-Tyr^7-D-Trp^8-Lys^9-Val^{10}-Phe^{11}] (**VI** in Table I), with potencies 50–100 times greater than that of somatostatin itself (Veber *et al.*, 1984). The role of a β-II′ conformation at D-Trp-Lys has been confirmed in numerous analogs of compound **V** (Mierke *et al.*, 1990; Van Binst and Tourwe, 1992) and in other disulfide-bridged octapeptides such as compound **VII** (Table I; Bauer, *et al.*, 1982). It is clear, however, that proximity of the D-Trp^8 and Lys^9 side chains is not sufficient for activity, based on the incorporation of proposed mimics of the *cis*-peptide bond into compound **V.** In several of these cases, the type II′ β turn is retained, but the activity of the analog is reduced (reviewed in Brady *et al.*, 1993). Conformationally constrained α- or β-methyl amino acids have been incorporated into peptide **V** and used to generate a revised model of its bioactive conformation in which the plane of the peptide backbone is bent in the middle, at residues Phe^7 and Thr^{10} (Huang *et al.*, 1992).

D. Angiotensin II

Conformational studies of a well-known pressor and myotropic agent, angiotensin II (Asp^1-Arg^2-Val^3-Tyr^4-Val/Ile^5-His^6-Pro^7-Phe^8, AT) have been extensive (e.g., see Duncia *et al.*, 1990, and references therein). These studies have yielded significantly different models of AT conformation(s) in

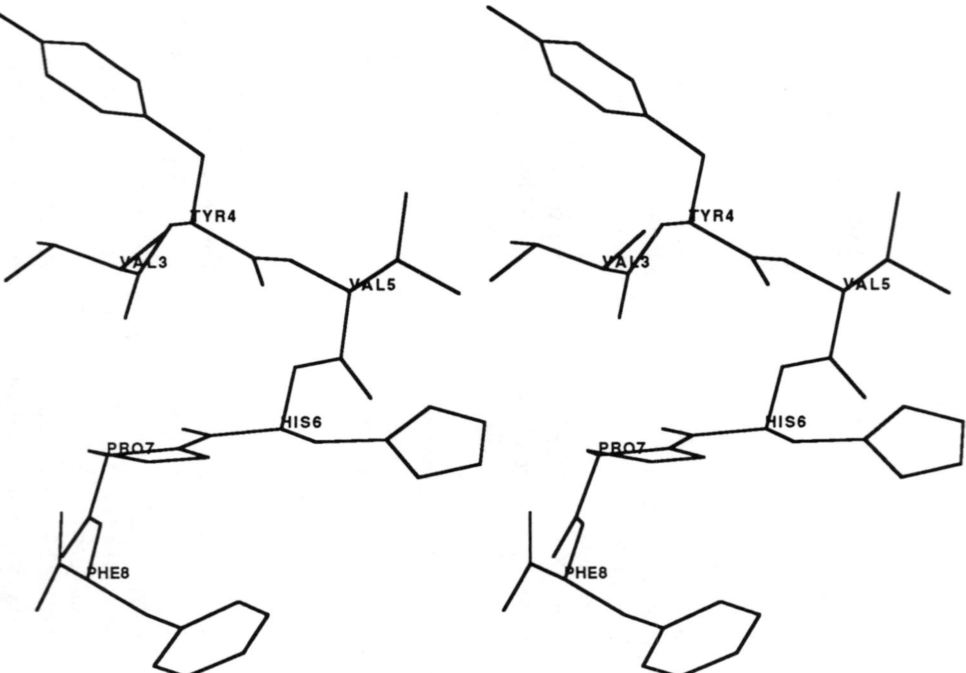

Figure 4 Stereoview of the proposed receptor-bound AT structure. Only the 3–8 fragment is shown. All hydrogen atoms are omitted.

solution and/or at the receptor surface. Different authors have suggested that AT in solution is present as a random coil (Paiva *et al.*, 1963), an α helix (Smeby *et al.*, 1962), a structure with an ionic–dipole interaction of the (His^6)NH$^+\cdots{}^-$OOC(Phe^8) type (Weinkam and Jorgensen, 1971), an inverse γ turn at the Tyr^4 residue or a β turn at the Tyr^4-Val^5 residues with a cis conformation of the Pro^7 residue (Printz *et al.*, 1972), the so-called cross-β-forms I (Devynck *et al.*, 1973) and II (Fermandjian *et al.*, 1976), a structure with clustering of the His, Tyr, and Phe aromatic rings (Matsoukas *et al.*, 1990b), or in state of dynamic equilibrium between two main forms with a β turn at either the Val^3-Tyr^4 or Tyr^4-Val^5 residues (Nikiforovich *et al.*, 1988). As a receptor-bound conformation of AT, different studies have proposed a β-III-like turn at the Tyr^4-Val^5 residues (Chipens *et al.*, 1979; Marshall and Nelson, 1986). Another model proposes proximity of the hydroxyl group of the Tyr^4 side chain and the C-terminal carboxyl (Turner *et al.*, 1991), whereas a more recent model points out the importance of a common spatial arrangement of the functional groups (discussed below) rather than any particular backbone structure (Nikiforovich and Marshall, 1993c).

One of the main reasons for such a variety of conformational models is the flexibility of angiotensin II. One way to distinguish among possible conformational models for AT, especially those for receptor-bound conformers, is to design conformationally constrained (e.g., cyclic) analogs retaining a high level of binding to specific receptors. Several attempts failed to do so, but one particular type of cyclic AT analog has been found to possess high affinity for specific receptor subtypes (Plucinska *et al.*, 1993; Spear *et al.*, 1990; Sugg *et al.*, 1989). These analogs have a disulfide bridge between positions 3 and 5 in the AT sequence, and are exemplified by cyclo[HCy/Cys^3,HCy^5]AT [IC$_{50}$ of 2.1 and 13 n*M* at rat uterine membrane, respectively (Spear *et al.*, 1990)] or cyclo[Sar^1,HCy/Cys^3,Mpt^5]AT [IC$_{50}$ of 0.65 and 0.82 n*M* at rabbit uterus, respectively (Plucinska *et al.*, 1993)]. The central fragment of each contains a 13- or 11-membered cyclic moiety, and in the case of cyclo[Sar^1,HCy^3,Mpt^5] analogs is certainly less flexible owing to the restricted torsional range (ϕ_5 about −60°) imposed by the mercaptoproline ring. Besides, the $C_\delta H_2$ group of the same ring creates an additional steric hindrance which limits the allowed values of the ψ_4 dihedral angle.

Computational studies revealed that conformations available to the set of active analogs are not consistent with a unique model for the backbone of the AT receptor-bound conformation(s). Whereas backbone conformations of cyclo[Sar^1,HCy^{3,5}]AT seem to be compatible with a γ turn or right-handed α-helical structure for Tyr^4, both of these conformers are sterically hindered for cyclo[Sar^1,Cys^3,Mpt^5]AT. Similarly, Tyr^4 backbone conformations for cyclo[Sar^1,Cys^3,Mpt^5]AT are, in turn, incompatible with those of bicyclo[Sar^1,Mpc^3,Cys^5]AT parallel and antiparallel dimers (see Plucinska *et al.*, 1993, for detailed discussion). To resolve this apparent contradiction, an alternate approach involving a search for a common spatial arrangement of functionally important (pharmacophoric) groups in low-energy conformers of AT and its active analogs, rather than comparison of dihedral angle values for the peptide backbone, has been used. One of the prerequisites for the successful use of this approach is *a priori* knowledge of which molecular groups are functionally important for recognition and activation. In the case of AT agonists, the four groups indispensable for optimal triggering of the biological response are the aromatic moieties of Tyr^4, His^6, and Phe^8 residues along with the C-terminal carboxyl (e.g., Regoli *et al.*, 1974).

The proposed spatial arrangement (Nikiforovich and Marshall, 1993c) of important functional groups in AT analogs (Fig. 4) is similar to a model used by Samanen *et al.* to develop potent nonpeptidic antagonists (Samanen *et al.*, 1992). However, in that model, the hydroxyl group of Tyr^4 is close to the C-terminal carboxyl, which is not the case in the model shown. An antibody-bound conformation of AT revealed by X-ray techniques (Garcia *et al.*, 1992) also shows some resemblance to this new model. At the same time, the side chains of Tyr^4 and His^6 are too far apart to be consistent with a model by Moore (1985), in which intramolecular interaction was suggested between them.

Comparison of proposed receptor-bound conformation(s) with those observed experimentally in solution is not straightforward and should be performed cautiously. However, it is noteworthy that the discussed model is in general agreement with NMR data on spin-labeled angiotensin in water (Nikiforovich *et al.*, 1987) that have been reconfirmed (Cushman *et al.*, 1992). On the other hand, clustering of all three aromatic rings, deduced using the NMR data for $[Sar^1]$AT in DMSO (Matsoukas *et al.*, 1990a,b), seems unlikely for this model.

The model shown in Fig. 4 suggested that the receptor-bound conformer of AT could be matched closely by a somewhat unusual analog that contains a Pro^5 residue but has D-Tyr^4 instead of Tyr^4 (Nikiforovich and Marshall, 1993c). The analog was synthesized, and tested for binding to rabbit aorta preparations. The IC_{50} was approximately 50 n*M*, which is equivalent to about 5% of the AT binding level [IC_{50} of 2.5 n*M* for AT (Plucinska *et al.*, 1993)]. It is noteworthy that [D-Tyr^4]AT had only 0.06% of the level of AT binding to tissue from rat ascending colon (Chipens *et al.*, 1976). Thus, the conformational studies led to a new and unusual AT analog on the basis of a proposed three-dimensional model for AT receptor recognition.

Acknowledgments

The authors thank the National Institutes of Health for support during the preparation of the manuscript (5 PO1 GM 24483, 5 RO1 GM48184, NO1-HD-3-3173), and Dr. Mark Smythe for assistance in preparing Fig. 3.

References

Alberg, D. G., and Schreiber, S. L. (1993). *Science* **262,** 248–250.

Allen, M. P., and Tildesley, D. J. (1989). "Computer Simulation of Liquids." Oxford Science Publ., Oxford.

Allured, V. S., Kelly, C. M., and Landis, C. R. (1991). *J. Am. Chem. Soc.* **113,** 1–12.

Altschuh, D., Vix, O., Rees, B., and Thierry, J.-C. (1992). *Science* **256,** 92–94.

Amir, D., and Haas, E. (1987). *Biochemistry* **26,** 2162–2175.

Appelt, K. (1993). *Perspect. Drug Discovery Des.* **1,** 23–48.

Aqvist, J., and Warshel, A. (1990). *J. Am. Chem. Soc.* **112,** 2860–2868.

Archer, S. J., Ellena, J. F., and Cafiso, D. S. (1991). *Biophys. J.* **60,** 389–398.

Arison, B. H., Hirschmann, R., and Weber, D. F. (1978). *Bioorg. Chem.* **7,** 447–451.

Arison, B. H., Hirschmann, R., Paleveda, W. J., Brady, S. F., and Veber, D. F. (1981). *Biochem. Biophys. Res. Commun.* **100,** 1148–1153.

Arrhenius, T., and Satterthwait, A. C. (1990). *In* "Peptides: Chemistry, Structure and Biology" (J. E. Rivier and G. R. Marshall, eds.), pp. 870–872. ESCOM, Leiden.

Arrondo, J. L. R., Muga, A., Castresana, J., and Goni, F. M. (1992). *Prog. Biophys. Mol. Biol.* **59,** 23–56.

Arseniev, A. S., Bystrov, V. F., Ivanov, V. T., and Ovchinnikov, Y. A. (1984). *FEBS Lett.* **165,** 51–56.

Arseniev, A. S., Barsukov, I. L., Bystrov, V. F., Lomize, A. L., and Ovchinnikov, Y. A. (1985a). *FEBS Lett.* **186,** 168–174.

Arseniev, A. S., Barsukov, K. L., and Bystrov, V. F. (1985b). *FEBS Lett.* **180,** 33–39.

Axelsen, P. H., Gratton, E., and Prendergast, F. G. (1991). *Biochemistry* **30,** 1173–1179.

Ball, J. B., and Alewood, P. F. (1990). *J. Mol. Recognit.* **3,** 55–64.

Ball, J. B., Hughes, R. A., Alewood, P. F., and Andrews, P. R. (1993). *Tetrahedron* **49,** 3467–3478.

Bandekar, J. (1992). *Biochim. Biophys. Acta* **1120,** 123–143.

Bano, M. C., Braco, L., and Abad, C. (1991). *Biochemistry* **30,** 886–894.

Bauer, W., Briner, U., Doepfner, W., Haller, R., Huguenin, R., Marbach, P., Petcher, T. J., and Pless, J. (1982). *Life Sci.* **31,** 1133–1140.

Bax, A., and Grezsiek, S. (1993). *Acc. Chem. Res.* **26,** 131–138.

Bayly, C. I., Cieplak, P., Cornell, W. D., and Kollman, P. A. (1993). *J. Phys. Chem.* **97,** 10269–10280.

Bechinger, B., Zasloff, M., and Opella, S. J. (1993). *Protein Sci.* **2,** 2077–2084.

Beechem, J. M., and Haas, E. (1989). *Biophys. J.* **55,** 1225–1236.

Betins, J., and Nikiforovich, G. V. (1984). *Bioorgan. Khim.* **10,** 1177–1182 (in Russian).

Beusen, D. D., Head, R. D., Clark, J. D., Hutton, W. C., Slomczynska, U., Zabrocki, J., Leplawy, M. T., and Marshall, G. R. (1993). *In* "Peptides 1992" (C. H. Schneider, ed.), pp. 79–80. ESCOM, Leiden.

Beusen, D. D., Christensen, A. M., Cohen, E. R., McDowell, L. M., Schmidt, A., and Schaefer, J. (1994). *In* "Peptides: Chemistry, and Biology" ((R. S. Hodges and J. A. Smith, eds.), pp. 760–762. ESCOM, Leiden.

Beveridge, D. L., and DiCapua, F. M. (1989). *In* "Computer Simulation of Biomolecular Systems" (S. van Gunsteren and P. K. Weiner, eds.), pp. 1–26. ESCOM, Leiden.

Bindal, R. D., and Marshall, G. R. (unpublished).

Blaney, J. M., and Dixon, J. S. (1994). *In* "Reviews in Computational Chemistry" (K. B. Lipkowitz and D. B. Boyd, eds.), pp. 299–335. VCH Publ., New York.

Blatt, E., and Sawyer, W. H. (1985). *Biochim. Biophys. Acta* **822,** 43–62.

Bohm, G., Buhr, R., and Jaenicke, R. (1992). *Protein Eng.* **5,** 191–195.

Bohm, H.-J. (1993). *J. Am. Chem. Soc.* **115,** 6152–6158.

Borel, J. F. (1989). *Transplant. Proc.* **21,** 810–815.

Bothner-By, A., Stephens, R. L., Lee, J.-M., Warren, C. D., and Jeanloz, R. W. (1984). *J. Am. Chem. Soc.* **106,** 811–813.

Bouchard, M., and Auger, M. (1993). *Biophys. J.* **65,** 2484–2492.

Bowen, J. P., and Allinger, N. L. (1991). *In* "Reviews in Computational Chemistry" (K. B. Lipkowitz and D. B. Boyd, eds.), pp. 81–98. VCH, New York.

Brady, S. F., Paleveda, W. J., Jr., Arison, B. H., Saperstein, R., Brady, E. J., Raynor, K., Reisine, T., Veber, D. F., and Freidinger, R. M. (1993). *Tetrahedron* **49,** 3449–3466.

Brahms, S., and Brahms, J. (1980). *J. Mol. Biol.* **138,** 149–178.

Braiman, M. S., and Rothschild, K. J. (1988). *Annu. Rev. Biophys. Biophys. Chem.* **17,** 541–570.

Brunne, R. M., van Gunsteren, W. F., Bruschweiler, R., and Ernst, R. R. (1993). *J. Am. Chem. Soc.* **115,** 4764–4768.

Bruschweiler, R., Blackledge, M., and Ernst, R. R. (1991). *J. Biomol. NMR* **1,** 3–11.

Buffington, L., Garsky, V., Massiot, G., Rivier, J., and Gibbons, W. A. (1980). *Biochem. Biophys. Res. Commun.* **93,** 376–384.

Burkert, U., and Allinger, N. L. (1982). "Molecular Mechanics." American Chemical Society, Washington, D. C.

Burt, S. K., and Greer, J. (1988). *Ann. Rep. Med. Chem.* **23,** 285–294.

Busath, D. D. (1993). *Annu. Rev. Physiol.* **55,** 473–501.

Byler, D. M., and Susi, H. (1986). *Biopolymers* **25,** 469–487.

Bystrov, V. F. (1976). *Prog. NMR Spectrosc.* **10,** 41–81.

Bystrov, V. F., and Arseniev, A. S. (1988). *Tetrahedron* **44,** 925–940.

Caldwell, J., Dang, L. X., and Kollman, P. A. (1990). *J. Am. Chem. Soc.* **112,** 9144–9147.

Callahan, J. F., Newlander, K. A., Burgess, J. L., Eggleston, D. S., Nichols, A., Wong, A., and Huffman, W. F. (1993). *Tetrahedron* **49,** 3479–3488.

Chapman, E. R., Alexander, K., Vorherr, T., Carafoli, E., and Storm, D. R. (1992). *Biochemistry* **31,** 12819–12825.

Chauhan, V. S., Sharma, A. K., Uma, K., Paul, P. K. C., and Balaram, P. (1987). *Int. J. Pept. Protein Res.* **29,** 126–133.

Chen, S., Chrusciel, R. A., Nakanishi, H., Raktabutr, A., Johnson, M. E., Sato, A., Weiner, D., Hoxie, J., Saragovi, H. U., Greene, M. I., and Kahn, M. (1992). *Proc. Natl. Acad. Sci. U.S.A.* **89,** 5872–5876.

Chipens, G., Ancan, J., Afanasyeva, G., Balodis, J., Indulen, J., Klusha, V., Kudryashova, V., Liepinsh, E., Makarova, N., and Mishlyakova, N. (1976). *In* "Peptides 1976" (A. Loffet, ed.), pp. 353–360. University of Brussels, Bruxelles.

Chipens, G. I., Ancan, Y. E., Nikiforovich, G. V., Balodis, Y. Y., and Makarova, N. A. (1979). *In* "Peptides 1978" (I. Z. Siemion and G. Kupryszewski, eds.), pp. 415–419. Wroclaw Univ. Press, Wroclaw, Poland.

Choi, C., and Elber, R. (1991). *J. Chem. Phys.* **94,** 751–760.

Christensen, A. M., and Schaefer, J. (1993). *Biochemistry* **32,** 2868–2873.

Chung, L. A., Lear, J. D., and DeGrado, W. F. (1992). *Biochemistry* **31,** 6608–6616.

Ciajolo, M. R., Tuzi, A., Pratesi, C. R., Fissi, A., and Pieroni, O. (1990). *Biopolymers* **30,** 911–920.

Clark, M., Cramer, R. D., III, and Van Opdenbosch, N. (1989). *J. Comput. Chem.* **10,** 982–1012.

Clore, G. M., and Gronenborn, A. M. (1982). *J. Magn. Reson.* **48,** 402–417.

Cramer, C. J. (1991). *J. Am. Chem. Soc.* **113,** 8552–8554.

Cramer, C. J., and Truhlar, D. G. (1991). *J. Am. Chem. Soc.* **113,** 8305–8311.

Cramer, C. J., and Truhlar, D. G. (1992a). *J. Comput. Chem.* **13,** 1089–1097.

Cramer, C. J., and Truhlar, D. G. (1992b). *Science* **256,** 213–217.

Creuzet, F., McDermott, A., Gebhard, R., van der Hoef, K., Spijker-Assink, M. B., Herzfeld, J., Lugtenburg, J., Levitt, M. H., and Griffin, R. G. (1991). *Science* **251,** 783–786.

Crippen, G. M., and Havel, T. F. (1988). "Distance Geometry and Molecular Conformation." Wiley, New York.

Cumming, D. A., and Carver, J. P. (1987). *Biochemistry* **26,** 6664–6676.

Cushman, J. A., Mishra, P. K., Bothner-By, A. A., and Khosla, M. S. (1992). *Biopolymers* **32,** 1163–1171.

Czerminski, R., and Elber, R. (1990). *Int. J. Quantum Chem.: Quantum Chem. Symp.* **24,** 167–186.

Daggett, V., and Levitt, M. (1992). *J. Mol. Biol.* **223,** 1121–38.

Dammkoehler, R. A., Karasek, S. F., Shands, E. F. B., and Marshall, G. R. (1989). *J. Comput. Aided Mol. Des.* **3,** 3–21.

Dang. L. X., Rice, J. E., Caldwell, J., and Kollman, P. A. (1991). *J. Am. Chem. Soc.* **113,** 2481–2486.

Davis, M. E., and McCammon, J. A. (1990). *Chem. Rev.* **90,** 509–521.

de Jong, E. A. M., Claesen, C. A. A., Daemen, C. J. M., Harmsen, B. J. M., Konings, R. N. H., Tesser, G. I., and Hilbers, C. W. (1988). *J. Magn. Reson.* **80,** 197–213.

Dellwo, M. J., and Wand, A. J. (1989). *J. Am. Chem. Soc.* **111,** 4571–4578.

Devynck, M.-A., Pernollet, M.-G., Meyer, P., Fermandjian, S., and Fromageot, P. (1973). *Nature (London) New Biol.* **245,** 55–58.

Dewar, M. J. S., Zoebisch, E. G., Healy, E. F., and Stewart, J. J. P. (1985). *J. Am. Chem. Soc.* **107,** 3902.

Di Blasio, B., Rossi, F., Benedetti, E., Pavone, V., Pedone, C., Temussi, P. A., Zanotti, G., and Tancredi, T. (1989). *J. Am. Chem. Soc.* **111,** 9089–9098.

Dudek, M., and Ponder, J. (1995). *J. Comput. Chem.* **16,** 791–816.

Duncia, J. V., Chiu, A. T., Carini, D. J., Gregory, G. B., Johnson, A. L., Price, W. A., Wells, G. J., Wong, P. C., Calabrese, J. C., and Timmermans, P. B. M. W. M. (1990). *J. Med. Chem.* **33,** 1312–1329.

Dykstra, C. E. (1993). *Chem. Rev.* **93,** 2339–2353.

Dyson, H. J., and Wright, P. E. (1991). *Annu. Rev. Biophys. Biophys. Chem.* **20,** 519–538.

Dzakula, Z., Westler, W. M., Edison, A. S., and Markley, J. L. (1992). *J. Am. Chem. Soc.* **114,** 6195–6199.

Elliott, J. F., Lin, Y., Mizel, S. B., Bleackley, R. C., Harnish, D. G., and Paetkau, V. (1984). *Science* **226,** 1439–1441.

Epps, D. E., Havel, H. A., Sawyer, T. K., Staples, D. J., Chung, N. N., Schiller, P. W., Hartrodt, B., and Barth, A. (1991). *Int. J. Pept. Protein Res.* **37,** 257–267.

Ernst, R. R., Bodenhausen, G., and Wokaun, A. (1988). "Principles of Nuclear Magnetic Resonance in One and Two Dimensions." Oxford Univ. Press, New York.

Esposito, G., Lesk, A. M., Molinari, H., Motta, A., Niccolai, N., and Pastore, A. (1992). *J. Mol. Biol.* **224,** 659–670.

Faerman, C. H., and Price, S. L. (1990). *J. Am. Chem. Soc.* **112,** 4915–4926.

Faulds, D., Goa, K. L., and Benfield, P. (1993). *Drugs* **45,** 953–1040.

Fermandjian, F., Lintner, K., Haar, W., Fromageot, P., Khosla, M. C., Smeby, R. R., and Bumpus, F. M. (1976). *In* "Peptides 1976" (A. Loffet, ed.), pp. 514–525. Univ. of Brussels, Bruxelles.

Fesik, S. W. (1991). *J. Med. Chem.* **34,** 2937–2945.

Fesik, S. W., and Olejniczak, E. T. (1987). *Magn. Reson. Chem.* **25,** 1046–1048.

Fesik, S. W., Luly, J. R., Erickson, J. W., and Abad-Zapatero, C. (1988). *Biochemistry* **27,** 8297–8301.

Fesik, S. W., Gampe, R. T., Jr., Holzman, T. F., Egan, D. A., Edalji, R., Luly, J. R., Simmer, R., Helfrich, R., Kishore, V., and Rich, D. H. (1990). *Science* **250,** 1406–1410.

Fesik, S. W., Gampe, R. T., Jr., Eaton, H. L., Gemmecker, G., Olejniczak, E. T., Neri, P., Holzman, T. F., Egan, D. A., Edalji, R., Simmer, R., Helfrich, R., Hochlowski, J., and Jackson, M. (1991a). *Biochemistry* **30,** 6574–6583.

Fesik, S. W., Gemmecker, G., Olejniczak, E. T., and Petros, A. M. (1991b). *J. Am. Chem. Soc.* **113,** 7080–7081.

Fischer, G., Wittmann-Liebold, B., Lang, K., Kiefhaber, T., and Schmid, F. X. (1989). *Nature (London)* **337,** 476–478.

Fox, R. O., Jr., and Richards, F. M. (1982). *Nature (London)* **300,** 325–330.

Franks, F. (1975). "Water: A Comprehensive Treatise." Plenum, New York.

Freidinger, R. M., Perlow, D. S., and Veber, D. F. (1982). *J. Org. Chem.* **47,** 104–109.

Galaktionov, S. G., Nikiforovich, G. V., Shenderovich, M. D., Chipens, G. I., and Vegner, R. E. (1976). *In* "Peptides 1976" (A. Loffet, ed.), pp. 617–624. Univ. of Brussels, Bruxelles.

Garbow, J. R., and McWherter, C. A. (1993). *J. Am. Chem. Soc.* **115,** 238–244.

Garcia, K. C., Desiderio, S. V., Ronco, P. M., Veroust, P. J., and Amzel, L. M. (1992). *Science* **257,** 528–531.

Gardner, B., Nakanishi, H., and Kahn, M. (1993). *Tetrahedron* **49,** 3433–3448.

Gasset, M., Baldwin, M. A., Lloyd, D. H., Gabriel, J. M., Holtzman, D. M., Choen, F., Fletterick, R., and Prusiner, S. B. (1992). *Proc. Natl. Acad. Sci. U.S.A.* **89,** 10940–10944.

Genin, M. J., and Johnson, R. L. (1992). *J. Am. Chem. Soc.* **114,** 8778–8783.

Gilson, M. K., Sharp, K. A., and Honig, B. H. (1987). *J. Comput. Chem.* **9,** 327–335.

Greenfield, N., and Fasman, G. D. (1969). *Biochemistry* **8,** 4108–4116.

Grinvald, A., Haas, E., and Steinberg, I. Z. (1972). *Proc. Natl. Acad. Sci. U.S.A.* **69,** 2273–2277.

Guenot, J., and Kollman, P. A. (1993). *J. Comput. Chem.* **14,** 295–311.

Haas, E., Wilchek, M., Katchalski-Katzir, E., and Steinberg, I. Z. (1975). *Proc. Natl. Acad. Sci. U.S.A.* **72,** 1807–1811.

Hagler, A. T., Huler, E., and Lifson, S. (1974). *J. Am. Chem. Soc.* **96,** 5319–5327.

Hallenga, K., Van Binst, G., Scarso, A., Michel, A., Knappenberg, M., Dremier, C., Brison, J., and Dirkx, J. (1980). *FEBS Lett.* **119,** 47–52.

Hancock, R. D. (1990). *Acc. Chem. Res.* **23,** 253–257.

Handschumacher, R. E., Harding, M. W., Rice, J., Drugge, R. J., and Speicher, D. W. (1984). *Science* **226,** 544–547.

Hansen, P. E., and Morgan, B. A. (1984). *In* "The Peptides Analysis, Synthesis, Biology" (S. Udenfriend and J. Meienhofer, eds.), pp. 269–321. Academic Press, New York.

Harding, M. W., and Handschumacher, R. E. (1988). *Transplantation* **46,** 29s–35s.

Harvey, S. C. (1989). *Proteins: Struct. Funct. Genet.* **5,** 78–92.

Havel, T. F. (1990). *Biopolymers* **29,** 1565–1585.

Havel, T. F. (1991). *Prog. Biophys. Mol. Biol.* **56,** 43–78.

Havel, T., and Wuthrich, K. (1984). *Bull. Math. Biol.* **46,** 673–698.

Hennessey, J. P. J., and Johnson, W. C. J. (1981). *Biochemistry* **20,** 1085–1094.

Hermans, J. (1993). *Curr. Opin. Struct. Biol.* **3,** 270–276.

Hillis, W. D., and Boghosian, B. M. (1993). *Science* **261,** 856–863.

Hinds, M. G., Richards, N. G. J., and Robinson, J. A. (1988). *J. Chem. Soc., Chem. Commun.,* 1447–1449.

Hinds, M. G., Welsh, J. H., Brennand, D. M., Fisher, J., Glennie, M. J., Richards, N. G. J., Turner, D. L., and Robinson, J. A. (1991). *J. Med. Chem.* **34,** 1777–1789.

Hladky, S. B., and Haydon, D. A. (1972). *Biochim. Biophys. Acta* **274,** 294–312.

Holl, S. M., Marshall, G. R., Beusen, D. D., Kociolek, K., Redlinski, A. S., Leplawy, M. T., McKay, R. A., Vega, S., and Schaefer, J. (1992). *J. Am. Chem. Soc.* **114,** 4830–4833.

Holladay, L. A., and Puett, D. (1976). *Proc. Natl. Acad. Sci. U.S.A.* **73,** 1199–1202.

Holzemann, G. (1991a). *Kontakte (Darmstadt),* 3–12.

Holzemann, G. (1991b). *Kontakte (Darmstadt),* 55–63.

Horn, A. S., and De Ranter, C. J., eds. (1984). "X-Ray Crystallography and Drug Action." Oxford Univ. Press (Clarendon), Oxford.

Hotchkiss, R. D., and Dubois, R. J. (1940). *J. Biol. Chem.* **132,** 791–792.

Hruby, V. J., and Nikiforovich, G. V. (1991). *In* "Molecular Conformation and Biological Interactions. G. N. Ramachandran Festschrift." (P. Balaram and S. Ramaseshan, eds.), pp. 429–445. Indian Academy of Sciences, Bangalore.

Hruby, V. J., Chow, M.-S., and Smith, D. D. (1990). *Annu. Rev. Pharmacol. Toxicol.* **30,** 501–534.

Huang, Z., He, Y.-B., Raynor, K., Tallent, M., Reisine, T., and Goodman, M. (1992). *J. Am. Chem. Soc.* **114,** 9390–9401.

Huston, S. E., and Marshall, G. R. (1994). *Biopolymers* **34,** 75–90.

IUPAC. (1979). "IUPAC Nomenclature of Organic Chemistry: Sections A, B, C, D, E, F, and H." Pergamon, Oxford.

Jacchieri, S. G., and Richards, N. G. (1993). *Biopolymers* **33,** 971–984.
Jackson, M., Mantsch, H. H., and Spencer, J. H. (1992). *Biochemistry* **31,** 7289–7293.
Jardetzky, O. (1980). *Biochim. Biophys. Acta* **621,** 227–232.
Jardetzky, O., and Roberts, G. C. K. (1981). "NMR in Molecular Biology." Academic Press, New York.
Johnson, W. C. J. (1988). *Annu. Rev. Biophys. Biophys. Chem.* **17,** 145–166.
Johnson, W. C. J. (1990). *Proteins: Struct. Funct. Genet.* **7,** 205–214.
Jorgensen, W. L., and Gao, J. (1988). *J. Am. Chem. Soc.* **110,** 4212–4216.
Jorgensen, W. L., Gao, J., and Ravimohan, C. (1985). *J. Phys. Chem.* **89,** 3470–3473.
Kaczmarek, K., Li., K.-M., Skeean, R., Dooley, D., Humblet, C., Lunney, E., and Marshall, G. R. (1994). *In* "Peptides: Chemistry and Biology" (R. S. Hodges and J. A. Smith, eds.), pp. 687–689. ESCOM, Leiden.
Kahn, M., Nakanishi, H., Chrusciel, R. A., Fitzpatrick, D., and Johnson, M. E. (1991). *J. Med. Chem.* **34,** 3395–3399.
Kelsh, L. P., Ellena, J. F., and Cafiso, D. S. (1992). *Biochemistry* **31,** 5136–5144.
Kemp, D. S. (1992). *In* "Medicinal Chemistry for the 21st Century" (C. G. Wermuth, ed.), pp. 259–277. IUPAC/Blackwell, London.
Kessler, H., Bernd, M., Kogler, H., Zarbock, J., Sorensen, O. W., Bodenhausen, G., and Ernst, R. R. (1983). *J. Am. Chem. Soc.* **105,** 6944–6952.
Kessler, H., Bats, J. W., Griesinger, C., Koll, S., Will, M., and Wagner, K. (1988a). *J. Am. Chem. Soc.* **110,** 1033–1049.
Kessler, H., Griesinger, C., Lautz, J., Muiller, A., van Gunsteren, W. F., and Berendsen, H. J. C. (1988b). *J. Am. Chem. Soc.* **110,** 3393–3396.
Kessler, H., Gehrke, M., Lautz, J., Kock, M., Seebach, D., and Thaler, A. (1990a). *Biochem. Pharmacol.* **40,** 169–173.
Kessler, H., Kock, M., Wein, T., and Gehrke, M. (1990b). *Helv. Chim. Acta* **73,** 1818–1832.
Ketchem, R. R., Hu, W., and Cross, T. A. (1993). *Science* **261,** 1457–1460.
Killian, J. A., Prasad, K. U., Hains, D., and Urry, D. W. (1988). *Biochemistry* **27,** 4848–4855.
Kock, M., Kessler, H., Seebach, D., and Thaler, A. (1992). *J. Am. Chem. Soc.* **114,** 2676–2686.
Koeppe, R. E. I., and Andersen, O. S. (1990). *Science* **30,** 1256–1259.
Kofron, J. L., Kuzmic, P., Kishore, V., Gemmecker, G., Fesik, S. W., and Rich, D. H. (1992). *J. Am. Chem. Soc.* **114,** 2670–2675.
Kolossvary, I., and Guida, W. C. (1993). *J. Am. Chem. Soc.* **115,** 2107–2119.
Kopple, K. D., Bean, J. W., Bhandary, K. K., Briand, J., D'Ambrosio, C. A., and Peishoff, C. E., (1993). *Biopolymers* **33,** 1093–1099.
Kreissler, M., Pesquer, M., Maigret, B., Fournie-Zaluski, M. C., and Roques, B. P. (1989). *J. Comput. Aided Mol. Des.* **3,** 85–94.
Krimm, S., and Bandekar, J. (1986). *Adv. Protein Chem.* **38,** 181–364.
Kronke, M., Leonard, W. J., Depper, J. M., Arya, S. K., and Wong-Staal, F. (1984). *Immunology* **81,** 5214–5218.
Ku, T. W., Ali, F. E., Barton, L. S., Bean, J. W., Bondinell, W. E., Burgess, J. L., Callahan, J. F., Calvo, R. R., Chen, L., Eggelston, D. S., Gleason, J. S., Huffman, W. F., Hwang, S. M., Jakas, D. R., Karash, C. B., Keenan, R. M., Kopple, K. D., Miller, W. H., Newlander, K. A., Nichols, A., Parker, M. F., Peishoff, C. E., Samanen, J. M., Uzinskas, I., and Venslavsky, J. W. (1993). *J. Am. Chem. Soc.* **115,** 8861–8862.
Kuliopulos, A., Westbrook, E. M., Talalay, P., and Mildvan, A. S. (1987). *Biochemistry* **26,** 3927–3937.
Kuntz, I. D., Thomason, J. F., and Oshiro, C. M. (1989). *In* "Methods in Enzymology" (N. J. Oppenheimer and T. L. James, eds.), Vol. 177, pp. 159–204. Academic Press, San Diego.
Kuwajima, S., and Warshel, A. (1990). *J. Phys. Chem.* **94,** 460–466.
Lakowicz, J. R., Laczko, G., Gryczynski, I., and Cherek, H. (1986). *J. Biol. Chem.* **261,** 2240–2245.
Landis, C., and Allured, V. S. (1991). *J. Am. Chem. Soc.* **113,** 9493–9499.
Langs, D. A. (1988). *Science* **241,** 188–191.

Langs, D. A. (1989). *Biopolymers* **28,** 259–266.

Lansbury, P. T., Jr. (1992). *Biochemistry* **31,** 6865–6870.

Lautz, J., Kessler, H., van Gunsteren, W. F., Weber, H.-P., and Wenger, R. M. (1990). *Biopolymers* **29,** 1669–1687.

Leach, A. R. (1993). *In* "Review in Computational Chemistry II" (K. B. Lipkowitz and D. B. Boyd, eds.), pp. 1–55. VCH Publ., New York.

Lee, J. P., Dunlap, B., and Rich, D. H. (1990). *Int. J. Pept. Protein Res.* **35,** 481–494.

Lepre, C. A., Cheng, J.-W., and Moore, J. M. (1993). *J. Am. Chem. Soc.* **115,** 4929–4930.

Lewis, P. N., Momany, F. A., and Scheraga, H. A. (1973). *Isr. J. Chem.* **11,** 121–152.

Lii, J.-H., and Allinger, N. L. (1991). *J. Comput. Chem.* **12,** 186–199.

LoGrasso, P. V., Moll III, F., and Cross, T. A. (1988). *Biochem. J.* **54,** 259–267.

Loncharich, R. J., and Brooks, B. R. (1989). *Proteins: Struct. Funct. Genet.* **6,** 32–45.

Loosli, H.-R., Kessler, H., Oschkinat, H., Weber, H.-P., and Petcher, T. J. (1985). *Helv. Chim. Acta* **68,** 682–704.

MacArthur, J. W., and Thornton, J. M. (1991). *J. Mol. Biol.* **218,** 397–412.

McCammon, J. A., and Harvey, S. C. (1987). "Dynamics of Protein and Nucleic Acids." Cambridge Univ. Press, Cambridge.

McDermott, A. E., Creuzet, F., Griffin, R. G., Zawadzke, L. E., Ye, Q.-Z., and Walsh, C. T. (1990). *Biochemistry* **29,** 5767–5775.

Madura, J. D., Davis, M. E., Gilson, M. K., Wade, R. C., Luty, B. A., and McCammon, J. A. (1994). *In* "Reviews in Computational Chemistry" (K. B. Lipkowitz and D. B. Boyd, eds.), pp. 229–267. VCH Publ., New York.

Mantsch, H. H., Perczel, A., Hollosi, M., and Fasman, G. D. (1993). *Biopolymers* **33,** 201–207.

Marchetto, R., Schreier, S., and Nakaie, C. R. (1993). *J. Am. Chem. Soc.* **115,** 11042–11043.

Marlow, G. E., Perkyns, J. S., and Pettit, B. M. (1993). *Chem. Rev.* **93,** 2503–2521.

Marshall, G. R., and Bosshard, H. E. (1972). *Circ. Res.* **31**(Suppl. 2), II-143–II-150.

Marshall, G. R., and Nelson, R. D. (1986). *In* "Peptide Chemistry 1985" (Y. Kiso, ed.), pp. 239–244. Protein Research Foundation, Osaka, Japan.

Marshall, G. R., Beusen, D. D., Kociolek, K., Redlinski, A. S., Leplawy, M. T., Pan, Y., and Schaefer, J. (1990a). *J. Am. Chem. Soc.* **112,** 963–966.

Marshall, G. R., Hodgkin, E. E., Langs, D. A., Smith, G. D., Zabrocki, J., and Leplawy, M. T. (1990b). *Proc. Natl. Acad. Sci. U.S.A.* **87,** 487–491.

Marshall, G. R., Beusen, D. D., and Nikiforovich, G. V. (1994). *In* "Peptides: Chemistry and Biology" (R. S. Hodges, and J. A. Smith, eds.), pp. 1105–1117. ESCOM, Leiden.

Matsoukas, J. M., Bigam, G., Zhou, N., and Moore, G. J. (1990a). *Peptides* **11,** 359–366.

Matsoukas, J. M., Yamdagni, R., and Moore, G. J. (1990b). *Peptides* **11,** 367–374.

Merutka, G., Morikis, D., Bruschweiler, R., and Wright, P. R. (1993). *Biochemistry* **32,** 13089–13097.

Metzler, W. J., Hare, D. R., and Pardi, A. (1989). *Biochemistry* **28,** 7045–7052.

Mierke, D. F., Pattaroni, C., Delaet, N., Toy, A., Goodman, M., Tancredi, T., Motta, A., Temussi, P. A., Moroder, L., Bovermann, G., and Wunsch, E. (1990). *Int. J. Pept. Protein Res.* **36,** 418–432.

Miick, S. M., Martinez, G. V., Fiori, W. R., Todd, A. P., and Millhauser, G. L. (1992). *Nature (London)* **359,** 653–655.

Miick, S. M., Casteel, K. M., and Millhauser, G. L. (1993). *Biochemistry* **32,** 8014–8021.

Mildvan, A. S. (1989). *FASEB J.* **3,** 1705–1714.

Millhauser, G. L. (1992). *Trends Biol. Sci.* **17,** 448–452.

Mirkin, N. G., and Krimm, S. (1991). *J. Am. Chem. Soc.* **113,** 9742–9747.

Misicka, A., Lipkowski, A. W., Horvath, R., Davis, P., Kramer, T. H., Yamamura, H. I., Porecca, F., and Hruby, V. J. (1992a). *In* "Peptides—1992" (C. Schneider, ed.), pp. 651–652. ESCOM, Leiden.

Misicka, A., Nikiforovich, G. V., Lipkowski, A. W., Horvath, R., Davis, P., Kramer, T. H., Yamamura, H. I., and Hruby, V. J. (1992b). *Bioorg. Med. Chem. Lett.* **2,** 547–552.

Monahan, M. W., Amoss, M. S., Anderson, H. A., and Vale, W. (1973). *Biochemistry* **12,** 4616–4620.

Moore, G. J. (1985). *Int. J. Pept. Protein Res.* **26,** 469–481.

Moskowitz, J. W., Schmidt, K. E., Wilson, S. R., and Cui, W. (1988). *Int. J. Quantum Chem.* **22,** 611–617.

Mueller, P., and Rudin, D. O. (1967). *Biochem. Biophys. Res. Commun.* **26,** 398–404.

Muller, N. (1992). *Trends Biol. Sci.* **17,** 459–463.

Nagai, U., Sato, K., Nakamura, R., and Kato, R. (1993). *Tetrahedron Lett.* **49,** 3577–3592.

Nakanishi, H., Chrusciel, R. A., Shen, R., Bertenshaw, S., Johnston, M. E., Rydel, T. J., Tulinsky, A., and Kahn, M. (1992). *Proc. Natl. Acad. Sci. U.S.A.* **89,** 1705–1709.

Nicholls, A., and Honig, B. (1991). *J. Comput. Chem.* **12,** 435–445.

Nikiforovich, G. V. (1986). *J. Mol. Structure—Theochem.* **134,** 315–340.

Nikiforovich, G. V., and Hruby, V. J. (1990). *Biochem. Biophys. Res. Commun.* **173,** 521–527.

Nikiforovich, G. V., and Hruby, V. J. (1993). *Biochem. Biophys. Res. Commun.* **194,** 9–16.

Nikiforovich, G. V., and Marshall, G. R. (1993a). *Int. J. Pept. Protein Res.* **42,** 171–180.

Nikiforovich, G. V., and Marshall, G. R. (1993b). *Int. J. Pept. Protein Res.* **42,** 181–193.

Nikiforovich, G. V., and Marshall, G. R. (1993c). *Biochem. Biophys. Res. Commun.* **195,** 222–228.

Nikiforovich, G. V., Vesterman, B., Betins, J., and Podins, L. (1987). *J. Biomol. Struct. Dyn.* **4,** 1119–1135.

Nikiforovich, G. V., Vesterman, B. G., and Betins, J. (1988). *Biophys. Chem.* **31,** 101–106.

Nikiforovich, G. V., Hruby, V. J., Prakash, O., and Gehrig, C. A. (1991). *Biopolymers* **31,** 941–955.

Nikiforovich, G. V., Sharma, S. D., Hadley, M. E., and Hruby, V. J. (1992). *In* "Peptides: Chemistry and Biology" (J. A. Smith and J. E. Rivier, eds.), pp. 389–392. ESCOM, Leiden.

Nikiforovich, G. V., Prakash, O., Gehrig, C. A., and Hruby, V. J. (1993). *J. Am. Chem. Soc.* **115,** 3399–3406.

Nilges, M., Clore, G. M., and Gronenborn, A. M. (1988). *FEBS Lett.* **229,** 317–324.

Noggle, J. H., and Schirmer, R. E. (1971). "The Nuclear Overhauser Effect." Academic Press, New York.

O'Connor, S. D., Smith, P. E., al-Obeidi, F., and Pettitt, B. M. (1992). *J. Med. Chem.* **35,** 2870–2881.

Ohmine, I., and Tanaka, H. (1993). *Chem. Rev.* **93,** 2545–2566.

Olatunij, O. L., and Premilat, S. (1987). *Int. J. Pept. Protein Res.* **29,** 1–8.

Opella, S. J., Stewart, P. L., and Valentine, K. G. (1987). *Q. Rev. Biophys.* **19,** 7–49.

Oshiro, C. M., Thomason, J., and Kuntz, I. D. (1991). *Biopolymers* **31,** 1049–1064.

Pachler, K. G. R. (1964). *Spectrochim. Acta* **20,** 581.

Paiva, T. B., Paiva, A. C. M., and Scheraga, H. A. (1963). *Biochemistry* **2,** 1327–1334.

Palmer III, A. G., Hochstrasser, R. A., Millar, D. P., Rance, M., and Wright, P. E. (1993). *J. Am. Chem. Soc.* **115,** 6333–6345.

Palmer, D. R., Pattaroni, C., Nunami, K., Chadha, R. K., Goodman, M., Wakamiya, T., Fukase, K., Horimoto, S., Kitazawa, M., Fujita, H., Kubo, A., and Shiba, T. (1992). *J. Am. Chem. Soc.* **114,** 5634–5642.

Palmo, K., Mirkin, N. G., Pietila, L.-O., and Krimm, S. (1993). *Macromolecules* **26,** 6831–6840.

Pan, Y., Gullion, T., and Schaefer, J. (1990). *J. Magn. Reson.* **90,** 330–340.

Partenskii, M. B., and Jordan, P. C. (1992). *Q. Rev. Biophys.* **25,** 477–510.

Pascal, S. M., and Cross, T. A. (1992). *J. Mol. Biol.* **226,** 1101–1109.

Pascal, S. M., and Cross, T. A. (1993). *J. Biomol. NMR* **3,** 495–513.

Pastor, R. W., Venable, R. M., and Karplus, M. (1988). *J. Chem. Phys.* **89,** 1112–1127.

Peersen, O. B., and Smith, S. O. (1993). *Concepts Magn. Reson.* **5,** 303–317.

Peersen, O. B., Yoshimura, S., Hojo, H., Aimoto, S., and Smith, S. O. (1992). *J. Am. Chem. Soc.* **114,** 4332–4335.

Peishoff, C. E., Dixon, J. S., and Kopple, K. D. (1990). *Biopolymers* **30,** 45–56.

Perczel, A., Angyan, J. G., Kajtar, M., Viviani, W., Rivail, J.-L., Marcoccia, M.-F., and Csizmadia, I. G. (1991a). *J. Am. Chem. Soc.* **113,** 6256–6265.

Perczel, A., Hollosi, M., Foxman, B. M., and Fasman, G. D. (1991b). *J. Am. Chem. Soc.* **113,** 9772–9784.
Perczel, A., Park, K., and Fasman, G. D. (1992). *Anal. Biochem.* **203,** 83–93.
Perczel, A., Hollosi, M., Sandor, P., and Fasman, G. D. (1993). *Int. J. Pept. Protein Res.* **41,** 223–236.
Petros, A. M., Luly, J. R., Liang, H., and Fesik, S. W. (1993). *J. Am. Chem. Soc.* **115,** 9920–9924.
Pflugl, G., Kallen, J., Schirmer, T., Jansonius, J. N., Zurini, M. G., and Walkinshaw, M. D. (1993). *Nature (London)* **361,** 91–94.
Pieroni, O., Fissi, A., Pratesi, C., Temussi, P. A., and Ciardelli, F. (1993). *Biopolymers* **33,** 1–10.
Pierotti, R. A. (1976). *Chem. Rev.* **76,** 717–726.
Pietrzynski, G., Rzeszotarska, B., and Kubica, Z. (1992). *Int. J. Pept. Protein Res.* **40,** 524–531.
Plucinska, K., Kataoka, T., Yodo, M., Cody, W. L., He, J. X., Humblet, C., Lu, G. H., Lunney, E., Major, T. C., Panek, R. L., Schelkun, P., Skeean, R., and Marshall, G. R. (1993). *J. Med. Chem.* **36,** 1902–1913.
Postma, J. P. M., Berendsen, H. J. C., and Haak, J. R. (1982). *Faraday Symp. Chem. Soc.* **17,** 55–67.
Pouny, Y., and Shai, Y. (1992). *Biochemistry* **31,** 9482–9490.
Pratt, L. R. (1985). *Annu. Rev. Phys. Chem.* **36,** 433–449.
Prestrelski, S. J., Byler, D. M., and Liebman, M. N. (1992). *Proteins: Struct. Funct. Genet.* **14,** 440–450.
Price, S. L., Faerman, C. H., and Murray, C. W. (1991). *J. Comput. Chem.* **12,** 1187–1197.
Price, S. L., Andrews, J. S., Murray, C. W., and Amos, R. D. (1992). *J. Am. Chem. Soc.* **114,** 8268–8276.
Printz, M. P., Nemethy, G., and Bleich, H. (1972). *Nature (London) New Biol* **237,** 135–40.
Provencher, S. W., and Glockner, J. (1981). *Biochemistry* **20,** 33–37.
Raleigh, D. P., Creuzet, R., Gupta, S. K. D., Levitt, M. H., and Griffin, R. G. (1989). *J. Am. Chem. Soc.* **111,** 4502–4503.
Ramachandran, G. N., and Sasisekharan, V. (1968). *Adv. Protein Chem.* **23,** 283–438.
Rao, B. G., and Singh, U. C. (1989). *J. Am. Chem. Soc.* **111,** 3125–3133.
Rao, S. N. (1992). *Pept. Res.* **5,** 148–155.
Rappe, A. K., and Goddard III, W. A. (1991). *J. Phys. Chem.* **95,** 3358–3363.
Rauhut, G., Clark, T., and Steinke, T. (1993). *J. Am. Chem. Soc.* **115,** 9174–9181.
Regoli, D., Park, W. K., and Rioux, F. (1974). *Pharm. Rev.* **26,** 69–123.
Richards, W. G., King, P. M., and Reynolds, C. A. (1987). *Protein Eng.* **2,** 319–327.
Ripka, W. C., DeLucca, G. V., Bach II, A. C., Pottorf, R. S., and Blaney, J. M. (1993a). *Tetrahedron* **49,** 3593–3608.
Ripka, W. C., Lucca, G. V. D., Bach II, A. C., Pottorf, R. S., and Blaney, J. M. (1993b). *Tetrahedron* **49,** 3609–3628.
Rivier, J., Brazeau, P., Vale, W., and Guillemin, R. (1975a). *J. Med. Chem.* **18,** 123–126.
Rivier, J., Brown, M., and Vale, W. (1975b). *Biochem. Biophys. Res. Commun.* **65,** 746–751.
Rivier, J., Kupryszewski, G., Varga, J., Porter, J., Rivier, C., Perrin, M., Hagler, A., Struthers, S., Corrigan, A., and Vale, W. (1988). *J. Med. Chem.* **31,** 677–682.
Rose, G. D., Gierarsch, L. M., and Smith, J. A. (1985). *Adv. Protein Chem.* **37,** 1–109.
Roux, B., and Karplus, M. (1993). *J. Am. Chem. Soc.* **115,** 3250–3262.
Roux, B., Bruschweiler, R., and Ernst, R. R. (1990). *Eur. J. Biochem.* **194,** 57–60.
Sagan, S., Amiche, M., Delfour, A., Camus, A., Mor, A., and Nicolas, P. (1989). *Biochem. Biophys. Res. Commun.* **163,** 726–732.
Samanen, J. M., Weinstock, J., Hempel, J. C., Keenan, R. M., Hill, D. T., Ohlstein, E. H., Weidley, E. F., Aiyar, N., and Edwards, R. (1992). *In* "Peptides: Chemistry and Biology" (J. A. Smith and J. E. Rivier, eds.), pp. 386–388. ESCOM, Leiden.
Sarges, R., and Witkop, B. (1964). *J. Am. Chem. Soc.* **86,** 1861–1862.
Sarges, R., and Witkop, B. (1965). *J. Am. Chem. Soc.* **87,** 2011–2020.
Schaefer, M., and Froemmel, C. (1990). *J. Mol. Biol.* **216,** 1045–1066.

Schafer, L., Newton, S. Q., Cao, M., Peeters, A., Van Alsenoy, C., Wolinski, K., and Momany, F. A. (1993). *J. Am. Chem. Soc.* **115,** 272–280.
Schiffer, C. A., Caldwell, J. W., Kollman, P. A., and Stroud, R. M. (1993). *Mol. Simul.* **10,** 121–149.
Schiller, P. W. (1991). *In* "Progress in Medicinal Chemistry" (G. P. Ellis and G. B. West, eds.), pp. 301–340. Elsevier, Amsterdam.
Schmidt, J. M., Bruschweiler, R., Ernst, R. R., Dunbrack, R. L., Jr., Joseph, D., and Karplus, M. (1993). *J. Am. Chem. Soc.* **115,** 8747–8756.
Schreiber, H., and Steinhauser, O. (1992a). *Biochemistry* **31,** 5856–5860.
Schreiber, H., and Steinhauser, O. (1992b). *Chem. Phys.* **168,** 75–89.
Schreiber, S. L. (1991). *Science* **251,** 283–7.
Schreiber, S. L. (1992). *Cell (Cambridge, Mass.)* **70,** 365–368.
Sharp, K. (1991). *J. Comput. Chem.* **12,** 454–468.
Sheridan, R. P., Nilakantan, R., Dixon, J. S., and Venkataraghavan, R. (1986). *J. Med. Chem.* **29,** 899–906.
Sherman, S. A., and Johnson, M. E. (1993). *Prog. Biophys. Mol. Biol.* **59,** 285–339.
Shimada, J., Kaneko, H., and Takada, T. (1993). *J. Comput. Chem.* **14,** 867–878.
Shin, Y.-K., Levinthal, C., Levinthal, F., and Hubbell, W. L. (1993). *Science* **259,** 960–963.
Shon, K. J., Kim, Y., Colnago, L. A., and Opella, S. J. (1991). *Science* **252,** 1303–1305.
Smeby, R. R., Arakawa, K., Bumpus, F. M., and Marsch, M. M. (1962). *Biochim. Biophys. Acta* **58,** 550–557.
Smith, J. A., and Pease, L. G. (1980). *Crit. Rev. Biochem.* **8,** 314–399.
Smith, P. E., and Pettit, B. M. (1991). *J. Chem. Phys.* **95,** 8430–8441.
Smith, S. O., and Peersen, O. B. (1992). *Annu. Rev. Biophys. Biomol. Struct.* **21,** 25–47.
Soman, K. V., Karimi, A., and Case, D. A. (1991). *Biopolymers* **31,** 1351–61.
Sorensen, R. A., Liau, W. B., Kesner, L., and Boyd, R. H. (1988). *Macromolecules* **21,** 200–208.
Sparks, S. W., Cole, H. B. R., Torchia, D. A., and Young, P. E. (1989). *Chem. Scr.* **29A,** 31–38.
Spear, K. L., Brown, M. S., Reinhard, E. J., McMahon, E. G., Olins, G. M., Palomo, M. A., and Patton, D. R. (1990). *J. Med. Chem.* **33,** 1935–1940.
Spencer, R. G. S., Halverson, K. J., Auger, J., McDermott, A. E., Griffin, R. G., and Lansbury, P. T., Jr. (1991). *Biochemistry* **30,** 10382–10387.
Sreerama, N., and Woody, R. W. (1993). *Anal. Biochem.* **209,** 32–44.
Still, W. C., Tempczyk, A., Hawley, R. C., and Hendrickson, T. (1990). *J. Am. Chem. Soc.* **112,** 6127–6129.
Stillinger, F. H. (1980). *Science* **209,** 451–457.
Stone, A. J. (1985). *Mol. Phys.* **56,** 1065–1082.
Stone, A. J., and Alderton, M. (1985). *Mol. Phys.* **56,** 1047–1064.
Stouten, P. F. W., Frommel, C., Nakamura, H., and Sander, C. (1993). *Mol. Simul.* **10,** 97–120.
Struthers, R. S., Tanaka, G., Koerber, S. C., Solmajer, T., Baniak, E. L., Gierasch, L. M., Vale, W., Rivier, J., and Hagler, A. T. (1990). *Proteins: Struct. Funct. Genet.* **8,** 295–304.
Stryer, L. (1978). *Annu. Rev. Biochem.* 47, 819–846.
Sugg, E. E., Cody, W. L., Abdel-Malek, Z., Hadley, M. E., and Hruby, V. J., (1986). *Biopolymers* **25,** 2029–2042.
Sugg, E. E., Dolan, C. A., Patchett, A. A., Chang, R. S. L., Faust, K. A., and Lotti, V. J. (1990). *In* "Peptides: Chemistry, Structure and Biology" (J. E. Rivier and G. R. Marshall, eds.), pp. 305–306. ESCOM, Leiden.
Surewicz, W. K., Mantsch, H. H., and Chapman, D. (1993). *Biochemistry* **32,** 389–394.
Tadesse, L., Nazarbaghi, R., and Walters, L. (1991). *J. Am. Chem. Soc.* **113,** 7036–7037.
Takahashi, N., Hayano, T., and Suzuki, M. (1989). *Nature (London)* **337,** 473–475.
Talluri, S., Montelione, G. T., van Duyne, G., Piela, L., Clardy, J., and Scheraga, H. A. (1987). *J. Am. Chem. Soc.* **109,** 4473–4477.
Tasaki, K., McDonald, S., and Brady, J. W. (1993). *J. Comput. Chem.* **14,** 278–284.
Tayar, N. E., Mark, A. E., Vallat, P., Brunne, R. M., Testa, B., and van Gunsteren, W. F. (1993). *J. Med. Chem.* **36,** 3757–3764.

Theriault, Y., Logan, T. M., Meadows, R., Yu, L., Olejniczak, E. T., Holzman, T. F., Simmer, R. L., and Fesik, S. W. (1993). *Nature (London)* **361,** 88–91.

Thompson, L. K., McDermott, A. E., Raap, J., van der Wielen, C. M., and Lugtenburg, J. (1992). *Biochemistry* **31,** 7931–7938.

Tobias, D. J., and Brooks., C.L., III. (1991). *Biochemistry* **30,** 6059–70.

Tobias, D. J., Sneddon, S. F., and Brooks, C.L., III. (1990). *J. Mol. Biol.* **216,** 783–96.

Tobias, D. J., Mertz, J. E., and Brooks, C.L., III. (1991). *Biochemistry* **30,** 6054–8.

Todd, A. P., Cong, J., Levinthal, F., Levinthal, C., and Hubbell, W. L. (1989). *Proteins* **6,** 294–305.

Torchia, D. A. (1984). *Annu. Rev. Biophys. Bioeng.* **13,** 125–44.

Torda, A. E., Scheek, R. M., and van Gunsteren, W. F. (1990). *J. Mol. Biol.* **214,** 223–235.

Tournois, H., Killian, J. A., Urry, D. W., Bokking, O. R., de Gier, J., and de Kruijff, B. (1987). *Biochim. Biophys. Acta* **905,** 222–226.

Tourwe, D., Toth, G., Lebl, M., Verschueren, K., Knapp, R. J., Davis, P., Van Binst, G., Yamamura, H. I., Burks, T. F., Kramer, T., and Hruby, V. J. (1992a). *In* "Peptides: Chemistry and Biology" (J. A. Smith and J. E. Rivier, eds.), pp. 307–308. ESCOM, Leiden.

Tourwe, D., Verschueren, K., Van Binst, G., Davis, P., Porecca, F., and Hruby, V. J. (1992b). *Bioorg. Med. Chem. Lett.* **2,** 1305–1308.

Turner, R. J., Matsoukas, J. M., and Moore, G. J. (1991). *Biochim. Biophys. Acta* **1065,** 21–28.

Urry, D. (1971). *Proc. Natl. Acad. Sci. U.S.A.* **68,** 672–676.

Urry, D. W., Goodall, M. C., Glickson, J. D., and Mayers, D. F. (1971). *Proc. Natl. Acad. Sci. U.S.A.* **68,** 1907–1911.

Urry, D. W., Long, M. M., Jacobs, M., and Trapane, T. L. (1982). *Proc. Natl. Acad. Sci. U.S.A.* **79,** 390–394.

Urry, D. W., Trapane, T. L., and Prasad, K. U. (1983). *Science* **221,** 1064–1067.

Vale, W., Brazeau, P., Rivier, C., Brown, M., Boss, B., Rivier, J., Burgus, R., Ling, N., and Guillermin, R. (1975). *Rec. Prog. Horm. Res.* **31,** 365–397.

Vale, W., Rivier, C., and Brown, M. (1977). *Annu. Rev. Physiol.* **39,** 473–527.

Van Binst, G., and Tourwe, D. (1992). *Pept. Res.* **5,** 8–14.

van Gunsteren, W. F., and Berendsen, H. J. C. (1990). *Angew. Chem., Int. Ed. Engl.* **29,** 992–1023.

van Schaik, R. C., van Gunsteren, W. F., and Berendsen, H. J. C. (1992). *J. Comput. Aided Mol. Des.* **6,** 1–24.

Veatch, W. T., Fossel, E. T., and Blout, E. R. (1974). *Biochemistry* **13,** 5249–5256.

Veber, D. F. (1982). *In* "Peptides: Synthesis–Structure–Function" (D. H. Rich and E. Gross, eds.), pp. 658–695. Pierce Chemical Company, Rockford, Illinois.

Veber, D. F. (1992). *In* "Peptides: Chemistry and Biology" (J. A. Smith and J. E. Rivier, eds.), pp. 3–14. ESCOM, Leiden.

Veber, D. F., Strachan, R. G., Bergstrand, S. J., Holly, F. W., Homnick, C. F., Hirschmann, R., Torchiana, M., and Saperstein, R. (1976). *J. Am. Chem. Soc.* **98,** 2367–2369.

Veber, D. F., Holly, R. W., Paleveda, W. J., Nutt, R. F., Bergstrand, S. J., Torchiana, M., Glitzer, M. S., Saperstein, R., and Hirschmann, R. (1978). *Proc. Natl. Acad. Sci. U.S.A.* **75,** 2636–2640.

Veber, D. F., Holly, R. W., Nutt, R. F., Bergstrand, S. J., Brady, S. F., Hirschmann, R., Glitzer, M. S., and Saperstein, R. (1979). *Nature (London)* **280,** 512–514.

Veber, D. F., Freidinger, R. M., Perlow, D. S., Palevada, J. W. J., Holly, F. W., Strachan, R. G., Nutt, R. F., Arison, B. H., Homnick, C., Randall, W. C., Glitzer, M. S., Saperstein, R., and Hirschmann, R. (1981). *Nature (London)* **292,** 55–58.

Veber, D. F., Saperstein, R., Nutt, R. F., Freidinger, R. M., Brady, S. F., Curley, P., Perlow, D. S., Paleveda, W. J., Colton, C. D., Zacchei, A. G., Tocco, D. J., Hoff, D. R., Vandlen, R. L., Gerich, J. E., Hall, L., Mandarino, L., Cordes, E. H., Anderson, P. S., and Hirschmann, R. (1984). *Life Sci.* **34,** 1371–1378.

Vedani, A., and Huhta, D. W. (1990). *J. Am. Chem. Soc.* **112,** 4759–4767.

Vega, M. C., Aleman, C., Giralt, E., and Perez, J. J. (1992). *J. Biomol. Struct. Dyn.* **10,** 1–13.

Vine, W., and Bowers, L. D. (1987). *Crit. Rev. Clin. Lab. Sci.* **25,** 275–311.
Wagner, G. (1993a). *Curr. Opin. Struct. Biol.* **3,** 748–754.
Wagner, G. (1993b). *J. Biomol. NMR* **3,** 375–385.
Wagner, G., Hyberts, S. G., and Havel, T. F. (1992). *Annu. Rev. Biophys. Biomol. Struct.* **21,** 167–198.
Wallace, B. A. (1986). *Biochem. J.* **49,** 295–306.
Wallace, B. A. (1990). *Annu. Rev. Biophys. Biophys. Chem.* **19,** 127–157.
Wallace, B. A. (1992). *Prog. Biophys. Mol. Biol.* **57,** 59–69.
Wallace, B. A., and Ravikumar, K. (1988). *Science* **241,** 182–187.
Walsh, C. T., Zydowsky, L. D., and McKeon, F. D. (1992). *J. Biol. Chem.* **267,** 13115–13118.
Ward, P., Ewan, G. B., Jordan, C. C., Ireland, S. J., Hagan, R. M., and Brown, J. R. (1990). *J. Med. Chem.* **33,** 1848–1851.
Weber, C., Weber, G., von Freyberg, B., Traber, R., Braun, W., Widmer, H., and Wuthrich, K. (1991). *Biochemistry* **30,** 6563–74.
Wei, J. A., Lin, Y. Z., Zhou, J. M., and Tsou, C. L. (1991). *Biochim. Biophys. Acta* **1080,** 29–33.
Weinkam, R. J., and Jorgensen, E. C. (1971). *J. Am. Chem. Soc.* **93,** 7033–7038.
Weinstein, S., Durkin, J. T., Veatch, W. R., and Blout, E. R. (1985). *Biochemistry* **24,** 4374–4382.
Wenger, R. (1986). *Transplant. Proc.* **18,** 213–18.
Whitlow, M., and Teeter, M. M. (1986). *J. Am. Chem. Soc.* **108,** 7163–7172.
Wilkes, B. C., and Schiller, P. W. (1992). *Int. J. Pept. Protein Res.* **40,** 249–254.
Wille, B., Franz, B., and Jung, G. (1989). *Biochim. Biophys. Acta* **986,** 47–60.
Williams, D. E. (1991). *In* "Reviews in Computational Chemistry" (K. B. Lipkowitz and D. B. Boyd, eds.), pp. 219–271. VCH Publ., New York.
Williams, R. J. P. (1989). *Eur. J. Biochem.* **183,** 479–497.
Wishart, D. S., Sykes, B. D., and Richards, F. M. (1991). *J. Mol. Biol.* **222,** 311–333.
Woody, R. (1985). *In* "The Peptides" (E. R. Blout, F. A. Bovey, M. Goodman, and N. Lotan, eds.), Vol. 7, pp. 15–114. Academic Press, New York.
Woolley, G. A., and Wallace, B. A. (1992). *J. Membr. Biol.* **129,** 109–136.
Wuthrich, K. (1986). "NMR of Proteins and Nucleic Acids." Wiley (Interscience), New York.
Yang, J. T., Wu, C.-S. C., and Martinez, H. M. (1986). *In* "Methods in Enzymology" (C. H. W. Hirs and S. N. Timasheff, eds.), Vol. 130, pp. 208–269. Academic Press, Orlando, Florida.
Zauhar, R. J. (1991). *J. Comput. Chem.* **12,** 575–583.
Zauhar, R. J., and Morgan, R. S. (1988). *J. Comput. Chem.* **9,** 171–187.
Zhang, Y. P., Lewis, R. N., Hodges, R. S., and McElhaney, R. N. (1992). *Biochemistry* **31,** 11572–11578.

6

Structure–Function Studies of Peptide Hormones: An Overview

Victor J. Hruby and Dinesh Patel
Department of Chemistry
University of Arizona
Tucson, Arizona 85721

I. Introduction

Many peptide hormones of biological importance are relatively small with respect to biological proteins, and as a result the degree of conformational diversity possible for these hormones in various biological environments is far greater due to the lack of a well-defined tertiary structure. This is further complicated by the knowledge that many of the peptide hormones are produced biosynthetically from larger precursor proteins, and when formed they may adopt numerous conformations at physiological temperature. Hence, the determination of which structural elements, conformations, and dynamic properties are of critical importance to biological activity, that is, the "bioactive conformations," becomes of overwhelming significance.

Many of the leading peptide research laboratories have been addressing this problem, with much emphasis being placed on developing a strategy that facilitates a systematic approach to peptide design. In general, these strategies must address three different aspects of peptide structure. First, one should consider which specific functional groups in the primary sequence (these are in most instances the side-chain groups on a peptide) are related to specific biological effects (this can be considered as the classic structure–function approach). Second, one must address the issues of backbone conformation (α helix, β turn, extended, pleated sheet, etc.) and topographical properties (side-chain conformations, surface properties, hydrophobic/hydrophilic moieties, surface charge distribution, and so forth) within the context of molecular recognition with a specific acceptor molecule or system (receptor, enzyme, antibody, etc.) and as it may relate to the transduction or activation process. Finally, one must understand the relationships between the conformational properties and dynamic behavior, and the measured biological activity. It is worth noting that a modern, well-designed bioassay is essentially a physical-chemistry experiment that can provide physical and/or chemical information on a variety of cellular level processes involved in the binding of a ligand to a receptor. Many scientists overlook the fact that the assays can provide a more sensitive probe to the structural, conformational, and dynamic requirements for biological activity for a peptide than can many of the more sophisticated biophysical experiments such as nuclear magnetic resonance (NMR) spectroscopy and X-ray crystallography.

This chapter attempts to provide a general overview of some of the considerations that go into the design of structure–function studies of peptide hormones. Increasingly, this approach can lead to a rational or semirational approach to ligand design and can provide insights into the physical-chemical and stereostructural properties that determine peptide hormone structure–biological relationships at a three-dimensional structural level. Emphasis is placed on general principles. It is impossible to be comprehensive in the space allotted, but we shall attempt to illustrate some of these principles by examples familiar to us. In many cases, numerous examples exist in the literature that could illustrate these principles, and we apologize to those whose excellent studies have not been referred to herein.

II. Mammalian Peptide Hormones

Attempting to define exactly what evidence is necessary to establish that a specific receptor exists for a hormone at a specific tissue or cell type is difficult. Generally, definitions based on putative physiological significance are used, but it must be kept in mind that the significance is based on historical observations rather than a detailed understanding of the overall significance of a particular observation of a biological effect. Within this context, the general criterion appears to be that the tissue is sensitive to stimulation or other biochemical changes at concentrations and conditions that are believed to be "physiologically relevant." This functional definition is often arrived at after a reasonably comprehensive series of biological, pharmacological, and physiological studies have been performed utilizing the purified peptide hormone, though often zeal for a particular effect will replace comprehensiveness.

A more pharmacological and biochemical perspective establishes a receptor on the basis of the ability of the endogenous peptide hormone, and several of its agonists and if possible competitive antagonists, to bind specifically and with high affinity to a membrane or whole-cell preparation of tissue that is believed to be physiologically relevant. A third, more chemically rigorous perspective insists that establishment of a particular receptor for a particular hormone at a particular cell or tissue can occur only if one isolates and determines the structure of a purified chemical entity (e.g., a membrane protein) that can specifically interact with (bind to) the hormone, and that can be reconstituted in a biologically relevant cell or tissue and display the appropriate biological activity in response to the hormone ligand.

All of these perspectives have their historical significance, and when properly pursued can provide important insights into the biological activity profile of a particular endogenous peptide hormone. For the purposes of this chapter we define a hormone–receptor interaction from the biological effect which must result from any such interaction, and in particular those direct effects that are an immediate result of the hormone–receptor interaction.

A hormone–receptor interaction involves at least three different states (Hruby, 1981a; Hruby and Hadley, 1986): (1) recognition or the hormone–receptor binding interaction, also known as the binding message (Schwyzer, 1977); (2) transduction or hormone–receptor interaction leading to activation of the biological response via a second messenger such as cyclic AMP or calcium; and (3) reversal or hormone–receptor dissociation or translocation which returns the cell to its basal state. The first two of these steps are well-recognized separate processes related to hormone–receptor interaction and can lead to different biological activities including full agonist activity, full competitive antagonist activity, and partial agonist and/or antagonist activities. In this regard it should be noted that there is much evidence, derived from several peptide hormone–receptor systems, that hormone–receptor interactions and structure–activity relationships are different for agonists and for competitive antagonists (e.g., Hruby, 1986, 1987).

The concept that reversal of the peptide hormone–receptor-mediated biological activities may depend on conformational structure properties of the hormone different from those used for binding and transduction is now also widely accepted. For example, it is believed that prolongation of biological activity (or reversal) may be a function of structural and conformational features of the hormone in the hormone–receptor complex after the transduction state for agonists or, in the case of antagonists, in a nonbiologically active (inhibitory) state, both of which are different from those required for binding and transduction. Thus, prolongation can be a receptor-related event irrespective of potency and efficacy.

The application of these ideas in topographical design are still under development, and thus in this chapter we emphasize those studies in which hormone–receptor interactions have been examined from the aspect of receptor recognition (binding, potency, etc.), transduction (agonists, antagonists, partial agonists/antagonists), and reversal (prolongation). Finally, it is important to consider both peptide hormones of the central nervous system and those found in the endocrine system due to the interrelationships of the two systems.

III. Use of Classic Structure–Function Studies: General Considerations

In general, peptide hormones and neurotransmitters are quite complex structures. Whereas they are made up of rather "simple" α-amino acid building blocks, because of the large variety of side-chain group functionalities that constitute α-amino acids, even a rather small peptide hormone of 5 to 10 residues can have available very complex chemical constituents including acid–base groups, lipophilic moieties, hydrophilic groups, hydrogen bond acceptors and donators, and aromatic and heteroaromatic groups. In general, the manner in which these functional groups arrange themselves in three-dimensional space are not obvious, and this aspect requires special considerations (see below). In addition, the chirality of each individual amino acid must be considered in relation to biological activity (in mammalian hormones and neurotransmitters nearly all amino acids are of the L configuration in the native hormone, except for glycine which can exist in D or L conformation space but is not itself chiral).

The early pioneers of peptide structure–biological activity studies including V. du Vigneaud, J. Rudinger, K. Hofmann, C. H. Li, R. Schwyzer, T. Wieland, M. Bodanszky, and H. Yajima, among others, took what is essentially a classic organic chemistry approach to structure–activity relationships. The simple questions they asked were primarily to determine which of the side-chain groups in particular biologically active peptides were important to biological activity. The methods that they used are still widely applicable and useful today. Beautiful and still quite useful expositions of the approaches used have been provided by Rudinger (1971, 1972) and Schwyzer (1977). We discuss here only some of the more general approaches used.

One of the classic systematic approaches used by du Vigneaud was replacement of the functional groups on each amino acid residue in a peptide one at a time by a hydrogen. Thus, aspartic acid (R = CH_2CO_2H) was replaced by alanine (R = CH_3), glutamic acid (R = $CH_2CH_2CO_2H$) by α-aminobutyric acid (R = CH_2CH_3), lysine (R = $CH_2CH_2CH_2CH_2NH_2$) by α-aminohexanoic acid (R = $CH_2CH_2CH_2CH_3$), serine (R = CH_2OH) by alanine (R = CH_3), and so forth. This led to the first superagonist analog [1-β-mercaptopropionic acid]oxytocin (deamino-oxytocin) (Ferrier *et al.*, 1965). Many similar studies have been made, and the methodology has now been generalized to an alanine scan. In this procedure each amino acid residue is replaced one at a time by an alanine residue, and the bioactivity of each compound examined. Alanine substitutions which lead to large decreases in potency (generally >100-fold) are assumed to be of critical importance to the biological recognition (binding) and/or transduction process. A caveat here is that one is making the assumption that such substitutions do not have a profound effect on conformation. Generally in small linear peptide hormones this will be the case, but in cyclic disulfide-containing peptides, elimination of the ring system may have profound conformational or dynamic effects that can strongly affect bioactivity.

Another widely used general screen of this type is the D-amino acid (or D-Ala) scan. In this case, each L-amino acid is replaced by the corresponding D-amino acid (one residue at a time) or by a D-Ala. Again, the bioactivity is evaluated for each analog, and for those analogs with profoundly decreased potency, the amino acid replaced is presumed to be critical for bioactivity. In this case, stereochemical relationships are changed that can add an extra degree of difficulty in interpretation of the results. Furthermore, D-amino acid residues can have profound effects on backbone conformation. For example, they can break up an α-helical segment. On the other hand, they can stabilize a β turn. If either of these conformations are important for receptor recognition (binding) or transduction, the former effect may greatly decrease potency even though that particular amino acid residue may not itself be critical to bioactivity. On the other hand, if a β turn is important to bioactivity, the use of a D-amino acid may in some cases actually increase biological potency. In addition, D-amino acid substitutions can greatly increase the stability of peptides to biodegradation. In many assays, especially *in vivo* assays, this stabilization may greatly affect bioactivity and yet have little or no significance for receptor recognition or transduction.

In these kinds of studies glycine residues present a particular challenge. Glycine, of course, has no α-substituent and hence is itself achiral. However, when incorporated into a peptide it can readily assume conformations in Ramachandran space (ϕ, ψ space) of an L- or D-amino acid residue. Thus, we suggest that at any place a glycine residue is found, both an L-alanine and a D-alanine residue should be examined, and for the sake of completeness perhaps an α-aminoisobutyric acid residue [-$NHC(CH_3)_2CO$-] might be examined.

A wide variety of other general approaches have been used to examine various aspects of the relationships of "chemical requirements" for specific functional groups in a peptide to its biological activity. Some examples include examining lipophilicity requirements by substituting a specific functional group for a more lipophilic group [e.g., a $CH_2CH(CH_3)_2$ by a $CH_2C(CH_3)_3$], an acid group by a hydrophilic group or a hydrophobic group (e.g., CO_2H by CH_2OH or CH_3), a basic group by a hydrophilic or a hydrophobic group (e.g., NH_2 by OH or CH_3), an aromatic group by a substituted aromatic group, e.g.,

$$\text{-CH}_2\text{-}\bigcirc \text{ to -CH}_2\text{-}\bigcirc\text{-NO}_2 \text{ or CH}_2\text{-}\bigcirc\text{-CH}_3,$$

a carboxyl terminal by a carboxamide terminal, and so forth. Special mention should be made of peptide bond (amide bond) replacements. Amide bond replacements have become widely used as a way to examine hydrogen bonding requirements of the backbone, the conformational requirements of the backbone, stabilization of the peptide against biodegradation, design of "transition state" inhibitors of proteases and esterases, and for other purposes. The general approaches to be used in this area were outlined in detail by Rudinger (1971), and a review and overview of this important area of research has been provided by Spatola (1983). The exact conformational and stereostructural consequences of such substitutions are still a matter of considerable speculation and intensive study. Such studies are sure to accelerate, as are the synthetic approaches to preparing the chiral molecules required for incorporation into peptides as true peptidomimetics, since so many peptide analogs already examined have provided critical insights into structure–biological activity relationships.

IV. Use of Conformational Constraints: General Considerations

Many developments toward a rational and systematic approach to peptide design have involved the use of conformational constraints. The basic idea in this approach is to modify the structure of one or more specific amino acid residues such as to restrict one or several of the ϕ, ψ , ω , and/or χ angles, and in so doing to provide either a specific local constraint (e.g., the amino acid residue becomes biased toward an α-helix or β-turn backbone conformation) or a more global constraint (e.g., a tetrapeptide segment of the peptide is biased toward a β turn by cyclization). This is an area of rapid development in which designer amino acids, designer peptides, pseudopeptides, and peptidomimetics are being designed, synthesized, and utilized to develop systematic strategies compatible with peptide and protein structure.

In our laboratory, emphasis has been placed on using constrained side-chain groups of amino acid residues (constrained, restricted, or fixed at the χ_1 and χ_2 torsional angles) as part of the design process. This has been of particular importance in peptide design projects in which a backbone conformational model or template has already been developed. Even in highly constrained cyclic polypeptides, there often is considerable conformational flexibility because the side-chain moieties may have considerable conformational freedom by rotation about the χ_1, χ_2, and other side-chain torsional angles. Careful evaluation of the preferred side chain conformations is a critical element in the biological activities of the peptide, and ultimately rational design requires that topographical design be an integral part of any peptide or protein design approach. Unfortunately this usually is not possible early on in most studies, and in such cases local and general global constraints are used, with the emphasis being placed on the introduction of rigidity into the peptide backbone.

The precise way in which conformational and topographical constraints can be incorporated into a conformationally flexible peptide hormone and still maintain the proper overall stereostructural, conformational, and dynamic properties of a peptide compatible with its biological activity is still a matter of considerable discussion and uncertainty. One can view this problem from two perspectives. On the one hand, one can aim to make a local or global restriction that will sufficiently "fix" the local or global conformation that it can be assumed to retain that conformational preference at the receptor, at least for initial recognition (because transduction may require conformational change, it is important that some structural element, e.g., a side-chain group(s), retain sufficient flexibility that it can adopt a different side-chain conformation). Alternatively, one can aim primarily to make accessible a limited range of conformational properties such that the peptide can more easily "fit" the receptor requirements (induced fit). In practice, given our present limited knowledge of peptide hormone–receptor interactions, neither approach is likely to be certain of success, and, in any case, the actual hormone–receptor interaction is probably more complex than either of the above views would provide. Thus, discussion here focuses on the general approach one takes, and the interested reader is directed to more comprehensive overviews of this important aspect of peptide hormone ligand design (Marshall, 1992, 1993; Rizo and Gierasch, 1992; Hruby, 1991, 1982a; Hruby *et al.*, 1991a; Kessler, 1982; Freidinger and Veber, 1984).

A. Local Constraints

Actually, the local constraints approach has been used from the earliest structure–activity studies, though the reasons for such efforts were not particularly a result of conformational considerations, but rather were done from the viewpoint of a functional group replacement. Examples include the use of *N*-methylamino acids such as sarcosine, dehydroamino acids such as $\Delta^{3,4}$-dehydroproline, and β-penicillamine (β,β-dimethylcystine, Pen). The latter amino acid is quite interesting in that it was first used by du Vigneaud as a replacement of Cys in oxytocin with the apparent object of increasing the hydrophobicity of the peptide. Interestingly, it was found that the analog obtained, [Pen^1]oxytocin, was an antagonist (Schulz and du Vigneaud, 1966; du Vigneaud *et al.*, 1964). Later it was shown by Hruby and co-workers (Meraldi *et al.*, 1977; Hruby *et al.*, 1978, 1979a) that in fact the penicillamine residue had both local conformation effects on the disulfide bridge (disulfide angle and chirality) and more global constraints in cyclic peptides, probably due to the geminal dimethyl substituents which can affect the peptide conformation via increased lipophilicity and especially transannular effects in medium-sized rings (Meraldi *et al.*, 1977; Hruby *et al.*, 1979b, 1980). The latter example points to the use of a constrained amino acid in already partially constrained cyclic peptide analogs or native peptides. Similarly, the use of an α-alkyl substituted amino acid such as α-aminoisobutyric acid was pointed out by Marshall *et al.* (1987) to have a considerable local restriction on the conformation preference of the residue for an α-helical structure.

Many other types of local constraints have been obtained by using specific N-substituted, α-substituted, and otherwise substituted amino acid residues in a peptide. In addition, intramolecular modifications involving side-chain to α-amino substitution, side-chain to adjacent amino acid backbone structures, and so forth, have been designed and synthesized. The uses of such constraints have been reviewed by Toniolo (1990) and by Hruby and Boteju (1993). A remarkable array of amino acids, amino acid derivatives and analogs, pseudodipeptides, and related compounds provide an ever increasing array of possibilities for local conformational constraints involving the peptide backbone ϕ and ψ angles. In addition, various amide bond replacements have been suggested as replacements for cis and trans peptide bonds (Spatola, 1983). Closely related to these developments has been the development of a variety of organic compounds which are meant to force and/or act as nucleation sites for various secondary structures, including β turns, γ turns, α helices, and β sheets. In many cases, these compounds are not designed to place the side-chain groups found in these structures in their correct stereochemical and conformational places. Hence, though these structures may induce appropriate secondary structures very well, they may fail as peptide mimetics in that side-chain groups critical to receptor recognition may be missing or in inappropriate three-dimensional space. These caveats make interpretation of biological activity in terms of conformational requirements difficult or misleading in these situations.

In addition to backbone constraints, local constraints in χ space also are possible and can be very illuminating. In most naturally occurring amino acids and in all of those most commonly found in natural proteins, the side-chain moieties have considerable flexibility in their torsional angles χ_1, χ_2, χ_3, etc. Thus, even in peptides with well-defined secondary structures, depending on the disposition of the side-chain groups, quite different topographical structures can occur. Thus, methods to bias, fix, or constrain side-chain moieties to favor particular χ_1, χ_2, and related side-chain torsion angles to a particular value [gauche(−) = $-60°$; gauche(+) = $+60°$ or trans = $\pm 180°$] are needed, especially where the perturbations of structure are compatible with standard peptide and protein secondary structure. Methods that have been used for such restrictions include β-alkyl substitution (methyl, ethyl, etc.), 2′, 6′ substitution on phenylalanine and tyrosine, aromatic to N^{α} covalent attachment either directly or through a methylene or other bridge, β-hydroxy groups, and γ-alkyl groups. These types of restrictions, particularly when examined for amino acid residues that are critical for the bioactivity of a hormone at a particular receptor, can profoundly affect the bioactivity not only because of the specific constraint they introduce, but because they introduce unique stereostructural and/or topographical relationships both in other key structural moieties in the hormone itself and in the way this structure presents itself in three-dimensional space to the receptor. The challenge, then, is to design unique amino acids that can provide new three-dimensional shapes compatible with protein–peptide recognition.

B. Global Constraints

Stabilization of particular secondary structures, especially those related to β turns and γ turns as well as an α helix, can often be best accomplished by a covalent constraint that is global in nature. Such secondary structures bring side-chain groups, distant from one another in sequence, close to one another in three-dimensional space. By making an appropriate covalent attachment between the two groups, an otherwise short-lived conformation can be stabilized to give a relatively more rigid, constrained structure that can possess a particular secondary structure. This strategy was first developed by globular proteins that often have within their structure disulfide bridges between two cysteine residues. It has been demonstrated in several cases that these disulfide bridges in protein substantially increase the stability of the protein to denaturation by heat, salt, or other perturbants. Thus, it is not surprising that peptide and protein chemists have turned to disulfide formation as a convenient method to stabilize secondary structure, with considerable success (Rizo and Gierasch, 1992; Hruby *et al.*, 1983, 1991a; Hruby, 1982a; Kessler, 1982). However, many failures also have occurred by this approach, and thus a few comments should be made regarding the proper choice of groups to be chosen for covalent attachment so as to maximize the possibility of obtaining a potent analog.

In general, when a peptide folds into a particular secondary structure (β turn, α helix, β sheet, etc.) it creates at least two distinct surfaces (for sake of argument, we define these as the front and back surface). In general, for interaction of the peptide with a biologically relevant receptor or other molecule, it will utilize primarily one of these surfaces for a binding interaction, and the other surface will remain completely or partially exposed to the surrounding solution or environment. In general, therefore, covalent stabilization should be made on the surface which is not utilized for molecular recognition (the ancillary surface). Alternatively, if the recognition surface is utilized, then an isosteric or pseudoisosteric replacement of the structural elements used for cyclization is ncessary. For example, we have utilized a covalent disulfide ($-CH_2SSCH_2-$) cyclization as a pseudoisosteric replacement for Met^4 ($-CH_2CH_2SCH_3$) and Gly^{10}(-H) residues to convert a linear potent hormone α-melanotropin into a superpotent cyclic hormone [$\overline{Cys^4,Cys^{10}}$]α-melanotropin (Sawyer *et al.*, 1982). This conformational constraint was designed to stabilize a putative β turn at the Phe^7 position.

Although the disulfide bridge is the most commonly used global conformational constraint, several other constraints not commonly found in proteins can be used for stabilizing a wide variety of secondary structures. Most commonly, disulfides have been used to help stabilize a variety of β-turn and γ-turn structures, but they have also been used to help stabilize α-helix structures and even β-sheet structures. In principle, a disulfide bridge can be used to help stabilize a loop structure as well. Finally, of course, it can be used as a way to maintain the proximity of groups believed to be close to one another in any secondary structure.

A wide variety of side chain to side chain cyclizations are possible using functional groups found on the common amino acids, and many others can be envisioned. In practice, very few of these have been explored in any detail. One exception is the cyclic lactam, in which the side-chain amino group of a lysine or ornithine residue is covalently linked to a side-chain carboxyl group on aspartic acid or glutamic acid. Stabilization of a variety of β-turn structures, an α-helix, etc., can be expected and has been observed. Because a lactam bridge is inherently more rigid than a disulfide bridge and has different stereoelectronic properties, one can anticipate that its uses will differ within the context of otherwise equivalent ring structures, though actual direct comparisons apparently have not been made. A variety of other covalent side chain to side chain attachments should be considered, including sulfides, ethers, esters, alkanes, alkynes, aromatics, alcohols, and other such moieties. It is interesting to note in this regard that, whereas a great deal of time has been spent on methods for macrocyclic synthesis by synthetic chemists, in general, much less time or effort has been given to studies of macrocyclic chemistry in the context of peptide or protein chemistry.

For smaller polypeptides (3 to 10 residues) cyclization of the C-terminal carboxyl to the N-terminal amino groups has been used widely. The habit of doing this came out of the fact that many peptide antibiotics such as gramicidin are cyclic peptides of this type, and such cyclic peptides serve as excellent models for β-turn and γ-turn structures (Hruby, 1974; Rose *et al.*, 1985; Rizo and Gierasch, 1992). However, when such cyclizations have been used for peptide hormones and neurotransmitters such as oxytocin, enkephalin, and others, they generally have led to larger decreases in potency, owing to the importance of C- and/or N-terminal groups for bioactivity. On the other hand, when bioactive partial sequences of peptide hormones and neurotransmitters are found that possess all of the major structural constituents for activity at a particular receptor, it is possible to construct an N- to C-terminal cyclic peptide which includes this structure and obtain analogs of high potency. This is elegantly exemplified by the studies of Veber and co-workers on the design of somatostatin analogs (Veber and Freidinger, 1984). These studies show how earlier fundamental studies on cyclic peptides, especially hexapeptides, provided a conformational context to design into a cyclic structure, a specific secondary structure (in this case a β turn) that was of great biological significance. The great advantage of cyclic peptides, especially cyclic pentapeptides and cyclic hexapeptides, is that they can provide an excellent secondary structure template from which to present side-chain moieties in specific stereochemical and topographical relationships for interaction with a receptor. Such considerations have been used, for example, by Kessler and co-workers in the design of an RGD-containing peptide for interactions with specific receptors (Kessler *et al.*, 1992). Clearly, the design of cyclic peptides as templates for the synthesis of specific pharmacophores is an intriguing prospect, and it is highly likely that continued efforts will be made in this direction.

In addition, of course one can consider backbone cyclization of a different type that may not involve the C- or N-terminal residues. For example, one could consider a covalent structure in which a structural moiety of some sort (most simply an alkyl chain) would physically connect the amide nitrogens of two amino acid residues *n* residues apart from one another in the polypeptide chain. A variety of possible linking groups can be considered, but thus far very little chemistry has been done along these lines.

Finally, one can consider a wide variety of methods by which side-chain groups are attached to the backbone of adjacent or quite distant amino acid residues to give a cyclic structure. This often is accomplished by covalently linking the N-terminal amino groups with a side-chain carboxyl group, or a C-terminal carboxyl group with a side-chain amino or hydroxyl group to form cyclic lactams and lactones. A considerable amount of work has been done with such cyclizations, and several highly successful applications have been made to bioactive peptides (e.g., see Al-Obeidi *et al.*, 1989a,b; Schiller and Geiger, 1984; Schiller, 1985a). Despite the successes using this approach, surprisingly little has been done to examine systematically the applications of such cyclizations for structure–activity studies. More importantly, relatively little has been done to examine the conformational consequences of such cyclizations. It would be anticipated that in medium-sized rings (8 to 20 members) many if not most of the topographical motifs used in protein and peptide molecular recognition could be obtained.

V. Selected Examples to Illustrate Approaches

We next illustrate some of the approaches which have led to insights into the stereochemical, conformational, topographical, and dynamic properties associated with the biological activities of a few peptide hormone neurotransmitters. We have not attempted to be comprehensive, but have chosen examples to illustrate some of the approaches that can be taken. Numerous other approaches are possible, and we apologize for omitting many that would have been equally illustrative or that have illustrated other important possibilities.

A. Gastrin and Related Peptides, Especially Cholecystokinin

The gastrin family of nonrelated peptides may be taken to include all peptides naturally containing the C-terminal tetrapeptide amide sequence of the gastrins, Trp-Met-Asp-Phe-NH_2. Although biologically active members of the family display diverse physiological activities, they nevertheless generally act on the same targets. It is important to distinguish between two extremes in the spectrum of activity: activities typical of gastrin itself such as gastrointestinal functions including secretion, motility, and adsorption are achieved at much higher doses than those functions stimulated within the gallbladder, typically cholecystokinin-related activity, which occur at much lower concentrations.

It was found that the minimal peptide sequence of human gastrin I [Pyr-Gly-Pro-Trp-Leu-$(Glu)^5$-Ala-Tyr-Gly-Trp-Met-Asp-Phe-NH_2] that is shown to have gastrinlike activity is (Ac-Trp-Met-Asp-Phe-NH_2). Numerous analogs have been synthesized in which single replacements in the C-terminal positions of the tetrapeptide have been made, including analogs in which two or more replacements have been made. These have been discussed elsewhere in earlier studies (Hruby *et al.*, 1990a; Gonzalez-Muniz *et al.*, 1990; Shiosaki *et al.*, 1990; Horwell *et al.*, 1990). The conclusions reached as to structure–function relations within the minimal fragment suggest that the interaction occurs over two distinct sites: first, binding sites involving Trp, Met, and Phe and, second, a functional site (involving Asp) which directly participates in the chemical events associated with the action of the hormone. The distinguishing feature of the binding site is that activity was still preserved when the amino acid residues were substituted, although decreases in binding potency were observed; the fact that activity is observed is evidence that the analog has activated the receptor. The substitution of the aspartic acid residue by other natural amino acids always resulted in inactivation. However, activity is preserved when the carboxyl group of the aspartyl residue is replaced by an unnatural residue (tetrazol-5-yl) which maintains the electronic character and location of the carboxyl group.

Cholecystokinin (CCK) has been long recognized as a gastrointestinal hormone mediating digestive functions and feeding behaviors. In 1975 a gastrinlike immunoreactive peptide was identified in the rat brain and subsequently identified as CCK. The appearance of CCK in both the gut and brain raises the issue of the evolutionary significance of separate pools of a peptide in two discrete biological systems. It has been shown that the C-terminal octapeptide portion of cholecystokinin, Asp-Tyr(SO_3H)-Met-Gly-Trp-Met-Asp-Phe-NH_2 (CCK-8) is as potent and efficient as the entire molecule. Several molecular forms of cholecystokinin have been identified in the brain, CCK-8 being the most abundant. CCK-8 is believed to function as a neuromodulator and/or a neurotransmitter. It also is known that both the mammalian brain and pancreas contain CCK receptors with different ligand specificities. The pancreatic CCK receptor (CCK-A) is highly selective for sulfated forms of CCK, whereas the brain receptor (CCK-B) interacts about equally with sulfated and unsulfated CCK-8, as well as with shorter C-terminal CCK fragments, including CCK-4 (Trp-Met-Asp-Phe-NH_2) which is the shortest fragment exhibiting high affinity for the central CCK receptor. Furthermore brain and pancreatic CCK receptors differ in size and subunit composition (Holladay *et al.*, 1992, and references therein), and their respective role in many CCK-induced pharmacological responses still creates some controversy. For this reason, it is of interest to design specific ligands for each class of receptors in order to investigate the physiological role of CCK. This is partly achieved with CCK-4, the C-terminal tetrapeptide of cholecystokinin which poorly binds to peripheral CCK receptors but which binds the central CCK receptors almost as well as CCK-8.

The conformation–bioactivity relationships for the two forms of the CCK receptor also differ. However, only relatively recently have models of receptor-bound conformations of CCK-8 been specifically proposed for the A and B receptor types (Nikiforovich and Hruby, 1993). The available data on possible CCK-8 or CCK-7 [CCK(27–33)] conformations are limited to those derived from fluorescence measurements (Schiller *et al.*, 1978), NMR (Durieux *et al.*, 1983; Gacel *et al.*, 1985; Hruby *et al.*, 1991b), a combination of both physical and computational energy minimization calculations (Fournie-Zaluski *et al.*, 1986), and energy calculations alone (Kreissler *et al.*, 1989; Pincus *et al.*, 1987; Coates and Knittle, 1990; Nikiforovich and Hruby, 1993). Several highly selective CCK analogs with *N*-methylamino residues have been synthesized in our laboratories (Hruby *et al.*, 1990a) and elsewhere (Lin *et al.*, 1990). These and findings from other studies open the possibility to propose several models of receptor-bound conformations of CCK-8 for the A and B type receptors by separate comparison of sets of low-energy structures of analogs with high affinity toward A or B receptors, respectively (Nikiforovich and Hruby, 1993).

Results from the study, which included four CCK-related compounds, indicated that it is indeed possible to suggest probable conformational factors that lead to receptor selectivity (or loss of it). The compounds chosen in this particular study were CCK-8 itself (having IC_{50} values for the A and B receptor types of 0.13 and 0.32 n*M*, respectively), desaminoTyr(SO_3)-Nle-Gly-Trp-Nle-Asp-Phe-NH_2 (IC_{50} of 0.77 and 0.50 n*M*), desaminoTyr(SO_3)-Nle-Gly-Trp-Nle-N-MeAsp-Phe-NH_2 (IC_{50} of 0.42 and 300 n*M*), and desaminoTyr(SO_3)-Nle-Gly-Trp-N-MeAsp-Phe-NH_2 (IC_{50} of 110 and 0.19 n*M*). Chosen for the diversity of selectivity, they demonstrated significant differences in their low-energy conformations and hence the probable conformational requirements of each receptor type.

Two models for CCK-8 were previously proposed (Gacel *et al.*, 1985, Fournie-Zaluski *et al.*, 1986) and have up to now been used most often for the design of CCK cyclic analogs. The calculated CCK-B conformer resembles the aforementioned model with respect to the presence of a β turn about the C-terminal Gly-Trp-Met-Asp sequence. However, the proposed CCK-A conformers does not bear any resemblance to the earlier proposed models, mainly because of a "shift" of the N-terminal β turn to the Tyr-Met-Gly-Trp fragment instead of the Asp-Tyr-Met-Gly sequence. This difference could explain the lack of affinity toward CCK-A receptors for conformationally constrained analogs such as X-Tyr(SO_3)-Nle-DLys-Trp-Nle-Asp-Phe-NH_2 (X = DAsp, γDGlu), which were designed with the purpose of stabilizing the N-terminal β turn of the Asp-Tyr-Nle-Glu sequence (Charpentier *et al.*, 1988). Preservation of the C-terminal tetrapeptide maintains the ability of the cyclic peptide to adopt the CCK-B required conformation, which in turn gives rise to the CCK-B selectivity of these compounds. Further cyclic analogs of the type

$$\begin{array}{l} \quad\ulcorner\!\!-CO(CH_2)_nCO-\!\!\urcorner \\ \text{X-Lys-Gly-Trp-Lys-Asp-Phe-NH}_2\ [\text{X} = \text{Ac-Tyr(SO}_3\text{), Ac-Tyr, Ac-H}] \end{array}$$

have also been shown to be CCK-B selective (Rodriquez *et al.*, 1990). In these cases the C^{α} atoms of the Lys residues are involved in the formation of a 24-membered macrocycle, such length perhaps being sufficient to close the ring without significant distortions from the CCK-B required C-terminal tetrapeptide sequence conformation.

The CCK-related analogs with substitutions of the conformationally flexible Gly residue by either L- or D-amino acids are also conformationally restricted. Substitution of the Gly^{29} residue by D-Ala results in a lower local energy minimum than if the corresponding L-amino acid is used. Indeed, the biological affinity of [DAla^{29}]CCK-8 remains of the same order as that of the parent compound (Gacel *et al.*, 1985; Fournie-Zaluski *et al.*, 1985), whereas that of the corresponding L-amino acid analog results in a severalfold decrease in affinity. These findings are also attributable in terms of preserving/changing the CCK-A conformer for corresponding CCK analogs; however, the loss of affinity toward CCK-B receptors for the same analogs appears not to be influenced by the same conformational transitions (Gacel *et al.*, 1984; Fournie-Zaluski *et al.*, 1985; Penke *et al.*, 1984). An LPro^{28} analog has been shown to have high affinity for both CCK receptors, which is in good agreement with a proposed ϕ value (~60°) for the Nle^{28} residue (Finchem *et al.*, 1992).

The biological data for conformationally restricted CCK analogs constitute substantial although indirect evidence for the CCK-A and -B conformers postulated. It is therefore necessary to examine the proposed A conformer with the structure of recently reported CCK-A selective tetrapeptides such as Boc-Trp-Lys(ε-NHCONH-*o*-MePhe)-Asp-Phe-NH_2 [IC_{50} values of 3.8 and 1500 n*M* respectively for the A and B CCK receptors (Shiosaki *et al.*, 1992)]. Calculated low-energy structures of tetrapeptides, when compared with the CCK-8 “A” conformation, show a high degree of similarity with root mean square (RMS) deviation values less than 1.0 Å. Furthermore, the aromatic rings of the Tyr(SO_3) residue and that of the Lye(εNHCONH-*o*-MePhe) residues occupy essentially the same spatial position. The proposed CCK-A and -B conformations of CCK-8 appear to elucidate the conformational factors that lead to receptor selectivity for several A- or B-selective CCK agonist analogs, including a nontraditional tetrapeptide. The latter case demonstrates that the topographical features of CCK agonists (i.e., the spatial arrangement of functionally important groups), rather than the conformational ones, are crucial for receptor binding. However, this need not necessarily be true for CCK antagonists and inhibitors, where other topographical features may indeed be important for binding (Gonzalez-Muniz *et al.*, 1990; Hruby, 1987; Hruby *et al.*, 1992).

Among the latest CCK-8 agonists are several bridge-containing analogs (Witte *et al.*, 1992) that impart conformational constraint by virtue of the cyclization from Lys^{28} to Lys^{31} residues. The two most interesting analogs are the C-terminal hexa- and heptapeptides of CCK-8, both containing the Lys-Lys cyclization; they represent among the most selective CCK-B receptor agonists to date (CCK-B/CCK-A > 10,000). The apparent selectivity must be attributed to the conformational constraint, as the analogous linear compounds do not demonstrate such selectivity. These results may give rise to insights into CCK-8 receptor requirements that differ considerably from those of the CCK-A receptor.

B. Oxytocin and Vasopressin

Oxytocin, cyclo(1–6) H-Cys-Tyr-Ile-Gln-Asn-Cys-Pro-Leu-Gly-NH_2 (OT), is a neurohypophyseal hormone that mediates uterine contraction and milk ejection, and it was the first peptide to be isolated, characterized, and synthesized by chemical methods (du Vigneaud *et al.*, 1953, 1954). Closely related to this peptide are the vasoconstrictive and antidiuretic hormones Lys- and Arg-vasopressin, cyclo(1–6) H-Cys-Tyr-Phe-Gln-Asn-Cys-Pro-(Lys or Arg)-Gly-NH_2 (LVP or AVP). Owing to their cyclic nature these peptide hormones have been the subject of extensive analog design based on various structural hypotheses and on conformational constraint (Hruby, 1981b, 1982a). Many analogs have been analyzed by theoretical and spectroscopic methods. The research on constrained oxytocin and vasopressin analogs illustrates how similar conformational elements can be required for the action of closely related peptides at different receptors and emphasizes the fact that not only different structural requirements, but also dynamic factors can govern the agonist or antagonist activity of analogs of a natural peptide.

Despite the disulfide bond constraint, oxytocin and vasopressin are still highly flexible, as shown by NMR in water (Brewster *et al.*, 1973; Brewster and Hruby, 1973), molecular dynamics simulations (Hagler *et al.*, 1985), and X-ray crystallography (Wood *et al.*, 1986); thus, additional constraints can be used to modulate activity and specificity. A "cooperative" conformational model based on NMR and structure–activity results predicts a proximity between the side chains in positions 5 and 8 along with turn conformations for residues 3–4 and 7–8 as important structural features for the antidiuretic activity of vasopressin (Walter *et al.*, 1977). On the basis of this model, a bicyclic analog with a 5–8 bridge, cyclo(1–6,5–8)[$Mpa^1Phe^2Asp^5$]LVP (Mpa = β-mercaptopropionic acid), has been synthesized and resulted in an antidiuretic antagonist. Although more potent monocyclic antidiuretic antagonists had been found earlier, they all were also found to be potent antagonists of the pressor activity of vasopressin, which was not found to be true of the bicyclic compounds (Skala *et al.*, 1984).

In the oxytocin series a "cooperative" model based on NMR and structure–activity studies was proposed by Walter, with turns in residues 3–4 and 7–8, and proximity between side chains of Tyr^2 and Asn^5 (Walter, 1977). The complementary dynamic model proposed by Hruby and co-workers predicted that conformational flexibility was necessary for agonist activity (Meraldi *et al.*, 1977). It was suggested that different modes of interaction with the receptor exist for oxytocin agonists and antagonists (Meraldi *et al.*, 1977; Hruby, 1981b, 1987). Analysis of several oxytocin antagonists with a penicillamine residue at position 1 has shown that the double methylation on the β carbon of this residue produces a higher degree of rigidity in the molecule (Meraldi *et al.*, 1977), giving further evidence that conformational flexibility may be necessary for signal transduction. The crystal structure of desamino-oxytocin (Wood *et al.*, 1986) has shown that, without steric constraints on the β carbon, more than one conformation is adopted by the disulfide bridge even in the solid state. This X-ray crystal structure has been used as a model to design a bicyclic analog, cyclo(1–6,4–8)[Mpa^1,Glu^4,Lys^8]OT, which is one of the most potent oxytocin antagonists in the uterine receptor (Hill *et al.*, 1990). The increased rigidity of these types of compounds is supported by NMR evidence, and the observation that the parent monocyclic analog lacking the lactam bridge acts as a weak agonist supports the correlation between rigidity and antagonist activity (Hill *et al.*, 1990). Antidiuretic antagonist activity of the bicyclic vasopressin analog mentioned above could also be due to increased rigidity.

The mobility of the side chains can also be an important factor for the agonist activity in oxytocin analogs, which is emphasized by the inhibitory effects observed in analogs with constrained side chains in positions 2 and 8 (Lebl *et al.*, 1990; Frîc *et al.*, 1990). It is worthwhile to note that the inhibition of vasopressor response to vasopressin is also produced by one of these oxytocin analogs containing a cycloleucine residue in position 8 (Frîc *et al.*, 1990). It is also important to point out that a new class of cyclic hexapeptide oxytocin antagonists have been isolated from *Streptomyces silvensis* (Pettibone *et al.*, 1989). These molecules contain a proline and two noncoded cyclic amino acid residues, which should restrict their conformation. Substitution studies on these hexapeptides have led to potent oxytocin antagonists with high receptor selectivity and enhanced solubility (Freidinger *et al.*, 1990; Bock *et al.*, 1990).

Three oxytocin analogs with $HSCH_2COOH$ in position 1 and β-homotyrosine, *O*-methyl-β-homotyrosine, or β-homophenylalanine have been synthesized (Lankiewich *et al.*, 1989) and their potencies and binding affinities determined. These small local modifications are also seen to enhance potency, again through improved solubility. Further analogs of oxytocin have included tetrahydroisoquinoline carboxylic acid (D- and L-Tic) in position 2 (Lebl *et al.*, 1990), and they have been found to be *in vitro* uterotonic inhibitors. Although the $DTic^2$ analog displayed increased inhibitory activity, its conformation has been shown to be similar to [$DPhe^2$]oxytocin, suggesting a conformation conducive to interaction with the receptor. Substitution by L-Tic gives rise to a different conformation, coinciding with poor receptor binding but antagonist activity.

Generally, substitution of the Asn^5 residue in oxytocin by almost any amino acid residue leads to a large decrease in agonist potency and efficacy

(Hruby and Smith, 1988; Lebl *et al.*, 1987a). However, very little has been done to examine the role of Asn^5 in oxytocin antagonists. The Walter "cooperative" model (Walter, 1977) proposes a key role for the Asn side chain in the induction of the biological message. More recent investigations have shown that a small degree of activity is maintained when aspartic acid is used in place of asparagine (Smith *et al.*, 1987). Little has been done to investigate the role of changes in the asparagine-5 residue in oxytocin antagonist analogs. Results of conformational analysis of Pen^1-containing oxytocin antagonists, which appear to be conformationally restricted owing to geminal dimethyl transannular effects, suggest increased rigidity at the 5 position (Meraldi *et al.*, 1977; Hruby and Mosberg, 1982a). This, combined with the hypothesis of the dynamic model that antagonists use different topographical features for receptor binding than agonists, and the finding that agonists and antagonists have different structure–activity relationships (Hruby, 1987; Lebl *et al.*, 1990; Hruby *et al.*, 1990b; Hruby and Mosberg, 1982b; Lebl, 1987), led to the hypothesis that interesting insights would be gained from investigations of 5-position changes within a potent antagonist compound, and this was confirmed by examination of several 5-position substituted analogs (Hill *et al.*, 1991).

Previous structure–activity studies (Chan *et al.*, 1987) have suggested that the oxytocin analog [Pen^1,DPen^2,Thr^4,Orn^8]OT is found to be a potent, long-lasting *in vitro* and *in vivo* antagonist. The introduction of D configuration aromatic amino acids into the 2 position of oxytocin and vasopressin is believed to bring about the change to antagonism. It has been suggested that the interaction of the aromatic side chain and sulfur in position 6 is important for oxytocin to adopt the correct agonist conformation (Lebl *et al.*, 1987b,c).

Antagonists of the vasopressor responses to AVP have been previously reported (Manning and Sawyer, 1989; Lâszlo *et al.*, 1991). Among the most widely used are [1-(β-mercapto-β,β-pentamethylenepropionic acid),2-*O*-methyltyrosine]AVP, or $d(CH_2)_5$Tyr(Me)AVP (Kruszynski *et al.*, 1980), and [1-deamino-penicillamine,2-*O*-methyltyrosine]AVP, or dPTyr(Me)AVP. These two molecules differ only in the nature of the alkyl substituents on the β carbon at position 1. The $d(CH_2)_5$Tyr(Me)AVP form is a highly potent antagonist of the vasopressor response to AVP ($pA_2 = 8.62$), whereas dPTyr(Me)AVP is a less potent V_1 antagonist ($pA_2 = 7.88$). (pA_2 is the negative logarithm of the molar concentration of antagonist required to reduce the response to an EC_{50} dose of drug to a response half that of the EC_{50} dose.) In attempts to increase the V_1-antagonistic potency and reduce the V_2 agonism (antidiuretic effects), Manning and co-workers (1992) focused on modifications, deletions, and substitutions at the C-terminal Gly-NH_2 position. Previous work (Manning *et al.*, 1984, 1987) had shown that deletion of C-terminal Gly or Gly-NH_2 from $d(CH_2)_5$Tyr(Me)AVP resulted in retention of the V_1-antagonistic component and almost total elimination of the V_2 agonism. It has also been shown (Manning *et al.*, 1984; Sawyer *et al.*, 1988) that the AVP V_2-selective antagonist $d(CH_2)_5$[DIle^2,Ile^4]AVP, and the nonselective AVP V_2/V_1 antagonists

$d(CH_2)_5$[DPhe^2,Ile^4]AVP, $d(CH_2)_5$[DTyr(Et)^2,Val^4]AVP, and $d(CH_2)_5$[DTyr-(Et)^2,Ile^4]AVP, can have the C-terminal Gly-NH_2 replaced with a variety of amino acid and non-amino acid substituents, with excellent retention of V_2 antagonism. The results of C-terminal Gly-NH_2 replacement by amino acid amides (Arg-NH_2, Ala-NH_2, Thr-NH_2, Tyr-NH_2, and Val-NH_2) suggest that the V_1 antagonism is well tolerated, with little or no gain in V_2 agonism. The effect was more pronounced, however, in dPTyr(Me)AVP(dP=desaminopenicillamine), analogs, and these may give rise to useful pharmacological tools for studies on the adrenocorticotropin (ACTH)-releasing effects (V_{1b} receptor) of AVP (Manning and Sawyer, 1988).

C. Glucagon

The peptide hormone glucagon is produced in the pancreas and is involved in carbohydrate metabolism (Farah, 1983). The "bihormonal hypothesis" for diabetes mellitus postulates glucagon to have an important role on the metabolic disorders of this disease (Ungar, 1978). Classic structure–activity studies have shown that residues 1–5 of glucagon are most critical for the transduction of the message into the cell, but that most of glucagon's 29 amino acids seem to be necessary for binding (Hruby, 1982b; Hruby *et al.*, 1986b). The complete three-dimensional structure of micelle-bound glucagon has been investigated by ^{1}H-NMR spectroscopy (Braun *et al.*, 1983). These studies indicate the presence of an amphiphilic helix through the sequence 17–27, which is oriented along the micelle–water interface, that residues Phe^6, Tyr^{10}, and Leu^{14} form a hydrophobic patch that interacts with the hydrophobic area of the micelle, and that residues 10–14 are in a helical orientation.

Although, for the most part small, linear peptide hormones are extended, flexible molecules in aqueous solution, they are thought to adopt a specific conformation when interacting with the receptor. This receptor conformation is inherent in the primary sequence which allows for only a limited number of secondary and tertiary structures. The conformation of larger polypeptides such as glucagon are dramatically influenced by changes in the amino acid sequence and by the introduction of constraints to residues of significant importance. The 19–27 sequence of glucagon readily forms an amphiphilic helix (Kaiser and Kezdy, 1983; Epand and Liepneks, 1983; Gysin and Schwyzer, 1984) which is considered to be important for the binding of glucagon to its receptor. Extension of this amphiphilic helix, which could have an important effect on the binding properties of glucagon, has been investigated in several ways. For example, the synthesis of [δ-(5-nitro-2-pyrimidyl)ornithine^{17,18}]glucagon (Epand and Liepneks, 1983) showed no significant changes in either the conformation or biological properties of the hormone. However, the exchange of Arg^{17,18} by Lys^{17,18} and Asp^{21} by Glu^{21} to increase the helical potential (Krstenansky *et al.*, 1988) of the sequence 17–27 resulted in the formation of a superagonist (Krstenansky *et al.*, 1986, 1987). Attempts to extend the helical region further toward the 10–13 residues, by substitution of Tyr^{10} and Tyr^{13} by

Phe, results in analogs with partial agonistic properties (Hruby *et al.*, 1986a). Analogs incorporating Ahx in positions 17 and 18 (Ahx is α-aminohexanoic acid) show a considerable loss in biological activity although the helical content of the amphiphilic region remains similar to that of the parent glucagon.

Following a combination of systematic modifications of the primary sequence in positions 3–5, a number of glucagon receptor antagonists were obtained (Hruby *et al.*, 1986a, 1993a; Gysin *et al.*, 1986, 1987). A combination of the residue sequence from the antagonists and that from the superagonist, therefore, should give a sequence with high affinity and very little or no efficacy. Increasing the hydrophobicity of the region Phe^{6},Tyr^{10},Lys^{14} by introducing a 3′,5′-diiodotyrosine into position 10 to give[DPhe^{4},Tyr^{5},3′,5′,-diiodo-Tyr^{10},Arg^{12},$Lys^{17,18}$,Glu^{21}]glucagon has little effect on the degree of observed helicity but produced an analog with only 50% of the binding affinity compared to the parent hormone. Further extension of the amphiphilic helical region by up to 4 residues is accomplished in the analog [Glu^{15},Lys^{18}]glucagon, thus resulting in a 5-fold increase in elipticity. The binding affinity of such analogs is reduced, however, possibly because of the interference of the message sequence by the increased size of the amphiphilic helical region. Results obtained from such sequence modifications suggest that the backbone conformation in the 9 to 12 region is crucial for the correct fit of the message sequence 1–5 to the active site of the receptor (Hruby *et al.*, 1993a). Results also suggest that the sequences 9–14 and 15–18 undergo a conformational change during the transition of the peptide from a membrane-bound state to a receptor-bound state. In such a model, glucagon would first interact nonspecifically with the cell membrane, following which the initial binding energy between the peptide and the membrane would be utilized to overcome the entropy requirements involved in the peptide–receptor interaction. In the case of glucagon, the conformation of the 10–14 and 15–18 residues may undergo a conformational transition during the second phase of the transduction process.

The discovery that the aspartic acid in position 9 was the locus of uncoupling and coupling resulted in the design and synthesis of potent glucagon antagonists (Unson *et al.*, 1990, 1991). Initial studies using semisynthetic [des-His^{1}]glucagon, and later with a totally synthetic analog, as well as the presence of the negatively charged functional group of Asp^{9}, led to speculations that an interaction of the negative Asp^{9} with positive His^{1} may constitute a part of the triggering mechanism at the molecular level (Unson *et al.*, 1987). Histidine at the amino terminus furnishes a positively charged α-amine and an additional positive charge through its imidazole side chain; hence, both [des-His^{1}]glucagon and [des-His^{1}]glucagon amide are recognized by the receptor, albeit with 1000-fold lower affinity. Replacement of His^{1} with singularly charged amino acids also resulted in a significant loss of binding affinity, whereas reversal of the charge had a similar effect. Replacement of histidine with other aromatic derivatives lacking an α-amino group (4-imidazoleacetic acid; 3-indoleacetic acid; N^{α}-2,4-difluorobenzoic acid) led to high-binding analogs that were all partial agonists. Blocking the free α-amino group has been reported to produce a partial agonist (Desbuquois, 1975), and similarly acylation, as in [N^{α}-2,4-difluorobenzoyl]glucagon amide, also produced a weak partial agonist. The loss of both the imidazole nitrogen and an aromatic ring as in [Pro^1]glucagon amide reduced binding affinity to merely 10% that of the parent hormone and abolished all cyclase activity. Interestingly, earlier it was shown that [N^{im}-2,4-dinitrophenylhistidine^1,homoarginine^{12}]glucagon was an antagonist (Bregman *et al.*, 1980).

Correct spatial orientation of the imidazole side chain is important for hydrogen bonding or an electrostatic interaction as has been shown by the report that [DHis^1]glucagon amide lost over 90% of its binding and showed a considerably reduced cyclase activity relative to the natural L-His^1 molecule (Hruby *et al.*, 1986b). Spatially constrained histidine analogs, namely, containing 4,5,6,7-tetrahydro-1H-imidazo[*c*]pyridine-6-carboxylic acid (Tip), a constrained fused ring, show both reduced binding affinity and cyclase activity (Zechel *et al.*, 1991). It is evident from the studies of Unson and co-workers (1993) that both positions 1 and 9 play an important role in transducing the hormonal signal. In the absence of an Asp carboxyl side chain at position 9, His^1-containing analogs retained weak agonist activity. To examine the role of Asp^9, 20 novel glucagon analogs were prepared (Unson *et al.*, 1990). Among these were analogs that included the deletion of position 9 or the replacement of Asp^9 by D-Asp, Asn, Glu, D-Glu, Gln, Glu-OMe, Gly, Nle, or Lys, each change being made in the presence or absence of His^1. Removal of the side chain functionality in position 9 (i.e., [Gly^9]glucagon amide) produced an analog with 32% of the receptor binding affinity, as compared to the parent hormone, but caused a 600-fold loss in cyclase activity, whereas deletion, as in the case of [des-Asp^9]glucagon, resulted in a 100-fold loss in cyclase activity but with only 50% loss of binding affinity. Replacement by L-amino acids in general caused a loss in activity, but again only a 50% loss in binding affinity was observed. The corresponding D-amino acid replacements, however, resulted in virtually 90% loss of binding affinity and consequently adversely affected the cyclase activity. It seems apparent that the side chain of position 9 is not directly involved in binding contacts but provides a major functional requirement for agonist activity. Aspartic acid is crucial at this position and cannot be replaced even by the closely related diacid glutamic acid.

D. Gonadotropin-Releasing Hormone

Mammalian gonadotropin-releasing hormone (GnRH) [or luteinizing hormone-releasing hormone (LHRH)] is a hypothalamic decapeptide, pGlu-His-Trp-Ser-Tyr-Gly-Leu-Arg-Pro-Gly-NH_2, that acts in the pituitary gland to stimulate the release of luteinizing hormone, and follicle-stimulating hormone, that in turn regulates ovulation/spermatogenesis (Matsuo *et al.*, 1971; Burgus *et al.*, 1972). Intense research efforts have been directed toward obtaining GnRH analogs with potential use as nonsteroidal contraceptives or as fertility agents (Karten and Rivier, 1986).

The development of cyclic GnRH antagonists (Struthers *et al.*, 1990) exemplifies how a tentative structural hypothesis can be pursued through the use of constraints and how conformational analysis of the constrained analogs by NMR and molecular dynamics can lead to the development of highly potent derivatives. As would be expected of a linear peptide of its size, GnRH is largely, if not exclusively, unstructured in solution (Chargy *et al.*, 1986). However, several folded conformations of GnRH have been suggested from empirical energy calculations, including the formation of a type II β turn about Gly^6-Leu^7 (Momany, 1976). This conformation could be favored by the presence of Gly in position $i + 1$ of the turn, as Gly can adopt the torsion angles characteristic of a D-amino acid, which is required for a residue in this position of the turn (Rose *et al.*, 1985). The relevance of a Gly^6-Leu^7 β-turn conformation for the biological activity of GnRH is supported by the increased potency of GnRH analogs in which D-residues have been incorporated into the 6 position (Monahan *et al.*, 1973). N-Methylation of Leu^7, which should be compatible with the turn conformation, produces yet a further increase in activity (Ling and Vale, 1975).

A potent GnRH agonist was produced by Freidinger *et al.* (1980) in which the type II′ β-turn conformation was constrained with an S-γ-lactam linking Gly^6 C$^\alpha$ and Leu^7 N. An approach between the N and C termini should be favored if a β turn around residues 6–7 is formed in the active conformation of GnRH, which suggests the possibility of cyclizing the molecule to stabilize the turn. Antagonist activity was found in one of the early cyclic GnRH analogs in which the residues 1 and 10 were constrained by an amide bond bridge, cyclo(1–10)[DPro^1,D*p*ClPhe^2,DTrp^3, DTrp^6,NMeLeu^7, βAla^{10}]GnRH (Rivier *et al.*, 1981). Using molecular dynamics simulation and energy minimization, only one well-defined conformational family was found for this peptide (Struthers *et al.*, 1990, 1985, 1984). The structure, which includes a type II β turn with residues 6–7 occupying corner positions, has since been confirmed by solution two-dimensional NMR spectroscopy (Baniak *et al.*, 1987). An interesting observation to arise from the suggested conformation was the proximity of residues 4 and 9, leading to the further possibility of bridging these residues. GnRH analogs with a bridge between residues 4 and 9 have been shown to be antagonists with similar activities to the parent cyclo(1–10) analog (Rivier *et al.*, 1986). These results are consistent with a retention of the structure of the cyclo(1–10) antagonist, but with some slight improvement the antagonist should be able to become one of higher affinity. Several such analogs have been reported (Struthers *et al.*, 1990; Rivier *et al.*, 1988), the most potent being cyclo(4–10)[Ac-DPro^1,D*p*FPhe^2,DTrp^3,Asp^4,DNal^6,DPro^{10}]GnRH. NMR and restrained molecular dynamics have shown this analog to adopt a β turn about residues 6 and 7 (Rizo *et al.*, 1992a,b). The close contacts between the side chains of Tyr^5 and Arg^8 and the N terminus suggested that new constraints could be introduced to obtain bicyclic GnRH antagonists.

Bicyclic (4–10,5–8) GnRH analogs have been demonstrated to be equipotent to the parent monocyclic cyclo(4–10) antagonists (Rivier *et al.*, 1990) and to have similar conformational behavior (Rizo *et al.*, 1993). It is also conceivable to form a tricyclic analog from the N termini to the side chain of residue 5 or 8, in order to obtain GnRH antagonists with increased potency; such targets may prove synthetically challenging, and none have yet been reported.

Among the latest GnRH antagonists are analogs containing basic unnatural amino acids. A report that Azaline B [Ac-DNal^1,DCpa^2,DPal^3,4Aph(Atz)^5, D4Aph(Atz)^6,ILys^8,DAla^{10}-NH_2] (Rivier *et al.*, 1992) was the most potent and longest acting GnRH antagonist has led several groups to attempt to further improve its activity. Analogs incorporating residues such as 3-aminophenylalanine (3Aph), 4-thiomorpholinomethylphenylalanine (Tmf), N^{α}-methyl-4-aminophenylalanine (NMe4Aph), and 4-isopropylaminophenylalanine (4IAph) at positions 5 (L isomer), 6 (D isomer), or 8 (L isomer) have been reported (Jiang *et al.*, 1993), with preliminary data suggesting that side-chain conformational restriction can also lead to potent analogs without the necessity for intramolecular cyclization.

Several agonists of GnRH are currently used in the treatment of prostate cancer, endometriosis, precocious puberty, and other indications which are testosterone or estrogen dependent (Karten and Rivier, 1986; Dutta, 1988; Garnick and Glode, 1984). All agonists are administered either subcutaneously, nasally, or as a depot, and so studies into the design of an orally active agonist have been conducted (Haviv *et al.*, 1993). Previously it has been demonstrated that substitution of NMe-Ser^4 into leuprolide ([DLeu^6-desGly-NH_2^{10},Pro-ethylamide^9]GnRH) can stabilize the molecule against enzymatic degradation (Garnick and Glode, 1984), as can modifications of positions 1, 2, or 3 (Haviv *et al.*, 1992a,b). In the latest study *N*-methyl substitutions of Nal^3, Ser^4, and Tyr^5 have for the first time converted the parent agonist into an antagonist, while rendering the 3–4 peptide bond completely stable to chymotrypsin degradation. Examination of the three-dimensional model of leuprolide when bound to the active site of chymotrypsin reveals that the NH groups of residues 3 and 5 are involved in hydrogen bonding interactions with the enzyme. N-Methylation of these positions not only disrupts the hydrogen bonding, but also sterically inhibits substrate fitting in the enzyme active site.

E. Somatostatin

A tetradecapeptide hormone released by the hypothalamus, somatostatin plays an important physiological role as a potent inhibitor of the release of several hormones (i.e., glucagon, growth hormone, insulin, gastrin) as well as a regulator of many other biological activities (Koerker *et al.*, 1975; Gerich *et al.*, 1975; Johansson *et al.*, 1981). The wide ranging physiological significance of somatostatin has led to substantial efforts to determine the underlying structural features responsible for the varied biological functions. Veber *et al.* (1981)

described an active analog of reduced size, cyclo[Pro-Phe-DTrp-Lys-Thr-Phe], on the basis of structure–function studies, conformational analysis, and molecular modeling. However, the tetrapeptide Phe^{7}-DTrp^{8}-Lys^{8}-Thr^{10} (superscript numbers indicate the position of the equivalent residue in the parent hormone) is postulated to be the biologically important sequence for receptor interaction, whereas the other residues in the parent hormone simply maintain the bioactive conformation.

Several cyclic analogs have been synthesized in which the bridging region has been modified through the incorporation of peptidomimetics. These compounds were synthesized on the premise that such analogs would allow for the screening of structural features believed to give rise to bioactivity. Veber's cyclic hexapeptide was employed as the reference peptide in the reported study. Cis/trans isomerization within the bridging linkage was tested by substituting Pro^{6} with L-thiazolidine-4-carboxylic acid residue (Thz: thioproline). Previous studies have shown that Thz residues exist primarily in the trans orientation (Goodman *et al.*, 1970; Benedetti *et al.*, 1976). However, because of a Curtius-type rearrangement during the cyclization, an ureido compound is also obtained. Retro-inverse modifications allow the importance of the bridging linkage to be probed with respect to biological activity, specific intramolecular hydrogen bonds, and main-chain and side-chain conformations. Changes in bridging linkages included inclusion of γSar^{6}–mPhe^{7}, γVal^{10}–mPhe^{11}, and γPhe^{11}–mAla^{6}. In the analog containing the γSar^{6}–mPhe^{7} modification the cis cyclization bond conformation is retained, and this is shown to play an important role in the bridging unit. Replacement of Thr by γVal allowed the examination of the combined effect of reversing the amide bond direction and replacing the secondary hydroxyl function at this position. Implementation of such a strategy into the original cyclic hexapeptide described by Veber also results in an analog with improved activity (Veber *et al.*, 1984).

Structural analysis of retro-inverse modified cyclic hexapeptides gives a unique picture of the important features introduced into the peptide (Goodman *et al.*, 1991). The analogs containing D-Trp and Lys result in an arc-like topology being adopted about the bridged linkage. The cis arrangement about the Phe^{11}–Pro^{6} amide bond is postulated to be required for maintaining an essentially planar backbone and in order to orientate the side chains of D-Trp, Lys, and Phe into correct spatial arrays. Replacement of Pro^{6} by (NMe)Ala gave rise to highly active analogs, in particular cyclo[(NMe)Ala^{6}-Phe^{7}-DTrp^{8}-Lys^{9}-Thr^{10}-Phe^{11}]somatostatin, which is shown to have similar conformational preferences as found for the original Pro^{6} analog (Pattaroni *et al.*, 1990). The two diastereomeric cyclic hexapeptides, namely, (*R*)-mAla and (*S*)-mAla analogs, were used primarily to probe the bridging structure. Additionally, analogs containing the (NMe)Ala substitution allowed investigation of the possibility of reduced cis/trans isomerization about the Phe^{11}–(NMe)Ala^{6} amide bond owing to the removal of the *N*-methyl group. ^{1}H-NMR studies indicate that the (*R*)-mAla analog assumes a type II′ β turn around the DTrp-Lys as found in the bioactive analogs; however, the (*R*)-mAla-containing analog adopts a trans conformation about the γ Phe^{11}–(*R*)-mAla^{6} amide bond and shows preferences of the $\chi_1 = g^+$ and g^- conformations for the D-Trp and γPhe^{11} side chains, respectively.

The difference in overall topology is illustrated by examination of the superposition of the two analogs while under molecular dynamic simulation (Diblasio *et al.*, 1992). Conformational space accessible to the (*S*)-mAla analog is unavailable to the inactive (*R*)-m-Ala analog. Two sets of distinct resonances are observed in the ^{1}H-NMR spectrum of the (*S*)-mAla analog. Rotating-frame nuclear Overhauser effect (NOE) spectroscopy (ROESY) experiments allowed observation of differences in chemical exchange for the two sets of resonances. Strong NOEs between α protons of the adjacent γPhe and (*S*)-mAla residues in the minor isomer enabled the assignment of the minor isomer as having a cis amide bond at γPhe–(*S*)-mAla. Integration of the one-dimensional resonances resulted in 64:36 percentage determination of the ratio of all-trans and all-cis amide bond-containing isomers. Relative stabilities of the cis and trans forms in cyclo[X^{6}-Phe^{7}-DTrp^{8}-Lys^{9}-Thr^{10}-Phe^{11}]somatostatin [X = Pro or (NMe)Ala] were determined by the energy difference of the three-bond steric interactions between Phe^{11} C^{α}–X C^{α} and Phe^{11} C^{α}–X *N*-alkyl substituent. In the cases of the (*S*)- and (*R*)-mAla analogs, the carbonyl oxygens play sterically similar roles in the cis–trans isomerization to the *N*-alkyl substituents in the Pro and (NMe)Ala analogs, and it is this effect that is pronounced in only the (*S*)-mAla analog (Elseviers *et al.*, 1988a). The preferred conformations of the bioactive cyclic somatostatin analogs contain cis amide bonds in the bridging linkages. Molecular dynamics simulation shows that the (*S*)-mAla substituted analog has considerable spatial homology with the Veber Pro^{6} analog, and consequently displays a corresponding activity.

More recently there has been much emphasis placed on producing reduced ring cyclic somatostatin analogs (Brady *et al.*, 1993). To this extent an eight-membered-Cys-Cys-unit has been incorporated in place of the -Phe^{11}-Pro^{6}-segment, thus achieving two aims: first, constraining the 11–6 amide bond into the cis geometry, as established for the earlier hexapeptide, and, second, positioning a disulfide in place of the position 11 phenyl group in order to act as a surrogate for the phenyl in receptor binding. This design principle relies on the knowledge that the role of the -Phe^{11}-Pro^{6}-dipeptide unit is considered largely as one of structural constraint, although an important component of the ligand–receptor interaction via the phenyl nucleus has been recognized (Veber *et al.*, 1984). It therefore seems probable that the 11–6 amide bond could be fixed into the cis geometry, as suggested by NMR for the cyclic hexapeptide (Veber *et al.*, 1984), in the eight-membered ring formed by the closure of two neighboring cysteine residues, with minimal perturbation of the stereoelectronic and physicochemical properties. Both *in vivo* and *in vitro* biological results suggest that the bridged bicyclic analogs retain essentially the full potencies of the corresponding monocyclic compounds. These findings strongly support the suggestion that a closed disulfide-linked dipeptide constitutes a constraint that can fix the conformation of a monocyclic species into its bioactive form. Other dipeptide bond surrogates have been reported with respect to amide bond replacement in somatostatin analogs (Morgan and Gainor, 1989; Toniolo, 1990; Elseviers *et al.*, 1988a,b). Many of these have been designed to enforce a type II′ β turn, and have been devised to simulate the trans amide bond typically encountered within reverse turns.

F. Angiotensin

The octapeptide angiotensin II (ANG-II), Asp-Arg-Val-Tyr-Ile-His-Pro-Phe, causes elevation of blood pressure in mammals by acting at receptors in a variety of target tissues. Two classes of angiotensin antagonists modified at positions 8 (type I) and 4 (type II), respectively, have been identified by structure–activity studies (Turker *et al.*, 1972; Scanlon *et al.*, 1984). Type I antagonists such as [Sar^1,Ile^8]ANG-II show protracted antagonist effects on isolated smooth muscle tissue (Matsoukas *et al.*, 1985, 1988), whereas type II antagonists such as [Sar^1,$MeTyr^4$]ANG-II (sarmesin) are reversible competitive antagonists at angiotensin receptors (Moore *et al.*, 1985). Type I antagonists are also characteristically different from sarmesin since the former retain significant antagonist activity when the N-terminal amino acid is deleted, whereas the latter is inactivated (Goghari *et al.*, 1986; Matsoukas *et al.*, 1988; Jorgensen *et al.*, 1971). Additionally [Sar^1,Ile^8]ANG-II is a potent inhibitor *in vivo*, whereas sarmesin is a relatively weak inhibitor in pressor assays (Scanlon *et al.*, 1984; Franklin and Moore, 1987).

It is known that positive charges at or near the N terminus of ANG-II are important for biological activity; for example, [$succinyl^1$]ANG-II and [Gly^2]ANG-II have biological activities of about 50 and 10%, respectively, suggesting that the charge on the guanidino group may be more important than the charge on the amino group for agonist activity. The situation for antagonists, however, is less well understood (Samanen *et al.*, 1988). An important study has attempted to delineate the contribution of the electrostatic interactions of the N-terminal region of antagonists with the angiotensin receptor (Moore *et al.*, 1991). These studies are interesting because recent findings suggest that type I angiotensin antagonists may interact with a different site on the angiotensin receptor. A series of closely related type I angiotensin antagonists were examined, which contain Ile^8 and residues impairing to varying degrees the positive charge at the N terminus and Arg residue position. These findings suggest that there may be two anionic binding sites on the smooth muscle receptor that interact with the N-terminal domain of the type I antagonist, and that ionic interaction of an analog with these sites has a profound influence on the resulting binding affinity of the peptide.

Furthermore, interactions at both anionic sites on the receptor may be important determinants of the prolonged inhibitory effect characteristics of [Sar^1,Ile^8]ANG-II and related analogs (Duncia *et al.*, 1990). The substitution of

Gly, Ala, or Nle for the Arg residue in [Sar^1,Ile^8]ANG-II reduced the pA_2 value by over one log unit (pA_2 is the negative logarithm of the molar concentration of antagonist required to reduce the response to an EC_{50} dose of ANG-II to a response half that of the EC_{50} dose). Similar results obtained for Gly-, Ala-, and Nle-substituted analogs suggest that the positive charge at the terminus on the Arg side chain is an important contributing factor, and that the long hydrophobic carbon side chain carrying the guanidino group contributes little to the binding affinity of these peptides (Magnan and Regoli, 1978). This is further demonstrated for [Sar^1,Nle^2,Ile^8]ANG-II, which contains the carbon side chain of Arg but lacks the terminal guanidino group, and correspondingly has a lower pA_2 value (6.7) than the analog completely lacking the side chain, namely, [Sar^1,Gly^2,Ile^8]ANG-II (7.0). The presence of Phe^2 in [Sar^1,Ile^8]ANG-II also contributes very little to the activity of the antagonist (pA_2 of 6.8); moreover, Pro^2 has a deleterious effect on antagonist potency. Following the same principle, incorporating Sar^2 similarly results in a greatly reduced pA_2 (5.3).

Introduction of a bend into the backbone by the presence of the secondary amino acids Pro or Sar in place of the Arg^2 is thought to severely disrupt the peptide–receptor binding. For heptapeptide analogs, deletion of the *N*-terminal amino acid of [Sar^1,Ile^8]ANG-II decreases the antagonist potency by half a log unit, suggesting that the N-terminal Sar residue contributes somewhat less than the Arg side chain to the overall binding affinity. The antagonist activity of the heptapeptide [des^1,Ile^8]ANG-II is greatly reduced when the Arg residue is replaced by Ala, Nle, Phe, or Pro but shows a less deleterious reduction in potency when the Arg residue is replaced by Gly or Sar. The lack of a side chain on the amino acid residue occupying the 2 position allows the heptapeptide to bind in a more favorable manner, thereby elevating the pA_2 value by almost one log unit. If the octapeptides are compared directly with the equivalent heptapeptides, then a trend is apparent when the Arg residue is replaced by an amino acid bearing a side chain. Hence, antagonist activity of Ala-, Nle-, and Phe-containing peptides, which is already reduced by more than one log unit for the octapeptides, is further reduced by another log unit when the N-terminal Sar residue is deleted. These observations confirm that both of the positive charges located on the Arg residue and the N terminus contribute substantially to the antagonist potency (Moore, 1985). Octapeptides and heptapeptides containing Gly^2 show similar pA_2 values, indicating that the deletion of the N-terminal Sar has no effect on the binding. These results suggest that [des^1,Gly^2,Ile^8]ANG-II has a different binding mode from the Ala-, Nle-, and Phe-containing heptapeptides, and further that [des^1,Gly^2,Ile^8]ANG-II may perhaps be able to adopt a similar binding conformation to the octapeptide in this particular case.

Conformational studies by Marshall and co-workers (Kataoka *et al.*, 1992) have shown that many backbone conformations are compatible with position 1–3 side-chain cyclizations of Cys or Hcy residues (Spear *et al.*, 1990). To further define the receptor-bound conformation, cyclic analogs of ANG-II have been reported in which mercaptoproline (4-trans, Mpt, and 4-cis, Mpc) has been incorporated into positions 3 and 5 in order to constrain the peptide backbone even further. Introduction of mercaptoproline in positions 3 and 5 provides more restricted bicyclic analogs and probes the likelihood of a bound conformation with a turn centered on residues 3–5. Compounds of the type Mpc^3-Tyr^4-Hcy^5 incorporating a disulfide-bridged substructure can accommodate type I and II β turns centered at residues 4–5, and they have been shown to have particularly high affinity for ANG-II receptors (Plucinska *et al.*, 1993). The inclusion of the cyclic disulfide and proline constraint of cyclo[Sar^1,Mpc^3, Hcy^5]ANG-II eliminates the formation of a type III′ β turn at residues 3 and 4. Additionally, analogs cyclo[Sar^1,Mpc^3,Hcy^5]ANG-II and cyclo[Sar^1,Cys^3, Hcy^5] ANG-II show extremely high activity and selectivity, implying that their conformational constraint is compatible with recognition and activation of the angiotensin II receptor. Interestingly, the central effects of angiotensin and its linear analogs do not differ substantially, whereas cyclic analogs have been reported to produce sedation or neuroleptic-like activity (Suirskis *et al.*, 1991, and references therein).

G. α-Melanotropin

α-Melanocyte-stimulating hormone (α-MSH), also called α-melantropin (Hadley, 1989), having the primary structure Ac-Ser^1-Tyr^2-Ser^3-Met^4-Glu^5-His^6-Phe^7-Arg^8-Trp^9-Gly^{10}-Lys^{11}-Pro^{12}-$Val^{13}NH_2$, has a long and storied history, being one of the first peptide hormones to be studied as a crude extract from the pituitary and to be isolated in pure form and structurally determined. It was early recognized to have essentially the same N-terminal tridecapeptide sequence as adrenocorticotropic hormone (ACTH, corticotropin), except that the N-terminal was acetylated in the case of α-MSH, indicating that the biosynthesis was different.

α-MSH is synthesized in the pituitary gland although it exerts its activity at peripheral receptors which are primarily responsible for its pigmentation properties. However α-MSH also appears to have several neurophysiological effects (O'Donohue and Jacobowitz, 1980). Research of α-MSH analogs has offered many examples of how the activity of a given peptide at different receptors can be modulated by diverse conformational features. Much of the research has been focused on the pigmentation activity of α-MSH, with an aim to obtain compounds with increased potency and prolonged duration of activity.

Structure–activity relationship studies on α-MSH began immediately with its isolation and structure determination in 1957 by Harris and Lerner, and since then several authors have written extensive reviews about these earlier studies (Hofmann, 1962; Ramachandran, 1973; Schwyzer, 1977; Medzihradszky, 1982; Hruby *et al.*, 1984). Nevertheless, many studies have been conducted since the early 1980s, and these form the basis of the current discussion.

A linear analog, [Nle^4,DPhe^7]α-MSH, with greatly increased potency and prolonged duration of activity in several bioassays, was obtained after the observation that heat–alkali-treated α-MSH underwent substantial racemization at position 7 (Sawyer *et al.*, 1980). On the basis of theoretical considerations and the activity enhancement produced by the D-Phe^7 substitution, a turn conformation within residues 5–9 was proposed to be important for melanotropic activity. This hypothesis has led to the synthesis of cyclic analogs, including cyclo(4–10)[Cys^4,Cys^{10}]α-MSH with extremely high potency in the frog skin assay, although the action of this compound had a shorter duration than [Nle^4,DPhe^7]α-MSH (Sawyer *et al.*, 1982; Knittel *et al.*, 1983). Both linear and cyclic analogs are substantially more potent in the frog skin assay than in the lizard skin assay, but only the linear compound is superpotent in a mouse melanoma tyrosinase assay, indicating that different features of the peptides are important for activity at different receptors (Sawyer *et al.*, 1982, 1983; Marwan *et al.*, 1985; Al-Obeidi *et al.*, 1989a,b).

Further research in the design of constrained α-MSH analogs has been directed toward obtaining long-acting compounds with high potency in the lizard skin and tyrosinase assay (Al-Obeidi *et al.*, 1989a,b), and potential clinical use in the treatment of pigmentary disorders and melanoma cancer (Al-Obeidi *et al.*, 1989a; Levine *et al.*, 1991). Molecular dynamics simulations of α-MSH and [Nle^4,DPhe^7]α-MSH yielded several low-energy structures displaying turn conformations about residues 6–9 and definite amphiphilicity, with polar and hydrophobic side chains being displayed on different sides of the structures (Al-Obeidi *et al.*, 1989a). Although side chains of Glu^5 and Lys^{11} generally were not close enough to form a salt bridge, it was observed that such an interaction could easily be formed if the Lys side chain was located in place of residue 10. A series of lactam bridges between residues 5 and 10 were designed, and one of the compounds, Ac-[Nle^4,Asp^5,DPhe^7,Lys^{10}]α-MSH(4–10)-NH_2, had a superpotent and prolonged activity in both the lizard skin and melanoma tyrosinase bioassays, whereas it was equipotent to α-MSH in the frog skin assay (Al-Obeidi *et al.*, 1989a,b). Therefore the cyclic (4–10) and (5–10) α-MSH analogs offer the opportunity to analyze additional details of the structural and topographical requirements for activity in the different α-MSH receptors, and further conformational studies of the analogs should provide new insights on these requirements (Al-Obeidi *et al.*, 1989a).

In an extensive report, Hruby and co-workers (1993b) demonstrate the multitude of analogs that can be or have been designed and synthesized, utilizing novel ideas of conformational and topographical control. The reader is directed to that paper and the references therein for further discussion. The increase in number and biological diversity of melanotropic receptor sites (Hadley *et al.*, 1993) and the cloning of 5 melanocortin receptors, will no doubt prompt the continuation of study in this particular field of peptide research.

H. δ-Opioid Peptides

Several years ago a series of cyclic, conformationally constrained enkephalins were designed that included [DPen²-DPen⁵]enkephalin (DPDPE) (Mosberg *et al.*, 1983), which has proved to be one the most highly δ-opioid receptor-selective ligands and has served as a standard ligand for the δ-opioid receptor (Hruby and Gehrig, 1989). More recently, alterative highly potent and selective ligands for the δ-opioid receptor have been isolated from frog skin extracts, namely, Tyr-DAla-Phe-Asp-Val-Val-Gly-NH_2 (deltorphin I) and Tyr-DMet-Phe-His-Leu-Met-Asp-NH_2 (dermenkephalin) (Erspamer *et al.*, 1989; Amiche *et al.*, 1989), although it has been shown that the N-terminal tetrapeptide sequence of dermenkephalin is μ-opioid selective (Amiche *et al.*, 1989). These findings give rise to the question of how two structurally different classes of compounds, the cyclic enkephalins and the linear deltorphin/dermenkephalin group, could both interact selectively with δ-opioid receptors. In view of the different structures and likely different conformations, it was deemed necessary to apply topographical considerations rather than similarities in backbone conformations to address this issue. From structure–activity studies and conformational analysis several independent groups had concluded that the high δ-opioid receptor selectivity of DPDPE and related analogs depended on a D configuration at the C^{α} of position 2, a Tyr^1 aromatic ring, and a Phe^4 aromatic ring (Schiller, 1985b; Hruby *et al.*, 1988; Wilkes and Schiller, 1990; Nikiforovich *et al.*, 1991; Heyl and Mosberg, 1992). In our laboratories NMR and computational studies (Hruby *et al.*, 1988), together with extensive structure–biological activity studies, provide strong support for these ideas.

To further investigate the topographical relationships between the DPDPE family of δ-selective ligands and the deltorphin/dermenkephalin family, a comprehensive comparative examination of these compounds using a combination of asymmetric synthesis of designer amino acids, analog design and synthesis, conformational analysis, computational examination of low-energy three-dimensional structures, topographical comparisons, and comprehensive binding and bioassay studies has been performed. A major component of these studies has been the development of a facile asymmetric synthesis of β-methyl-substituted analogs of phenylalanine and tyrosine (Dharanipragada *et al.*, 1989). The initial synthetic approach to these amino acid derivatives was to use precursors to the relevant amino acids in which the β-carbon stereochemistries were preset and the α-carbon stereochemistry could then be synthetically introduced in conjunction with a chiral auxiliary. Recently developed methodologies allow both α and β stereochemical control through use of only a single chiral auxiliary. Methyl group substitution at the 2′ and 6′ positions of the aromatic rings also have been achieved in some of the novel Phe and Tyr mimetics. The simple β-methylphenylalanine and β-methyltyrosine analogs have been incorporated into DPDPE, and results from studies indicate that, for the molecular recognition process, gauche(−)χ_1 values for the Phe^4 and Tyr^1 side chains are preferred. This places the Tyr^1 and Phe^4 side-chain groups somewhat farther apart than originally suggested, but still remaining on the same lipophilic face for interaction with the δ-opioid receptor (Hruby *et al.*, 1991d).

To obtain further insights into the topographical similarities between the cyclic DPDPE series and the deltorphins, extensive conformational search procedures in conjunction with molecular mechanics calculations have been performed (Nikiforovich *et al.*, 1991). These began with comprehensive examination of the low-energy conformations accessible to dermenkephalin in which certain classes of low-energy conformations of this molecule and DPDPE topographically matched in three-dimensional space around the C$^\alpha$ carbon of the second residue and the Tyr^1 and Phe^3 (Phe^4) aromatic rings. A more comprehensive study including examination of some β-MePhe^4 substituted analogs of DPDPE gave rise to similar findings. Further calculations involving all four β-MePhe-substituted DPDPE analogs (two compounds having high potency and two having low potency at the δ-opioid receptor, with respect to DPDPE), as well as a re-examination of DPDPE conformation based on NMR studies (Hruby *et al.*, 1991c), have yielded additional insights. Utilizing all the data, a unique pharmacophore for the δ-opioid receptor has been developed. Preliminary results suggest that the pharmacophore consists of the C$^\alpha$ carbon of residue 2, the two aromatic rings located in three-dimensional space on the same face of the 13-membered disulfide-containing ring within 6 to 10 Å, and the α-amino group and the C-terminal carboxylate on the same surface but on opposite edges of the lipophilic surface.

Linear opioid peptides such as the deltorphins are highly flexible molecules and need to be conformationally restricted in order to obtain insight into their bioactive conformation. Various types of cyclizations of opioid peptides resulted in a number of compounds with high selectivity toward μ or δ receptors (Hruby and Gehrig, 1989; Schiller *et al.*, 1992a). Conformational restriction of phenylalanine residues in opioid peptides has previously been shown to have drastic effects on receptor selectivity and signal transduction. Thus, substitution of Phe^3 in the relatively nonselective cyclic dermorphin analog Tyr-DOrn-Phe-Glu-NH_2 with 2-aminoindan-2-carboxylic acid (Aic) or 2-aminotetralin-2-carboxylic acid (Atc) produced μ-selective agonists (Schiller *et al.*, 1992a), and opioid tri- and tetrapeptide analogs containing a tetrahydroisoquinoline-3-carboxylic acid (Tic) residue in the 2 position of the peptide sequence turn out to be potent and very selective δ antagonists (Schiller *et al.*, 1993). Novel deltorphin analogs incorporating Aic or Atc in the 2 position and Tic in position of residue 3 have been reported (Schiller *et al.*, 1992b). Cyclic deltorphin analogs were obtained through side chain to side chain cyclizations of Orn (or Lys) and Asp (or Glu) which had been substituted in various positions of the peptide sequence. Both Aic^3 and Atc^3 analogs displayed extraordinary δ receptor affinity and δ selectivity; substitution of a Tic residue in position 2 of deltorphin-related peptides did in one case produce a highly selective δ antagonist, and in other cases produced partial δ agonists. In this particular series of analogs, conformational restriction through cyclization did not produce analogs displaying a high degree of δ selectivity, primarily owing to the significant loss of receptor binding affinity.

Interestingly, comparison of the tetrapeptide antagonist Tyr-Tic-Phe-Phe-OH (TIPP) (Schiller *et al.*, 1992b), with the deltorphin-related antagonist Tyr-Tic-Phe-Phe-Leu-Nle-Asp-NH_2 showed a marked decrease in δ receptor affinity, selectivity, and antagonistic potency. Carboxyl-terminal extensions of the μ-selective tetrapeptide analog Tyr-DOrn-Phe-Asp-NH_2 with the C-terminal tripeptide sequence of the δ agonist [DAla^2]deltorphin I, reduces the μ affinity but drastically improves the δ affinity. This interesting observation suggests that a difference exists in the mode of binding between antagonists and agonists.

VI. Summary and Conclusions

Scientific advances since the 1960s have made peptides, proteins, and derivatives such as glycoproteins and lipoproteins major targets for the treatment and cure of diseases. Thus, the development of a rational approach to the design of peptides and proteins with specific structural, conformational, stereochemical, topographical, and dynamic properties has become of central importance to chemistry, biology, and medicinal applications of these compounds. In view of the tremendous diversity of structures and conformations possible in even quite simple peptides and peptidomimetics, it is especially important that systematic, rational, and predictive physical/chemical and biological approaches be developed which can quickly identify important structural leads for a particular biological target and in particular pharmacological and physiological contexts. It seems obvious that this will require highly interdisciplinary efforts by chemists, biophysicists, and biologists who have a common goal of understanding the details of the physical/chemical basis for the biological effect(s) under consideration. Thus, in our opinion, it will be critical that such problems be addressed as soon as possible after they become identified by a multidisciplinary approach.

In Table I, we outline some of the elements of such an interdisciplinary approach whether one is interested in the basic science of the system or in some specific application in medicine. As can be seen from Table I, state-of-the-art methodology needs to be applied to these problems from every perspective. Hence, a strong interdisciplinary approach will always be required. From this perspective, the training of chemists, biologists, and biophysicists to participate in such scientific problems with a continued high degree of creativity and productivity in their specialty and at the same time directing their research in ways useful to the larger problem will be critical.

There are several aspects of this that strike us as being of particular importance at this time, based on our current perspective, though we realize,

Table 1 Elements of an Interdisciplinary Approach to Peptide and Protein Structure–Biological Activity Relationships

I. Chemistry
  A. Rapid, highly efficient synthesis of complex peptides, peptidomimetics, pseudopeptides, and nonpeptide peptidomimetics
  B. Asymmetric synthesis of specialized amino acids, peptidomimetics, scaffolds, templates, and biocompatible polymers
  C. Macrocyclic chemistry
  D. Design, synthesis, and analysis of peptides, peptidomimetics, etc., with specific conformational, topographical, and dynamic properties
  E. Rapid methods for conformational analysis
  F. Computational chemistry suitable for complex structures with predictive power
  G. Highly sensitive analytical methodology for detection, structural analysis, and conformational determination
  H. High efficient methods of purification and separation of complex peptide and peptidomimetic structures

II. Biology
  A. Highly efficient binding assays with minimal nonspecific binding
  B. Highly efficient *in vitro* bioassays including second messengers, ion channels, enzymes, etc.
  C. Highly efficient *in vitro* assay with specific biological end points related to biological and/or medical effects
  D. Appropriate cell lines specific for biological effects under investigation
  E. Cell lines appropriate for transfecting specific membrane-bound receptors
  F. Highly efficient, reproducible, and stable molecular biology methodology for site-specific mutagenesis for receptors and ligands
  G. Development of cellular or cell-free systems suitable for performing reproducible thermodynamic and kinetic studies
  H. Animal models for specific diseases; cellular systems and nonbiological methods suitable as models for human disease

III. Biophysics
  A. Continued development of spectroscopic methods for obtaining structural, conformational, and dynamic properties of peptides, proteins, glycoproteins, etc.
    1. Nuclear magnetic resonance spectroscopy
      A. Solution methods
      B. Solid-phase methods
    2. Circular dichroism
    3. Ultraviolet/visible spectroscopy
    4. Fluorescence spectroscopy
    5. Infrared spectroscopy
    6. Raman spectroscopy
  B. X-Ray crystallography of complex structures including methodology to obtain crystals suitable for X-ray analysis
  C. Computational methods: force field algorithms, dynamic analysis, intermolecular interactions, conformational predictions, etc.
  D. Development of FACS, STM, lasers, and other high technologies for studying complex biological systems

even as we write, that new, more pressing needs will exist. On the chemistry side, although "simple" peptide chemistry has been developed into one of the most efficient areas of synthetic organic chemistry, quite the opposite is true for peptidomimetics. This area is very fertile for synthetic chemists with knowledge of peptide and protein chemistry to develop asymmetric synthetic methodology, macrocyclic synthetic methodology, etc., for use in the design and synthesis of peptidomimetics and nonpeptide peptide mimetics. Similar comments could be made about scaffolds, templates, and polymers suitable for use with protein and peptide synthesis and structure. On the biological side, there is a need for biologists to understand that a good bioassay, especially a binding assay and many second messenger and *in vitro* bioassays, is essentially a chemical experiment. In the case of assays involving peptides and proteins this is especially important to realize. From our reading of literature it appears that many binding assays with peptides and proteins are not run under equilibrium conditions. In view of the slow on and off rates, several hours of incubation are often required to reach equilibrium, but many binding assays are run only for 1 or 2 hr or less. Under these circumstances when using a radioligand displacement assay, it is not valid to convert an EC_{50} value to a K_i value, though this is done all the time. Unfortunately, many editors and reviewers insist on such conversions. As biologists are increasingly using chemical and physical methods to study systems, they should be aware of the laws of chemistry and physics that apply. On the other hand, it is very important for chemists and physicists who study biological activity to realize the limitations and complexities that are inherent in the biological system under investigation and that the biologist faces every day. All of this, of course, points to the need and the advantage of close collaboration in these areas.

It is very clear that investigations in the area of peptide and protein hormone and neurotransmitter research has entered a new era. With the cloning and stable transfection of many of the hormones and neurotransmitters and many of the receptors for these molecules, it will be possible to study these systems at a molecular level heretofore unavailable. We have no doubt that the challenges this provides will be met, and we look forward with excitement and anticipation to participating in this area of science.

Acknowledgments

We thank Shubh Sharma, Nathan Collins, and Carrie Haskell-Luevano for useful comments. The financial support of the U.S. Public Health Service, the National Institute of Drug Abuse, and the National Science Foundation for the research carried out in our laboratory is gratefully acknowledged.

References

Al-Obeidi, F., Castrucci, A. M., Hadley, M. E., and Hruby, V. J. (1989a). *J. Med. Chem.* **32,** 3413–3416.

Al-Obeidi, F., Hadley, M. E., Pettitt, B. M., and Hruby, V. J. (1989b). *J. Am. Chem. Soc.* **111,** 3413–3416.

Amiche, M., Sagan, S., Mor, A., Delfour, A., and Nicolas, P. (1989). *Mol. Pharmacol.* **35,** 770–779.

Baniak, E. L., Rivier, J. E., Struthers, R. S., Hagler, A. T., and Gierasch, L. M. (1987). *Biochemistry* **26,** 2642–2651.

Benedetti, E., Christensen, A., Gilon, C., Fuller, W., and Goodman, M. (1976). *Biopolymers* **15,** 2523–2534.

Bock, M. G., Dipardo, R. M., Williams, P. D., Pettibone, D. J., and Clineschmidt, B. V. (1990). *J. Med. Chem.* **33,** 2321–2323.

Brady, S. F., Paleveda, W. J., Jr., Arison, B. H., Saperstein, R., Brady, E. J., Raynor, K., Reisen, T., Veber, D. F., and Freidinger, R. M. (1993). *Tetrahedron* **49,** 3449–3466.

Braun, W., Wider, G., Lee, K. H., and Wüthrich, K. (1983). *J. Mol. Biol.* **169,** 921–948.

Bregman, M. D., Trevedi, D., and Hruby, V. J. (1980). *J. Biol. Chem.* **255,** 11725–11733.

Brewster, A. I. R., and Hruby, V. J. (1973). *Proc. Natl. Acad. Sci. U.S.A.* **70,** 3806–3809.

Brewster, A. I. R., Hruby, V. J., Glasel, A. J., and Tonelli, A. E. (1973). *Biochemistry* **12,** 5294–5304.

Burgus, R., Butcher, M., Amoss, M., Ling, N., and Monahan, M. (1972). *Proc. Natl. Acad. Sci. U.S.A.* **69,** 278–282.

Chan, W. Y., Hruby, V. J., Rockway, T. W., and Hlavacek, J. (1987). *J. Pharmacol. Exp. Therp.* **239,** 84–87.

Chargy, K. V. R., Srivastrava, S., Hosur, R. V., Roy, K. B., and Govil, G. (1986). *Eur. J. Biochem.* **158,** 323–332.

Charpentier, B., Pelaprat, D., Durieux, C., Dor, A., Reibaud, M., Blanchard, J. C., and Roques, B. P. (1988). *Proc. Natl. Acad. Sci. U.S.A.* **85,** 1968–1972.

Coates, E. A., and Knittle, J. J. (1990). *Quant. Struct. Act. Rel.* **9,** 94–101.

Desbuquois, B. (1975). *Eur. J. Biochem.* **60,** 335–347.

Dharanipragada, R., Nicolás, E., Toth, G., and Hruby, V. J. (1989). *Tetrahedron Letters* **30,** 6841–6844.

Dharanipragada, R., Van Hulle, K., Bannister, A., Bear, S., Kennedy, L., and Hruby, V. J. (1992). *Tetrahedron* **48,** 4733–4748.

Diblasio, B., Rossie, F., Beneditti, E., Pavone, V., Saviano, M., Pedone, C., Zanotti, G., and Tancredi, T. (1992). *J. Am. Chem. Soc.* **114,** 8277–8283.

Duncia, J. V., Chiu, A. T., Carini, D. J., Gregory, G. B., Johnsos, A. C., Price, W. A., Wells, G. J., Wong, P. C., Calabrese, J. C., and Timmermanns, P. B. (1990). *J. Med. Chem.* **33,** 1312–1329.

Durieux, C., Belleneny, J., Lalelemanol, J. Y., Roques, B. P., and Fournie-Zaluskie, M. C. (1983). *Biochem. Biophys. Res. Commun.* **114,** 705–712.

Dutta, A. S. (1988). *Drugs Future* **13,** 43–57.

du Vigneaud, V., Ressler, C., Swan, J. M., Katsoyannis, P. G., and Gordon, S. (1953). *J. Am. Chem. Soc.* **75,** 4879–4880.

du Vigneaud, V., Ressler, C., Swan, J. M., Roberts, C. W., and Katsoyannis, P. G. (1954). *J. Am. Chem. Soc.* **76,** 3115–3121.

du Vigneaud, V., Denning, G. S., Jr., Drabarek, S., and Chan, W. Y. (1964). *J. Biol. Chem.* **239,** 472–478.

Elseviers, M., Van der Auwera, L., Pepermans, H., Tourwé, D., and Van Binst, G. (1988a). *Biochem. Biophys. Res. Commun.* **154,** 515–521.

Elseviers, M., Van der Auwera, L., Pepermans, H., Tourwé, D., and Van Binst, G. (1988b). *In* "Peptide Chemistry, 1987" (T. Shiba and S. Sakakibara, eds.), pp. 607–610. Protein Research Foundation, Osaka, Japan.

Epand, R. M., and Liepneks, J. J. (1983). *J. Biol. Chem.* **258,** 203–207.

Erspamer, V., Melchiorri, P., Falconieri-Erspamer, G., Negri, L., Corsi, R., Severini, C., Barra, D., Simmaco, M., and Kreil, G. (1989). *Proc. Natl. Acad. Sci. U.S.A.* **86,** 5188–5192.

Farah, A. E. (1983). *Pharmacol. Rev.* **35,** 181–217.

Ferrier, B. M., Jarvis, D., and du Vigneaud (1965). *J. Biol. Chem.* **240,** 4264–4273.

Finchem, C. I., Horwell, D. C., Ratcliff, G. S., and Rees, D. C. (1992). *Bioorg. Med. Chem. Lett.* **2,** 403–406.

Fournie-Zaluski, M. C., Belleney, J., Durieux, C., Gacel, G., Roques, B. P., Begue, D., Menant, I., Lux, B., and Gerand, D. (1985). *Ann. N.Y. Acad. Sci.* **448,** 598–600.

Fournie-Zaluski, M. C., Belleney, J., Lux, B., Duriex, C., Gerand, D., Gacel, G., Mougret B., and Roques, B. P. (1986). *Biochemistry* **25,** 3778–3787.

Franklin, K. J., and Moore, G. J. (1987). *J. Med. Chem.* **31,** 1418–1421.

Freidinger, R. M., and Veber, D. F. (1984). *In* "Conformationally Directed Drug Design" (J. A. Vida and M. Gordon, eds.), ACS Monograph Series, Vol. 251, pp. 169–187. American Chemical Society, Washington, D.C.

Freidinger, R. M., Veber, D. F., Perlow, D. S., Brooks, J. R., and Saperstein, R. (1980). *Science* **210,** 656–658.

Freidinger, R. M., Williams, P. D., Tung, R. D., Bock, M. G., and Pettibone, D. J. (1990). *J. Med. Chem.* **33,** 1843–1845.

Frîc, I., Hdavacek, J., Rockway, T. W., Chen, W. Y., and Hruby, V. J. (1990). *J. Protein Chem.* **9,** 9–15.

Gacel, G., Morinsuron, M. P., Champagnat, J., Denauitsaubie, M., and Roques, B. P. (1984). *Eur. J. Pharmacol.* **98,** 214–247.

Gacel, G., Durieux, C., Fellion, E., Fournie-Zaluski, M. C., Begue, B., Menant, I., Rossignol, P., and Roques, B. P. (1985). *In* "Peptides 1984: Proceedings of the 18th European Peptide Symposium" (U. Ragnarsson, ed.), pp. 385–386. Almquist and Wiksell, Stockholm.

Garnick, M. B., and Glode, M. (1984). *J. Med. Chem.* **311,** 1281–1286.

Gerich, J. E., Lovinger, R., and Grodsky, G. M. (1975). *Endocrinology (Baltimore)* **96,** 749–754.

Goghari, M. H., Franklin, K. J., and Moore, G. J. (1986). *J. Med. Chem.* **29,** 1121–1124.

Gonzalez-Muniz, R., Bergeron, E., Marseigne, I., Durieux, C., and Roques, B. P. (1990). *J. Med. Chem.* **33,** 3199–3204.

Goodman, M., Su, K. C., and Niu, G. C.-C. (1970). *J. Am. Chem. Soc.* **92,** 5219–5222.

Goodman, M., Yamazaki, T., Huang, Z., Spencer, J. R., and Said-Nejad, J. D. (1991). *In* "Peptides, Chemistry and Biology. Proceedings of the 12th American Peptide Symposium" (J. A. Smith and J. E. Rivier, eds.), pp. 39–42. ESCOM, Leiden.

Gysin, B., and Schwyzer, R. (1984). *Biochemistry* **23,** 1811–1818.

Gysin, B., Trivedi, D. B., Johnson, D. G., and Hruby, V. J. (1986). *Biochemistry* **25,** 8278–8284.

Gysin, B., Johnson, D. G., Trivedi, D. B., and Hruby, V. J. (1987). *J. Med Chem.* **30,** 1409–1415.

Hadley, M. E. (1989). "The Melanotropic Peptides," Vols. 1–3. CRC Press, Boca Raton, Florida.

Hadley, M. E., Sharma, S. D., Hruby, V. J., Levine, N., and Dorr, R. T. (1993). *Annu. N.Y. Acad. Sci.* **680,** 424–439.

Hagler, A. T., Osguthrope, D. J., Dauber-Osguthrope, P., and Hemple, J. (1985). *Science* **227,** 1309–1315.

Harris, J. I., and Lerner, A. B. (1957). *Nature (London)* **179,** 1346–1348.

Haviv, F., Fitzpatrick, T. D., Nichols, C. J., Bush, E. N., Diaz, G., Nellans, H. N., Hoffman, D. J., Ghanbari, H., Johnson, E. S., Love, S., Cybulski, V., Nguyen, A., and Greer, J. (1992a). *In* "Peptides: Chemistry and Biology. Proceedings of the 12th American Peptide Symposium" (J. A. Smith and J. E. Rivier, eds.), pp. 54–56. ESCOM, Leiden.

Haviv, F., Fitzpatrick, T. D., Nichols, C. J., Swenson, R. E., Bush, E. N., Diaz, G., Nguyen, A. T., Nellans, H. N., Hoffman, D. J., Ghanbari, H., Johnson, E. S., Love, S., Cybulski, V., and Greer, J. (1992b). *J. Med. Chem.* **35,** 3890–3894.

Haviv, F., Fitzpatrick, T. D., Swenson, R. E., Nichols, C. J., Mort, N. A., Bush, E. N., Diaz, G., Bammert, G., Nguyen, A., Rhutasel, N. S., Nellans, H. N., Hoffman, D. J., Johnson, E. S., and Greer, J. (1993). *J. Med. Chem.* **36,** 363–369.

Heyl, D. L., and Mosberg, H. I. (1992). *Int. J. Pept. Protein Res.* **39,** 450–457.

Hill, P. S., Smith, D. D., Slaninova, J., and Hruby, V. J. (1990). *J. Am. Chem. Soc.* **112,** 3110–3113.

Hill, P. S., Chan, W. Y., and Hruby, V. J. (1991). *Int. J. Peptide Protein Res.* **38,** 32–37.

Hofmann, K (1962). *Annu. Rev. Biochem.* **31,** 213–246.

Holladay, M. W., Bennett, M. J., Tufano, M. D., Lin, C. W., Asin, K. E., Witte, D. G., Miller, T. R., Bianchi, B. R., Nikkel, A. L., Bednarz, L., and Nadzan, A. M. (1992). *J. Med. Chem.* **35,** 2919–2928.

Horwell, D. C., Birchmore, B., Boden, P. R., Higginbottom, M., Ho, Y. P., Hughes, J., Hunter, J. C., and Richardson, R. S. (1990). *Eur. J. Med. Chem.* **33,** 2950–62.

Hruby, V. J. (1974). *In* "Chemistry and Biology of Amino Acids, Peptides and Proteins" (B. Weinstein, ed.), Vol. 3, pp. 1–188. Dekker, New York.

Hruby, V. J. (1981a). *In* "Perspectives in Peptide Chemistry" (A. Eberle, R. Geiger, and T. Weilend, eds.), pp. 207–220. Karger, Basel.

Hruby, V. J. (1981b). *Top. Mol. Pharmacol.* **1,** 99–126.

Hruby, V. J. (1982a). *Life Sci.* **31,** 189–199.

Hruby, V. J. (1982b). *Mol. Cell. Biochem.* **44,** 49–64.

Hruby, V. J. (1986). *In* "Biochemical Actions of Hormones" (G. Litwack, ed.), Vol. 13, pp. 191–241. Academic Press, New York.

Hruby, V. J. (1987). *Trends Pharmacol. Sci.* **8,** 336–339.

Hruby, V. J. (1991). *In* "Peptides, Peptoids and Proteins" (P. D. Garzon, W. A. Colburn, and M. Mokotoff, eds.), pp. 3–13. Harvey Whitney, Cincinnati, Ohio.

Hruby, V. J., and Boteju, L. K. (1995). *In* "Encyclopedia of Molecular Biology." in press.

Hruby, V. J., and Gehrig, C. A. (1989). *J. Med. Chem.* **9,** 343–410.

Hruby, V. J., and Hadley, M. E. (1986). *In* "Design and Synthesis of Organic Molecules Based on Molecular Recognition" (G. Van Binst, ed.), pp. 269–289. Springer-Verlag, Heidelberg.

Hruby, V. J., and Mosberg, H. I. (1982a). *In* "Hormone Antagonists" (M. K. Agarwal, ed.), pp. 433–473. deGruyter, Berlin.

Hruby, V. J., and Mosberg, H. I. (1982b). *Peptides* (N.Y.) **3,** 329–336.

Hruby, V. J., and Smith, C. W. (1988). *In* "The Peptides" (C. W. Smith, ed.), Vol. 8, pp. 77–207. Academic Press, New York.

Hruby, V. J., Deb, K. K., Fox, J., Bjarnason, J., and Tu, A. T. (1978). *J. Biol. Chem.* **253,** 6060–6067.

Hruby, V. J., Deb, K. K., Yamamoto, D. M., Hadley, M. E., and Chen, W. Y. (1979a). *J. Med. Chem.* **22,** 7–12.

Hruby, V. J., Upson, D. A., Yamamoto, D. M., Smith, C. W., and Walter, R. (1979b). *J. Am. Chem. Soc.* **101,** 2717–2721.

Hruby, V. J., Mosberg, H. I., Hadley, M. E., Chan, W. Y., and Powell, A. M. (1980). *Int. J. Pept. Protein Res.* **16,** 377–381.

Hruby, V. J., Mosberg, H. I., Sawyer, T. K., Knittle, J. J., Rockway, T. W., Ormberg, J., Darmen, P., Chen, W. Y., and Hadley, M. E. (1983). *Biopolymers* **22,** 517–530.

Hruby, V. J., Wilkes, B. C., Cody, W. L., and Sawyer, T. K. (1984). *Pept. Protein Rev.* **3,** 1–64.

Hruby, V. J., Krstenansky, J. L., Gysin, B., Pelton, J. L., Trivedi, D., and McKee, R. (1986a). *Biopolymers* **25,** 5135–5155.

Hruby, V. J., Krstenansky, J. L., McKee, R., and Pelton, J. T. (1986b). *In* "Hormonal Control of Glucogenesis" (N. Kraus-Friedman, ed.), Vol. 2: Signal Transduction, pp. 3–20. CRC Press, Boca Raton, Florida.

Hruby, V. J., Kao, L.-F., Pettitt, M., and Karplus, M. (1988). *J. Am. Chem. Soc.* **110,** 3351–3359.

Hruby, V. J., Fang, S., Knapp, R., Kazmierski, W., Lui, G., and Yamamura, H. I. (1990a). *Int. J. Pept. Protein Res.* **35,** 566–573.

Hruby, V. J., Kazmierski, W., Kawasaki, A. M., and Matsunaga, T. (1990b). *In* "Peptide Pharmaceutics: Approaches to the Design of Novel Drugs" (D. J. Ward, ed.), pp. 135–184. Open Univ. Press, London.

Hruby, V. J., Al-Obeidi, F., and Kazmierski, W. (1991a). *Biochem. J. Rev.,* 159–172.

Hruby, V. J., Fang, S., Toth, G., Jiao, D., Matsunaga, T. O., Collins, N., Knapp, R., and Yamamura, H. I. (1991b). *In* "Peptides 1990, Proceedings of the 21st European Peptide Symposium" (E. Giralt and D. Andreu, eds.), pp. 707–709. ESCOM, Leiden.

Hruby, V. J., Prakash, O., Kazmierski, W., Gehrig, C., and Matsunaga, T. O. (1991c). *In* "Emerging Technologies and New Directions in Drug Abuse Research" (S. Rao, A. M. Rapaka, and M. J. Kuhar, eds.), NIDA Research Monograph 112, pp. 198–217. NIDA, Rockville, Maryland.

Hruby, V. J., Toth, G., Gehrug, C. A., Kao, L.-F., Knapp, R., Lui, R. K., Yamamura, H. I., Kramer, T. H., Davis, P., and Burkes, T. F. (1991d). *J. Med. Chem.* **39,** 1823–1830.

Hruby, V. J., Fang, S., Nikiforovich, G. V., Knapp, R., Jiao, D., and Yamamura, H. I. (1992). *In* "Peptides: Chemistry and Biology. Proceedings 12th American Peptide Symposium" (J. A. Smith and J. E. Rivier, eds.), pp. 142–143. ESCOM, Leiden.

Hruby, V. J., Gysin, B., Trivedi, D., and Johnson, G. (1993a). *Life Sci.* **52,** 845–844.

Hruby, V. J., Sharma, S. D., Toth, K., Jaw, J. Y., Al-Obeidi, F., Sawyer, T. K., and Hadley, M. C. (1993b). *Annu. N.Y. Acad. Sci.* **680,** 51–63.

Jiang, G., Porter, J., Rivier, C., Corrigan, A., Vale, W., and Rivier, J. (1993). *In* "Peptides: Chemistry, Structure and Biology. Proceedings of the 13th American Peptide Symposium" (R. S. Hodges and J. A. Smith, eds.), pp. 403–408, ESCOM, Leiden.

Johansson, C., Wilsen, O., Efendic, S., and Uvnan-Wallensten, K. (1981). *Digestion* **22,** 126–137.

Jorgensen, E. C., Wunridge, G. C., and Lee, T. C. (1971). *J. Med. Chem.* **14,** 631–636.

Kaiser, E. T., and Kezdy, F. J. (1983). *Proc. Natl. Acad. Sci. U.S.A.* **80,** 1137–1143.

Karten, M. J., and Rivier, J. E. (1986). *Endocr. Rev.* **7,** 44–66.

Kataoka, T., Beisen, D. D., Clark, J. D., Yodo, M., and Marshall, G. R. (1992). *Biopolymers* **32,** 1519–1533.

Kessler, H. (1982). *Angew. Chem., Int. Ed. Engl.* **21,** 521–523.

Kessler, H., Muller, G., Gurrath, M., and Timpl, R. (1992). *Angew. Chem., Int. Ed. Engl.* **31,** 326–328.

Knittel, J. J., Sawyer, T. K., Hruby, V. J., and Hadley, M. E. (1983). *J. Med. Chem.* **26,** 125–129.

Koerker, D. J., Harker, L. A., and Goodnew, C. J. (1975). *N. Eng. J. Med.* **295,** 476–479.

Kreissler, M., Perquer, M., Maigret, B., Fournier-Zaluski, M. C., and Roques, B. P. (1989). *J. Comput. Aided Mol. Des.* **3,** 85–94.

Krstenansky, J. L., Trivedi, D., Johnson, D., and Hruby, V. J. (1986). *J. Am. Chem. Soc.* **108,** 1696–1698.

Krstenansky, J. L., Trivedi, D., Johnson, D., and Hruby, V. J. (1987). *J. Med. Chem.* **30,** 1409–1445.

Krstenansky, J. L., Zechel, C., Trivedi, D., and Hruby, V. J. (1988). *Int. J. Pept. Protein Res.* **32,** 468–478.

Kruszynski, M., Lammek, B., Manning, M., Seto, J., Haldar, J., and Sawyer, W. H. (1980). *J. Med. Chem.* **23,** 364–368.

Lankiewich, L., Glanz, D., Gronka, Z., Slaninov, J., Barth, T., and Fahrenholz, F. (1989). *Bull. Pol. Acad. Sci. Chem.* **37,** 45–49.

Lâszlo, F. A., Lâszlo, F., Jr., and de Wied, D. (1991). *Pharmacol. Rev.* **43,** 73–108.

Lebl, M. (1987). *In* "CRC Handbook of Neurohypophyseal Hormone Analogs" (K. Jost, M. Lebl, and F. Brtnik, eds.), Vol. 2(1), pp. 17–34. CRC Press, Boca Raton, Florida.

Lebl, M., Frîc, I., Sugg, E. E., Cody, W. L., and Hruby, V. J. (1987a). *In* "Peptides" (D. Theodoropoulos, ed.), pp. 341–344. de Gruyter, Berlin.

Lebl, M., Jost, K., and Brtnik, F. (1987b). *In* "CRC Handbook of Neurohypophyseal Hormone Analogs" (K. Jost, M. Lebl, and F. Brtnik, eds.), Vol. 2(2), pp. 127–167. CRC Press, Boca Raton, Florida.

Lebl, M., Sugg, E. E., and Hruby, V. J. (1987c). *Int. J. Pept. Protein Res.* **29,** 40–45.

Lebl, M., Hill, P., Kazmierski, W., Kárászová, L., Slaninová, J., Frîc, I., and Hruby, V. J. (1990). *Int. J. Pept. Protein Res.* **36,** 321–330.

Levine, N., Sheftel, S. N., Eytan, T., Dorr, R. T., Hadley, M. E., Weinrach, J. C., Ertl, G. A., Toth, K., and Hruby, V. J. (1991). *J. Am. Med. Assoc.* **266,** 2730–2736.

Lin, C. W., Holladay, M. W., Witte, D. G., Miller, T. R., Wolfran, C. A. W., Bianchi, B. R., Bennett, M. J., and Nadzan, A. M. (1990). *Am. J. Physiol.* **285,** G648–G651.

Ling, N., and Vale, W. (1975). *Biochem. Biophys. Res. Commun.* **63,** 801–806.

Magnan, J., and Regoli, D. (1978). *Can. J. Physiol. Pharmacol.* **56,** 39–45.

Manning, M., and Sawyer, W. H. (1988). *In* "Peptides" (B. Penle and A. Torok, eds.), pp. 297–309. de Gruyter, Berlin.

Manning, M., and Sawyer, W. H. (1989). *J. Lab. Clin. Med.* **114,** 617–632.

Manning, M., Olma, A., Klis, W., Kolodziejczyk, A., Nawrocka, E., Misicka, A., Seto, J., and Sawyer, W. H. (1984). *Nature (London)* **308,** 652–653.

Manning, M., Misicka, A., Olma, A., Klis, W. A., Bankowski, K., Nawrocka, E., Kruszynski, M., Kolodziejczyk, A., Cheng, L.-L., Seto, J., Wo, N. C., and Sawyer, W. H. (1987). *J. Med. Chem.* **30,** 2245–2252.

Manning, M., Stoev, S., Bankowski, K., Misicka, A., and Lammek, B. (1992). *J. Med. Chem.* **35,** 382–388.

Marshall, G. R. (1992). *Biopolymers* **32,** 1519–1532.

Marshall, G. R. (1993). *Tetrahedron* **49,** 3547–3558.

Marshall, G. R., Needleman, P., and Hsieh, K. H. (1987). *J. Med. Chem.* **30,** 1097–1100.

Marwan, M. M., Malek, Z. A. A., Kreutzfeld, K. L., Hadley, M. E., Wilkes, B. C., and Hruby, V. J. (1985). *Mol. Cell. Endocrinol.* **41,** 171–177.

Matsoukas, J. M., Goghari, M. H., Scanlon, M. N., Franklin, K. J., and Moore, G. J. (1985). *J. Med. Chem.* **28,** 780–783.

Matsoukas, J., Cordopatis, P., Belte, V., Goghari, M. H., Ganter, R. C., Franklin, K. J., and Moore, G. J. (1988). *J. Med. Chem.* **33,** 1418–1421.

Matsuo, H., Baba, Y., Nair, R. M. G., Arimura, A., and Schally, A. V. (1971). *Biochem. Biophys. Res. Commun.* **43,** 1334–1339.

Medzihradszky, Y. (1982). *Med. Res. Rev.* **2,** 247–270.

Meraldi, J. P., Hruby, V. J., and Brewster, A. I. R. (1977). *Proc. Natl. Acad. Sci. U.S.A.* **74,** 1373–1377.

Momany, F. A. (1976). *J. Am. Chem. Soc.* **98,** 2990–2996.

Monahan, M. V., Amos, M. S., Anderson, H. A., and Vale, W. (1973). *Biochemistry* **12,** 4616–4620.

Moore, G. J. (1985). *Int. J. Pept. Protein Res.* **26,** 469–481.

Moore, G. J., Franklin, K. J., Nystrom, D. M., and Goghari, M. H. (1985). *Can. J. Physiol. Pharmacol.* **63,** 966–971.

Moore, G. J., Ganter, R. C., Goghari, M. H., and Franklin, K. J. (1991). *Int. J. Pept. Protein Res.* **38,** 1–7.

Morgan, B. A., and Gainor, J. A. (1989). *Annu. Rep. Med. Chem.* **24,** 243–252.

Mosberg, H. I., Hurst, R., Hruby, V. J., Gee, K., Yamamura, H. I., Galligan, J. J., and Burks, T. F. (1983). *Proc. Natl. Acad. Sci. U.S.A.* **80,** 5871–5874.

Nikiforovich, G., and Hruby, V. J. (1993). *Biochem. Biophys. Res. Commun.* **194,** 9–16.

Nikiforovich, G., Hruby, V. J., Prakash, O., and Gehrig, C. A. (1991). *Biopolymers* **31,** 941–955.

O'Donohue, T. L., and Jacobowitz, D. M. (1980). *In* "Polypeptide Hormones" (R. F. Beers and E. G. Bassett, eds.), pp. 203–222. Raven, New York.

Pattaroni, C., Lucietto, P., Goodman, M., Yamamoto, G., Vale, W., Moroder, L., Gazerro, L., Gohring, W., Schmied, B., and Wünsch, E. (1990). *Int. J. Pept. Protein Res.* **36,** 401–417.

Penke, B., Hajnal, F., Lunovics, J., Holzinger, G., Kadar, T., and Teledgy, G. (1984). *J. Med. Chem.* **27,** 845–849.

Pettibone, D. J., Clineschmidt, B. V., Anderson, P. S., Freidinger, R. M., and Lundell, G. F. (1989). *Endocrinology (Baltimore)* **125,** 217–222.

Pincus, M. R., Carty, R. P., Chen, J., Lubowsky, J., Avitable, M., Shah, D., Scheraga, H. A., and Murphy, R. B. (1987). *Proc. Natl. Acad. Sci. U.S.A.* **84,** 4821–4825.

Plucinska, K., Kataoka, T., Yodo, M., Cody, W. L., He, J. X., Humblet, C., Lu, G. H., Lunney, E., Major, T. C., Panek, R. L., Schelkun, P., Skeean, R., and Marshall, G. R. (1993). *J. Med. Chem.* **36,** 1902–1913.

Ramachandran, J. (1973). *In* "Hormonal Proteins and Peptides" (C. H. Li, ed.), pp. 1–28. Academic Press, New York.

Rivier, J., Rivier, C., Perrin, M., Porter, J., and Vale, W. (1981). *In* "LHRH Peptides as Female and Male Contraception" (G. I. Zatuchni, J. D. Shelton, and J. J. Seiarra, eds.), pp. 13–23. Harper & Row, Philadelphia, Pennsylvania.

Rivier, J., Varga, J., Porter, J., Perrin, M., and Hass, Y. (1986). *In* "Peptides: Structure and Function" (C. M. Deber, V. J. Hruby, and K. D. Kopple, eds.), pp. 541–544. Pierce Chemical Company, Rockford, Illinois.

Rivier, J., Krupryszewski, G., Varga, J., Porter, J., and Rivier, C. (1988). *J. Med. Chem.* **31,** 677–682.

Rivier, J., Rivier, C., Vale, W., Koerber, S., Corrgan, A., Porter, J., Gierasch, J. M., and Hagler, A. T. (1990). *In* "Peptides: Chemistry, Structure and Biology" (E. J. Rivier and G. R. Marshall, eds.), pp. 33–37. ESCOM, Leiden.

Rivier, J., Porter, J., Hoegar, C., Theobald, P., and Craig, A. G. (1992). *J. Med. Chem.* **35,** 4270–4278.

Rizo, G., and Gierasch, L. M. (1992). *Annu. Rev. Biochem.* **61,** 387–418.

Rizo, J., Koerber, S. C., Bienstock, R. J., Rivier, J. E., Gierasch, L. M., and Hagler, A. T. (1992a). *J. Am. Chem. Soc.* **114,** 2852–2859.

Rizo, J., Koerber, S. C., Bienstock, R. J., Rivier, J. E., Hagler, A. T., and Gierasch, L. M. (1992b). *J. Am. Chem. Soc.* **114,** 2860–2871.

Rizo, J., Sullan, B., Breslaw, J., Koerber, S. C., Rivier, J. E., Hagler, A. T., and Gierasch, L. M. (1993). *J. Cell. Biochem.* **17C**(Suppl.), 234–239.

Rodriquez, M., Lignon, M. F., Gales, M. C., Ambland, M., and Martinez, J. (1990). *Mol. Pharmacol.* **38,** 333–341.

Rose, G. D., Gierasch, L. M., and Smith, J. A. (1985). *Adv. Protein Chem.* **37,** 1–109.

Rudinger, J. (1971). *In* "Drug Design" (E. J. Ariens, ed.), Vol. 2, pp. 318–419. Academic Press, New York.

Rudinger, J. (1972). *In* "Endocrinology 1971: Proceedings of the 3rd International Symposium" (S. Taylor, ed.), pp. 201–222. William Heinemann Medical Books, London.

Samanen, J., Brandeis, E., Narindray, D., Adams, W., Cash, Y., Yellen, T., and Regoli, D. (1988). *J. Med. Chem.* **31,** 342–373.

Sawyer, T. K., Sanfilippo, P. J., Hruby, V. J., Engel, M. H., and Heward, C. B. (1980). *Proc. Natl. Acad. Sci. U.S.A.* **77,** 5745–5758.

Sawyer, T. K., Hruby, V. J., Darmen, P. S., and Hadley, M. E. (1982). *Proc. Natl. Acad. Sci. U.S.A.* **79,** 1751–1755.

Sawyer, T. K., Hruby, V. J., Hadley, M. E., and Engel, M. H. (1983). *Am. Zool.* **23,** 529–540.

Sawyer, W. H., Bankowski, K., Misicka, A., Nawrocka, E., Krusazynski, M., Stoev, S., Klis, W. A., Przybylski, J., and Manning, M. (1988). *Peptides (N.Y.)* **9,** 157–163.

Scanlon, M. N., Matsoukas, J. M., Franklin, K. J., and Moore, G. J. (1984). *Life Sci.* **34,** 317–321.

Schiller, P. W. (1985a). *Biochem. Biophys. Res. Commun.* **127,** 558–564.

Schiller, P. W. (1985b). *J. Med. Chem.* **28,** 1766–1773.

Schiller, P. W. (1991). *In* "Progress in Medicinal Chemistry" (G. P. Ellis and G. B. West, eds.), Vol. 28, pp. 301–340. Elsevier, Amsterdam.

Schiller, P. W., and Geiger, R. (1984). *J. Protein Chem.* **2,** 279–287.

Schiller, P. W., Natarajan, S., and Bodanszky, M. (1978). *Int. J. Pept. Protein Res.* **12,** 139–142.

Schiller, P. W., Weltrowska, G., Nguyen, T. M., Wilkes, B. C., Chung, N. N., and Lemiex, C. (1992a). *J. Med. Chem.* **35,** 3956–3961.

Schiller, P. W., Nguyen, T. M.-D., Weltrowska, G., Wilkes, B. C., Marsden, B. J., Lemieux, C., and Chung, N. N. (1992b) *Proc. Natl. Acad. Sci. U.S.A.* **89,** 11871–11875.

Schiller, P. W., Wilkes, B. C., Marsden, B. J., Nguyen, T. M., Weltrowska, G., Lemieux, C., and Chung, C. (1993). *J. Cell. Biochem.* **17C**(Suppl.), 206–209.

Schulz, H., and du Vigneaud, V. (1966). *J. Med. Chem.* **9,** 647–650.

Schwyzer, R. (1977). *Annu. N.Y. Acad. Sci.* **297,** 3–26.

Shiosaki, K., Lin, C. W., Kopeca, H., Craig, R., Blanchi, B., Miller, T., Witle, D., and Nadzan, A. M. (1990). *J. Med. Chem.* **33,** 2950–2963.

Shiosaki, K., Lin, C. W., Kopeca, H., Craig, R., Blanchi, B., Miller, T., Witle, D., Stashko, M., and Nadzan, A. M. (1992). *J. Med. Chem.* **35,** 2007–2014.

Skala, G., Smith, C. W., Taylor, C. J., and Ludens, J. H. (1984). *Science* **226,** 443–445.

Smith, C. W., Suguna, K., Padlan, E. A., Carlson, W. D., and Davis, D. (1987). *Proc. Natl. Acad. Sci. U.S.A.* **84,** 7009–7013.

Spatola, A. F. (1983). *In* "Chemistry and Biochemistry of Amino Acids, Peptides and Proteins" (B. Weinstein, ed.), Vol.7, pp. 267–357. Dekker, New York.

Spear, K. L., Brown, M. S., Reinhard, E. J., McMahon, E. G., Olins, G. M., Palomo, M. A., and Patton, D. R. (1990). *J. Med. Chem.* **33,** 1935–1940.

Struthers, R. S., Rivier, J., and Hagler, A. T. (1984). *In* "Conformationally Directed Drug Design" (J. A. Vida and M. Gordon, eds.), pp. 239–262. American Chemical Society, Washington, D. C.

Struthers, R. S., Rivier, J., and Hagler, A. T. (1985). *Annu. N.Y. Acad. Sci.* **439,** 81–96.

Struthers, R. S., Tanaka, G., Koerber, S. C., Solmajer, T., and Baniak, E. L. (1990). *Proteins* **8,** 295–304.

Suirskis, S., Muceniece, R., Klusa, V., Ancans, J., Georgier, V., and Getova, D. (1991). *Central Peripheral Peptide Regulation,* 120–128.

Toniolo, C. (1990). *Int. J. Pept. Protein Res.* **35,** 287–300.

Turker, R. K., Hall, M. N., Yamamoto, M., Sweet, C. S., and Bumpus, F. M. (1972). *Science* **177,** 1203–1205.

Ungar, R. H. (1978). *Metabolism* **27,** 1691–1706.

Unson, C. G., Andreu, D., Gurzenda, E. M., and Merrifield, R. B. (1987). *Proc. Natl. Acad. Sci. U.S.A.* **84,** 4083–4087.

Unson, C. G., Iwasa, K., Gurzenda, E. M., Durrah, T. L., and Merrifield, R. B. (1990). *In* "Peptides, Chemistry, Structure, and Biology, Proceedings of the 11th American Peptide Symposium" (J. E. Rivier and G. R. Marshall, eds.), pp. 203–204. ESCOM, Leiden.

Unson, C. G., Macdonald, D., Ray, K., Durrah, T. L., and Merrifield, R. B. (1991). *J. Biol. Chem.* **266,** 2763–2766.

Unson, C. G., Macdonald, D., and Merrifield, R. B. (1993). *Arch. Biochem. Biophys.* **300,** 747–750.

Veber, D. F., and Friedinger, R. M. (1984). *In* "Conformationally Directed Drug Design" (J. A. Vida and M. Gordon, eds.), ACS Symposium Series No. 251, pp. 169–187. American Chemical Society, Washington, D. C.

Veber, D. F., Freidinger, R. M., Perlow, D. S., Paleveder, W. J., Holly, F. W., Strachean, R. G., Nutt, R. F., Arison, B. J., Homnick, C., Randall, W. C., Glitzer, M. S., and Saperstein, R. (1981). *Nature (London)* **292,** 13–15.

Veber, D. F., Saperstein, R., Nutt, R. F., Freidinger, R. M., Brady, S. F., Curley, P., Perlow, D. S., Paleveda, W. J., Colton, C. D., Zacchei, A. G., Tocco, D. J., Hoff, D. R., Vandlen, D. R., Gerich, J. E., Hall, L., Mandarino, L., Cordes, E. H., Anderson, P. S., and Hirschmann, R. (1984). *Life Sci.* **34,** 1371–1378.

Walter, R. (1977). *Fed. Proc.* **36,** 1872–1878.

Walter, R., Smith, C. W., Mehta, P. K., Boonjarern, S., Arruda, J. A. L., and Kurtzman, N. A. (1977). *In* "Disturbances in Body Fluid Osmolarity" (T. E. Andreoli, J. J. Grantham, and F. C. Rector, eds.), pp. 1–36. American Physiology Society, Bethesda, Maryland.

Wilkes, B. C., and Schiller, P. W. (1990). *Biopolymers* **29,** 89–96.

Witte, D. G., Nadzan, A. M., Martinez, J., Rodriguez, M., and Lin, C. W. (1992). *Peptides* **13,** 1227–1232.

Wood, S. P., Tickle, I. J., Treharne, A. M., Pitts, J. E., and Mascarenhas, Y. (1986). *Science* **232,** 633–636.

Zechel, C., Trivedi, D., and Hruby, V. J. (1991). *Int. J. Pept. Protein Res.* **38,** 131–138.

7

Neuropeptides: Peptide and Nonpeptide Analogs

Andrzej W. Lipkowski
Medical Research Centre
Polish Academy of Sciences
00-784 Warsaw, Poland
and Industrial Chemistry Research Institute
01-793 Warsaw, Poland

Daniel B. Carr
Departments of Anesthesia and Medicine
New England Medical Center and Tufts University School of Medicine
Boston, Massachusetts 02111

Continuing advances in the isolation, chemical characterization, and definition of the roles of neuropeptides in homeostasis reflect not only the evolution of medical and biological scientific techniques, but also progressive refinement of our concepts of the function of living systems. In this chapter, we survey structure–activity relationships of several prototypical classes of neuropeptides. For each of the neuropeptide families discussed, knowledge of these relationships has been exploited to generate both peptide and nonpeptide analogs that have biological activity and therapeutic potential. Our motivation to discuss neuropeptides for which both peptide and nonpeptide analogs have been prepared is that in addition to the intrinsic biological importance of these peptides, their analysis has extended available knowledge of hormone chemistry and in some cases has generated new hypotheses of hormone–receptor interactions.

Peptides: Synthesis, Structures, and Applications

These classes of neuropeptides include the endogenous opioids, the tachykinins, angiotensin II, oxytocin, vasopressin, and cholecystokinin.

I. Biosynthesis of Neuropeptides

Neuropeptides are initially synthesized as large, inactive protein precursors, which undergo proteolytic processing to yield biologically active peptides. All precursors possess a signal peptide, required for vectorial transport across the membranes of the endoplasmic reticulum. The signal peptide consists predominantly of hydrophobic amino acids, but no specific sequence of amino acids is required for recognition by the signal protease. On the other hand, the secondary structure of the signal peptide (typically a β turn) is important for determining the exact site of cleavage (Inouye *et al.*, 1985). The eukaryotic endoplasmic reticulum protease that cleaves the signal peptide is an integral membrane protein (Wolfe and Wickner, 1984) with strong sequence similarity to the signal protease isolated from bacteria (Rice and Wickner, 1985).

Numerous precursor proteins have been characterized (Andrews *et al.*, 1987), including three precursor proteins for opioids (Hollt, 1983), and others for tachykinins, thyrotropin-releasing hormone (TRH), corticotropin-releasing hormone (CRH), and gonadotropin-releasing hormone (GnRH). Prior to their cleavage to generate active end products, propeptides may be modified by glycosylation, amidation, or acetylation and thereby differentiated for further metabolism. Within propeptides, biologically active sequences such as those for the opioids are flanked by pairs of basic amino acids (Lys-Lys, Lys-Arg, Arg-Lys, or Arg-Arg) or single basic amino acids, often followed by proline. Cleavage of propeptides at such basic sites by one or more trypsin-like endopeptidases (Gainer *et al.*, 1985), termed prohormone-converting enzyme(s), initiates the generation of neuropeptides. Despite their functional similarities, these enzymes are distinct from either trypsin or cathepsin B (Loh *et al.*, 1984). Interestingly, prohormone-converting enzymes do not cleave small synthetic substrates (Chang *et al.*, 1982), suggesting a strong conformational dependence of the reactions. This steric dependency may underlie the relative resistance to cleavage of some basic amino acid pairs within neuropeptides (e.g., dynorphins) liberated at this stage of processing.

Basic residues at the amino terminus of the resultant tryptic products are postulated to be removed by an aminopeptidase (Gainer *et al.*, 1984; Lipkowski and Misicka, 1989). Basic residues at the carboxyl termini have been shown to be removed by carboxypeptidases, of which the best characterized is carboxypeptidase H (CPH) (Hook, 1988; Hook and Affolter, 1988) (designated EC 3.4.17.10). When CPH action exposes a glycine residue at the carboxyl terminus, transformation of the peptide into an amide may occur (Bradbury and Smyth, 1985). Considering that amidated peptides are relatively resistant to carboxypeptidases, it is not surprising that about half of the bioactive neuropeptides possess carboxyl-terminal amide groups. α-Amidation is catalyzed by peptidylglycine α-amidating monooxygenase (PAM). This reaction is stimulated by the presence of a reduced cofactor, ascorbic acid (Eipper *et al.*, 1983). The above-mentioned enzymatic processes are examples drawn from a broad constellation of enzymes that together determine the final peptide products formed by processing and degradation of neuropeptides.

In addition to proteolytic cleavage, posttranslational processing can modify the peptide precursor by glycosylation, sulfation, and so forth (Martinez and Potier, 1986; Carr *et al.*, 1991) thereby altering the nature and biological activity of the gene product (Mains *et al.*, 1990; Marx, 1991). All enzymatic systems are very sensitive to local biochemical conditions. Posttranslational processing can generate simultaneously a number of active peptides from one precursor protein (Gainer *et al.*, 1985). A good illustration is proopiomelanocortin (POMC) (Mains and Eipper, 1979), the common precursor for β-endorphin, adrenocorticotropic hormone (ACTH), melanocyte-stimulating hormone (MSH), corticotropin-like intermediate lobe peptide (CLIP) (Chastrette *et al.*, 1990), and related compounds. Individual propeptides have historically been conceptualized, isolated, and sequenced as precursors for particular neuopeptides, each possessing a well-defined biological function. After sequencing, recognition of other peptides generated simultaneously has often produced fruitful insights as to the functions of these previously unrecognized sibling peptides, for example, calcitonin gene-related peptide (CGRP).

Neuropeptides are stored in intracellular vesicles from which they are released to the extracellular space. Neuropeptides released into the bloodstream function as neurohormones, but those released into the synaptic cleft serve as neuromodulators (to inhibit the action of excitatory neurotransmitters) or neuromediators (to prolong the action of neurotransmitters). In the extracellular space, neuropeptides enter both pharmacodynamic and pharmacokinetic cascades. Thus, neuropeptides interact with receptors on target cells to generate a biological signal within these cells, and serve as substrates for a variety of elimination reactions (enzymatic degradation, urinary excretion, etc.). The equilibrium between active regulation of cellular function through receptor interactions, and their own catabolism, is a critical element in the homeostasis of neuropeptides.

II. Neuropeptide Receptors

Neuropeptides act on target cells by interacting with selective receptors. Over a 2-year period, these neuropeptide receptors have been cloned and characterized: opioid δ (Evans *et al.*, 1992; Kieffer *et al.*, 1992); opioid μ (Chen *et al.*, 1993); opioid κ (Yasuda *et al.*, 1993); tachykinins (Takeda *et al.*, 1991; Nakanishi, 1991); bradykinin (Hess *et al.*, 1992; Eggericks *et al.*, 1992; McIntyre *et al.*, 1993); angiotensin (Sakmar *et al.*, 1989; Iwai *et al.*, 1991; Bergsma *et al.*, 1991); neuropeptide Y1 (Eva *et al.*, 1990); neuropeptide Y3 (Rimland *et al.*, 1991); vasoactive intestinal peptide (VIP) (Sreedharan *et al.*, 1991); and somatostatin (Bell and Reisine, 1993). All characterized receptors of neuropeptides share the common familial features of seven transmembrane protein domains and coupling to G proteins that serve as intracellular effectors (Humblet and Mirzadegan, 1992).

III. Opioids

Compared to our knowledge of other neuropeptides, that of the opioid system is unique. The structure of fentanyl (**1**) (Janssen, 1962) and the initial results of studies on structure–activity relationships (SARs) of morphine alkaloids were known for over a decade before the isolation of endogenous opioid peptides and the discovery of multiple opioid receptors in the 1970s. Modification of the benzomorphan skeleton of morphine (**2**) had produced numerous analogs (Janssen and Tollenaere, 1979; Lenz *et al.*, 1986), including antagonists, with differing spectra of biological activities. Therefore, the discovery of endogenous peptide ligands (Hughes *et al.*, 1975) and characterization of four major types (μ, δ, κ, σ) of opioid receptors (Pasternak, 1988; Knapp *et al.*, 1990) did not initiate SAR studies (as was true for other neuropeptide systems) but were themselves the result of longstanding research.

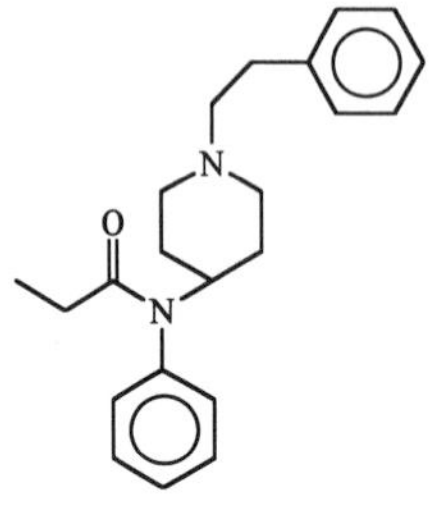

1. Fentanyl 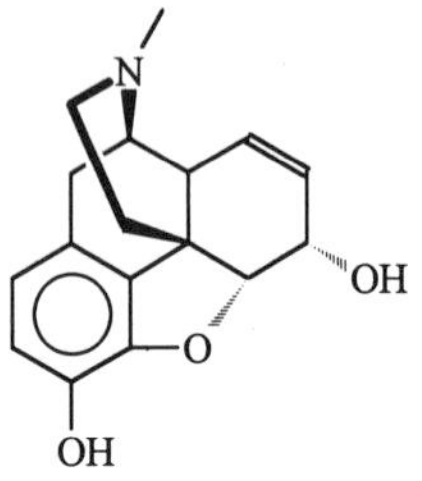 **2. Morphine** 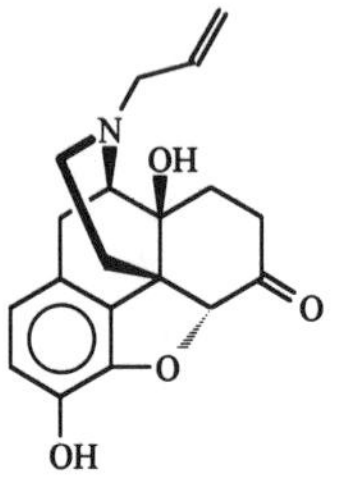 **3. Naloxone** 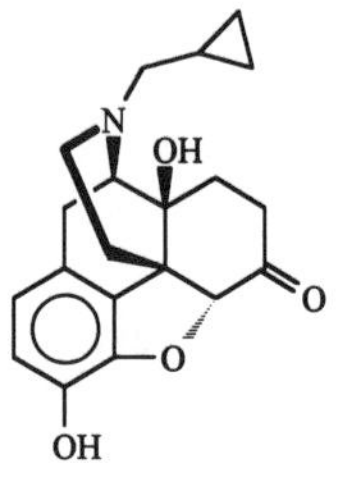 **4. Naltrexone**

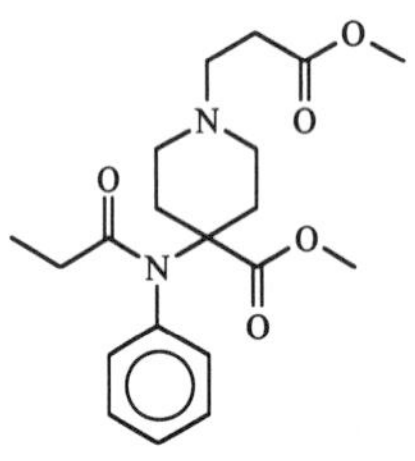

7. Remifentanil 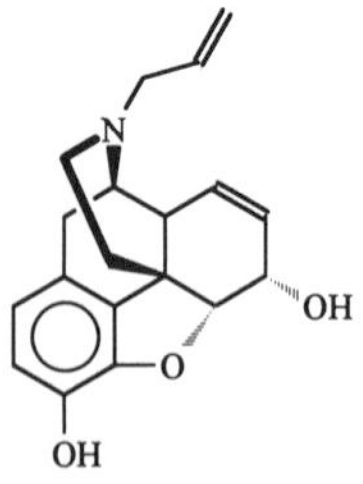 **5. Nalorphine** 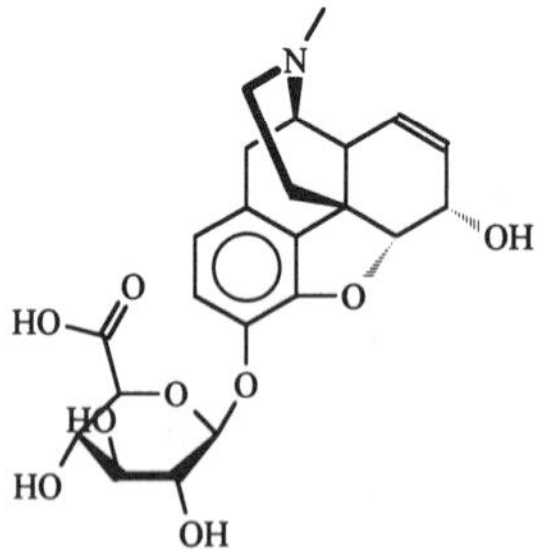 **6. Morphine-3-glucuronide**

Considering that two complementary families of ligands for opioid SAR studies—endogenous flexible peptides and exogenous rigid nonpeptide alkaloids—have been available since the 1970s, it is surprising that early hopes for fast results on three-dimensional structural requirements for opioid ligand–receptor interactions, as well as topographical correlations between peptide and nonpeptide ligands, have been so slowly attained (Hudson *et al.,* 1979; Gorin and Marshall, 1977; Schiller, 1984; Michel *et al.,* 1991). Even today, only general rules regarding structural comparisons are known. For example, the *N*-terminal tyrosine of opioid peptides is necessary for binding to opioid receptors. The obvious inference is that the tyramine element of tyrosine in opioid peptides and morphine alkaloids are topographically related. The primary amine group of tyrosine can be transformed into a secondary or tertiary amine, as in alkaloids, with preservation of opioid receptor affinity (Hansen and Morgan, 1984; Hruby and Gehrig, 1989). The tertiary amine group of morphine can be further transformed to a quaternary amine salt that has persistent opioid receptor affinity (Iorio and Frigeni, 1984).

Rigid opioid alkaloids have a well-defined conformation and are ready at all times to interact with receptors. In contrast, the flexible tyrosine at the amino terminus of opioid peptides is not by itself recognized by opioid receptors. Therefore, all opioid peptide analogs require additional elements to support their binding to receptors. This requirement illustrates the proposal by Schwyzer that natural peptide ligands are a combination of two types of elements: a "message" and an "address" (Schwyzer, 1977). The opioid peptide message, an N-terminal dipeptide, is required to activate the peptide–receptor complex, and is directly related to the tyramine moiety of opiate alkaloids. Nevertheless, this flexible dipeptide sequence has too low an affinity for opioid receptors to make an effective complex with the receptor binding site. Additional, specific address sequences that mediate recognition of opioid receptor types are present in endogenous compounds. Opioid peptide address sequences are composed of two elements: (1)...-Gly-Phe-..., or...-Phe-..., which are present in the majority of natural opioid peptides and confer a high affinity for opioid receptors in general, and (2) C-terminal peptide fragments which differentiate the affinity of peptides between particular receptor types (Lipkowski, 1987).

The considerable independence of ongoing SAR research programs for peptide versus nonpeptide opioids reflects a lack of well-defined structural correlation between the two structural families. Morphine (**2**), the prototypical opioid, has relatively high affinity for μ receptors (Kosterlitz *et al.,* 1986). Normorphine, the desmethyl analog of morphine, also has high μ-receptor selectivity (Magnan *et al.,* 1982). Alkylation of the amino group with a larger, lipophilic alkyl group results in antagonist or partial agonist compounds. Naloxone (**3**) naltrexone (**4**), and nalorphine (**5**) are the most commonly used opioid antagonists. All three compounds have selectivity for the μ-receptor type, but their affinities for δ and κ receptors are high enough that they can reverse the effects of agonist ligands of these receptors. *In vivo,* morphine-related compounds are cleared mostly through biotransformation into 3-glucuronides (e.g., **6**) Lenz *et al.,* 1986).

Two other families of nonpeptide opioids also have high selectivity for the μ receptor. The first are phenylpiperidine derivatives such as fentanyl (Janssen,1982) and sufentanil. These compounds are metabolized mainly by *N*-dealkylation (Goromaru *et al.,* 1981). Relatively minor modifications within the

Tyr-Gly-Gly-Phe-Met
11. Met-enkephalin

Tyr-Gly-Gly-Phe-Leu
12. Leu-enkephalin

Tyr-Gly-Gly-Phe-Leu-Arg-Arg-Ile-Arg-Pro-Lys-Leu-Lys-Trp-Asp-Asn-Gln
31. Dynorphin A

Tyr-D-Ala-Phe-Gly-Tyr-Pro-Ser-NH_2
9. Dermorphin

Tyr-D-Ala-Phe-Asp-Val-Val-Gly-NH_2
20. Deltorphin I

Tyr-D-Ala-Phe-Glu-Val-Val-Gly-NH_2
21. Deltorphin II

Tyr-D-Met-Phe-His-Leu-Met-Asp-NH_2
22. Dermenkephalin

Tyr-Pro-Phe-Pro-Gly-Pro-Ile
10. β-casomorphin

fentanyl family have generated compounds with differing durations of action, allowing their use to match particular clinical circumstances. One compound in this series, remifentanil (**7**), is an ultra-short-acting μ-opioid agonist now under evaluation for same-day surgical procedures. An ester linkage in the chemical structure of remifentanil renders this compound susceptible to rapid metabolism by blood and tissue esterases. The primary metabolic pathway of remifentanil is deesterification to form a carboxylic acid metabolite which has one-thousandth the potency of the parent compound (Westmoreland *et al.,* 1993). On the other hand, the long duration of action of other μ-receptor-selective diphenylpropylamine ligands, such as methadone or L-α-acteylmethadol (LAAM), allows their use in single daily doses for maintenance treatment of drug dependence.

The endogenous agonist ligand for the μ-receptor type is probably β-endorphin (Li, 1986), but its affinities for μ and δ receptors are almost equal (Kosterlitz *et al.,* 1986). Dermorphin (**9**) isolated from frog skin, is the most μ selective natural opioid peptide, but it also possesses some affinity for the δ-receptor type (Marastoni *et al.,* 1987). Natural human and bovine β-casomorphins (**10**), α-casein derived peptides, have been also characterized as μ-opioid receptor ligands, but with relatively low potency (Koch and Brantl, 1990). Modifications of dermorphin, β-casomorphin, and Leu- or Met-enkephalins (**11, 12**) have all resulted in very selective μ-opioid receptor ligands. Thus, DALDA (**13**) (Schiller *et al.,* 1989), PLO17 (**14**) (Chang *et al.,* 1983; Shook *et al.,* 1987), and DAMGO (DAGO, **15**) (Handa *et al.,* 1981) are the most common analogs used as μ-receptor-selective ligands. Surprisingly, the peptide antagonists with high μ-receptor selectivity were discovered not from SAR studies of the endogenous opioid peptides, but instead through modifications of somatostatin, which itself has low affinity for opioid receptors. SAR studies of somatostatin in conjunction with conformational and topographical considerations resulted in the analog CTOP (**16**), the prototype of a series of antagonists with high selectivity for the μ receptor (Pelton *et al.*, 1985; Kazmierski *et al.*, 1988).

Tyr-D-Arg-Phe-Lys-NH_2
13. DALDA

Tyr-Pro-(N-Me)Phe-D-Pro-NH_2
14. PLO17

Tyr-D-Ala-Gly-Phe-NH-CH_2-CH_2-OH
15. DAMGO

D-Phe-Cys-Tyr-D-Trp-Orn-Thr-Pen-Thr-NH_2
16. CTOP

Tyr-D-Ala-Gly-Phe-D-Leu
17. DADLE

Tyr-D-Ser-Gly-Phe-Leu-Thr
18. DSLET

Tyr-D-Pen-Gly-Phe-D-Pen
19. DPDPE

$(CH_2$=CH-CH_2-)=Tyr-Aib-Aib-Phe-Leu
23. ICI174864

Tyr-Tic-Phe-Phe
Tic, L-1,2,3,4-tetrahydroisoquinoline-3-carboxylic acid
24. TIPP

Tyr-Gly-Gly-Phe-Leu-Arg-Arg-Cys-Arg-Pro-Lys-Leu-Cys-NH_2
32.

The δ agonists with the highest selectivity have enkephalin-derived peptide structures. Early on, DADLE (**17**) was used as a δ-selective ligand because it had 3- to 10-fold higher selectivity for the δ receptor than the μ receptor, and was more stable than the native compounds Leu- or Met-enkephalin (Kosterlitz *et al.*, 1986). Further modifications of the parent enkephalin sequence resulted in two groups of peptide analogs. The first group, a series of linear analogs, included DSLET (**18**) (Gacel *et al.*, 1981); the second, all rigid cyclic analogs, included DPDPE (**19**) (Mosberg *et al.*, 1983). In binding assays, the affinities of these analogs for δ over μ receptors are over 100-fold, whereas the selectivities for δ versus κ receptors are greater than 1000-fold. Further analogs having either linear (Gacel *et al.*, 1988; Delay-Goyet *et al.*, 1991) or cyclic (Toth *et al.*, 1992) structures displayed higher selectivity. Deltorphins (**20, 21**) and dermenkephalin (**22**), analogs of dermorphin with high δ selectivity, have been isolated from amphibian skin (Kreil *et al.*, 1989; Amiche *et al.*, 1989). Deltorphins and dermenkephalin have common topographical features responsible for δ selectivity (Misicka *et al.*, 1992a), which can be stabilized into respective cyclic analogs (Misicka *et al.*, 1993).

The enkephalin analog ICI 174864 (**23**) has been found to show considerable δ selectivity but only moderate potency as an antagonist (Cotton *et al.*, 1984). The synthesis of TIPP (**24**) analogs has offered significant progress in the search for selective peptide-derived δ antagonists. TIPP itself has nanomolar antagonist affinity to and extraordinary selectivity for the δ receptor (Schiller *et al.*, 1992). Pseudopeptide analogs of TIPP containing a reduced bond between Tic^2 and Phe^3 displayed subnanomolar δ receptor affinity with selectivity for δ versus μ receptors of over 10,000-fold (Schiller *et al.*, 1993).

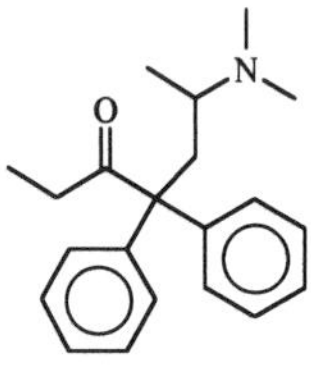

8. Methadone 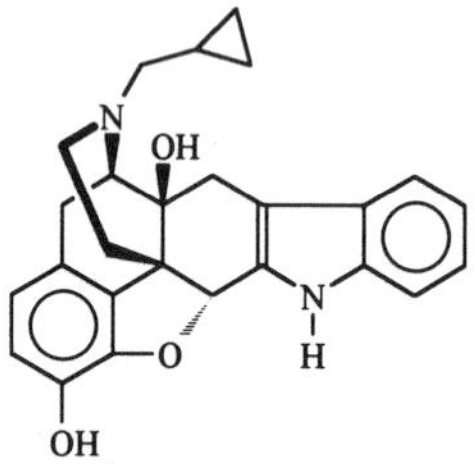 **25. NTI** 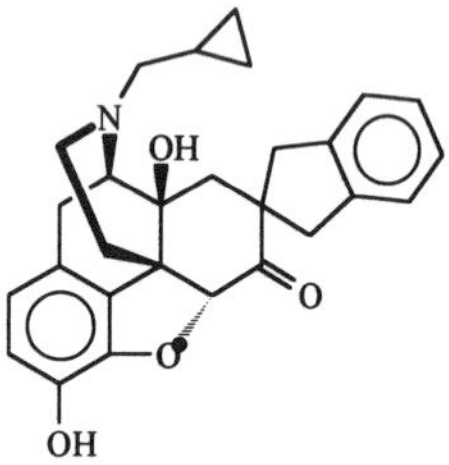 **26. SIOM**

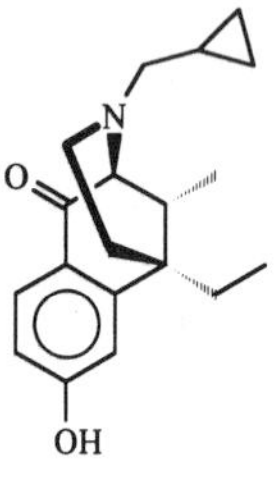

27. EKC 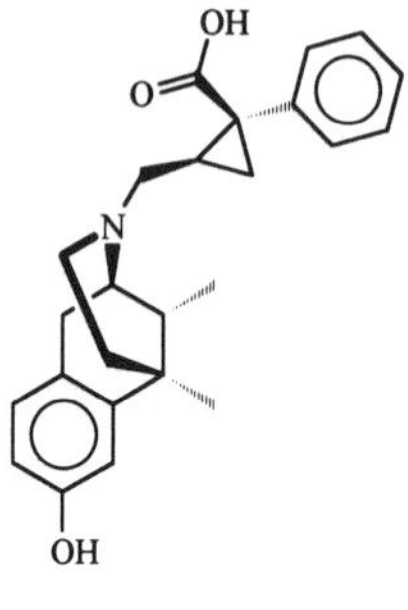 **28.** 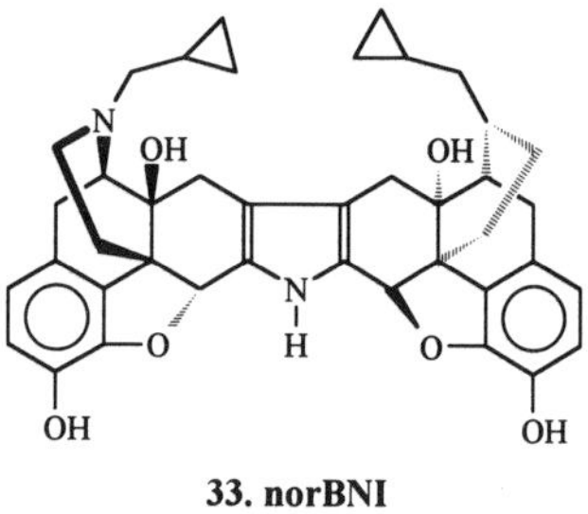 **33. norBNI**

The discovery of alkaloid analogs with selectivity for the δ receptor type has a very short history. The structure and biological activities of the first nonpeptide δ antagonist, naltrindole (NTI, **25**) were reported 7 years ago (Portoghese *et al.*, 1988). NTI is a naltrexone analog with a conformationally restricted aromatic system fused to the C-6 and C-7 carbons. The NTI analog in which an *N*-phenethyl group replaces cyclopropylmethyl shows δ-agonist properties (Portoghese *et al.*, 1992). Both NTI and this analog have proved useful for structural comparisons of δ-opioid peptide and alkaloid ligands. Comparison of the native enkephalins (Portoghese *et al.*, 1988), the cyclic analog DPDPE, and the linear analog deltorphin with NTI analogs led to the conclusion that Phe^4 in DPDPE and enkephalin is related to the indole ring of NTI, but Phe^3 of deltorphin is related to the phenethyl group (Misicka *et al.*, 1992b). Syntheses and subsequent biological evaluation of new NTI analogs have indicated that, for agonist properties, the slightly different topographical location of the aromatic ring induced by spiro substitution in position 7 of oxymorphone (SIOM, **26**) is preferred (Portoghese *et al.,* 1993).

Ketazocine, the benzomorphan derivative, was used in pioneering experiments that inferred the existence of the κ receptor (Martin *et al.,* 1976). Analogs of ketazocine such as ethylketazocine (EKC, **27**) or bremazocine (Romer *et al.,* 1980) have long been used as κ-receptor markers. However, in spite of their selective κ-agonist properties, they also display high affinities to μ- and δ-receptor types (Tam, 1985). Synthesis of a phenyl carboxy ester analog of normetazocine (**28**) is a significant advance in the development of κ-selective ligands based on the benzomorphan skeleton (Ronsisvalle *et al.,* 1993). The discovery of 1,2-aminoamide derivatives, such as U-50,488 (**29**) (Szmuszkovicz and Von Voigtlander, 1982), was a further step forward in SAR studies of κ-receptor-selective ligands. Not only has U-50,488 become the current κ-agonist reference ligand, but it has also served as the structural starting point for the design of more selective and potent κ agonists. For example, incorporation of a spiro ether group into the cyclohexane ring gives U-62,066 (spiradoline, **30**) (Peters *et al.,* 1987) and U-69,593. U-69,593 is widely used in receptor binding assays because it is commercially available in tritiated form (Lahti *et al.,* 1985). Another highly selective κ-agonist, PD117302, is also structurally related to U-50,488 (Halfpenny *et al.,* 1989).

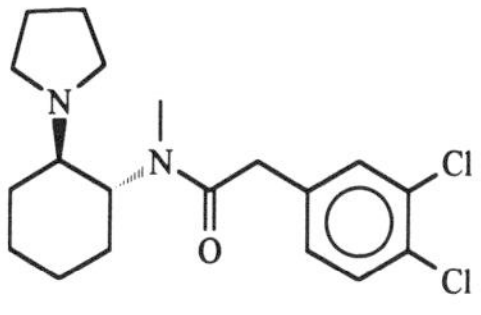

29. U50,488

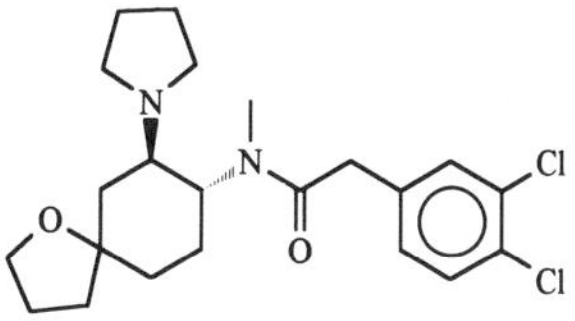

30. Spiradoline

Dynorphin (DYN) related peptides are endogenous agonist ligands for the κ receptor (Chavkin and Goldstein, 1981). From the same propeptide, prodynorphin, are generated peptides of differing lengths and with different receptor selectivites. DYN(1–8), DYN(1–13), and DYN(1–17) (**31**) are distributed in the central nervous system of vertebrates in physiologically significant concentrations (Cuello, 1983). Nonpeptide κ-receptor ligands, resembling opioids with selectivity to other receptor types, possess one amino group necessary for receptor binding. SAR studies of dynorphin peptide analogs have shown that the basic residues Arg^{7}, Lys^{11}, and Lys^{13} are important for high κ-receptor selectivity and/or potency (James, 1986). Successful substitutions of lipophilic residues at position 8 with D-amino acids (Yoshino *et al.,* 1990) or position 10 with D-Pro (Gairin *et al.,* 1984) suggest that reverse turns are present in the active conformations of DYN-related peptides. Introduction of such reverse turns formed the basis for the design of cyclic DYN analogs (Kawasaki *et al.,* 1993) of high potency (e.g., **32**), although with κ/μ selectivity comparable to that of native compounds. Nevertheless, these compounds showed some differences in κ-receptor subtype binding.

Peptide analogs with κ-antagonist potency have not yet been discovered. The most potent and selective antagonist for the κ receptor is norbinaltorphamine (norBNI, **33**) (Portoghese *et al.*, 1987), a bivalent ligand derived from naltrexone (Lipkowski *et al.*, 1986a). Despite the bivalent structure, probably only a single pharmacophore within the bivalent ligand plays the role of the opioid message; the second is probably responsible for topographical positioning of the amino group that serves as an essential part of the basic dynorphin address. Indeed, hybridization of naloxone with the dynorphin address sequence (e.g., **34**) (Lipkowski *et al.*, 1986b, 1988) or attachment of the address sequence to NTI (e.g., **35**) (Olmsted *et al.*, 1993) produces analogs with high κ selectivity.

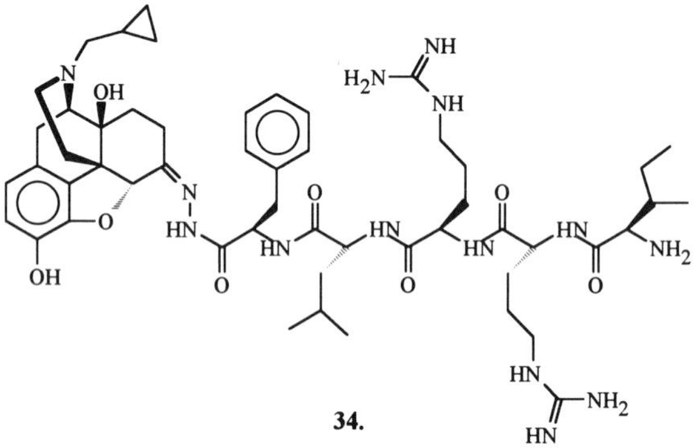

34.

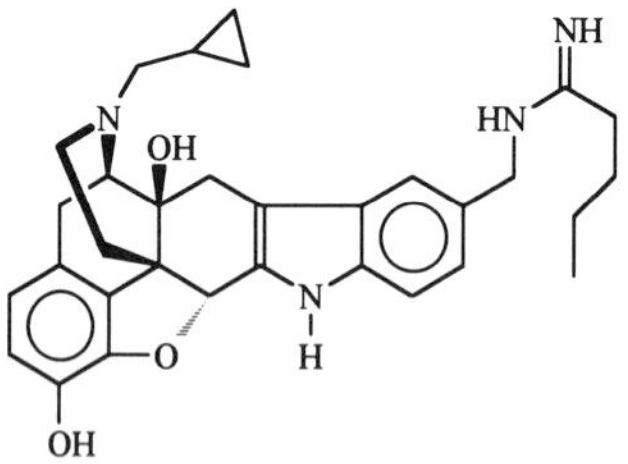

35.

The division of opioid peptide receptors into μ, δ, and κ types is widely used in current SAR studies of opioid analogs. However, there has emerged growing evidence of the existence of a much more complicated mosaic of opioid receptors. Two μ-receptor subtypes (Wolozin and Pasternak, 1981; Pasternak and Wood, 1986), two subtypes of δ receptors (Rothman *et al.*, 1984; Porreca *et al.*, 1992), and three κ receptor subtypes (Clark *et al.*, 1989) have been suggested. These subtypes, not all of which were proposed on the basis of uniform criteria, remain to be better characterized. Accordingly, SAR studies of opioid ligands with respect to such receptor subtypes are still rudimentary.

IV. Tachykinins

The endogenous mammalian tachykinins substance P (SP, **36**), neurokinin A (NKA, **37**) and neurokinin B (NKB, **38**), exert a variety of effects in the central nervous system as well as in the periphery, where they act on a number of smooth muscle and glandular tissues (Maggio, 1988). All members of the tachykinin family share close sequence similarity at the carboxyl terminus (Erspamer, 1981), a region required to evoke the biological effects characteristic of this group of neuropeptides (Konecka *et al.*, 1981; Lipkowski *et al.*, 1980; Kitagawa *et al.*, 1979). Tachykinins interact specifically with at least three neurokinin receptor types, termed NK_1, NK_2, and NK_3 (Guard and Watson, 1991). The N-terminal portions of tachykinins serve as address sequences that confer receptor selectivity. Thus, substance P expresses selectivity to the NK_1 receptor, neurokinin A to NK_2, and neurokinin B to NK_3. Neurokinin receptor subtypes were identified on the basis of rank ordering of potencies of tachykinin agonists and desensitization experiments (Lee *et al.*, 1982). All three NK receptors have been cloned and shown to be homologous members of the G-protein-coupled, seven-transmembrane helix receptor superfamily.

Arg-Pro-Lys-Pro-Gln-Gln-Phe-Phe-Gly-Leu-Met-NH_2

36. Substance P

His-Lys-Thr-Asp-Ser-Phe-Val-Gly-Leu-Met-NH_2

37. Neurokinin A

Asp-Met-His-Asp-Phe-Phe-Val-Gly-Leu-Met-NH_2

38. Neurokinin B

To probe the relation between receptor selectivity and biological effect, it is vital to develop highly selective and metabolically stable agonists and antagonists. In the case of NK_1 receptors (originally called SP-P receptors), the initial starting point was the endogenous peptide, substance P. Methyl esterification of the carboxyl-terminal amide of SP preserved affinity to the NK_1 receptor while decreasing by approximately 99.9% the affinity to the NK_2 (previously named SP-E) and NK_3 receptors (previously termed SP-N) (Watson *et al.*, 1983). The carboxyl-terminal hexapeptide analog of SP, [Glu^6]SP(6–11) (**39**) is equipotent with the native undecapeptide SP at the NK_1 receptor, and it has been widely used as a parent compound in SAR studies (Sandberg *et al.*, 1981; Lipkowski *et al.*, 1981a). The major disadvantage of using the analog is its low water solubility, the problem that can be overcome by adsorbing it to dextran (Lipkowski *et al.*, 1981b). Replacement of the Gly^9 residue with Pro preserved high affinity to the NK_1 receptor while reducing by 300-fold the affinity to other receptors (Laufer *et al.*, 1986). This analog, referred to as septide (**40**), is widely used as a selective NK_1 agonist. To improve resistance to enzymatic breakdown as well as to increase aqueous solubility, the sanktide analog, δAva[Pro^9, *N*-MeLeu^{10}] SP(7–11) has been synthesized (Hagan *et al.*, 1991a).

<Glu-Phe-Phe-Gly-Leu-Met-NH_2
39.

<Glu-Phe-Phe-Pro-Leu-Met-NH_2
40. Septide

Asp-Lys-Phe-Val-Gly-(N-Me)Leu-Nle-NH_2
41.

HOOC-CH_2-CH_2-CO-Asp-Phe-(N-Me)Phe-Gly-Leu-Met-NH_2
43. Senktide

Arg-D-Pro-Lys-Pro-Gln-Gln-D-Phe-Phe-D-Trp-Leu-Met-NH_2
44.

Elimination of the three *N*-terminal amino acid residues from neurokinin A did not alter the affinity or selectivity of the parent compound for the NK_2 receptor (Appell *et al.*, 1992). This minimal sequence has been used for further modifications. Replacement of Gly^8 with βAla resulted in [βAla^8]neurokinin A(4–10), a potent and selective compound. Interestingly Met^{10} can be successfully replaced with Nle; Ser^5 with Lys (Rovero *et al.*, 1989); and Leu^9 with *N*-MeLeu, resulting in a stable and selective agonist (**41**) now employed as a standard reagent. Incorporation of the γ-lactam unit between Gly^8 and Leu^9 to yield the analog [Lys^3,Gly^8,(*R*-γ-lactam)Leu^9]neurokinin A(3–10) (**42**) resulted in the most potent, selective, and enzymatically stable compound of this series, useful for *in vivo* NK_2 receptor studies (Overton *et al.*, 1992).

Replacement of Val^7 in the complete neurokinin B sequence with imino acid residues such as Pro or *N*-MePhe yielded a more selective agonist for NK_3 receptors (Lavielle *et al.*, 1988). Nevertheless, the most potent and selective compound for the NK_3 receptor is a carboxyl-terminal hexapeptide fragment analog of substance P, succinyl-[Asp^6,$MePhe^8$]SP(6–11), referred to as senktide (**43**). This compound has at least 10,000-fold higher affinity to the NK_3 receptor than to the NK_1 or NK_2 receptors (Wormser *et al.*, 1986).

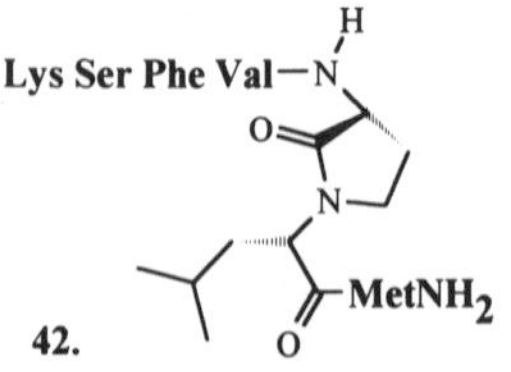

Lys Ser Phe Val—N(H) … $MetNH_2$
42.

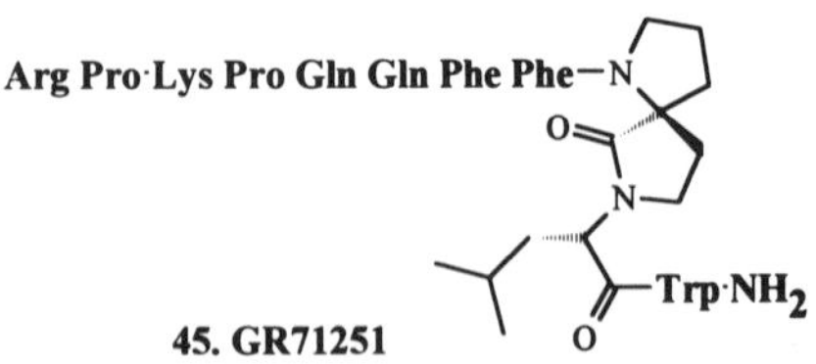

Arg Pro·Lys Pro Gln Gln Phe Phe—N … Trp·NH_2
45. GR71251

The first tachykinin antagonists were developed over 10 years ago by replacing Phe^7 and Gly^9 in substance P or SP(4–11) with D-aromatic acid residues, and changing the chirality of Pro^2 or Pro^4 (Folkers *et al.*, 1982). Such antagonists (e.g., **44**) have low affinity and poor selectivity but were widely applied in early tachykinin receptor studies (Regoli *et al.*, 1985; Vaught, 1988) and served as a basis for subsequent development of selective NK_1 receptor antagonists. Incorporation of spirolactams into linear peptides homologous to substance P has yielded both selective agonists and antagonists for NK_1 receptors (Ward *et al.*, 1990). Agonists were produced when an (*R*)-spirolactam was used. Incorporation of the (*S*)-spirolactam to yield [DPro^9,(spiro-γ-lactam)Leu^{10},Trp^{11}]SP (GR71251, **45**) resulted in antagonist activity almost 1000-fold more selective for NK_1 than for NK_2 and NK_3 receptors. An analog of physalaemin with an identical modification, GR82334, was successfully used as an NK_1-selective antagonist *in vivo* (Hagan *et al.*, 1991b; Birch *et al.*, 1992). The search for crucial elements in peptide NK_1 antagonists resulted in the development of very short (dipeptide) ligands. The first of this series was FR113680 (**46**) (Morimoto *et al.*,

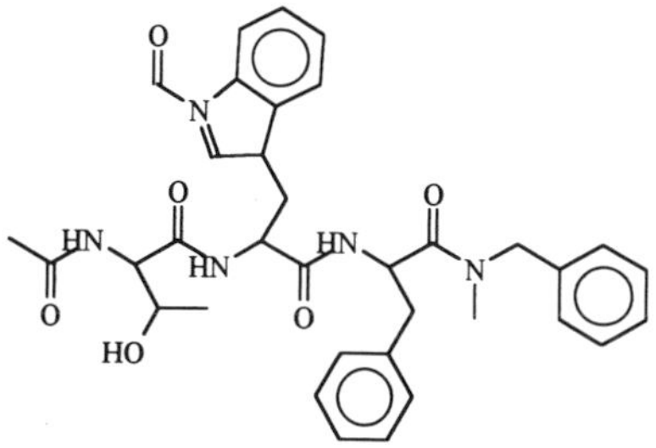

46. FR113680

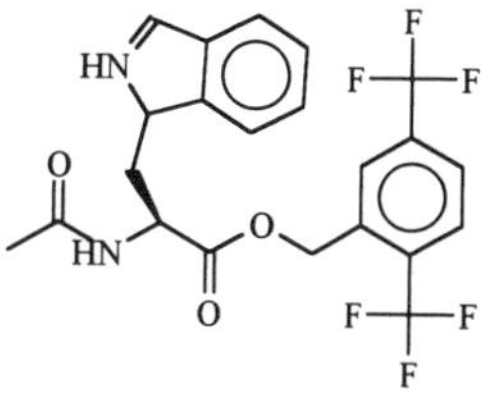

47.

1992a; Hagiwara *et al.*, 1992). N-Terminal replacement (e.g., *N*-acetylthreonine or N^{in}-formyl) can result in compounds with greater water solubility and oral bioavailability (Kucharczyk *et al.*, 1993). Further structural analysis of this series of analogs resulted in topographically related indolylcarbonyl di- and tripeptides (Hagiwara *et al.*, 1993). Interestingly, screening of peptide libraries to identify novel compounds with NK_1 antagonist properties led to the finding that *N*-Et-TrpOBzl is a weak NK_1 receptor inhibitor. Modification of this molecule resulted in **47,** with over 2000-fold increased affinity for the NK_1 receptor than the parent compound (MacLeod *et al.*, 1993).

Asp-Tyr-D-Trp-Val-D-Trp-D-Trp-Lys-NH_2
48. MEN 10,376

Ac-Leu-Met-Gln-Trp-Phe-Gly-NH_2
49. L-659,874

Ac-Leu-Asp-Gln-Trp-Phe-Gly-NH_2
50. R396

Asp-Ser-Phe-Trp-β-Ala-Leu-Met-NH_2
52.

Random amino acid substitutions within neurokinin A produced a series of linear peptide NK_2 antagonists (Maggi *et al.*, 1990), of which MEN 10,376 (**48**) is the most selective (Maggi *et al.*, 1991). Another series of selective NK_2 antagonists was based on the linear peptide sequence L-659,847 (**49**) (Williams *et al.*, 1988); replacement of Met^2 with Asp (R396, **50**) increases affinity and selectivity (Maggi *et al.*, 1990). Differences in the pharmacological profiles of MEN 10,376 and R396 suggest species differences between NK_2 receptors or the existence of subclasses of NK_2 receptors (Maggi *et al.*, 1992; Henderson *et al.*, 1992). Mixed NK_1/NK_2 antagonistic properties have been found for a macrocyclic peptide FK224 (**51**) derived from a natural product isolated from the fermentation broth of *Streptomyces violaceoniger* (Morimoto *et al.*, 1992b).

To date, no selective peptide antagonists for the NK_3 receptor have been described. Proposed analogs [DPro^2,DTrp^{6,8},Nle^{10}]NKB (Jacoby *et al.*, 1986) or [Gly^6]NKB (Hashimoto *et al.*, 1987) express low antagonist affinity as well as low selectivity for NK_3 receptors. Improved affinity has been reported with the neurokinin A-derived analog [Trp^7,β-Ala^8]NKA(4–10) (**52**) (Drapeau *et al.*, 1990). However, this compound also displays significant affinity to NK_1 and NK_2 receptors.

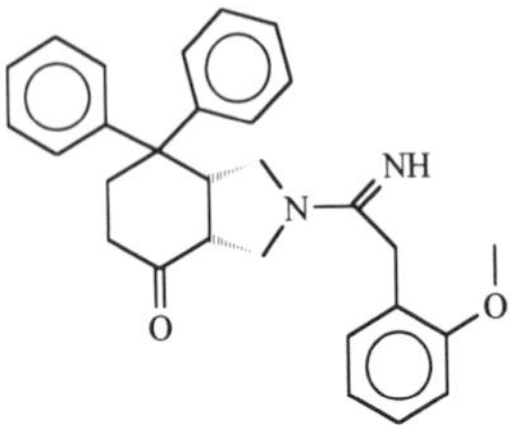

53. RP67580 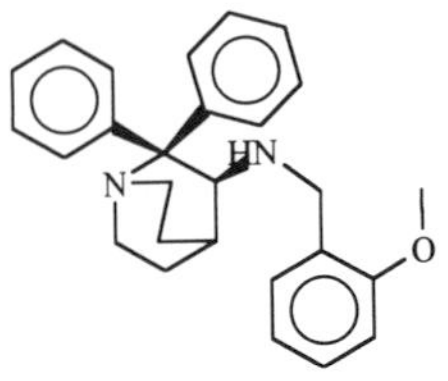 **54. CP96345** 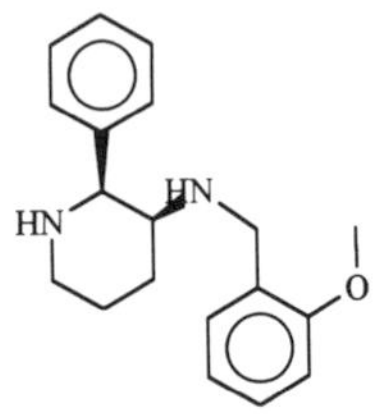 **55. CP99994**

A high throughput screening strategy led to discovery of the selective nonpeptide ligands RP67580 (**53**) (Garret *et al.*, 1991) and CP96345 (**54**) (Snider *et al.*, 1991), the first nonpeptide antagonists for tachykinin receptors. Both were very selective for the NK_1 receptor. SAR studies based on the structure of CP96345 led to the discovery of the highly potent and NK_1 receptor selective (+)-(2*S*,3*R*)-3-[(2-methoxybenzyl)amino]-2-phenylpiperidine, CP99994 (**55**) (Rosen *et al.*, 1993). Other NK_1-receptor-selective antagonists (WIN51,708, **56**; WIN62,577, **57**) were discovered through random screening of library compounds (Venepalli *et al.*, 1992). The pharmacological profiles of the two compounds illustrate the species diversity of the NK_1 receptor. Both compounds possess nanomolar affinity to the guinea pig brain NK_1 receptor but are practically inactive for the human brain NK_1 receptor (Appell *et al.*, 1992). A similar screening approach to ligand development led to the discovery of the very potent NK_2 receptor antagonist SR48968 (**58**) (Advenier *et al.*, 1992; Emonds-Alt *et al.*, 1993; Hale *et al.*, 1993).

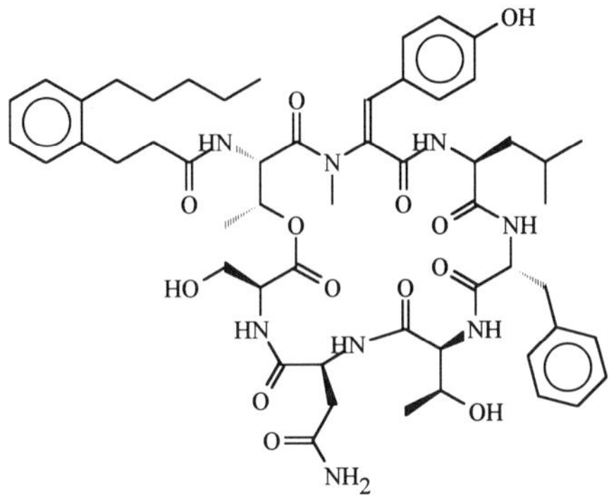

51. FK224

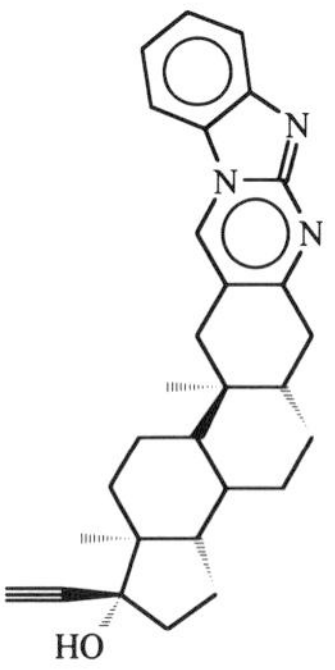

56. WIN51,708

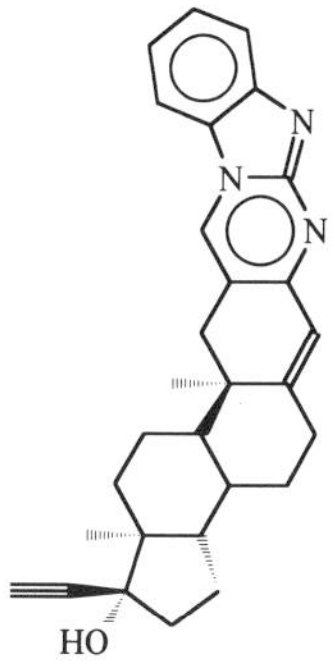

57. WIN62,577

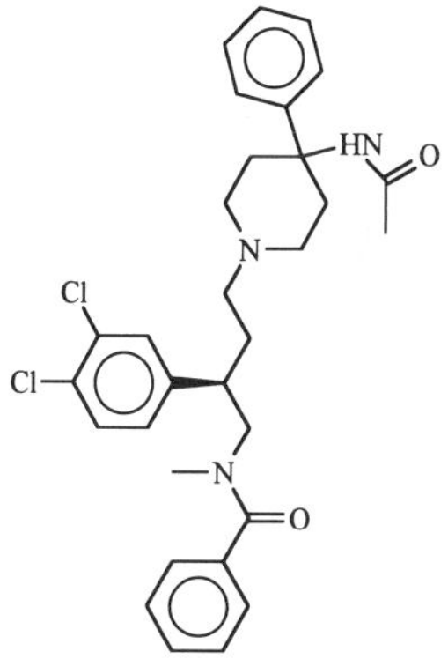

58. SR48968

V. Angiotensin II

The renin–angiotensin system plays a crucial role in the hormonal regulation of blood pressure and fluid volume homeostasis (Timmermans *et al.,* 1993; Rubin and Levin, 1994). Drugs designed to interfere with this system are effective for the treatment of hypertension and congestive heart failure. Captopril and enalapril, clinically useful angiotensin-converting enzyme (ACE) inhibitors, both block the conversion of angiotensin I (AI, **59**) to the potent vasoconstrictor angiotensin II (AII, **60**). However, the application of such compounds is limited by side effects related to inhibition of the metabolism of other neuropeptides (e.g., bradykinin) Lindgren and Andersson, 1989). Specific inhibition of the terminal step in the renin-angiotensin cascade, namely, the action of AII at its receptor, offers potentially more selective, and hence better tolerated, antihypertensive drugs

(Timmermans *et al.*, 1991). Therefore, SAR studies of AII have focused on developing selective antagonists for its receptor. One of the AII receptor subtypes, AT_1, is a particular focus of selective antagonist development because AT_1 exclusively mediates the hypertensive effects of AII. The physiological role of a second AII receptor subtype, AT_2, has not yet been fully elucidated, but appears to involve actions within the central nervous system (CNS).

Asp-Arg-Val-Tyr-Ile-His-Pro-Phe-His-Leu
59. Angiotensin I

Asp-Arg-Val-Tyr-Ile-His-Pro-Phe
60. Angiotensin II

Structure–activity relationship studies of ATII analogs have reached a consensus that the amino acid at position 8 is crucial in determining whether the entire sequence is an agonist or antagonist. Replacement of the native phenylalanine at position 8 of AII either by the D enantiomer (Samanen *et al.*, 1988) or a nonaromatic, lipophilic L-amino acid (Ile, Ala) (Khosla *et al.*, 1973) yields potent AII antagonists. Replacement of Asp^1 with Sar enhances potency in series of both agonists or antagonists. Although peptidic antagonists such as saralasin [Sar^1,Ala^8]AII (**61**) (Moore and Fulton, 1984) were shown to be potent and specific antagonists of AII at its receptor, the short duration of action and partial agonist properties limit their utility. Recently synthesized AII linear analogs are promising; for example, sarmesin (**62**), a saralasin analog, has been proposed as a selective antagonist without agonist activity (Hondrelis *et al.*, 1991).

Sar-Arg-Val-Tyr-Ile-His-Pro-Ala
61. Saralasin

Sar-Arg-Val-Tyr(Me)-Ile-His-Pro-Phe
62. Sarmesin

Sar-Arg-Hcy-Tyr-Hcy-His-Pro-Phe
63.

Sar-Arg-Hcy-Tyr-Mpt-His-Pro-Phe
Mpt, *trans*-4-mercaptoproline
64.

Sar-Arg-Cys-Tyr-Mpt-His-Pro-Phe
65.

Substitution of His in position 6 with Phe(4-NH_2) results in a selective ligand for the AT_2 receptor (Whitebread *et al.*, 1989). Elimination of the *N*-terminal tripeptides of AII (yielding Ac-Tyr-Val-His-Pro-Ile, CGP-37065) does not affect the affinity to the AT_2 receptor but dramatically decreases affinity to AT_1 receptor

(Greenlee and Siegl, 1991). Dimerization of analogs of AII at the carboxyl termini (CGP-37534) leads to a substantial loss of binding potency, confirming the importance of the carboxyl-terminal carboxylic acid (de Gasparo *et al.,* 1991).

Because AII, a linear peptide, is capable of adopting a large number of conformations (Nikiforovich *et al.,* 1987; Marchionini *et al.,* 1983), one objective has been to obtain a more constrained structure of AII that retains a high degree of biological activity. The introduction of a constrained amino acid (e.g., 5,5-dimethylthiazolidine-4-carboxylic acid, or cycloleucine) produced analogs with higher activity in both agonist and antagonist series (Samanen *et al.,* 1991). Conformational restriction by means of a disulfide bridge between substituted homocysteine or cysteine residues in position 3 and 5 resulted in potent agonists and antagonists (Spear *et al.,* 1990; Sugg *et al.,* 1990), consistent with the prediction of a γ or β turn in the center of the AII peptide chain in its active conformation. Further constraints of this type employing 4-*trans*-mercaptoproline (Mpt) in position 5 resulted in analogs with affinity to AT_2 receptors exceeding that of AII. None of the constrained analogs has thus far been reported to display particular selectivity for AT_1 versus AT_2 receptors. Nevertheless, the moderate differences in receptor selectivity between different cyclic analogs may imply differences in the conformational requirements for interaction with the two receptors. Whereas AII and saralasin have almost equal affinity to both AT_1 and AT_2 receptors, cyclo[Sar,$Hcy^{3,5}$]AII (**63**) has 60 times higher affinity to AT_1; cyclo[Sar,Hcy^3,Mpt^5]AII (**64**) has equal affinity to AT_1 and AT_2; and cyclo [Sar,Cys^3,Mpt^5]AII (**65**) has 20 times lower affinity to AT_1 than to AT_2 (Plucinska *et al.,* 1993).

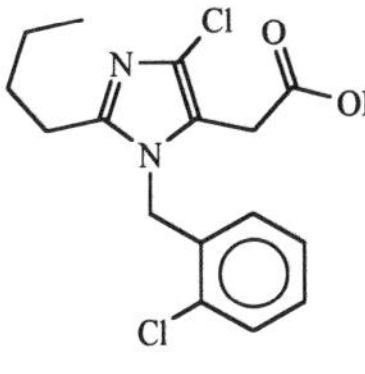

66. S8307

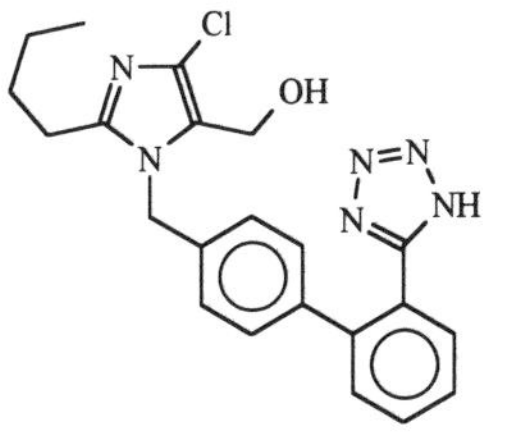

67. Losartan

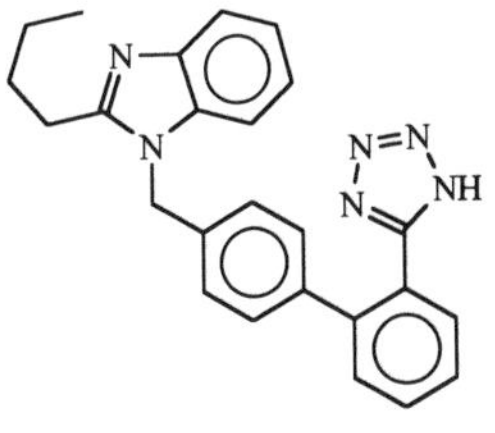

68.

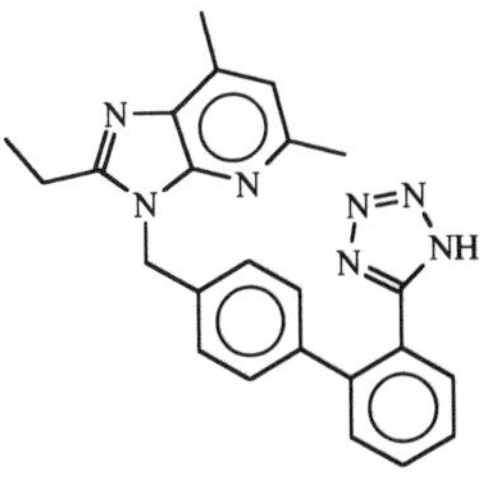

69.

The design of nonpeptide ligands for AT receptors started with patents of Furakawa *et al.* (1982a,b), who described two vasodilator imidazole-5-acetic derivatives (S8307, **66**; S8308) that antagonized AII effects. Both compounds were selective to AT receptors although they had very low affinity (Wong *et al.,* 1988). This basic structure was extensively modified by a group from DuPont who linked the imidazole to a biphenylyltetrazole moiety, leading to the discovery of the orally active DuP 753 (Losartan, **67**) which is now under evaluation in clinical trials (Duncia *et al.*, 1992). Losartan has high affinity for the AT_1 receptor. Losartan has served as the basis for most subsequent SAR studies of nonpeptide antagonists of AT_1 receptors. The imidazole group can be replaced with imidazolinone, mercaptotriazoles (Bandurco *et al.,* 1993), benzimidazoles (**68**), nicotinic acid (Winn *et al.,* 1993), imidazo[4,5*b*]pyridine (**69**) (Mantlo *et al.,* 1991), quinoline, 1,5-naphthyridine (Allott *et al.,* 1993), triazolones (Olins *et al.,* 1992), pyrrolidin-2-ones (Murray *et al.*, 1993), and cycloheptimidazolone (Bovy *et al.,* 1991). Biphenyl can be replaced by a similar group, for example, phenyl-1*H*-pyrrole (Bovy *et al.,* 1993) or biphenylmethoxy (Bradbury *et al.,* 1993). Interestingly, replacement of the tetrazole with a carboxyl group resulted in a significant loss of affinity to the AT_1 receptor subtype.

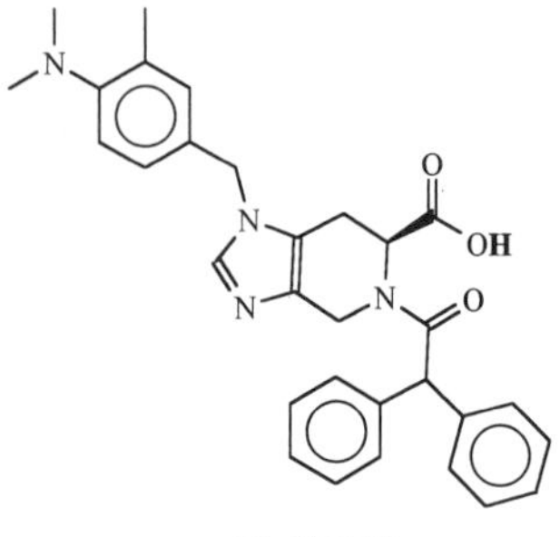

70. 123319

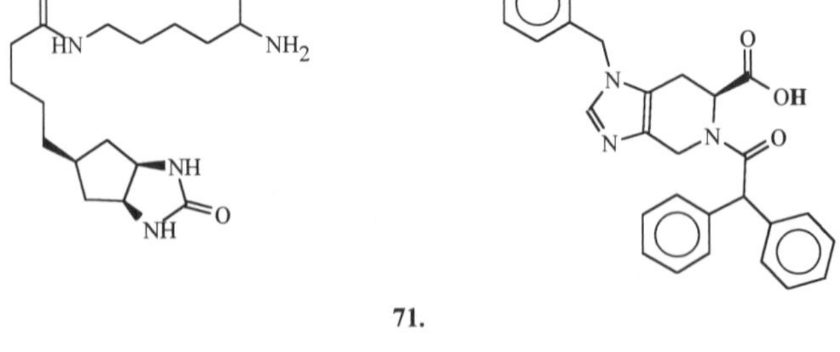

71.

The independently discovered compound PD 123319 (**70**) expressed high selectivity to the AT_2 receptor subtype (Dudley *et al.,* 1990). The *N*-dimethyl group in PD 123319 can be replaced not only by a primary amino group or by a methoxyl group, but also with a bulky fluorescent group or biotin (**71**) for use in affinity studies (Hodges *et al.*, 1993). Because replacements at this site by bulky lipophilic or basic groups do not lead to a loss of receptor affinity, it is likely that an empty space surrounds this position in the ligand–AT_2 receptor complex. Radiolabeled Losartan (Rivero *et al.*, 1993) and PD 123319 are widely used as selective ligands for the study of AT_1 and AT_2 receptors, respectively.

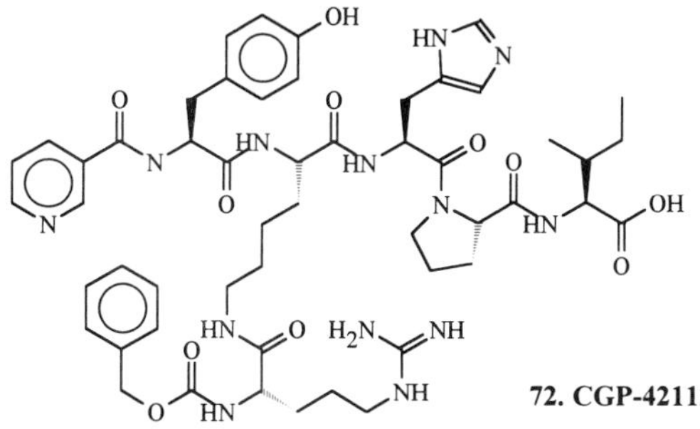

72. CGP-42112A

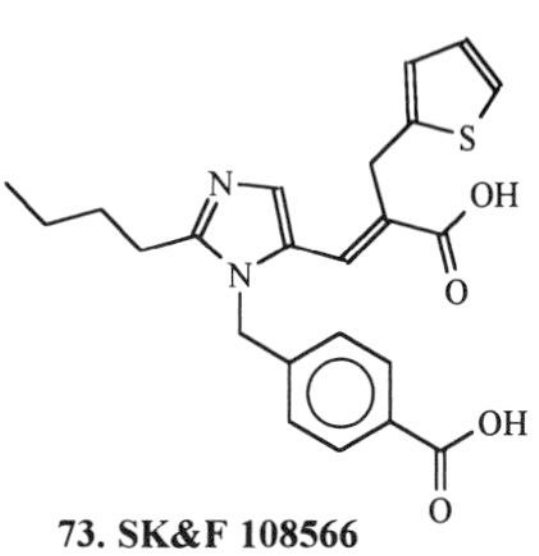

73. SK&F 108566

Studies of peptidomimetic compounds results in the synthesis of CGP-42112A (**72**), a ligand with high selectivity for the AT_2 receptor (Speth and Kim, 1990). The use of a molecular model of AII as a template for modifying the weak benzylimidazole AII antagonist S8307 by the addition of selected binding groups from AII resulted in the synthesis of SKF 108566 (**73**), an AII antagonist which has 40,000-fold greater affinity to AT_1 receptors than the parent compound (Weinstock *et al.*, 1991). Following the development of SKF 108566, highly potent peptide AII analogs appeared in which substituted amino acid residues in positions 3 and 5 formed a disulfide bridge. The previously proposed model of AII, which reflected work with the nonpeptide antagonist, was not consistent with the high potencies of the cyclic analogs, since the side chains in positions 3 and 5 in that model point in opposite directions, preventing formation of a disulfide bridge. Further study of possible steric similarities of the active elements of the nonpeptide antagonist and the cyclic peptide analog allowed the SmithKline group to propose a structure for the active peptide conformer related to the nonpeptide ligand (Samanen *et al.*, 1993). This sequence of events well illustrates how neuropeptide SAR studies and nonpeptide drug development may interact.

VI. Oxytocin and Vasopressin

Four types of neurohypophyseal hormone receptors have been characterized, three for vasopressin (**74**) (V_{1a}, V_{1b}, V_2) and one for oxytocin (OT, **75**) (Jard *et al.*, 1987; Manning and Sawyer, 1993). The neurohypophyseal hormones exert vasopressor actions through interactions with V_{1a} receptors, whereas V_{1b} receptors mediate the ACTH-releasing effect of vasopressin on the anterior pituitary. The

V_2 receptors, present in the renal tubule and collecting duct, mediate the antidiuretic response. OT receptors, present in the uterus and the mammary gland, mediate uterine contraction and milk let down. For simplicity, in many papers the affinity to V_{1b} receptors is omitted and the V_{1a} receptor is defined as V_1 (vascular) (Thibonnier, 1988).

Cys-Tyr-Phe-Gln-Asn-Cys-Pro-Arg-Gly-NH_2
74. Arg-vasopressin

Cys-Tyr-Ile-Gln-Asn-Cys-Pro-Leu-Gly-NH_2
75. Oxytocin

OT and vasopressin (VP) differ only at positions 3 and 8. Because of their sequence and structural similarities, they are able to interact with all four neurohypophyseal receptor types but with different selectivity. Since the original synthesis of OT (du Vigneaud *et al.,* 1954b) and VP (du Vigneaud *et al.,* 1954a) by du Vigneaud and colleagues, many selective agonists have been synthesized and applied as pharmacological probes and for clinical therapy (Sawyer and Manning, 1989). V_2 antagonists have potential therapeutic value to treat the syndrome of inappropriate antidiuretic hormone (ADH) secretion (Manning and Sawyer, 1989), and OT antagonists could be useful to suppress premature labor (Akerlund *et al.*, 1987). Most selective agonist analogs of OT and VP result from systematic modification of the native sequences of these neurohypophyseal hormones. In contrast, OT and VP antagonists fall into three classes: (1) cyclic or linear analogs of parent hormones: (2) peptides with origins distinct from group (1); and (3) nonpeptides. The primary study of OT and VP analogs fostered the dogma that cyclic peptides are necessary for agonist interactions with OT and VP receptors (Drabarek and Lipkowski, 1971; Lipkowski and Drabarek, 1972). Indeed, at present all widely available selective agonists of neurohypophyseal hormones are cyclic. Modification of cyclic parent structures resulted in analogs with antagonist properties (**76**) (Flouret *et al.,* 1993). Therefore, it was a surprise (Manning *et al.,* 1987) to identify linear analogs of parent hormones that possessed antagonist properties for V_2 (**77**), V_{1a} (**78**) (Manning *et al.*, 1990), and OT receptors (Manning *et al.*, 1993).

Pmp-D-Trp-Ile-Gln-Asn-D-Cys-Pro-Arg-Gly-NH_2
Pmp, β,β-pentamethylene-β-mercaptopropionyl
76.

Pa-D-Tyr(Et)-Phe-Val-Asn-Nva-Pro-Arg-Arg-NH_2
Pa, propionyl
77.

Phaa-D-Tyr(Et)-Phe-Gln-Asn-Lys-Pro-Arg-NH_2
Phaa, phenylacetyl
78.

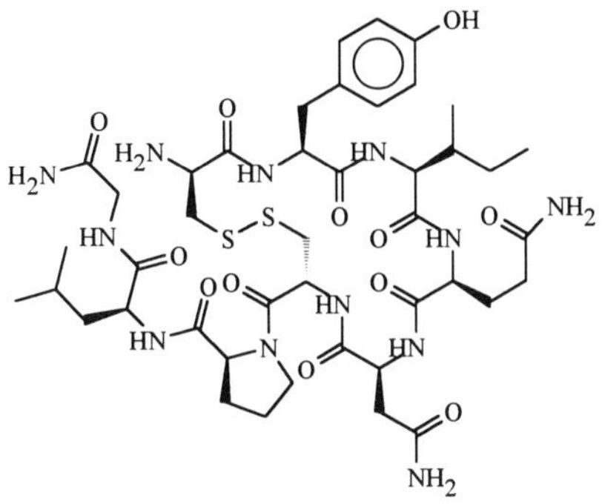

75. Oxytocin

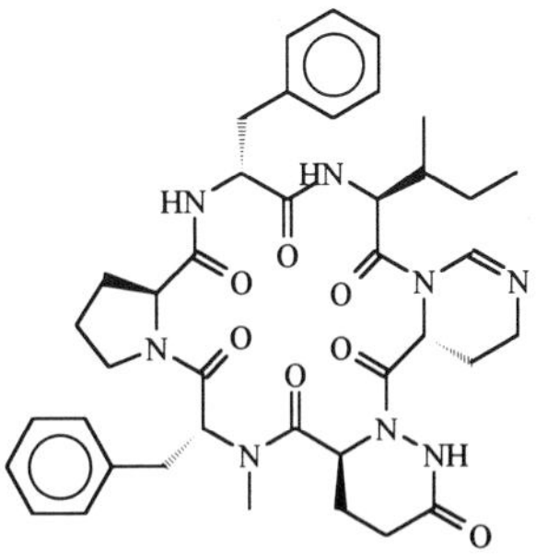

79.

A new structural class of OT antagonists, derived from a microbial source, was reported by Pettibone *et al.* (1989). Further modification of the original structure resulted in a highly selective OT antagonist (**79**) (Bock *et al.*, 1990). Pharmacological screening of small, nonpeptide molecules has resulted in the discovery of ligands for the three types of receptors: OT (**80,** Salituro *et al.*, 1993; **81,** Evans *et al.*, 1993), V_1 (**82,** Ogawa *et al.*, 1993), and V_2 (**83,** Yamamura *et al.*, 1992).

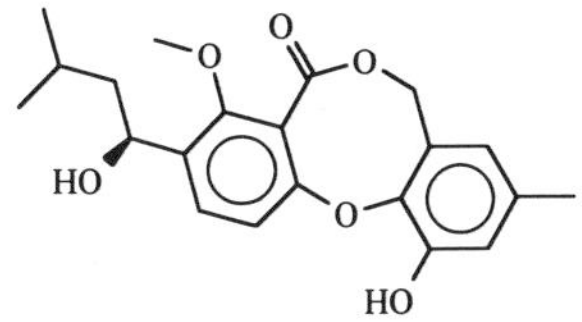

80.

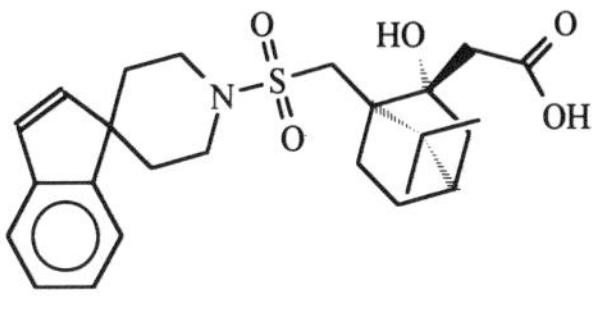

81. L-366,509

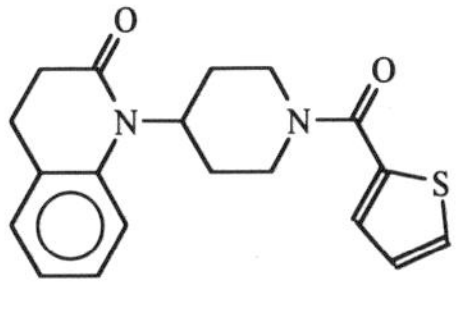

82. OPC-18549

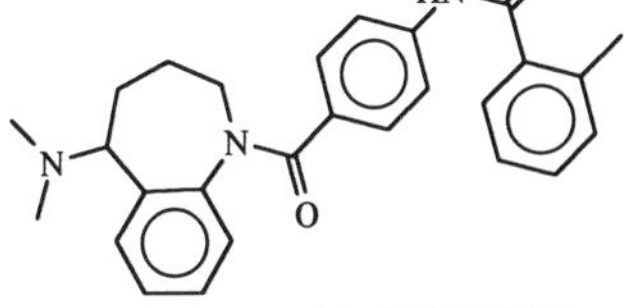

83. OPC-31260

VII. Cholecystokinin

Cholecystokinin (CCK, **84**) is a member of the brain–gut family of peptides. It mediates various peripheral biological functions including gallbladder contraction and pancreatic enzyme secretion (Jensen *et al.*, 1989). In addition, CCK has been implicated in central processes including analgesia (Wiertelak *et*

Lys-Ala-Pro-Ser-Gly-Arg-Met-Ser-Ile-Val-Lys-Asn-Leu-Gln-Asn-Leu-Asp-Pro-Ser-His-Arg-Ile-Ser-Asp-Arg-Asp-Tyr(SO_3H)-Met-Gly-Trp-Met-Asp-Phe-NH_2
84. CCK

Asp-Tyr(SO_3H)-Met-Gly-Trp-Met-Asp-Phe-NH_2
85. $CCK_{26\text{-}33}$

al., 1992), appetite regulation, and modulation of dopaminergic pathways (Morley, 1982; Figlewich *et al.*, 1989). Receptors for CCK are usually classified into two types: CCK-A (alimentary), found predominantly in peripheral tissues such as the pancreas and gallbladder, and CCK-B (brain), localized in the central nervous system (Steigerwalt and Williams, 1984). The gastrin receptor, cloned from gastric tissues (Kopin *et al.*, 1992), has a ligand-binding profile similar to that of the CCK-B receptor and has been identified with it (Dourish and Hill, 1987; Moran *et al.*, 1986). Nevertheless, some authors identify the peripheral gastrin receptor as CCK-B_1 and consider that the central CCK receptor, termed CCK-B_2, is a second subtype of the CCK-B receptor (Makovec *et al.*, 1992).

The CCK-A and CCK-B receptors can be differentiated by their relative affinities for CCK and its fragments. Whereas CCK and CCK(26–33) (CCK-8, **85**) bind potently to both the peripheral and brain CCK receptors, Boc-CCK-4, the carboxyl-terminal tetrapeptide, exhibits high affinity only for the CCK-B receptor. The desulfated octapeptide (CCK-8-DS) has significantly less potency for the pancreatic (CCK-A) receptor. Several CCK-8 and CCK-4 derivatives containing *N*-methylnorleucine at positions 28 and/or 31 are highly selective for the CCK-B receptor (Hruby *et al.*, 1990). Therefore, one approach to the design of further CCK peptide agonists has been based on the modification of Boc($Nle^{28,31}$)CCK-8 (BDNL, **86**). Interestingly, replacement of Met^{28} with

Boc-Asp-Tyr(SO_3H)-Nle-Gly-Trp-Nle-Asp-Phe-NH_2
86. BDNL

HOOC-CH_2-CH_2-CO-Tyr(SO_3H)-Met-Gly-Trp-Met-Asp-FA
FA, β-phenethylamine
87. Ge 410

(2*R*,3*R*)β-MePhe in CCK-8 practically eliminates affinity for both CCK-A and CCK-B receptors, but substitution of (2*R*,3*S*)β-MePhe in the same position results in high affinity for the CCK-B receptor, with no significant affinity to the CCK-A receptor (Hruby *et al.*, 1993). Receptor binding profiles and agonistic properties of CCK peptide analogs can be modulated by modifying the side chains of CCK-8 and CCK-4 related peptides. Replacing Nle^{31} with Orn led to full agonists. When the amine function in the side chain of Orn^{31} was benzyloxycarbonylated, the resulting peptides were mixed antagonists (Gonzalez-Muniz *et al.*, 1990). Introduction of Phe in position 31 of BDNL increased its affinity for the CCK-B receptor (Roques *et al.*, 1993) but decreased the CCK-B receptor affinity of respective tetrapeptide analogs (Corringer *et al.*, 1993). Interestingly, alkylation of the carboxyl-terminal amide of BDNL-related agonists transforms them into antagonist compounds. BDNL analogs in which the carboxyl-terminal -PheNH_2 or Asp-PheNH_2 is replaced with 2-phenylethyl alcohol (Fulcrand *et al.*, 1988) or 3-amino-phenylheptanoic acid (Amblard *et al.*, 1993) retain their affinities as agonists at peripheral CCK receptors. On the other hand, the succinyl-related compound (Ge 410, **87**) is a potent antagonist at CCK-A receptor (Boomgaarden *et al.*, 1992).

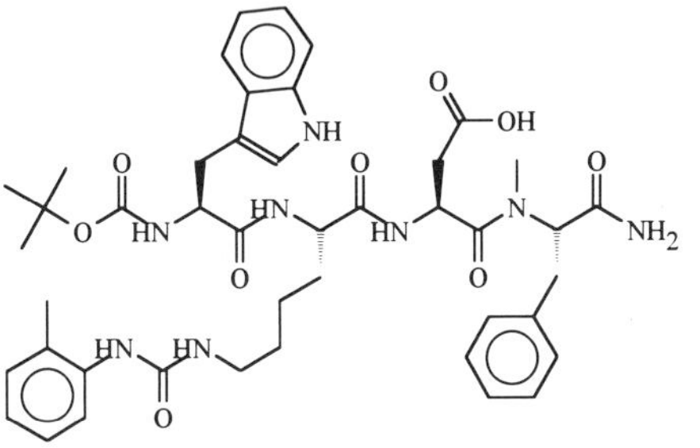

88. A-71623

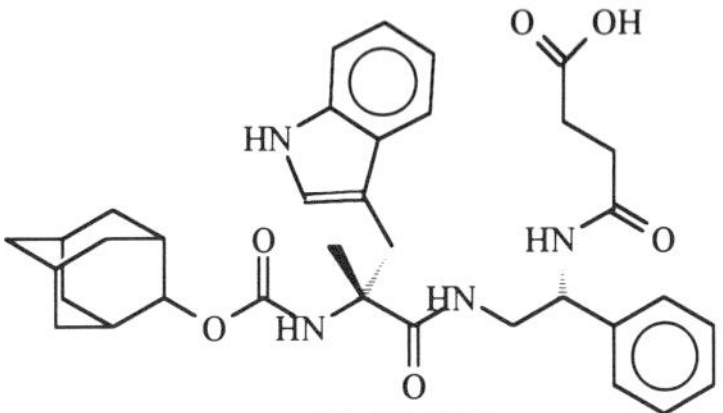

89. CI-988

A template approach has yielded agonist ligands for the CCK-A receptor. Modified tetrapeptides, such as A-71623 (**88**), are the shortest fragments possessing agonist activity that bind with high affinity to the CCK-A receptor (Shiosaki *et al.*, 1990, 1991). In this series of modified tetrapeptide CCK-A agonists, the necessary elements are aromatic groups connected with the ε-side group of Lys^{31} through an amide, thioamide, or urea (Shiosaki *et al.*, 1993). The aromatic group in this series of agonists is probably related to phenol of Tyr^{27} in the native sequence of CCK. These tetrapeptide compounds, unlike longer peptide analogs of CCK, do not require an acidic moiety for potency. Further, in contrast to CCK-8, these tetrapeptides are roughly 1000-fold selective for the CCK-A receptor. For example, A-71623 binds to the pancreatic (CCK-A) receptor with an IC_{50} of 3.7 n*M* and to the cerebral cortical (CCK-B) receptor with an IC_{50} of 4500 n*M*.

A similar template strategy has been successfully applied to develop "dipeptoids" that exhibit antagonistic properties for both CCK-A and CCK-B receptors. First, SAR analysis of CCK-8 resulted in characterization of the topographically important groups for interaction with these receptors. The combination of optimal groups in a dipeptoid resulted in CI-988 (**89**), a high-affinity selective antagonist for the CCK-B receptor (Horwell *et al.*, 1991). Further modifications of CI-988 produced compounds selective for CCK-A or having "mixed selectivity" for CCK-A and CCK-B (Boden *et al.*, 1993). It is interesting that two enantiomers within this series have opposite selectivity (Higginbottom *et al.*, 1993).

All those new agonists and antagonists have been constructed on the basis of SAR studies of biologically active CCK fragments. Parallel, practically independent studies have produced a number of nonpeptide ligands for both CCK-A and

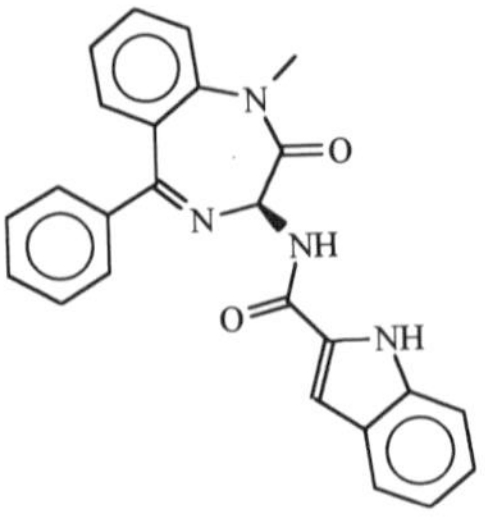

90. MK-329 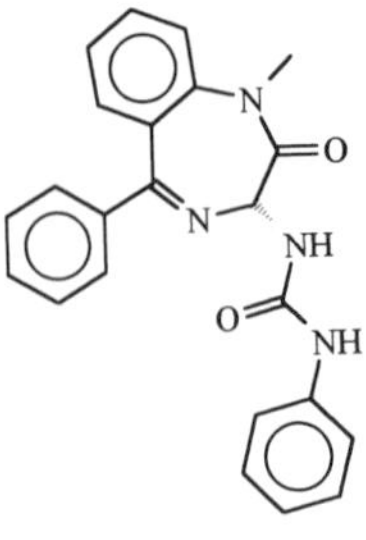 **91. L-365,260**

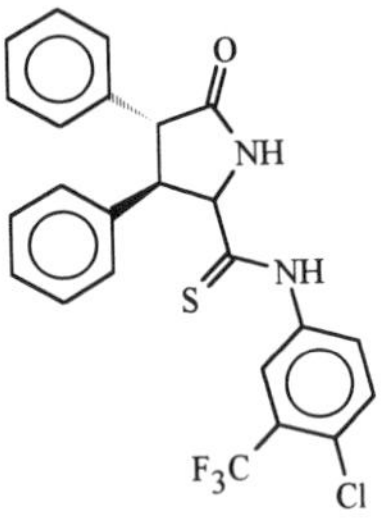

92. LY219057 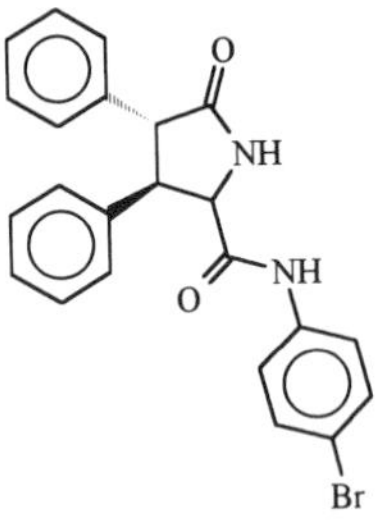 **93. LY288513**

CCK-B receptors. Among the first classes of CCK nonpeptide antagonists to be reported were 1,4-benzodiazepines such as MK-329 (**90**) (Evans *et al.*, 1986), selective for CCK-A, and L-365,260 (**91**) (Bock *et al.*, 1989), selective for CCK-B. Various benzodiazepine analogs have been proposed as nonpeptide CCK receptor ligands (Bock *et al.*, 1993). Other nonpeptide CCK receptor ligands are derived from diphenylpyrazolidinones; these include LY219057 (**92**) (Howbert *et al.*, 1991), selective for CCK-A and LY288513 (**93**) (Rasmussen *et al.*, 1991), selective for CCK-B receptors. As for the dipeptoid antagonists described above, the receptor selectivity of diphenylpyrazolidinone-related analogs depends strongly on their chirality. Selectivity for the CCK-B receptor appears to correlate most strongly with (4*S*,5*R*) stereochemistry at the phenyl groups, whereas selectivity for CCK-A is associated with (4*R*,5*S*) stereochemistry. Although the chemical origins of benzodiazepine and diphenylpyrazolidinone classes of CCK receptor antagonists are different, the predicted low-energy conformations show remarkable topographical homology between the two structures (Howbert *et al.*, 1993).

VIII. Conclusions

We live in a time of revolutionary progress in neuropeptide research. A number of endogenous peptides generated by neurons have been newly recognized and isolated. Cloning and characterization of neuropeptide receptors are now almost routine. Physicochemical and mathematical methods have allowed modeling of the active conformations of neuropeptide ligands during their interaction with receptors. Together, these powerful methods have set the stage for the creation of new potent and selective peptide analogs as well as peptide mimetics.

Because of their potential ease of large-scale synthesis, relative stability, and bioavailability through the oral route, nonpeptide ligands have become the goal of many medicinal chemists currently engaged in SAR studies of neuropeptides. As surveyed above, however, most currently available nonpeptide ligands for neuropeptide receptors have been found through large-scale biological screening of available organic compounds. The design of rigid nonpeptide ligands starting from SAR studies of flexible peptides must contend with differences in the mechanisms of interaction of rigid versus flexible ligands and their receptors. Peptides and other flexible ligands bind to receptors through a "zipper" mechanism (Burgen *et al.,* 1975), whereas rigid nonpeptide molecules appear to bind in an "induced-fit" fashion. Additionally, the possibility that nonpeptide ligands bind to other sites of the receptor than do peptide ligands must be taken into account (Schwartz *et al.,* 1993).

Perhaps for the above reasons, programs to deliberately synthesize novel peptomimetic ligands have, to date, generated only a few nonpeptide compounds (Olson *et al.,* 1993). Further, although a few successes in the generation of nonpeptide ligands have followed modeling of the active conformation(s) of neuropeptides during their receptor interactions, the refinement of active nonpeptide ligands often proceeds through SAR studies that focus on nonpeptide ligands (Freidinger, 1984). Results obtained from SAR studies of nonpeptide ligands have often, independently from SAR studies of peptides, extended knowledge of the topography and chemical requirements of active pockets within receptors, thereby advancing SAR studies of neuropeptide conformation and ultimately fostering the creation of new peptide analogs. The complementarity of both peptide and nonpeptide SAR studies is likely to result in advances in the design of both classes of compounds. Thus, whereas medicinal chemists are presently fascinated by the prospect of nonpeptide ligands for neuropeptide receptors, in our opinion the pharmacokinetic and pharmacodynamic properties of peptides, together with increasing clinical application of sophisticated means to target drug delivery, will continue to favor peptide analogs as potential drugs (Lipkowski *et al.,* 1994).

References

Advenier, C., Emonds-Alt, X., Vilain, P., Goulaouic, P., Proietto, V., Van Broeck, D., Naline, E., Neliat, G., Le Fur, G., and Breliore, J.-C. (1992). *Br. J. Pharmacol.* **105,** 77P.

Akerlund, M., Stromberg, P., Hauksson, A., Andersen, L. F., Lyndrup, J., and Melin, P. (1987). *Br. J. Obstet. Gynaecol.* **94,** 1040–1045.

Allott, C. P., Bradbury, R. H., Dennis, M., Fisher, E., Luke, R. W. A., Major, J. S., Oldham, A. A., Pearce, R. J., Reid, A. C., Roberts, D. A., Rudge, D. A., and Russell, S. T. (1993). *Bioorg. Med. Chem. Lett.* **3,** 899–904.

Amblard, M., Rodriguez, M., Lignon, M.-F., Galas, M.-C., Bernad, N., Artis-Noel, A.-M., Hauad, L., Laur, J., Califano, J.-C., Aumelas, A., and Martinez, J. (1993). *J. Med. Chem.* **36,** 3021–3028.

Amiche, M., Sagan, S. Mor, A., Delfour, A., and Nicolas, P. (1989). *Mol. Pharmacol.* **35,** 774–779.

Andrews, P. C., Brayton, K., and Dixon, J. E. (1987). *Experientia* **43,** 784–790.

Appell, K. C., Fragale, B. J., Loscig, J., Singh, S., and Tomczuk, B. E. (1992). *Mol. Pharmacol.* **41,** 772–778.

Bandurco, V. T., Murray, W. V., Gill, A., Addo, M., Lewis, J., Wachter, M. P., Hadden, S., Underwood, D. C., and Cheung, W.-M. (1993). *Bioorg. Med. Chem. Lett.* **3,** 375–379.

Battey, J. F., Way, J. M., Corjay, M. H., Shapira, H., Kusano, K., Harkins, R., Wu, J. M., Slattery, T., Mann, E., and Feldman, R. I. (1991). *Proc. Natl. Acad. Sci. U.S.A.* **88,** 395–399.

Bell, G. I., and Reisine, T. (1993). *Trends Neurosci.* **16,** 34–38, and references cited therein.

Bergsma, D. J., Ellis, C., Kumar, C., Nuthulaganti, P., Kersten, H., Eishourbagy, N., Griffin, E., Stadel, J. M., and Ayiar, N. (1992). *Biochem. Biophys. Res. Commun.* **183,** 989–995.

Birch, P. J., Beresford, I. J. M., Rogers, H., Hagan, R. M., Bailey, F., Hayes, A. G., Harrison, S. M., and Ireland, S. J. (1992). *Br. J. Pharmacol.* **105,** 134P.

Bock, M. G., DiPardo, R. M., Evans, B. E., Rittle, K. E., Whitter, W. L., Veber, D. F., Anderson, P. S., and Freidinger, R. M. (1989). *J. Med. Chem.* **32,** 13–16.

Bock, M. G., DiPardo, R. M., Williams, P. D., Pettibone, D. J., Clineschmidt, B. V., Ball, R. G., Veber, D. F., and Freidinger, R. M. (1990). *J. Med. Chem.* **33,** 2321–2323.

Bock, M. G., DiPardo, R. M., Veber, D. F., Chang, R. S. L., Lotti, V. J., Freedman, S. B., and Freidinger, R. M. (1993). *Bioorg. Med. Chem. Lett.* **3,** 871–874.

Boden, P. R., Higginbottom, M., Hills, D. R., Horwell, D. C., Hughes, J., Rees, D. C., Roberts, E., Singh, L., Suman-Chauhan, N., and Woodruff, G. N. (1993). *J. Med. Chem.* **36,** 552–565.

Boomgaarden, M., Henklein, P., Morgenstern, R., Sohr, R., Ott, T., and Martinez, J. (1992). *Eur. J. Med. Chem.* **27,** 955–959.

Bovy, P. R., O'Neal, J., Collins, J. T., Olins, G. M., Corpus, V. M., Burrows, S. D., McMahon, E. G., Palomo, M., and Koehler, K. (1991). *Med. Chem. Res.* **1,** 86–94.

Bovy, P. R., Reitz, D. B., Collins, J. T., Chamberlain, T. S., Olins, G. M., Corpus, V. M., McMahon, E. G., Palomo, M. A., Koepke, J. P., Smits, G. J., McGraw, D. E., and Gaw, J. F. (1993). *J. Med. Chem.* **36,** 101–110.

Bradbury, A. F., and Smyth, D. G. (1985). *In* "Biogenetics of Neurohormonal Peptides" (R. Hakanson and J. Thorell, eds.), pp. 171–186. Academic Press, London.

Bradbury, R. H., Allott, C. P., Dennis, M., Girdwood, A., Kenny, P. W., Major, J. S., Oldham, A. A., Ratcliffe, A. H., Rivett, J. E., Roberts, D. A., and Robins, P. J. (1993). *J. Med. Chem.* **36,** 1245–1254.

Burgen, A. S. V., Roberts, G. C. K., and Feeney, J. (1975). *Nature (London)* **253,** 753–755.

Carr, D. B., Lipkowski, A. W., and Silbert, B. S. (1991). *In* "Opioids in Anesthesia II" (F. G. Estafanous, ed.), pp. 3–19. Butterworth-Heinemann, Boston, and references cited therein.

Chang, K.-J., Wei, E. T., Killian, A., and Chang, J.-K. (1983). *J. Pharm. Exp. Ther.* **227,** 403–408.

Chang, T. L., Gainer, H., Russell, J. T., *et al.* (1982). *Endocrinology (Baltimore)* **11,** 1607–1614.

Chastrette, N., Cespuglio, R., and Jouvet, M. (1990). *Neuropeptides* **15,** 61–74.

Chavkin, C., and Goldstein, A. (1981). *Proc. Natl. Acad. Sci. U.S.A.* **78,** 6543–6547.

Chen, Y., Mestek, A., Liu, J., Hurley, J. A., and Yu, L. (1993). *Mol. Pharmacol.* **44,** 8–12.

Clark, J. A., Liu, L., Price, M., Hersh, B., Edelson, M., and Pasternak, G. W. (1989). *J. Pharm. Exp. Ther.* **251,** 461–468.

Corringer, P. J., Weng, J. H., Ducos, B., Durieux, C., Boudeau, P., Bohme, A., and Roques, B. P. (1993). *J. Med. Chem.* **36,** 166–172.

Cotton, R., Giles, M. G., Miller, L., Shaw, J. S., and Timms, D. (1984). *Eur. J. Pharmacol.* **97,** 331–332.

Cuello, A. C. (1983). *Br. Math. Bull.* **39,** 11–16.

de Gasparo, M., Whitebread, S., Kamber, B., Criscione, L., Thomann, H., Riniker, B., and Andreatta, R. (1991). *J. Recept. Res.* **11,** 247–253.

Delay-Goyet, P., Ruiz-Gayo, M., Baamonde, A., Gacel, G., Morgat, J.-L., and Roques, B. P. (1991). *Pharmacol. Biochem. Behav.* **38,** 155–162.

Dourish, C. T., and Hill, D. R. (1987). *Trends Pharmacol. Sci.* **8,** 207–210.

Drabarek, S., and Lipkowski, A. W. (1971). *Acta Physiol. Pol.* **22**(Suppl. 3), 737–754.

Drapeau, G., Rouissi, N., Nantel, F., Rhaleb, N.-E., Tousignant, C., and Regoli, D. (1990). *Regul. Pept.* **31,** 125–135.

Dudley, D. T., Panek, R. L., Major, T. C., Lu, G. H., Bruns, R. F., Klinkefus, B. A., Hodges, J. C., and Weishaar, R. E. (1990). *Mol. Pharmacol.* **38,** 370–377.

Duncia, J. V., Carini, D. J., Chiu, A. T., Johnson, A. L., Price, W. A., Wong, P. C., Wexler, R. R., and Timmermans, P. B. M. W. M. (1992). *Med. Res. Rev.* **12,** 149–191.

du Vigneaud, V., Gish, D. T., and Katsoyannis, P. G. (1954a). *J. Am. Chem. Soc.* **76,** 4751–4752.

du Vigneaud, V., Ressler, C., Swan, J. M., Katosoyannis, P. G., and Roberts, C. W. (1954b). *J. Am. Chem. Soc.* **76,** 3115–3121.

Eggericks, D., Raspe, E., Bertrand, D., Vassart, G., and Parmentier, M. (1992). *Biochem. Biophys. Res. Commun.* **187,** 1306–1313.

Eipper, B. A., Mains, R. E., and Glembotski, C. C. (1983). *Proc. Natl. Acad. Sci. U.S.A.* **80,** 5144–5148.

Emonds-Alt, X., Proietto, V., Broeck, D. V., Vilain, P., Advenier, C., Neliat, G., Le Fur, G., and Breliere, J.-C. (1993). *Bioorg. Med. Chem. Lett.* **3,** 925–930.

Erspamer, V. (1981). *Trends Neurosci.* **4,** 267–269, and references cited therein.

Eva, C., Keinanen, K., Monyer, H., Seeburg, P., and Sprengel, R. (1990). *FEBS Lett.* **271,** 81–84.

Evans, B. E., Bock, M. G., Rittle, K. E., DiParo, R. M., Whitter, W. L., Veber, D. F., Anderson, P. S., and Freidinger, R. M. (1986). *Proc. Natl. Acad. Sci. U.S.A.* **83,** 4918–4922.

Evans, B. E., Lundell, G. F., Gilbert, K. F., Bock, M. G., Rittle, K. E., Carroll, L. A., Williams, P. D., Pawluczyk, J. M., Leighton, J. L., Young, M. B., Erb, J. M., Hobbs, D. W., Gould, N. P., DiPardo, R. M., Hoffman, J. B., Perlow, D. S., Whitter, W. L., Veber, D. F., Pettibone, D. J., Clineschmidt, V., Anderson, P. S., and Freidinger, R. M. (1993). *J. Med. Chem.* **36,** 3993–4005.

Evans, C. J., Keith, D. E., Jr., Morrison, H., Magendzo, K., and Edwards, R. H., (1992). *Science* **258,** 1952–1955.

Figlewich, D. P., Sipols, A. J., Porte, D., Jr., Woods, S. C., and Liddle, R. A. (1989). *Am. J. Physiol.* **256,** R1313.

Flouret, G., Majewski, T., and Brieher, W. (1993). *J. Med. Chem.* **36,** 747–749.

Folkers, K., Horig, J., Rampold, G., Lane, P., Rosell, S., and Bjorkroth, U. (1982). *Acta Chem. Scand.* **B36,** 389–395.

Freidinger, R. M. (1984). *Trends Pharmacol. Sci.* **10,** 270–274, and references cited therein.

Fulcrand, P., Rodriguez, M., Galas, M. C., Lignon, M. F., Laur, J., Aumelas, A., and Martinez, J. (1988). *Int. J. Pept. Protein Res.* **32,** 384–395.

Furakawa, Y., Kishimoto, S., and Nisikawa, K. (1982a). *U.S. Patent* 4,340,598.

Furakawa, Y., Kishimoto, S., and Nisikawa, K. (1982b). *U.S. Patent* 4,355,040.

Gacel, G., Fournie-Zaluski, M.-C., Fellion, E., and Roques, B. P. (1981). *J. Med. Chem.* **24,** 1119–1124.

Gacel, G., Dauge, V., Breuze, P., Delay-Goyet, P., and Roques, B. P. (1988). *J. Med. Chem.* **31,** 1891–1897.

Gainer, H., Russell, J. T., and Loh, Y. P. (1984). *FEBS Lett.* **175,** 135–139.

Gainer, H., Russell, J. T., and Loh, Y. P. (1985). *Neuroendocrinology* **40,** 171–184.

Gairin, J. E., Gouarderes, C., Mazarguil, H., Alvinerie, P., and Cros, J. (1984). *Eur. J. Pharmacol.* **106,** 457–458.

Garret, C., Carruette, A., Fardin, V., Maoussaoui, S., Peyronel, J.-F., Blanchard, J.-C., and Laduron, P. M. (1991). *Proc. Natl. Acad. Sci. U.S.A.* **88,** 10208–10212.

Gonzalez-Muniz, R., Bergeron, F., Marseigne, I., Durieux, C., and Roques, B. P. (1990). *J. Med. Chem.* **33,** 3199–3204.

Gorin, F. A., and Marshall, G. R. (1977). *Proc. Natl. Acad. Sci. U.S.A.* **75,** 5179–5183.

Goromaru, T., Furuta, T., Baba, S., Yoshimura, N., Miyawaki, T., Sameshima, T., and Miyao, (1981). *J. Anesthesiol.* **55,** A173.

Greenlee, W. J., and Siegl, P. K. S. (1991). *Ann. Rep. Med. Chem.* **26,** 63–72, and references cited therein.

Guard, S., and Watson, S. P. (1991). *Neurochem. Int.* **18,** 149–165.

Hagan, R. M., Ireland, S. J., Bailey, F., McBride, C., Jordan, C. C., and Ward, P. (1991a). *Br. J. Pharmacol.* **102,** 168P.

Hagan, R. M., Ireland, S. J., Jordan, C. C., Beresford, I. J. M., Deal, M. J., and Ward, P. (1991b). *Neuropeptides* **19,** 127–135.

Hagiwara, D., Miyake, H., Morimoto, H., Murai, M., Fujii, T., and Matsuo, M. (1992). *J. Med. Chem.* **35,** 3184–3192.

Hagiwara, D., Miyake, H., Murano, K., Morimoto, H., Murai, M., Fujii, T., Nakanishi, I., and Matsuo, M. (1993). *J. Med. Chem.* **36,** 2266–2278.

Hale, J. J., Finke, P. E., and MacCoss, M. (1993). *Bioorg. Med. Chem. Lett.* **3,** 319–322.

Halfpenny, P. R., Hill, R. G., Horwell, D. C., Hughes, J., Hunter, J. C., Johnson, S., and Rees, D. C. (1989). *J. Med. Chem.* **32,** 1620–1626.

Handa, B. K., Land, A. C., Lord, J. A., Morgan, B. A., Rance, M. J., and Smith, C. F. (1981). *Eur. J. Pharmacol.* **70,** 531–540.

Hansen, P. E., and Morgan, B. A. (1984). *In* "The Peptides" (S. Udenfriend and J. Meienhofer, eds.), Vol. 6, pp. 269–321. Academic Press, and references cited therein.

Hasimoto, T., Uchida, Y., Naminohira, S., and Sakai, T. (1987). Jpn. J. Pharmacol. **45,** 570–573.

Henderson, A. K., Yamamura, H. I., Maggi, C. A., Buck, S. H., Van Giersbergen, P. L. M., and Roeske, W. R. (1992). *Eur. J. Pharmacol.* **225,** 175–178.

Hess, J., Borkowski, J., Young, G., Strader, C., and Ransom, R. (1992). *Biochem. Biophys. Res. Commun.* **184,** 260–268.

Higginbottom, M., Horwell, D. C., and Roberts, E. (1993). *Bioorg. Med. Chem. Lett.* **3,** 881–884.

Hodges, J. C., Edmunds, J. J., Nordblom, G. D., and Lu, G. H. (1993). *Bioorg. Med. Chem. Lett.* **3,** 905–908.

Hollt, V. (1983). *Trends Neurosci.* **6,** 24–26.

Hondrelis, J.,. Matsoukas, J., Cordopatis, P., Ganter, R. C., Franklin, K. J., and Moore, G. J. (1991). *Int. J. Pept. Protein Res.* **37,** 21–26.

Hook, V. Y. H. (1988). *Cell. Mol. Neurobiol.* **8,** 49–55.

Hook, V. Y. H., and Affolter, H. U. (1988). *FEBS Lett.* **238,** 338–342.

Horwell, D. C., Hughes, J., Hunter, J. C., Pritchard, M. C., Richardson, R. S., Roberts, E., and Woodruff, G. N. (1991). *J. Med. Chem.* **34,** 404–414.

Howbert, J. J., Mason, N. R., Bruns, R. F., Netterville, L. A., Schwartz, G. J., and Moran, T. H. (1991). *Soc. Neurosci. Abstr.* **17,** 491.

Howbert, J. J., Lobb, K. L., Britton, T. C., Mason, N. R., and Bruns, R. F. (1993). *Bioorg. Med. Chem. Lett.* **3,** 875–880.

Hruby, V. J. (1981). *In* "Topics in Molecular Pharmacology" (A. S. V. Burgen and G. C. K. Roberts, eds.), pp. 99–126. Elsevier North-Holland Publ., Amsterdam.

Hruby, V. J., and Gehrig, C. A. (1989). *Med. Res. Rev.* **9,** 343–401, and references cited therein.

Hruby, V. J., Fang, S., Knapp, R., Kazmierski, W., Lui, G. K., and Yamamura, H. I. (1990). *In* "Peptides: Chemistry, Structure and Biology" (J. E. Rivier and G. E. Marshall, eds.), p. 53 (Proceedings of the 11th American Peptide Symposium). ESCOM, Leiden.

Hruby, V. J., Sharma, S. D., Fang, S., Misicka, A., Lipkowski, A., Knapp, R. J., Yamamura, H. I., and Porreca Hadley, M. E. (1993). *In* "Peptide Chemistry 1992" (N. Yanaihara, ed.), pp. 461–465. (Proceedings of the 2nd Japan Symposium on Peptide Chemistry). ESCOM, Leiden, and references cited therein.

Huang, G. C., Page, M. J., Roberts, A. J., Malik, A. N., Spence, H., McGregor, A. M., and Banga, J. P. (1990). *FEBS Lett.* **264,** 193–197.

Hudson, D., Kenner, G. W., Sharpe, R., and Szelke, M. (1979). *Int. J. Pept. Protein Res.* **14,** 177–185.

Hughes, J., Smith, T. W., Kosterlitz, H. W., Fothergill, L. A., Morgan, B. A., and Morris, H. R. (1975). *Nature (London)* **258,** 577–579.

Humblet, C., and Mirzadegan, T. (1992). *Ann. Rep. Med. Chem.* **27,** 291–300.

Inouye, M., Inouye, S., Pollitt, S., *et al.* (1985). *In* "Protein Transport and Secretion" (M. J. Gething, ed.), p. 54. Cold Spring Harbor Laboratory, Cold Spring Harbor, New York.

Iorio, M. A., and Frigeni, V. (1984). *Eur. J. Med. Chem.* **19,** 301–303.

Iwai, N., Yamano, Y., Chaki, S., Konishi, F., Bardhan, S., Tibbetts, C., Sasaki, K., Hasegawa, M., Matsuda, Y., and Inagami, T. (1991). *Biochem. Biophys. Res. Commun.* **177,** 299–304.

Jacoby, H. I., Lopez, I., Wright, D., and Vaught, J. L. (1986). *Life Sci.* **39,** 1995–2003.

James, I. F. (1986). *In* "Opioid Receptors for the Dynorphin Peptides" (R. S. Rapaka and R. L. Hawks, eds.), pp. 192–208. NIDA Res. Monograph Ser. 70, NIDA, Rockville, Maryland, and references cited therein.

Janssen, P. A. J. (1962). *Br. J. Anaesth.* **34,** 260–265.

Janssen, P. A. J. (1982). *Acta Anesthesiol. Scand.* **26,** 262–268.

Janssen, P. A. J., and Tollenaere, J. P. (1979). *Adv. Biochem. Psychopharmacol.* **20,** 103–129, and references cited therein.

Jard, S., Barberis, C., Audigier, S., and Tribollet, E. (1987). *Prog. Brain Res.* **72,** 173–187.

Jensen, R. T., Wank, S. A., Rowley, W. H., Sato, S., and Gardner, J. D. (1989). *Trends Pharmacol. Sci.* **10,** 418–423.

Kawasaki, A. M., Knapp, R. J., Kramer, T. H., Walton, A., Wire, W. S., Hashimoto, S., Yamamura, H. I., Porreca, F., Burks, T. F., and Hruby, V. J. (1993). *J. Med. Chem.* **36,** 750–757.

Kazmierski, W. M., Wire, W. S., Lui, G. K., Knapp, R. J., Shook, J. E., Burks, T. F., Yamamura, H. I., and Hruby, V. J. (1988). *J. Med. Chem.* **31,** 2170–2177.

Khosla, M. C., Hall, M. M., Smeby, R. R., and Bumpus, F. M. (1973). *J. Med. Chem.* **16,** 1184–1185.

Kieffer, B. L., Befort, K., Gaveriaux-Ruff, C., and Hirth, C. G. (1992). *Proc. Natl. Acad. Sci. U.S.A.* **89,** 12048–12052.

Kitagawa, K., Ujita, K., Kiso, Y., Akita, T., Nakata, Y., Nakamoto, N., Segawa, T., and Yajima, H. (1979). *Chem. Pharm. Bull.* **27,** 48–57.

Knapp, R. J., Hawkins, K. N., Lui, G. K., Shook, J. E., Heyman, J. S., Porreca, F., Hruby, V. J., Burks, T. F., and Yamamura, H. I. (1990). *Adv. Pain Res. Ther.* **14,** 45–85.

Koch, G., and Brantl, V. (1990). *In* "β-Casomorphins and Related Peptides" (F. Nyberg and V. Brantl, eds.), pp. 43–52. Fyris-Tryck AB, Uppsala, Sweden.

Konecka, A. M., Sadowski, B., Sroczynska, I., Lipkowski, A. W., and Drabarek, S. (1981). *Gen. Pharmacol.* **12,** 119–121.

Kopin, A. S., Lee, Y. M., McBride, E. W., Miller, L. J., Lu, M., Lin, H. Y., Kolakowski, L. F., Jr., and Beinborn, M. (1992). *Proc. Natl. Acad. Sci. U.S.A.* **89,** 3605–3609.

Kosterlitz, H. W., Corbett, A. D., Gillan, M. G. C., McKnight, A. T., Paterson, S. J., and Robson, L. E. (1986). *In* "Opioid Peptides: Molecular Pharmacology, Biosynthesis, and Analysis" (R. S. Rapaka and R. L. Hawks, eds.), pp. 223–236. NIDA Res. Monograph Ser. 70, NIDA, Rockville, Maryland.

Kreil, G., Barra, D., Simmaco, M., Erspamer, V., Falconieri-Erspamer, G., Negri, L., Severini, C., Corsi, R., and Melchiorri, P. (1989). *Eur. J. Pharmacol.* **162,** 123–128.

Kucharczyk, N., Thurieau, C., Paladino, J., Morris, A. D., Bonnet, J., Canet, E., Krause, J. E., Regoli, D., Couture, R., and Fauchere, J.-L. (1993). *J. Med. Chem.* **36,** 1654–1661.

Lahti, R. A., Mickelson, M. M., McCall, J. M., and Von Voigtlander, P. F. (1985). *Eur. J. Pharmacol.* **109,** 281–284.

Laufer, R., Gilon, C., Chorev, M., and Selinger, Z. (1986). *J. Med. Chem.* **29,** 1284–1288.

Lavielle, S., Chassaing, G., Loeuillet, D., Robilliard, P., Marquet, A., Viret, J., Beaujouan, J.-C., Torrens, Y., Saffroy, M., Petitet, F., Dietl, M., and Glowinski, J. (1988). *Regul. Pept.* **22,** 108–114.

Lee, C. M., Iversen, L. L., Hanley, M. R., and Sandberg, B. E. B. (1982). *Naunyn-Schmiedeberg's Arch. Pharmacol.* **318,** 281–287.

Lenz, G. R., Evans, S. M., Walters, D. E., Hopfinger, A. J., and Hammond, D. L. (1986). "Opiates."

Academic Press, Orlando, Florida, and references cited therein.
Li, C. H. (1986). *In* "Opioid Peptides: Molecular Pharmacology, Biosynthesis, and Analysis" (R. S. Rapaka and R. L. Hawks, eds.), pp. 109–127. NIDA Res. Monograph Ser. 70, NIDA, Rockville, Maryland, and references cited therein.
Lindgren, B. R., and Andersson, R. G. G. (1989). *Med. Toxicol. Adverse Drug. Exp.* **4,** 369–380.
Lipkowski, A. W. (1987). *Pol. J. Pharmacol. Pharm.* **39,** 585–596.
Lipkowski, A. W., and Drabarek, S. (1972). *Wiad. Chem.* **26,** 159–177.
Lipkowski, A. W., and Misicka, A. (1989). *Acta Physiol. Pol.* **40**(Suppl. 34), 29–43, and references cited therein.
Lipkowski, A. W., Majewski, T., and Drabarek, S. (1980). *Pol. J. Chem.* **54,** 1707–1712.
Lipkowski, A. W., Drabarek, S., Majewski, T., Konecka, A. M., and Sadowski, B. (1981a). *Experienta* **37,** 49–50.
Lipkowski, A. W., Majewski, T., Drabarek, S., and Osipiak, B. (1981b). *In* "Peptides 1980" (K. Brunfeldt, eds.), pp. 530–533. Scriptor, Copenhagen.
Lipkowski, A. W., Nagase, H., and Portoghese, P. S. (1986a). *Tetrahedron Lett.* **27,** 4257–4260.
Lipkowski, A. W., Tam, S. W., and Portoghese, P. S. (1986b). *J. Med. Chem.* **29,** 1222–1225.
Lipkowski, A. W., Misicka, A., Portoghese, P. S., and Tam, S. W. (1988). *In* "Peptide Chemistry 1987" (T. Shiba and S. Sakakibara, eds.), Proc. 1st Japan Symp. Peptide Chem., pp. 709–712. Protein Research Foundation Osaka, Japan.
Lipkowski, A. W., Misicka, A., Hruby, V. J., and Carr, D. B. (1994). *Pol. J. Chem.* **68,** 907–912.
Loh, Y. P., Brownstein, M. J., and Gainer, H. (1984). *Annu. Rev. Neurosci.* **7,** 189–222.
McIntyre, P., Phillips, E., Skidmore, E., Brown, M., and Webb, M. (1993). *Mol. Pharmacol.* **44,** 346–355.
MacLeod, A. M., Merchant, K. J., Cascieri, M. A., Sadowski, S., Ber, E., Swain, C. J., and Baker, R. (1993). *J. Med. Chem.* **36,** 2044–2045.
Maggi, C. A., Patacchini, R., Giuliani, S., Rovero, P., Dion, S., Regoli, D., Giachetti, A., and Meli, A. (1990). *Br. J. Pharmacol.* **100,** 588–592.
Maggi, C. A., Giulaini, S., Ballati, L., Lecci, A., Manzini, S., Patacchini, R., Renzetti, A. R., Rovero, P., Quartara, L., and Giachetti, A. (1991). *J. Pharm. Exp. Ther.* **257,** 1172–1178.
Maggi, C. A., Eglezos, A., Quartara, L., Patacchini, R., and Giachetti, A. (1992). *Regul. Pept.* **37,** 85–93.
Maggio, J. E. (1988). *Annu. Rev. Neurosci.* **11,** 13–28.
Magnan, J., Paterson, S. J., Tavani, A., and Kosterlitz, H. W. (1982). *Naunyn-Schmiedeberg's Arch. Pharmacol.* **319,** 197–198.
Mains, R. E., and Eipper, B. A. (1979). *J. Biol. Chem.* **254,** 7885–7894.
Mains, R. E., Dickerson, I. M., May, V., Stoffers, D. A., Perkins, S. N., Ouafik, L. H., Husten, E. J., and Eipper, B. A. (1990). *Front. Neuroendocrinol.* **11,** 52–89.
Makovec, F., Peris, W., Revel, L., Giovannetti, R., Mennuni, L., and Rovati, L. C. (1992). *J. Med. Chem.* **35,** 28–38.
Manning, M., and Sawyer, W. H. (1989). *J. Lab. Clin. Med.* **114,** 617–632.
Manning, M., and Sawyer, W. H. (1993). *J. Rec. Res.* **13,** 195–214.
Manning, M., Przybylski, J. P., Olma, A., Klis, W. A., Kruszynski, M., Wo, N. C., Pelton, G. H., and Sawyer, W. H. (1987). *Nature (London)* **329,** 839–840.
Manning, M., Stoev, S., Kolodziejczyk, A., Klis, W. A., Kruszynski, M., Misicka, A., Olma, A., Wo, N. C., and Sawyer, W. H. (1990). *J. Med. Chem.* **33,** 3079–3086.
Manning, M., Stoev, S., Chan, W. Y., and Sawyer, W. H. (1993). *Ann. N.Y. Acad. Sci.* **689,** 219–232.
Mantlo, N. B., Chakravarty, P. K., Ondeyka, D. L., Siegl, P. K. S., Chang, R. S., Lotti, V. J., Faust, K. A., Chen, T. B., Schorn, T. W., Sweet, C. S., Emmert, S. E., Patchett, A. A., and Greenlee, W. J. (1991). *J. Med. Chem.* **34,** 2919–2922.
Marastoni, M., Salvadori, S., Tomatis, R., Borea, P. A., and Bertelli, G. (1987). *Il Farmaco* **42,** 125–131.
Marchionini, C., Maigret, B., and Premilat, S. (1983). *Biochem. Biophys. Res. Commun.* **112,**

339–246.

Martin, W. R., Eades, C. G., Thompson, J. A., Huppler, R. E., and Gilbert, P. E. (1976). *J. Pharm. Exp. Ther.* **197,** 517–532.

Martinez, J., and Potier, P. (1986). *Trends Pharmacol. Sci.* **7,** 139–147.

Marx, J. (1991). *Science* **252,** 779–780.

Michel, A., Villeneuve, G., and DiMaio, J. (1991). *J. Comput. Aided Mol. Des.* **5,** 553–569.

Misicka, A., Lipkowski, A. W., Horvath, R., Davis, P., Kramer, T. H., Yamamura, H. I., and Hruby, V. J. (1992a). *Life Sci.* **51,** 1025–1032.

Misicka, A., Lipkowski, A. W., Nikiforovich, G. V., Kazmierski, W. M., Knapp, R. J., Yamamura, H. I., and Hruby, V. J. (1992b). *In* "Peptides: Chemistry and Biology (J. A. Smith and J. E. Rivier, eds.), pp. 140–141 (Proc. 12th Am. Peptide Symp.). ESCOM, Leiden.

Misicka, A., Lipkowski, A. W., Horvath, R., Davis, P., Yamamura, H. I., Porreca, F., and Hruby, V. J. (1993). *In* "Peptides 1992" (C. H. Schneider and A. N. Eberle, eds.), Proc. 22th Eur. Peptide Symp., pp. 651–652. ESCOM, Leiden.

Moore, A. F., and Fulton, R. W. (1984). *Drug. Dev. Rev.* **4,** 331–249.

Moran, T. H., Robinson, P. H., Goodrich, M. S., and McHugh, P. R. (1986). *Brain Res.* **362,** 175–179.

Morimoto, H., Murai, M., Maeda, Y., Hagiwara, D., Miyake, H., Matsuo, M., and Fujii, T. (1992a). *Br. J. Pharmacol.* **106,** 123–126.

Morimoto, H., Murai, M., Maeda, Y., Yamaoka, M., Nishikawa, M., Kiyotoh, S., and Fujii, T. (1992b). *J. Pharm. Exp. Ther.* **262,** 398–402.

Morley, J. E. (1982). *Life Sci.* **30,** 479–493.

Mosberg, H. I., Hurst, R., Hruby, V. J., Gee, K., Yamamura, H. I., Galligan, J. J., and Burks, T. F. (1983). *Proc. Natl. Acad. Sci. U.S.A.* **80,** 5871–5874.

Murray, W. V., Lalan, P., Gill, A., Addo, M. F., Lewis, J. M., Lee, D. K. H., Wachter, M. P., Rampulla, R., and Underwood, D. C. (1993). *Bioorg. Med. Chem. Lett.* **3,** 369–374.

Nakanishi, S. (1991). *Annu. Rev. Neurosci.* **14,** 123–136.

Nikiforovich, G. V., Vesterman, B., Betins, J., and Podins, L. (1987). *J. Biomol. Struct. Dyn.* **4,** 1119–1135.

Ogawa, H., Yamamura, Y., Miyamoto, H., Kondo, K., Yamashita, H., Nakaya, K., Chihara, T., Mori, T., Tominaga, M., and Yabuuchi, Y. (1993). *J. Med. Chem.* **36,** 2011–2017.

Olins, G. M., Corpus, V. M., McMahon, E. G., Palomo, M. A., Schub, J. R., Blehm, D. J., Huang, H. C., Reitz, D. B., Manning, R. E., and Blaine, E. H. (1992). *J. Pharm. Exp. Ther.* **261,** 1037–1043.

Olmsted, S. L., Takemori, A. E., and Portoghese, P. S. (1993). *J. Med. Chem.* **36,** 179–180.

Olson, G. L., Bolin, D. R., Bonner, M. P., Bos, M., Cook, C. M., Fry, D. C., Graves, B. J., Hatada, M., Hill, D. E., Kahn, M., Madison, V. S., Rusiecki, V. K., Sarabu, R., Sepinwall, J., Vincent, G. P., and Voss, M. E. (1993). *J. Med. Chem.* **36,** 3039–3049.

Overton, P., Elliot, P. J., Hagan, R. M., and Clark, D. (1992). *Eur. J. Pharmacol.* **213,** 165–166.

Pasternak, G. W. (1988). "The Opiate Receptors." Humana, Clifton, New Jersey, and references cited therein.

Pasternak, G. W., and Wood, P. L. (1986). *Life Sci.* **38,** 1889–1898.

Pelton, J. T., Gulya, K., Hruby, V. J., Duckles, S. P., and Yamamura, H. I. (1985). *Proc. Natl. Acad. Sci. U.S.A.* **82,** 236–239.

Peters, G. R., Ward, N. J., Antal, E. G., Lai, P. Y., and DeMaar, E. E. (1987). *J. Pharm. Exp. Ther.* **240,** 128–134.

Pettibone, D. J., Clineschmidt, B. V., Anderson, P. S., Freidinger, R. M., Lundell, G. F., Koupal, L. R., Schwartz, C. D., Williamson, J. M., Goetz, M. A., Hensens, O. D., Liesch, J. M., and Springer, J. P. (1989). *Endocrinology (Baltimore)* **225,** 217–222.

Plucinska, K., Kataoka, T., Yodo, M., Cody, W. L., He, J. X., Humblet, C., Lunney, E., Major, T. C., Panek, R. L., Schelkun, P., Skeean, R., and Marshall, G. R. (1993). *J. Med. Chem.* **36,**

1902–1913.

Porreca, F., Takemori, A. E., Portoghese, P. S., Sultana, M., Bowen, W. D., and Mosberg, H. I. (1992). *J. Pharm. Exp. Ther.* **263,** 147–152.

Portoghese, P. S., Lipkowski, A. W., and Takemori, A. E. (1987). *Life Sci.* **40,** 1287–1292.

Portoghese, P. S., Sultana, M., Nagase, H., and Takemori, E. (1988). *J. Med. Chem.* **31,** 281–282.

Portoghese, P. S., Larson, D. L., Sultana, M., and Takemori, A. E. (1992). *J. Med. Chem.* **35,** 4325–4329.

Portoghese, P. S., Moe, S. T., and Takemori, A. E. (1993). *J. Med. Chem.* **36,** 2572–2574.

Rasmussen, K., Stockton, M. E., Czachura, J. F., and Howbert, J. J. (1991). *Eur. J. Pharmacol.* **209,** 135–138.

Regoli, D., Mizrahi, J., D'Orleans-Juste, P., Dion, S., Drapeau, G., and Echer, E. (1985). *Eur. J. Pharmacol.* **109,** 121–125.

Rice, M. C., and Wickner, W. T. (1985). *In* "Protein Transport and Secretion" (M. J. Gething, ed.), p. 44. Cold Spring Harbor Laboratory, Cold Spring Harbor, New York.

Rimland, J., Xin, W., Sweetnam, P., Saijol, K., Nestler, J., and Duman, R. (1991). *Mol. Pharmacol.* **40,** 869–875.

Rivero, R. A., Chakravarty, P. K., Chen, R., Greenlee, W. J., Rosegay, A., and Simpson, R. (1993). *Bioorg. Med. Chem. Lett.* **3,** 557–560.

Romer, D., Buscher, H. H., Hill, R. C., Maurer, R., Petcher, T. J., Welle, H. B. A., Bakel, H. C. C. K., and Akkerman, A. M. (1980). *Life Sci.* **27,** 971–977.

Ronsisvalle, G., Pasqiunucci, L., Pappalardo, M. S., Vittorio, F., Fronza, G., Romagnoli, C., Pistacchio, E., Spampinato, S., and Ferri, S. (1993). *J. Med. Chem.* **36,** 1860–1865.

Roques, B. P., Corringer, P.-J., Derrien, M., Dauge, V., and Durieux, C. (1993). *Bioorg. Med. Chem. Lett.* **3,** 847–850.

Rosen, T., Seeger, T. F., McLean, S., Desai, M. C., Guarino, K. J., Bryce, D., Pratt, K., and Heym, J. (1993). *J. Med. Chem.* **36,** 3197–3201.

Rothman, R. B., Bowen, W. D., Bykov, V., Schumacher, U. K., Pert, C. B., Jacobson, A. E., Burke, T. R., Jr., and Rice, K. C. (1984). *Neuropeptides* **4,** 201–215.

Rovero, P., Rhaleb, N.-E., Dion, S., Rouissi, N., Tousignant, C., Telemaque, S., Drapeau, G., and Regoli, D. (1989). *Neuropeptides* **13,** 263–270.

Rubin, S. A., and Levin, E. R. (1994). *J. Clin. Endocrinol. Metab.* **78,** 6–10.

Sakmar, T. P., Franke, R. R., and Khorana, H. G. (1989). *Proc. Natl. Acad. Sci. U.S.A.* **86,** 8309–8313.

Salituro, G. M., Pettibone, D. J., Clineschmidt, B. V., Williamson, J. M., and Zink, D. L. (1993). *Bioorg. Med. Chem. Lett.* **3,** 337–340.

Samanen, J., Narindray, D., Adams, W., Jr., Cash, T., Yellin, and Regoli, D. (1988). *J. Med. Chem.* **31,** 510–516.

Samanen, J., Cash, T., Narindray, D., Brandeis, E., Adams, W. J., Weideman, H., and Yellin, T. (1991). *J. Med. Chem.* **34,** 3036–3043.

Samanen, J. M., Peishoff, C. E., Keenan, R. M., and Weinstock, J. (1993). *Bioorg. Med. Chem. Lett.* **3,** 909–914.

Sandberg, B. E. B., Lee, C. M., Hanley, M. R., and Iversen, L. L. (1981). *Eur. J. Biochem.* **114,** 329–337.

Sawyer, W. H., and Manning, M. (1989). trends Endocrinol. Metab., 48–50.

Schiller, P. W. (1984). *In* "The Peptides" (S. Udenfriend and J. Meienhofer, eds.), Vol. 6, pp. 219–268. Academic Press, New York, and references cited therein.

Schiller, P. W., Nguyen, T. M.-D., Chung, N. N., and Lemieux, C. (1989). *J. Med. Chem.* **32,** 698–704.

Schiller, P. W., Nguyen, T. M.-D., Weltrowska, G., Wilkes, B. C., Marsden, B. J., Lemieux, C., and Chung, N. N. (1992). *Proc. Natl. Acad. Sci. U.S.A.* **89,** 11871–11875.

Schiller, P. W., Weltrowska, G., Nguyen, T. M.-D., Wilkes, B. C., Chung, N. N., and Lemieux, C. (1993). *J. Med. Chem.* **36,** 3182–3187.

Schwartz, T. W., Gether, U., Schambye, H. T., Nielsen, S. M., Elling, C., Jensen, C., Zoffmann, S.,

and Hjorth, S. (1993). *Abstr. 206th ACS Natl. Meeting, Chicago, August 1993*, 263.
Schwyzer, R. (1977). *Annu. N.Y. Acad. Sci.* **247,** 3–26.
Shiosaki, K., Lin, C. W., Kopecka, H., Craig, R., Wagenaar, F. L., Bianchi, B., Miller, T., Witte, D., and Nadzan, A. M. (1990). *J. Med. Chem.* **33,** 2950–2952.
Shiosaki, K., Lin, C. W., Kopecka, H., Tufano, M. D., Bianchi, B. R., Miller, T. R., Witte, D. G., and Nadzan, A. M. (1991). *J. Med. Chem.* **34,** 2837–2842.
Shiosaki, K., Lin, C. w., Leanna, M. R., Morton, H. E., Miller, T. R., Witte, D., Stashko, M., and Nadzan, A. M. (1993). *Bioorg. Med. Chem. Lett.* **3,** 855–860.
Shook, J. E., Pelton, J. T., Lemcke, P. K., Porreca, F., Hruby, V. J., and Burks, T. F. (1987). *J. Pharm. Exp. Ther.* **242,** 1–9.
Snider, R. M., Constantine, J. W., Lowe, J. A., Longo, K. P., Lebel, W. S., Woody, H. A., Drozda, S. E., Desai, M. C., Vinick, F. J., Spencer, R. W., and Hess, H. J. (1991). *Science* **251,** 435–439.
Spear, K. L., Brown, M. S., Reinhard, E. J., McMahon, E. G., Olins, G. M., Palomo, M. A., and Patton, D. R., (1990). *J. Med. Chem.* **33,** 1935–1940.
Speth, R. C., and Kim, K. H. (1990). *Biochem. Biophys. Res. Commun.* **269,** 997–1006.
Sreedharan, S. P., Robichon, A., Peterson, K. E., and Goetzl, E. J. (1991). *Proc. Natl. Acad. Sci. U.S.A.* **88,** 4986–4990.
Steigerwalt, R. W., and Williams, J. A. (1984). *Regul. Pept.* **8,** 51–59.
Sugg, E. E., Dolan, C. A., Patchett, A. A., Chang, R. S. L., Faust, K. A., and Lotti, V. J. (1990). *In* "Peptides: Chemistry and Biology" (J. E. Rivier and G. R. Marshall, eds.), pp. 305–306 (Proceedings of the 11th American Peptide Symposium). ESCOM, Leiden.
Szmuszkovicz, J., and Von Voigtlander, P. F. (1982). *J. Med. Chem.* **25,** 1126–1127.
Takeda, Y., Chou, K. B., Takeda, J., Sachais, B. S., and Krause, J. E. (1991). *Biochem. Biophys. Res. Commun.* **179,** 1232–1240.
Tam, S. W. (1985). *Eur. J. Pharmacol.* **109,** 33–41.
Thibonnier, M. (1988). *Kidney Int.* **34**(Suppl. 26), S48-S51, and references cited therein.
Timmermans, P. B. M. W. M., Wong, P. C., Chiu, A. T., and Herblin, W. F. (1991). *Trends Pharmacol. Sci.* **12,** 55–62, and references cited therein.
Timmermans, P. B. M. W. M., Wong, P. C., Chiu, A. T., Herblin, W. F., Benfield, P., Carini, D. J., Lee, R. J., Wexler, R. R., Saye, J. A. M., and Smith, R. D. (1993). *Pharmacol. Rev.* **45,** 205–251, and references cited therein.
Toth, G., Russell, K. C., Landis, G., Kramer, T. H., Fang, L., Knapp, R., Davis, P., Burks, T. F., Yamamura, H. I., and Hruby, V. J. (1992). *J. Med. Chem.* **35,** 2384–2391.
Vaught, J. L. (1988). *Life Sci.* **43,** 1419–1431.
Venepalli, B. R., Aimone, I. D., Appell, K. C., Bell, M. R., Dority, J. A., Goswami, R., Hall, P. L., Kumar, V., Lawrence, K. B., Logan, M. E., Scensny, P. M., Seelye, J. A., Tomczuk, B. E., and Yanni, J. M. (1992). *J. Med. Chem.* **35,** 374–378.
Ward, P., Ewan, G. B., Jordan, C. C., Ireland, S. J., Hagan, R. M., and Brown, J. R. (1990). *J. Med. Chem.* **33,** 1848–1851.
Watson, S. P., Sandberg, B. E. B., Hanley, M. R., and Iversen, L. L. (1983). *Eur. J. Pharmacol.* **87,** 77–84.
Weinstock, J., Keenan, R. M., Samanen, J., Hempel, J., Finkelstein, J. A., Franz, R. G., Gaitanopoulos, D. E., Girard, G. R., Gleason, J. G., Hill, D. T., Morgan, T. M., Peishoff, C. E., Aiyar, N., Brooks, D. P., Fredrickson, T. A., Ohlstein, E. H., Ruffolo, R. R., Jr., Stack, E. J., Sulpizio, A. C., Weidley, E. F., and Edwards, R. M. (1991). *J. Med. Chem.* **34,** 1514–1517.
Westmoreland, C. L., Hoke, J. F., Sebel, P. S., Hug, C. C., and Muir, K. T. (1993). *Anesthesiology* **79,** 893–903.
Whitebread, S., Mele, M., Kamber, B., and de Gasparo, M. (1989). *Biochem. Biophys. Res. Commun.* **163,** 284–291.
Wiertelak, E. P., Maier, S. F., and Watkins, L. R. (1992). *Science* **256,** 830–833.
Williams, B. J., Curtis, N. R., McKnight, A. T., Maguire, J., Foster, A., and Tridgett, R. (1988).

Regul. Pept. **22,** 189–195.

Winn, M., De, B., Zydowsky, T. M., Altenbach, R. J., Basha, F. Z., Boyd, S. A., Brune, M. E., Buckner, S. A., Crowell, D., Drizin, I., Hancock, A. A., Jae, H. S., Kester, J. A., Lee, J. Y., Mantei, R. A., Marsh, K. C., Novosad, E. I., Oheim, K. W., Rosenberg, S. H., Shiosaki, K., Sorensen, B. K., Spina, K., Sullivan, G. M., Tasker, A. S., von Geldern, T. W., Warner, R. B., Opgenorth, T. J., Kerkman, D. J., and DeBernardis, J. F. (1993). *J. Med. Chem.* **36,** 2676–2688.

Wolfe, P. B., and Wickner, W. T. (1984). *Cell (Cambridge, Mass.)* **36,** 1067–1072.

Wolozin, B. L., and Pasternak, G. W. (1981). *Proc. Natl. Acad. Sci. U.S.A.* **78,** 6181–6185.

Wong, P. C., Chiu, A. T., Price, W. A., Thoolen, M. J. M. C., Carini, D. J., Johnson, A. L., Taber, R. I., and Timmermans, P. B. M. W. M. (1988). *J. Pharm. Exp. Ther.* **247,** 1–7.

Wormser, U., Laufer, R., Hart, Y., Chorev, M., Gilon, C., and Selinger, Z. (1986). *EMBO J.* **5,** 2805–2808.

Yamamura, Y., Ogawa, H., Yamashita, H., Chihara, T., Myamoto, H., Nakamura, S., Onogawa, T., Yamashita, T., Hosokawa, T., Mori, T>, Tominaga, M., and Yabuuchi, Y. (1992). *Br. J. Pharmacol.* **105,** 787–791.

Yasuda, K., Raynor, K., Kong, H., Breder, C. D., Takeda, J., Reisine, T., and Bell, G. I. (1993). *Proc. Natl. Acad. Sci. U.S.A.* **90,** 6736–6740.

Yoshino, H., Kaneko, T., Arakawa, Y., Nakazawa, T., Yamatsu, K., and Tachibana, S. (1990). *Chem. Pharm. Bull.* **38,** 404–406.

Yoshida, H., Kakuchi, J., Guo, D. F., Furuta, H., Iwai, N., van der Meer-de Jong, R., Inagami, T., and Ichikawa, I. (1992). *Biochem. Biophys. Res. Commun.* **186,** 1042–1049.

8

Reversible Inhibitors of Serine Proteinases

Naturally Occurring Miniproteins, Semisynthetic Variants, Recombinant Homologs, and Synthetic Peptides

Herbert R. Wenzel and Harald Tschesche
Universität Bielefeld
Fakultät für Chemie/Biochemie I
D-33501 Bielefeld, Germany

Peptides: Synthesis, Structures, and Applications

I. Introduction

Proteinases are known to play a crucial role in the normal functioning of biological systems (Holzer and Tschesche, 1979). They are involved in important processes such as food digestion, blood coagulation and fibrinolysis, blood pressure regulation, fertilization, phagocytosis, and complement immune reactions. Certain proteinases release peptide hormones and neuromodulators from inactive precursors or degrade message-transmitting peptides, thus initiating or terminating a variety of biological responses.

Regarding the great number of potential peptide or protein substrates in an organism, it is clear that besides a more or less pronounced cleavage specificity a tight regulation of proteolytic activity by endogenous proteinase inhibitors is mandatory (Barrett and Salvesen, 1986). Endogenous inhibitors always appear to be proteins; small nonproteinaceous compounds are produced only by microorganisms (Umezawa, 1972). The number of proteinaceous inhibitors isolated and characterized so far is large and growing steadily. The vast majority are directed toward members of the serine proteinases, one of the four classes of endopeptidases. Serine proteinase inhibitors can be grouped into at least sixteen different inhibitor families on the basis of sequence similarity, topological relationships between the disulfide bridges, and the location of the binding site for the cognate proteinase (Laskowski and Kato, 1980; Laskowski, 1986; Bode and Huber, 1992).

Inhibitors belonging to four of these families were chosen as a basis for discussion of the different strategies for inhibitor engineering in this chapter. These families are the bovine pancreatic trypsin inhibitor family, the pancreatic secretory trypsin inhibitor family (including ovomucoid third domains), the squash seed inhibitors, and the potato inhibitor 1 family (barley proteinase inhibitor 2 and eglin c).

The current interest of many research groups in naturally occurring proteinase inhibitors has mainly emerged for two reasons. On the one hand, these proteins are ideally suited to study general aspects of protein conformation or folding (Creighton, 1978; Kim and Baldwin, 1990) and of protein–protein interaction (Laskowski, 1980; Laskowski *et al.*, 1983). On the other hand, detailed knowledge of the structure and reactivity of the inhibitors is indispensable for a thorough understanding of the controlling functions they exercise in a variety of fundamental physiological proteolytic processes, some of which have been mentioned above.

Moreover, several pathological situations have been related to the action of proteolytic enzymes. Pulmonary emphysema, for example, is a disease characterized by a decrease in physiological lung function due to the loss of elastic recoil (Mittman, 1972; Taylor and Mittman, 1987; Janoff, 1985, 1988). A proteinase/proteinase inhibitor imbalance hypothesis has been proposed as a model for the development of the disease: It suggests that either an increase in destructive proteolytic activity, mainly due to the release of proteinases from leukocytes, or a decrease in protective serum inhibitors, brought about, for example, by an environmental factor such as cigarette smoke or related to a hereditary deficiency, can lead to emphysema. The final stages involve the irreversible destruction of lung connective tissue. A similar model may apply to other related connective tissue diseases such as rheumatoid arthritis (Barrett, 1978; Baici *et al.*, 1982; Burkhardt *et al.*, 1987).

The possible use of exogenous natural or synthetic proteinase inhibitors to control proteolytic activities and thereby regulate certain disease states has been recognized for a long time; however, only since the 1980s have major successes actually been achieved (Rich, 1990; Grant *et al.*, 1993). Problems encountered with this therapeutic approach may be due to the reactivity, specificity, stability, bioavailability, toxicity, or immunogenicity of the inhibitors envisaged (Powers and Bengali, 1989; Travis and Fritz, 1991).

It is the purpose of this chapter to discuss, with only a few typical serine proteinase/inhibitor systems as examples, some approaches that might lead to satisfactory therapeutic compounds. As background information, the catalytic enzyme mechanism and basic concepts for the binding of substrates and proteinaceous inhibitors of proteinases are briefly reviewed first.

II. Serine Proteinase–Inhibitor Interactions

Serine proteinases are a class of proteolytic enzymes whose central catalytic machinery is composed of three invariant residues, an aspartic acid, a histidine, and a uniquely reactive serine, the last giving rise to their name. As a result of intensive study since the 1950s their mode of operation is now well understood in terms of the spatial atomic structures (Kraut, 1977; Steitz and Shulman, 1982; Bode and Huber, 1986). Nevertheless, this textbook knowledge is still subject to refinements and additions (Warshel *et al.*, 1989; Dutler and Bizzozero, 1989).

Most small protein inhibitors of serine proteinases interact with the cognate enzymes according to a common standard or canonical mechanism, which was mainly established by Laskowski and co-workers (Laskowski and Sealock, 1971; Laskowski and Kato, 1980).

A. Catalytic Mechanism of Hydrolysis

Hydrolysis of a peptide bond by a serine proteinase occurs by a two-step displacement with an amine produced first, followed by production of a carboxylic acid. The catalysis starts with the nucleophilic attack of the hydroxyl group in the side chain of a serine residue (S195 in chymotrypsin) on the partially positive carbonyl carbon of the amide bond in the substrate, thus forming a tetrahedral intermediate. This step is favored by the close proximity of the imidazole moiety of a histidine (H57 in chymotrypsin), which activates the hydroxyl group. The third component of the catalytic system or triad, the carboxyl group of an aspartic acid (D102 in chymotrypsin), is apparently necessary to position the imidazole moiety properly and to aid in proton transfer. Two backbone amide protons of the enzyme forming the so-called oxyanion hole are available for hydrogen bonding to the developing oxygen anion from the carbonyl group of the peptide bond undergoing attack. The tetrahedral intermediate then collapses to form the amine derived from the carboxyl end of the substrate and an ester intermediate between the serine hydroxyl group and the acyl portion of the substrate. Hydrolysis of the ester, which proceeds through a second tetrahedral intermediate, liberates the second product, the carboxylic acid derived from the amino end of the substrate.

All steps of the catalytic process are reversible. Thus, under proper experimental conditions, the reverse reaction, peptide bond formation, can be brought about and used for synthetic purposes (Chaiken *et al.*, 1982; Kullmann, 1985; Schellenberger and Jakubke, 1991).

B. Substrate Specificity

The system of notation introduced by Schechter and Berger (1967) has become generally accepted to describe the interaction of proteinases and their substrates or inhibitors. The peptide segment of the substrate or inhibitor, which is in contact with the active site of the enzyme, is designated, from the amino to the carboxyl terminus, $\ldots$-P_3-P_2-P_1-P_1'-P_2'-P_3'-$\ldots$, where the scissile bond is between P_1 and P_1'. The subsites of the proteinase, which accommodate the single residues, are termed S_3-S_2-S_1-S_1'-S_2'-S_3' by analogy. It has been found that for most serine proteinases the residue P_1 interacting with the S_1 subsite is the primary determinant in binding and subsequent cleavage. The preference for certain P_1 side chains can be rationalized at the molecular level by the complementary S_1 structure, which is also often called specificity pocket.

Trypsin (EC 3.4.21.4) preferentially cleaves peptides with lysine or arginine in position P_1. This can be attributed to a favorable ionic interaction between the positively charged side chains of the substrate and a negatively charged aspartic acid at the bottom of the specificity pocket of the enzyme.

Chymotrypsin (EC 3.4.21.1) is able to accommodate bulky hydrophobic side chains in its S_1 subsite, which lacks charged residues. It typically cleaves peptide bonds following phenylalanine, tyrosine, tryptophan, and leucine residues. Subtilisin (EC 3.4.21.14), a bacterial serine proteinase, has a spatial structure completely different from that of chymotrypsin. The active site groups of the two enzymes are, however, essentially identical, and their three-dimensional positions are nearly indistinguishable (convergent evolution). Subtilisin shares a preferential specificity for large P_1 residues with the mammalian enzyme.

Pancreatic elastase (EC 3.4.21.36) can digest native elastin, the elastic, fibrous protein in connective tissue. Amino acids with small side chains such as alanine are preferred in position P_1. A homologous elastase is found in azurophil granules of leukocytes (EC 3.4.21.37). Owing to a larger specificity pocket it prefers valine or leucine at P_1. Cathepsin G (EC 3.4.21.20), a proteinase with chymotrypsin-like cleavage specificity, is found in the same granules. Both enzymes are involved in pathological tissue destruction and are responsible, at least in part, for lung emphysema, arthritis, and inflammatory conditions.

This array of six proteinases provides the screening system for the search of potent and specific inhibitors by using the different approaches described in the next four sections. In special cases additional, highly specific enzymes are included: the four trypsin-like proteinases kallikrein from porcine pancreas (EC 3.4.21.35), kallikrein from human plasma (EC 3.4.21.34), endoproteinase Lys-C from *Lysobacter enzymogenes* (EC 3.4.21.—), and endoproteinase Arg-C from murine submaxillary glands (EC 3.4.21.—) and two enzymes which prefer glutamic acid as the P_1 residue, namely, *Staphylococcus aureus* V8 proteinase (EC 3.4.21.19) (Drapeau *et al.*, 1972) and *Streptomyces griseus* proteinase (EC 3.4.21.—) (Yoshida *et al.*, 1988).

C. Standard Mechanism of Inhibition

A large fraction of all small proteinaceous inhibitors (with the possible exception of members of the potato inhibitor 1 family lacking cystine bridges) interact with cognate enzymes according to the following minimum mechanism (Finkenstadt *et al.*, 1974):

$$E + I \underset{k_{off}}{\overset{k_{on}}{\rightleftharpoons}} EI \underset{k^*_{on}}{\overset{k^*_{off}}{\rightleftharpoons}} E + I^*,$$

where E is the proteinase, I and I* are the native single-chain inhibitor and the so-called modified inhibitor specifically cleaved at its reactive site peptide bond P_1–P_1', and EI is the stable 1:1 complex. The different rate constants are designated as k. The equilibrium hydrolysis constant defined as $K_{Hyd} = [I^*]/[I]$ is typically close to unity at physiological pH values, but increases sharply with increasing or decreasing pH (Niekamp *et al.*, 1969; Siekmann *et al.*, 1988a; Ardelt and Laskowski, 1991).

When judging by typical numerical values for the ratio k^*_{off}/K_D, where $K_D = k_{off}/k_{on}$, the inhibitors paradoxically appear to be good substrates. Both k^*_{off} and K_D values are, however, many orders of magnitude lower than the values for normal substrates, so that stable complexes are formed. [Values for the system bovine trypsin/aprotinin are as follows: $K_D \approx 10^{-13}$ *M*, $k^*_{off} \approx 10^{-10}$ sec^{-1} (Quast *et al.*, 1978).]

III. Naturally Occurring Proteinase Inhibitors

Protein inhibitors of serine proteinases are of widespread occurrence in nature. They have been isolated from numerous tissues of animals, from many plants—legume seeds, cereals, and tubers are especially rich sources—and from microorganisms (Tschesche, 1974; Eguchi, 1993; Ryan, 1981; Hiromi *et al.*, 1985). The vast collection of sequence data is continuously being supplemented by spatial atomic structures elucidated by X-ray or two-dimensional nuclear magnetic resonance (NMR) methods. Thus, for eleven of the sixteen inhibitor families at least one representative spatial structure is known to date (Read and James, 1986; Bode and Huber, 1991, 1992).

A. Bovine Pancreatic Trypsin Inhibitor Family

The Kunitz trypsin inhibitor from bovine pancreas, or aprotinin, the prototype of the bovine pancreatic trypsin inhibitor family, is probably the most extensively studied protein (Fritz and Wunderer, 1983; Gebhard *et al.*, 1986). Thus, it is not surprising that numerous reviews and fundamental papers are available, which cover all facets of research on aprotinin: historical aspects and chemical modifications (Tschesche, 1989), crystal structures derived from X-ray and neutron diffraction data (Wlodawer *et al.*, 1984, 1987), solution structures studied by NMR (Wagner, 1983; Wagner *et al.*, 1987; Berndt *et al.*, 1992), X-ray structures of trypsin and kallikrein complexes (Huber *et al.*, 1974; Chen and Bode, 1983), thermodynamic stability (Moses and Hinz, 1983), folding pathway (Goldenberg, 1992; Creighton, 1992; Weissman and Kim, 1992), calculation of conformation and molecular dynamics (Vasquez and Scheraga, 1988a,b; Swaminathan *et al.*, 1990; Hao *et al.*, 1993), immunohistochemical localization (Fritz *et al.*, 1979; Fiorucci *et al.*, 1989; Businaro *et al.*, 1989), biosynthesis, processing, and evolution (Creighton and Charles, 1987; Creighton *et al.*, 1993), and established and perspective therapeutic use (Fritz and Wunderer, 1983; Royston, 1990; Westaby, 1993).

The single polypeptide chain of aprotinin consists of 58 amino acid residues, which are cross-linked by three disulfide bridges (Fig. 1). It is folded in such a way that a pear-shaped molecular structure results. The chain is mainly organized in an extended amino-terminal segment, a reactive site loop, a three-

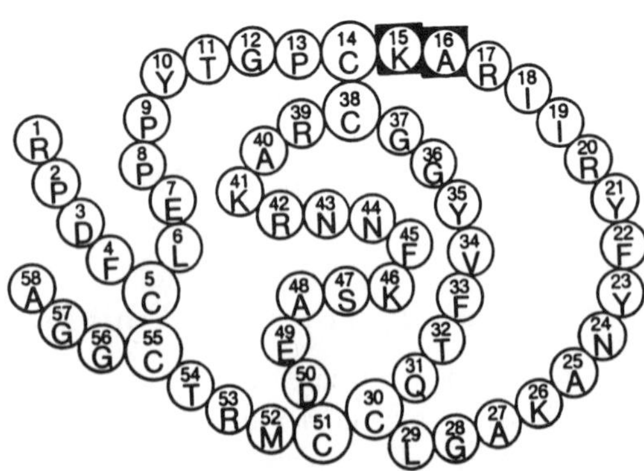

Figure 1 Primary structure of the bovine pancreatic trypsin inhibitor. The P_1 and P_1' residues are shown in squares (K15–A16).

stranded β-pleated sheet, and a short carboxyl-terminal α-helical segment (Fig. 2). The highly dipolar character of the basic aprotinin (isoelectric point close to 10.5) results from a concentration of the negatively charged groups (D3,

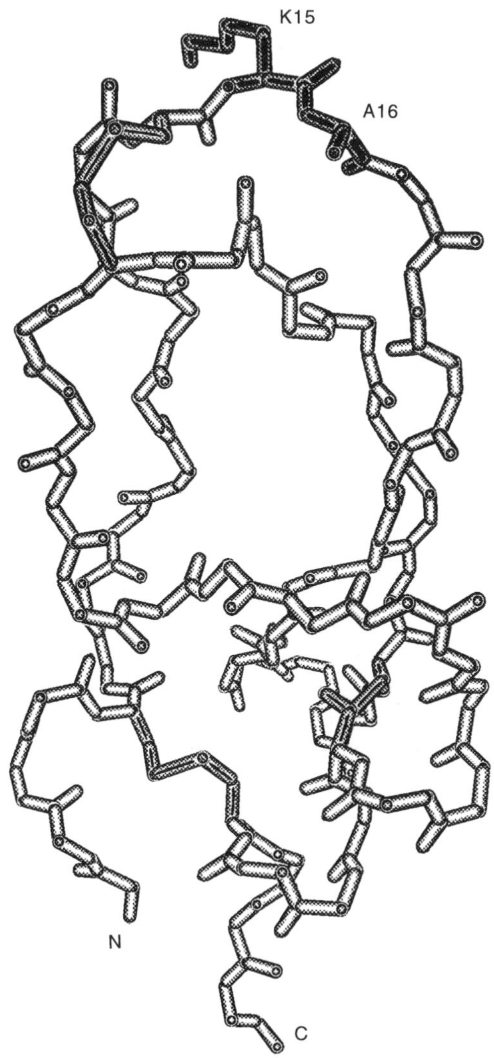

Figure 2 Spatial backbone structure of the bovine pancreatic trypsin inhibitor. The reactive site residues K15 and A16, including their side chains and the three disulfide bridges, are drawn with solid bonds; N denotes the amino and C the carboxyl terminus. The drawings of protein structures in this and the following figures were made with the program MOBY 1.5 (Springer-Verlag, Heidelberg).

E7, E49, D50, A58) at the bottom of the pear-shaped molecule. The conformation of the reactive site loop at the top around the scissile P_1–P_1' peptide bond K15–A16 is complementary to proteinase binding sites, thus allowing an unconstrained, tight interaction with several enzymes.

Aprotinin is an excellent inhibitor of bovine trypsin, mainly because of the positively charged side chain of its K15 residue, which is accommodated in the specificity pocket of the proteinase ($K_D = 6 \times 10^{-14}$ *M*). For the same reason, relatively strong complexes are formed with porcine pancreatic kallikrein ($K_D = 5 \times 10^{-10}$ *M*), human trypsin ($K_D = 2.0 \times 10^{-8}$ *M*), human plasma kallikrein ($K_D = 4.5 \times 10^{-8}$ *M*), endoproteinase Lys-C ($K_D = 6.7 \times 10^{-10}$ *M*), and, surprisingly, when taking into account its substrate specificity, also with bovine chymotrypsin ($K_D = 2 \times 10^{-8}$ *M*). Inhibition of endoproteinase Arg-C ($K_D = 1.8 \times 10^{-6}$ *M*), human leukocyte elastase ($K_D = 3 \times 10^{-6}$ *M*), and human cathepsin G ($K_D = 4 \times 10^{-6}$ *M*) is very weak; porcine pancreatic elastase is not inhibited at all.

Depending on breed, bovine organs contain, besides aprotinin, several Kunitz-type inhibitors in varying concentrations. One was shown to have an additional N-terminal pyroglutamic acid (Z-1) (Siekmann *et al.*, 1987). Its occurrence can be explained by a different proteolytic processing of the aprotinin precursor which, according to the aprotinin gene structure, has Q adjacent to R1 (Anderson and Kingston, 1983).

Other homologs are true isoinhibitors, like the following three with the indicated replacements[1]: T32P, A48S aprotinin, I18M, T32P aprotinin, and Z-1, I18M, T32P, R42S aprotinin (Siekmann *et al.*, 1988b). The most thoroughly investigated aprotinin isoinhibitor, also called bovine spleen inhibitor, has seven replacements: R17K, I18M, L29F, Q31E, R39K, R42S, K46R (Fioretti *et al.*, 1985; Wenzel *et al.*, 1985). It can also be isolated in three differently processed forms with short N- or C-terminal extensions (Barra *et al.*, 1987, 1991). Lower levels of the spleen inhibitor compared to aprotinin are found in bovine tissues. This can be correlated with differences in translational efficiency of the corresponding mRNAs (Gambacurta *et al.*, 1993). All these natural aprotinin homologous inhibitors display very similar activities toward serine proteinases in agreement with their great structural similarity, including the reactive site loop (Siekmann *et al.*, 1988b; Fioretti *et al.*, 1989).

Among the almost 50 Kunitz-family sequences currently tabulated in sequence databases, several other small potent trypsin inhibitors can be found. They were isolated from bovine serum, cow colostrum, snails, sea anemones, and certain snake venoms. All of them have a basic P_1 residue, either K or R, but differ considerably in their ability to inhibit porcine pancreatic kallikrein. To account for this it was surmised that additional basic residues were required in positions P_2', P_4', or P_{24}' for efficient kallikrein inhibition (Fritz and Wunderer, 1983).

[1] Amino acid replacements are designated by the original residue (wild type) using the one-letter code for the 20 coded amino acids, followed by the residue number and the new (mutant) residue, written out for noncoded amino acids.

With the X-ray structure determination of the kallikrein–aprotinin complex it became evident, however, that only R17 (P_2') forms favorable interactions with residues of the proteinase binding site (Chen and Bode, 1983).

Kunitz structures with their characteristic disulfide pattern can also be found as constituents of larger multidomain proteins. The long-known inter-α-trypsin inhibitor with its two closely related Kunitz domains (Ødum, 1990) has been joined by another plasma protein, the tissue factor pathway inhibitor, which is probably essential for a normal hemostatic balance and has three Kunitz domains (Lindahl *et al.*, 1992).

It has been found that the C-terminal domain in the human type VI collagen α3 chain is also of the Kunitz structure (Chu *et al.*, 1990). The most spectacular example, however, is finding Kunitz domain inserts in some of the precursors of the amyloid β-proteins characteristic of Alzheimer's disease (Ponte *et al.*, 1988; Tanzi *et al.*, 1988; Kitaguchi *et al.*, 1988). Whereas it was clearly shown in several cases that these domains are active as inhibitors of typical pancreatic serine proteinases, the real target enzymes and their physiological role are still a matter of speculation (Nakamura *et al.*, 1992; Kido *et al.*, 1993).

B. Pancreatic Secretory Trypsin Inhibitor Family

The group of homologous pancreatic secretory inhibitors of trypsin, also called the Kazal family, is included in this review for two reasons. On one hand, small Kazal inhibitors have been found in all vertebrates examined including humans. This is in contrast to the Kunitz family where, for example, the claimed identification of aprotinin homologs in human serum (Fioretti *et al.*, 1987, 1993b) needs to be confirmed. Thus, the human pancreatic trypsin inhibitor is a promising starting point for the engineering of useful therapeutic proteins with a minimum likelihood of an immunogenic reaction.

On the other hand, a huge set of homologous Kazal inhibitors with a great variability in the contact region is easily available from avian ovomucoids. Such inhibitors have been the basis for a sequence to reactivity algorithm developed and still being refined by Laskowski. Ovomucoids are abundant glycoproteins, which account for about 10% of the protein content of avian egg whites (Feeney, 1970). Their polypeptide chains consist of three homologous Kazal domains that can, in favorable cases, be chemically or enzymatically cleaved from one another without any intradomain nicking (Kato *et al.*, 1987). The production of third domains by limited proteolysis using *Staphylococcus aureus* proteinase V8, thermolysin, or pancreatic elastase proved most efficacious. Thus, since the 1980s ovomucoid third domains from 153 species of birds have been isolated and sequenced (Laskowski *et al.*, 1987a, 1990; Apostol *et al.*, 1993).

X-Ray crystal structures were determined for two of these single Kazal domains, namely, those from Japanese quail (Weber *et al.*, 1981; Papamokos *et al.*, 1982) and from silver pheasant (Bode *et al.*, 1985). They contain 56 amino acid residues cross-linked by three disulfide bridges (Fig. 3). A three-stranded

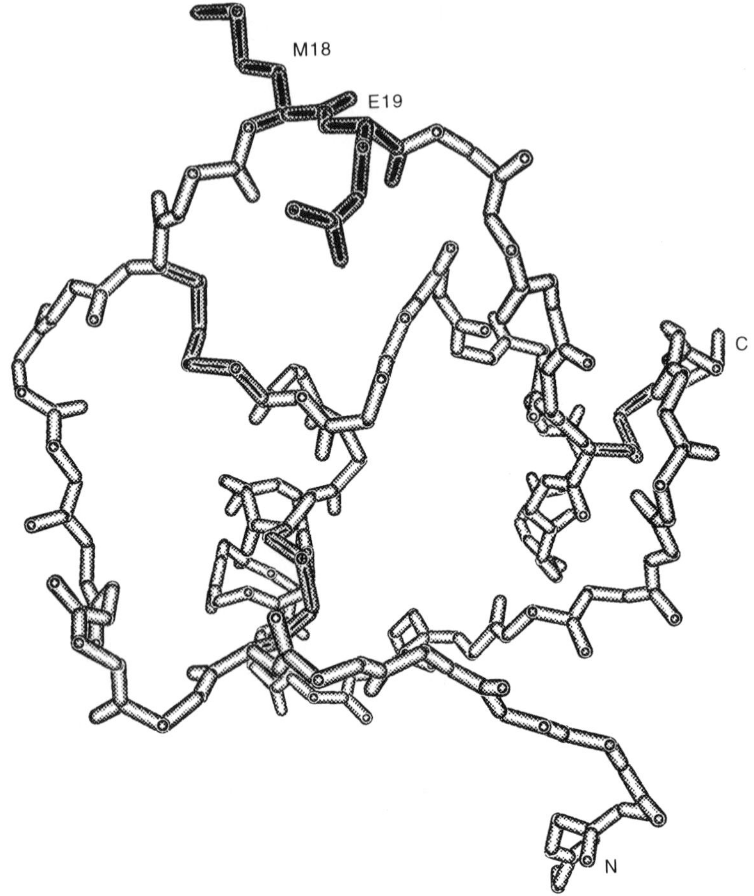

Figure 3 Spatial backbone structure of the silver pheasant ovomucoid third domain. The reactive site residues M18 and E19, including their side chains and the three disulfide bridges, are drawn with solid bonds; N denotes the amino and C the carboxyl terminus.

antiparallel β sheet and an α helix approximately parallel to the β strands form the nucleus of the molecule, to which the amino-terminal segment, the reactive site loop, and a segment connecting the α helix with the carboxyl-terminal β strand are attached.

Spatial structural data derived from X-ray diffraction are also available for three ovomucoid third domain–serine proteinase complexes (Read *et al.*, 1983; Bode *et al.*, 1986a; Fujinaga *et al.*, 1987). They all show, in accordance with the Kunitz inhibitors, that the reactive site loops associate with the residues of the catalytic center in a manner similar to that of productively bound substrates. A characteristic feature is an antiparallel β structure mediated by hydrogen bonds between the P_3-P_2-P_1 segment of the inhibitor and opposite backbone components of the proteinase. The oxygen of the catalytic triad serine is in "sub-van der Waals" contact with the carbonyl carbon of the P_1 residue. The adjacent carboxyl-terminal inhibitor segment interacts through another hydrogen bond at P_2'.

An X-ray structure analysis has been accomplished for two modified ovomucoid third domains I*, with cleaved reactive site peptide bond P_1–P_1' (Musil *et al.*, 1991). Larger deviations from the intact single-chain inhibitors were only observed for the flexible first six amino-terminal residues and the reactive site loop. The remainder of the molecule essentially keeps its shape and conformation, and also the new termini are not very far apart from one another. These results are compatible with the standard mechanism of inhibition and with the generally small equilibrium hydrolysis constant K_{Hyd}.

We now return to the Kazal sequences, which have originated from the massive effort of Laskowski and co-workers. Alignment of the 92 unique ovomucoid third domain sequences from the selected 153 species of birds shows that the structurally important residues are strongly conserved. On the other hand, those residues in contact with the cognate enzymes are not conserved but, instead, are by far the most variable ones in the molecule. Thus, ovomucoids seem to be an example of positive Darwinian selection for functional diversity in the evolution of proteins (Laskowski *et al.*, 1987b).

The dissociation constants K_D for the complexes of virtually all of the Kazal inhibitors with five proteinases, namely, bovine α-chymotrypsin, porcine pancreatic elastase, subtilisin, and the *Streptomyces griseus* proteinases A and B, were measured, and the smallest K_D values were of the order of 10^{-12} M (Empie and Laskowski, 1982; Park, 1985). The following generalizations can be drawn from the results: P_1 is predominant in the interaction with all serine proteinases except for subtilisin which, in fact, has an unusual S_1 structure. Changes in residues other than P_1 may exert large effects on K_D provided that the residues make contact with the enzyme; moreover, they often exert considerable differential effects on the different enzymes, as the same change can render the inhibitor stronger for one enzyme and weaker for another. Changes in surface residues, which do not contact the enzyme, are virtually without effect. The ultimate goal, which is being approached, is to state a sequence to reactivity algorithm, which is a set of rules allowing the prediction of the K_D values for various enzyme complexes from the amino acid sequence of the inhibitor alone (Laskowski, 1980).

Single-domain Kazal inhibitors with a clear antitrypsin specificity are found in the pancreatic juice of vertebrates, where they are secreted with the proenzymes. Their physiological function is to prevent premature activation of the proenzymes by inhibiting trace amounts of active trypsin. A special feature of this inhibition is that the enzymatic activity reappears from the proteinase–inhibitor complex as incubation proceeds. This phenomenon, called temporary inhibition, has been most thoroughly investigated for the porcine inhibitor, where it could be attributed to the cleavage of the R44–Q45 inhibitor bond (Tschesche *et al.*, 1974). The corresponding inactivation of the human inhibitor could be correlated with the enzymatic deletion of the dipeptide K43–R44 (Kikuchi *et al.*, 1989). In both cases the "nominal point of rupture" is located opposite to the reactive site. At least four forms of the pancreatic secretory trypsin inhibitor are present in human pancreatic juice. Besides the major native inhibitor, two deamidated species with reduced activity and a fully active derivative lacking five amino-terminal residues have been purified and characterized (Kikuchi *et al.*, 1985).

C. Squash Seed Inhibitor Family

The term "miniproteins" is certainly appropriate for the members of the squash seed inhibitor family. With peptide chains comprising from 28 to 32 amino acid residues, they are the smallest naturally occurring inhibitors of serine proteinases described so far and possibly even the smallest rigid proteins.

The first member was independently isolated from squash or pumpkin seeds as a trypsin inhibitor (Polanowski *et al.*, 1980; Wilusz *et al.*, 1983) and as an inhibitor for trypsin and β-factor XIIa (Hojima *et al.*, 1982) during an extensive screening of plant materials for serine proteinase inhibitors (Hojima *et al.*, 1983). Since then, more than two dozen miniinhibitors have been extracted from a variety of seeds, where they often occur in the form of isoinhibitors, and have, in many cases, been sequenced. Typical sources are the seeds of zucchini, pumpkin, summer squash, cucumber, gourd, melon, and red bryory (Joubert, 1984; Wieczorek *et al.*, 1985; Otlewski *et al.*, 1987; Favel *et al.*, 1989a; Nishino *et al.*, 1992; Ling *et al.*, 1993).

The complete three-dimensional structure of the inhibitor from the seeds of squash (*Cucurbita maxima*) in aqueous solution was determined by nuclear magnetic resonance (Holak *et al.*, 1989a; Nilges *et al.*, 1991) (Fig. 4). A comparison of this solution structure with the X-ray structure of the inhibitor determined from its complex with bovine trypsin (Bode *et al.*, 1989) revealed a high degree of similarity in terms of global folding and secondary structure (Holak *et al.*, 1989b). The polypeptide chain is rigidly fixed by three disulfide bridges, giving rise to a striking, knotted structure (Le-Nguyen *et al.*, 1990). The binding loop with the reactive site residues R5 and I6 is in a conformation similar to that observed for the Kunitz and Kazal inhibitors. The crystal structure of the complex between porcine trypsin and the inhibitor from the seeds of bitter gourd (*Momordica charantia*) has also been determined, corroborating the earlier observations (Huang *et al.*, 1992, 1993).

Nearly all the squash seed family inhibitors have either K or R at the P_1 position and are excellent trypsin inhibitors, with K_D values in the range of 10^{-9} to 10^{-12} *M* (Otlewski and Zbyryt, 1994). The only exception is an isoinhibitor from bitter gourd seeds, which has L at P_1 and is effective toward porcine pancreatic elastase, with a K_D of 3×10^{-7} *M* (Hara *et al.*, 1989). Two squash seed inhibitors with K and R as the P_1 residue were tested against an array of nine serine proteinases (Otlewski *et al.*, 1990) with the following results. Bovine chymotrypsin is not inhibited, inhibition is very weak for porcine pancreatic elastase, subtilisin, and *Streptomyces griseus* proteinase B. Of special interest

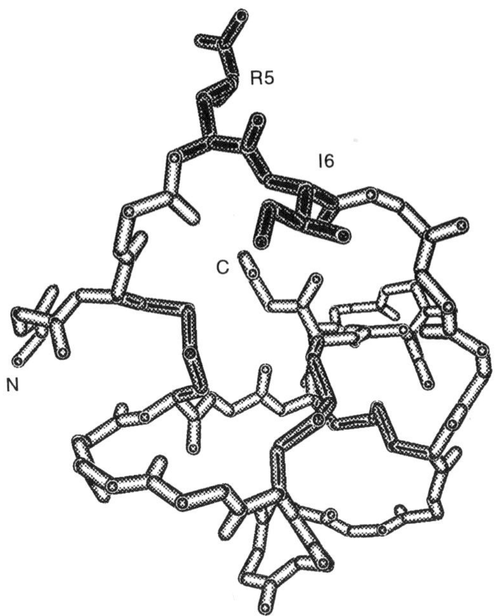

Figure 4 Spatial backbone structure of the squash seed trypsin inhibitor. The reactive site residues R5 and I6, including their side chains and the three disulfide bridges, are drawn with solid bonds; N and C denote the amino and carboxyl terminus, respectively.

are the human enzymes: Inhibition of thrombin is hardly detectable, plasma kallikrein exhibits a clear P_1 discrimination with a K_D of 5×10^{-4} M when P_1 is K and 7.7×10^{-7} M when P_1 is R, and plasmin is effectively inhibited by both isoinhibitors with a K_D of 3.5×10^{-9} M when P_1 is K and 2.6×10^{-9} M when P_1 is R. Highly unexpected is the strong inhibition of cathepsin G ($K_D = 7.7 \times 10^{-9}$ M for P_1 = K, $K_D = 5.9 \times 10^{-9}$ M for P_1 = R), because this proteinase shows chymotrypsin-like specificity and is not inhibited by, for example, aprotinin.

D. Potato Inhibitor 1 Family

The potato inhibitor 1 family includes two well-characterized members which, unlike the vast majority of proteinase inhibitors, do not contain stabilizing disulfide bridges, namely, eglin c, an inhibitor from the leech *Hirudo medicinalis,* and barley serine proteinase inhibitor 2 (Svendsen *et al.*, 1982). The latter has served as an excellent system for studying protein folding and stability (Jackson and Fersht, 1991a,b); the high interest in eglin c mainly arose because of its pharmacological potential (Braun and Schnebli, 1986).

The parent form of serine proteinase inhibitor 2 comprising 83 amino acid residues is usually isolated from barley together with several shorter forms lacking up to 18 residues from the amino terminus. The inhibitor is obviously subject to multiple hydrolytic cleavages during purification, or it already exists in the grain as a ragged N-terminus polypeptide (Svendsen *et al.*, 1980). Three-dimensional structures of shortened forms were determined from X-ray diffraction data (McPhalen and James, 1987) (Fig. 5) and from NMR measurements (Clore *et al.*, 1987a). Both methods yielded molecular models, which are very similar in terms of overall shape, polypeptide fold, and secondary structure (Clore *et al.*, 1987b). A four-stranded mixed parallel and antiparallel β sheet forms the core of the molecule. On one side it is flanked by an α helix of 3.6 turns and on the other side by the reactive site loop in an extended conformation,

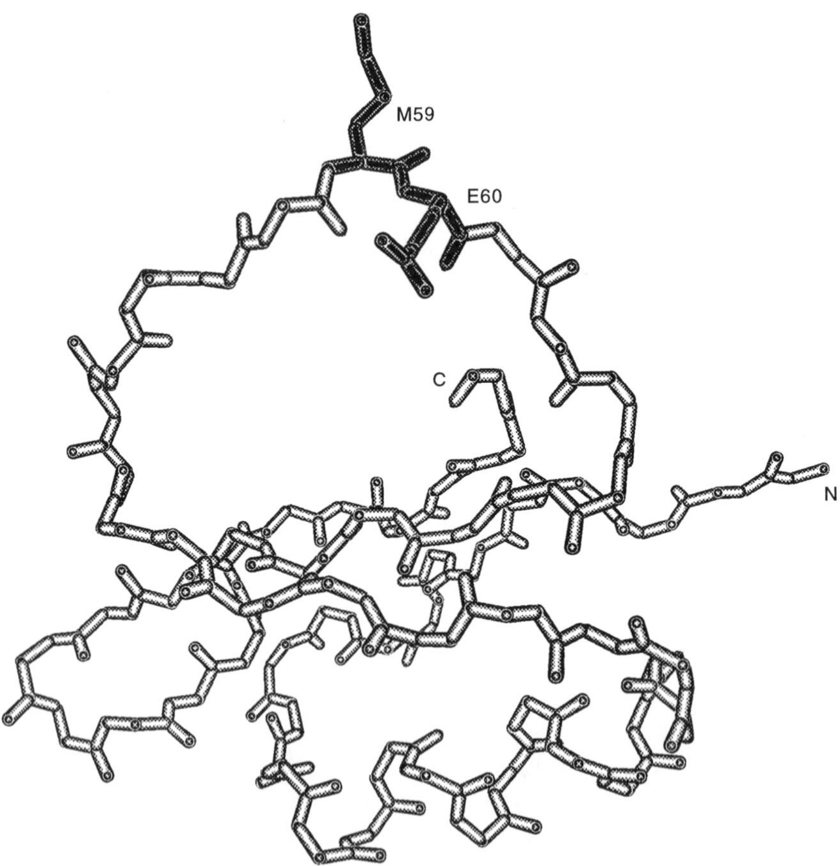

Figure 5 Spatial backbone structure of the barley serine proteinase inhibitor 2. The reactive site residues M59 and E60, including their side chains, are drawn with solid bonds; N denotes the amino terminus and C the carboxyl terminus.

which is stabilized by a network of hydrogen bonds and electrostatic interactions. Most of these contacts are provided by two arginine side chains extending to the loop from the β sheet.

In accordance with the M59 and E60 reactive site residues at P_1–P_1', the barley inhibitor is active against chymotrypsin ($K_D = 1.6 \times 10^{-9}$ *M*), pancreatic elastase ($K_D = 3.0 \times 10^{-8}$ *M*), and different subtilisins ($K_D \approx 10^{-11}$ *M*), but trypsin is not inhibited. The X-ray crystal structure of one complex was also determined, namely, that of barley serine proteinase inhibitor 2 and subtilisin Novo (McPhalen *et al.*, 1985). It reveals a conformation around the reactive site similar to that found for the other inhibitor families already discussed.

Among the variety of proteinase inhibitors which can be isolated from leeches (Seemüller *et al.*, 1986), there are also two members of the potato inhibitor 1 family: eglin c, which has been used for all structural investigations cited below, and eglin b, which has the single substitution Y35H (Chang *et al.*, 1985). Both isoinhibitors comprise 70 amino acid residues and exhibit, despite the lack of cystine stabilization, a high resistance against denaturation by heat or acid.

X-Ray crystal structures of free eglin c have become available, both for the native single-chain form and for the modified form nicked at the reactive site peptide bond L45–E46. The polypeptide chain fold of native eglin c is similar to that of barley serine proteinase inhibitor 2 (Hipler *et al.*, 1992). On enzymatic P_1–P_1' cleavage, which unintentionally took place when a cocrystallization of eglin with subtilisin DY was intended, the rigid well-defined core with the α-helix and the four-stranded β sheet maintains its structure. The new termini with the adjacent residues of the former reactive site loop, however, are subjected to a complete reorientation (Betzel *et al.*, 1993).

The structures of complexes formed by eglin c with the following proteinases have been determined by X-ray diffraction analysis: subtilisin Carlsberg (Bode *et al.*, 1986b, 1987; McPhalen and James, 1988), thermitase, a thermostable member of the subtilisin family (Gros *et al.*, 1989a,b), subtilisin Novo (Heinz *et al.*, 1991), and bovine α-chymotrypsin (Frigerio *et al.*, 1992). The paper by McPhalen and James (1988) also contains a detailed structure comparison of the subtilisin Carlsberg–eglin c and the subtilisin Novo–barley proteinase inhibitor 2 complexes, and the conclusion was that most of the enzyme–inhibitor interactions are maintained.

Eglin c strongly inhibits bovine α-chymotrypsin ($K_D = 7 \times 10^{-10}$ *M*), subtilisin ($K_D = 1.2 \times 10^{-10}$ *M*), and, most notably, both human leukocyte elastase ($K_D = 2 \times 10^{-10}$ *M*) and cathepsin G ($K_D = 2.8 \times 10^{-10}$ *M*); eglin b has an identical inhibitory specificity (Seemüller *et al.*, 1986; Braun *et al.*, 1987). The high potency of eglin c against human enzymes, which are thought to be a major factor in connective tissue damage in inflammation, combined with a high selectivity—no interaction, for example, with the blood-borne proteinases of the clotting, fibrinolysis, and complement cascades could be observed—gave rise to an ambitious drug development project (Schnebli *et al.*, 1985), which is discussed in Section V.

IV. Semisynthetic Proteinase Inhibitors

The terms semisynthesis or partial synthesis as applied to peptides or proteins refer to the use of a fragment of a natural polypeptide as a prefabricated intermediate in a synthesis (Offord, 1980, 1990). Besides total synthesis discussed in Section VI, the recombinant approach discussed in Section V, and side-chain transformation, it is an established method for the preparation of protein homologs. Although recombinant techniques have become the most often used for that purpose and should be considered as the first option, the chemical techniques have more to offer than mere historical interest, which is especially true in the field of proteinase inhibitors. For example, they permit the introduction of noncoded amino acids and amino acid analogs in a broad sense.

It has been shown that it is feasible to extend the recombinant methods to "noncoded exchanges" in certain cases by using synthetic acylated suppressor tRNAs (Anthony-Cahill *et al.*, 1989; Offord, 1991). A number of technical obstacles must be overcome, however, before this new strategy can yield sufficient material for most investigations.

A. Aprotinin Homologs Mutated at P_1

After it had become evident that P_1 dominates the specificity of serine proteinase inhibitors, it was of course tempting to change this amino acid residue and, thereby, the inhibition spectrum. The first successful method to this end, the so-called enzymatic mutation, which was originally developed for the soybean trypsin inhibitor (Sealock and Laskowski, 1969) and later adopted for aprotinin (Jering and Tschesche, 1974, 1976b), is outlined in Fig. 6 and comprises the following steps: enzymatic cleavage of aprotinin (**1**) at the reactive site to yield *seco*-aprotinin (**2**) (Jering and Tschesche, 1976a; Estell *et al.*, 1980); removal of the original P_1 residue K15 by carboxypeptidase B, providing des-K15-*seco*-

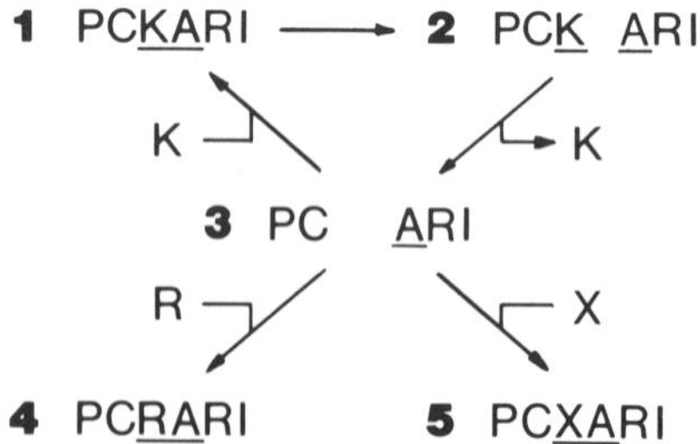

Figure 6 Enzymatic mutations of the aprotinin P_1 residue. Only residues P13 to I18 of the reactive site region are shown, P_1 and P_1' are underlined.

aprotinin (**3**); coupling of arginine to C14 by carboxypeptidase B to yield *seco*-K15R-aprotinin in minute amounts, which are trapped by complex formation with pancreatic kallikrein, thus shifting the carboxypeptidase-catalyzed reaction toward synthesis; and dissociation of the complex at acidic pH to yield K15R aprotinin (**4**) as the main product. It is essential to use kallikrein in the trapping reaction, because trypsin unintentionally led to a product lacking R39, and also when ^{13}C-labeled lysine was introduced for NMR experiments (Richarz *et al.*, 1979). When using carboxypeptidase A and chymotrypsin, aromatic amino acids can be incorporated to yield K15X aprotinin (**5**; X = F, W).

The K15R replacement results in a substantially better inhibition for several proteinases. The complex dissociation constants decrease by a factor of about 20 for human trypsin ($K_D = 1.2 \times 10^{-9}$ *M*) and endoproteinase Arg-C ($K_D = 9.4 \times 10^{-8}$ *M*) and of almost 80 for human plasma kallikrein ($K_D = 5.9 \times 10^{-10}$ *M*) (Tschesche *et al.*, 1989). Most notably, human coagulation factor XIa, although not inhibited by natural aprotinin, has a K_D of 3.4×10^{-8} *M* for the K15R homolog (Scott *et al.*, 1987). On the other hand, the highly specific endoproteinase Lys-C with a subnanomolar K_D for aprotinin is not affected by K15R aprotinin even when incubated with a 100-fold excess of the inhibitor.

K15F and K15W aprotinin exhibit pronounced antichymotryptic activity. The dissociation constant for the complexes with bovine chymotrypsin is decreased by one order of magnitude ($K_D = 2.8 \times 10^{-9}$ *M*). When tested against human cathepsin G, however, K15F aprotinin is devoid of appreciable inhibitory activity.

There have been several attempts to extend the elegant "enzymatic mutation" to the introduction of aliphatic amino acids with alkyl side chains in order to render aprotinin to be, for example, an efficient elastase inhibitor. However, so far most efforts have remained unsuccessful, mainly because a proper enzyme system is lacking (methionine, ornithine, and homolysine could be introduced in rather low yields). We have therefore developed more general strategies, partly analogous to that described above, partly with more chemically based routes involving coupling reagents and reversible protection, which allow for replacement of K15 by virtually any other amino acid (Wenzel *et al.*, 1986; Tschesche *et al.*, 1987; Beckmann *et al.*, 1988a, 1989a). Briefly, the reaction sequences consist of the following seven or eight steps (Fig. 7).

Reactive site cleavage of aprotinin (**1**) is accomplished as in the enzymatic mutation scheme (above). Complete esterification of *seco*-aprotinin (**2**) at its six carboxyl groups (cf. Fig. 1) is promoted with acidified methanol to yield the hexamethyl ester (**6**). Specific saponification of the K15 methyl ester with endoproteinase Lys-C in the presence of dioxane as organic cosolvent yields the pentamethyl ester (**7**). Removal of K15 by carboxypeptidase B produces the key intermediate (**8**) with its single free carboxyl group, from which the reaction pathway branches out. In the more general route (**8** → **10** → **11** → **12** → **13**), the new P_1 amino acid X as the *tert*-butyl ester is coupled to C14 with a water-soluble carbodiimide. Specific hydrolysis of the *tert*-butyl ester group with trifluoroacetic acid is followed by a second carbodiimide condensation joining

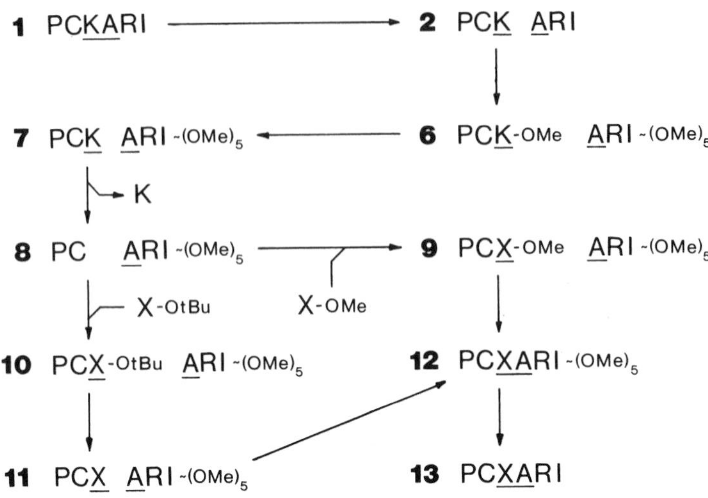

Figure 7 General chemical methods for the exchange of the aprotinin P_1 residue. Residues P13 to I18 are shown, P_1 and P_1' are underlined, and the methyl ester formation indicated by $(OMe)_5$ takes place at D3, E7, E49, D50, and A58.

X15 and A16. Alkaline saponification of all methyl ester groups yields the homologous K15X aprotinins. In the shorter second route (**8 → 9 → 12 → 13**), X as the methyl ester is coupled to C14. Resynthesis of the reactive site peptide bond X–A depends on the availability of a suitable proteinase and can be achieved, for example, with trypsin (X = R) or chymotrypsin (X = M, norleucine, E-5-methyl ester). Saponification of the five ester groups, six in the last case, again yields K15X aprotinins.

Using these "chemical mutations," about 20 P_1 substitutions have been performed. Table I lists 10 of the homologs obtained, which are also homologous in a chemical sense with regard to the structure of their X15 alkyl side chains. The dissociation constants for complexes with the two elastases certainly reflect the spatial structures of the complementary specificity pockets and are in agreement with known substrate preferences and X-ray data.

Aprotinin homologs with noncoded P_1 residues having a linear side chain of medium length (2-aminobutyric acid, norvaline, norleucine) are especially strong inhibitors of porcine pancreatic elastase, whereas homologs with branched residues have a significantly lower affinity toward this proteinase. Human leukocyte elastase, on the other hand, is obviously able to accommodate bulky branched side chains in its specificity pocket. K15V aprotinin is the most powerful inhibitor of the series for this elastase when tested with synthetic chromogenic substrates, followed by the homologs with norvaline, 2-aminobutyric acid, and isoleucine at position 15. It is known that leukocyte elastase exhibits a high affinity to its natural substrate, elastin, which renders inhibition of the adsorbed pro-

Table I Dissociation Constants of Elastase Complexes with Lysine15X Aprotinin Homologs

X	Side chain	K_D (n*M*) of complex with elastase from	
		Porcine pancreas	Human leukocytes
Glycine		>100	7.0
Alanine		28	2.5
2-Aminobutyric acid		1.1	0.3
Norvaline		0.4	0.2
Norleucine		2.6	2.7
Valine		57	0.1
Leucine		19	2.9
Isoleucine		>100	0.4
tert-Leucine		>1000	15
Neopentylglycine	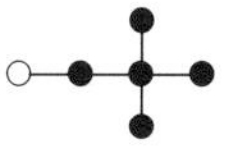	>100	67

teinase very difficult. In contrast to the physiological inhibitor, α_1-proteinase inhibitor (Reilly and Travis, 1980), K15V aprotinin is also able to entirely stop the elastin breakdown by the prebound enzyme (Beckmann *et al.*, 1988b).

K15M aprotinin is an efficient inhibitor of the elastase from human leukocytes ($K_D = 2.9 \times 10^{-9}$ *M*) and, to a lesser degree, of that from porcine pancreas ($K_D = 7.6 \times 10^{-8}$ *M*) (Beckmann *et al.*, 1989a). On oxidation of M15 to the sulfoxide, the K_D value for the leukocyte enzyme increases by three orders of magnitude, and the affinity for the pancreatic enzyme is totally abolished (Beckmann *et al.*, 1989b). These drastic effects are strikingly similar to those observed with the oxidation of α_1-proteinase inhibitor, which has also a P_1 methionine residue (Beatty *et al.*, 1980). The oxidation and concomitant inactivation of this natural inhibitor are thought to be responsible, for example, for the increased risk of emphysema among smokers.

Trypsin is appreciably inhibited by only two of the homologs listed in Table I, namely, those containing norvaline and norleucine, which have unbranched side chains of medium length, the K_D values being 1.3×10^{-7} and 2.0×10^{-8} *M*, respectively. Empirical free energy calculations based on X-ray crystallographic structures yielded values that are within an order of magnitude agreement with the experimental data (Krystek *et al.*, 1993). The inhibitory activity reported for K15G aprotinin derivatives from preliminary experiments (Wenzel and Tschesche, 1981) could not be confirmed.

K15E aprotinin was tested for inhibition of the glutamate-specific proteinases from *Staphylococcus aureus* and *Streptomyces griseus*. No decrease in enzymatic activity was detectable in either case.

B. Aprotinin Homologs Mutated at P′ Positions

Although the inhibitory specificity of an aprotinin homolog is dominated by the P_1 amino acid residue, it can certainly also be influenced by the adjacent P′ residues A16, R17, and I18 (cf. Fig. 1), which are all situated within the contact region. Thus, additional fine-tuning of the proteinase–inhibitor interaction can be expected from variations of these residues. Several semisynthetic strategies for the required replacements have been exploited. They are often parallel to the reactions described in the previous section. A simple enzymatic method, somewhat reminiscent of DNA cassette mutagenesis, is outlined in Fig. 8 (Groeger *et al.*, 1991a; Tschesche *et al.*, 1991).

After the obligatory cleavage of aprotinin (**1**) at its reactive site, A16 and R17 are successively removed from *seco*-aprotinin (**2**) by the action of aminopeptidase K to yield the nicked derivatives **14** and **15**, respectively. From an incubation mixture of des-(A16-R17)-*seco*-aprotinin (**15**) with the dipeptide AR and equimolar amounts of trypsin in the presence of 80% 1,4-butanediol, native aprotinin (**1**) can be isolated after some hours. The proteinase obviously closes two peptide bonds in the order R–I before K–A and traps the single chain product by complex formation. When using AK as dipeptide, R17K aprotinin (**17**) is obtained; the dipeptides XR (X = G, L) open the way to A16X homologs.

Because of the pronounced specificity of trypsin, this enzymatic fragment substitution is not generally applicable; it requires a carboxyl-terminal R or K in

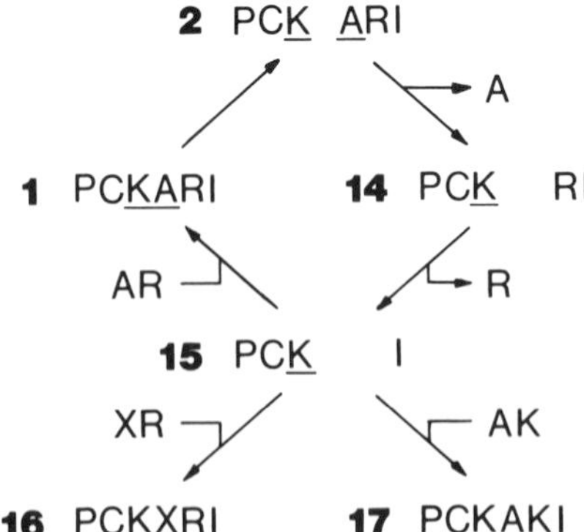

Figure 8 Enzymatic mutations of the aprotinin P_1' and P_2' residues. Only residues P13 to I18 of the reactive site region are shown, and P_1 and P_1' are underlined.

the dipeptides. Other dipeptides or single amino acids can be incorporated, however, by chemical coupling procedures and reversible protection with ester groups (Groeger *et al.*, 1991b; Wenzel *et al.*, 1989a). Key intermediates in the reaction sequences are the pentamethyl esters of **14** or **15** with a single free carboxyl group at K15, to which *tert*-butyl esters of amino acids or dipeptides can be coupled. Selective hydrolysis of the *tert*-butyl ester group enables a second residue to be coupled or the peptide chain to be closed. R17A aprotinin can be prepared in this way.

The P_2' substitutions R17K and R17A are only of minor to moderate significance for the inhibition of bovine trypsin, bovine chymotrypsin, and porcine kallikrein. The same holds true for the P_1' substitution A16G. A16L aprotinin, however, has considerably increased K_D values when compared with the native inhibitor: trypsin, about 10^6-fold; chymotrypsin, 120-fold; kallikrein, 60-fold.

A combination of the enzymatic and chemical methods outlined in this and the previous section allows the semisynthesis of K15V, A16S, R17I aprotinin with the optimum P_1 residue for leukocyte elastase and P_1' and P_2' residues identical to the corresponding residues in the potent natural α_1-proteinase inhibitor. Instead of the improved elastase inhibitor expected, however, an increase in the K_D value by a factor of 40 was observed in comparison with K15V aprotinin (Beckmann *et al.*, 1988b; Wenzel *et al.*, 1989b).

C. Backbone Variants of Aprotinin

The rational design of inhibitors as potential drugs requires detailed knowledge of all aspects of the interaction of the inhibitors with proteinases. This primarily includes not only the arrangement of amino acid side chains within the contact region discussed above but also the underlying peptide backbone structure. We have therefore embarked on systematic variations of the elements making up the aprotinin backbone around the reactive site (Groeger *et al.*, 1993).

The "reference compound" in the first series of variants is A16G, R17G aprotinin with the link -NH-CH_2-**CO-NH**-CH_2-CO- between residues K15 and I18. Despite the loss of two side chains, this homolog remains an excellent permanent trypsin inhibitor ($K_D = 2 \times 10^{-12}$ *M*). A complete or partial replacement of the G16–G17 amide bond with methylene groups is feasible by semisynthetic introduction of urethane-protected ω-amino carboxylic acid *N*-hydroxysuccinimide esters. These are in a first step coupled to the I18 amino group of des-(A16-R17)-*seco*-aprotinin (**15**) (Fig. 8). Although this inhibitor fragment possesses a total of six potential sites for acylation, the reaction can be carried out with a high degree of selectivity at pH 4.75; about 25% of the product is monoacylated at I18. After cleavage of the N-protecting groups, the second amide bond, that to K15, is closed by using trypsin. The three ω-amino acid residues substituting for GG, their structures, and the dissociation constants of the aprotinin homolog–trypsin complexes are as follows:

5-Aminolevulinic acid, -NH-CH_2-**CO-CH$_2$**-CH_2-CO-, $K_D = 5 \times 10^{-11}$ *M*;
N-Aminoethylglycine, -NH-CH_2-**CH$_2$-NH**-CH_2-CO-, $K_D = 7 \times 10^{-9}$ *M*;
5-Aminovaleric acid, -NH-CH_2-**CH$_2$-CH$_2$**-CH_2-CO-, $K_D = 2 \times 10^{-9}$ *M*.

The first homolog is still a strong and permanent inhibitor. As the keto function is involved in neither intra- nor intermolecular hydrogen bonding, the activity is probably due to the retained backbone rigidity in this region. The latter two homologs prove to be much weaker, and, furthermore, temporary trypsin inhibitors, which are slowly degraded and inactivated by the enzyme. A reason for this switch to a substrate may lie in the enhanced flexibility of the binding loops.

Current studies are aimed at introducing ester and hydroxyethylene moieties at different positions of the aprotinin backbone (Groeger *et al.*, 1994). Reports indicate that genetic engineering might supplement these peptide chemical methods to also allow the incorporation of novel backbone structures into proteins (Ellmann *et al.*, 1992). Successful replacements include α,α-disubstituted amino acids, *N*-alkyl amino acids, and an isoelectronic analog of alanine, lactic acid.

D. Ovomucoid Third Domains with Changes at P_1

The methods outlined above comprise the exchange of up to 3 amino acid residues out of 58, so that the term semisynthesis in a literal sense appears somewhat exaggerated. However, it also proved practicable to substitute large fragments of proteinase inhibitors, for example, the complete P strand, as was comprehensively demonstrated for members of the Kazal family. Ovomucoid third domains of 56 residues (cf. Fig. 3) from two species of birds are enzymatically cleaved at their reactive site between residues 18 and 19. After reduction of the three disulfide bridges, two peptides can be isolated in each case, the amino-terminal 1–18 fragment and the carboxyl-terminal 19–56 fragment. The chains are then converted to mixed disulfides with glutathione. On incubation of the short fragment of one species with the long fragment of the other, interchain disulfide bridges are formed. Enzymatic closure of the reactive site peptide bond finally yields a hybrid third domain with a new primary structure (Wieczorek and Laskowski, 1983).

This strategy can be extended to the recombination of a natural, large fragment 19–56 with a synthetic, short fragment 1–18 or, more conveniently, with a 6–18 fragment, as the five amino-terminal residues have essentially no influence on the enzyme–inhibitor association (Wieczorek *et al.*, 1987). Thus, the way is open for many interesting substitutions within the P strand, independent of their occurrence in natural ovomucoids and also including noncoded amino acids.

For obvious reasons, the main interest is again concentrated on the P_1 residue. Thus, a collection of 13 homologs of turkey ovomucoid third domain has been introduced (Bigler *et al.*, 1993), which display the following L18X substitutions: X = A, S, V, M, Q, E, K, and F, and X = α-aminobutyric acid, norvaline, norleucine, α-aminoheptanoic acid, and homoserine. The last substitution was effected by CNBr cleavage of a natural inhibitor with M18 followed by closure of the reactive site peptide bond.

The equilibrium dissociation constants for the complexes of these variants with a set of six serine proteinases were determined, and the substitution yielding the smallest value is given in parentheses: bovine chymotrypsin (L18F, $K_D = 4.8 \times 10^{-13}$ *M*), porcine pancreatic elastase (L18α-aminobutyric acid, $K_D = 3.0 \times 10^{-12}$ *M*), subtilisin Carlsberg (L18norvaline, $K_D = 7.1 \times 10^{-12}$ *M*), *Streptomyces griseus* proteinase A (L18α-aminoheptanoic acid, $K_D = 2.1 \times 10^{-12}$ *M*), *Streptomyces griseus* proteinase B (wild type L18, $K_D = 1.8 \times 10^{-11}$ *M*), human leukocyte elastase (L18α-aminobutyric acid, $K_D = 7.1 \times 10^{-11}$ *M*). Discussing all the 81 constants measured and the tendencies, for example, within a homologous series, would be beyond the scope of this review, and the reader is referred to the original paper (Bigler *et al.*, 1993). It suffices here to mention that for the two elastases a reasonable agreement exists between the conclusions drawn in this study and those drawn in the study with the corresponding aprotinin homologs.

A striking difference between the two inhibitor families becomes apparent, however, when testing the glutamate-specific proteinases. L18E turkey ovomucoid third domain is a powerful inhibitor of the enzyme from *Streptomyces griseus* ($K_D = 1.8 \times 10^{-11}$ *M*) (Komiyama *et al.*, 1991), whereas K15E aprotinin is at least five orders of magnitude weaker. Neither homolog inhibits *Staphylococcus aureus* V8 proteinase.

V. Proteinase Inhibitors by Genetic Engineering

A substantial proportion of the numerous papers published since the 1980s that illustrate the dramatic advances in protein engineering using the technology of recombinant DNA is concerned with proteinase inhibitors. Studies have been aimed at unraveling folding pathways, understanding the thermodynamic stability of proteins, tailoring specific inhibitors against a given proteinase, providing mutant proteins for structure determination, or even producing kilogram quantities of an inhibitor for *in vivo* pharmacological experiments. Examples are selected from three families of serine proteinase inhibitors.

A. Recombinant Aprotinin and Homologs

Several different expression systems for the production of recombinant Kunitz inhibitors in *Escherichia coli* have been described. Some of these rely on secretion of the protein into the periplasmic space and direct recovery of native, correctly folded homologs (Marks *et al.*, 1986; Goldenberg, 1988; Goldenberg *et al.*, 1993). Expression levels in secretion systems vary by several orders of magnitude depending on the mutation in the inhibitor gene; thus, it may turn out to be impossible to obtain sufficient material, for example, for structural studies. Methods for the intracellular production of homologs in inclusion bodies are often superior in this respect. Such approaches sometimes require, however, modification of the aprotinin sequence to avoid internal nicking, for example, at M52 when the fusion protein is to be cleaved with CNBr (von Wilcken-Bergmann *et al.*, 1986; Auerswald *et al.*, 1987; Altman *et al.*, 1991). Expression of synthetic genes in the yeast *Saccharomyces cerevisiae* resulting in secretion of correctly processed aprotinin and homologs thereof into the culture medium has also been described (Norris *et al.*, 1990; Berndt *et al.*, 1993; Barthel and Kula, 1993).

The most detailed and informative pathway of folding, namely, reductive unfolding and oxidative refolding, which has been elucidated thus far, is that of aprotinin (Creighton, 1990). During the 1970s, decisive experiments of isolating and characterizing partially disulfide-bonded intermediates trapped with SH-alkylating reagents had already been carried out (Creighton, 1978). A further insight into the forces that guide the aprotinin folding pathway has become possible with the rise in recombinant methods. They provide, for example, analogs of one- or two-disulfide intermediates, in which the cysteine residues not involved in disulfide bridges are replaced by alanine or serine residues (Staley and Kim, 1992; Kosen *et al.*, 1992; Darby and Creighton, 1993). The results of kinetic measurements could be correlated with those from complementary conformational analysis either by NMR (van Mierlo *et al.*, 1993) or X-ray diffraction (Eigenbrot *et al.*, 1990).

All two-disulfide mutants of aprotinin investigated thus far display two very slow folding reactions, which have characteristics of proline isomerization. They were thought to occur because the nonnative cis form of two prolines significantly destablizes the folded state of the protein. It was then shown that P8Q, C30V, C51A aprotinin does not exhibit the slowest folding reaction, suggesting that the cis form of P8 is the most destabilizing of the four prolines (Hurle *et al.*, 1991). Another interesting mutant is G36S aprotinin, which, owing to its enlarged side chain, lacks one of the four tightly bound internal water molecules of aprotinin. This results in a slightly reduced thermal stability and in weaker complexes with trypsin and plasma kallikrein (Berndt *et al.*, 1993).

From the very beginning of the production of recombinant Kunitz inhibitors, the intentional change in inhibitory specificity by site-directed mutagenesis within the reactive site region was a main objective. The first P_1 homologs K15I aprotinin (von Wilken-Bergmann *et al.*, 1986) and K15R, M52E aprotinin (Auerswald *et al.*, 1988) exhibit inhibition characteristics comparable to those of the semisynthetically prepared proteins. Moreover, the latter homolog, at a concentration of 50 μM, clearly reduces human immunodeficiency virus type 1 (HIV-1) replication in H9 cells (Auerswald *et al.*, 1991). This effect parallels that of the Kunitz inhibitor rat trypstatin at a still lower concentration (Hattori *et al.*, 1989) and will be the subject of further investigation.

Neither aprotinin (Fioretti *et al.*, 1993a) nor any of the P_1 homologs prepared so far shows appreciable inhibitory activity against cathepsin G. When testing a selection of 12 recombinant inhibitors with additional mutations in the

P_1' and P_2' positions, however, activity against this human leukocyte proteinase can be detected. The most effective homologs are K15L, R17F, M52E aprotinin and K15L, R17Y, M52E aprotinin ($K_D = 10^{-8}$ *M*) followed by K15L, A16G, R17L, M52E aprotinin and K15F, A16G, R17F, M52E aprotinin ($K_D = 10^{-7}$ *M*). Obviously, besides a hydrophobic residue at P_1 another at P_2' is mandatory for strong cathepsin G inhibition. All four proteins are excellent chymotrypsin inhibitors, with K_D values in the subnanomolar range (Brinkmann *et al.*, 1991).

All of the potent inhibitors of serine proteinases described above have emerged by using one of two classic strategies: screening of the large but, nevertheless, limited and incomplete set of naturally occurring proteins, such as the ovomucoids, or changing the specificity of a known inhibitor in one or several cycles of refinement. That means starting with a more or less well-founded guess as to the result of a substitution, performing the substitution, testing the new activity, and entering into another cycle to further improve the inhibitory properties if necessary. With regard to the vast number of possible changes even for the few amino acid residues in the contact region, it is clear that this route is extremely time-consuming, laborious, and often unsuitable to end up with the optimum sequence. What would be needed is a library of homologs with random mutations and a screening method allowing rapid selection of the most potent inhibitors. This is exactly what the new technique of peptide display on filamentous phages has to offer (Marks *et al.*, 1992; Cesareni, 1992). Its first application to proteinase inhibitors yielded aprotinin homologs with the highest affinity toward leukocyte elastase reported so far. The procedure is as follows (Roberts *et al.*, 1992a,b).

Prior to construction of the library the mutations R39M, A40G, K41N, and R42G are introduced, because they enhance the binding of aprotinin homologs to leukocyte elastase. A variegated oligonucleotide is then used to mutate the DNA encoding residues adjacent to the reactive site. The following amino acids are possible at the different positions: P_1 = V, L, F, I, M; P_1' = G, A; P_2' = V, L, F, I, M; P_3' = F, S, I, T; P_4' = K, S, Q, P, T. The complete Kunitz domains are now displayed on the surface of bacteriophage M13 as a fusion to the gene III product. Fractionation of the phages with their potential 1000 aprotinin homologous sequences is achieved by adsorbing them to immobilized leukocyte elastase followed by desorption with buffers of decreasing pH, with the phages displaying the most potent inhibitors being desorbed last. DNA sequencing of 20 clones reveals that only V or I appears at P_1 of the strong elastase inhibitors, consistent with the results of semisynthesis, F and M are favored at P_2', and, rather surprisingly, only F is observed at P_3'. The encoding genes for the three strongest inhibitors are expressed in *E. coli* and the dissociation constants of the complexes with leukocyte elastase determined: K15I, R17F, I18F, I19P aprotinin ($K_D = 1.0 \times 10^{-12}$ *M*), K15V, R17M, I18F, I19P aprotinin ($K_D = 2.7 \times 10^{-12}$ *M*), and K15V, A16G, R17F, I18F, I19S aprotinin ($K_D = 2.8 \times 10^{-12}$ *M*); substitutions at positions 39 to 42 are as specified above. Thus, the elastase inhibition by aprotinin homologs with a single substitution at P_1 can be improved by at least two orders of magnitude when additional exchanges are performed within the P′ strand. This phage display strategy will certainly be applied to engineer high-affinity inhibitors for a wide variety of target proteinases.

B. Recombinant Kazal Inhibitors

Evidence has been accumulated that human pancreatic secretory trypsin inhibitor, besides its role in preventing premature activation of zymogens in the pancreas and pancreatic duct, also exhibits growth factor activity for endothelial cells (McKeehan *et al.*, 1986). Given these important physiological functions and potential therapeutic application, it is not surprising that an efficient production of the recombinant protein was already envisaged in the mid 1980s. Since then, several groups have reported on the secretory production of this Kazal inhibitor in heterologous systems such as *E. coli* (Kanamori *et al.*, 1988; Maywald *et al.*, 1988), *Bacillus subtilis* (Nakayama *et al.*, 1992), and yeast (Izumoto *et al.*, 1987).

One group chose the human pancreatic inhibitor as a starting structure for a wide-ranging protein design project, which aimed at the construction of a highly specific and effective inhibitor of human leukocyte elastase (Collins *et al.*, 1989, 1990). Several design cycles involving gene cloning and expression, investigation of properties and structure of the new proteins, and suggestion of new variants led to increasingly better inhibitors. The conclusions drawn from the results obtained from over 24 homologs primarily corroborate the straightforward rules on the crucial role of the P_1 residue: The native trypsin inhibitor with P_1 being K is completely inactive toward chymotrypsin and leukocyte elastase, P_1 being Y is correlated with strong chymotrypsin and weak elastase inhibition, P_1 being V vice versa (most specific elastase inhibitor is the K18V, I19E, D21R homolog, $K_D = 1.5 \times 10^{-11}\,M$), and P_1 being L is the optimal choice when both enzymes are to be effectively inhibited (most potent elastase inhibitor is the K18L, I19E, D21F homolog, $K_D = 5.2 \times 10^{-12}\,M$).

The three-dimensional structures of the complexes between bovine chymotrypsinogen A and the following inhibitors were solved: K18Y, I19E, D21R homolog and K18L, I19E, D21R homolog (Hecht *et al.*, 1991). The structure of the former inhibitor was also determined in the uncomplexed form (Hecht *et al.*, 1992).

An *E. coli* expression system for ovomucoid third domain variants was also established (Lu *et al.*, 1993). Thus, it can be anticipated that the basis for the sequence to reactivity algorithm will be substantially extended in the future. A first notable result is the engineering of a moderate inhibitor for furin, a member of the subtilisin-related, highly specific proprotein covertases (Steiner *et al.*, 1992): A15R, T17K, L18R turkey ovomucoid third domain matched to the substrate specificity of furin, which cleaves after the consensus sequence RXKR, clearly forms a complex with this enzyme ($K_D = 9.1 \times 10^{-8}\,M$). The L18R substitution alone causes only a very weak inhibition ($K_D = 2.2 \times 10^{-3}\,M$).

C. Recombinant Members of Potato Inhibitor 1 Family

The gene encoding for the barley serine proteinase inhibitor 2 has been cloned and can be express at high levels in *E. coli* (Longstaff *et al.*, 1990). Three mutant proteins with new P_1 residues have also been prepared and tested against a set of proteinases: Trypsin, for which the wild-type protein functions only as a substrate, is effectively inhibited by the M59K homolog ($K_D = 5.6 \times 10^{-9}$ *M*). The M59Y substitution lowers the dissociation constant for the chymotrypsin complex by a factor of 3, whereas the M59A substitution unexpectedly raises the value for the pancrease elastase complex by a factor of 20.

Site-directed mutagenesis has also allowed production of systematic cavity-creating mutations in the β-sheet/α-helix core of the inhibitor. Substitutions such as I39V, V66A, I76A, or L68A result in changes in the free energy of unfolding measured by both guanidinium chloride-induced denaturation and differential scanning calorimetry. The average change for deleting one methylene group amounts to 1.3 ± 0.5 kcal/mol (Jackson *et al.*, 1993a,b).

Detailed dynamic studies by ^{13}C NMR on the effects of mutations within the barley inhibitor have been reported. Highly accurate relaxation data could be obtained by the introduction of a specifically ^{13}C-enriched phenylalanine or tryptophan residue into the recombinant protein (Leatherbarrow and Matthews, 1992; Matthews *et al.*, 1993).

As mentioned in the introduction, several proteinaceous inhibitors have been envisaged as potential therapeutic compounds for disease states, which are characterized by overshooting proteinase activities. Of all the projects started, the eglin project was undoubtedly developed furthest. To obtain large quantities of eglin c, the gene was chemically synthesized and cloned in *E. coli* (Rink *et al.*, 1984). The resulting bacterial strain expresses the gene at high levels; indeed, eglin c comprises 20% of the bacterial protein produced, and about 400 mg of the inhibitor can be isolated from 1 liter of culture medium. Recombinant eglin c has the same amino acid sequence as the naturally occurring inhibitor, but the amino terminus is blocked by an acetyl group. With respect to enzyme inhibitory properties, stability, and recognition by a number of monoclonal antibodies, the proteins derived from either leech or *E. coli* are indistinguishable. Wide-ranging studies investigating animal models of disease, pharmacokinetics, and toxicology were carried out with encouraging results. Nevertheless, the development of eglin c as a drug had to be abandoned because of hypersensitivity reactions in humans (Schnebli and Liersch, 1989).

The experience gained during this project is, of course, also reflected in numerous experiments and results of general interest (Heinz *et al.*, 1989, 1991, 1992). As with the barley inhibitor, the specificity of eglin c can be changed by a substitution at P_1: L45R eglin c is an excellent trypsin inhibitor ($K_D = 2.5 \times 10^{-11}$ *M*). Surprisingly, the double P_1, P_1' mutant L45R, D46S eglin c is a substrate of trypsin and various other serine proteinases, although NMR studies show virtually the same conformation as for the wild-type inhibitor. There are indications, however, that in the double mutant the internal rigidity of the binding loop is significantly weakened due to the loss of the intraresidue hydrogen bond between the backbone amide proton P_1' and its own side chain.

The homolog T44P eglin c was constructed to examine the role of a proline residue at position P_2, which is frequently found in serine proteinase inhibitors. The mutant remains a potent elastase inhibitor but no longer inhibits subtilisin. The substitutions R51K and R53K concern the crucial hydrogen bonding network between the core of the molecule and residues in the binding loop, respectively. Both mutants show a decrease in inhibitory potential.

VI. Chemically Synthesized Proteinase Inhibitors and Fragments

Extraordinary progress in the chemical synthesis of peptides and proteins has been achieved since the late 1980s. The key methodological improvements have been the development of automated, highly optimized, solid-phase peptide synthesis and of high-resolution chromatographic and electrophoretic techniques for purification and analysis (Kent, 1988; Bayer, 1991). Thus, the total chemical synthesis of all the proteinase inhibitors discussed so far, having up to about 80 residues, is clearly within the capabilities of current methods. This holds especially true for the small squash seed inhibitors. In this connection, the question arises of to which minimum size proteinaceous inhibitors can be reduced while their activity is maintained. The problem is most conveniently addressed by chemical synthesis, not least because it allows the incorporation of non-genetically encoded structures (e.g., D-amino acids) in a completely general fashion.

A. Squash Seed Inhibitor Homologs

The first member of the squash seed inhibitor family of miniproteins to be chemically synthesized is the *Cucurbita maxima* trypsin inhibitor III. The linear 29-residue peptide was assembled using the solid-phase procedure (Boc protection, DCC coupling), oxidized to yield the three disulfide bridges, and purified by affinity chromatography on immobilized anhydrotrypsin. The product proved to be identical with the natural inhibitor according to all criteria applied (Kupryszewski *et al.*, 1986). Shortly afterward, the homologous 28-residue *Ecballium elaterium* trypsin inhibitor II was also synthesized by the solid-phase technique (Boc protection, BOP coupling) (Le-Nguyen *et al.*, 1989) in quantities large enough for structural studies using NMR, which led to the correct disulfide bridge assignment (Heitz *et al.*, 1989).

Encouraged by these successful syntheses, several groups embarked on synthesizing homologs in order to facilitate the correct folding and thus improve the yields or to change the inhibitory specificity (Rolka *et al.*, 1989, 1991; McWherter *et al.*, 1989; Favel *et al.*, 1989b). Most of the results from monosubstitutions at P_1 are in line with the general rules deduced from the other inhibitor families and can be summarized as follows. Inhibition of chymotrypsin and cathepsin G is achieved with P_1 being L, A, F, and M; these homologs also retain some antitryptic activity. The relatively low affinity of the R5F homolog of the *Cucurbita maxima* inhibitor III to chymotrypsin ($K_D \approx 10^{-6}$ *M*) can be increased by five orders of magnitude when the binding loop is approximated to that of turkey ovomucoid third domain, yielding the V2G, P4T, R5F, I6E, L7Y, M8R homolog ($K_D = 1.7 \times 10^{-12}$ *M*). Pancreatic elastase accepts A ($K_D = 2.5 \times 10^{-8}$ *M*) and to a lesser degree norleucine and V as P_1 residues. Inhibition of the leukocyte enzyme is observed with V, I, G, L, A, F, and M. Dissociation constants were only determined for the R5V homolog–leukocyte elastase complex and found to be in the nanomolar range. An additional substitution to yield the R5V, I6D homolog, which shares its P_1' residue with eglin c, does not improve the antielastase activity.

More drastic changes of the *Cucurbita maxima* inhibitor (cf. Fig. 4) aiming at reducing the size of this miniprotein and simplifying its structure by removal of disulfide bridges or certain side chains have been reported (Rolka *et al*., 1991, 1992; Rozycki *et al*., 1993). Shortening the polypeptide chain at the amino terminus (i.e., elimination of R1 and V2) and at the carboxyl terminus (i.e., elimination of G29) does not influence the activity of the trypsin inhibitor. This is rather surprising because, according to crystallographic data, V2 and G29 are in van der Waals contact with the enzyme. A further shortening is not readily possible, as C3 and C28 are the outermost disulfide bridgeheads in the sequence, and all three cystine cross-links are essential for effective trypsin inhibition. The substitution of each of them with a pair of glycine and alanine residues increases the dissociation constant by six to seven orders of magnitude. Equally deleterious are the introductions of glycine at positions 17 or 19 and valine at position 27, which were, among others, performed to enhance the stability of a β turn or to match the structure more closely with that of other native inhibitors. Thus, despite the available X-ray and NMR structures, there is still a long way to go toward understanding all contributions of single residues to structure and inhibitory activity.

B. Peptides Based on Aprotinin

Successful chemical syntheses of the entire bovine pancreatic trypsin inhibitor were already reported in the early 1970s. They were carried out both by the Merrifield method (Noda *et al*., 1971; Tan and Kaiser, 1976) and by fragment condensation (Yajima and Kiso, 1974). From today's point of view, some doubt as to the homogeneity of the isolated proteins is certainly understandable.

The current methodological status is nicely exemplified by a recent synthesis of aprotinin and two homologs with three or two (C5–C55, C30–C51) disulfide bridges replaced by pairs of α-aminobutyric acid residues. The linear sequences were assembled by stepwise Fmoc solid-phase peptide synthesis and the final products characterized by amino acid analysis, sequencing, ion electrospray mass spectrometry, analytical high-performance liquid chromatography (HPLC), capillary zone electrophoresis, and circular dichroism (CD) spectroscopy. Synthetic aprotinin and the natural inhibitor were indistinguishable by all criteria applied (Ferrer *et al.*, 1992). Thus, reliable techniques are also available to chemically prepare aprotinin and homologs thereof. With the latter, however, problems of correct folding, as they are amply known with homologs prepared by recombinant methods, may be encountered.

We now return to the problem of size reduction and structure simplification, which was already touched on in the last section. In the trypsin–aprotinin complex only 12 of the 58 amino acids of the inhibitor are within van der Waals distance of enzyme residues. As is readily seen from the ball-and-stick model in Fig. 9, these residues are situated in the slender portion of aprotinin, clustered in two strands which are distant in the linear sequence but held together in a close antiparallel conformation in the folded molecule. This portion is stabilized by a disulfide bridge, C14–C38, several internal hydrogen bonds, and a scaffold supplied by the remainder of the molecule.

Several attempts have been made to synthesize small fragments of aprotinin, which should retain as much inhibitory potency as possible. The peptides, which are compiled in Table II, comprise some or all of the 12 contact residues and are often circular to compensate for the loss of rigidity brought about by the elimination of the aprotinin core.

The first two model compounds, **18** and **19** are extremely weak competitive trypsin inhibitors, although they already contain the aprotinin reactive site K–A bond and the adjacent disulfide bridge. The inhibition is only slightly improved when the peptide bond between the two fragments is closed (**20**) or when the disulfide-bridged peptide chain is enlarged by the two residues R and F (**21**) (Weber and Schmid, 1975, 1976; Wiejak and Rzeszotarska, 1974). The circular hexapeptide **22** (Wiejak and Rzeszotarska, 1975) and the highly rigid bicyclic hexapeptide **23** (Siemion *et al.*, 1973) do not show any appreciable antitryptic activity.

Dissociation constants in the micromolar range are only attained with peptide **24,** comprising all 12 aprotinin contact residues (Tan and Kaiser, 1977). Virtually the same inhibition is achieved with the shorter peptide **25** containing the bridging dipeptide fragment -D-phenylalanyl-prolyl-, which is surmised to direct tight β turns in peptidic structures. The fact that the reduced linear peptide **25** with acetamidomethyl groups on the two cysteines is still a trypsin inhibitor ($K_D = 5 \times 10^{-5}$ *M*) may indicate that a favorable aprotinin-like conformation is produced at least in the bound state (Kitchell and Dyckes, 1982). Peptide **26** was designed to mimic the aprotinin contact region by joining the ends of the two antiparallel strands using a cystine moiety and a glycine residue, respectively. It is, however, devoid of any inhibitory activity, probably due to a flexible conformation favored by the deletion of the disulfide bridge adjacent to the putative reactive site.

It should be noted that all inhibitory peptides are subject to a more or less rapid degradation by the proteinase. Thus, it appears that, although only a small

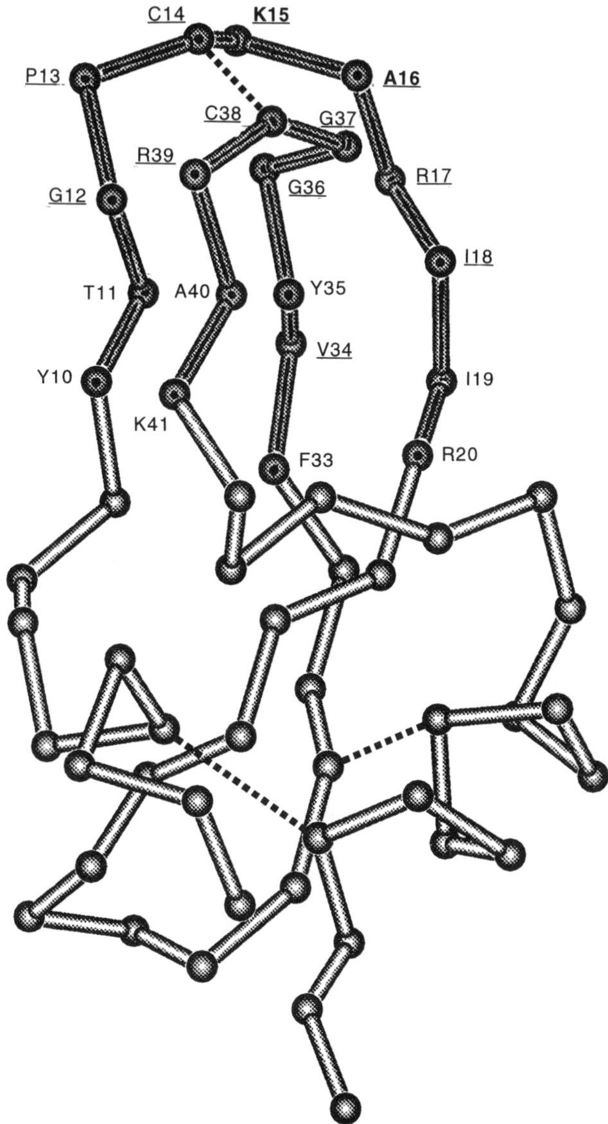

Figure 9 Projection of the α-carbon atoms of the bovine pancreatic trypsin inhibitor. The amino acid residues taken as basis for the construction of inhibitory peptides (Table II) are drawn with solid lines and designated by their one letter code and position number, those in close contact with the enzyme in the trypsin–inhibitor complex are underlined. The dotted lines represent the three disulfide bridges.

Table II Inhibitory Activity of Peptides Mimicking the Aprotinin Contact Region

No.	Sequence (putative P_1P_1' in bold)	Corresponding aprotinin fragments (cf. Fig. 9)	K_D (trypsin complex)
18	C **K A** G G C	C14-. . .-A16, G36-. . .-C38	2.3×10^{-2} *M*
19	G P C **K A** G G C	G12-. . .-A16, G36-. . .-C38	4.9×10^{-3} *M*
20	C **K A** G G C-OMe	C14-. . .-A16-G36-. . .-C38	2.1×10^{-3} *M*
21	G P C **K A** R F G G C	G12-. . .-R17-F-G36-. . .-C38	4.7×10^{-4} *M*
22	**K A** R Y G G	K15-. . .-R17-Y35-. . .-G37	No inhibition
23	P C **K G** C R	P13-. . .-K15-G37-. . .-R39	No inhibition
24	G P C **K A** R I I R F V Y G G C R A	G12-. . .-R20-F33-. . .-A40	1.8×10^{-6} *M*
25	Ac-P C **K A** R I DF P Y G G C R-NH_2	P13-. . .-I18-DF-P-Y35-. . .-R39	2.1×10^{-6} *M*
26	C V Y G G A R A K G Y T G P A **K A** R I I C	C-V34-. . .-G37-A-R39-. . .-K41-G-Y10-. . .-P13-A-K15-. . .-I19-C	No inhibition

portion of aprotinin directly interacts with trypsin, the remainder of the molecule imposing rigidity is essential for the strong interaction. A vast gap of about eight orders of magnitude separates the K_D value of the most efficient synthetic peptide from that of the natural inhibitor.

This conclusion is corroborated by a recent study of linear acetylated peptide amides based on the aprotinin reactive site region (Deshpande *et al.*, 1991). The series starts with Ac-GP-NH_2 corresponding to G12 and P13 of aprotinin, followed by Ac-GPα-aminobutyramide in which the unnatural residue substitutes for C14. The chain is then elongated step by step by K, A, R, I, and I, following the aprotinin sequence. Weak inhibition of porcine pancreatic kallikrein is only found with the penta-, hexa-, hepta-, and octapeptide; the dissociation constant of the last ($K_D = 1.2 \times 10^{-4}$ *M*) exceeds that of aprotinin by more than five orders of magnitude. Again, rigidity is of utmost importance: NMR measurements of the octapeptide confirm the assumed absence of a stable solution conformation.

C. Peptides Derived from the Potato Inhibitor 1 Family

The leech inhibitor eglin c with its 70 amino acid residues is one of the scarce examples of proteins synthesized by conventional solution methods (Okada and Tsuboi, 1991a,b). Several of the relatively short peptides, which were to be assembled later on to yield the entire eglin c, were tested against α-chymotrypsin, leukocyte elastase, and cathepsin G. None of them exhibits K_D values below 10^{-5} *M* (Okada *et al.*, 1989). Even the constant of a large peptide comprising residues 22 to 70 with the reactive site L45–E46 and thus 70% of the total sequence exceeds 10^{-6} *M*.

The switch from weak to very strong inhibition is only observed when up to seven amino-terminal residues are missing. The synthetic peptide 8–70 (Okada *et al.*, 1990) as well as the truncated eglins 7–70 and 5–70, prepared by degradation with the exopeptidase cathepsin C (Dodt *et al.*, 1987), exhibit at least nanomolar dissociation constants, as does the natural inhibitor.

When comparing the backbone structure of aprotinin (Fig. 2) with that of barley serine proteinase inhibitor 2 (Fig. 5), it is evident that the latter, because of its extended reactive site loop, preferentially lends itself to the design of small peptide-based inhibitors. Indeed, an 18-residue peptide comprising the reactive site M59–E60 and most of the other loop residues was synthesized and forced to adopt a cyclic structure by introduction of a disulfide bridge. Virtually identical inhibitory properties were claimed for the synthetic peptide and the natural 83-residue inhibitor (Leatherbarrow and Salacinski, 1991). The results, however, could not be reproduced. The cyclic peptide appears to be devoid of any appreciable inhibitory activity.

VII. Concluding Remarks and Outlook

Extensive screening of natural sources has yielded an ample array of small proteinaceous inhibitors of serine proteinases. Peptide sequencing, which has become more and more routine for molecules of that size, allows for grouping them into several families, four of which have been dealt with in this review as instructive examples. It can be anticipated that the number of new structures will increase considerably in future, not least because of the modern fast and highly selective methods of purification. It will also increase, however, because of extensive use of synthetic strategies, either total synthesis, semisynthesis, or recombinant techniques. This particularly applies when the specificity of a given inhibitor is to be changed. X-Ray and NMR structural data now available for several free inhibitors and their proteinase complexes are of invaluable help with regard to a rational procedure.

A persistent challenge for many groups involved in proteinase inhibitor research has been the reduction and simplification of these molecules to the absolutely necessary structural elements. At first glance, this task does not seem too complicated, because the backbone conformation of the crucial amino acid residues P_3 to P_3' is highly conserved throughout the inhibitor families (Wibley and Barlow, 1992) and the remainder of the molecules look like an optional template to arrange the short peptide strands properly. What has up to now been achieved, however, is the mere truncation of some inhibitors by a few residues at the amino or carboxyl terminus. All efforts to mimic the contact regions by short rigid peptides have obviously been somewhat naive and have essentially failed to maintain the high inhibitory potency.

The more surprising is the report on so-called pepzymes (Atassi and Manshouri, 1993), cyclic 29-residue peptides in which important active site residues of trypsin or chymotrypsin are linked by glycine spacers. The claim that these compounds exhibit virtually the same activity and specificity as the respective proteinase has been met with some skepticism. We hold these findings to be highly improbable and believe that they urgently require independent confirmation.

Two strategies for the production of tailor-made proteinaceous or peptide inhibitors, which in a way duplicate evolution, are of great promise for the future. One of them, the phage display system discussed above, has already led to the most effective aprotinin homologous elastase inhibitor, and the other makes use of synthetic peptide combinatory libraries (Jung and Beck-Sickinger, 1992). By a systematic screening of specifically designed libraries of hexapeptide and dodecapeptide mixtures, the sequence -AKIYR- (P_2 to P_3') was identified as the essential motif for trypsin inhibition (Eichler and Houghten, 1993). It is conceivable that in future studies this or other sequences found could be structurally modified or incorporated into suitable molecular scaffolds to eventually obtain therapeutically useful compounds.

Acknowledgments

This review is dedicated to Professor Dr. Hans Fritz, Munich, on the occasion of his 60th birthday. The work in our laboratory has been supported by the Deutsche Forschungsgemeinschaft (Ts 8/24 and SFB, B08). Thanks are due to Ms. G. Delany for linguistic advice.

References

Altman, J. D., Henner, D., Nilsson, B., Anderson, S., and Kuntz, I. D. (1991). *Protein Eng.* **4,** 593–600.

Anderson, S., and Kingston, I. B. (1983). *Proc. Natl. Acad. Sci. U.S.A.* **80,** 6838–6842.

Anthony-Cahill, S. J., Griffith, M. C., Noren, C. J., Suich, D. J., and Schultz, P. G. (1989). *Trends Biochem. Sci.* **14,** 400–403.

Apostol, I., Giletto, A., Komiyama, T., Zhang, W., and Laskowski, M., Jr. (1993). *J. Protein Chem.* **12,** 419–433.

Ardelt, W., and Laskowski, M., Jr. (1991). *J. Mol. Biol.* **220,** 1041–1053.

Atassi, M. Z., and Manshouri, T. (1993). *Proc. Natl. Acad. Sci. U.S.A.* **90,** 8282–8286.

Auerswald, E.-A., Schröder, W., and Kotick, M. (1987). *Biol. Chem. Hoppe-Seyler* **368,** 1413–1425.

Auerswald, E.-A., Hörlein, D., Reinhardt, G., Schröder, W., and Schnabel, E. (1988). *Biol. Chem. Hoppe-Seyler* **369,** 27–35.

Auerswald, E. A., Schubert, A., Dolinar, M., Gürtler, L., and Deinhardt, F. (1991). *Biomed. Biochim. Acta* **50,** 697–700.

Baici, A., Salgam, P., Cohen, G., Fehr, K., and Böni, A. (1982). *Rheumatol. Int.* **2,** 11–16.

Barra, D., Simmaco, M., Bossa, F., Fioretti, E., Angeletti, M., and Ascoli, F. (1987). *J. Biol. Chem.* **262,** 13916–13919.

Barra, D., Fioretti, E., Angeletti, M., Maras, B., Bossa, F., and Ascoli, F. (1991). *Biochim. Biophys. Acta* **1076,** 143–147.

Barrett, A. J. (1978). *Agents Actions* **8,** 11–18.

Barrett, A. J., and Salvesen, G. (eds.) (1986). "Proteinase Inhibitors." Elsevier, Amsterdam.

Barthel, T., and Kula, M.-R. (1993). *Biotechnol. Bioeng.* **42,** 1331–1336.

Bayer, E. (1991). *Angew. Chem., Int. Ed. Engl.* **30,** 113–129.

Beatty, K., Bieth, J., and Travis, J. (1980). *J. Biol. Chem.* **255,** 3931–3934.

Beckmann, J., Mehlich, A., Schröder, W., Wenzel, H. R., and Tschesche, H. (1988a). *Eur. J. Biochem.* **176,** 675–682.

Beckmann, J., Mehlich, A., Wenzel, H. R., and Tschesche, H. (1988b). *Adv. Exp. Med. Biol.* **240,** 107–114.

Beckmann, J., Mehlich, A., Schröder, W., Wenzel, H. R., and Tschesche, H. (1989a). *J. Protein Chem.* **8,** 101–113.

Beckmann, J., Mehlich, A., Feldmann, A., Wenzel, H. R., and Tschesche, H. (1989b). *In* "Intracellular Proteolysis Mechanisms and Regulations" (N. Katunuma and E. Kominami, eds.), pp. 361–368. Japan Scientific Societies Press, Tokyo.

Berndt, K. D., Güntert, P., Orbons, L. P. M., and Wüthrich, K. (1992). *J. Mol. Biol.* **227,** 757–775.

Berndt, K. D., Beunink, J., Schröder, W., and Wüthrich, K. (1993). *Biochemistry* **32,** 4564–4570.

Betzel, C., Dauter, Z., Genov, N., Lamzin, V., Navaza, J., Schnebli, H. P., Visanji, M., and Wilson, K. S. (1993). *FEBS Lett.* **317,** 185–188.

Bigler, T. L., Lu, W., Park, S. J., Tashiro, M., Wieczorek, M., Wynn, R., and Laskowski, M., Jr. (1993). *Protein Sci.* **2,** 786–799.

Bode, W., and Huber, R. (1986). *In* "Molecular and Cellular Basis of Digestion" (P. Desnuelle, H. Sjöström, and O. Noren, eds.), pp. 213–234. Elsevier, Amsterdam.

Bode, W., and Huber, R. (1991). *Curr. Opin. Struct. Biol.* **1,** 45–52.

Bode, W., and Huber, R. (1992). *Eur. J. Biochem.* **204,** 433–451.

Bode, W., Epp, O., Huber, R., Laskowski, M., Jr., and Ardelt, W. (1985). *Eur. J. Biochem.* **147,** 387–395.
Bode, W., Wei, A.-Z., Huber, R., Meyer, E., Travis, J., and Neumann, S. (1986a). *EMBO J.* **5,** 2453–2458.
Bode, W., Papamokos, E., Musil, D., Seemueller, U., and Fritz, H. (1986b). *EMBO J.* **5,** 813–818.
Bode, W., Papamokos, E., and Musil, D. (1987). *Eur. J. Biochem.* **166,** 673–692.
Bode, W., Greyling, H. J., Huber, R., Otlewski, J., and Wilusz, T. (1989). *FEBS Lett.* **242,** 285–292.
Braun, N. J., and Schnebli, H. P. (1986). *Eur. J. Respir. Dis.* **69**(Suppl. 146), 541–547.
Braun, N. J., Bodmer, J. L., Virca, G. D., Metz-Virca, G., Maschler, R., Bieth, J. G., and Schnebli, H. P. (1987). *Biol. Chem. Hoppe-Seyler* **368,** 299–308.
Brinkmann, T., Schnierer, S., and Tschesche, H. (1991). *Eur. J. Biochem.* **202,** 95–99.
Burkhardt, H., Kasten, M., Rauls, S., and Rehkopf, E. (1987). *Rheumatol. Int.* **7,** 133–138.
Businaro, R., Fioretti, E., Fumagalli, L., De Renzis, G., Fiorucci, G., and Ascoli, F. (1989). *Histochemistry* **93,** 69–74.
Cesareni, G. (1992). *FEBS Lett.* **307,** 66–70.
Chaiken, I. M., Komoriya, A., Ohno, M., and Widmer, F. (1982). *Appl. Biochem. Biotechnol.* **7,** 385–399.
Chang, J.-Y., Knecht, R., Maschler, R., and Seemüller, U. (1985). *Biol. Chem. Hoppe-Seyler* **366,** 281–286.
Chen, Z., and Bode, W. (1983). *J. Mol. Biol.* **164,** 283–311.
Chu, M.-L., Zhang, R.-Z., Pan, T.-C., Stokes, D., Conway, D., Kuo, H.-J., Glanville, R., Mayer, U., Mann, K., Deutzmann, R., and Timpl, R. (1990). *EMBO J.* **9,** 385–393.
Clore, G. M., Gronenborn, A. M., Kjaer, M., and Poulsen, F. M. (1987a). *Protein Eng.* **1,** 305–311.
Clore, G. M., Gronenborn, A. M., James, M. N. G., Kjaer, M., McPhalen, C. A., and Poulsen, F. M. (1987b). *Protein Eng.* **1,** 313–318.
Collins, J., Szardenings, M., Maywald, F., Fritz, H., Bruns, W., Reinhardt, G., Schnabel, E., Schröder, W., Blöcker, H., Reichelt, J., and Schomburg, D. (1989). *In* "Advances in Protein Design" (H. Blöcker, J. Collins, R. D. Schmid, and D. Schomburg, eds.), pp. 201–210. VCH, Weinheim.
Collins, J., Szardenings, M., Maywald, F., Blöcker, H., Frank, R., Hecht, H.-J., Vasel, B., Schomburg, D., Fink, E., and Fritz, H. (1990). *Biol. Chem. Hoppe-Seyler* **371**(Suppl.), 29–36.
Creighton, T. E. (1978). *Prog. Biophys. Mol. Biol.* **33,** 231–297.
Creighton, T. E. (1990). *Biochem. J.* **270,** 1–16.
Creighton, T. E. (1992). *Science* **256,** 111–112.
Creighton, T. E., and Charles, I. G. (1987). *Cold Spring Harbor Symp. Quant. Biol.* **52,** 511–519.
Creighton, T. E., Bagley, C. J., Cooper, L., Darby, N. J., Freedman, R. B., Kemmink, J., and Sheikh, A. (1993). *J. Mol. Biol.* **232,** 1176–1196.
Darby, N. J., and Creighton, T. E. (1993). *J. Mol. Biol.* **232,** 873–896.
Deshpande, M. S., Boylan, J., Hamilton, J. A., and Burton, J. (1991). *Int. J. Pept. Protein Res.* **37,** 536–543.
Dodt, J., Seemüller, U., and Fritz, H. (1987). *Biol. Chem. Hoppe-Seyler* **368,** 1447–1453.
Drapeau, G. R., Boily, Y., and Houmard, J. (1972). *J. Biol. Chem.* **247,** 6720–6726.
Dutler, H., and Bizzozero, S. A. (1989). *Acc. Chem. Res.* **22,** 322–327.
Eguchi, M. (1993). *Comp. Biochem. Physiol.* **105B,** 449–456.
Eichler, J., and Houghten, R. A. (1993). *Biochemistry* **32,** 11035–11041.
Eigenbrot, C., Randal, M., and Kossiakoff, A. A. (1990). *Protein Eng.* **3,** 591–598.
Ellman, J. A., Mendel, D., and Schultz, P. G. (1992). *Science* **255,** 197–200.
Empie, M. W., and Laskowski, M., Jr. (1982). *Biochemistry* **21,** 2274–2284.
Estell, D. A., Wilson, K. A., and Laskowski, M., Jr. (1980). *Biochemistry* **19,** 131–137.
Favel, A., Mattras, H., Coletti-Previero, M. A., Zwilling, R., Robinson, E. A., and Castro, B. (1989a). *Int. J. Pept. Protein Res.* **33,** 202–208.
Favel, A., Le-Nguyen, D., Coletti-Previero, M. A., and Castro, B. (1989b). *Biochem. Biophys. Res. Commun.* **162,** 79–82.
Feeney, R. E. (1970). *In* "Proceeding of the International Research Conference on Proteinase Inhibitors" (H. Fritz and H. Tschesche, eds.), pp. 189–195. de Gruyter, Berlin and New York.

Ferrer, M., Woodward, C., and Barany, G. (1992). *Int. J. Pept. Protein Res.* **40,** 194–207.

Finkenstadt, W. R., Hamid, M. A., Mattis, J. A., Schrode, J., Sealock, R. W., Wang, D., and Laskowski, M., Jr. (1974). *In* "Proteinase Inhibitors" (H. Fritz, H. Tschesche, L. J. Greene, and E. Truscheit, eds.), pp. 389–411. Springer-Verlag, Berlin.

Fioretti, E., Iacopino, G., Angeletti, M., Barra, D., Bossa, F., and Ascoli, F. (1985). *J. Biol. Chem.* **260,** 11451–11455.

Fioretti, E., Angeletti, M., Citro, G., Barra, D., and Ascoli, F. (1987). *J. Biol. Chem.* **262,** 3586–3589.

Fioretti, E., Angeletti, M., Passeri, D., and Ascoli, F. (1989). *J. Protein Chem.* **8,** 51–60.

Fioretti, E., Angeletti, M., Coletta, M., Ascenzi, P., Bolognesi, M., Menegatti, E., Rizzi, M., and Ascoli, F. (1993a). *J. Enzyme Inhibition* **7,** 57–64.

Fioretti, E., Eleuteri, A. M., Angeletti, M., and Ascoli, F. (1993b). *J. Chromatogr. Biomed. Appl.* **617,** 308–312.

Fiorucci, L., De Renzis, G., Businaro, R., Fumagalli, L., Fioretti, E., Giardina, B., and Ascoli, F. (1989). *Histochem. J.* **21,** 721–730.

Frigerio, F., Coda, A., Pugliese, L., Lionetti, C., Menegatti, E., Amiconi, G., Schnebli, H. P., Ascenzi, P., and Bolognesi, M. (1992). *J. Mol. Biol.* **225,** 107–123.

Fritz, H., Kruck, J., Rüsse, I., and Liebich, H. G. (1979). *Hoppe-Seyler's Z. Physiol. Chem.* **360,** 437–444.

Fritz, H., and Wunderer, G. (1983). *Arzneim.-Forsch./Drug Res.* **33,** 479–494.

Fujinaga, M., Sielecki, A. R., Read, R. J., Ardelt, W., Laskowski, M., Jr., and James, M. N. G. (1987). *J. Mol. Biol.* **195,** 397–418.

Gambacurta, A., Piro, M. C., and Ascoli, F. (1993). *Biochim. Biophys. Acta* **1174,** 267–273.

Gebhard, W., Tschesche, H., and Fritz, H. (1986). *In* "Proteinase Inhibitors" (A. J. Barrett and G. Salvesen, eds.), pp. 375–388. Elsevier, Amsterdam.

Goldenberg, D. P. (1988). *Biochemistry* **27,** 3481–2489.

Goldenberg, D. P. (1992). *Trends Biochem. Sci.* **17,** 257–261.

Goldenberg, D. P., Bekeart, L. S., Laheru, D. A., and Zhou, J. D. (1993). *Biochemistry* **32,** 2835–2844.

Grant, S. K., Meek, T. D., Metcalf, B. W., and Petteway, S. R., Jr. (1993). *In* "Biomedical Applications of Biotechnology" (W. V. Williams and D. B. Weiner, eds.), Vol. 1, pp. 325–353. Technomic Publ., Lancaster, Pennsylvania.

Groeger, C., Wenzel, H. R., and Tschesche, H. (1991a). *J. Protein Chem.* **10,** 245–251.

Groeger, C., Wenzel, H. R., and Tschesche, H. (1991b). *J. Protein Chem.* **10,** 527–533.

Groeger, C., Wenzel, H. R., and Tschesche, H. (1993). *Angew. Chem., Int. Ed. Engl.* **32,** 898–900.

Groeger, C., Wenzel, H. R., and Tschesche, H. (1994). *Int. J. Pept. Protein Res.* **44,** 166–172.

Gros, P., Fujinaga, M., Dijkstra, B. W., Kalk, K. H., and Hol, W. G. J. (1989a). *Acta Crystallogr. Sect. B Struct. Sci.* **B45,** 488–499.

Gros, P., Betzel, C., Dauter, Z., Wilson, K. S., and Hol, W. G. J. (1989b). *J. Mol. Biol.* **210,** 347–367.

Hao, M.-H., Pincus, M. R., Rackovsky, S., and Scheraga, H. A. (1993). *Biochemistry* **32,** 9614–9631.

Hara, S., Makino, J., and Ikenaka, T. (1989). *J. Biochem. (Tokyo)* **105,** 88–92.

Hattori, T., Koito, A., Takatsuki, K., Kido, H., and Katunuma, N. (1989). *FEBS Lett.* **248,** 48–52.

Hecht, H. J., Szardenings, M., Collins, J., and Schomburg, D. (1991). *J. Mol. Biol.* **220,** 711–722.

Hecht, H. J., Szardenings, M., Collins, J., and Schomburg, D. (1992). *J. Mol. Biol.* **225,** 1095–1103.

Heinz, D. W., Liersch, M., and Grütter, M. G. (1989). *J. Mol. Biol.* **207,** 641–642.

Heinz, D. W., Priestle, J. P., Rahuel, J., Wilson, K. S., and Grütter, M. G. (1991). *J. Mol. Biol.* **217,** 353–371.

Heinz, D. W., Hyberts, S. G., Peng, J. W., Priestle, J. P., Wagner, G., and Grütter, M. G. (1992). *Biochemistry* **31,** 8755–8766.

Heitz, A., Chiche, L., Le-Nguyen, D., and Castro, B. (1989). *Biochemistry* **28,** 2392–2398.

Hipler, K., Priestley, J. P., Rahuel, J., and Grütter, M. G. (1992). *FEBS Lett.* **309,** 139–145.

Hiromi, K., Akasaka, K., Mitsui, Y., Tonomura, B., and Murao, S. (eds.) (1985). "Protein Protease Inhibitor—The Case of *Streptomyces* Subtilisin Inhibitor (SSI)." Elsevier, Amsterdam.

Hojima, Y., Pierce, J. V., and Pisano, J. J. (1982). *Biochemistry* **21,** 3741–3746.
Hojima, Y., Pisano, J. J., and Cochrane, C. G. (1983). *Biochem. Pharmacol.* **32,** 985–990.
Holak, T. A., Gondol, D., Otlewski, J., and Wilusz, T. (1989a). *J. Mol. Biol.* **210,** 635–648.
Holak, T. A., Bode, W., Huber, R., Otlewski, J., and Wilusz, T. (1989b). *J. Mol. Biol.* **210,** 649–654.
Holzer, H., and Tschesche, H. (eds.) (1979). "Biological Functions of Proteinases." Springer-Verlag, Berlin.
Huang, Q., Liu, S., Tang, Y., Zeng, F., and Qian, R. (1992). *FEBS Lett.* **297,** 143–146.
Huang, Q., Liu, S., and Tang, Y. (1993). *J. Mol. Biol.* **229,** 1022–1036.
Huber, R., Kukla, D., Bode, W., Schwager, P., Bartels, K., Deisenhofer, J., and Steigemann, W. (1974). *J. Mol. Biol.* **89,** 73–101.
Hurle, M. R., Anderson, S., and Kuntz, I. D. (1991). *Protein Eng.* **4,** 451–455.
Izumoto, Y., Sato, T., Yamamoto, T., Yoshida, N., Kikuchi, N., Ogawa, M., and Matsubara, K. (1987). *Gene* **59,** 151–159.
Jackson, S. E., and Fersht, A. R. (1991a). *Biochemistry* **30,** 10428–10435.
Jackson, S. E., and Fersht, A. R. (1991b). *Biochemistry* **30,** 10436–10443.
Jackson, S. E., Moracci, M., el Masry, N., Johnson, C. M, and Fersht, A. R. (1993a). *Biochemistry* **32,** 11259–11269.
Jackson, S. E., el Masry, N., and Fersht, A. R. (1993b). *Biochemistry* **32,** 11270–11278.
Janoff, A. (1985). *Am. Rev. Respir. Dis.* **132,** 417–433.
Janoff, A. (1988). *In* "Inflammation: Basic Principles and Clinical Correlates" (J. I. Gallin, I. M. Goldstein, and R. Snyderman, eds.), pp. 803–814. Raven, New York.
Jering, H., and Tschesche, H. (1974). *Angew. Chem., Int. Ed. Engl.* **13,** 662–663.
Jering, H., and Tschesche, H. (1976a). *Eur. J. Biochem.* **61,** 443–452.
Jering, H., and Tschesche, H. (1976b). *Eur. J. Biochem.* **61,** 453–463.
Joubert, F. J. (1984). *Phytochemistry* **23,** 1401–1406.
Jung, G., and Beck-Sickinger, A. G. (1992). *Angew. Chem., Int. Ed. Engl.* **31,** 367–383.
Kanamori, T., Mizushima, S., Shimizu, Y., Morishita, H., Kubota, H., Nii, A., Ogino, H., Nagase, Y., Kisaragi, M., and Nobuhara, M. (1988). *Gene* **66,** 295–300.
Kato, I., Schrode, J., Kohr, W. J., and Laskowski, M., Jr. (1987). *Biochemistry* **26,** 193–201.
Kent, S. B. H. (1988). *Annu. Rev. Biochem.* 57, 957–989.
Kido, H., Takeda, M., Wakabayashi, H., Tanaka, S., Nishimura, N., Takenaka, M., and Okada, M. (1993). *Gerontology* **39**(Suppl. 1), 30–37.
Kikuchi, N., Nagata, K., Yoshida, N., and Ogawa, M. (1985). *J. Biochem. (Tokyo)* **98,** 687–694.
Kikuchi, N., Nagata, K., Shin, M., Mitsushima, K., Teraoka, H., and Yoshida, N. (1989). *J. Biochem. (Tokyo)* **106,** 1059–1063.
Kim, P. S., and Baldwin, R. L. (1990). *Annu. Rev. Biochem.* 59, 631–660.
Kitaguchi, N., Takahashi, Y., Tokushima, Y., Shiojiri, S., and Ito, H. (1988). *Nature (London)* **331,** 530–532.
Kitchell, J. P., and Dyckes, D. F. (1982). *Biochim. Biophys. Acta* **701,** 149–152.
Komiyama, T., Bigler, T. L., Yoshida, N., Noda, K., and Laskowski, M., Jr. (1991). *J. Biol. Chem.* **266,** 10727–10730.
Kosen, P. A., Marks, C. B., Falick, A. M., Anderson, S., and Kuntz, I. D. (1992). *Biochemistry* **31,** 5705–5717.
Kraut, J. (1977). *Annu. Rev. Biochem.* 46, 331–358.
Krystek, S., Stouch, T., and Novotny, J. (1993). *J. Mol. Biol.* **234,** 661–679.
Kullmann, W. (1985). *J. Protein Chem.* **4,** 1–22.
Kupryszewski, G., Ragnarsson, U., Rolka, K., and Wilusz, T. (1986). *Int. J. Pept. Protein Res.* **27,** 245–250.
Laskowski, M., Jr. (1980). *Biochem. Pharmacol.* **29,** 2089–2094.
Laskowski, M., Jr. (1986). *In* "Nutritional and Toxicological Significance of Enzyme Inhibitors in Foods" (M. Friedman, ed.), pp. 1–17. Plenum, New York.
Laskowski, M., Jr., and Sealock. R. W. (1971). *In* "The Enzymes" (P. D. Boyer, ed.), 3rd Ed., Vol. 3, pp. 375–473. Academic Press, New York.

Laskowski, M., Jr., and Kato, I. (1980). *Annu. Rev. Biochem.* 49, 593–626.

Laskowski, M., Jr., Tashiro, M., Empie, M. W., Park, S. J., Kato, I., Ardelt, W., and Wieczorek, M. (1983). *In* "Proteinase Inhibitors: Medical and Biological Aspects" (N. Katunuma, H. Umezawa, and H. Holzer, eds.), pp. 55–68. Japan Scientific Societies Press, Tokyo/Springer-Verlag, Berlin.

Laskowski, M., Jr., Kato, I., Ardelt, W., Cook, J., Denton, A., Empie, M. W., Kohr, W. J., Park, S. J., Parks, K., Schatzley, B. L., Schoenberger, O. L., Tashiro, M., Vichot, G., Whatley, H. E, Wieczorek, A., and Wieczorek, M. (1987a). *Biochemistry* **26,** 202–221.

Laskowski, M., Jr., Kato, I., Kohr, W. J., Park, S. J., Tashiro, M., and Whatley, H. E. (1987b). *Cold Spring Harbor Symp. Quant. Biol.* **52,** 545–553.

Laskowski, M., Jr., Apostol, I., Ardelt, W., Cook, J., Giletto, A., Kella, C. A., Lu, W., Park, S. J., Qasim, M. A., Whatley, H. E., Wieczorek, A., and Wynn, R. (1990). *J. Protein Chem.* **9,** 715–725.

Leatherbarrow, R. J., and Matthews, S. J. (1992). *Magn. Res. Chem.* **30,** 1255–1260.

Leatherbarrow, R. J., and Salacinski, H. J. (1991). *Biochemistry* **30,** 10717–10721.

Le-Nguyen, D., Nalis, D., and Castro, B. (1989). *Int. J. Pept. Protein Res.* **34,** 492–497.

Le-Nguyen, D., Heitz, A., Chiche, L., Castro, B., Boigegrain, R. A., Favel, A., and Coletti-Previero, M. A. (1990). *Biochimie* **72,** 431–435.

Lindahl, A. K., Sandset, P. M., and Abildgaard, U. (1992). *Blood Coagul. Fibrinol.* **3,** 439–449.

Ling, M.-H., Qi, H.-Y., and Chi, C.-W. (1993). *J. Biol. Chem.* **268,** 810–814.

Longstaff, C., Campbell, A. F., and Fersht, A. R. (1990). *Biochemistry* **29,** 7339–7347.

Lu, W., Zhang, W., Molloy, S. S., Thomas, G., Ryan, K., Chiang, Y. W., Anderson, S., and Laskowski, M., Jr. (1993). *J. Biol. Chem.* **268,** 14583–14585.

McKeehan, W. L., Sakagami, Y., Hoshi, H., and McKeehan, K. A. (1986). *J. Biol. Chem.* **261,** 5378–5383.

McPhalen, C. A., and James, M. N. G. (1987). *Biochemistry* **26,** 261–269.

McPhalen, C. A., and James, M. N. G. (1988). *Biochemistry* **27,** 6582–6598.

McPhalen, C. A., Svendsen, I., Jonassen, I., and James, M. N. G. (1985). *Proc. Natl. Acad. Sci. U.S.A.* **82,** 7242–7246.

McWherter, C. A., Walkenhorst, W. F., Campbell, E. J., and Glover, G. I. (1989). *Biochemistry* **28,** 5708–5714.

Marks, C. B., Vasser, M., Ng, P., Henzel, W., and Anderson, S. (1986). *J. Biol. Chem.* **261,** 7115–7118.

Marks, J. D., Hoogenboom, H. R., Griffiths, A. D., and Winter, G. (1992). *J. Biol. Chem.* **267,** 16007–16010.

Matthews, S. J., Jandu, S. K., and Leatherbarrow, R. J. (1993). *Biochemistry* **32,** 657–662.

Maywald, F., Böldicke, T., Gross, G., Frank, R., Blöcker, H., Meyerhans, A., Schwellnus, K., Ebbers, J., Bruns, W., Reinhardt, G., Schnabel, E., Schröder, W., Fritz, H., and Collins, J. (1988). *Gene* **68,** 357–369.

Mittman, C. (ed.) (1972). "Pulmonary Emphysema and Proteolysis." Academic Press, New York.

Moses, E., and Hinz, H.-J. (1983). *J. Mol. Biol.* **170,** 765–776.

Musil, D., Bode, W., Huber, R., Laskowski, M., Jr., Lin, T. Y., and Ardelt, W. (1991). *J. Mol. Biol.* **220,** 739–755.

Nakamura, S., Suenaga, T., Akiguchi, I., Kimura, J., Nakamura, S., Tokushima, Y., Kitaguchi, N., Takahashi, Y., and Shiojiri, S. (1992). *Acta Neuropathol.* **84,** 244–249.

Nakayama, A., Kobayashi, H., Ando, K., Hori, M., Ohnishi, T., and Honjo, M. (1992). *J. Biotechnol.* **23,** 225–229.

Niekamp, C. W., Hixson, H. F., Jr., and Laskowski, M., Jr. (1969). *Biochemistry* **8,** 16–22.

Nilges, M., Habazettl, J., Brünger, A. T., and Holak, T. A. (1991). *J. Mol. Biol.* **219,** 499–510.

Nishino, J., Takano, R., Kamei-Hayashi, K., Minakata, H., Nomoto, K., and Hara, S. (1992). *Biosci. Biotech. Biochem.* **56,** 1241–1246.

Noda, K., Terada, S., Mitsuyasu, N., Waki, M., Kato, T., and Izumiya, N. (1971). *Naturwissenschaften* **58,** 147–148.

Norris, K., Norris, F., Bjorn, S. E., Diers, I., and Petersen, L. C. (1990). *Biol. Chem. Hoppe-Seyler* **371**(Suppl.), 37–42.

Ødum, L. (1990). *Int. J. Biochem.* **22,** 925–930.

Offord, R. E. (1980). "Semisynthetic Proteins." Wiley, Chichester and New York.

Offord, R. E. (1990). *In* "Protein Design and the Development of New Therapeutics and Vaccines" (J. B. Hook and G. Poste, eds.), pp. 253–282. Plenum, New York.

Offord, R. (1991). *Protein Eng.* **4,** 709–710.

Okada, Y., and Tsuboi, S. (1991a). *J. Chem. Soc., Perkin Trans. 1,* 3315–3319.

Okada, Y., and Tsuboi, S. (1991b). *J. Chem. Soc., Perkin Trans. 1,* 3321–3328.

Okada, Y., Tsuboi, S., Tsuda, Y., Nakabayashi, K., Nagamatsu, Y., and Yamamoto, J. (1989). *Biochem. Biophys. Res. Commun.* **161,** 272–275.

Okada, Y., Tsuboi, S., Tsuda, Y., Nagamatsu, Y., and Yamamoto, J. (1990). *FEBS Lett.* **272,** 113–116.

Otlewski, J., and Zbyryt, T. (1994). *Biochemistry* **33,** 200–207.

Otlewski, J., Whatley, H., Polanowski, A., and Wilusz, T. (1987). *Biol. Chem. Hoppe-Seyler* **368,** 1505–1507.

Otlewski, J., Zbyryt, T., Krokoszynska, I., and Wilusz, T. (1990). *Biol. Chem. Hoppe-Seyler* **371,** 589–594.

Papamokos, E., Weber, E., Bode, W., Huber, R., Empie, M. W., Kato, I., and Laskowski, M., Jr. (1982). *J. Mol. Biol.* **158,** 515–537.

Park, S. J. (1985). Ph.D. Thesis, Purdue University, West Lafayette, Indiana.

Polanowski, A., Wilusz, T., Nienartowicz, B., Cieslar, E., Slominska, A., and Nowak, K. (1980). *Acta Biochim. Pol.* **27,** 371–382.

Ponte, P., Gonzalez-De Whitt, P., Schilling, J., Miller, J., Hsu, D., Greenberg, P., Davis, K., Wallace, W., Lieberburg, I., Fuller, F., and Cordell, B. (1988). *Nature (London)* **331,** 525–527.

Powers, J. C., and Bengali, Z. H. (1987). *J. Enzyme Inhibition* **1,** 311–319.

Quast, U., Engel, J., Steffen, E., Tschesche, H., and Kupfer, S. (1978). *Biochemistry* **17,** 1675–1682.

Read, R. J., and James, M. N. G. (1986). *In* "Proteinase Inhibitors" (A. J. Barrett and G. Salvesen, eds.), pp. 301–336. Elsevier, Amsterdam.

Read, R. J., Fujinaga, M., Sielecki, A. R., and James, M. N. G. (1983). *Biochemistry* **22,** 4420–4433.

Reilly, C. F., and Travis, J. (1980). *Biochim. Biophys. Acta* **621,** 147–157.

Rich, D. H. (1990). *In* "Comprehensive Medicinal Chemistry, The Rational Design, Mechanistic Study and Therapeutic Application of Chemical Compounds" (C. Hansch, ed.), Vol. 2, pp. 391–441. Pergamon, Oxford.

Richarz, R., Tschesche, H., and Wüthrich, K. (1979). *Eur. J. Biochem.* **102,** 563–571.

Rink, H., Liersch, M., Sieber, P., and Meyer, F. (1984). *Nucleic Acids Res.* **12,** 6369–6387.

Roberts, B. L., Markland, W., Ley, A. C., Kent, R. B., White, D. W., Guterman, S. K., and Ladner, R. C. (1992a). *Proc. Natl. Acad. Sci. U.S.A.* **89,** 2429–2433.

Roberts, B. L., Markland, W., Siranosian, K., Saxena, M. J., Guterman, S. K., and Ladner, R. C. (1992b). *Gene* **121,** 9–15.

Rolka, K., Kupryszewski, G., Ragnarsson, U., Otlewski, J., Wilusz, T., and Polanowski, A. (1989). *Biol. Chem. Hoppe-Seyler* **370,** 499–502.

Rolka, K., Kupryszewski, G., Ragnarsson, U., Otlewski, J., Krokoszynska, I., and Wilusz, T. (1991). *Biol. Chem. Hoppe-Seyler* **372,** 63–68.

Rolka, K., Kupryszewski, G., Rozycki, J., Ragnarsson, U., Zbyryt, T., and Otlewski, J. (1992). *Biol. Chem. Hoppe-Seyler* **373,** 1055–1060.

Royston, D. (1990). *Blood Coagul. Fibrinol.* **1,** 55–69.

Rozycki, J., Kupryszewski, G., Rolka, K., Ragnarsson, U., Zbyryt, T., Krokoszynska, I., and Otlewski, J. (1993). *Biol. Chem. Hoppe-Seyler* **374,** 851–854.

Ryan, C. A. (1981). *In* "The Biochemistry of Plants" (A. Marcus, ed.), Vol. 6, pp. 351–370. Academic Press, New York.

Schechter, I., and Berger, A. (1967). *Biochem. Biophys. Res. Commun.* **27,** 157–162.

Schellenberger, V., and Jakubke, H.-D. (1991). *Angew. Chem., Int. Ed. Engl.* **30,** 1437–1449.

Schnebli, H. P., and Liersch, M. H. (1989). *In* "Elastin and Elastases" (L. Robert and W. Hornebeck, eds.), pp. 137–143. CRC Press, Boca Raton, Florida.

Schnebli, H. P., Seemüller, U., Fritz, H., Maschler, R., Liersch, M., Virca, G. D., Bodmer, J. L., Snider, G. L., Lucey, E. C., and Stone, P. G. (1985). *Eur. J. Respir. Dis.* **66**(Suppl. 139), 66–70.

Scott, C. F., Wenzel, H. R., Tschesche, H., and Colman, R. W. (1987). *Blood* **69,** 1431–1436.

Sealock, R. W., and Laskowski, M., Jr. (1969). *Biochemistry* **8,** 3703–3710.

Seemüller, U., Dodt, J., Fink, E., and Fritz, H. (1986). *In* "Proteinase Inhibitors" (A. J. Barrett and G. Salvesen, eds.), pp. 337–359. Elsevier, Amsterdam.

Siekmann, J., Wenzel, H. R., Schröder, W., Schutt, H., Truscheit, E., Arens, A., Rauenbusch, E., Chazin, W. J., Wüthrich, K., and Tschesche, H. (1987). *Biol. Chem. Hoppe-Seyler* **368,** 1589–1596.

Siekmann, J., Wenzel, H. R., Matuszak, E., von Goldammer, E., and Tschesche, H. (1988a). *J. Protein Chem.* **7,** 633–640.

Siekmann, J., Wenzel, H. R., Schröder, W., and Tschesche, H. (1988b). *Biol. Chem. Hoppe-Seyler* **369,** 157–163.

Siemion, I. Z., Konopinska, D., Wiejak, S., Rzeszotarska, B., and Najbar, Z. (1973). *In* "Peptides 1972" (H. Hanson and H. D. Jakubke, eds.), pp. 210–213. North-Holland Publ., Amsterdam.

Staley, J. P., and Kim, P. S. (1992). *Proc. Natl. Acad. Sci. U.S.A.* **89,** 1519–1523.

Steiner, D. F., Smeekens, S. P., Ohagi, S., and Chan, S. J. (1992). *J. Biol. Chem.* **267,** 23435–23438.

Steitz, T. A., and Shulman, R. G. (1982). *Annu. Rev. Biophys. Bioeng.* **11,** 419–444.

Svendsen, I., Jonassen, I., Hejgaard, J., and Boisen, S. (1980). *Carlsberg Res. Commun.* **45,** 389–395.

Svendsen, I., Boisen, S., and Hejgaard, J. (1982). *Carlsberg Res. Commun.* **47,** 45–53.

Swaminathan, S., Ravishanker, G., Beveridge, D. L., Lavery, R., Etchebest, C., and Sklenar, H. (1990). *Proteins* **8,** 179–193.

Tan, N. H., and Kaiser, E. T. (1976). *J. Org. Chem.* **41,** 2787–2793.

Tan, N. H., and Kaiser, E. T. (1977). *Biochemistry* **16,** 1531–1541.

Tanzi, R. E., McClatchey, A. I., Lamperti, E. D., Villa-Komaroff, L., Gusella, J. F., and Neve, R. L. (1988). *Nature (London)* **331,** 528–530.

Taylor, J. C., and Mittman, C. (eds.) (1987). "Pulmonary Emphysema and Proteolysis: 1986." Academic Press, Orlando, Florida.

Travis, J., and Fritz, H. (1991). *Am. Rev. Respir. Dis.* **143,** 1412–1415.

Tschesche, H. (1974). *Angew. Chem., Int. Ed. Engl.* **13,** 10–28.

Tschesche, H. (1989). *In* "The Kallikrein–Kinin System in Health and Disease" (H. Fritz, I. Schmidt, and G. Dietze, eds.), pp. 237–260. Limbach-Verlag, Braunschweig.

Tschesche, H., Reidel, G., and Schneider, M. (1974). *In* "Proteinase Inhibitors" (H. Fritz, H. Tschesche, L. J. Greene, and E. Truscheit, eds.), pp. 235–242. Springer-Verlag, Berlin.

Tschesche, H., Beckmann, J., Mehlich, A., Schnabel, E., Truscheit, E., and Wenzel, H. R. (1987). *Biochim. Biophys. Acta* **913,** 97–101.

Tschesche, H., Beckmann, J., Mehlich, A., Feldmann, A., Wenzel, H. R., Scott, C. F., and Colman, R. W. (1989). *Adv. Exp. Med. Biol.* **247B,** 15–21.

Tschesche, H., Groeger, C., and Wenzel, H. R. (1991). *Biomed. Biochim. Acta* **50,** S175–S180.

Umezawa, H. (1972). "Enzyme Inhibitors of Microbial Orgin." Univ. Park Press, Baltimore, Maryland.

van Mierlo, C. P. M., Darby, N. J., Keeler, J., Neuhaus, D., and Creighton, T. E. (1993). *J. Mol. Biol.* **229,** 1125–1146.

Vasquez, M., and Scheraga, H. A. (1988a). *J. Biomol. Struct. Dyn.* **5,** 705–755.

Vasquez, M., and Scheraga, H. A. (1988b). *J. Biomol. Struct. Dyn.* **5,** 757–784.

von Wilcken-Bergmann, B., Tils, D., Satorius, J., Auerswald, E. A., Schröder, W., and Müller-Hill, B. (1986). *EMBO J.* **5,** 3219–3225.

Wagner, G. (1983). *Q. Rev. Biophys.* **16,** 1–57.

Wagner, G., Braun, W., Havel, T. F., Schaumann, T., Go, N., and Wüthrich, K. (1987). *J. Mol. Biol.* **196,** 611–639.

Warshel, A., Naray-Szabo, G., Sussman, F., and Hwang, J.-K. (1989). *Biochemistry* **28,** 3629–3637.

Weber, E., Papamokos, E., Bode, W., Huber, R., Kato, I., and Laskowski, M., Jr. (1981). *J. Mol. Biol.* **149,** 109–123.

Weber, U., and Schmid, H. (1975). *Hoppe Seyler's Z. Physiol. Chem.* **356,** 1505–1515.

Weber, U., and Schmid, H. (1976). *Hoppe Seyler's Z. Physiol. Chem.* **357,** 1359–1363.

Weissman, J. S., and Kim, P. S. (1992). *Science* **256,** 112–114.

Wenzel, H. R., and Tschesche, H. (1981). *Angew. Chem., Int. Ed. Engl.* **20,** 295–296.

Wenzel, H. R., Beckmann, J., Mehlich, A., Siekmann, J., Tschesche, H., and Schutt, H. (1985). *In* "Modern Methods in Protein Chemistry" (H. Tschesche, ed.), Vol. 2, pp. 173–184. de Gruyter, Berlin.

Wenzel, H. R., Beckmann, J., Mehlich, A., Schnabel, E., and Tschesche, H. (1986). *In* "Chemistry of Peptides and Proteins" (W. Voelter, E. Bayer, Y. A. Ovchinnikov, and V. T. Ivanov, eds.), Vol. 3, pp. 105–117. de Gruyter, Berlin.

Wenzel, H. R., Beckmann, J., Feldmann, A., Mehlich, A., Siekmann, J., and Tschesche, H. (1989a). *In* "Peptides 1988" (G. Jung and E. Bayer, eds.), pp. 381–383. de Gruyter, Berlin.

Wenzel, H. R., Mehlich, A., Beckmann, J., and Tschesche, H. (1989b). *In* "Chemistry of Peptides and Proteins" (W. A. König and W. Voelter, eds.), Vol. 4, pp. 197–204. Attempto-Verlag, Tübingen.

Westaby, S. (1993). *Ann. Thorac. Surg.* **55,** 1033–1041.

Wibley, K. S., and Barlow, D. J. (1992). *J. Enzyme Inhibition* **5,** 331–338.

Wieczorek, M., and Laskowski, M., Jr. (1983). *Biochemistry* **22,** 2630–2636.

Wieczorek, M., Otlewski, J., Cook, J., Parks, K., Leluk, J., Wilimowska-Pelc, A., Polanowski, A., Wilusz, T., and Laskowski, M., Jr. (1985). *Biochem. Biophys. Res. Commun.* **126,** 646–652.

Wieczorek, M., Park, S. J., and Laskowski, M., Jr. (1987). *Biochem. Biophys. Res. Commun.* **144,** 499–504.

Wiejak, S., and Rzeszotarska, B. (1974). *Roczniki Chem.* **48,** 2207–2215.

Wiejak, S., and Rzeszotarska, B. (1975). *Roczniki Chem.* **49,** 1105–1112.

Wilusz, T., Wieczorek, M., Polanowski, A., Denton, A., Cook, J., and Laskowski, M., Jr. (1983). *Hoppe Seyler's Z. Physiol. Chem.* **364,** 93–95.

Wlodawer, A., Walter, J., Huber, R., and Sjölin, L. (1984). *J. Mol. Biol.* **180,** 301–329.

Wlodawer, A., Deisenhofer, J., and Huber, R. (1987). *J. Mol. Biol.* **193,** 145–156.

Yajima, H., and Kiso, Y. (1974). *Chem. Pharm. Bull.* **22,** 1087–1094.

Yoshida, N., Tsuruyama, S., Nagata, K., Hirayama, K., Noda, K., and Makisumi, S. (1988). *J. Biochem. (Tokyo)* **104,** 451–456.

9

Design of Polypeptides

Bernd Gutte and Stephan Klauser
Biochemisches Institut der Universität Zürich
CH-8057 Zürich, Switzerland

I. Introduction

A. Approaches to Polypeptide Design

A growing number of researchers are trying to design novel functional polypeptides despite the fact that natural proteins are abundant. A single eukaryotic cell, for example, has more than 10,000 different proteins with almost as many different activities, which provokes the question whether polypeptide design is significant at all. Considering the immediate need for virus repressors, enzyme inhibitors, pesticide-degrading polypeptides, etc., in order to cope with new epidemics, increasing concentrations of anthropogenic toxins, and other health and environmental problems, the above question must be answered in the affirmative. But are we able to "invent" effective repressors, inhibitors, and enzymes? At present, the gap between rationale and experimental results of design is still considerable. However, if polypeptide design were performed as rigorously as shown in Fig. 1 and if structural and functional modules of natural proteins were utilized whenever appropriate, tailor-made polypeptides with biomedical activities could

Peptides: Synthesis, Structures, and Applications

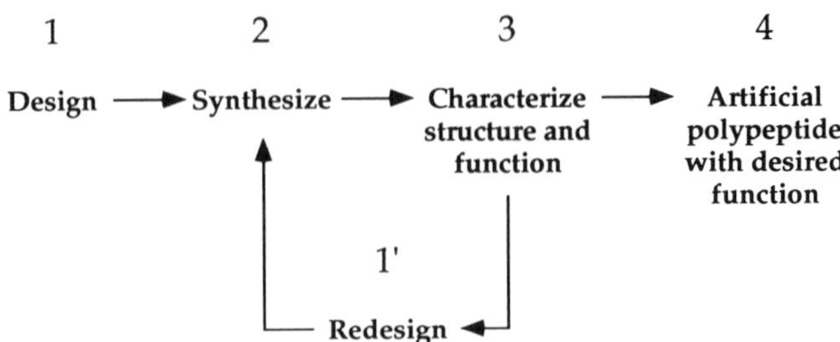

Figure 1 Optimal procedure for designing polypeptides. Ideally, the designer should be in command of analytical tools such as nuclear magnetic resonance spectroscopy and X-ray crystallography.

become available relatively soon. Circling through steps 1′, 2, and 3 of Fig. 1 would also provide a wealth of data on the relationship between amino acid sequence and folding of a polypeptide chain and, last not least, on folding pathways.

Basically, there are three approaches to design artificial polypeptides:

1. Modification of natural proteins or peptides by replacement, deletion, and insertion of amino acid residues and shuffling of structural modules
2. Rational or *de novo* design which yields sequences that inherently show little if any similarity with those of natural products
3. Construction of hybrid polypeptides from designed and authentic (natural) structural elements.

Investigators have used these approaches to construct stable scaffolds for subsequent functionalization (Pessi *et al.*, 1993), to mimic or modify the structure and function of natural proteins (Hahn *et al.*, 1990), and to design novel ion channels (Lear *et al.*, 1988), enzymes (Johnsson *et al.*, 1993), and repressors (Hehlgans *et al.*, 1993) that possess unique mechanisms and specificities.

In this chapter, we discuss mainly trends in polypeptide design based on a survey of the literature and apologize for not citing and reviewing all relevant references. Furthermore, we deal only with products that are accessible by peptide synthesis; constructs such as peptide-linked variable domains of antibodies (Huston *et al.*, 1988; Bird *et al.*, 1988; Marasco *et al.*, 1993) are too large (~240 residues) and are examples of protein design or protein engineering. Earlier reviews of polypeptide design include those by Moser *et al.* (1985), DeGrado (1988), and Richardson and Richardson (1989).

B. The Beginnings of Polypeptide Design

Polypeptide design starts in the mind with positive thinking, that is, the strong belief that it can be done. As for the practical aspect, polypeptide design has always depended on the availability of efficient synthesis (Merrifield, 1963, Chapter 3, this volume) and structure prediction methods (Chou and Fasman, 1978; Levitt, 1978; Benner and Gerloff, 1991; Rost and Sander, 1993).

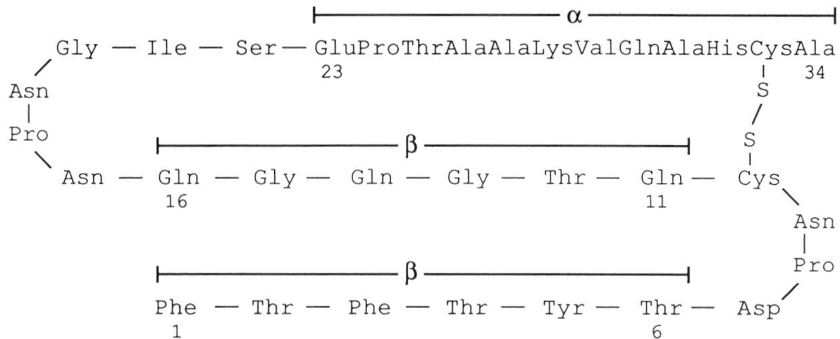

Figure 2 Designed 34-residue polypeptide with weak RNase A-like activity (Gutte *et al.*, 1979). Labeling experiments using iodo[2–^{14}C]acetic acid showed that the mechanism of action of the artificial nuclease was probably different from that of RNase A.

As early as 1973, a decapeptide with lysozyme like activity was designed (Chakravarty *et al.*, 1973) on the basis of the X-ray structure of the active site of lysozyme (Phillips, 1966). The sequence of the peptide was Glu-Phe-Ala-Ala-Glu-Glu-Ala-Ala-Ser-Phe. At relatively high concentration or on interaction with (polymeric) substrates, it may form an amphiphilic α helix with Glu^1, Phe^2, Glu^5, Glu^6, and Ser^9 on the hydrophilic side, Ala^3, Ala^4, Ala^7, Ala^8, and Phe^{10} on the hydrophobic side, and the side-chain carboxyl group of the active-site glutamic acid (here supposedly Glu^6) sandwiched between the aromatic rings of Phe^2 and Phe^{10}. In the years since, the amphiphilic α helix has become the favorite structural element employed in polypeptide design.

Our interest in polypeptide design was triggered by the finding that synthetic cut-and-paste analogs of bovine ribonuclease A (70 and 63 residues, respectively; Gutte, 1975, 1977) and porcine proinsulin (46 residues; Kullmann and Gutte, 1975) retained considerable biological activity. Moreover, the truncated ribonucleases had the same substrate specificity as the natural 124-residue enzyme and cross-reacted with anti-ribonuclease A antibodies. The results suggested that polypeptides accessible by chemical synthesis can have enzymatic properties and that the sequences of stable structural motifs of natural proteins could perhaps be used as building blocks with a predetermined, committed fold for polypeptide design.

In 1979 we designed a 34-residue nucleic acid-binding peptide on the basis of structure prediction and model building around the trinucleotide GAA (Gutte *et al.*, 1979) (Fig. 2). This peptide did not have the proposed sequence specificity but, by serendipity, cleaved RNA with high preference at the 3′ end of C. The dissociation constant (K_D) of the complex of dimeric peptide and 2′-CMP (2′-cytidine monophosphate) was 5×10^{-6} M. Titration of the dimeric peptide with 2′-CMP gave an extrinsic Cotton effect in the near UV (Jaenicke *et al.*, 1980), indicating interactions between nucleotide and aromatic side chains of the peptide as postulated in the model (Gutte *et al.*, 1979).

a 198 E H L S T L S E K A K P A L E D L R Q G L L 219

b 1 P K L E E L K E K L K E L L E K L K E K L A 22

Figure 3 Sequence of (a) residues 198–219 of human apolipoprotein A-I (Osborne and Brewer, 1977), composed of 12 different amino acids, and (b) designed 22-residue model peptide (Fukushima *et al.*, 1979) containing only 5 different amino acids. Identical residues are boxed.

Another early example of polypeptide design is that of Fukushima *et al.* (1979). These authors constructed a simplified model peptide of the homologous, helical 22-amino acid residue repeats of plasma apolipoprotein A-I (Osborne and Brewer, 1977) (Fig. 3) which bound to phospholipid bilayer vesicles in a manner similar to that of the intact apolipoprotein. It was concluded that the lipid binding capacity of apolipoprotein A-I resided largely in its helical segments and that this capacity could be initiated by peptides of high amphiphilic helix-forming potential.

C. Tools of Polypeptide Design

Polypeptide design may be as simple as drawing a helical wheel (Schiffer and Edmundson, 1967) if one wants to create the sequence of an amphiphilic α helix (Fukushima *et al.*, 1979; Johnsson *et al.*, 1993). On the other hand, knowledge-based design of a zinc binding site onto an antibody variable domain-derived scaffold (Pessi *et al.*, 1993) was performed by computer modeling and required the coordinates of the X-ray structures of both the zinc enzyme carbonic anhydrase B (Kannan *et al.*, 1975) and the Fab fragment of monoclonal antibody McPC 603 (Segal *et al.*, 1974). For some time already, sequences of designed peptides have been refined by energy minimization using computer programs. In our laboratory, designed structures including polypeptide–DNA complexes were always verified by model building (Gutte *et al.*, 1979; Moser *et al.*, 1983; Kullmann, 1984; Hehlgans *et al.*, 1990, 1993; Moser, 1992) and in some cases by computer graphics using the coordinates of structurally related proteins (Feldmann, 1983; Moser, 1992).

II. Recent Design Activities

A. The Amphiphilic α Helix

The reason why the amphiphilic α helix has become the most frequently used structural element for polypeptide design may be 2-fold: first, it seems to be easy to create and, second, it is widespread in nature, suggesting a great variety of model studies. As amphiphilic α helices have an inherent tendency to aggregate regardless of the nature of the solvent, the art of design in this case consists mainly in the control of the aggregation.

1. Four-Helix Bundles

a. Natural Four-Helix Bundle Proteins Four-helix bundles are a recurring structural motif in proteins (Argos *et al.*, 1977; Weber and Salemme, 1980). In the majority of cases they are made of sequentially connected amphiphilic α helices, and they have a characteristic left-handed twist. Four-helix bundles occur in proteins as diverse as apoferritin (Banyard *et al.*, 1978), *Escherichia coli* cytochrome b_{562} (Mathews *et al.*, 1979), and tobacco mosaic virus coat protein (Bloomer *et al.*, 1978; Stubbs *et al.*, 1977). The packing of the helices in the bundles produces a central cavity which can accommodate metals or heme.

An example of a protein forming a stable intermolecular four-helix bundle is ROP (repressor of primer) (Twigg and Sherrat, 1980; Tomizawa, 1990). Each of the two identical subunits of ROP contributes a two-helix coiled coil to assemble the four-helix bundle (Banner *et al.*, 1987; Eberle *et al.*, 1990).

b. Designed Unsupported Four-Helix Bundles In a series of papers, DeGrado and co-workers (Ho and DeGrado, 1987; Regan and DeGrado, 1988; Osterhout *et al.*, 1992) described an incremental approach to the design of a four-helix bundle protein. They developed a fundamental 16-residue sequence called α_1B which had high propensity to form an amphiphilic α helix (Fig. 4a). Four-helix bundles with increasing stability were obtained by cooperative tetramerization of α_1B, dimerization of α_1B–loop–α_1B, and folding of α_1B–loop–α_1B–loop–α_1B–loop–α_1B (Regan and DeGrado, 1988) (Fig. 4b). The driving force for self-assembly of a helical bundle was the clustering of the leucine side chains of the four helices in the interior of the structure. Nuclear magnetic resonance (NMR) studies confirmed the overall helical structure of α_1B but did not allow conclusions to be drawn about the exact arrangement of the helices in the tetramer $(\alpha_1\text{B})_4$ because useful interhelical nuclear Overhauser effects (NOEs) could not be identified because of a substantial overlap of the side-chain resonances (Osterhout *et al.*, 1992).

More recently, DeGrado and co-workers designed two heme-binding polypeptides whose sequences were closely related to that of α_1B–Pro-Arg-Arg–α_1B (Choma *et al.*, 1994) (Fig. 5a) and which were intended to dimerize into four-helix bundles. The structure of these peptides was stabilized by disulfide formation, and binding of heme in the core of the bundle was mediated mainly by bishistidine side-chain ligation of the heme iron as indicated by electron paramagnetic resonance (EPR) spectroscopy (Fig. 5b). The dissociation constants of the two peptide–heme complexes were 7×10^{-7} and $<1 \times 10^{-7}$ *M*, respectively. Finally, Robertson *et al.* (1994) reported the design of a multiheme protein. A 31-residue helical peptide possessing essential residues of the *b*-heme-binding B and D helices of cytochrome bc_1 (e.g., two histidines 14 residues apart) and most of the sequence of α_1B (Fig. 5c) assembled into a four-helix unit and bound four

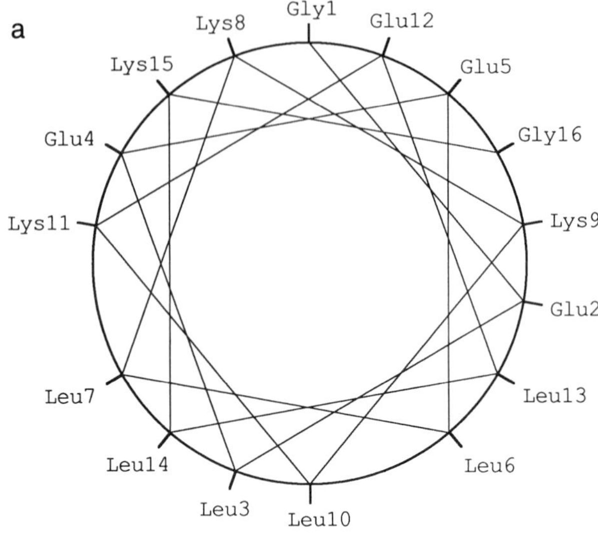

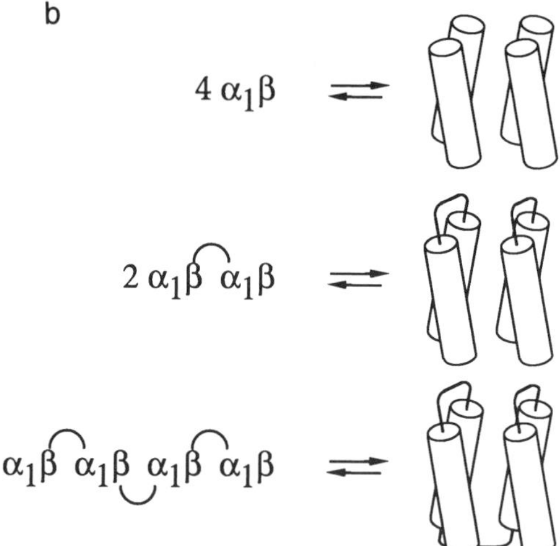

Figure 4 Incremental approach of DeGrado and co-workers (Ho and DeGrado, 1987; Regan and DeGrado, 1988; Osterhout *et al.*, 1992) to the design of a four-helix bundle protein. (a) Helical wheel representation of the 16-residue amphiphilic helix α_1B. (b) Three different modes of four-helix bundle formation. Cylinders are helices, and each of the loops linking two helices has the sequence -Pro-Arg-Arg-.

a

helix loop

I Ac-C-GELEELLKKAKELLKG-PRR-

GEVEEHLKKVKELLKG-NH2

helix

II Ac-C-GKLLEKVKKLHEEVEG-PRR-

GKLLEKAKKLLEELEG-NH2

b

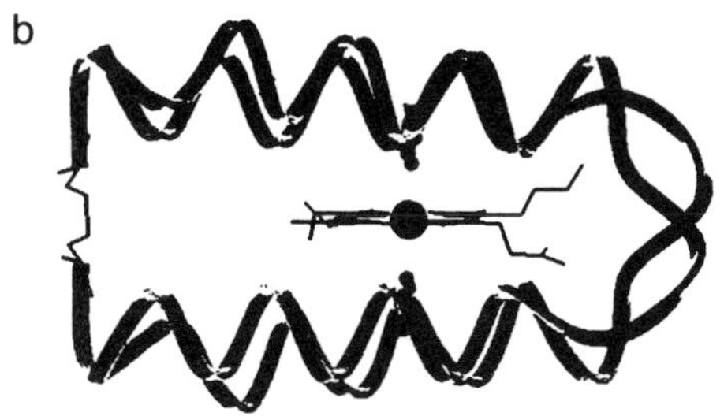

c

Figure 5 (a) Two 36-residue polypeptides designed to form a four-helix bundle and to bind heme after dimerization (Choma *et al.*, 1994). In peptide **I**, the deviations from the original α_1B sequence are marked by an asterisk. The sequence of peptide **II** is the reverse of that of peptide **I.** (b) Energy-minimized model of the disulfide dimer of peptide **II** with bound heme liganded by the side chains of the two histidine residues. The two loops of the dimeric peptide are on the right-hand side; the disulfide bond is to the left of the bundle. Reprinted with permission from Choma *et al.*, 1994. Copyright 1994 American Chemical Society. (c) Designed 31-residue helical peptide that, after disulfide formation and dimerization of the resulting 62-residue polypeptide, bound four heme groups as indicated (Robertson *et al.*, 1994). α_1B sequences are boxed.

hemes after disulfide formation and dimerization of the resulting 62-residue two-helix peptide.

Replacing some of the nonspecific hydrophobic interactions of the leucine side chains in the interior of a single-chain four-helix bundle with more specific metal–ligand bonds induced a more nativelike state of the designed polypeptide (Handel *et al.*, 1993). In the experiment, two leucines and one lysine of two of the helices were replaced with histidines to construct a Zn^{2+} binding site. Four-helix bundles with one or two Zn^{2+} ligation sites were prepared.

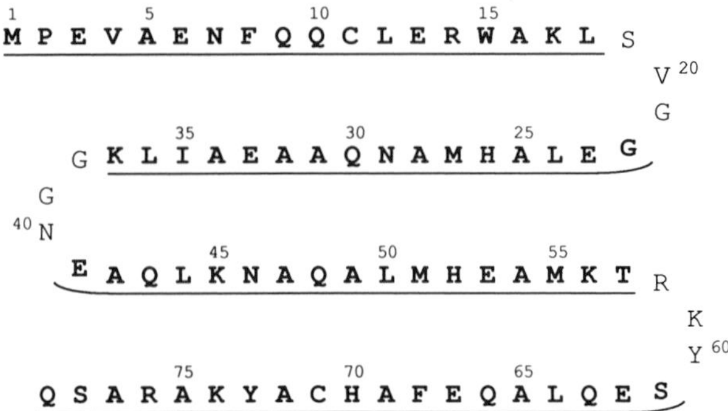

Figure 6 Proposed secondary structure (helices underlined) of a 79-residue polypeptide designed to fold into an antiparallel four-helix bundle (Hecht *et al.*, 1990). The disulfide bond formed between Cys-11 and Cys-71 stabilizes the (postulated) three-dimensional structure of the artificial protein.

Another approach to constructing an antiparallel four-helix bundle was chosen by Hecht *et al.* (1990). Using 19 of the 20 naturally occurring amino acids, these authors designed four different amphiphilic α helices which they linked by short loops to give a single-chain 79-residue polypeptide. The choice of amino acids for the various positions of the helices was based on statistical analysis of amino acid sequences of natural helices. To stabilize the proposed structure by a disulfide bond, cysteines were incorporated in positions 11 (helix 1) and 71 (helix 4) (Fig. 6). After helix and turn propensities along the sequence of the designed polypeptide had been confirmed by various secondary structure prediction programs, the gene of the 79-residue peptide was synthesized and, with the aid of the bacteriophage T7 system (Studier and Moffatt, 1986; Rosenberg *et al.*, 1987), cloned and expressed in *E. coli*. The recombinant product was isolated from insoluble inclusion bodies, purified, and by various physico-chemical properties found to fold approximately into the designed three-dimensional structure. The authors emphasized that the purpose of this work was controlling the formation of tertiary structure.

That designed antiparallel four-helix bundles are not fictitious was demonstrated most clearly by Schafmeister *et al.* (1993). These authors constructed a 24-residue amphipathic α-helical peptide (Fig. 7a) to act as a detergent for transmembrane proteins. In the crystal, the peptide alone was shown by X-ray analysis to form an antiparallel four-helix bundle (Fig. 7b) held together mainly by interdigitating leucine side chains in the core of the structure, as predicted.

2. *Three-Helix Bundles*

A designed 15-residue amphiphilic α-helical peptide with a 2,2′-bipyridine-5-carboxy-modified N terminus (Fig. 8a) assembled in the presence of transition metal ions into a 45-residue parallel triple-helical coiled coil (Ghadiri *et*

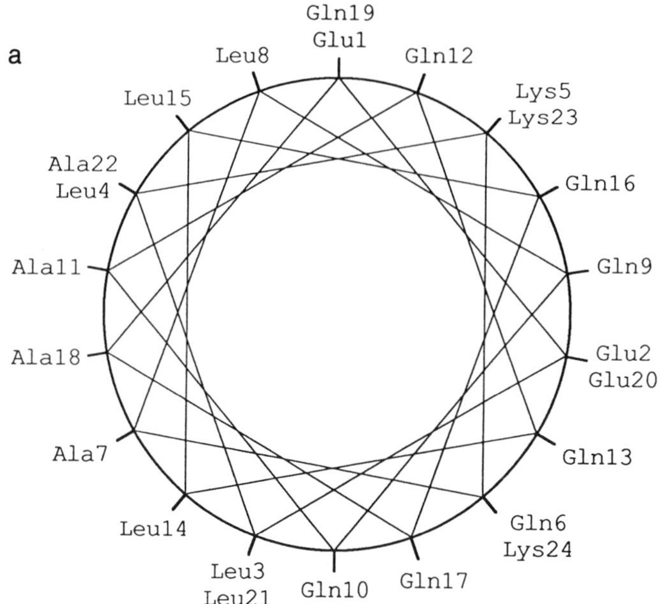

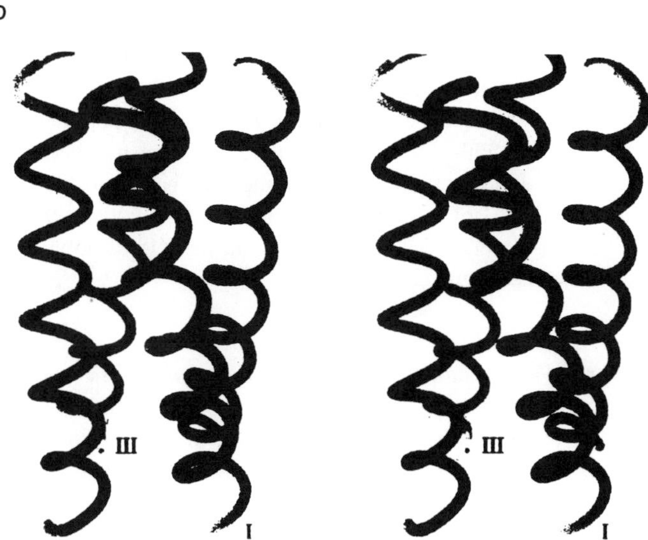

Figure 7 (a) Helical wheel representation of a designed 24-residue amphiphilic α helix (Schafmeister *et al.*, 1993) that was found by X-ray crystallography to form an antiparallel four-helix bundle (b). Part b is reprinted with permission from Schafmeister, C. E., Miercke, L. J. W., and Stroud, R. M. (1993). *Science* **262,** 734–738. Copyright 1993 American Association for the Advancement of Science.

a [bipyridyl]–C(=O)– GELAQKLEQALQKLA-NH_2 (residues numbered 5, 10, 15)

b EWEALEKKLAALESKLQALEKKLEALEHG (residues numbered 1, 10, 20, 29)

Figure 8 Designed peptides that form triple-stranded α-helical bundles. (a) Assembly of the pentadecapeptide of Ghadiri *et al.* (1992) into the presumptive parallel triple-helical bundle may be initiated by the complexation of a transition metal ion by the bipyridyl moieties of three peptide chains. (b) The 29-residue peptide of Lovejoy *et al.* (1993), designed to form a double-stranded parallel coiled coil, assembled into a triple-stranded α-helical bundle as shown by X-ray crystallography. Residues corresponding to positions a and d of the heptad repeats of leucine zippers are underlined.

al., 1992). This controlled trimerization was driven by the known tris-chelation of transition metal ions with bidentate ligands such as 2,2′-bipyridyl and by the sequence of the designed peptide, which allowed sterically favorable interhelical packing of the hydrophobic side chains in the core of the structure as well as intrahelical and interhelical electrostatic interactions on the solvent-exposed hydrophilic surface.

Lovejoy *et al.* (1993) determined the X-ray crystal structure of a synthetic 29-residue peptide (Fig. 8b) that assembled into a triple-stranded α-helical bundle. Two of the helices of the bundle ran parallel, and the third ran antiparallel. Equilibrium sedimentation experiments indicated that the peptide was trimeric also in 0.15 *M* sodium chloride solution. Interestingly, it had been designed to form a double-stranded parallel coiled coil and was based on the tropomyosin model sequence $(KLEALEG)_5$ contained in a 43-residue construct of Hodges *et al.* (1981). Sooner or later, everybody working in the field of polypeptide design will experience similar surprises.

3. *Two-Stranded α-Helical Coiled Coils*

a. Natural Two-Stranded α-Helical Coiled Coils The two-stranded α-helical coiled coil structural motif consists of two parallel amphipathic α helices held together by interchain hydrophobic bonds. It may occur in more than 200 proteins (Lupas *et al.*, 1991) and is characterized by a repeating seven-residue sequence pattern whose positions are termed a to g (McLachlan and Stewart, 1975). In position a, Leu, Ile, and Met occur most frequently, whereas in position d Leu is strongly preferred (Lupas *et al.*, 1991). Of particular interest have been leucine zipper coiled coil proteins (Landschulz *et al.*, 1988) such as the eukaryotic transcriptional activators GCN4 (Ellenberger *et al.*, 1992), Jun homodimer (Junius *et al.*, 1993), and Fos–Jun heterodimer (O'Shea *et al.*, 1992).

b. Designed Two-Stranded α-Helical Coiled Coils The first model coiled-coil protein was designed by Hodges *et al.* (1981). Later, Engel *et al.*

(1991) incorporated four of the seven-residue repeats of this protein with slightly modified sequence [(RIEAIEA)$_4$] into a 39-residue construct which also contained a potential integrin binding site (-Arg-Gly-Asp-) and C-terminal Cys. The 78-residue dimer of the peptide formed through interchain hydrophobic interactions of eight pairs of isoleucine residues was stabilized by a disulfide bridge between the C-terminal cysteines of the two chains, as demonstrated by heat and urea denaturation. Circular dichroism (CD) spectroscopy indicated that both the covalent and the noncovalent dimer contained stable α-helical coiled coils.

Further studies by Hodges and co-workers (Zhou *et al.*, 1992, 1993) on model coiled coil proteins showed that any single Leu to Ala substitution in positions a and d of the repeating heptads (except at the ends of the two chains) and interchain disulfide bonds replacing leucine residues in nonterminal positions a destabilized or disrupted the structure of two-stranded α-helical coiled coils. Interchain disulfide bonds replacing leucine residues in positions d, however, did not perturb the coiled coil structure and made the largest contribution to coiled coil stability.

4. Carrier-Supported Association of α Helices and β Strands

We have seen that the helices in α-helical bundles are held together mainly by interchain noncovalent hydrophobic interactions and that additional stabilization may be provided by disulfide bonds. One would expect that designed peptide chains anchored at one end on a suitable carrier would also readily associate in the desired manner to form a stable bundle. This notion was successfully put into reality by several research groups using mostly short linear or branched peptides as carriers. To prepare a complex, two-domain protein, Mutter *et al.* (1989) first synthesized the resin-bound carrier peptide K_B-K_F-K_B-K_F-P-G-K_F-K_B-K_F-K_B-ε-aminocaproic acid-G with alternating Boc (B) and Fmoc (F) protection of the ε-amino groups of the lysine residues. After removal of the four Boc protecting groups, 15 coupling cycles were performed to build a four-helix bundle domain; subsequent cleavage of the four Fmoc groups allowed the synthesis, on the same peptide template, of a second, β-barrel-like domain (nine residues per strand, Fig. 9). Removal of the protecting groups of the α helices and β strands and cleavage from the resin yielded the free two-domain "template-assembled synthetic protein" that, at a peptide concentration of 0.1 m*M*, showed a CD spectrum typical of α + β proteins. Individual, not template-bound peptide chains of the artificial two-domain protein, however, had only low tendency to form stable secondary structures. The versatility and potential of template-based assembly of proteins were pointed out by Mutter and Vuilleumier (1989).

Hahn *et al.* (1990) used the resin-bound branched tripeptide amide

```
    ──┐
   —Lys─┐
 –Lys—Orn-NH2,
 ───┘
```

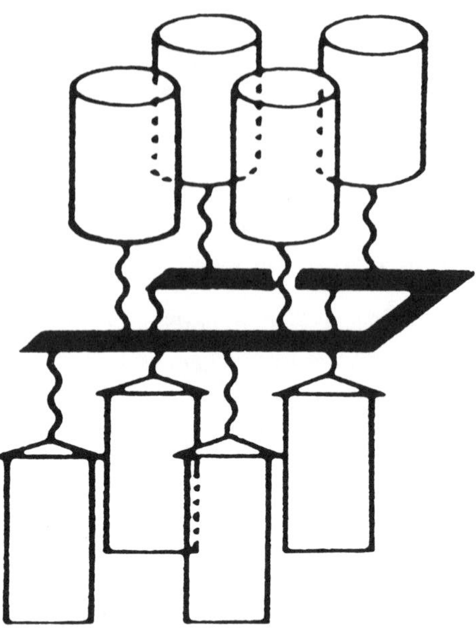

Figure 9 Template-assembled two-domain protein constructed by Mutter *et al.* (1989). Heavy bars represent the template (a reverse-turn-forming dodecapeptide), cylinders are amphiphilic α helices, and broad arrows are amphiphilic β strands. From Mutter, M., Hersperger, R., Gubernator, K., and Müller, K. (1989). *Proteins Struct. Funct. Genet.* **5,** 13–21. Copyright © 1989 John Wiley & Sons, Inc. Reprinted by permission of John Wiley & Sons, Inc.

reminiscent of Tam's carriers for multiple peptide synthesis (Tam, 1988), to prepare in one operation an artificial four-helix protein with chymotrypsin-like esterase activity which had been designed by computer modeling. The design was refined using energy minimization algorithms. The four helices were amphiphilic and contained 15, 17, 18, and 20 residues with N-terminal Ser, Glu, His, and Asp, respectively. After deprotection and cleavage from the resin, the tripeptide-anchored helices could associate in a manner that brought the N-terminal serine, histidine, and aspartic acid residues into the same relative positions as in the active site of chymotrypsin and produced a binding pocket for small substrates and an oxyanion hole for stabilizing the transition state of the reaction (Blow, 1976). The 73-residue chymotrypsin model was 75% helical as shown by CD spectroscopy in 50 m*M* NaCl, hydrolyzed acetyltyrosine ethyl ester (values of k_{cat} and k_{cat}/K_m were $\frac{1}{4500}$ and $\frac{1}{6600}$ of those of natural chymotrypsin, respectively), reacted irreversibly with phenylmethylsulfonyl fluoride at the hydroxyl group of the active-site serine residue, and did not cleave the trypsin substrate benzoylarginine ethyl ester.

Nonpeptide carriers for multiple-chain proteins have also been described. Sasaki and Kaiser (1989) coupled the designed prefabricated pentadecapeptide amide AEQLLQEAEQLLQEL-NH_2 to the tetrahydroxysuccinimide active ester of coproporphyrin I, yielding a porphyrin-anchored four-helix protein. CD studies showed that the anchored peptide chains had an α-helical content of approximately 70%, whereas the free peptide amide was disordered. The design was such that the four porphyrin-linked amphiphilic helices could form a hydrophobic pocket for substrate binding above the porphyrin ring. The authors reported that the Fe(III) complex of the construct had weak aniline hydroxylase activity (k_{cat} 0.02 min^{-1}). They did not say whether other compounds were tested as substrates. In light of the considerable aniline hydroxylase activity of hemin in the presence of a reducing agent and oxygen (Sakurai and Ogawa, 1975), it is surprising that Fe(III)–coproporphyrin I should be inactive against this substrate in the same conditions.

Obviously, the carrier-bound multichain proteins described above cannot be formed *in vivo*. This may be a disadvantage with respect to potential pharmaceutical applications.

5. *Transmembrane α-Helical Bundles*

Transmembrane α-helical coiled coils are inside-out versions of α-helical coiled coils soluble in aqueous buffers; that is, the positions of the hydrophobic and the hydrophilic residues of the latter have been reversed, resulting in an apolar exterior and a polar interior of the helical bundle. If the design is such that the side chains in the interior do not pack tightly, ion channels of various sizes can be obtained. Lear *et al.* (1988) synthesized a 21-residue model peptide that contained serine in positions a and d of each heptad repeat [$(LSLLLSL)_3$-NH_2] and could span the hydrophobic part of a phospholipid bilayer. The peptide produced channels that conducted protons, but not alkali metal ions, supporting the proposed formation of transmembrane four-helix bundles. If one of the leucines of each heptad was replaced by serine [$(LSSLLSL)_3$-NH_2], the channels formed were wider and conducted protons, alkali metal ions, and organic cations up to a diameter of 8 Å. Computer modeling showed that, in order to accommodate the side chains of all three serine residues of such a heptad in the hydrophilic interior, a circular arrangement of six or more α helices would be required.

The helical nature of both peptides in phospholipid membranes was demonstrated by CD and infrared (IR) spectroscopy, and a model for the voltage dependence of ion channel formation and of the frequency of the channel openings was derived from fluorescence studies using a series of 21-residue peptides in which the residues of the middle heptad had been replaced, one at a time, by tryptophan (Chung *et al.*, 1992). To summarize, application of a voltage gradient at the membrane most likely moved the peptides from a near-surface to a channel-forming transmembrane position due to the alignment of the helical dipole moments with the electric field (Baumann and Mueller, 1974). The fluorescence studies also confirmed that the membrane-bound peptides were α-helical.

When a very similar peptide [$(LSLBLSL)_3$-NH_2, where B is α-aminoisobutyric acid] was coupled to the four carboxyl groups of a tetrakis (3-carboxyphenyl)porphyrin template, more defined, nearly voltage-independent four-helix proton channels with stabilized conducting states were formed

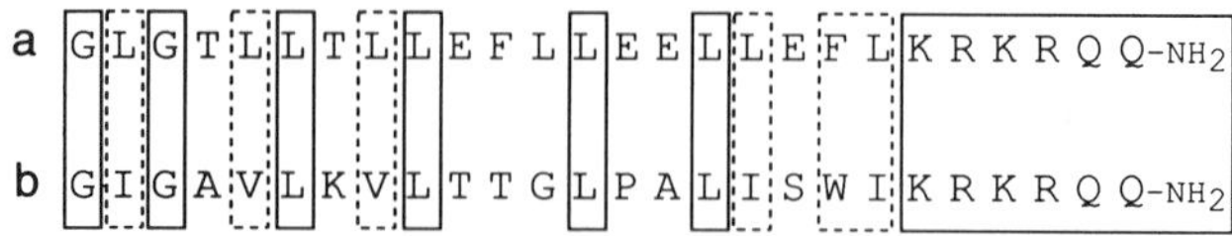

Figure 10 (a) A 26-residue amphiphilic α-helical peptide with net charge 0 and hemolytic activity at acidic pH (Moser, 1992) derived from (b) bee venom melittin (net charge +5). Solid boxes denote conserved residues; dashed boxes, conservative exchanges.

(Åkerfeldt *et al.*, 1992). Preliminary X-ray crystallographic studies suggested that the same peptide was also tetrameric in unanchored form when crystallized from a 2,2,2-trifluoroethanol–water mixture (Lovejoy *et al.*, 1992).

Artificial membrane-spanning ion channels or pores have also been prepared by other authors. Moser (1992) designed an analog of bee venom melittin (Fig. 10) that folded into a stable amphiphilic α helix and formed hemolytic pores in erythrocyte membranes at pH 5 but not at pH 7.5. To obtain the observed pores (2.3 to 2.9 nm inner diameter as revealed by transmission electron microscopy), circular aggregation of 18 to 20 helical peptides would be required. Alternatively, the melittin analog could bind to band 3 or other erythrocyte membrane proteins and induce pore-forming aggregation of those proteins (Clague and Cherry, 1988). CD spectroscopy showed that the designed peptide was nearly 100% α-helical both as an aggregate in aqueous buffer and as a monomer in methanol (Moser, 1992). The helical nature of the monomer in methanolic solution was confirmed nicely by high-resolution two-dimensional ^{1}H-NMR spectroscopy and molecular dynamics simulations (Klaus and Moser, 1992).

The template-assembled multichain proteins of Mutter and co-workers (Mutter and Vuilleumier, 1989) also lend themselves strongly to studies of the formation of artificial transmembrane ion channels. Grove *et al.* (1993) used the cyclic template

Ac - Cys – Lys – Ala – Lys – Pro
| |
H_2N - Cys – Lys – Ala – Lys – Gly

as an aid in assembly and folding of four identical polypeptide chains which were linked to the ε-amino groups of the four lysine residues and could form amphiphilic α helices (Glu^{18}-Leu-Leu-Glu^{15}-Ala-Leu-Glu-Lys^{11}-Ala-Leu-Lys-Glu-Ala-Leu-Ala-Lys-Leu-Gly). Membrane-spanning hydrophilic cation-selective channels with distinct open and closed states were obtained when the four peptide chains were 15 and 18 residues long, respectively; 11-residue chains were too short to span the hydrophobic core of the lipid bilayer, as shown by the single broad band of the current histogram of the transmembrane conductance. CD studies showed that the helix content of the constructs in trifluoroethanolic solution increased with increasing length of the peptides. Membrane channels with well-defined conductance properties were also formed by the free octadecapeptide but

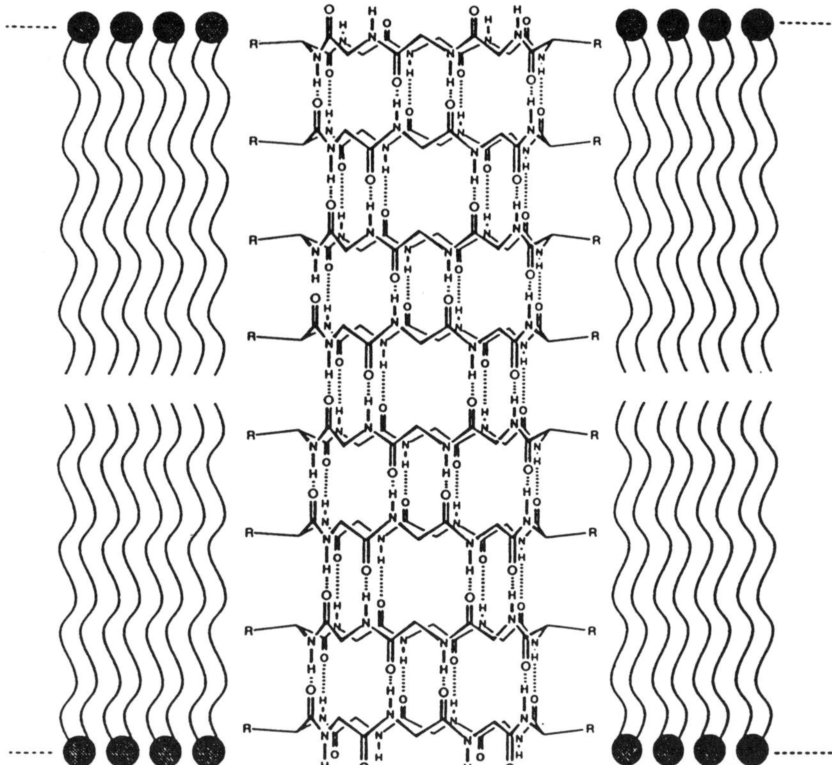

Figure 11 Artificial transmembrane ion channel with an inner diameter of 7.5 Å embedded in a lipid bilayer. The channel is formed by hydrogen-bonded, antiparallel stacking of a cyclic octapeptide. Reprinted with permission from *Nature,* Ghadiri *et al.*, 1994. Copyright 1994 Macmillan Magazines Limited.

not the free pentadecapeptide, reflecting perhaps differences in the aggregation properties of the two peptides in lipid bilayers. The selectivity of the channels for cations, calculated from the reversal potential, was approximately 90%.

Ghadiri *et al.* (1994) has reported the design of artificial transmembrane ion channels that had an inner diameter of approximately 7.5 Å and showed single-channel conductances of 55 pS in 500 m*M* NaCl (1.8×10^7 ions/sec) and 65 pS in 500 m*M* KCl (2.2×10^7 ions/sec) independent of applied voltages of 10 to 100 mV. These channels were formed by a tubular hydrogen-bonded, antiparallel stacking of probably eight molecules of the cyclic octapeptide cyclo[Trp-DLeu-Trp-DLeu-Trp-DLeu-Gln-DLeu] (Fig. 11). The ion channels described by Ghadiri *et al.* (1994) may be more uniform than those produced by a transmembrane α-helical bundle because, at a given peptide concentration, the latter may exist in an equilibrium with different aggregation states, resulting in the formation of channels with different sizes.

6. *Other Designed α-Helical Peptides*

a. An Amphiphilic α-Helical Peptide with Enzymatic Activity Johnsson *et al.* (1993) designed a tetradecapeptide (R-Leu-Ala-Lys-Leu-Leu-Lys-Ala-Leu-Ala-Lys-Leu-Leu-Lys-Lys-NH_2, were R is H or CH_3CO-) that, especially in the N-terminally acetylated form, had the expected helix-forming propensity, as shown by CD and two-dimensional NMR spectroscopy, and was able to catalyze the decarboxylation of oxaloacetate. The helicity was a function of the concentrations of both peptide and trifluoroethanol and correlated with the catalytic activity; the values of k_{cat} and k_{cat}/K_m for the peptide with the acetylated N-terminal amino group in 5% trifluoroethanol were 0.0068 sec^{-1} and 0.21 sec^{-1} M^{-1}, respectively. A $Leu^4 \rightarrow$ Gly, $Leu^8 \rightarrow$ Pro, $Leu^{11} \rightarrow$ Gly triple mutant of the peptide had a random coil conformation at all concentrations and only 13% of the activity of the original peptide. The proposed intermediates of the decarboxylation reaction, namely, protonated Schiff bases with substrate and product, could be trapped by reduction and identified. The authors did not mention whether other α-ketodicarboxylic acids were substrates of the artificial oxaloacetate decarboxylase. Both versions of the designed 14-residue peptide aggregated in a concentration-dependent manner to give stable four-helix bundles or larger complexes.

b. A Helical Heparin-Binding Peptide Heparin–protein interactions are important in regulating hemostasis (Rosenberg, 1977) and many other biological processes. On the basis of the structure of heparin-binding domains of natural proteins, Ferran *et al.* (1992) designed an α-helical 19-residue peptide containing six arginine residues evenly distributed over two-thirds of the helix surface (succinyl-Ala-Glu-Ala-Ala-Ala-Arg-Ala-Ala-Ala-Arg-Arg-Ala-Ala-Arg-Arg-Ala-Ala-Ala-Arg-NH_2). In the presence of heparin, the helix content of the peptide was 100%, and its apparent thermal stability increased by about 1 kcal/mol; the melting point for the helix-to-coil transition of the heparin–peptide complex was 50°C and that of the peptide alone, 25°C. The heparin–peptide complex, with a K_D around 3 μM, also dissociated under conditions of high ionic strength.

B. Mixed Secondary Structures

Goraj *et al.* (1990) derived a "unit peptide" from naturally occurring α/β-barrel proteins which contained the following folding information:

turn 1	β	turn 2	α
DARS	GLVVYL	GKRPDSG	TARELLRHLVAEG.

The synthetic gene encoding the peptide was cloned and expressed in *E. coli* either as the monomer or as direct repeats. CD measurements and urea-gradient gel electrophoresis (Creighton, 1979) indicated that the octamer of the unit peptide was most compact and had the predicted amount of α-helical structure. No mention was made by the authors of the β-sheet and turn contents of the α/β-barrel protein models.

The 73-residue polypeptide of Fedorov *et al.* (1992) was designed to fold into two consecutive identical αββ units. The synthetic gene for the peptide was cloned and expressed in a cell-free translation system. The crude product was shown to have a compact, stable structure as judged by size-exclusion chromatography, urea-gradient gel electrophoresis, and limited proteolysis.

C. β Sheets

1. Artificial Polypeptides That Bind the Insecticide DDT

Although the design of a β sheet is basically not more difficult than the design of an α helix, special attention has to be directed to the control or prevention of aggregation of designed β-sheet polypeptides. We have constructed a DDT-binding peptide (DBP, Fig. 12) (Moser *et al.*, 1983) which even in 8 *M* urea and at low peptide concentration (~20 μ*M*) retained approximately 50% of the proposed β structure as demonstrated by CD spectroscopy (Klauser *et al.*, 1991). The dissociation constant K_D of the peptide–DDT complex in 50 m*M* Tris-HCl, pH 7.0/ethanol (1:1) was 0.8 μ*M*. Among the many natural proteins tested, β-casein had the highest and bovine serum albumin the lowest affinity for DDT; the corresponding K_D values were 55 μ*M* (Hehlgans, 1987) and >1 m*M* (Moser *et al.*, 1983). Aromatic residues in positions 14 and 16 of DBP seemed to be crucial for tight DDT binding because the replacement of Phe^{14} by Val and His^{16} by Thr weakened the interaction with DDT 30- and 80-fold, respectively. The mutual exchange of Phe and Ile in positions 3 and 4, however, had only a minor effect on DDT binding. Other hydrophobic ligands such as dioxin were clearly bound less strongly.

Mixtures of DBP, hemin, and various reducing agents were able to degrade DDT partially with cytochrome *P*-450-like activity (Langen *et al.*, 1989), whereby the role of DBP seemed to be that of a mediator of the solubility and thus reactivity of DDT. When the reducing agent was cysteine, DDT degradation was approximately 3×10^5 times faster (k_{cat} of 0.03 min^{-1} based on the hemin concentration) than in the uncatalyzed reaction, and the major products formed were three water-soluble, nontoxic conjugates of DDT metabolites with cysteine that had lost two or three of the five chlorine atoms of DDT per molecule and whose structures were elucidated by gas chromatography/mass spectrometry (Langen *et al.*, 1989). DDT metabolite–amino acid conjugates are also formed *in vivo* and have been isolated from rats (Pinto *et al.*, 1965). Most likely, glutathione *S*-transferase-catalyzed addition of glutathione (Meister and Anderson, 1983) also plays a role in DDT metabolism in animals and humans. Interestingly, a β-galactosidase–DBP fusion protein expressed from the recombinant plasmid pUR291-DBP in *E. coli* JM 109 was 25 times more potent than an equimolar amount of free DBP in stimulating the degradation of DDT by a mixture of hemin and excess cysteine (Moser *et al.*, 1987).

To provide a binding site for the prosthetic group of the degradation reaction next to the proposed DDT binding site, a slightly modified version of DBP

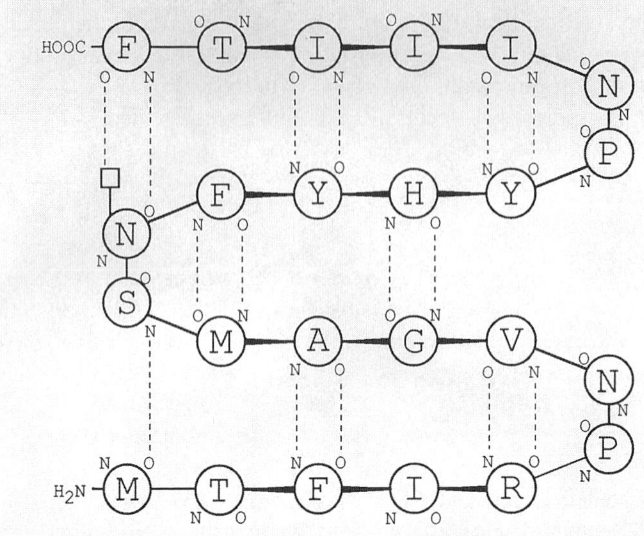

Figure 12 Proposed structure of a designed 24-residue DDT-binding peptide (Moser *et al.*, 1983; Klauser *et al.*, 1991). The space-filling model shows the complex with the DDT ligand. The two rectangular *p*-chlorophenyl groups of DDT (in the middle of the picture) are sandwiched between the aromatic side chains of Phe^{14}, His^{16}, and Phe^{3}.

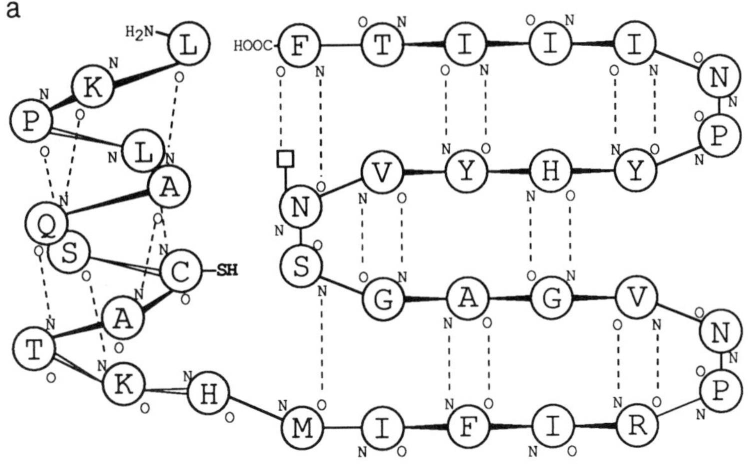

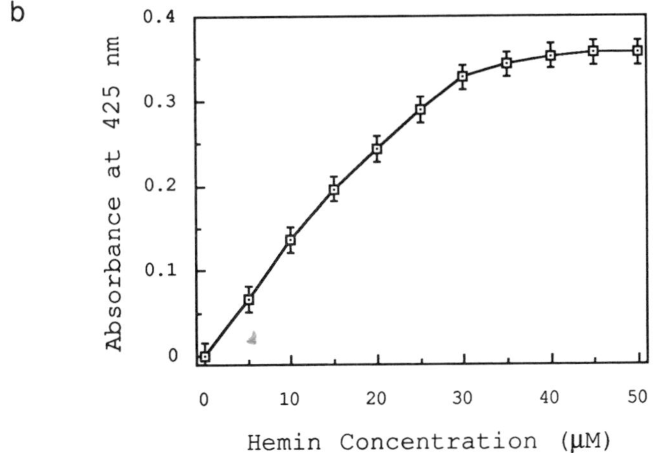

Figure 13 (a) DDT-binding peptide extended at the N terminus by the heme-binding F-helix of myoglobin (Langen, 1988). (b) Hemin saturation curve of the F-helix–DBP fusion protein.

was extended at the N terminus by the F-helix of myoglobin (Fig. 13a; Langen, 1988). Because the fifth ligand of the heme iron in cytochrome *P*-450 enzymes is usually cysteine, we replaced the proximal histidine of the F-helix with this amino acid. Although the resulting 36-residue F-helix–DBP fusion protein bound hemin 80 times more strongly (K_D 6×10^{-6} *M*) than did the 24-residue DBP, the DDT degradation reactions were not accelerated further. Probably, orientation and distance of "active site"-bound hemin and DDT were not optimal

Figure 14 Model of the designed synthetic 61-residue "minibody" (Pessi *et al.*, 1993) derived from the variable domain of the heavy chain of monoclonal antibody McPC603 (Segal *et al.*, 1974). Three histidine residues properly placed in the hypervariable loops 1 and 2 mimicked the Zn^{2+} binding site of carbonic anhydrase B (Kannan *et al.*, 1975). Reprinted with permission from *Nature*, Pessi *et al.*, 1993. Copyright 1993 Macmillan Magazines Limited.

for catalysis. A hemin titration curve of the F-helix–DBP fusion protein is shown in Fig. 13b.

During our work on the artificial DDT-binding peptide we began to wonder what a DDT binding site designed by nature would look like. To answer this question, we prepared polyclonal and monoclonal anti-DDT antibodies (Bürgisser *et al.*, 1990), sequenced the V_H and the V_L domains of an immunoglobulin M (IgM)-type monoclonal antibody, and derived a structure of its variable region by computer modeling using INSIGHT/DISCOVER (Frey, 1994; Frey *et al.*, 1995). The modeling was based on the X-ray structures of two Fab antibody fragments that showed high sequence similarity with the H- and L-chains of our anti-DDT antibody, respectively (Padlan *et al.*, 1989; Stanfield *et al.*, 1990). Manual docking with DDT identified a binding site whose main features agreed closely with those of the proposed DDT binding site of our artificial 24-residue polypeptide (Frey,

1994; Frey *et al.*, 1995). On the basis of our work leading to the preparation of anti-DDT antibodies, a commercial kit for DDT analyses was developed.

2. *The Minibody*

Pessi *et al.* (1993) have designed, synthesized, and characterized a 61-residue metal-binding all-β protein called the minibody (Fig. 14). This protein contains three β strands from each of the two β sheets and the hypervariable loops 1 and 2 of a V_H domain of the monoclonal antibody McPC603 (Segal *et al.*, 1974). To engineer a zinc binding site, Pessi *et al.* (1993) transferred the zinc ligation motif of carbonic anhydrase B (Kannan *et al.*, 1975) onto the hypervariable loops of the minibody by replacing each amino acid in the three positions crucial for ligation with histidine. The resulting sequence was synthesized by the solid-phase method. The product gave a CD spectrum typical of β-sheet structure and bound Zn(II) with a K_D of approximately 10^{-6} *M*. The minibody seems to be well suited for further functionalization studies.

3. *Betabellin*

In pursuing the goal of constructing nativelike novel proteins (Hecht *et al.*, 1990), J. S. and D. C. Richardson together with B. W. Erickson tried to design betabellin, a β sandwich consisting of two identical four-stranded antiparallel β sheets stabilized by a cross-sheet disulfide bond. After several rounds of redesign and synthesis, a 64-residue product with satisfactory solubility and CD and laser Raman spectroscopic properties was obtained (Daniels *et al.*, 1988; Richardson and Richardson, 1989).

D. Sequence-Specific DNA-Binding Polypeptides

1. *Principles of Protein–DNA Recognition*

Life is regulated by manifold specific and nonspecific DNA–protein interactions. The proteins involved in these interactions comprise DNA polymerases, auxiliary proteins of DNA replication and other cell cycle proteins, RNA polymerases and transcription factors, repressors, intracellular hormone receptors, histones, and high mobility group and other scaffold proteins, among others.

Proteins achieve sequence-specific DNA recognition by means of structural motifs that usually interact with bases in the major groove and with the backbone of the DNA double helix. The nature of these structural motifs (mainly helix–turn–helix, helix–loop–helix, zinc finger–helix, leucine zipper coiled coil, β sheet) and the principles of sequence-specific DNA binding have been reviewed by Harrison (1991), Pabo and Sauer (1992), and Phillips (1994). However, the repertoire of DNA-recognizing motifs is still growing as demonstrated by the discovery of the "Zn ribbon" in eukaryotic transcriptional elongation factor TFIIS which combines a Cys_4 zinc binding site with a three-stranded β sheet (Qian *et al.*, 1993). It is also interesting to see that the number of proteins that recognize DNA by antiparallel β sheets is steadily increasing

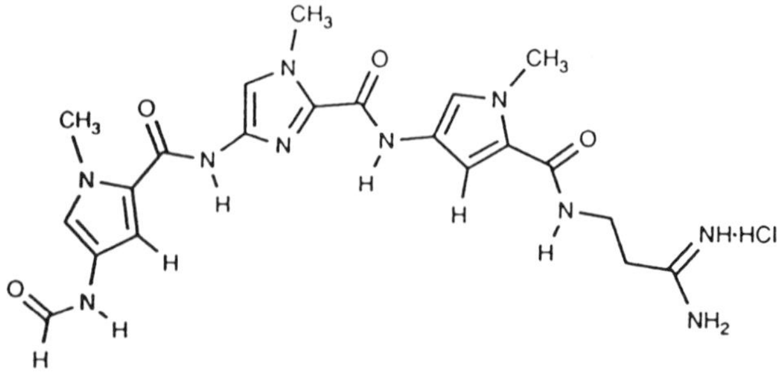

Figure 15 Minor-groove sequence-specific DNA-binding imidazole analog of antibiotic distamycin A. Reprinted with permission from Dwyer *et al.*, 1992. Copyright 1992 American Chemical Society.

(Raumann *et al.*, 1994). The great variety in the specific interactions between DNA and protein sequences suggests that there is no unique amino acid–DNA base recognition code.

2. *Minor Groove Sequence-Specific DNA-Binding Peptide Model*

Using two-dimensional NMR spectroscopy, Dwyer *et al.* (1992) showed that an imidazole analog of the antibiotic distamycin A (Fig. 15) bound with high affinity and cooperativity to AAGTT in the minor groove of synthetic double-stranded d(CGCAAGTTGGC). The stoichiometry of the ligand–DNA complex was 2:1, and the DNA-bound ligand molecules were oriented head to tail. When the central GC base pair of the above oligodeoxyribonucleotide was replaced with an AT base pair, no specific complex detectable by NMR spectroscopy was formed.

3. *Major Groove Sequence-Specific DNA-Binding Polypeptides: From Yeast to HIV*

a. Basic Region–Leucine Zipper Model Peptides Talanian *et al.* (1990) found that the dimerizing leucine zipper moiety of the yeast transcriptional activator GCN4 could be replaced by a disulfide bridge without affecting the DNA binding specificity of a peptide comprising the basic region of GCN4 (residues 222 to 252; Hinnebusch, 1984), as judged by DNase I footprinting at 4°C. This showed that the GCN4 basic region was sufficient for specific binding to the GCN4 recognition element

5′–ATGACTCAT–3′
×
3′–TACTGAGTA–5′.

CD difference spectroscopy indicated that the disulfide dimer of the basic region peptide was helical when bound to a 20-bp target DNA. The dissociation con-

	1	11	21
GCN4	ALKRARNTEA	ARRSRARKLQ	RMKQLEDKVE
MBR-CC	EARRARNREA	AARSRARRAE	KLKALEEKLK

	31	41	51
GCN4	ELLSKNYHLE	NEVARLKKLV	GER
MBR-CC	ALEEKLKALE	EKLKALEEKL	KALG

Figure 16 Sequences of the basic region–leucine zipper of GCN4 and a minimal basic region–coiled coil (MBR-CC) designed by O'Neil *et al.* (1990).

stant of the complex at 4°C was approximately 10^{-8} *M*; binding of the reduced, monomeric peptide was much weaker. Later, the same authors (Talanian *et al.*, 1992) showed that only 15 residues of the GCN4 basic region (positions 231 to 245) were involved in specific DNA binding. Dimerization of a DNA-binding peptide via formation of a disulfide bond was reported also by other authors (Gutte *et al.*, 1980).

Using the same helical basic region–leucine zipper model system for sequence-specific DNA binding, O'Neil *et al.* (1990) designed two peptides which were identical to natural leucine zipper proteins only at those positions that seemed to be crucial for DNA recognition and peptide dimerization (Fig. 16). The designed peptides formed sequence-specific DNA-binding dimers, thus confirming the assumption that their basic regions were α-helical when bound to DNA.

b. HIV Enhancer-Binding Polypeptides Derived from Bacteriophage 434 Repressor A project was undertaken to study specific DNA–protein interactions with respect to constructing human immunodeficiency virus (HIV) enhancer-binding polypeptides. Such polypeptides might be able to displace cellular enhancer-binding proteins, such as NF-κB (Grimm and Baeuerle, 1993) and EBP-1 (Wu *et al.* 1988), and thus inhibit HIV long terminal repeat (LTR)-controlled transcription.

Numerous approaches have been undertaken to interfere with the "life cycle" of HIV-1 *in vitro* and also *in vivo*. Investigators have found the following:

1. 3′-Azido-3′-deoxythymidine (AZT; Dickson and Macilwain, 1993), 2′,5′-oligoadenylates (Sobol *et al.*, 1993), and the 3′-deoxy derivatives of the latter (Müller *et al.*, 1991) inhibited HIV-1 reverse transcriptase
2. Sequence-specific peptidyl-chloromethylketones blocked the furin-mediated cleavage activation of HIV-1 envelope glycoprotein gp160 (Hallenberger *et al.*, 1992)
3. An octameric, branching peptide (Tam, 1988) representing the principal neutralizing determinant of HIV-$1_{\text{III B}}$ envelope glycoprotein gp120 elicited long-term high-titer virus-neutralizing activity in guinea pigs (Wang *et al.*, 1991)
4. A monoclonal antibody directed to the same region of HIV-1_{IIIB} gp120 and administered intravenously to chimpanzees was able to prevent HIV-1 infection in these animals (Emini *et al.*, 1992)

5. Antibodies to a central portion of gp120 inhibited fusion of HIV-infected cells in culture (Rusche *et al.*, 1988)

6. A human monoclonal antibody recognizing the sequence Glu-Leu-Asp-Lys-Trp-Ala within the external domain of the gp41 portion of the HIV-1 envelope glycoprotein neutralized a broad range of lymphoid cell culture-adapted HIV-1 variants and HIV-1 primary isolates (Conley *et al.*, 1994)

7. A 30-residue peptide of gp41 (residues 637 to 666, net charge –5, not toxic at a concentration of ~800 μM) inhibited infection of MT-2 cells by both homologous and heterologous HIV-1 isolates completely at concentrations above 2 μM by preventing the interaction between the fusogenic domain of gp41 and relevant cell membrane components (Jiang *et al.*, 1993)

8. A dimeric "CD4 immunoadhesin" consisting of the first two extracellular, N-terminal domains of the HIV receptor CD4 and the Fc portion of an IgG_1 heavy chain possessed properties of both recombinant soluble CD4 (i.e., gp120 binding and HIV blocking) and IgG (i.e., long plasma half-life and Fc receptor binding) and mediated antibody-dependent cell-mediated cytotoxicity toward HIV-infected cells (Byrn *et al.*, 1990)

9. Oligodeoxyribonucleotide phosphorothioates, 28 nucleotides long and complementary to different regions of HIV-1 RNA ("antisense oligonucleotides"), inhibited replication of the virus in cell culture in a sequence-specific manner at 1 μM concentration, whereas mismatched or random oligonucleotide phosphorothioates only delayed HIV-1 replication (Lisziewicz *et al.*, 1992)

10. A hairpin ribozyme designed to cleave HIV-1 RNA between nucleotides +111 and +112 (C↓GUC), and transfected into HeLa cells in various expression vectors together with DNA constructs from a variety of HIV-1 strains, inhibited HIV-1 expression as indicated by the p24 antigen levels and the reduced Tat activity (Ojwang *et al.*, 1992; Yu *et al.*, 1993; see also Joseph and Burke, 1993, for optimization of an anti-HIV hairpin ribozyme by *in vitro* selection)

11. Stimulation of HIV-1 and HIV-2 transcription by Tat as well as NF-κB- and Tat-induced superactivation of HIV-1 LTR could be inhibited in cell culture by nontoxic concentrations of the antiviral drugs Ro5-3335 (Hsu *et al.*, 1991) and pentoxifylline and Ro24-7429 (Biswas *et al.*, 1993), respectively, with the latter two compounds acting in a cooperative manner

12. A human anti-HIV-1 gp120 single-chain antibody, designed for expression in eukaryotic cells and composed of heavy chain leader sequence and heavy and light chain variable regions connected by a 15-residue linker peptide, bound to intracellular envelope protein and inhibited both processing of the envelope precursor and syncytia formation (Marasco *et al.*, 1993)

13. Inhibitors of HIV-1 protease prevented formation of infectious virus particles (Meek, 1992; Wlodawer and Erickson, 1993)

As far as these drugs have been applied in clinical trials, none has shown significant long-term anti-HIV activity or relief from symptoms of the disease. The reasons for this are manifold, for example, rapid mutation of the relevant regions of

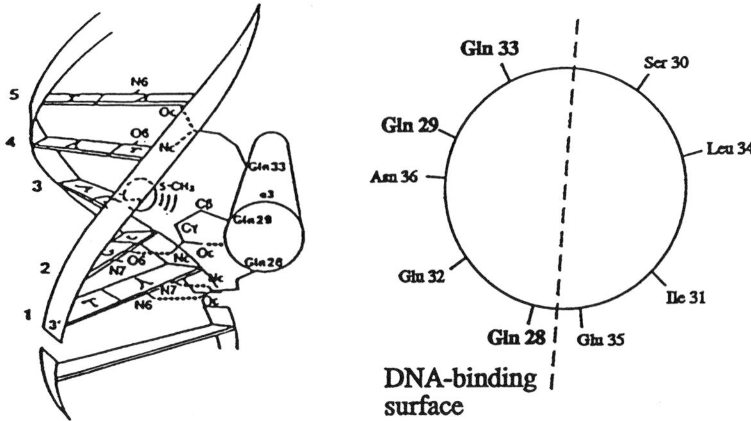

Sequence of the designed 42-residue HIV enhancer-binding peptide derived from 434 repressor:

GKTKRPRFGKTKRPRVGQQSIEQLENGKTKRPRFGKTKRPRY

Figure 17 Design of artificial HIV enhancer-binding peptide R42. *(Left)* Structure of the complex formed between the recognition helix (cylinder, residues 28–36) of the repressor and the ACAAG or ACAAT sequences of the operators of bacteriophage 434. Reprinted with permission from *Nature,* Anderson *et al.,* 1987. Copyright 1987 Macmillan Magazines Limited. The same (ACAAG) or similar sequences (ACTTT) occur in the enhancer region of HIV LTR. *(Right)* Helical wheel representation of the sequence of the recognition helix. The specificity of binding is determined mainly by the three glutamine residues. *(Bottom)* Sequence of R42; underlined residues are identical with residues 28–44 of 434 repressor.

the genes for envelope protein and reverse transcriptase, lack of stability in the circulation or toxicity of some of the drugs, and failure of many of the anti-HIV agents mentioned above to enter cells freely. The search for efficient anti-HIV drugs therefore continues.

Because the enhancer region (positions –104 to –80) of the long terminal repeat of HIV-1 (Ratner *et al.,* 1985; Wain-Hobson *et al.,* 1985) seems to be genetically stable, we tried to design enhancer-binding, inhibitory polypeptides using repressor–operator complexes of bacteriophage 434 as a model system (Anderson *et al.,* 1987; Aggarwal *et al.,* 1988; Shimon and Harrison, 1993). This choice was based on sequence similarities between the 434 operators and the HIV enhancer region. We found that a 42-residue construct (R42) (Fig. 17), comprising the recognition helix and four copies of a positively charged segment of 434 repressor, bound strongly to HIV-1 enhancer DNA (Hehlgans *et al.,* 1990, 1993; Gutte and Klauser, 1991). The specificity of the interaction was demonstrated by competitive band shift assays (Fig. 18a) (Brenz Verca, 1992; Hiltpold, 1994), inhibition of *in vitro* transcription of HIV enhancer-containing

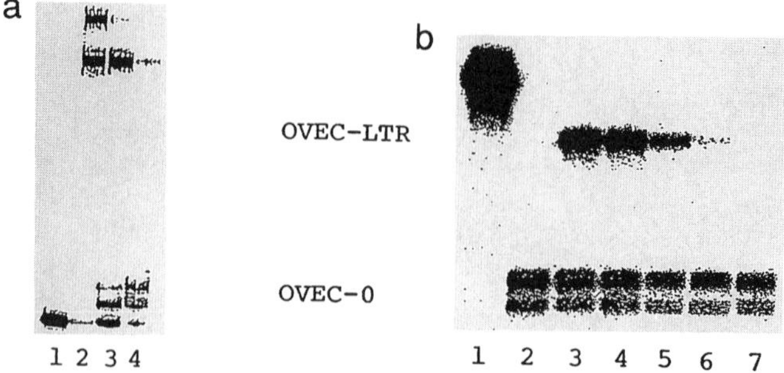

Figure 18 *In vitro* activity of designed HIV enhancer-binding peptide R42. (a) Gel electrophoretic analysis of the interaction of R42 and p50 (DNA-binding subunit of transcription factor NF-κB) with the HIV enhancers, which have two NF-κB binding sites (Brenz Verca, 1992; Hiltpold, 1994). Lane 1 contained 70-bp enhancer DNA; lane 2, enhancer DNA plus 0.4 pmol p50; lane 3, enhancer DNA plus 0.4 pmol p50 and 0.5 pmol R42; lane 4, enhancer DNA plus 0.4 pmol p50 and 1.0 pmol R42. (b) Effect of R42 on *in vitro* transcription in HeLa cell nuclear extract (Caderas *et al.*, manuscript in preparation). Plasmid OVEC-0 was a gift from Prof. W. Schaffner (Westin *et al.*, 1987); plasmid OVEC-LTR contained an HIV LTR insert (–109 to –45 including the HIV enhancers). Lane 1 contained oligodeoxyribonucleotide (93-mer) used for hybridization/nuclease S_1 digestion of the transcription products; lanes 2–7, products of *in vitro* transcription after hybridization with 93-mer and nuclease S_1 digestion: lanes 2 and 3, no R42 added; lane 4, 5 pmol R42; lane 5, 10 pmol R42; lane 6, 20 pmol R42; lane 7, 250 pmol R42.

plasmids in HeLa cell extracts in the presence of R42 (Fig. 18b) (Hehlgans *et al.*, 1993; Caderas *et al.*, manuscript in preparation), and DNase I footprinting of complexes of R42 with 57- and 70-bp HIV enhancer DNA, respectively (Hehlgans *et al.*, 1993; Städler *et al.*, in press). Detailed structural characterization of one of the R42–HIV enhancer DNA complexes is under way.

Redesign of R42 in such a way that the resulting product recognizes a nearly symmetric dimer of the target sequence ACAAG at the 5′ end of HIV enhancer may increase the binding specificity of an artificial repressor greatly. To test the *in vivo* activity of HIV enhancer-binding, inhibitory polypeptides, the synthetic genes of these peptides must be introduced into cells via recombinant plasmids and viruses and then expressed (Brochier *et al.*, 1991).

E. Miscellaneous Design Approaches

This overview of strategies for designing novel polypeptides should not end without mentioning the work of Balaram and co-workers on a modular approach to the construction of synthetic protein mimics (e.g., Karle *et al.*, 1989; Srinivasan *et al.*, 1991). In addition, Atassi and co-workers devised the technique of "surface-simulation" synthesis (Atassi, 1986).

III. Polypeptide Design: *Quo Vadis?*

As indicated in Section I, polypeptide design has great potential to make interesting and useful contributions to basic research, biotechnology, and medicine. However, comparing early and recent work in the field (e.g., Gutte *et al.*, 1979; Johnsson *et al.*, 1993) shows very clearly that progress has been slow. Successful *de novo* design is still hampered by limited knowledge of the relationship between amino acid sequence and folding of a polypeptide chain. There have been advances in following protein folding pathways using site-directed mutagenesis (e.g., Clarke and Fersht, 1993; Jackson *et al.*, 1993) and hydrogen exchange labeling with subsequent analysis of the proton labeling pattern by two-dimensional NMR spectroscopy (Roder *et al.*, 1988). However, at present, and probably for a long time to come, the results of these studies will only tell how but not why a given sequence folds into a distinct, unique conformation. It seems that radically new methods are needed for solving the enigma of protein folding. Until such methods are available, we should make the best of the present knowledge of protein folding and try to design some useful peptides such as dioxin-binding peptides, stable and tight-binding enzyme inhibitors, or transmembrane channels of various sizes.

A promising approach to designing novel proteins is based on the use of structural and functional modules or domains of natural proteins. A number of interesting products have been obtained in this way already, for example, the HIV enhancer-binding polypeptide derived from the bacteriophage 434 repressor (Hehlgans *et al.*, 1993) and a modified, metal-binding IgG heavy-chain variable domain (Pessi *et al.*, 1993), and many more will follow.

We should be aware that designed, biologically active polypeptides will have to compete with artificial, intracellularly acting antibodies (Marasco *et al.*, 1993), peptide mimetics (Hirschmann *et al.*, 1993; Smythe and von Itzstein, 1994), peptide libraries (Scott and Smith, 1990; Devlin *et al.*, 1990; Lam *et al.*, 1991; Houghten *et al.*, 1991), RNA libraries (Bartel and Szostak, 1993; Benner, 1993), ribozymes (Cech, 1988; Haseloff and Gerlach, 1988; Yu *et al.*, 1993), antisense oligodeoxyribonucleotides (Lisziewicz *et al.*, 1992; Offensperger *et al.*, 1993; Stein and Cheng, 1993), and other compounds. We should also be aware that the application of designed peptides or their genes in humans will raise bioethical questions.

In spite of these limitations and obstacles, polypeptide design seems to be attractive as judged by the growing number of people working in this area. For them, Shakespeare's famous line from *Hamlet* (Act III, Scene 1) could be modified to read, "To design or not to design, that is the question."

Acknowledgments

We thank the Deutsche Forschungsgemeinschaft, the Stiftung Volkswagenwerk, the Kanton of Zürich, the Schweizerische Nationalfonds, the Bundesamt für Gesundheitswesen in collaboration with the Stiftung zur Förderung der Aids-Forschung in der Schweiz, the Roche Research Foundation, and the Stiftung für wissenschaftliche Forschung an der Universität Zürich for financial support of our part of the work discussed in this chapter.

References

Aggarwal, A. K., Rodgers, D. W., Drottar, M., Ptashne, M., and Harrison, S. C. (1988). *Science* **242,** 899–907.

Åkerfeldt, K. S., Kim, R. M., Camac, D., Groves, J. T., Lear, J. D., and DeGrado, W. F. (1992). *J. Am. Chem. Soc.* **114,** 9656–9657.

Anderson, J. E., Ptashne, M., and Harrison, S. C. (1987). *Nature (London)* **326,** 846–852.

Argos, P., Rossmann, M. G., and Johnson, J. E. (1977). *Biochem. Biophys. Res. Commun.* **75,** 83–86.

Atassi, M. Z. (1986). *In* "Protein Engineering, Applications in Science, Medicine and Industry" (M. Inouye and R. Sarma, eds.), pp. 125–153. Academic Press, Orlando, Florida.

Banner, D. W., Kokkinidis, M., and Tsernoglou, D. (1987). *J. Mol. Biol.* **196,** 657–675.

Banyard, S. H., Stammers, D. K., and Harrison, P. M. (1978). *Nature (London)* **271,** 282–284.

Bartel, D. P., and Szostak, J. W. (1993). *Science* **261,** 1411–1418.

Baumann, G., and Mueller, P. (1974). *J. Supramol. Struct.* **2,** 538–557.

Benner, S. A. (1993). *Science* **261,** 1402–1403.

Benner, S. A., and Gerloff, D. (1991). *Adv. Enzyme Regul.* **31,** 121–181.

Bird, R. E., Hardman, K. D., Jacobson, J. W., Johnson, S., Kaufman, B. M., Lee, S. M., Lee, T., Pope, S. H., Riordan, G. S., and Whitlow, M. (1988). *Science* **242,** 423–426.

Biswas, D. K., Ahlers, C. M., Dezube, B. J., and Pardee, A. B. (1993). *Proc. Natl. Acad. Sci. U.S.A.* **90,** 11044–11048.

Bloomer, A. C., Champness, J. N., Bricogne, G., Staden, R., and Klug, A. (1978). *Nature (London)* **276,** 362–368.

Blow, D. M. (1976). *Acc. Chem. Res.* **9,** 145–152.

Brenz Verca, S. (1992). Diplomarbeit, Universität Zürich.

Brochier, B., Kieny, M. P., Costy, F., Coppens, P., Bauduin, B., Lecocq, J. P., Languet, B., Chappuis, G., Desmettre, P., Afiademanyo, K., Libois, R., and Pastoret, P.-P. (1991). *Nature (London)* **354,** 520–522.

Bürgisser, D., Frey, S., Gutte, B., and Klauser, S. (1990). *Biochem Biophys. Res. Commun.* **166,** 1228–1236.

Byrn, R. A., Mordenti, J., Lucas, C., Smith, D., Marsters, S. A., Johnson, J. S., Cossum, P., Chamow, S. M., Wurm, F. M., Gregory, T., Groopman, J. E., and Capon, D. J. (1990). *Nature (London)* **344,** 667–670.

Caderas *et al.* (manuscript in preparation).

Cech, T. R. (1988). *J. Am. Med. Assoc.* **260,** 3030–3034.

Chakravarty, P. K., Mathur, K. B., and Dhar, M. M. (1973). *Experientia* **29,** 786–788.

Choma, C. T., Lear, J. D., Nelson, M. J., Dutton, P. L., Robertson, D. E., and DeGrado, W. F. (1994). *J. Am. Chem. Soc.* **116,** 856–865.

Chou, P. Y., and Fasman, G. D. (1978). *Adv. Enzymol. Relat. Areas Mol. Biol.* **47,** 45–148.

Chung, L. A., Lear, J. D., and DeGrado, W. F. (1992). *Biochemistry* **31,** 6608–6616.

Clague, M. J., and Cherry, R. J. (1988). *Biochem. J.* **252,** 791–794.

Clarke, J., and Fersht, A. R. (1993). *Biochemistry* **32,** 4322–4329.

Conley, A. J., Kessler, J. A., Boots, L. J., Tung, J.-S., Arnold, B. A., Keller, P. M., Shaw, A. R., and Emini, E. A. (1994). *Proc. Natl. Acad. Sci. U.S.A.* **91,** 3348–3352.

Creighton, T. E. (1979). *J. Mol. Biol.* **129,** 235–264.

Daniels, S. B., Williams, R. W., Richardson, J. S., Richardson, D. C., and Erickson, B. W. (1988) *FASEB J.* **2,** A1543.

DeGrado, W. F. (1988). *Adv. Protein Chem.* **39,** 51–124.

Devlin, J. J., Panganiban, L. C., and Devlin, P. E. (1990). *Science* **249,** 404–406.
Dickson, D., and Macilwain, C. (1993). *Nature (London)* **362,** 483.
Dwyer, T. J., Geierstanger, B. H., Bathini, Y., Lown, J. W., and Wemmer, D. E. (1992). *J. Am. Chem. Soc.* **114,** 5911–5919.
Eberle, W., Klaus, W., Cesareni, G., Sander, C., and Roesch, P. (1990). *Biochemistry* **29,** 7402–7407.
Ellenberger, T. E., Brandl, C. J., Struhl, K., and Harrison, S. C. (1992). *Cell (Cambridge, Mass.)* **71,** 1223–1237.
Emini, E. A., Schleif, W. A., Nunberg, J. H., Conley, A. J., Eda, Y., Tokiyoshi, S., Putney, S. D., Matsushita, S., Cobb, K. E., Jett, C. M., Eichberg, J. W., and Murthy, K. K. (1992). *Nature (London)* **355,** 728–730.
Engel, M., Williams, R. W., and Erickson, B. W. (1991). *Biochemistry* **30,** 3161–3169.
Fedorov, A. N., Dolgikh, D. A., Chemeris, V. V., Chernov, B. K., Finkelstein, A. V., Schulga, A. A., Alakhov, Y. B., Kirpichnikov, M. P., and Ptitsyn, O. B. (1992). *J. Mol. Biol.* **225,** 927–931.
Feldmann, R. J. (1983). *In* "Computer Applications in Chemistry" (S. R. Heller and R. Potenzone, Jr., eds.), pp. 9–18. Elsevier, Amsterdam.
Ferran, D. S., Sobel, M., and Harris, R. B. (1992). *Biochemistry* **31,** 5010–5016.
Frey, S. (1994). Thesis, Universität Zürich.
Frey, S., Klauser, S., Gutte, B., and Honegger, A. (1995). *J. Biol. Chem.* in press.
Fukushima, D., Kupferberg, J. P., Yokoyama, S., Kroon, D. J., Kaiser, E. T., and Kézdy, F. J. (1979). *J. Am. Chem. Soc.* **101,** 3703–3704.
Ghadiri, M. R., Soares, C., and Choi, C. (1992). *J. Am Chem. Soc.* **114,** 825–831.
Ghadiri, M. R., Granja, J. R., and Buehler, L. K. (1994). *Nature (London)* **369,** 301–304.
Goraj, K., Renard, A., and Martial, J. A. (1990). *Protein Eng.* **3,** 259–266.
Grimm, S., and Baeuerle, P. A. (1993). *Biochem. J.* **290,** 297–308.
Grove, A., Mutter, M., Rivier, J. E., and Montal, M. (1993). *J. Am. Chem. Soc.* **115,** 5919–5924.
Gutte, B. (1975). *J. Biol. Chem.* **250,** 889–904.
Gutte, B. (1977). *J. Biol. Chem.* **252,** 663–670.
Gutte, B., and Klauser, S. (1991). *In* "Molecular Conformation and Biological Interactions" (P. Balaram and S. Ramaseshan, eds.), pp. 447–455. Indian Academy of Sciences, Bangalore.
Gutte, B., Däumigen, M., and Wittschieber, E. (1979). *Nature (London)* **281,** 650–655.
Gutte, B., Schindler, S., Standar, F., and Wittschieber, E. (1980). *Biochem. Biophys. Res. Commun.* **95,** 1071–1079.
Hahn, K. W., Klis, W. A., and Stewart, J. M. (1990). *Science* **248,** 1544–1547.
Hallenberger, S., Bosch, V., Angliker, H., Shaw, E., Klenk, H.-D., and Garten, W. (1992). *Nature (London)* **360,** 358–361.
Handel, T. M., Williams, S. A., and DeGrado, W. F. (1993). *Science* **261,** 879–885.
Harrison, S. C. (1991). *Nature (London)* **353,** 715–719.
Haseloff, J., and Gerlach, W. L. (1988). *Nature (London)* **334,** 585–591.
Hecht, M. H., Richardson, J. S., Richardson, D. C., and Ogden, R. C. (1990). *Science* **249,** 884–891.
Hehlgans, T. (1987). Diplomarbeit, Universität Zürich.
Hehlgans, T., Stolz, M., Cui, T., Mohajeri, H., Salgam, P., Leiser, A., Klauser, S., Mertz, R., and Gutte, B. (1990). *Experientia* **46,** A3.
Hehlgans, T., Stolz, M., Klauser, S., Cui, T., Salgam, P., Brenz Verca, S., Widmann, M., Leiser, A.,, Städler, K., and Gutte, B. (1993). *FEBS Lett.* **315,** 51–55.
Hiltpold, A. (1994). Diplomarbeit, Universität Zürich.
Hinnebusch, A. G. (1984). *Proc. Natl. Acad. Sci. U.S.A.* **81,** 6442–6446.
Hirschmann, R., Nicolaou, K. C., Pietranico, S., Leahy, E. M., Salvino, J., Arison, B., Cichy, M. A., Spoors, P. G., Shakespeare, W. C., Sprengeler, P. A., Hamley, P., Smith III, A. B., Reisine, T., Raynor, K., Maechler, L., Donaldson, C., Vale, W., Freidinger, R. M., Cascieri, M. R., and Strader, C. D. (1993). *J. Am. Chem. Soc.* **115,** 12550–12568.
Ho, S. P., and DeGrado, W. F. (1987). *J. Am. Chem. Soc.* **109,** 6751–6758.
Hodges, R. S., Saund, A. K., Chong, P. C. S., St.-Pierre, S. A., and Reid, R. E. (1981). *J. Biol. Chem.* **256,** 1214–1224.

Houghten, R. A., Pinilla, C., Blondelle, S. E., Appel, J. R., Dooley, C. T., and Cuervo, J. H. (1991). *Nature (London)* **354,** 84–86.

Hsu, M.-C., Schutt, A. D., Holly, M., Slice, L. W., Sherman, M. I., Richman, D. D., Potash, M. J., and Volsky, D. J. (1991). *Science* **254,** 1799–1802.

Huston, J. S., Levinson, D., Mudgett-Hunter, M., Tai, M. S., Novotny, J., Margolies, M. N., Ridge, R. J., Bruccoleri, R. E., Haber, E., Crea, R., and Oppermann, H. (1988). *Proc. Natl. Acad. Sci. U.S.A.* **85,** 5879–5883.

Jackson, S. E., el Masry, N., and Fersht, A. R. (1993). *Biochemistry* **32,** 11270–11278.

Jaenicke, R., Gutte, B., Glatter, U., Strassburger, W., and Wollmer, A. (1980). *FEBS Lett.* **114,** 161–164.

Jiang, S., Lin, K., Strick, N., and Neurath, A. R. (1993). *Nature (London)* **365,** 113.

Johnsson, K., Allemann, R. K., Widmer, H., and Benner, S. A. (1993). *Nature (London)* **365,** 530–532.

Joseph, S., and Burke, J. M. (1993). *J. Biol. Chem.* **268,** 24515–24518.

Junius, F. K., Weiss, A. S., and King, G. F. (1993). *Eur. J. Biochem.* **214,** 415–424.

Kannan, K. K., Notstrand, B., Fridborg, K., Lövgren, S., Ohlsson, A., and Petef, M. (1975). *Proc. Natl. Acad. Sci. U.S.A.* **72,** 51–55.

Karle, I. L., Flippen-Anderson, J. L., Uma, K., and Balaram, P. (1989). *Biochemistry* **28,** 6696–6701.

Klaus, W., and Moser, R. (1992). *Protein Eng.* **5,** 333–341.

Klauser, S., Gantner, D., Salgam, P., and Gutte, B. (1991). *Biochem. Biophys. Res. Commun.* **179,** 1212–1219.

Kullmann, W. (1984). *J. Med. Chem.* **27,** 106–115.

Kullmann, W., and Gutte, B. (1975). *Biochem. Soc. Trans.* **3,** 899–902.

Lam, K. S., Salmon, S. E., Hersh, E. M., Hruby, V. J., Kazmierski, W. M., and Knapp, R. J. (1991). *Nature (London)* **354,** 82–84.

Landschulz, W. H., Johnson, P. F., and McKnight, S. L. (1988). *Science* **240,** 1759–1764.

Langen, H. (1988). Thesis, Universität Zürich.

Langen, H., Epprecht, T., Linden, M., Hehlgans, T., Gutte, B., and Buser, H.-R. (1989). *Eur. J. Biochem.* **182,** 727–735.

Lear, J. D., Wasserman, Z. R., and DeGrado, W. F. (1988). *Science* **240,** 1177–1181.

Levitt, M. (1978). *Biochemistry* **17,** 4277–4285.

Lisziewicz, J., Sun, D., Klotman, M., Agrawal, S., Zamecnik, P., and Gallo, R. (1992). *Proc. Natl. Acad. Sci. U.S.A.* **89,** 11209–11213.

Lovejoy, B., Åkerfeldt, K. S., DeGrado, W. F., and Eisenberg, D. (1992). *Protein Sci.* **1,** 1073–1077.

Lovejoy, B., Choe, S., Cascio, D., McRorie, D. K., DeGrado, W. F., and Eisenberg, D. (1993). *Science* **259,** 1288–1293.

Lupas, A., van Dyke, M., and Stock, J. (1991). *Science* **252,** 1162–1164.

McLachlan, A. D., and Stewart, M. (1975). *J. Mol. Biol.* **98,** 293–304.

Marasco, W. A., Haseltine, W. A., and Chen, S. Y. (1993). *Proc. Natl. Acad. Sci. U.S.A.* **90,** 7889–7893.

Mathews, F. S., Bethge, P. H., and Czerwinski, E. W. (1979). *J. Biol. Chem.* **254,** 1699–1706.

Meek, T. D. (1992). *J. Enzyme Inhib.* **6,** 65–98.

Meister, A., and Anderson, M. E. (1983). *Annu. Rev. Biochem.* **52,** 711–760.

Merrifield, R. B. (1963). *J. Am. Chem. Soc.* **85,** 2149–2154.

Moser, R. (1992). *Protein Eng.* **5,** 323–331.

Moser, R., Thomas, R. M., and Gutte, B. (1983). *FEBS Lett.* **157,** 247–251.

Moser, R., Klauser, S., Leist, T., Langen, H., Epprecht, T., and Gutte, B. (1985). *Angew. Chem., Int. Ed. Engl.* **24,** 719–727.

Moser, R., Frey, S., Münger, K., Hehlgans, T., Klauser, S., Langen, H., Winnacker, E.-L., Mertz, R., and Gutte, B. (1987). *Protein Eng.* **1,** 339–343.

Müller, W. E. G., Weiler, B. E., Charubala, R., Pfleiderer, W., Leserman, L., Sobol, R. W., Suhadolnik, R. J., and Schröder, H. C. (1991). *Biochemistry* **30,** 2027–2033.

Mutter, M., and Vuilleumier, S. (1989). *Angew. Chem.* **101,** 551–571.

Mutter, M., Hersperger, R., Gubernator, K., and Müller, K. (1989). *Proteins Struct. Funct. Genet.* **5,** 13–21.

Offensperger, W. B., Offensperger, S., Walter, E., Teubner, K., Igloi, G., Blum, H. E., and Gerok, W. (1993). *EMBO J.* **12,** 1257–1262.

Ojwang, J. O., Hampel, A., Looney, D. J., Wong-Staal, F., and Rappaport, J. (1992). *Proc. Natl. Acad. Sci. U.S.A.* **89,** 10802–10806.

O'Neil, K. T., Hoess, R. H., and DeGrado, W. F. (1990). *Science* **249,** 774–778.

Osborne, J. C., Jr., and Brewer, H. B., Jr. (1977). *Adv. Protein Chem.* **31,** 253–337.

O'Shea, E. K., Rutkowski, R., and Kim, P. S. (1992). *Cell (Cambridge, Mass.)* **68,** 699–708.

Osterhout, J. J., Jr., Handel, T., Na, G., Toumadje, A., Long, R. C., Connolly, P. J., Hoch, J. C., Johnson, W. C., Jr., Live, D., and DeGrado, W. F. (1992). *J. Am. Chem. Soc.* **114,** 331–337.

Pabo, C. O., and Sauer, R. T. (1992). *Annu. Rev. Biochem.* **61,** 1053–1095.

Padlan, E. A., Silverton, E. W., Sheriff, S., Cohen, G. H., Smith-Gill, S. J., and Davies, D. R. (1989). *Proc. Natl. Acad. Sci. U.S.A.* **86,** 5938–5942.

Pessi, A., Bianchi, E., Crameri, A., Venturini, S., Tramontano, A., and Sollazzo, M. (1993). *Nature (London)* **362,** 367–369.

Phillips, D. C. (1966). *Sci. Am.* **215,** 78–90.

Phillips, S. E. V. (1994). *Annu. Rev. Biophys. Biomol. Struct.* **23,** 671–701.

Pinto, J. D., Camien, M. N., and Dunn, M. S. (1965). *J. Biol. Chem.* **240,** 2148–2154.

Qian, X., Jeon, C. J., Yoon, H. S., Agarwal, K., and Weiss, M. A. (1993). *Nature (London)* **365,** 277–279.

Ratner, L., Haseltine, W., Patarca, R., Livak, K. J., Starcich, B., Josephs, S. F., Doran, E. R., Rafalski, J. A., Whitehorn, E. A., Baumeister, K., Ivanoff, L., Petteway, S. R., Jr., Pearson, M. L., Lautenberger, J. A., Papas, T. S., Ghrayeb, J., Chang, N. T., Gallo, R. C., and Wong-Staal, F. (1985). *Nature (London)* **313,** 277–284.

Raumann, B. E., Rould, M. A., Pabo, C. O., and Sauer, R. T. (1994). *Nature (London)* **367,** 754–757.

Regan, L., and DeGrado, W. F. (1988). *Science* **241,** 976–978.

Richardson, J. S., and Richardson, D. C. (1989). *Trends Biochem. Sci.* **14,** 304–309.

Robertson, D. E., Farid, R. S., Moser, C. C., Urbauer, J. L., Mulholland, S. E., Pidikiti, R., Lear, J. D., Wand, A. J., DeGrado, W. F., and Dutton, P. L. (1994). *Nature (London)* **368,** 425–432.

Roder, H., Elöve, G. A., and Englander, S. W. (1988). *Nature (London)* **335,** 700–704.

Rosenberg, A. H., Lade, B. N., Chui, D., Lin, S., Dunn, J. J., and Studier, F. W. (1987). *Gene* **56,** 125–135.

Rosenberg, R. D. (1977). *Semin. Hematol.* **14,** 427–440.

Rost, B., and Sander, C. (1993). *J. Mol. Biol.* **232,** 584–599.

Rusche, J. R., Javaherian, K., McDanal, C., Petro, J., Lynn, D. L., Grimaila, R., Langlois, A., Gallo, R. C., Arthur, L. O., Fischinger, P. J., Bolognesi, D. P., Putney, S. D., and Matthews, T. J. (1988). *Proc. Natl. Acad. Sci. U.S.A.* **85,** 3198–3202.

Sakurai, H., and Ogawa, S. (1975). *Biochem. Pharmacol.* **24,** 1257–1260.

Sasaki, T., and Kaiser, E. T. (1989). *J. Am. Chem. Soc.* **111,** 380–381.

Schafmeister, C. E., Miercke, L. J. W., and Stroud, R. M. (1993). *Science* **262,** 734–738.

Schiffer, M., and Edmundson, A. B. (1967). *Biophys. J.* **7,** 121.

Scott, J. K., and Smith, G. P. (1990). *Science* **249,** 386–390.

Segal, D. M., Padlan, E. A., Cohen, G. H., Rudikoff, S., Potter, M., and Davies, D. R. (1974). *Proc. Natl. Acad. Sci. U.S.A.* **71,** 4298–4302.

Shimon, L. J. W., and Harrison, S. C. (1993). *J. Mol. Biol.* **232,** 826–838.

Smythe, M. L., and von Itzstein, M. (1994). *J. Am. Chem. Soc.* **116,** 2725–2733.

Sobol, R. W., Fisher, W. L., Reichenbach, N. L., Kumar, A., Beard, W. A., Wilson, S. H., Charubala, R., Pfleiderer, W., and Suhadolnik, R. J. (1993). *Biochemistry* **32,** 12112–12118.

Srinivasan, N., Sowdhamini, R., Ramakrishnan, C., and Balaram, P. (1991). *In* "Molecular Conformation and Biological Interactions" (P. Balaram and S. Ramaseshan, eds.), pp. 59–73. Indian Academy of Sciences, Bangalore.

Städler, K., Liu, N., Trotman, L., Hiltpold, A., Caderas, G., Klauser, S., Hehlgans, T., and Gutte, B. (1995). *Int. J. Peptide Protein Res.*, in press.

Stanfield, R. L., Fieser, T. M., Lerner, R. A., and Wilson, I. A. (1990). *Science* **248,** 712–719.

Stein, C. A., and Cheng, Y.-C. (1993). *Science* **261,** 1004–1012.

Stubbs, G., Warren, S., and Holmes, K. (1977). *Nature (London)* **267,** 216–221.

Studier, F. W., and Moffatt, B. A. (1986). *J. Mol. Biol.* **189,** 113–130.

Talanian, R. V., McKnight, C. J., and Kim, P. S. (1990). *Science* **249,** 769–771.

Talanian, R. V., McKnight, C. J., Rutkowski, R., and Kim, P. S. (1992). *Biochemistry* **31,** 6871–6875.

Tam, J. P. (1988). *Proc. Natl. Acad. Sci. U.S.A.* **85,** 5409–5413.

Tomizawa, J.-I. (1990). *J. Mol. Biol.* **212,** 695–708.

Twigg, A. J., and Sherrat, D. (1980). *Nature (London)* **283,** 216–218.

Wain-Hobson, S., Sonigo, P., Danos, O., Cole, S., and Alizon, M. (1985). *Cell (Cambridge, Mass.)* **40,** 9–17.

Wang, C. Y., Looney, D. J., Li, M. L., Walfield, A. M., Ye, J., Hosein, B., Tam, J. P., and Wong-Staal, F. (1991). *Science* **254,** 285–288.

Weber, P. C., and Salemme, F. R. (1980). *Nature (London)* **287,** 82–84.

Westin, G., Gerster, T., Müller, M. M., Schaffner, G., and Schaffner, W. (1987). *Nucleic Acids Res.* **15,** 6787–6798.

Wlodawer, A., and Erickson, J. W. (1993). *Annu. Rev. Biochem.* **62,** 543–585.

Wu, F. K., Garcia, J. A., Harrich, D., and Gaynor, R. B. (1988). *EMBO J.* **7,** 2117–2130.

Yu, M., Ojwang, J., Yamada, O., Hampel, A., Rappaport, J., Looney, D., and Wong-Staal, F. (1993). *Proc. Natl. Acad. Sci. U.S.A.* **90,** 6340–6344.

Zhou, N. E., Kay, C. M., and Hodges, R. S. (1992). *J. Biol. Chem.* **267,** 2664–2670.

Zhou, N. E., Kay, C. M., and Hodges, R. S. (1993). *Biochemistry* **32,** 3178–3187.

10

Soluble Chemical Combinatorial Libraries: Current Capabilities and Future Possibilities

Richard A. Houghten
Torrey Pines Institute for Molecular Studies
and Houghten Pharmaceuticals, Inc.
San Diego, California 92121

Peptides: Synthesis, Structures, and Applications

I. Introduction

Pharmaceutical drug development, in virtually all instances, begins with the identification of an initial compound (often termed a "lead" compound) that has an activity of interest. Although the activity of interest is expressed, this activity may be low, however, and/or may be combined with undesirable properties such as unacceptable toxicity, poor availability *in vivo,* and unsuitable half-life. In the past, attempts to improve the drug discovery process have relied on the synthesis and testing of large numbers of individual analogs of a lead compound. Structure–activity relationship studies of a molecule, that is, the molecular changes that may increase or decrease its biological activity, are time-consuming and expensive. In most instances, the sheer number of analogs required, and the time and resources required to make them, have been the limiting factors in such studies.

The process involved in the development of new diagnostically or therapeutically useful compounds, as with all pharmaceutical drug development, requires the synthesis and screening of hundreds to thousands of analogs of an original active lead structure. Peptides have increasingly emerged as useful tools in all areas of biomedical research, and as effective immunodiagnostic and therapeutic agents. Methodological developments have made it possible for a modestly equipped laboratory to prepare, using a variety of approaches (Geysen *et al.,* 1984; Houghten, 1985; Frank and Döring, 1988; Eichler *et al.,* 1990), thousands of peptides per year at a fraction of the cost per peptide of earlier methods (Merrifield, 1963; Atherton *et al.,* 1989). Automated synthesizers, which can synthesize 50–100 peptides simultaneously, have also been described (Schnorrenberg and Gerhardt, 1989; Gausephol *et al.,* 1990).

In spite of these advances, however, which greatly increase the availability of individual peptides, synthesis of the immense number of possible analogs of even short peptides (i.e., there are 64×10^6 L-amino acid substitution analogs of a hexapeptide) is impractical. The introduction of combinatorial libraries made up of millions of individual peptide sequences prepared by chemical or molecular biological approaches is at the forefront of a revolution in drug discovery and basic research involving peptides. Such libraries offer a fundamental, practical advance in the study of interactions between peptides and their biochemical or pharmacological targets. They have been used, for example, to study antibody–antigen interactions (Scotto and Craig, 1994; Cwirla *et al.,* 1990; Felicia *et al.,* 1991; Houghten *et al.,* 1991, 1992b; Lam *et al.,* 1991; Pinilla *et al.,* 1992a,b); to develop enzyme inhibitors (Eichler and Houghten, 1993a,b; Owens *et al.,* 1991) and novel antimicrobials (Houghten *et al.,* 1991, 1992a,b; Blondelle *et al.,* 1994b); and to identify biologically active peptides (Houghten and Dooley, 1993a,b; Hortin *et al.,* 1992) or to engineer novel properties in antibodies (Mark *et al.,* 1991; Huse *et al.,* 1989; Barbas *et al.,* 1991).

The peptide libraries presented thus far fall into three broad categories, the difference being the manner in which the sequences are synthesized and/or presented: (1) synthetic chemical approaches, in which mixtures of peptides are synthesized in a manner which permits their use directly in solution since they are not bound to an insoluble pin, resin bead, phage particle, etc. (Houghten *et al.*, 1991, 1992b; Pinilla *et al.*, 1992a,b; Blondelle *et al.*, 1994b; Houghten and Dooley, 1993a,b; Owens *et al.*, 1991; Hortin *et al.*, 1992; Zuckerman *et al.*, 1992; Blake and Litzi-Davis, 1992); (2) synthetic chemistry approaches in which peptide mixtures are presented on a solid support, that is, plastic pins (Geysen *et al.*, 1986), individual resin beads (Lam *et al.*, 1991), or cotton supports (Eichler and Houghten, 1993a,b); and (3) the molecular biology approach, in which peptides or proteins are presented on the surface of filamentous phage particle or plasmids (Scott and Craig, 1994; Cwirla *et al.*, 1990; Felicia *et al.*, 1991; Mark *et al.*, 1991; Huse *et al.*, 1989; Barbas *et al.*, 1991; Devlin *et al.*, 1990; Cull *et al.*, 1992). This chapter illustrates the preparation of soluble combinatorial libraries made up of peptides for a variety of applications.

A. Synthetic Chemical Approaches

1. Soluble Non-Support-Bound Chemical Combinatorial Libraries

Libraries that are bound to a solid support as described below have the inherent limitation that a soluble, purified receptor (antibody, enzyme, etc.) must be available for these approaches to be of use. The majority of pharmacologically relevant bioassays, however, do not utilize soluble receptors, but rather receptors that are membrane-bound or available only as a small component of a complex mixture. Also, most functional assays that rely on affecting a particular interaction or activity (bacterial growth assays requiring functional whole tissue, etc.) will not typically be amenable to use of a peptide on a phage, pin, bead, cotton, etc. Soluble synthetic combinatorial libraries made up of peptides (SPCLs) (Houghten *et al.*, 1991, 1992a,b; Pinilla *et al.*, 1992a,b; Blondelle *et al.*, 1994b; Houghten and Dooley, 1993a,b; Owens *et al.*, 1991; Hortin *et al.*, 1992; Zuckerman *et al.*, 1992; Blake and Litzi-Davis, 1992) produce a peptide chemical diversity that can be screened in any existing *in vitro* or *in vivo* assay system. Figure 1 illustrates the concept of the SPCL approach. The defined positions (O) can be located in any position of the sequence.

As presented in 1991, the first practical SPCL (Houghten *et al.*, 1991) consisted of six-residue peptide sequences having either acetylated (Ac) or nonacetylated N termini, and amidated C termini. The first two amino acids in each peptide chain were individually and specifically defined, whereas the last four amino acids consisted of equimolar, or close to equimolar, mixtures of 19 of the 20 natural L-amino acids. Cysteine was omitted from the mixture positions of the two SPCLs but included in the defined positions. These libraries can be generally represented by the formulas Ac-O_1O_2XXXX-NH_2 and O_1O_2XXXX-NH_2 with O_1 and O_2 equal to AA, AC, AD, etc. through YV, YW, YY, for a total of 400 combinations (20^2), and each X position representing an equimolar mixture of the 19 amino acids. Four mixture positions result in a total

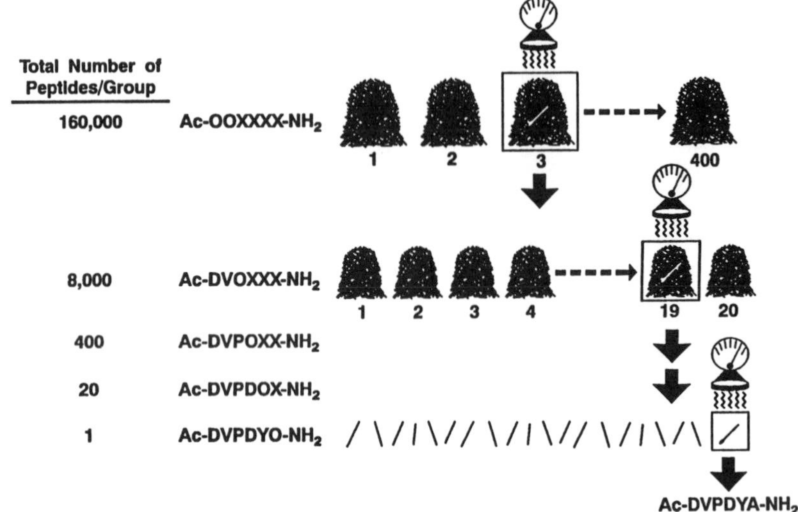

Figure 1 Synthetic peptide combinatorial libraries. The iterative process can be likened to finding a needle in a haystack (with the metal detector being analogous to the bioassay). The initial "haystack" is divided into 400 groups of 160,000 "straws," each with only one of the groups containing the "needle." The needle is followed in ever-decreasing haystack sizes, until a single needle (peptide sequence) is identified among 19 straws. Reprinted from Houghten, R. A. (1993), **9,** 235–239, with permission from Elsevier Trends Journals.

of 130,321 combinations (19^4). Each of the 400 different peptide mixtures that make up each of these libraries thus consists of 130,321 individual hexamers, which in total represent 52,321,400 peptides.

Preparation of the libraries involves minimal manipulation following the final cleavage from the resin. Only extraction and lyophilization are necessary before use. Each non-support-bound peptide mixture is typically used at 1.0 mg/ml. Therefore, if one assumes that the average molecular weight of Ac-O_1O_2XXXX-NH_2 is 785, then a mixture of 130,321 peptides at a total final concentration of 1.0 mg/ml yields a concentration of every peptide within each mixture of 7.67 ng/ml (9.8 n*M*). Once the library of mixtures is screened in an assay of interest, the remaining mixture positions are defined through an iterative enhancement and selection process in order to identify the most active sequences by an iterative synthesis.

Another approach, termed a positional scanning SPCL (PS-SPCL) (Pinilla *et al.*, 1992a; Dooley and Houghten, 1993), enables specific sequence information to be determined in as short as a single day. Each of the individual peptide sublibraries (one for each position of the library) which make up the PS-SPCL is composed of 18 different peptide mixtures in which a single position is defined (represented as O) with one of 18 of the 20 natural L-amino acids (cysteine and tryptophan were omitted); the remaining five positions of the six-residue sequence are composed of mixtures (represented as X) of the same 18 amino acids. The six different peptide sublibraries vary only in the location of the defined amino acids. They are represented as Ac-O_1XXXXX-NH_2, Ac-XO_2XXXX-NH_2, Ac-XXO_3XXX-NH_2, Ac-XXXO_4XX-NH_2, Ac-XXXXO_5X-NH_2, and Ac-XXXXXO_6-NH_2 (108 peptide mixtures in total). As each peptide mixture represents 1,889,568 (18^5) individual sequences, each of the six positional peptide sublibraries contains in total 34,012,224 (18 x1,889,568) different hexamers. Although each of the six positional SPCLs can be examined independently, this set of 108 peptide mixtures, when used in concert, yields a PC-SPCL.

The above approaches have been used to study monoclonal antibody interactions (Houghten *et al.*, 1991, 1992b; Appel *et al.*, 1992; Pinilla *et al.*, 1992b, 1993, 1994), to develop novel opioid peptides in radioreceptor assays (Houghten and Dooley, 1993a,b; Dooley *et al.*, 1993), to develop new antimicrobial peptides in microdilution assays (Houghten *et al.*, 1991, 1992a,b, 1993; Blondelle *et al.*, 1994b; Blondelle and Houghten, 1994), and to identify inhibitors of the bee toxin melittin (Blondelle *et al.*, 1993a, 1994; Blondelle and Houghten, 1994). This approach was also successfully used in plaque assays to develop antiviral peptides, as well as for the development of potent blood pressure and heart rate modulators in direct *in vivo* intravenous studies in rats (Weber *et al.*, 1994).

2. Support-Bound Combinatorial Libraries

A number of approaches have been presented which enable one to generate a diversity of peptides attached to a solid support by synthetic chemistry methods. The first of these to appear was the synthesis of mixtures of peptides on plastic pins (Geysen *et al.*, 1986). The synthesis utilized mixtures of incoming amino acids and resulted in each pin containing hundreds of thousands of octapeptides, each with two positions individually defined. The pins making up the defined mixture sets, when exposed to a soluble receptor such as a monoclonal antibody, were found to yield differential binding. The most effective defined mixture was then enhanced by synthesizing a new set of mixtures on the pins, now with a third position individually defined and the others remaining as mixtures. In this manner, it was thought that one could define individual sequences which bound specifically to discontinuous determinants recognized by monoclonal antibodies. This approach, while conceptually sound, has met with only limited success (Geysen *et al.*, 1986; Geysen and Mason, 1993).

The first practical illustration of immobilized combinatorial libraries was initially presented as the "one resin bead/one peptide" approach, in which 1–3 million individual pentapeptides were generated, each on a separate resin bead. The bead-attached peptides were accessible to water-soluble receptors (Lam *et al.*, 1991). The receptor, now bound to a specific peptide attached to a single bead, is visualized by standard colorimetric approaches to differentiate it from the other beads in the library. The resulting visible beads are removed from the millions other beads using microforceps, and sequenced using standard Edman degradation microsequencing approaches (Lam *et al.*, 1991). The one bead/one peptide approach has been used successfully in the study of monoclonal antibody–antigen interactions (Lam *et al.*, 1991) and to study biotin–streptavidin interactions (Lam and Lebl, 1992). It has also been successfully extended to enable a range of other chemical moieties to be identified using a separate, encoded peptide microsequencing approach (Nikolaiev *et al.*, 1993).

Using cotton as solid support, peptide mixtures were prepared by standard solid-phase chemical synthesis methods (Eichler and Houghten, 1993a,b). As with the pin approach described above, mixtures of amino acids were used in the successive coupling steps to generate a hexapeptide library. As with the bead method, this approach requires a soluble receptor, although in the case of cotton the libraries can also be successfully removed from the support. The use of this approach for the development of a new trypsin inhibitor has been reported (Eichler and Houghten, 1993a,b).

B. Molecular Biological Approaches

The phage library approach, which has been the most commonly reported, centers on the expression of a small number (reviewed by Scott and Craig, 1994) to as many as thousands (Felicia *et al.*, 1991) of copies of the same peptide sequence expressed on the end of filamentous bacteriophage particles. Initial examples of this approach involved six-residue peptides, with later examples extending the peptide length to 20 residues. A library made up of a mixture of phage particles carrying different peptide sequences is produced by preparing a large number (i.e., millions) of oligonucleotides (all possible combinations being the goal) and inserting the random oligonucleotides into the gene of the coat protein. Ultimately, the individual peptides expressed following insertion into the gene are expressed at the tip, or entire surface, of the phage. The peptide–phage particles that bind to a purified receptor which has been immobilized (typically a monoclonal antibody) can be enriched in a process referred to as "biopanning." The peptide–phage particles that do not bind, or are weakly bound, are first washed away, followed by elution of the bound peptide–phage particles. The eluted phage particles are then expressed in *Escherichia coli,* and the process is repeated two or more times to remove weakly or nonspecifically bound sequences. Thus, in panning for a specific peptide–phage, one uses a repetitive process that enriches the amount of the desired sequence with each panning step.

The specific peptide sequence associated with the phage is determined by sequencing the coat protein coding region of the viral DNA. The central features of this approach are the generation of a very large diversity of individual peptides attached to phage particles and the ability to determine which of the peptides have specifically bound to an immobilized, purified receptor. The strength of the method derives from the wide availability of molecular biology techniques and the ability to generate peptide or protein sequences beyond the capabilities of chemical synthetic approaches. The limitations of the approach are the inability to use any but the 20 genetically coded amino acids and the insoluble nature of the phage particle, which limits the range of assays that can be used.

All of the library approaches presented thus far are more rapid than traditional methods for the identification of active compounds by at least three orders of magnitude, and each has particular advantages and disadvantages in specific applications. A set of eight broadly applicable criteria by which the various methods can be compared and contrasted has been presented (Houghten, 1993a). The utility of SPCLs and PS-SPCLs is illustrated here for applications ranging from simple enzyme-linked immunosorbent assay (ELISA) determination of linear antigenic determinants of peptides to the use of SPCLs in direct *in vivo* studies.

II. Classes and Applications of Soluble Combinatorial Libraries of Peptides

A. Classes of Soluble Synthetic Combinatorial Libraries of Peptides

A range of different soluble SPCLs have been prepared and successfully used to determine potent peptide lead sequences. To illustrate the various library types, as well as the utility of the library approach in general, a number of different libraries and their uses are shown here. The first is the positional scanning SPCL (i.e., Ac-OXXXXX-NH_2, Ac-XOXXXX-NH_2, etc.) for the rapid (i.e., overnight) determination of linear antigenic determinants. A sizing library (Ac-OXX-NH_2 $\rightarrow$ Ac-OXXXXXXXX-NH_2) and a standard SPCL (Ac-OOXXXX-NH_2) are used to illustrate the determination of potent, previously unknown opioid peptide antagonists in a radioreceptor binding assay. The use of a tetrapeptide library composed of 58 different amino acids illustrates the identification of potent antibacterial peptides. A peptide library synthesized on cotton illustrates the identification of trypsin inhibitors. The use of a library composed entirely of D-amino acids (Ac-ooxxxx-NH_2) illustrates the determination of potent antiviral peptides. A modified peptide library (per-N-methylated hexapeptide PS-SPCL) illustrates the development of potent anti-staphylococcal peptides which are stable to enzyme hydrolysis. Finally, a preliminary study illustrating the potential of peptide libraries for direct *in vivo* studies for the identification of peptides that affect blood pressure and/or heart rate is discussed. The libraries described range in number from 361 peptides per mixture (Ac-OXX-NH_2) to 893 $\times 10^9$ (Ac-OXXXXXXXX-NH_2), and they include all of the commonly occurring L-amino acids, their D-amino acid counterparts, 19 different unnatural amino acids, as well as amide-modified libraries.

B. Preparation of Soluble Synthetic Combinatorial Libraries of Peptides

1. Dual Defined Synthetic Combinatorial Libraries of Peptides

The 400 peptide mixtures making up each of the two SPCLs (OOXXXX-NH_2) and Ac-OOXXXX-NH_2), as well as the peptide mixtures synthesized for the iterative steps, were prepared using methylbenzhydrylamine (MBHA) polystyrene resin and standard *tert*-Boc chemistry in combination with simultaneous multiple peptide synthesis (SMPS) (Houghten, 1985). Peptide mixture resins were prepared using a process termed divide, couple, and recombine (DCR) (Furka *et al.*, 1991; Houghten *et al.*, 1991; Lam *et al.*, 1991). Briefly, 19 porous polypropylene packets, each containing 18.6 mmol (20 g) MBHA resin, were coupled to each of the protected *N*-α-*t*-Boc natural L-amino acids; cysteine was excluded from the mixture positions because either its presence would require the continued presence of a reducing agent in the assay being used or, in the absence of a reducing agent, one would expect generation of difficult to define disulfide aggregates in the library. All coupling reactions proceeded to completion (>99.5%) as assessed by Kaiser's ninhydrin test (Kaiser *et al.*, 1970) or Gisin's picric acid procedure (Gisin, 1972). The resulting resins (typically 100–500 mg) from each packet were then combined and thoroughly mixed (there are ~ 4000 beads/mg resin). The resin mixture was divided into 19 portions of equal weight which were placed back into porous polypropylene packets, followed by *N*-α-*t*-Boc protecting group removal and neutralization. Each resin packet was then reacted once again with solutions of individual activated amino acids, and the resins mixed thoroughly to yield 361 dipeptide combinations (19^2). The above two cycles were repeated twice more, yielding a final mixture of 130,321 protected tetrapeptide resins (19^4; Fig. 2). The above peptide resins (XXXX-resin) were divided into 400 aliquots and placed in numbered, porous, polypropylene packets. Synthesis of the two defined positions and N-acetylation were then carried out using SMPS (Houghten, 1985). These libraries contain of all of the possible combinations of the amino acids used for a six-amino acid peptide (thus the term "combinatorial" in SPCL). The side-chain protecting groups of the peptide mixtures were removed and the peptides cleaved from their respective resins using the low–high hydrogen fluoride method (Houghten *et al.*, 1986; Tam *et al.*, 1983). Peptide mixtures were extracted with water or dilute acetic acid and the solutions lyophilized.

2. Positional Scanning Synthetic Combinatorial Libraries of Peptides

In contrast to the physical method described above, the PS-SPCLs were assembled by the chemical addition of mixtures of amino acids to the growing peptide chains (Houghten *et al.*, 1988; Pinilla *et al.*, 1992a; Dooley and Houghten, 1993). Other than the use of mixtures of activated amino acids, the general synthesis procedures were carried out as described above. The amino

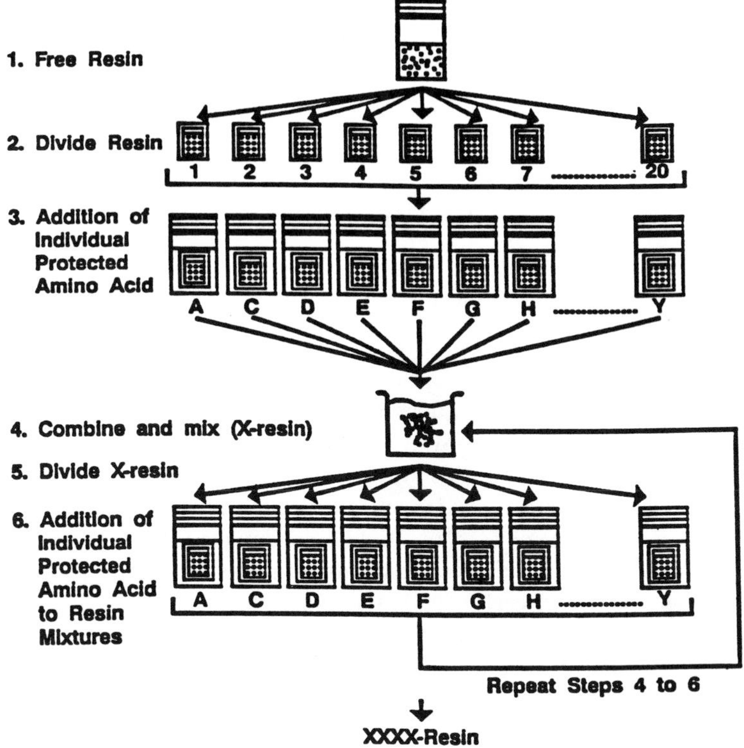

Figure 2 Preparation of equimolar multi-peptide-resin mixture. Reprinted from Houghten *et al.* (1992b), with permission.

acids were present in concentrations which yielded an approximately equimolar coupling of each amino acid. The ratio of concentrations of the individual amino acids found to yield approximately equimolar coupling was determined empirically by repeated amino acid analysis of various test mixtures. Final equimolarity (Houghten and Dooley, 1993a,b) was determined by amino acid analysis, relative to standard mixtures prepared by the DCR method (Houghten *et al.,* 1991). Coupling completion was determined as described above. Side-chain deprotection and cleavage from the resin support were achieved using low HF (Tam *et al.,* 1983) and high HF (Houghten *et al.,* 1986) procedures. The 120 peptide mixtures (i.e., OXXXXX, XOXXXX, etc.) were individually extracted with water, lyophilized, and resuspended in water at a final concentration of 10 mg/ml.

This PS-SPCL format is not limited to the naturally occurring L-amino acids, nor is it limited to a length of six residues. A decapeptide PS-SPCL composed of more than 4×10^{12} sequences has been successfully used to identify an antigenic determinant using ELISA and opioid sequences using a radioreceptor assay (Pinilla *et al.,* 1994; Houghten, 1993b). Also, a library derived from the chemical transformation of an existing N-acetylated L-amino acid PS-SPCL was prepared and used to identify new, permethylated compounds which exhibit potent antimicrobial activity against gram-positive bacteria (Ostresh *et al.*, 1994).

III. Applications of Soluble Peptide Libraries

A. Identification of Linear Peptide Antigenic Determinants

The determination of linear peptide antigenic determinants can be accomplished using all of the library approaches described in the literature. The ease and speed of the determination therefore become the deciding factors in such determinations. The use of PS-SPCLs permits, in many instances, the identification of linear determinants in as little as a single day through the screening of only 120 separate mixtures in a single ELISA format. A competitive ELISA was used as the screening assay. Briefly, an antigenic peptide (Ac-YPYD-VPDYASLRS-NH_2) was adsorbed to microtiter plates (100 pmol/well). After a blocking step, 25 μl of each peptide mixture (6 m*M* final concentration) was added to each well followed by the addition of 25 μl of monoclonal antibody (MAb) at a previously determined dilution. Percent inhibition of antibody binding by each peptide mixture was determined relative to the 100% binding of the MAb to the antigenic peptide on the plate. The concentration of peptide mixture necessary to inhibit 50% antibody binding (IC_{50}) was determined by serially diluting the peptide mixture prior to the addition of the MAb.

The screening of an acetylated hexapeptide PS-SPCL, which was carried out in a single day using three microtiter plates, enabled the determination of the most effective amino acid residues at each position (Fig. 3). Although not every position was found to yield a single specific amino acid, the information obtained (aspartic acid specific to the first and fourth positions, tyrosine specific to the fifth position, and either alanine or serine specific to the sixth position) permits the ready identification of -DVPDYA- as the determinant recognized in the peptide sequence used to generate this MAb (the immunogenic peptide was HCDGFQNEKWDLFVERSKAFSNCYPYDVPDYALSRS, which represents residues 75–110 of the hemagglutinin of influenza virus, Wilson *et al.*, 1984).

If more than one amino acid is effective at a given position, or if the sequence of the immunogen used to generate the MAb is not known, then representative peptides can be synthesized in order to identify the most active peptide. In the study described here, 12 individual peptides representing the combinations of the most effective peptide mixtures (>40% inhibition) found for YPYDVPDYASLRS/MAb 17/9 were synthesized. The most active amino acids at this threshold level at each position were as follows: aspartic acid for position 1; isoleucine and valine for position 2; glutamic acid, lysine, and proline for position 3; aspartic acid for position 4; tyrosine for position 5; and

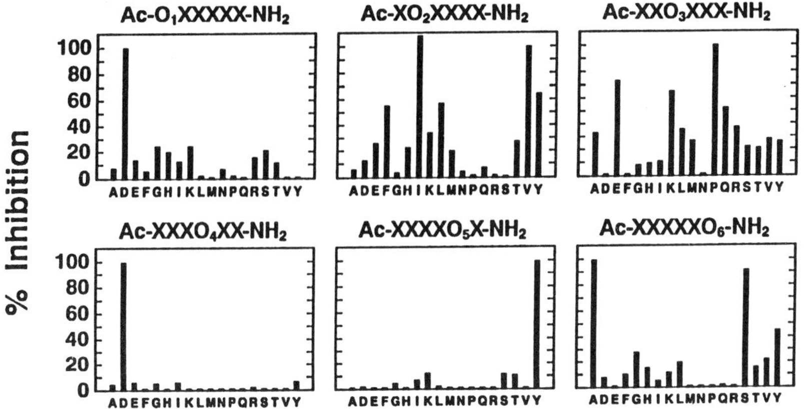

Figure 3 Screening of PS-SPCL Ac-YPYDVPDYASLRS-NH_2/MAb 17/9.

alanine and serine for position 6 (the positional combinations therefore yield $1 \times 2 \times 3 \times 1 \times 1 \times 2 = 12$ peptides). Of the 12 individual peptides prepared, Ac-DVPDYA-NH_2 along with five other peptides resulting from single or double substitutions at the second, third, and sixth redundant positions of the antigenic determinant were found to have the lowest IC_{50} values (Table 1).

B. Identification of Opioid Peptides

The enkephalins (i.e., YGGFM-OH and YGGFL-OH) were the first natural ligands found for the opioid receptors (Hughes *et al.*, 1975). They bind to three known receptor subclasses (μ, δ, and κ) with differing affinities (Schiller, 1990). For this study, an analog of Met-enkephalin, ^{3}H-labeled [D Ala^2,MePhe^4, Gly-ol^5]enkephalin (DAMGO) was used, which is known to bind specifically to the μ receptor. Because the enkephalins are not N-acetylated, a nonacetylated SPCL was used in an initial test study. Each of the 400 different peptide mixtures of the SPCL (O_1O_2XXXX-NH_2) was assayed to determine its ability to inhibit binding of [^{3}H]DAMGO to membrane-bound receptors in crude rat brain homogenates. Following the initial screening, IC_{50} values were determined for each of the most effective peptide mixtures. YGXXXX-NH_2, with an IC_{50} of 3452 n*M*, was found to be the most effective inhibiting peptide mixture. The remaining four mixture positions in YGXXXX-NH_2 were iteratively chosen as described earlier (Houghten *et al.*, 1991; Houghten and Dooley, 1993a,b). It was found that the first five residues of the two peptide mixtures exactly matched the sequences of the naturally occurring Met- and Leu-enkephalins (Hughes *et al.*, 1975).

To determine if an N-acetylated peptide would be able to function as an effective agonist of antagonist of opioid receptors (none were known from previous studies), a sizing library made up of a series of individual, N-acetylated

Table I IC_{50} Values for Individual Peptides Derived from Positional Scanning Synthetic Peptide Combinatorial Library Screening against Monoclonal Antibody 17/9

Sequence	IC_{50} (n*M*)
Ac-DVPDYA-NH_2	1.9
Ac-DIPDYA-NH_2	2.2
Ac-DVEDYA-NH_2	2.3
Ac-DVPDYS-NH_2	3.6
Ac-DIEDYA-NH_2	3.7
Ac-DVEDYS-NH_2	4.4
Ac-DVQDYA-NH_2	5.2
Ac-DIEDYS-NH_2	7.5
Ac-DIPDYS-NH_2	7.5
Ac-DVQDYS-NH_2	26.3
Ac-DIQDYS-NH_2	29.2
Ac-DIQDYA-NH_2	33.4

libraries ranging from three to eight residues in length was used to determine the optimal length associated with a potential N-acetylated peptide. The results found when the sizing library was examined in the same radioreceptor binding assay (Ac-OXX-NH_2 to Ac-OXXXXXXX-NH_2) are shown in Fig. 4. No activity was seen until a length of six amino acids was reached. From six to eight amino acids, arginine was highly active at the N terminus. On screening an acetylated SPCL composed of hexapeptides, Ac-RWXXXX-NH_2 and Ac-RFXXXX-NH_2 were found to be the most active mixtures. We therefore carried out the standard iterative enhancement process for Ac-RFXXXX-NH_2. The final 20 defined peptides are shown in Table II. These peptides were found to be potent opioid antagonists (Dooley *et al.*, 1993). A similar acetylated hexapeptide library composed of D-amino acids was also screened in the receptor binding assay. The most active peptide found on completion of the iterative process, Ac-rfwink-NH_2, is an agonist with high selectivity for the μ receptor. Furthermore, this peptide produces analgesia in mice following intraperitoneal administration (Dooley *et al.*, 1994).

C. Development of New Antimicrobial Peptides Containing Mixtures of L-, D-, and Unnatural Amino Acids

The repertoire of amino acids incorporated into SPCLs was increased in two other SPCLs. These libraries, one N-acetylated and the other not, were prepared by the DCR approach. Each was composed of amidated C-terminal

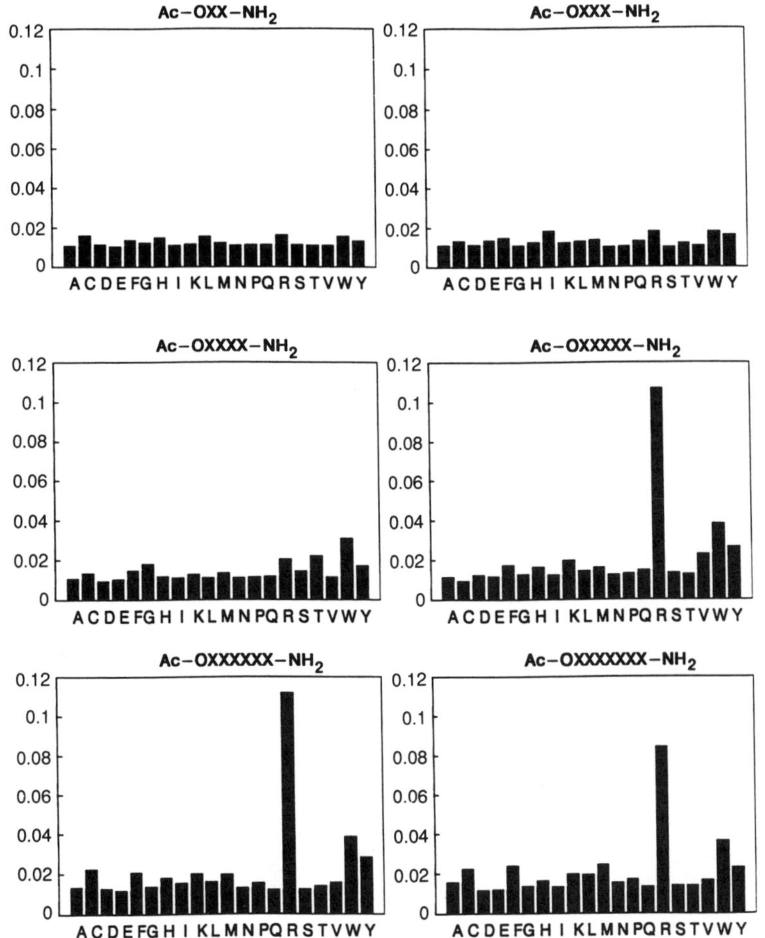

Figure 4 Varying length and diversity libraries. The libraries vary in length from three residues (Ac-OXX-NH_2; 361 peptides per mixture) to eight residues (Ac-OXXXXXXX-NH_2; 893×10^6 peptides per mixture). The results are expressed as the reciprocal of the IC_{50} of [$^3H_2$]DAMGO binding to crude rat brain homogenates.

tetrapeptides having the first position defined as one of 58 different amino acids (composed of 20 L-, 19 D-, and the 19 unnatural amino acids listed in Table III). The remaining three positions were equimolar mixtures of 56 different L-, D-, and unnatural amino acids (19 L-, 18 D-, and 19 unnatural amino acids; L- and D-cysteine were omitted from the mixture positions). The two SPCLs can be represented by the general formula Ac-UZZZ-NH_2 and UZZZ-NH_2. Each SPCL is therefore composed of 58 peptide mixtures; each peptide mixture contains 175,616 (56^3) tetrapeptides, for a total of 10,185,728 (58 × 175,616) individual tetrapeptides per SPCL. Each peptide mixture was initially assayed at a final concentration of 2.5 mg/ml. Thus, at 2.5 mg/ml, every peptide within each

Table II Inhibition of [^{3}H]DAMGO by Peptide Mixtures Making Up an Acetylated Synthetic Peptide Combinatorial Library

Ac-RFMWMO-NH$_2$	IC$_{50}$ ± SE (n*M*)
Ac-RFMWMK-NH$_2$	5 ± 3
Ac-RFMWMT-NH$_2$	5 ± 1
Ac-RFMWMR-NH$_2$	6 ± 1
Ac-RFMWMS-NH$_2$	7 ± 3
Ac-RFMWMG-NH$_2$	10 ± 1
Ac-RFMWMN-NH$_2$	11 ± 3
Ac-RFMWMQ-NH$_2$	12 ± 3
Ac-RFMWMA-NH$_2$	15 ± 4
Ac-RFMWMV-NH$_2$	17 ± 7
Ac-RFMWMY-NH$_2$	17 ± 4
Ac-RFMWMM-NH$_2$	18 ± 4
Ac-RFMWML-NH$_2$	20 ± 3
Ac-RFMWMH-NH$_2$	20 ± 5
Ac-RFMWMP-NH$_2$	24 ± 7
Ac-RFMWMW-NH$_2$	46 ± 13
Ac-RFMWMI-NH$_2$	67 ± 27
Ac-RFMWMF-NH$_2$	68 ± 20
Ac-RFMWMC-NH$_2$	79 ± 14
Ac-RFMWME-NH$_2$	96 ± 9
Ac-RFMWMD-NH$_2$	109 ± 16

mixture is present at 14.2 ng/ml (26 n*M* if one assumes an average molecular weight of 540). The ease of synthesis of such SPCLs readily expands the chemical diversity of peptides that can be screened in a given assay.

The two tetramer SPCLs were screened for inhibition of *E. coli* grown in a microdilution assay. The IC$_{50}$ values were determined for each of the peptide mixtures. Overall, the non-N-acetylated peptide mixtures were found to be more active than the corresponding N-acetylated ones. The peptide mixture (αFmoc-εlys)ZZZ-NH$_2$ was found to exhibit the highest activity (IC$_{50}$ of 179 μg/ml). This peptide mixture was selected for the iterative selection and identification process. Thus, new synthetic peptide mixtures were derived from the most active preceding peptide mixtures by successively defining each position, thus permitting the identification of optimal tetramer antimicrobial peptides. Several peptide mixtures have been carried through this process, and a single example

Table III Unnatural Amino Acids Included in the Two Tetramer Synthetic Peptide Combinatorial Libraries

Abbreviation in sequences	Protected amino acid used during synthesis
Bala	Boc-ß-alanine
aABA	Boc-L-α-aminobutyric acid
gABA	Boc-L-γ-aminobutyric acid
aAIB	Boc-L-α-aminoisobutyric acid
eAca	Boc-L-ε-aminocaproic acid
7aHa	Boc-7-aminoheptanoic acid
bAsp	Boc-L-aspartic acid (α-Bzl)
gGlu	Boc-L-glutamic acid (α-Bzl)
Cys[ACM]	Boc-cysteine (acetamidomethyl)
εlys	*N*-ε-Box-*N*-α-CBZ-L-lysine
αFmoc-εlys	*N*-ε-Boc-*N*-α-Fmoc-L-lysine
$MetO_2$	Boc-L-methionine sulfone
Nle	Boc-L-norleucine
Nve	Boc-L-norvaline
Orn	*N*-α-Boc-*N*-δ-CBZ-L-ornithine
dOrn	*N*-δ-Boc-*N*-α-CBZ-L-ornithine
NO_2F	Boc-*p*-nitro-L-phenylalanine
Hyp	Boc-hydroxyproline (Bzl)
Thiopro	Boc-L-thioproline

of the final active tetrapeptide sequences, (αFmoc-εlys)WKU-NH_2, is illustrated in Table IV. (αFmoc-εlys)WKW-NH_2 was found to show the highest activity against *E. coli* of this series of tetrapeptides. A number of tetrapeptides in the series are expected to be more stable to proteolysis than the hexapeptides found in earlier studies (Houghten *et al.*, 1991, 1992a,b, 1993).

D. Development of Enzyme Inhibitors

Enzyme inhibitors have a wide range of research and therapeutic uses. The applicability of using peptide libraries for the identification of peptide enzyme inhibitors was therefore investigated using trypsin as the model enzyme. A specifically designed L-amino acid library of hexapeptide mixtures was synthesized on cotton carriers and screened. The hexapeptide library initially screened in this study was designed to represent all possible reactive sites for trypsin inhibitors with the P1 residue (lysine or arginine) in every position

Table IV Antimicrobial Activity against *Escherichia coli* of (αFmoc-εlys)WKU-NH_2

Peptide	MIC[a] (μg/ml)
(αFmoc-εlys)WKW-NH_2	16–31
(αFmoc-εlys)WKw-NH_2	31–62
(αFmoc-εlys)WKL-NH_2	31–62
(αFmoc-εlys)WK(NO_2F)-NH_2	31–62
(αFmoc-εlys)WKF-NH_2	31–62
(αFmoc-εlys)WKl-NH_2	31–62
(αFmoc-εlys)WKc-NH_2	62–125
(αFmoc-εlys)WKi-NH_2	31–62
(αFmoc-εlys)WKM-NH_2	31–62
(αFmoc-εlys)WK(Nle)-NH_2	31–62
(αFmoc-εlys)WKY-NH_2	31–62
(αFmoc-εlys)WKC-NH_2	31–62
(αFmoc-εlys)WKI-NH_2	31–62
(αFmoc-εlys)WKV-NH_2	62–125
(αFmoc-εlys)WK(Nve)-NH_2	62–125

[a]Minimum inhibitory concentration (MIC).

of the sequence except the last. This library was made up of 10 groups of peptide mixtures, represented by the following formulas: Ac-KOXXXX, Ac-XKOXXX, Ac-XXKOXX, Ac-XXXKOX, Ac-XXXXKO, Ac-ROXXXX, Ac-XROXXX, Ac-SSROSS, Ac-XXXROS, and Ac-XXXXRO. Each of the groups consisted of 20 different peptide mixtures representing a total number of 2,606,420 individual peptides. In each peptide mixture, four positions (X) represent a mixture of 19 of the 20 genetically coded amino acids (cysteine excluded), whereas O represents those positions occupied by individual amino acids (A through Y). The mixtures were synthesized using cotton as solid-phase peptide synthesis support (Eichler *et al.*, 1991; Eichler and Houghten, 1993b).

Trypsin was chosen for study because numerous polypeptide and protein inhibitors of trypsin from plant and animal sources are known, ranging in length from 30 to several hundred amino acids. The inhibitors are highly specific, limited proteolysis substrates for their target enzymes (Laskowski and Kato, 1980). Their amino acid sequences contain at least one peptide bond called the reactive site (-P1-P1′-) (Ozawa and Laskowski, 1966), which specifically interacts with the active site of the cognate enzyme, and is identical to the potential cleavage site. Trypsin inhibitors are known to have lysine or arginine at the P1 position. In contrast to normal substrates, the dissociation constant of an enzyme–inhibitor complex is extremely low, resulting in very slow hydrolysis. It was anticipated that short peptides having a reactive site in their sequence would therefore be capable of inhibiting trypsin.

The results of the initial screening of the peptide library are shown in Fig. 5. Two peptide mixtures, Ac-XKIXXX and Ac-XXKIXX, were found to be the most effective inhibitors of the tryptic hydrolysis of *N*α-benzoyl-D,L-arginine-*p*-nitroanilide. By iteratively defining the X positions adjacent to the lysine in Ac-XKIXXX, 20 new mixtures with three X positions (Ac-OKIXXX) were synthesized and screened. Ac-AKIXXX was found to be the strongest trypsin inhibitor. On defining the remaining positions in a like manner, it was found that the most active inhibitor was Ac-AKIYRP-NH_2, with an IC_{50} of 46 μ*M*.

E. Development of New Antiviral Compounds

Preliminary screening using a standard plaque assay with the D305 strain of herpes simplex virus 1, in which each of the 400 peptide mixtures making up an all-D-amino acid SPCL (Ac-ooxxxx-NH_2) was added prior to the addition of a virus, revealed a number of mixtures to be active (≥80% reduction of plaques). Seventy of the 400 mixtures gave plaque inhibition between 95 and 100%. These mixtures were titrated using four serial 2-fold dilutions starting at 400 μg/ml. The iterative process was carried out for one of the more active mixtures, Ac-kwxxxx-NH_2. On completion, the iterative process yielded peptides made up entirely of D-amino acids that inhibited plaque formation at 5 μg/ml.

F. Libraries from Libraries: Modified Peptide Libraries for the Development of Anti-Staphylococcal Peptides

The physicochemical properties of peptides are primarily derived from the amide bonds that make up peptide chains. We reasoned that if the character of the -CONH- moiety could be changed, then this would result in an overall change in the chemical property of peptides. A range of possibilities included N-alkylation and/or reduction or modification (Ostresh *et al.*, 1994). If such changes could be made to an entire library, a much wider range of chemical diversities would be possible.

Modified peptidomimetic libraries have now been prepared which greatly extend the range and repertoire of chemical diversity. An example of this approach is illustrated here. A peptidomimetic library was prepared by the permethylation of the amides in an existing resin-bound PS-SPCL, with the protected peptides remaining attached to the solid-phase peptide support used in synthesis. Following removal from the solid-phase synthesis resin (Fig. 6), libraries of this type, which lack the typical -CONH- amide bonds of peptides, can be used in solution with existing assay systems for the identification of individual permethylated peptides having activity. Such permethylated peptides are expected to have very different physical, chemical, and biological properties when contrasted with existing -CONH- based peptides.

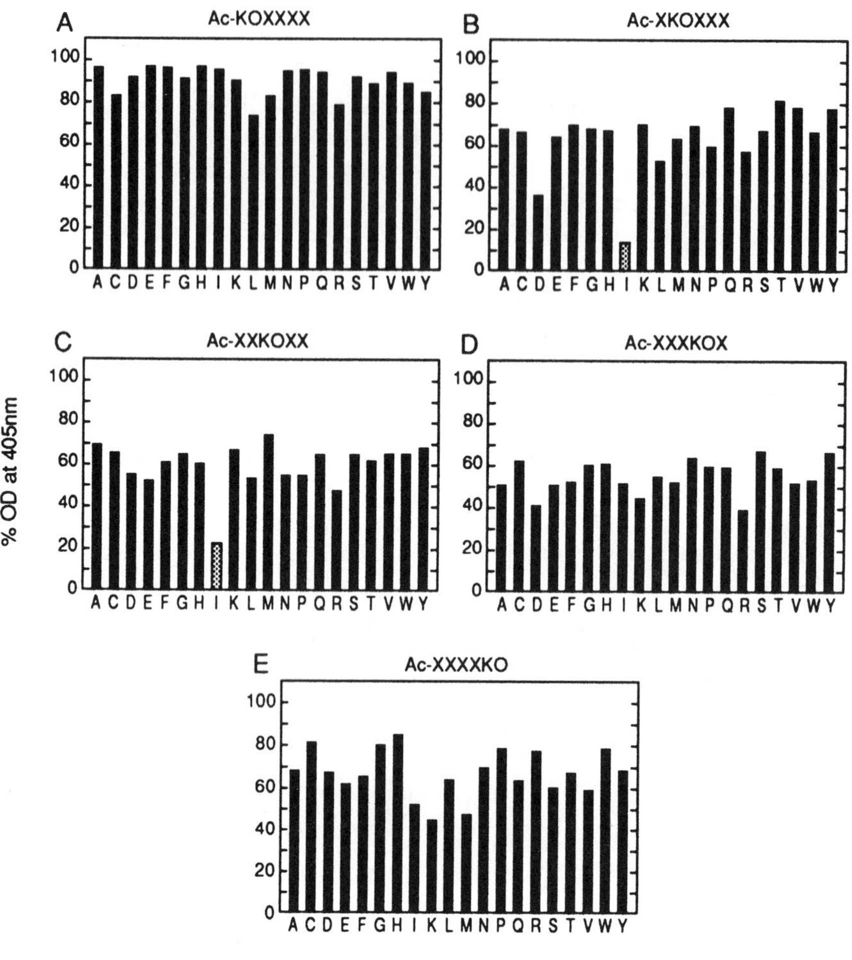

Figure 5 Initial screening of the hexapeptide library for trypsin-inhibiting peptide mixtures (0.05 mg/ml trypsin). Each graph represents one of the 10 groups of peptide mixtures making up the library. Each bar represents trypsin inhibition by a peptide mixture defined in the O position with one of the 20 amino acids. Hatched bars indicate peptide mixtures that were chosen to be carried through the iterative process. Reprinted with permission from Eichler and Houghten (1993b). Copyright 1993 American Chemical Society.

The utility of the permethylated library is shown here in which a permethylated PS-SPCL was screened for anti-staphylococcal activity. Following cleavage from the solid support, it was used along with a standard microdilution assay to identify unique individual permethylated peptides with potent antimicrobial activity against *Staphylococcus aureus*. Because phenylalanine was active at every position, a series of permethylated phenylalanine-containing peptides was then synthesized, which ranged in length from a single phenylalanine

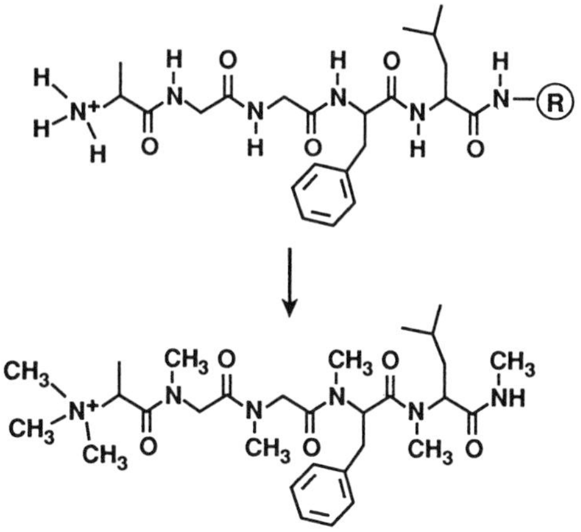

Figure 6 Per-N-methylation of resin-bound peptides: AGGFL-NH-resin.

amide to its hexapeptide form. This series was prepared in order to determine if the repetitive appearances of phenylalanine found at all six positions in the PS-SPCL were due to an individual hexapeptide sequence or to a frame-shifting fit by a shorter peptide (i.e., di-, tri-, tetraphenylalanine). As can be seen (Table V), the greatest activity found (minimum inhibitory concentration of 3.9 μg/ml) was associated with the hexapeptide sequence. The generation of libraries through related chemical modifications using ethyl bromide, benzyl bromide, allyl bromide, etc., have been successfully carried out, as have libraries of polyfunctional amines generated by the reduction of amide carbonyls to methylenes using SPCLs as starting materials (Ostresh *et al.*, 1994).

G. Direct *in Vivo* Studies

The direct *in vivo* study of pharmaceutically active substances requires a specific and readily quantifiable measure. Blood pressure and heart rate are such measures. The 400 L-amino acid peptide mixtures making up OOXXXX-NH_2 were used to study their effect following direct intravenous administration of 250- to 300-g Sprague-Dawley rats. Owing to the rapid clearance rate found for these peptide mixtures, a new mixture could be injected every 10 min. Rats were anesthetized and underwent acute cannulation procedures. After the peptide mixtures were administered intravenously, heart rates and blood pressures were monitored with a Grass recorder. An example of the results obtained is shown in Fig. 7, which illustrates the results for the 20 mixtures making up MOXXXX-NH_2. Two features are of note: first, the majority of mixtures having the second amino acid defined had no significant effect on blood pressure or heart rate; second, the effects of two amino acids lysine and threonine, in the

Table V Antimicrobial Activity against *Staphylococcus aureus* of Per-N-methylated Peptides

Sequence	IC_{50} (μg/ml)	MIC (μg/ml)
Pm-[F-NH_2]	>500	>500
Pm-[FF-NH_2]	>500	>500
Pm-[FFF-NH_2]	288	500
Pm-[FFFF-NH_2]	116	250
Pm-[FFFFF-NH_2	19	31.25
Pm-[FFFFFF-NH_2]	2.5	3.91
Ac-FFFFFF-NH_2	>500	>500
Oxacillin	0.04	0.125
Erythromycin	0.13	0.5

second position, are markedly different. Thus, whereas threonine in the second position results in a profound drop in blood pressure and heart rate, lysine in this position results solely in a drop in blood pressure. It was found that any combination of an increase or decrease of blood pressure and/or heart rate could be obtained from the remaining 380 mixtures (Weber *et al.*, 1994). We are currently pursuing selected mixtures to define individual peptides having potent *in vivo* effects on heart rate and/or blood pressure.

IV. Conclusion

The availability of immense libraries of chemically diverse compounds that can be used to identify highly active peptide and nonpeptide libraries is at the forefront of a revolution in basic research and drug discovery. In contrast to the expectations of "rational drug design," which enables analogs to be designed on the basis of detailed understanding of molecular interactions, chemical library diversity enables the direct *de novo* discovery of lead compounds, as well as the enhancement of the activity of existing compounds. Currently available libraries permit the identification of the most active sequences from hundreds of millions of peptides or modified peptides. We have successfully utilized the concepts briefly reviewed here for the preparation and screening of small molecule organic, as well as heterocyclic chemistry-based libraries using soluble combinatorial libraries as starting materials. Such organic chemical libraries can be prepared using known pharmacophores such as diazepams and piperizines. Peptide libraries are also being constructed which are cyclic and disulfide-linked in nature. These developments, as well as others not yet presented, will broaden the utility of libraries to enable the full repertoire of

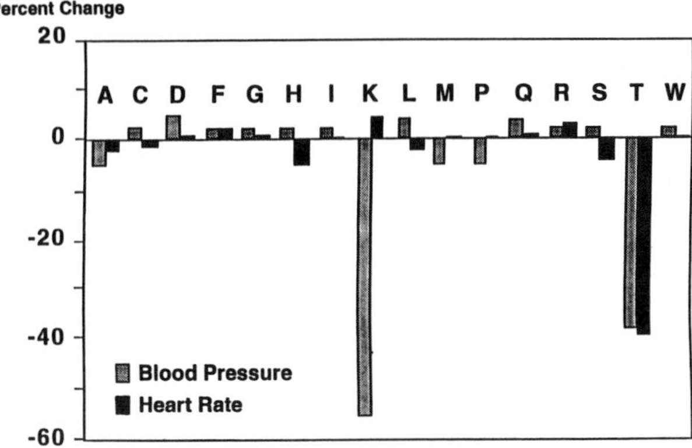

Figure 7 Cardiovascular responses to MOXXXX.

existing molecular architectures to be screened. It can be anticipated that libraries will assume an ever-increasing role in drug discovery and basic research.

Acknowledgments

The examples reviewed here to illustrate the utility of peptide SCLs and PS-SCLs were carried out in collaboration with Jon Appel and Clemencia Pinilla (ELISA-based antigen–antibody studies), Sylvie Blondelle (antimicrobial studies), Colette Dooley (opioid radioreceptor assays), Jutta Eichler (trypsin inhibition studies), John Ostresh (permethylation studies), Patti Weber (HSV-1 studies), and Ron Tuttle (*in vivo* studies). I also thank Eileen Silva for assistance in preparing the manuscript. The specific examples described in this review (P.W., R.T.) were funded by Houghten Pharmaceuticals, Inc., San Diego.

References

Appel, J., Pinilla, C., and Houghten, R. A. (1992). *Immunomethods* **1,** 17–23.

Atherton, E., Hübscher, W., Sheppard, R. C., and Woolley, V. (1989). *Hoppe-Seyler's Z. Physiol. Chem.* **362,** 833–839.

Barbas, C. F., Kang, A. S., Lerner, R. A., and Benkovic, S. J. (1991). *Proc. Natl. Acad. Sci. U.S.A.* **88,** 7978–7982.

Blake, J., and Litzi-Davis, L. (1992). *Bioconjugate Chem.* **3,** 510–513.

Blondelle, S. E., and Houghten, R. A. (1994). *In* "Techniques in Protein Chemistry V" (G. W. Crabb, ed.), pp. 509–516. Academic Press, San Diego.

Blondelle, S. E., Simpkins, L. R., and Houghten, R. A. (1994a). *In* "Innovations and Perspectives in Solid Phase Synthesis: Peptides, Proteins, and Nucleic Acids" (R. Epton, ed.), pp. 163–168. Mayflower Worldwide Limited, Birmingham.

Blondelle, S. E., Takahashi, E., Weber, P. A., and Houghten, R. A. (1994b). *Antimicrob. Agents and Chemother.* **38,** 2280–2286.

Cull, M. G., Miller, J. F., and Schatz, P. J. (1992). *Proc. Natl. Acad. Sci. U.S.A.* **89,** 1865–1869.

Cwirla, S. E., Peters, E. A., Barrett, R. W., and Dower, W. J. (1990). *Proc. Natl. Acad. Sci. U.S.A.* **87,** 6378–6382.

Devlin, J. J., Panganiban, L. C., and Devlin, P. E. (1990). *Science* **249,** 404–406.

Dooley, C. T., and Houghten, R. A. (1993). *Life Sci.* **52,** 1509–1517.

Dooley, C. T., Chung, N. N., Schiller, P. W., and Houghten, R. A. (1993) *Proc. Natl. Acad. Sci. U.S.A.* **90,** 10811–10815.

Dooley, C. T., Chung, N. N., Wilkes, B. C., Schiller, P. W., Bidlack, J. M., Pasternak, G. W., and Houghten, R. A (1994). *Science* **266,** 2019–2022.

Eichler, J., and Houghten, R. A. (1993a). *In* "Proceedings of the 22nd European Peptide Symposium" (C. H. Schneider and A. N. Eberle, eds.), pp. 320–321. ESCOM, Leiden.

Eichler, J., and Houghten, R. A. (1993b). *Biochemistry* **32,** 11035–11041.

Eichler, J., Bienert, M., Sepetov, N. F. Śtolba, P., Krchňák, Smékal, O., Gut, V., and Lebl, M. (1990). *In* "Innovation and Perspective in Solid Phase Synthesis" (R. Epton, ed.), pp. 337–343, SPCC (U.K.), Oxford.

Eichler, J., Bienert, M., Stierandová, A., and Lebl, M. (1991). *Pep. Res.* **4,** 296–307.

Felicia, F., Castagnoli, L., Mussacchio, A., Jappelli, R., and Cesarnei, G. (1991). *J. Mol. Biol.* **222,** 301–310.

Frank, R., and Döring, R. (1988). *Tetrahedron* **44,** 6031–3040.

Furka, A., Sebestyen, F., Asgedom, M., and Dibo, G. (1991). *Int. J. Pept. Protein Res.* **37,** 487–493.

Gausephol, H., Kraft, M., Boulin, C., and Frank, R. W. (1990). *In* "Proceedings of the 11th American Peptide Symposium" (J. E. Rivier and G. R. Marshall, eds.), pp. 1003–1004. ESCOM, Leiden.

Geysen, H. M., and Mason, T. J. (1986). *Biol. Med. Chem. Lett.* **3,** 397–404.

Geysen, H. M., Meloen, R. H., and Bartelin, S. J. (1984). *Proc. Natl. Acad. Sci. U.S.A.* **81,** 3998–4002.

Geysen, H. M., Rodda, S. J., and Maston, T. J. (1986). *Mol. Immunol.* **23,** 709–715.

Gisin, B. F. (1972). *Anal. Chim. Acta* **58,** 248–249.

Hortin, G. L., Staatz, W. D., and Santoro, S. A. (1992). *Biochem. Int.* **26,** 731–738.

Houghten, R. A. (1985). *Proc. Natl. Acad. Sci. U.S.A.* **82,** 5131–5135.

Houghten, R. A. (1993a). *Trends Genet.* **9,** 235–239.

Houghten, R. A. (1993b). *Gene* **137,** 7–11.

Houghten, R. A. and Dooley, C. T. (1993a). *Bioorg. Med. Chem. Lett.* **3,** 405–412.

Houghton, R. A. and Dooley, C. T. (1993b). *In* "Peptide Chemistry 1992: Proceedings of the 2nd Japan Symposium on Peptide Chemistry." (N. Yanaihara, ed.), pp. 11–13. ESCOM, Leiden.

Houghten, R. A., Bray, M. K., De Graw, S. T., and Kirby, C. J. (1986). *Int. J. Pept. Protein Res.* **27,** 673–678.

Houghten, R. A., Appel, J. R., and Pinilla, C. (1988). *In* "Peptide Chemistry 1987" (T. Shiba and S. Sakakibara, eds.), pp. 769–774. Protein Research Foundation, Osaka, Japan.

Houghten, R. A., Pinilla, C., Blondelle, S. E., Appel, J. R., Dooley, C. T., and Cuervo, J. H. (1991). *Nature (London)* **354,** 84–86.

Houghten, R. A., Blondelle, S. E., and Cuervo, J. H. (1992a). *In* "Innovation and Perspectives in Solid Phase Synthesis: Peptides, Polypeptides and Oligonucleotides" (R. Epton, ed.), pp. 237–239. Intercept Limited, Andover, England.

Houghten, R. A., Appel, J. R., Blondelle, S. E., Cuervo, J. H., Dooley, C. T., and Pinilla, C. (1992b). *BioTechniques* **13,** 412–421.

Houghten, R. A., Dinh, K. T., Burcin, D. E., and Blondelle, S. E. (1993). *In* "Techniques in Protein Chemistry IV" (R. H. Angeletti, ed.), pp. 249–256. Academic Press, Orlando, Florida.

Hughes, J., Smith, T. W., Kosterlitz, H. W., Fothergill, L. A., Morgan, B. A., and Morris, H. R. (1975). *Nature (London)* **258,** 577–579.

Huse, W. D., Sastry, L., Iverson, S. A., Kang, A. S., Alting, M., and Burton, D. R. (1989). *Science* **246,** 1275–1281.

Kaiser, E. T., Colescott, R. L., Blossinger, C. D., and Cook, P. I. (1970). *Anal. Biochem.* **34,** 595–598.

Lam, K. S., and Lebl, M. (1992). *Immunomethods* **1,** 11–15.

Lam, K. S., Salmon, S. E., Hersh, E. M., Hruby, V. J., Kazmierski, W. M., and Knapp, R. J. (1991). *Nature (London)* **354,** 82–84.

Laskowski, M., Jr., and Kato, I. (1980). *Annu. Rev. Biochem.* **49,** 593–626.

Mark, J. D., Hoogenboom, H. R., Bonnert, T. P., McCafferety, J., Griffiths, A. D., and Winter, G. (1991). *J. Mol. Biol.* **222,** 581–597.

Merrifield, R. B. (1963). *J. Am. Chem. Soc.* **85,** 2149–2154.

Nikolaiev, V., Stierandová, A., Krchnák, V., Seligmann, B., Lam, K. S., Salmon, S. E., and Lebl, M. (1993). *Pept. Res.* **6,** 161–170.

Ostresh, J. M., Husar, G. M., Blondelle, S. E., Dörner, B., Weber, P. A., and Houghten, R. A. (1994). *Proc. Natl. Acad. Sci. U.S.A.* **91,** 11138–11142.

Owens, R. A., Gesellchen, P. D., Houchins, B. J., and DiMarchi, R. D. (1991). *Biochem. Biophys. Res. Commun.* **181,** 402–408.

Ozawa, K., and Laskowski, M., Jr. (1966). *J. Biol. Chem.* **241,** 3955–3961.

Pinilla, C., Appel, J. R., Blanc, P., and Houghten, R. A. (1992a). *BioTechniques* **13,** 901–905.

Pinilla, C., Appel, J. R., and Houghten, R. A. (1992b). *In* "Vaccines 92" (R. A. Lerner, H. Ginsberg, R. M. Chanock, and F. Brown, eds.), pp. 25–28. Cold Spring Harbor Laboratory, Cold Spring Harbor, New York.

Pinilla, C., Appel, J. R., and Houghten, R. A. (1993). *Gene* **128,** 71–76.

Pinilla, C., Appel, J. R., and Houghten, R. A. (1994). *Biochem. J.* **301,** 847–853.

Schiller, P. W. (1990). *In* "Progress in Medicinal Chemistry" (G. P. Ellis and G. B. West, eds.), pp. 301–340. Elsevier Science Publishers, U.K.

Schnorrenberg, G., and Gerhardt, H. (1989). *Tetrahedron* **45,** 7759–7764.

Scott, J. K., and Craig, L. (1994). *Curr. Opin. Biotechnol.* **5,** 40–48.

Tam, J. P., Heath, W. F., and Merrifield, R. B. (1983). *J. Am. Chem. Soc.* **105,** 6442–6455.

Weber, P., Blanc, P., Bloyer, R., and Houghten, R. A. (1994). In "Recent Advances in Chemotherapy" (J. Einhorn, C. R. Nord, and S. R. Norrby, eds.), pp. 649–650. American Society for Microbiology, Washington, D.C.

Wilson, I., Niman, H. L., Houghten, R. A., Cherenson, A. R., Connolly, M. L., and Lerner, R. A. (1984). *Cell (Cambridge, Mass.)* **37,** 767–778.

Zuckermann, R. N., Kerr, J. M., Siani, M. A., Banville, S. C., and Santi, D. V. (1992). *Proc. Natl. Acad. Sci. U.S.A.* **89,** 4505–4509.

11

Epitope Mapping with Peptides

Hans Rudolf Bosshard
Biochemisches Institut der Universität Zürich
CH-8057 Zürich, Switzerland

I. Introduction

The immune system has the capacity to respond to and distinguish between an immense number of different proteins: Proteins form the most abundant and

Peptides: Synthesis, Structures, and Applications

diverse class of antigens. The specificity by which an antibody recognizes and binds to a protein has been a major subject of immunochemical research. Every protein–antibody complex provides a fine example of a highly specific and singular event of macromolecular recognition. There are several different means to investigate the structural elements of recognition, called epitopes. Among these, epitope mapping with peptides is the most widely used. Improvements in chemical peptide synthesis and use of imaginative techniques to construct peptide libraries through the simultaneous and rapid synthesis of large numbers of individual peptides have given the field new impetus.

In this chapter, we first discuss the nature of epitopes, which are operational entities that defeat attempts at a universal definition (Section II). Using peptides to map epitopes of proteins rests on the assumption that peptides can simulate some of the features of the "true" epitope recognized by an antiprotein antibody. The essence of the mapping procedure is to find peptides that cross-react with an antibody directed against an intact protein of undisturbed three-dimensional structure. In most cases, the cross-reactive peptide has a sequence that corresponds to a partial sequence of the protein, but this is not a prerequisite for cross-reaction. Mapping epitopes of proteins with the help of synthetic peptides is the main theme of Section IV.

From epitope mapping with peptides it is a small step to antibody production with peptides (Section V). The objective is to obtain antipeptide antibodies that cross-react with a protein. Again, the approach is based on a probable similarity between the structural features of a peptide and parts of a protein. The approach led to the successful production of several virus-neutralizing antibodies and even to some peptide-based vaccines.

This is not a comprehensive review about the use of peptides to map epitopes, a Sisyphean task. We wish to assess critically the strengths and weaknesses of the peptide approach. The examples shown, of which several are from our own work, merely illustrate principles and are hardly representative of all the many diverse and inventive applications of peptides to epitope mapping. We apologize to investigators who do not find their work cited. Issues at the heart of the current discussion are antibody specificity, antibody cross-reactivity, and the nature of epitopes. These topics have been reviewed by Berzofsky and Schechter (1981), Van Regenmortel (1989), Arnon and Van Regenmortel (1992), and Greenspan (1992).

II. The Nature of Epitopes

An epitope or antigenic determinant[1] is defined as "the portion of an antigen that makes contact with a particular antibody or T cell receptor" *(Macmillan Dictionary of Immunology)*. Epitopes recognized by T cells differ fundamentally from

[1]"Antigenic determinant" is a more precise term, but "epitope" is in common usage and preferred for the sake of brevity.

epitopes recognized by antibodies or B cells. T cells recognize protein antigens in a fragmented form on the surface of cells expressing MHC proteins, called antigen-presenting cells. MHC proteins, the products of the major histocompatibility gene complex, are integral membrane glycoproteins that bind peptides and present them as T-cell epitopes to T-cell receptors. In contrast, soluble antibodies and B cells, the latter through antibody molecules embedded in their membrane, recognize an enormous variety of antigens in solution or on surfaces: native and unfolded proteins, peptides, nucleic acids, polysaccharides, small organic molecules such as steroids, etc. Moreover, B-cell epitopes are recognized directly by antibodies and not by way of presentation through another protein molecule.

The spatial folding of B-cell epitopes is important and often decisive for binding to antibodies. This is the main reason why B-cell epitopes are far more complex than T-cell epitopes. This chapter focuses on mapping of B-cell epitopes of proteins and peptides. We shall see shortly that the above simple definition of an epitope becomes arguable as soon as we try to define a B-cell epitope more precisely. First, we briefly summarize the simpler T-cell epitopes.

A. T-Cell Epitopes

T-cell epitopes help to provoke the cellular branch of the immune response. T-cell epitopes are short flexible peptides produced by intracellular processing of foreign proteins. The peptides are presented to the T-cell receptor through either a class I or a class II MHC antigen.[2] MHC class I antigens present peptides to CD8 (mainly cytotoxic) T cells, and MHC class II antigens to CD4 (predominantly helper) T cells. Class I antigens are found on most cells, whereas class II antigens occur on macrophages and B lymphocytes.

Research advances have provided a clear and congruent picture of the nature of T-cell epitopes and of the way these epitopes are recognized by and bound to MHC antigens. Less is known about how a T-cell receptor recognizes the peptide–MHC antigen complex. T-cell epitopes correspond to unique segments of the polypeptide chain of a protein antigen. Their nature depends largely on the assortment and specificity of the processing enzymes (Goldberg and Rock, 1992). T-cell epitopes have been analyzed in detail for many proteins. One way of mapping T-cell epitopes is to synthesize the presumed epitopes and use them to stimulate T-cell clones (Gammon *et al.*, 1987; Fox *et al.*, 1987). Peptides that stimulate best are found by methodical changes of the length and sequence of the peptide. For example, in pigeon and moth cytochrome *c*, the main helper T-cell epitope is confined to a consecutive segment of 8–10 residues close to the C-terminal end of the polypeptide chain (Fox *et al.*, 1987). In contrast to the B-cell epitopes of a protein, the number of T-cell epitopes is restricted; a few sequences dominate the T-cell response to a particular protein.

[2]The products of the MHC gene cluster are called antigens to indicate their role in the discrimination between self and nonself.

It has become possible to release, purify, and sequence the minute amounts of peptides bound to isolated MHC antigens (Hunt *et al.*, 1992a,b; Falk *et al.*, 1991a,b). Studies confirmed that T-cell epitopes are indeed short and flexible peptides. Those presented by MHC class I antigens to CD8 T cells are of very uniform length, 8 to 10 residues. The reason is apparent from the crystal structure of peptide–MHC class I complexes (Bjorkman *et al.*, 1987; Garret *et al.*, 1989). The peptide binds to a narrow groove on the MHC molecule. The groove is blocked at either end and imposes restrictions on the size of peptides it can accommodate, with longer peptides bulging out (Guo *et al.*, 1992).

The structure of a peptide complex with a class II MHC molecule is similar to the complex with a class I MHC molecule (Brown *et al.*, 1993). There is, however, one main difference: the class II peptide binding groove accommodates longer peptides—averaging 15–18 residues—as the groove allows peptides to protrude from either end. Studies of peptides eluted from class II molecules directly established the size heterogeneity (Demotz *et al.*, 1989; Chicz *et al.*, 1992).

An individual organism has only a few different MHC molecules displayed on the cell surface, at most six class I molecules for humans. How can such few molecules bind with sufficient affinity to hundreds or perhaps thousands of different peptides, which when presented to the T-cell receptors stimulate a response against a myriad foreign antigens? This puzzle seems to have been solved. Peptide binding to the MHC protein is governed by only 2 or 3 anchor residues whose side chains fit to complementary polymorphic pockets of the groove of the MHC molecule (Fremont *et al.*, 1992; Matsumura *et al.*, 1992; Guo *et al.*, 1992; Silver *et al.*, 1992; Brown *et al.*, 1993). Other binding interactions with the MHC molecule involve preferentially main-chain atoms, leaving other side chains except those of the anchor residues open to variation. In this manner, many peptides with different sequences can bind to the same MHC molecule. Looked at in another way, the peptide–MHC complex is a natural peptide library whose function it is to present to T-cell receptors as many different peptides as possible. Such promiscuity of peptide binding permits presentation of virtually any T-cell epitope through a restricted set of MHC antigens to an almost infinite number of different T-cell receptors.

It is not yet well understood how the peptide complexes with MHC class I and class II antigens set in motion the helper and killer T-cell response, respectively, and how the peptide–MHC complexes contribute to self–nonself discrimination. Such understanding seems difficult before we know more of the three-dimensional structures of not only T-cell receptors but also their several accessory molecules and the ternary and higher order complexes formed among these molecules when T cells contact with antigen-presenting cells.

In concluding this brief review of T-cell epitopes, we note that synthetic peptides have been instrumental in steering research. It would have been impossible to attain today's knowledge had we not benefitted enormously from the ease with which large numbers of peptides can be synthesized and tested as T-cell epitopes. Equally important was much progress in high-performance liquid chromatography, capillary electrophoresis, and mass spectrometry, techniques of decisive importance to the analysis of T-cell epitope peptides isolated from MHC molecules.

B. B-Cell Epitopes

We restrict our discussion to B-cell epitopes on proteins and peptides; epitopes on other biopolymers and haptenic groups are not covered. The first important generalization about B-cell epitopes is that they are directed against three-dimensional features of the molecular surface of proteins and polypeptides. A particular three-dimensional topology is a hallmark of B-cell epitopes as opposed to T-cell epitopes. It is likely that any amino acid residue accessible from the surface of a protein can be part of one or another B-cell epitope (Benjamin *et al.,* 1984). Hence, proteins may contain very many different epitopes although, for steric reasons, only a limited number of antibodies can bind to the antigen at any one time. The B-cell response is stereospecific, being much weaker against D enantiomers of peptides (Gill *et al.,* 1963) and proteins (Dintzis *et al.,* 1993), possibly because D-enantiomeric proteins are not efficiently processed to yield peptides for T-cell help.

Earlier, proteins were believed to have a well-defined antigenic structure characterized by a limited number of epitopes. Insight into the complex nature of the B-cell immune response and its regulation, and into the specificity of monoclonal antibodies to proteins, made it clear that a protein does not possess a defined antigenic structure. It is not possible to define the "complete antigenic structure" of a protein, contrary to what has been claimed by some investigators (Atassi and Lee, 1978; Atassi, 1984). The antigenicity of a protein is both a property of the protein topography as well as of regulatory mechanisms of the host immune system, including tolerance to structures resembling the host's own proteins, the specificity of T-cell help, and idiotypic networks (Benjamin *et al.,* 1984; Berzofsky, 1985). Immunodominant sites, that is, sites to which most but not all antibodies of the immune response are directed, are not an intrinsic property per se of the protein. As pointed out by others before, epitopes do not exist in their own right but only by virtue of a connection with the complementary antibody binding site, the so-called paratope (Berzofsky, 1985; Van Regenmortel, 1986, 1989). Hence, an epitope is a relational concept, and the definition of an epitope is necessarily operational (Van Regenmortel, 1986). In other words, the definition of a particular epitope depends to a large extent on the molecular geometry and chemical nature of the corresponding paratope and, perhaps more importantly, on the experimental approach chosen to map the epitope.

This state of affairs can be illustrated by the example of the first protein–antibody complex whose structure was solved by X-ray crystallography (Amit *et al.,* 1986). In this complex, 16 residues of lysozyme contact 17 residues of a monoclonal Fab (antibody fragment) to lysozyme. The epitope extends over 750 $Å^2$ of lysozyme surface. In contrast, epitope mapping with a series of sequence-related avian lysozymes indicates that only a few residues are important to binding of lysozyme to antilysozyme monoclonal antibodies. Mutation of very few residues can reduce radically the association constant of the lysozyme–antibody complex (Harper *et al.,* 1987). In one case, a single Arg to Lys substitution reduced by two orders of magnitude the affinity of lysozyme for a monoclonal antibody (Smith-Gill *et al.*, 1982). Theoretical calculations based on the crystal structures of two complexes of lysozyme with Fab fragments demonstrated that, of the many residues that define the epitope in the crystal, only a few actually contribute to the stability of the complex (Novotny *et al.*, 1989). Based on their calculations, Novotny *et al.* distinguished between an energetic epitope and a passive epitope. The energetic epitope encompasses those residues that contribute to the energetics of binding. The passive epitope provides only surface complementarity around the residues that form the energetic epitope. That only a few of the interactions seen in the crystal structure play a major role in stabilizing the antigen–antibody complex was confirmed for binding of influenza virus neuraminidase to a monoclonal antibody. Nineteen residues of the neuraminidase contact 17 residues of the antibody in the crystal, but site-specific mutation of only 3 residues totally abolishes binding (Air *et al.*, 1990; Nuss *et al.*, 1993).

A more generally applicable operational distinction of epitopes is that between a *contact epitope* and a *functional epitope*. The contact epitope relates to information obtained from the three-dimensional structure of the antigen–-antibody complex; the functional epitope relates to information from noncrystallographic mapping procedures, including epitope mapping with peptides. A contact epitope is represented by a matching fit between large complementary surface areas of antigen and antibody, as seen in several X-ray structures of protein–antibody complexes (Davies and Padlan, 1990; Wilson and Stanfield, 1994; Braden and Poljak, 1995). Contact epitopes cover several hundred square angstroms of molecular surface. The functional epitope defines residues which seem significant for antibody binding and whose mutation may reduce or totally abolish binding. The functional epitope may comprise as few as 2 to 3 residues, as in the examples of lysozyme and neuraminidase mentioned above. It is not possible to deduce the contact epitope from the functional epitope. Similarly, the contact epitope does not by itself reveal the functional epitope. On the part of the antibody, one may also differentiate two types of paratopes: a functional paratope and a contact paratope. This follows from thermodynamic analysis of the complementarity determining regions of a monoclonal antibody (Kelley and O'Connell, 1993).

The dual nature of an epitope as revealed by crystallography and noncrystallographic mapping techniques reflects two different models of molecular recognition. Viewed in this way, the difficulty in defining the nature of epitopes is shifted to the level of an epistemological difficulty: how are we to model reality by the restricted experimental means at our disposal? These limits will be kept in mind when discussing epitope mapping by peptides.

Here we must mention the long-known conceptual classification of B-cell epitopes into *sequential* and *conformational* (Sela *et al.*, 1967; Sela, 1969; Atassi and Smith, 1978). An epitope is called sequential or continuous if it can be represented by a series of contiguous residues of a polypeptide chain. An antibody is said to recognize a sequential epitope if it reacts with a short, flexible peptide or with the denatured, unfolded polypeptide chain. A conformational epitope, also called discontinuous or topographic or assembled, is built from noncontiguous parts of the amino acid sequence through folding of the polypeptide chain in the native protein. An antibody is said to recognize a conformational epitope if it reacts with a native protein and not with the unfolded polypeptide chain, or if it reacts with a peptide of unique conformation, for example, a helix, but not with a random coil peptide.

The distinction between sequential and conformational epitopes is somewhat arbitrary and may be misleading. Because every paratope has a well-defined three-dimensional structure, the interaction between paratope and epitope is always a fit of structures in three-dimensional space. This applies to an epitope on a well-ordered globular protein as well as to an epitope on a short flexible peptide. In the latter case, the peptide also must adopt a unique conformation when binding to the antibody; hence, a continuous epitope is also "conformational." The binding conformation either preexists or is induced by the paratope (see Sections V,B and V,C).

In the case of antibodies to native proteins, it has been argued that most or perhaps all epitopes are discontinuous (Barlow *et al.*, 1986). Because of the large size of a typical contact epitope in an antigen–antibody crystal, it is indeed unlikely for an antibody to bind exclusively to a contiguous stretch of the polypeptide chain and not also to contact residues apart in sequence but close in space. Space-filling models of proteins show few linear stretches longer than 4 to 5 residues in direct peptide linkage accessible on the molecule surface. This is not to say that an antibody directed against an assembled epitope on the surface of a protein may not cross-react also with a peptide corresponding to a segment of the protein.

Concluding our overview of the nature of B-cell epitopes, we emphasize once more the sheer difficulty to give a general definition of "epitope." A pragmatic recourse to operational definitions may not please the purist, but operational definitions can be helpful in answering questions about the character of a particular antigen–antibody interaction.

III. Overview of Noncrystallographic Mapping Procedures

Epitope mapping with peptides must be viewed in the context of other noncrystallographic mapping procedures. Several reviews have been written (Jemmerson and Paterson, 1986a; Horsfall *et al.*, 1991; Paterson, 1992). A possible classification is shown in Table I. Methods 1 and 2 are our main subject and are discussed in detail in Sections IV and V, respectively. Method 3, cross-reactivity studies with homologous proteins, consists of comparing the reaction of an antibody with sequence-related proteins. The method is also known as fine-specificity analysis. Well-studied examples are epitopes for antibodies to *c*-type

Table I Classification of Epitope Mapping Procedures

1. Cross-reactivity studies between proteins and peptides using antiprotein antibodies
2. Cross-reactivity studies between proteins and peptides using antipeptide antibodies
3. Cross-reactivity studies between homologous proteins
4. Scanning mutational studies
5. Differential accessibility of protein surfaces to chemical modification or proteolysis in the presence and absence of antibodies
6. Studies using NMR analysis of antigen–antibody complexes

cytochromes, myoglobins, lysozymes, and ribonucleases (reviewed by Benjamin *et al.,* 1984). If an antibody discriminates between two related proteins, one assumes that the epitope contains one or more of the residues in which the two proteins differ. The above-mentioned epitope for a monoclonal antibody to lysozyme, which discriminates between Arg and Lys at a single position, was analyzed in this way. The antibody could distinguish between the lysozymes of hen and quail, which differ by the Arg to Lys substitution at position 68 (Smith-Gill *et al.,* 1982). We found a monoclonal antibody to horse cytochrome *c* (antibody 2.61) that cross-reacts strongly with beef cytochrome *c* but not with eight other, mostly mammalian *c*-type cytochromes. Glu-92 was the only residue that was replaced in all the cytochromes that did not bind to antibody 2.61, and Glu-92 was therefore assigned to the epitope (Burnens *et al.,* 1987).

Epitope mapping by studying the cross-reactivity with homologous proteins is most successful if the three-dimensional structure of the antigenic protein is known. Before the first crystal structures of complexes of antigens and antibodies were known, cross-reactivity studies between homologous proteins provided the best information about epitopes formed through the folding of the polypeptide chain of a native protein.

Method 4 (Table I) is related to method 3 and uses homologous proteins prepared by recombinant DNA techniques for comparative immune reactions. Scanning mutational studies can systematically elaborate the functional interactions in folded proteins (Wells, 1991). The requirements for binding of 21 monoclonal antibodies to human growth hormone was mapped by homolog- and alanine-scanning mutagenesis (Jin *et al.,* 1992).

Method 5 measures the differential accessibility of residues on the surface of proteins in the presence and absence of an antibody. We have pioneered this method, termed differential chemical modification, which is useful in analyzing protein–protein interactions in general (Bosshard, 1979, 1995). The method is also aimed primarily at epitopes on native proteins and is most effective if the results can be related to the crystal structure of the protein antigen. In one variant of the method, the free antigen and the antigen–antibody complex are chemically modified with a group-specific reagent, for example, by acetylation of lysine side-chain amino groups with acetic anhydride (Burnens *et al.,* 1987; Oertle *et al.,*

1989; Saad and Bosshard, 1990). With the help of dual isotope labeling, the degree of chemical modification of each modified group is deduced from an isotope ratio. In this way, we could add Lys-60 and Lys-90 to the epitope for antibody 2.61 against horse cytochrome *c*, Glu-92 having been assigned to the epitope by fine-specificity analysis (Burnens *et al.*, 1987). The side chains of the three residues are near one another on the surface of the cytochrome *c* molecule. The epitope exists only on the native molecule; antibody 2.61 does not react with unfolded apocytochrome *c*. Analysis of chemically modified residues has also been achieved by mass spectrometric peptide mapping (Suckau *et al.*, 1990, 1992).

In another variant of method 5, the free protein and the protein–antibody complex are digested with a protease and the freed peptides analyzed. Because antibodies are quite resistant to proteolysis, mainly peptides of the antigen are obtained. The area of the epitope is deduced from differences in the pattern of proteolysis of free and antibody-bound protein antigen (Jemmerson and Paterson, 1986b).

Finally, method 6 (Table I) uses hydrogen–deuterium exchange labeling and two-dimensional ^{1}H NMR to detect surface areas protected by the paratope in the antigen–antibody complex (Paterson *et al.*, 1990; Paterson, 1992; Zinn-Justin *et al.*, 1993). Consistent and complementary results were obtained by the NMR method and the differential protection technique for an epitope on cytochrome *c*. Whereas the differential protection technique assigned a few particular residues to the epitope and excluded others (Oertle *et al.*, 1989), the isotope-exchange technique revealed the surface area of the epitope in broad outline (Paterson *et al.*, 1990).

IV. Cross-Reactivity Studies between Proteins and Peptides Using Antiprotein Antibodies

In this section, we discuss the use of peptides to map and interpret the antigenic structure of whole proteins. There are hundreds, perhaps thousands of studies in which the cross-reaction of peptides with antibodies to proteins have been followed. Micromolar amounts of peptide are mostly sufficient for the tests, and solid-phase peptide synthesis can provide pure peptides for hundreds of assays. The limitation is not the amount of material but the speed at which different peptides can be synthesized. Here, automated instruments for the simultaneous synthesis of many peptides are very helpful (Schnorrenberg and Gerhardt, 1989).

One of the earliest applications of peptides to epitope mapping was the fractionation of antisera to native proteins on affinity columns with peptides as affinity ligands. Using the loop peptide of hen egg white lysozyme (residues 64–83 obtained by pepsin digestion of lysozyme), Arnon and Sela (1969) isolated a fraction of antibodies from rabbit antilysozyme serum which reacts with the isolated loop peptide as well as (more strongly) with the intact native protein. The loop peptide contains a disulfide bond between Cys-64 and Cys-80. The affinity-purified antibodies react only with the loop peptide and not with the open-chain form in which the two cysteines are reduced and carboxymethylated. Subsequently, other antiprotein antisera were affinity-fractionated to obtain monospecific antibodies. Another early example was the isolation of antibodies reactive with the carboxyl-terminal third of staphylococcal nuclease (Sachs *et al.*, 1972). In this example, the peptide used for affinity fractionation did not contain an internal disulfide bridge. To explain binding of antibodies to the peptide, it was postulated that the peptide adopts a conformation similar to that of the carboxyl-terminal segment of intact native nuclease. It is not always the case that antisera can be fractionated on peptide-affinity columns. From rabbit antiserum to cytochrome *c*, no antibodies could be isolated that bind to CNBr fragments of this protein (Jemmerson and Margoliash, 1979).

The early success of affinity purification of antibodies was seminal to the future development of epitope mapping with peptides as well as to the idea of using peptides to elicit protein-reactive antibodies (Section V). Affinity purification of monospecific antibodies on peptides indicated that structural features of a protein can be simulated by synthetic peptides. Among the many conformations a peptide can adopt in solution, one or a few may be recognized by an antiprotein antibody. Alternatively, some linear peptides may already exhibit conformational preferences in solution related to conformational features of the protein whose epitopes are to be mapped (see Sections V,B and V,C). There is evidence for preferred solution conformations of some short peptides (Dyson and Wright, 1991; Waltho *et al.*, 1993; Saberwal and Nagaraj, 1993).[3] It should be considered also that identical short peptide sequences can have different conformations in different proteins (Wilson *et al.*, 1985) and that the same peptide may cross-react with unrelated antiprotein antibodies. Peptides may also weakly cross-react with antibodies directed to discontinuous epitopes. In this case, synthetic peptides may mimic a portion of the discontinuous epitope and react weakly with the antiprotein antibody. Sometimes a combination of peptides can be used to simulate discontinuous epitopes (Cooper *et al.*, 1989).

A. Importance of the Immunoassay Format

There are many ways to quantify the cross-reaction of peptides and proteins with an antibody. Precipitation reactions or anaphylaxis tests were used in early studies. These assays guarantee that antigen and antibody react in solution under conditions where the danger of a change of conformation of either molecule is small. The tests are cumbersome and have been replaced by radioimmunoassays (RIA) and enzyme immunoassays (EIA).[4] In the simplest EIA format, the solid-phase antibody-capture EIA, the peptide or protein is adsorbed directly to the microtiter plate and

[3]See also Chapter 4 in this volume by J. M. Scholtz and R. L. Baldwin.

[4]Experimental details of immunoassay procedures are found, for example, in Harlow and Lane (1988); theoretical aspects of assay procedures are discussed by Butler (1991).

then tested on the plate for reaction with antiprotein antibodies. The problem with this assay is that the protein may unfold on the solid surface (Soderquist and Walton, 1980; Kennel, 1982; Butler *et al.,* 1987; Schwab and Bosshard, 1992). Short peptides may not be easily accessible to the antibody molecule and often do not adsorb well to plastic, or the amount of material adsorbed varies from peptide to peptide.

Peptides are more easily accessible if adsorbed through a carrier protein, for example, BSA (bovine serum albumin) or polylysine, which binds very strongly to plastic. The EIA plate is coated with the peptide–carrier conjugate and incubated with the antibody. There are many coupling procedures to link peptides to carriers (Maelicke *et al.,* 1989, and references therein). The structure and conformational properties of a peptide may change if coupling occurs through side-chain functional groups of the peptide. Coupling through the N- or C-terminal residue interferes least and is the method of choice. We have found N-terminal biotinylation of peptides followed by adsorption to a streptavidin-coated plate to provide a convenient and efficient method to fix peptides to EIA plates. The biotin group is separated from the peptide by a spacer (Savoca *et al.,* 1991).

Not only peptides but also proteins can be fixed to the plate through a carrier or another antibody to lessen the danger of unfolding (Smith and Wilson, 1986). However, fixation to a solid surface by whatever means will always restrict and alter the conformational states a peptide or protein can adopt. Therefore, any cross-reaction of an antibody with a peptide or protein on a solid surface differs from the analogous cross-reaction in solution (Pesce and Michael, 1992). For this reason, it is absolutely essential to apply solution-phase conformation-sensitive immunoassays if one wishes to examine whether a peptide and a native protein compete for the paratope of an antibody that is directed against the native protein (Goldberg and Djavadi-Ohaniance, 1993). The Farr-type RIA is still one of the best assays for this purpose (Farr, 1958). A constant amount of ^{125}I-labeled protein and increasing amounts of nonradioactive competitor peptide are incubated with the antibody in solution, until equilibrium is reached. Thereafter, the antigen–antibody complex is precipitated by ammonium sulfate or polyethylene glycol. (The free antigen must remain in solution during the precipitation step.) The concentration of the antibody should be sufficient to precipitate 50% to 70% of the radioactive tracer antigen in the absence of the nonradioactive competing antigen.

We have developed a nonradioactive alternative to the Farr assay, the protein A antibody-capture EIA (PACE) (Ngai *et al.,* 1993). In this assay, monoclonal or polyclonal antibodies are first allowed to equilibrate with a biotinylated antigen in solution. Afterward, the antigen–antibody complex (and free antibody) is captured to the microtiter plate through protein A, which binds to the Fc part of most IgG types. Captured antigen–antibody complex is detected by streptavidin–alkaline phosphatase conjugate and *p*-nitrophenyl phosphate as substrate; the streptavidin moiety attaches to the biotinylated antigen. A competition assay is accomplished by coincubation of biotinylated and nonbiotinylated antigens before capture to the protein A-coated plate. PACE combines the advantages of a solution-phase immunoassay with the ease of a solid-phase EIA.

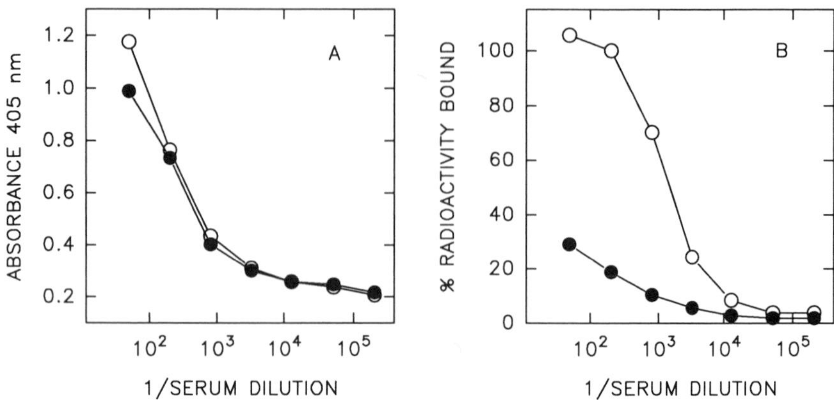

Figure 1 Reaction of rabbit antiserum to native yeast iso-1 cytochrome *c* with native cytochrome *c* (open circles) and apocytochrome *c* (filled circles) in solid-phase and solution-phase immunoassays. **(A)** Antibody-capture EIA. Native and apocytochrome *c* were adsorbed directly to the microtiter plate and reacted with anti-native cytochrome *c* antiserum. **(B)** Solution-phase EIA. ^{125}I-Labeled native and apocytochrome *c* were incubated in solution with antiserum to native cytochrome *c*, followed by precipitation of the antigen–antibody complex with polyethylene glycol and counting of the radioactivity of the precipitate. See Schwab and Bosshard (1992) for experimental details. Reprinted from Schwab and Bosshard (1992) with permission of Elsevier Science Publishers.

B. Different Assays Yield Different Results with the Same Peptides

To illustrate the danger of unfolding of a protein on a solid surface, we studied the reaction of rabbit antibody against native cytochrome *c* with native and apocytochrome *c* in a solid-phase EIA and a solution-phase RIA (Fig. 1). Apocytochrome *c* is the random coil form of cytochrome *c* (Stellwagen *et al.*, 1972; Fisher *et al.*, 1973). Antibodies could not distinguish between native and apocytochrome *c* in the solid-phase EIA, whereas the reaction with the native protein was clearly specific in the solution-phase RIA. The solution-phase assay reveals a subpopulation of antibodies directed against unfolded cytochrome *c*.

A comparison of assay formats was reported also for rabbit antisera to lactate dehydrogenase (Hogrefe *et al.*, 1989). Peptides corresponding to surface-exposed and flexible segments of the enzyme bound strongly to anti-lactate dehydrogenase antibodies in a solid-phase EIA (peptides adsorbed directly to microtiter plate), but not in a competition RIA performed in solution with ^{125}I-labeled enzyme as the tracer antigen. In this example, the antibodies, which were induced by immunization with a seemingly native protein, reacted with peptides on a solid surface but not with the same peptides in solution.

Table II represents another example of how the assay format may affect the cross-reaction with synthetic peptides. Rabbit antisera to native FNR (ferredoxin:NADP$^+$ oxidoreductase from spinach) and denatured FNR were tested with nine different peptides representing parts of the FNR sequence. The

Table II Influence of Immunoassay Format on Reaction of Peptides with Antibodies[a,b]

Peptide sequence[c]	EIA[d]		Streptavidin EIA[e]		PACE[f]	
	Antibodies to native FNR	Antibodies to denatured FNR	Antibodies to native FNR	Antibodies to denatured FNR	Antibodies to native FNR	Antibodies to denatured FNR
1–20	+	+	+	+	+	+
30–39	–	+	+	+	–	+
81–89	–	–	(+)	+	–	(+)
102–109	–	+	–	(+)	–	+
120–129	+	+	+	+	–	+
173–181	–	+	–	+	–	+
186–195	–	+	–	(+)	–	+
220–226	–	+	–	(+)	–	(+)
235–245	–	(+)	–	+	–	+

[a]IgG fractions of rabbit antisera to native ferredoxin:$NADP^+$ oxidoreductase (FNR) and to denatured FNR were tested with a series of peptides corresponding to various sequences of FNR (I. Jelesarov and H. R. Bosshard, unpublished experiments).

[b]The + sign indicates a strong reaction; (+) indicates a weak but significant reaction as compared to the background reaction, which was measured with preimmune serum; the – sign indicates no reaction above background level.

[c]The peptides correspond to peaks of surface accessibility except peptide 173–181, which is completely buried. Accessibility was measured according to Connolly (1983) and averaged over 9 residues. The N-terminal 18 residues of FNR are invisible in the FNR structure and are mobile and disordered (Karplus *et al.*, 1991).

[d]Peptide adsorbed directly to EIA plate and reacted with antibody. Detection with alkaline phosphatase conjugate of goat anti-rabbit IgG.

[e]Peptide bound to plate through streptavidin. Biotin group attached to peptide through triglycine spacer (Savoca *et al.*, 1991). Detection as in EIA.[d]

[f]Biotinylated peptide preincubated with antibody in solution. Biotin–antigen–antibody complex captured to protein A, which is coated to EIA plate. Detection by alkaline phosphatase conjugate of streptavidin (Ngai *et al.*, 1993).

peptides selected are well accessible on the surface of the molecule and should be good candidates for cross-reaction with anti-FNR antibodies. Three assay formats were tested. In the first, the peptides were coated directly to the EIA plate. In the second, peptides with an N-terminal biotinyl-Gly-Gly-Gly extension were bound to the plate through streptavidin. The third assay was a PACE in which binding of peptides to antibodies takes place in solution, as described above.

When denatured FNR was used to produce the antibodies, virtually all peptides cross-reacted in all three assays. Because denatured FNR is a random coil, antibodies to any segment of the polypeptide chain may be found after immunization with denatured FNR. Therefore, the reaction with antibodies to denatured FNR serves as positive control for the reaction with antibodies to native FNR. Indeed, antibodies to native FNR recognize only a few peptides. Only the N-terminal peptide 1–20 was recognized in the solution-phase assay as well as in the two solid-phase assays. The positive reaction of peptide 1–20 in all three assays is not surprising as the first 18 residues of FNR are disordered and mobile in the FNR crystal (Karplus *et al.*, 1991). Antibodies to this segment may have similar specificities whether produced in response to native or denatured FNR. Peptide 30–39 reacted with anti-native FNR antibodies only in the streptavidin EIA; peptide 120–129 reacted in the EIA and the streptavidin EIA. Such differential reactivity of the same peptides in different assay formats is not untypical but is difficult to explain. The observation emphasizes that a cross-reaction with a peptide is not a yes-or-no event but depends critically on the assay format. The observation also raises a question: Are the cross-reactive antibodies indeed directed against the native protein? A reaction with a peptide in a solid-phase assay can indicate that, even if native protein was used for immunization, the cross-reacting antibodies are directed against the unfolded protein. From this it follows that we have to distinguish between genuine and apparent cross-reaction.

C. Genuine and Apparent Cross-Reaction of Antibodies to Proteins

Native preparations of proteins may always contain some unfolded isoforms. Emulsification of a protein in adjuvant will assist unfolding (Torensma, 1993). The reaction of antilysozyme antisera with denatured and carboxymethylated lysozyme was ascribed many years ago to contamination of the immunogen with denatured protein (Scibienski, 1973). Any cross-reaction of a peptide with an antiprotein antibody raises the question of whether the cross-reaction is genuine or only apparent. The cross-reaction is genuine if the cross-reacting antibody is specific for an epitope on the native protein. In this case, the peptide does indeed mimic an epitope on the native protein. On the other hand, the cross-reaction with the peptide is only apparent if the cross-reacting antibody was in fact produced against an unfolded form of the protein. The occurrence of apparent cross-reactivity by antiprotein antibodies was demonstrated unequivocally by the isolation of monoclonal antibodies that react exclusively with the denatured protein even though the animals were immunized with the native protein (Friguet *et al.*, 1984; Merrill *et al.*, 1988).

There is no easy way to distinguish between a genuine and an apparent cross-reaction. This is unfortunate as epitope mapping with peptides is mostly directed at discovering genuine cross-reactions. It was shown that a fraction of the antibodies to sperm whale myoglobin react preferentially with myoglobin peptides and only weakly with native myoglobin (Lando and Reichlin, 1982). These antibodies seem to have been induced by unfolded myoglobin. The peptide-reactive antibody fraction was detected by a competition RIA. The antiserum was incubated in solution with a constant amount of ^{125}I-labeled tracer antigen and variable amounts of nonlabeled competitor antigen. When the tracer was ^{125}I-labeled myoglobin, 10^4-fold excess of peptide was required to achieve the same level of inhibition as with native, nonradioactive myoglobin. However, if the tracer was ^{125}I-labeled peptide, nonradioactive peptide competed better than nonradioactive native myoglobin. This experimental protocol exposed different antibody populations reactive with different antigenic forms of myoglobin.

We have taken up this type of analysis in the experiments shown in Fig. 2. A rabbit antiserum to yeast iso-1 cytochrome *c* reacted very specifically with the native cytochrome in a competition immunoassay in which native cytochrome *c* was the tracer antigen. This is seen in Fig. 2A: native cytochrome *c* is a good competitor, whereas unfolded, random coil apocytochrome *c* hardly competes in the reaction of antibodies with labeled tracer cytochrome *c* (see legend to Fig. 2 for details of the immunoassay). However, when the labeled tracer antigen was apocytochrome *c*, apocytochrome *c* turned out to be a much stronger competitor than native cytochrome *c* (Fig. 2B). This clearly indicates that the anti-cytochrome *c* serum contains an antibody subpopulation specific for native cytochrome *c* and another subpopulation specific for unfolded cytochrome *c*. The cross-reaction with apocytochrome *c* is only an apparent cross-reaction as the cross-reactive antibodies are antibodies specific for the unfolded protein. There is no detectable genuine cross-reaction in this example. Genuine cross-reaction would have occurred if native cytochrome *c* had been a stronger competitor than apocytochrome *c* also in the experiment in which labeled apocytochrome *c* was used as tracer antigen.

An opposite result was obtained when a similar experiment was performed with antiserum to the random coil apocytochrome *c* (Fig. 2C,D). Now the cross-reaction of antibodies to apocytochrome *c* with native cytochrome *c* was genuine: competition by apocytochrome *c* was stronger than competition by native cytochrome *c*, irrespective of whether the tracer antigen was labeled apocytochrome *c* (Fig. 2C) or labeled native cytochrome *c* (Fig. 2D). This result had been expected because immunization with apocytochrome *c* can never induce antibodies that react preferentially with native cytochrome *c*, as apocytochrome *c* cannot refold to native cytochrome *c* without a covalently attached heme. The

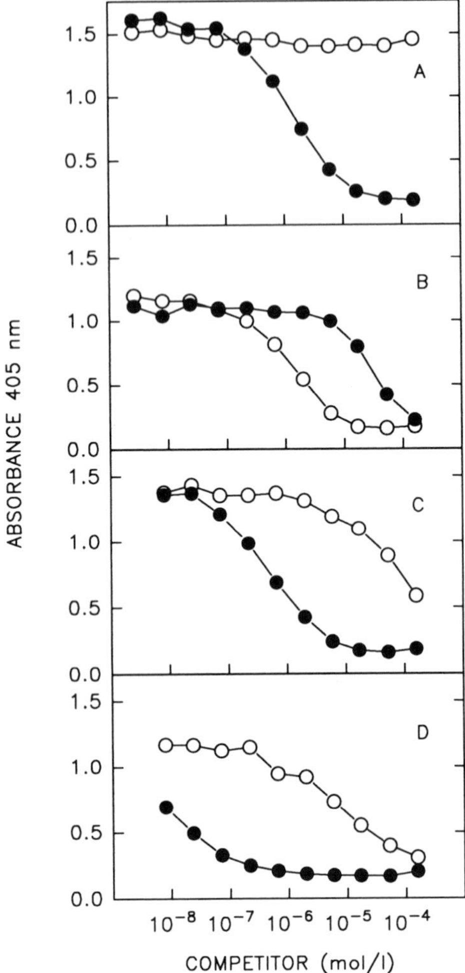

Figure 2 Distinction between genuine and apparent cross-reaction of rabbit antibodies to native yeast iso-1 cytochrome *c* and to the homologous random coil apocytochrome *c* (data from Leder *et al.*, 1994). The antibody reaction was followed by a protein A antibody-capture EIA (PACE; see Ngai *et al.*, 1993, for details of this conformation-sensitive immunoassay). **(A)** Antiserum, produced by immunizing a rabbit with native yeast iso-1 cytochrome *c*, was reacted with biotin-labeled native yeast iso-1 cytochrome *c* (tracer antigen) in the presence of increasing amounts of either native cytochrome *c* (filled circles) or apocytochrome *c* (open circles). After equilibrium between tracer antigen and competitor was established in solution, the biotin-labeled antigen–antibody complex was adsorbed to a protein A-coated microtiter plate and detected with streptavidin–alkaline phosphatase conjugate and *p*-nitrophenyl phosphate as the substrate. In this experiment, apocytochrome *c* does not compete with binding of antibodies to native cytochrome *c*. **(B)** The same antiserum was reacted with biotin-labeled apocytochrome (tracer antigen) in the presence of increasing amounts of either native cytochrome *c* (filled cycles) or apocytochrome *c* (open circles). In this experiment, apocytochrome *c* is a stronger competitor than native cytochrome *c*. A subpopulation of antibodies that bind apocytochrome *c* better than native cytochrome *c* is detected. This subpopulation exhibits apparent cross-reactivity with

cross-reaction of the anti-apocytochrome *c* antibodies with native cytochrome *c* is, thus, a genuine cross-reaction.[5]

D. Mapping of Epitopes of Monoclonal Antibodies

Some of the problems encountered with mapping polyclonal antibodies do not exist for monoclonal antibodies, which are directed to a single epitope. Valid results are obtained if a monoclonal antibody recognizes a contiguous stretch of a polypeptide chain, as often is the case with antibodies to N- or C-terminal segments of a protein. In this case, one may map the functional epitope to within a few residues in the polypeptide chain. An example published by Hodges *et al.* (1988) is shown in Fig. 3. Epitopes for two monoclonal antibodies to bovine rhodopsin, rho 3A6 and rho 1D4, were mapped to a few residues near to the C terminus of rhodopsin. Peptides corresponding to the C-terminal 12 and 18 residues were synthesized and each position was replaced by Ala, Gly, or Gln, as indicated in Fig. 3. Twenty-seven peptides were synthesized by the classic Merrifield solid-phase method, purified, and compared with rhodopsin for binding to the two monoclonal antibodies. Only residues 3', 5', and 7' are very critical for binding to antibody rho 1D4, and residues 8', 10', and 11' for binding to antibody rho 3A6. The critical residues are spaced by residues whose replacement changes the affinity to a lesser degree: residues 1', 4', and 6' for antibody rho 1D4, residue 9' for antibody rho 3A6 (see Fig. 3). It is a common observation that, within a linear sequence, some positions are more important than others for binding to the antibody, a phenomenon dubbed "discontinuous linear determinants" (Appel *et al.*, 1990) and explicable by the spatial restrictions imposed by the paratope of the antibody.

Not only are some residues more important than others for binding, but these critical residues also must be presented in a proper context. Removal of apparently unimportant residues may abolish binding. In the case of antibody rho 1D4, the heptapeptide 1'–7' is not bound even though this peptide contains all

apocytochrome *c*.[5] **(C)** Antiserum, produced by immunizing a rabbit with random coil yeast iso-1 apocytochrome *c*, was reacted with biotin-labeled apocytochrome *c* (tracer antigen) in the presence of increasing amounts of either apocytochrome *c* (filled circles) or native cytochrome *c* (open circles). **(D)** The same antiserum to apocytochrome *c* was reacted with biotin-labeled native cytochrome *c* (tracer antigen) in the presence of increasing amounts of either native cytochrome *c* (open circles) or apocytochrome *c* (filled circles). In the experiments in C and D, apocytochrome *c* is a stronger competitor than native cytochrome *c*, irrespective of whether biotinylated native cytochrome *c* (D) or biotinylated apocytochrome *c* (C) is used as tracer antigen. The cross-reaction with native cytochrome *c* of the antibodies to apocytochrome *c* is therefore genuine.[5] (Reprinted from Leder *et al.* (1994) with permission of the *European Journal of Biochemistry*.)

[5]Some competition by apocytochrome *c* contaminating the native protein cannot be excluded.

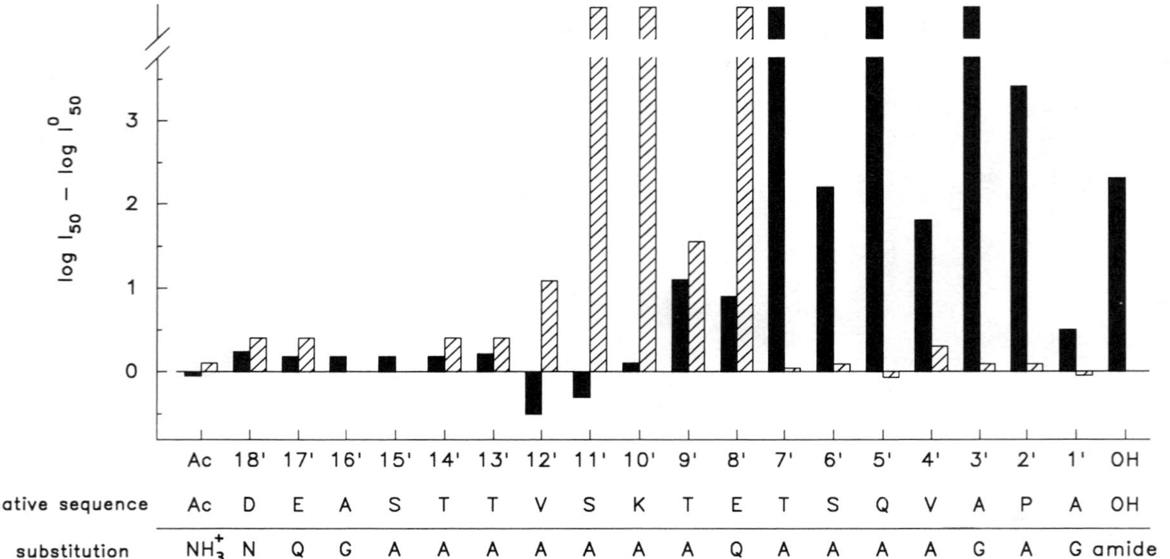

Figure 3 Effect of single amino acid substitutions on the reaction of two monoclonal antibodies to bovine rhodopsin with C-terminal peptides of rhodopsin (data from Hodges *et al.*, 1988). Peptides corresponding to the C-terminal 12 and 18 residues, respectively, were synthesized. Each position was systematically replaced by Ala or Gln or Gly, as indicated. The reaction of the peptide with the native sequence was compared with the reaction of peptides with single amino acid substitutions. I°_{50} is the concentration of free native peptide necessary to achieve 50% inhibition of binding of the antibody to rhodopsin. I_{50} is the concentration of the mutant peptide necessary to achieve 50% inhibition of binding of the antibody to rhodopsin. Positive values of $\log I_{50} - \log I^{\circ}_{50}$ indicate a weaker reaction with the mutant peptide. Bars ending above the break in the ordinate indicate that I_{50} could not be determined because the corresponding mutation produced a peptide that did not react with the antibody. *Hatched bars:* Antibody rho 3A6. *Filled bars:* Antibody rho 1D4. Amino acids are expressed in the one-letter code and are numbered following the direction of the chemical synthesis from the carboxyl ends as 1′ to 18′. (Reprinted from Hodges *et al.* (1988) with permission of the American Society for Biochemistry & Molecular Biology.)

the residues that are critical according to the data shown in Fig. 3 (Hodges *et al.*, 1988). This reminds us of the concepts of functional and contact epitopes. To allow for a good fit, the residues that actively contribute to binding can do so only if other residues enable the functional residues to adopt an optimal binding conformation. From a thermodynamic point of view, replacement of a critical residue leads to an unfavorable change of the free energy of binding, whereas changing of an apparently uncritical residue leaves the free energy of binding unchanged. To give an example, the Gln to Ala replacement in position 5' results in a positive $\Delta\Delta G$ (weaker binding to the mutant than to the native sequence). $\Delta\Delta G$ is approximately zero for the Ala to Lys replacement in position 10'. However, $\Delta\Delta G \approx 0$ may result from enthalpy/entropy compensation; the residue may contribute to binding by a negative enthalpy term ($-\Delta\Delta H$), which is cancelled by a positive entropy term ($+\Delta\Delta S$). Attempts at dissecting the antibody binding reaction into its thermodynamic components are beginning (Sigurskjold *et al.*, 1990; Kelley *et al.*, 1992; Braden and Poljak, 1995).

Most monoclonal antibodies produced in response to immunization with a native protein do not cross-react with peptides. In a rigorous study, Jemmerson (1987) showed that the vast majority of 575 monoclonal antibodies to rat cytochrome *c* do not react with the three CNBr fragments of this protein. When the conformation of the epitope is critical for binding to the antibody, mapping with peptides becomes difficult. Quite often, the reaction depends on a unique folding of the polypeptide chain yet one still finds peptides that cross-react weakly. Two examples are shown in Fig. 4 (Saad and Bosshard, 1990). Monoclonal antibodies 30.6 and 55.1 are directed against native bacterial cytochrome c_2. Antibody 30.6 is highly specific for the native conformation of the protein. By differential acetylation of lysine residues of cytochrome c_2 (method 5 in Table I) we found that antibody 30.6 protects residues Lys-27, Lys-88, and Lys-90 from reaction with acetic anhydride, indicating that these three residues, which are clustered in the three-dimensional structure of cytochrome c_2, contribute to the epitope. However, the epitope cannot be mimicked by peptide 9–37 (containing protected Lys-27) nor by peptide 81–100 (containing protected Lys-88 and Lys-90), as shown in Fig. 4A. Monoclonal antibody 55.1 is directed to an epitope close to the N terminus of the protein. This epitope could be simulated in part by peptides 1–8 and 9–37, and also by random coil apocytochrome c_2. However, also in the case of antibody 55.1, the native cytochrome c_2 is still a far better antigen than the two peptides or the random coil protein. Clearly, conformational features of the native protein contribute to the epitope of both antibodies, although the epitope of antibody 55.1 may be designated continuous according to the conventional operational definition (Section II,B). Sometimes even epitopes on a denatured protein cannot be simulated adequately by sequential peptides because binding of the denatured protein to the antibody is conformationally constrained (Saad *et al.*, 1988).

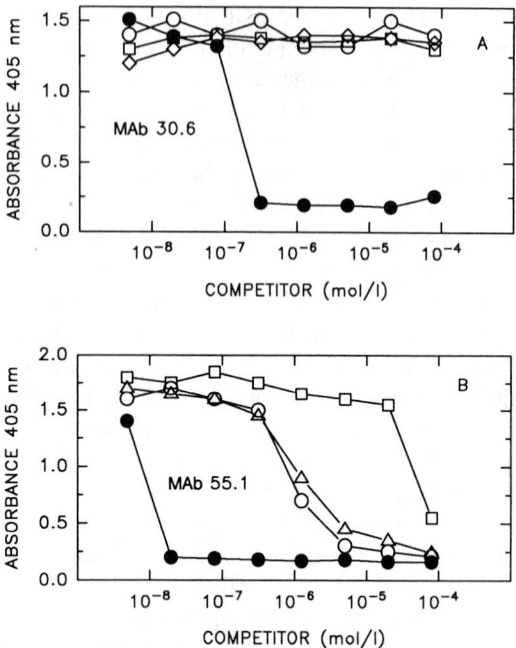

Figure 4 Reaction of peptides and denatured protein with monoclonal antibodies to cytochrome c_2 (data from Saad and Bosshard, 1990). The monoclonal antibodies 30.6 (A) and 55.1 (B) were incubated with increasing amounts of competing antigen (concentration on abscissa) before being added to a microtiter plate coated with cytochrome c_2. The amount of antibody bound to the plate was determined by a second antibody coupled to alkaline phosphatase and using *p*-nitrophenyl phosphate as chromogen. The competitor antigens are indicated by the following symbols: ●, cytochrome c_2; ○, apocytochrome c_2; Δ, peptide 1–8; □, peptide 9–37; ◊, peptide 81–100. (Reprinted from Saad and Bosshard (1990) with permission of the *European Journal of Biochemistry*.)

E. Mapping of Epitopes with Peptide Libraries

New and ingenious methods to rapidly and simultaneously synthesize large numbers of peptides in small amounts have revolutionized mapping procedures. Peptide libraries containing millions of peptides can be produced by chemical methods or recombinant DNA techniques (recent review by Jung and Beck-Sickinger, 1992). Peptide libraries allow one to map sequential epitopes at a resolution unheard of before.

1. Multiple Peptide Synthesis

Multiple peptide synthesis is the name given to any method capable of simultaneous synthesis of large numbers of peptides.[6] The tea-bag method of

[6]Techniques of multiple peptide synthesis and their applications are reviewed in Chapter 10 of this volume by R. A. Houghten.

Houghten (1985) and the pin-technology of Geysen (Geysen *et al.*, 1984, 1987a) were the first widely used approaches at multiple peptide synthesis. In the tea-bag method, the polymeric resin on which the synthesis is accomplished is sealed in a porous polypropylene net, the tea bag, and all the steps that are identical, such as deprotection and washing, are performed on the combined tea bags in a common vessel. The bags are separated only for individual coupling reactions. A representative example of the application of the tea-bag method to the synthesis of peptide families is found in Appel *et al.* (1990).

The Geysen method, also called pepscan method, produces only nanomole amounts of peptides covalently attached to polyethylene pins (Geysen *et al.*, 1987a). The pins fit into the wells of a standard 96-well microtiter plate, and all the synthesis steps are performed on the pins immersed in the wells. The peptide is not cleaved from the support after completion of synthesis. The polymer-bound peptides are used directly in an EIA in which the pins are incubated with the antibody, washed, incubated with an enzyme-coupled second antibody, and processed with substrate. Antibody and second antibody can be removed from the pins, which can be reused many times. In a variation of the method, peptides can be cleaved from the pins (Maeji *et al.*, 1990).

Thousands of different peptides have been synthesized on derivatized glass beads, which then were tested with monoclonal antibodies. The EIA was performed on the beads, which were so small that the EIA color was seen only under the microscope (Lam *et al.*, 1991). Similar miniaturization can be achieved by the synthesis of large numbers of peptides on chips. The techniques of semiconductor-based photolithography and solid-phase synthesis are combined to obtain ordered, high-density arrays of peptides on a solid surface (Fodor *et al.*, 1991, 1993).

An obvious difficulty with synthetic peptide libraries is that individual peptides cannot be analyzed; synthesis errors will not be detected. This disadvantage is partly offset by the possibility of performing statistical analyses of the large data sets obtained from mapping hundreds or thousands of peptides. For example, an antibody may react with 5 hexapeptides that cover in an overlapping manner a stretch of 10 residues of a polypeptide chain. The fact that the antibody recognizes the five consecutive, overlapping peptides validates the result even if the reactive peptides are contaminated with some peptides of faulty sequence.

2. *Mapping Epitopes with Overlapping Peptides (Pepscan)*

The method of Geysen and co-workers (1987a) made it possible to map systematically the entire sequence of a protein with overlapping peptides. Once a peptide reactive with an antibody has been identified, each position of the peptide can be substituted by any of the other 19 naturally occurring amino acids, or with nonnatural amino acids, to test the contribution of each sequence position to the antibody reaction (replaceability matrix analysis). In the example of the monoclonal antibodies to rhodopsin shown in Fig. 3, the pepscan method would

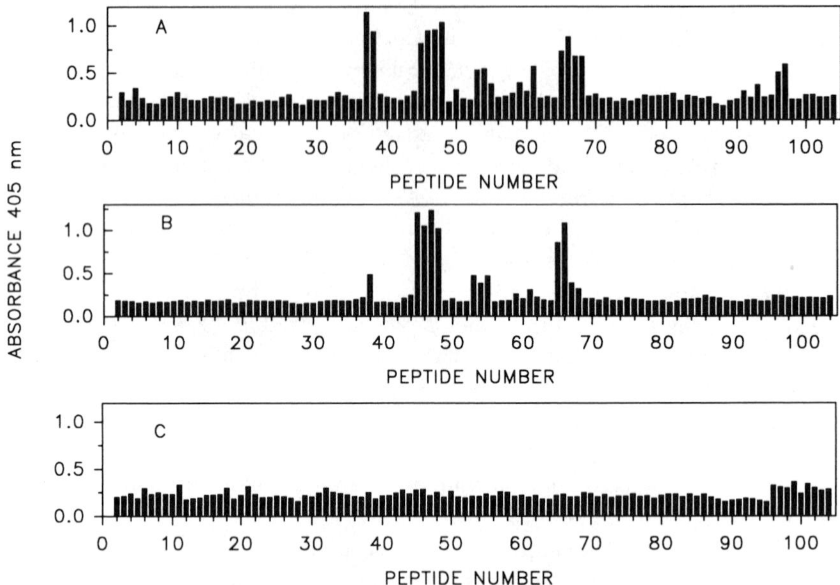

Figure 5 Epitope mapping by the pepscan method (data from Schwab *et al.*, 1993). Rabbit antiserum to yeast iso-1 cytochrome *c* was reacted with 103 overlapping hexapeptides covering the 108-residue sequence of the protein. Each bar represents a hexapeptide coupled to a plastic pin. The numbers on the abscissa indicate the sequence number of the first residue of the hexapeptide in the sequence of cytochrome *c*. The height of each bar corresponds to the color of the EIA reaction and is a measure of the reactivity of the peptide with the antiserum. **(A)** Reaction with whole antiserum. **(B)** Reaction with subpopulation of serum bound to apocytochrome *c* affinity column. **(C)** Reaction with subpopulation of serum not bound to apocytochrome *c* affinity column. (Reprinted from Schwab *et al.* (1993) with permission of Cambridge University Press.)

allow one to replace every position by all the 19 other natural amino acids. With the pepscan method it became possible to analyze the "chemistry of antibody binding to a protein," the title given to the analysis of myohemerythrin by all possible overlapping penta- and hexapeptides (Geysen *et al.*, 1987b). The results suggested that the entire myohemerythrin surface is antigenic but that some reactive peptides cluster in surface areas of high local mobility and convex surface shape. Although no peptides reacted with all seven antisera tested, some peptides were recognized by most sera and were thought to belong to immunodominant regions. A great many epitope mapping studies have since been performed by the pepscan and tea-bag methods. From a compilation of the results obtained by pepscan for 103 epitopes within 63 antigenic peptides, it was deduced that on average only about four or five residues are required to determine specificity and provide binding energy to a sequential epitope (Geysen *et al.*, 1988).

3. Determining Whether Results with Short Synthetic Peptides Are Representative of the Antigenic Structure of Native Protein

Figure 5A, shows an example of a pepscan analysis performed with antiserum to yeast cytochrome *c*. The serum was tested with the 103 hexapeptides that cover the 108-residue sequence of yeast cytochrome *c* (Schwab *et al.*, 1993). The reactive hexapeptides are clustered in a few segments of the polypeptide chain. Is the pattern seen in Fig. 5A representative of the antigenic structure of yeast cytochrome *c?* By operational definition, only sequential epitopes can be revealed by epitope mapping with sequential overlapping peptides. Still, it seems possible that some of the reactive peptides might represent parts of discontinuous epitopes of cytochrome *c*.

To test if hexapeptides cross-react with antibodies to discontinuous epitopes of cytochrome *c*, the antiserum was fractionated on an apocytochrome *c* affinity column. Because apocytochrome *c* is a random coil, antibodies specific for discontinuous epitopes on native cytochrome *c* will not bind to apocytochrome *c*. Figure 5B, obtained with antibodies bound to apocytochrome *c*, shows virtually the same pattern as Fig. 5A, which means that antibodies cross-reacting with apocytochrome *c* are, in essence, responsible for the entire epitope binding pattern observed with the 103 hexapeptides. In support, antibodies that do not bind to apocytochrome *c* also do not react with any of the 103 hexapeptides (Fig. 5C). This results makes it very unlikely that hexapeptides contain elements of assembled, conformation-dependent epitopes of native cytochrome *c*.

Results such as those shown in Fig. 5 were obtained for three other antisera to native cytochrome *c* (Schwab *et al.*, 1993). The antibody fractions that bound to the apocytochrome *c* affinity column always accounted for only 1–2% of the cytochrome *c*-specific antibodies in an anti-cytochrome *c* serum (Schwab, 1991; Schwab *et al.*, 1993). This fraction is most likely composed of antibodies directed primarily against the unfolded protein, as discussed in Section IV,C. At least in the case of the small globular and conformationally stable protein cytochrome *c*, we have to conclude that epitope mapping by pepscan opens only a very narrow window of the antigenicity of the protein. Perhaps the situation differs for larger, less compact proteins with flexible segments and loops. However, such proteins will also tend to unfold more easily, and more antibodies to unfolded forms may be produced on immunization with a less compact and more flexible protein. Epitope mapping with peptides will be biased toward these antibodies.

An observation on the species specificity of anti-cytochrome *c* antibodies further illustrates the problems that may be encountered with the peptide approach (Leder and Bosshard, 1994). Antibodies to cytochrome *c* can distinguish between closely sequence-related cytochromes *c* (Benjamin *et al.*, 1984). Because the three-dimensional structure of the polypeptide chain is virtually identical among eukaryotic cytochromes *c*, species specificity must arise from antibodies that can discriminate against amino acid substitutions within a common polypeptide folding pattern. The question arises if the specificity is observed at the level of the three-dimensional structure and/or at the level of the primary structure. In the latter case, species specificity can be demonstrated by the peptide approach. Using rabbit sera to horse cytochrome *c*, it was shown that discrimination against the host's own cytochrome *c* (six amino acid changes) occurs exclusively at the three-dimensional level and not between peptides with sequences typical for horse and rabbit cytochrome *c*. Furthermore, deliberate immunization with horse apocytochrome *c* produced antibodies that could not discriminate efficiently between sequence-related apocytochromes *c* (Leder and Bosshard, 1994). B-cell tolerance to the host's own protein seems to be restricted to the intact, native protein and, therefore, the high species specificity of anti-cytochrome *c* antibodies could not be detected by the peptide approach.

The strength of pepscan and other related mapping procedures lies in the analysis of sequential epitopes as recognized by monoclonal antibodies. However, even if one concedes that pepscan and related methods can reveal only the subset of sequential epitopes of a protein, one has to keep in mind that many sequential epitopes do not exist on the native protein but only on its unfolded or partially unfolded isoforms (Jemmerson, 1987).[7] In our opinion, and perhaps contrary to earlier expectations (Geysen *et al.*, 1987b), epitope mapping with large numbers of overlapping sequential peptides is of little use to acquire a comprehensive picture of the antigenicity of whole proteins.

F. Phage-Displayed Peptide Libraries

Vast libraries of random peptides can be displayed on the surface of bacteriophages (Scott and Smith, 1990; Devlin *et al.*, 1990; reviewed by Lane and Stephen, 1993). Phages containing the random peptides inserted into a surface-exposed protein are screened by affinity-binding to antibody ("biopanning"). Affinity-purified phage are then amplified and rescreened. The yield of affinity-purified phage increases with every screening round if binding is specific. After several rounds of screening, amino acid sequences of peptide inserts are determined by sequencing the corresponding coding region in the phage DNA.

Phage-displayed peptide libraries are used mainly to screen epitopes of monoclonal antibodies. They also may have the potential to identify ligands of antibodies in polyclonal sera whether or not the antigen is known. For example, autoimmune sera may be used to screen phage libraries for peptides characteristic for specific au-

[7]Some investigators estimate that about 10% of monoclonal antibodies to a protein recognize a sequential epitope (Pellequer *et al.*, 1991). This figure may be too high if one considers only antibodies specific for the protein with a preserved native conformation and excludes antibodies induced by (partly) unfolded protein.

toimmune states. The main problem of such applications will be to ascertain the specificity of the reaction between an antibody and the peptide displayed by the phage.

Studies with antibodies to known epitopes indicate that residues which make a major contribution to the binding energy are easily identified and cannot be replaced by other residues. For example, two monoclonal antibodies known to bind the sequence Asp-Phe-Leu-Glu-Lys-Ile, a sequence of myohemerythrin, selected hexapeptide sequences containing the motif Asp-Phe-Leu-Glu from a hexapeptide library comprising 23 million clones (Scott and Smith, 1990). In position 6, selection for Ile, Val, and Leu was observed, but neither antibody selected for Lys in position 5. A monoclonal antibody directed against the N-terminal sequence Tyr-Gly-Gly-Phe of natural opioid peptides was used to screen a hexapeptide library comprising 300 million clones (Cwirla *et al.*, 1990). Screening with 5 n*M* antibody selected peptides of which 30% had the N-terminal sequence Tyr-Gly-Gly. When selection was repeated using only 50 p*M* Fab, 68% of the isolated clones had the Tyr-Gly-Gly-Phe sequence. The conditions of the screening procedures are very critical.

G. Epitopes and Mimotopes

Epitope mapping with phage-displayed libraries revealed another important phenomenon of molecular recognition: Affinity screening may uncover sequences that are completely unrelated to the original epitope of the antibody. Christian *et al.* (1992) prepared a library of 4×10^8 decapeptides, which they screened with a monoclonal antibody raised against a decapeptide of the V3 loop of protein gp120 of HIV-1 (human immunodeficiency virus type 1). The sequences selected from the decapeptide library contained the common motifs Xaa-Ser-Thr-Arg-Xaa-Met and Xaa-Cys-Cys-Arg-Xaa, where Xaa is any amino acid. These motifs are not related to the original gp120 epitope sequence Arg-Ala-Phe-His-Thr-Thr-Gly-Arg-Ile-Ile to which the antibody was produced. That unrelated sequences can be affinity-selected implies that an entirely different structure can be accommodated by the antibody. The unrelated peptide is a mimotope or mimetic peptide. In another example, dodecapeptides with the consensus sequence Tyr-Pro-Tyr were found to bind to concanavalin A with a similar affinity as the entirely unrelated methyl α-D-mannopyranoside, a known ligand of concanavalin A (Oldenburg *et al.*, 1992).

It has long been known that an antibody can bind to completely unrelated molecules and also bind some molecules more strongly than the one which has induced the antibody (Mäkalä, 1965; Underwood, 1985). The biological explanation for this heterospecificity is that in clonal selection of a B cell, a moderate fit and affinity between B-cell receptor and immunogen might suffice to trigger B-cell maturation and antibody production. Molecules that bind the antibody more tightly than the original B-cell-stimulating immunogen may always exist, but they were difficult to identify before millions of peptides could be screened with the help of peptide libraries.

Mimotopes are also seen in studies with chemical peptide libraries. We found the sequence Thr-Val-Trp-Tyr-Ala-Cys to be recognized by anti-cytochrome *c* antibodies although this sequence does not appear in the yeast iso-1 cytochrome *c* used for immunization (Savoca *et al.*, 1991). Multiple peptide synthesis has facilitated the deliberate search for mimotopes. A monoclonal antibody to rabies virus was found to bind weakly to the dipeptide LPhe-LHis (Rodda *et al.*, 1986). The dipeptide is only a very weak mimotope of the (unknown) epitope of the antibody. To find a stronger mimotope, all the tripeptides LPhe-LHis-DLXaa were synthesized and tested. Strongest binding was found with LPhe-LHis-DLeu. An additional round of mimotope search gave LPhe-LHis-DLeu-DMet as the best tetrapeptide mimotope, which bound 1250 times more strongly than the original dipeptide (Geysen *et al.*, 1987a). The development of methods for the synthesis of unnatural biopolymers composed of building blocks other than amino acids may provide for the deliberate search for mimotopes with novel properties (Cho *et al.*, 1993). This could lead to results that may also be of importance to drug design.

V. Cross-Reactivity Studies between Proteins and Peptides Using Antipeptide Antibodies

In 1963, F. A. Anderer published a paper entitled "Preparation and properties of an artificial antigen related to tobacco mosaic virus." In this study, rabbits were immunized with peptides of the C-terminal amino acid sequence of tobacco mosaic virus coat protein attached to BSA. The antisera thus obtained cross-reacted with tobacco mosaic virus and neutralized the virus (Anderer, 1963; Anderer and Schlumberger, 1965). After Anderer's work followed many successful applications of synthetic peptides to produce virus-neutralizing antibodies. Such "synthetic vaccines" are the best proof for the existence of genuine cross-reactions between antipeptide antibodies and intact proteins (Arnon, 1987; Vajda *et al.*, 1990; Arnon and Van Regenmortel, 1992). Antibodies to peptides are also often employed in expression cloning. A peptide corresponding to a sequence of the cloned protein is used as immunogen, and the ensuing antibodies are then used to visualize the expressed protein on nitrocellulose filters and other polymeric matrices. Commercial companies offer to prepare antibodies to peptides with a guarantee for cross-reaction with the corresponding protein.

A. Possibility of Order–Disorder Paradox

In view of our discussion of the scarcity of antiprotein antibodies cross-reacting with peptides in a genuine way (Sections IV,C and IV,E), the many reports about successful cross-reactions of antipeptide antibodies with proteins seem paradoxical. Dyson *et al.* (1988) put forward the question of how an antibody raised against a highly disordered structure, the peptide, can recognize a highly ordered structure, the protein. They called this the "order–disorder paradox." We believe this paradox is unfounded. The reason is that the problem of unfolding of proteins in solid-phase immunoassays has been overlooked. Most of the cross-reactions of antipeptide antibodies with proteins were and still are observed with proteins adsorbed to a solid surface, as in a Western blot or any other solid-phase antibody-capture EIA. Indeed, Spangler (1991) has proved for several monoclonal antipeptide antibodies, which before were thought to cross-react in a genuine way with the corresponding intact protein, that the cross-reaction was only due to unfolding of the protein during the assay procedure and that no cross-reaction occurred when the experiment was repeated with both the antipeptide antibody and the protein in solution. In the crystal structure of a myohemerythrin peptide complexed to the Fab fragment of an antipeptide antibody, the peptide has a different conformation than the homologous section of myohemerythrin (Stanfield *et al.*, 1990). This agrees with Spangler's observation that the same antibody does not bind to intact myohemerythrin in solution but only on a solid surface.

The high incidence of protein-reactive antipeptide antibodies can thus be explained by the nature of the immunoassays used to detect cross-reactions. Fortunately, partial unfolding of proteins on a solid surface is unimportant in many immunochemical applications of antipeptide antibodies. For example, the detection of gene products on nitrocellulose will usually succeed with antipeptide antibodies directed against the unfolded protein. Only if antipeptide antibodies are used as conformation-specific probes must care be taken to show that the cross-section occurs with the correct form of the protein. An example is the application of antipeptide antibodies to probe the accessibility and inside/outside orientation of folds and loops in membrane proteins (Maelicke *et al.*, 1989).

B. Cross-Reactions of Antipeptide Antibodies with Proteins Can Have Different Origins

Genuine cross-reactions of antipeptide antibodies with native proteins may be as infrequent as the converse genuine cross-reaction of antiprotein antibodies with peptides. The two situations are not quite equivalent for the following reason. The paratope of an antibody cannot undergo a large conformational adaptation when binding to the epitope; this at least follows from the limited number of structures of protein–antibody and peptide–antibody complex available today (Davies and Padlan, 1990; Garcia *et al.*, 1992; Rini *et al.*, 1993). Conformational adaptations are more likely to occur if the antigen is a peptide. A historical example is antiserum to apomyoglobin which binds myoglobin but distorts the myoglobin molecule to such an extend that the heme is squeezed out of the molecule (Crumpton, 1966). If a peptide contains residues that contribute strongly to binding but are not arrayed optimally to fit the paratope, then a conformational adaptation of the peptide may lead to cross-reaction with the antiprotein antibody. In the opposite case of an antipeptide antibody, the reaction with the protein may require a conformational adaptation of the protein. This will be energetically more costly than the adaptation of a peptide, except perhaps in flexible and not well-ordered domains of the protein.

These considerations lead to two principal ways in which a peptide can induce cross-reactive antibodies. In the first case, the peptide possesses a preexisting and predominant conformation, for example, a β turn, which simulates a surface feature of the three-dimensional structure of the protein. Included are also disordered protein domains, which are the easiest to simulate by a peptide. A high percentage of protein-reactive antipeptide antibodies are expected to arise in this first case. In the second case, only very few out of many conformational isomers of the peptide simulate a surface feature of the protein, and protein-reactive antibodies will arise only rarely. The two origins of protein-reactive antibodies mark two opposite possibilities with a range of possibilities, including models of partial conformational similarity combined with conformational adaptations.

Different origins of protein-reactive antipeptide antibodies are illustrated by the experiments of Fig. 6. In Fig. 6A, antibodies to peptide 1–20 of FNR were tested with FNR in a competition PACE (Section IV,A). In one experiment, peptide 1–20 was used as tracer antigen (solid curves), and FNR (open circles) and nonlabeled peptide 1–20 (closed circles) were tested as competitors. FNR was recognized by the antibodies to peptide 1–20, although, as expected, FNR was a weaker competitor than peptide 1–20. Actually, the competition curve with FNR levels off at half-height. This indicates that only a fraction of the antipeptide antibodies cross-reacts with the protein and is prevented by FNR from binding to the peptide. When FNR was the tracer antigen (Fig. 6A, dashed curves), FNR was again a weaker competitor than peptide 1–20. It follows that antibodies to peptide 1–20 cross-react in a genuine way with the native FNR in the conformation-sensitive solution-phase assay, yet the peptide is clearly a better antigen than the protein. The number of cross-reactive antibodies was quite large as judged from the competition curve with FNR (Fig. 6A, solid line with open circles). In view of the good cross-reaction of peptide 1–20 with antibodies to intact FNR (Table II), a complementary cross-reaction of antibodies to peptide 1–20 with intact FNR could have been expected, even more so as the N-terminal 18 residues of FNR are disordered in the crystal (Karplus *et al.*, 1991).

The situation differs drastically for antibodies to peptide 30–39. This peptide, found to cross-react with anti-FNR antibodies in the streptavidin EIA (Table II), produced antibodies which do not react with FNR in a competition assay using labeled peptide 30–39 as tracer antigen (solid curve with open circles in Fig. 6B). On the basis of this assay alone, peptide 30–39 seems unable to produce FNR-reactive antibodies. However, when we changed to labeled FNR as the tracer antigen, FNR became a cross-reacting antigen, even a stronger antigen than peptide 30–39 itself (compare dashed curves in Fig. 6B). This surprising result is explained in the following way. The antiserum to peptide 30–39 contains a tiny fraction of antibodies that cross-react with FNR. These antibodies were

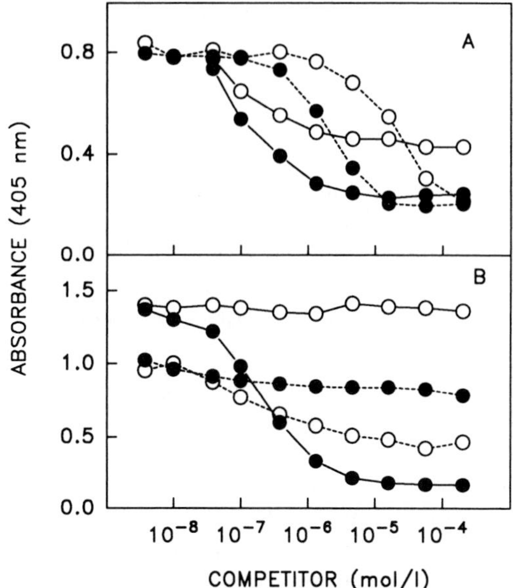

Figure 6 Cross-reaction of antipeptide antibodies with intact protein (data from Leder *et al.*, 1994). Rabbit antibodies to peptides representing the sequences 1–20 and 30–39 of ferredoxin:NADP$^+$ oxidoreductase (FNR) were tested for reaction with FNR in a protein A antibody-capture EIA (PACE; see the legend to Fig. 2 and Section IV,A for details of the assay). **(A, solid lines)** Antiserum to peptide 1–20 was reacted with biotin-labeled peptide 1–20 in the presence of increasing amounts of either peptide 1–20 (filled circles) or FNR (open circles). FNR cross-reacts with the antipeptide antibodies. **(A, dashed lines)** Antiserum to peptide 1–20 was reacted with biotin-labeled FNR in the presence of increasing amounts of peptide 1–20 (filled circles) or FNR (open circles). The experimental confirms cross-reaction of antipeptide antibodies with FNR. **(B, solid lines)** Antiserum to peptide 30–39 was reacted with biotin-labeled peptide 30–39 in the presence of increasing amounts of peptide 30–39 (filled circles) or FNR (open circles). No cross-reaction of antibodies to peptide 30–39 with FNR is seen in this experiment. **(B, dashed lines)** Antiserum to peptide 30–39 was reacted with biotin-labeled FNR in the presence of peptide 30–39 (filled circles) or FNR (open circles). Change of the tracer antigen to biotinylated FNR uncovers a small subpopulation of antibodies to peptide 30–39 that react with FNR more strongly than with the majority conformation(s) of peptide 30–39. (Reprinted from Leder *et al.* (1994) with permission of the *European Journal of Biochemistry.*)

produced in response to a conformational isomer of peptide 30–39 that simulates the conformation of segment 30–39 of intact FNR. Because this conformational isomer of peptide 30–39 is very rare, the antibodies to it are also rare. They could be detected only with labeled FNR as the tracer antigen. That peptide 30–39 was a weaker competitor than intact FNR in the assay in which labeled FNR was the tracer, is in full agreement with the assumption that the FNR-like conformation of the peptide is infrequent among the many different conformations peptide 30–39 may adopt in solution.

C. Antibodies to Peptides with Restrained Conformations

As we have seen, peptides with restrained conformations are better suited to elicit protein-reactive antibodies. This has been demonstrated in numerous cases. One of the first was the production of antibodies to the loop peptide of lysozyme. Anti-loop peptide antibodies cross-reacted to a significant degree with the intact native lysozyme molecule (Arnon and Sela, 1969; Arnon *et al.*, 1971). Cyclization of a peptide is a simple way to restrict its conformational flexibility, and introduction of disulfide bonds has been used very often to try to increase the chances of obtaining protein-reactive antibodies. However, success is often equivocal (Jemmerson and Hutchinson, 1990; Joisson *et al.*, 1993). A monoclonal antibody that neutralizes HIV-1 was obtained by immunization with a disulfide-linked 16-residue peptide corresponding to the hypervariable loop region V3 of gp120 of HIV-1 (White-Scharf *et al.*, 1993). In the crystal of this peptide with antipeptide monoclonal antibody, the first 9 residues contact a 25-Å-long nonpolar groove, and the conformation of the bound peptide is in partial agreement with the structure predicted for the V3 loop of gp120 (Rini *et al.*, 1993). In this as in several other examples, the peptide does indeed at least partially simulate the corresponding conformation in the intact protein.

It is increasingly being realized that many peptides show marked conformational preferences in aqueous solutio (Dyson and Wright, 1991). Once methods to predict conformations from sequences become more advanced, it should become feasible to design peptides having a high probability to induce protein-reactive antibodies. An illustrative example of how restricting the conformational flexibility of a peptide can increase antibody binding was reported by Hinds *et al.* (1991). These authors prepared monoclonal antibodies to a synthetic 19-residue peptide of the HA1 chain of influenza virus hemagglutinin. Antibody DB19/1 recognized the first 4 or 5 residues of the peptide Ac-Tyr-Pro-Tyr-Asp-Val-Pro-Asp-Tyr-Ala, which corresponds to the first 9 residues of the 19-residue peptide used to elicit the antibody. In the intact hemagglutinin as well as in the isolated peptide, the sequence Ac-Tyr-Pro-Tyr-Asp forms a β turn, as shown by nuclear magnetic resonance (NMR) spectroscopy and computer modeling studies (Fig. 7, peptide **1**). However, peptide **1** is not rigid but still populates other conformations. The β-turn conformation of peptide **1** could be stabilized by introduction of a methyl group in position 2 of the proline side chain (peptide **2** of Fig. 7). Binding of antibody DB19/1 to peptide **2** was 50 times stronger than that to peptide **1.** First, this observation proves that the antibody is directed to a β-turn conformation of the hemagglutinin peptide. Second, the results clearly show how binding can be strengthened significantly if the conformation of the antigen is restrained through substitution of *(S)*-α-methylproline for Pro. At least part of the more favorable free energy of binding can be accounted for by a reduced entropy loss when peptide **2** binds to the antibody (Hinds *et al.*, 1991).

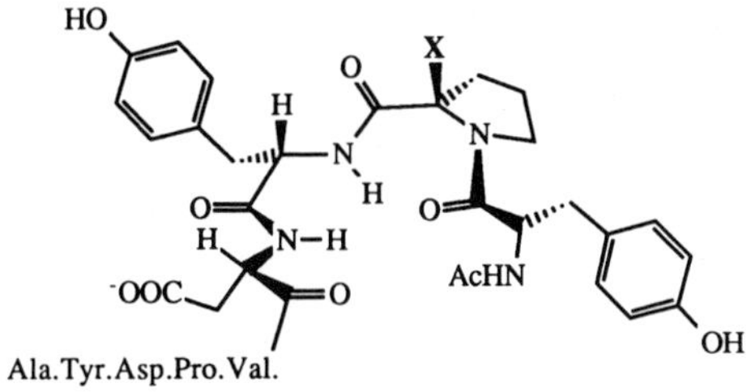

1: X = H $K_d = (1.8 \pm 0.2) \times 10^{-7}$ M
2: X = Me $K_d = (3.6 \pm 0.5) \times 10^{-9}$ M

Figure 7 The β-turn conformation of peptide Ac-Tyr-Pro-Tyr-Asp-Val-Pro-Asp-Tyr-Ala (compound **1**) is stabilized by a *(S)*-α-methylproline residue in position 2 (compound **2**), resulting in stronger binding (lower K_d) of monoclonal antibody DB19/1 (data from Hinds *et al.*, 1991).

D. Can We Predict Antigenic Peptides?

The large body of information about epitopes on proteins of known three-dimensional structure combined with results from epitope mapping with peptides have been used to develop prediction methods for antigenic sites of proteins. The prediction algorithms can be valuable to design peptides having a higher probability of inducing protein-reactive antibodies. The methods are restricted to the prediction of sequential epitopes. Parameters such as hydrophilicity and hydrophobicity, surface accessibility, mobility (from temperature factors of crystal structures) and preponderance for certain secondary structure features have been correlated with antigenicity. Empirical rules have been established which are based on propensity scales for the 20 natural amino acids, for example, scales of hydrophilicity and preponderance for β turns.

Assets and limitations of the many prediction methods have been reviewed (Pellequer *et al.*, 1991). Success rates of correct predictions range from 50 to 70%, but the evaluation of success rates itself is a matter of debate. On the whole, we believe that a good success rate is again at least partly due to the conformation-insensitive immunoassay procedures. For example, the peptides of Table II were selected for maximum surface accessibility in the three-dimensional structure of FNR. Peptides 1–20, 30–39, 120–129, 186–195, and 220–226 induced antibodies that cross-reacted with intact FNR in a solid-phase EIA, but only peptides 1–20 and 30–39 induced antibodies that reacted with FNR in a solution-phase assay (Leder *et al.*, 1994). As mentioned, in many applications the antibody need not cross-react with the conformationally intact protein, and prediction methods may be of little relevance to the design of peptides.

VI. Concluding Remarks

Epitope mapping with peptides is a well-established and extensively applied procedure. It has gained value from the new possibilities of multiple peptide synthesis. Most protein epitopes identified today are actually represented only by cross-reacting peptides, and we can safely say that the peptide approach is the single most important source of information about epitopes. However, a peptide never can reproduce exactly the epitope against which the antibody has been induced through clonal selection and activation of a B cell. Thus, a cross-reacting peptide is always a mimotope, and epitope mapping with peptides is searching for the best mimotopes. Mimotopes need not be chemically related to the epitope, an observation exploited in modern drug design to which epitope mapping provides inspiration.

It is no trivial matter to verify whether a peptide simulates features of a protein epitope. Verification depends strongly on how we define and assay cross-reaction between peptides and proteins. The cross-reaction is hardly ever a yes-or-no event. Depending on the immunoassay format, a peptide may or may not react with an antiprotein antibody. True mimicry of surface features of proteins by peptides may be less frequent than assumed by many who use peptides for epitope mapping, the reason being that, on a protein of undisturbed three-dimensional structure, sequential epitopes are rare. Yet only sequential epitopes can be mimicked with a high grade of success. In this regard it is interesting to note that, already half a century ago, Karl Landsteiner commented: "Antisera to native protein, in general, react much less, if at all, with the denatured products" (Landsteiner, 1945).

There is a complementary relationship between epitope mapping with peptides and X-ray analysis of antigen-antibody complexes. Crystallography provides a very detailed static picture of an epitope, from which it is difficult to appraise the dynamic and functional aspects of antibody recognition. In contrast, epitope mapping with peptides contributes only fragmentary information about the size and bounds of an epitope yet can expose the functionally critical residues. Together, the two approaches disclose the functional and structural facets of an epitope.

The use of peptides to produce protein-reactive antibodies is another important application of synthetic peptides. Success depends greatly on what structural aspects are to be simulated by peptides. It is relatively easy to obtain protein-reactive antipeptide antibodies to visualize partly unfolded proteins on a solid surface. It is much more demanding to produce antipeptide antibodies able to recognize an ordered and unique structural element of an intact native protein. There are several ways to improve the yield of protein-reactive antibodies and to synthesize peptides for active immunization.[8]

[8]Approaches to active immunization using synthetic peptides are discussed in Chapter 12, this volume, by J. P. Tam.

Because of the many difficulties and uncertainties intrinsic to the study of epitopes of native proteins with peptides and to producing antipeptide antibodies reactive with a native protein, some investigators have doubted that peptides are at all useful for these purposes (Laver *et al.*, 1990). We do not share this view. With the necessary precautions, the peptide approach can provide valid information about epitopes, even though peptides cannot disclose the majority of epitopes on a native protein as these are mostly discontinuous. Synthetic peptides must not be used as the only tools for epitope mapping, but we believe that epitope mapping with peptides will remain as a simple and perhaps the single most important approach.

Acknowledgments

The author thanks Professor John A. Robinson for a critical reading of the manuscript and for providing Fig. 7. Work performed in the author's laboratory was supported by the Swiss National Science Foundation, the Hartmann Müller-Stiftung, the Stiftung für Wissenschaftliche Forschung und Jubiäumsspende der Universität Zürich, and by the Kanton of Zürich.

References

Air, G. M., Laver, W. G., and Webster, R. G. (1990). *J. Virol.* **64,** 5797–5803.

Amit, A. G., Mariuzza, R. A., Phillips, S. E. V., and Poljak, R. J. (1986). *Science* **233,** 747–753.

Anderer, F. A. (1963). *Biochim. Biophys. Acta* **71,** 246–248.

Anderer, F. A., and Schlumberger, H. D. (1965). *Biochim. Biophys. Acta* **97,** 503–509.

Appel, J. R., Pinilla, C., Niman, H., and Houghten, R. (1990). *J. Immunol.* **144,** 976–983.

Arnon, R. (1987). "Synthetic Vaccines." CRC Press, Boca Raton, Florida.

Arnon, R., and Sela, M. (1969). *Proc. Natl. Acad. Sci. U.S.A.* **62,** 163–170.

Arnon, R., and Van Regenmortel, M. H. V. (1992). *FASEB J.* **6,** 3265–3274.

Arnon, R., Maron, E., Sela, M., and Anfinsen, C. B. (1971). *Proc. Natl. Acad. Sci. U.S.A.* **68,** 1450–1455.

Atassi, M. Z. (1984), *Eur. J. Biochem.* **145,** 1–20.

Atassi, M. A., and Lee, C. L. (1978). *Biochem. J.* **171,** 429–434.

Atassi, M. Z., and Smith, J. A. (1978). *Immunochemistry* **15,** 609–610.

Barlow, D. J., Edwards, M. S., and Thornton, J. M. (1986). *Nature (London)* **322,** 747–748.

Benjamin, D. C., Berzofsky, J. A., East, I. J., Gurd, F. R. N., Hannum, C., Leach, S. J., Margoliash, E., Michael, J. G., Miller, A., Prager, E. M., Reichlin, M., Sercarz, E., Smith-Gill, S. J., Todd, P. E., and Wilson, A. C. (1984). *Annu. Rev. Immunol.* **2,** 67–101.

Berzofsky, J. A. (1985). *Science* **229,** 932–940.

Berzofsky, J. A., and Schechter, A. N. (1981). *Mol. Immunol.* **18,** 751–763.

Bjorkman, P. J., Saper, M. A., Samraoui, B., Bennett, W. S., Strominger, J. L., and Wiley, D. C. (1987). *Nature (London)* **329,** 506–512.

Bosshard, H. R. (1979). *Methods Biochem. Anal.* **25,** 273–301.

Bosshard, H. R. (1995). *In* "Cytochrome *c:* A multidisciplinary approach" (A. G. Mauk and R. A. Scott, eds.), Chapter 10, in press. Univ. Science Books, Mill Valley, California.

Braden, B. C., and Poljak, R. J. (1995). *FASEB J.* **9,** 9–16.

Brown, J. H., Jardetzky, T. S., Gorga, J. C., Stern, L. J., Urban, R. G., Strominger, J. L., and Wiley, D. C. (1993). *Nature (London)* **364,** 33–39.

Burnens, A., Demotz, S., Corradin, G., Binz, H., and Bosshard, H. R. (1987). *Science* **235,** 780–783.

Butler, J. E. (1991). "Immunochemistry of Solid-Phase Immunoassays." CRC Press, Boca Raton, Florida.

Butler, J. E., Peterman, J. H., Suter, M., and Dierks, S. E. (1987). *Fed. Proc.* **46,** 2548–2556.

Chicz, R. M., Urban, R. G., Lane, W. S., Gorga, J. C., Stern, L. J., Vignali, D. A. A., and Strominger, J. L. (1992). *Nature (London)* **358,** 764–768.

Cho, C. Y., Moran, E. J., Cherry, S. R., Stephans, J. C., Fodor, S. P. A., Adams, C. L., Sundaram, A., Jacobs, J. W., and Schultz, P. G. (1993). *Science* **261,** 1303–1305.

Christian, R. B., Zuckermann, R. N., Kerr, J. M., Wang, L., and Malcolm, B. A. (1992). *J. Mol. Biol.* **227,** 711–718.

Connolly, M. L. (1983). *Science* **221,** 709–713.

Cooper, H. M., Klinman, N. R., and Paterson, Y. (1989). *Eur. J. Immunol.* **19,** 315–322.

Crumpton, M. J. (1966). *Biochem. J.* **100,** 223–232.

Cwirla, S. E., Peters, E. A., Barrett, R. W., and Dower, W. J. (1990). *Proc. Natl. Acad. Sci. U.S.A.* **87,** 6378–6382.

Davies, D. R., and Padlan, E. A. (1990). *Annu. Rev. Biochem.* **59,** 439–473.

Demotz, S., Grey, H. M., Appella, E., and Sette, A. (1989). *Nature (London)* **342,** 682–684.

Devlin, J. J., Panganiban, L. C., and Devlin, P. E. (1990). *Science* **249,** 404–406.

Dintzis, H. M., Symer, D. E., Dintzis, R. Z., Zawadzke, L. E., and Berg, J. M. (1993). *Proteins* **16,** 306–308.

Dyson, H. J., and Wright, P. E. (1991). *Annu. Rev. Biophys. Biophys. Chem.* **20,** 519–538.

Dyson, H. J., Lerner, R. A., and Wright, P. E. (1988). *Annu. Rev. Biophys. Biophys. Chem.* **17,** 305–324.

Falk, K., Rötzschke, O., Deres, K., Metzger, J., Jung, G., and Rammensee, H. G. (1991a). *J. Exp. Med.* **174,** 425–434.

Falk, K., Rötzschke, O., Stevanovic, S., Jung, G., and Rammensee, H. G. (1991b). *Nature (London)* **351,** 290–296.

Farr, R. S. (1958). *J. Infect. Dis.* **103,** 239–262.

Fisher, W. R., Taniuchi, H., and Anfinsen, C. B. (1973). *J. Biol. Chem.* **248,** 3188–3195.

Fodor, S. P. A., Read, J. L., Pirrung, M. C., Stryer, L., Lu, A. T., and Solas, D. (1991). *Science* **251,** 767–773.

Fodor, P. A., Rava, R. P., Huang, X. C., Pease, A. C., Holes, C. P., and Adams, C. L. (1993). *Nature (London)* **364,** 555–556.

Fox, B. S., Chen, C., Fraga, E., French, C. A., Singh, B., and Schwartz, R. H. (1987). *J. Immunol.* **139,** 1578–1588.

Fremont, D. H., Matsumura, M., Stura, E. A., Peterson, P. A., and Wilson, I. A. (1992). *Science* **257,** 919–927.

Friguet, B., Djavadi-Ohaniance, L., and Goldberg, M. E. (1984). *Mol. Immunol.* **21,** 673–677.

Gammon, G., Shastri, N., Cogswell, J., Wilbur, S., Sadegh-Nasseri, S., Krzych, U., Miller, A., and Sercarz, E. (1987). *Immunol. Rev.* **98,** 53–73.

Garcia, K. C., Ronco, P. M., Verroust, P. J., Brünger, A. T., and Amzel, L. M. (1992). *Science* **257,** 502–507.

Garrett, T. P. J., Saper, M. A., Bjorkman, P. J., Strominger, J. L., and Wiley, D. C. (1989). *Nature (London)* **342,** 692–696.

Geysen, H. M., Meloen, R. B., and Barteling, S. J. (1984). *Proc. Natl. Acad. Sci. U.S.A.* **81,** 3998–4002.

Geysen, H. M., Rodda, S. J., Mason, T. J., Tribbick, G., and Schoofs, P. G. (1987a). *J. Immunol. Methods* **102,** 259–274.

Geysen, H. M., Tainer, J. A., Rodda, S. J., Mason, T. J., Alexander, H., Getzoff, E. D., and Lerner, R. A. (1987b). *Science* **235,** 1184–1190.

Geysen, H. M., Mason, T. J., and Rodda, S. J. (1988). *J. Mol. Rec.* **1,** 32–41.

Gill III, T. J., Gould, H. J., and Doty, P. (1963). *Nature (London)* **197,** 746–747.

Goldberg, A. L., and Rock, K. L. (1992). *Nature (London)* **357,** 375–379.

Goldberg, M. E., and Djavadi-Ohaniance, L. (1993). *Curr. Opin. Immunol.* **5,** 278–281.

Greenspan, N. S. (1992). *Bull. Inst. Pasteur 90,* 267–279.

Guo, H. C., Jardetzky, T. S., Garrett, T. P. J., Lane, W. S., Strominger, J. L., and Wiley, D. C. (1992). *Nature (London)* **360,** 364–366.

Harlow, E., and Lane, D. (1988). "Antibodies, A Laboratory Manual." Cold Spring Harbor Laboratories, Cold Spring Harbor, New York.

Harper, M., Lema, F., Boulot, G., and Poljak, R. J. (1987). *Mol. Immunol.* **24,** 97–108.

Hinds, M. G., Welsh, J. H., Brennand, D. M., Fisher, J., Glennie, M. J., Richards, N. G. J., Turner, D. L., and Robinson, J. A. (1991). *J. Med. Chem.* **34,** 1777–1789.

Hodges, R. S., Heaton, R. J., Parker, J. M. R., Molday, L., and Molday, R. S. (1988). *J. Biol. Chem. J. Biol. Chem.* **263,** 11768–11775.

Hogrefe, H. H., Kaumaya, P. T. P., and Goldberg, E. (1989). *J. Biol. Chem.* **264,** 10513–10519.

Horsfall, A. C., Hay, F. C., Soltys, A. J., and Jones, M. G. (1991). *Immunol. Today* **12,** 211–213.

Houghten, R. A. (1985). *Proc. Natl. Acad. Sci. U.S.A.* **82,** 5131–5135.

Hunt, D. F., Henderson, R. A., Shabanowitz, J., Sakaguchi, K., Michel, H., Sevilir, N., Cox, A. L., Appella, E., and Engelhard, V. H. (1992a). *Science* **255,** 1261–1263.

Hunt, D. F., Michel, H., Dickinson, T. A., Shabanowitz, J., Cox, A. L., Sakaguchi, K., Appella, E., Grey, H. M., and Sette, A. 1992b). *Science* **256,** 1817–1820.

Jemmerson, R. (1987). *Proc. Natl. Acad. Sci. U.S.A.* **84,** 9180–9184.

Jemmerson, R., and Hutchinson, R. M. (1990). *Eur. J. Immunol.* **20,** 579–585.

Jemmerson, R., and Margoliash, E. (1979). *J. Biol. Chem.* **254,** 12706–12716.

Jemmerson, R., and Paterson, Y. (1986a). *BioTechniques* **4,** 18–31.

Jemmerson, R., and Paterson, Y. (1986b). *Science* **232,** 1001–1004.

Jin, L., Fendly, B. M., and Wells, J. A. (1992). *J. Mol. Biol.* **226,** 851–865.

Joisson, C., Kuster, F., Plaué, S., and Van Regenmortel, M. H. V. (1993). *Arch. Virol.* **128,** 299–317.

Jung, G., and Beck-Sickinger, A. G. (1992). *Angew. Chem., Int. Ed. Engl.* **31,** 367–383.

Karplus, P. A., Daniels, M. J., and Herriott, J. R. (1991). *Science* **251,** 60–66.

Kelley, R. F., and O'Connell, M. P. (1993). *Biochemistry* **32,** 6828–6835.

Kelley, R. F., O'Connell, M. P., Carter, P., Presta, L., Eigenbrot, C., Covarrubias, M., Snedecor, B., Bourell, J. H., and Vetterlein, D. (1992). *Biochemistry* **31,** 5434–5441.

Kennel, J. J. (1982). *J. Immunol. Methods* **55,** 1–12.

Lam, K. S., Salmon, S. E., Hersh, E. M., Hruby, V. J., Kazmierski, W. M., and Knapp, R. J. (1991) *Nature (London)* **354,** 82–84.

Lando, G., and Reichlin, M. (1982). *J. Immunol.* **129,** 212–216.

Landsteiner, K. (1945). "The Specificity of Serological Reactions." Harvard Univ. Press, Cambridge, Massachusetts.

Lane, D. P., and Stephen, C. W. (1993). *Curr. Opin. Immunol.* **5,** 268–271.

Laver, W. G., Air, G. M., Webster, R. G., and Smith-Gill, S. J. (1990). *Cell (Cambridge, Mass.)* **61,** 553–556.

Leder, L., and Bosshard, H. R. (1994). *Biochimie* **76,** 465–470.

Leder, L., Wendt, H., Schwab, C., Jelesarov, I., Bornhauser, S., Ackermann, F., and Bosshard, H. R. (1994). *Eur. J. Biochem.* **219,** 73–81.

Mäkalä, O. (1965). *J. Immunol.* **95,** 378–386.

Maeji, N. J., Bray, A. M., and Geysen, H. M. (1990). *J. Immunol. Methods* **134,** 23–33.

Maelicke, A., Pluemer-Wilk, R., Fels, G., Spencer, S. R., Englehard, M., Veltel, D., and Conti-Tronconi, B. M. (1989). *Biochemistry* **28,** 1396–1405.

Matsumara, M., Fremont, D. H. Peterson, P. A., and Wilson, I. A. (1992). *Science* **257,** 927–934.

Merrill, G. A., Horowitz, P. M., Bowman, S., Bentley, K., and Klebe, R. (1988). *J. Biol. Chem.* **36,** 19324–19330.

Ngai, P. K., Ackermann, F., Wendt, H., Savoca, R., and Bosshard, H. R. (1993). *J. Immunol. Methods* **158,** 267–276.

Novotny, J., Bruccoleri, R. E., and Saul, F. A. (1989). *Biochemistry* **28,** 4735–4749.

Nuss, J. M., Whitaker, P. B., and Air, G. M. (1993). *Proteins* **15,** 121–132.

Oertle, M., Immergluck, K., Paterson, Y., and Bosshard, H. R. (1989). *Eur. J. Biochem.* **182,** 699–704.

Oldenburg, K. R., Loganathan, D., Goldstein, I. J., Schultz, P. G., and Gallop, M. A. (1992). *Proc. Natl. Acad. Sci. U.S.A.* **89,** 5393–5397.

Paterson, J. (1992). *Nature (London)* **356,** 456–457.

Paterson, Y., Englander, S. W., and Roder, H. (1990). *Science* **249,** 755–759.

Pelleguer, J. L., Westhof, E., and Van Regenmortel, M. H. V. (1991). *In* "Methods in Enzymology" (J. L. Langone, ed.), Vol. 203, pp. 176–201. Academic Press, San Diego.
Pesce, A. J., and Michael, J. G. (1992). *J. Immunol. Methods* **150,** 111–119.
Rini, J. M., Stanfield, R. L., Stura, E. A., Salinas, P. A., Profy, A. T., and Wilson, I. A. (1993). *Proc. Natl. Acad. Sci. U.S.A.* **90,** 6325–6329.
Rodda, S. J., Geysen, H. M., Mason, T. J., and Tribbick, G. (1986). *Protides Biol. Fluids* **34,** 91–93.
Saad, B., and Bosshard, H. R. (1990). *Eur. J. Biochem.* **187,** 425–430.
Saad, B., Corradin, G., and Bosshard, H. R. (1988). *Eur. J. Biochem.* **178,** 219–224.
Saberwal, G., and Nagaraj, R. (1993). *J. Biol. Chem.* **268,** 14081–14089.
Sachs, D. H., Schechter, A. N., Eastlake, A., and Anfinsen, C. B. (1972). *J. Immunol.* **109,** 1300–1310.
Savoca, R., Schwab, C., and Bosshard, H. R. (1991). *J. Immunol. Methods* **141,** 245–252.
Schnorrenberg, G., and Gerhardt, H. (1989). *Tetrahedron* **45,** 7759–7764.
Schwab, C. (1991). Vergleich der Immunantworten gegen die native und denaturierte Form eines Proteinantigens. Ph.D. Thesis, Universität Zürich, Zürich.
Schwab, C., and Bosshard, H. R. (1992). *J. Immunol. Methods* **147,** 125–134.
Schwab, C., Twardek, A., Lo, T. P., Brayer, G. D., and Bosshard, H. R. (1993). *Protein Sci.* **2,** 175–182.
Scibienski, R. J. (1973). *J. Immunol.* **111,** 114–120.
Scott, J. K., and Smith, G. P. (1990). *Science* **249,** 386–390.
Sela, M. (1969) *Science* **166,** 1365–1374.
Sela, M., Schechter, B., Schechter, I., and Borek, F. (1967). *Cold Spring Harbor Symp. Quant. Biol.* **32,** 537–545.
Sigurskjold, B. W., Altman, E., and Bundle, D. R. (1990). *Eur. J. Biochem.* **197,** 239–246.
Silver, M. L., Guo, H.-C., Strominger, J. L., and Wiley, D. C. (1992). *Nature (London)* **360,** 367–369.
Smith, A. D., and Wilson, J. E. (1986). *J. Immunol. Methods* **94,** 31–35.
Smith-Gill, S. J., Wilson, A. C., Potter, M., Prager, E. M., Feldmann, R. J., and Mainhart, C. R. (1982). *J. Immunol.* **128,** 314–322.
Soderquist, M. E., and Walton, A. G. (1980). *J. Colloid. Interface Sci.* **75,** 386–397.
Spangler, B. D. (1991). *J. Immunol.* **146,** 1591–1595.
Stanfield, R. L., Fieser, T. M., Lerner, R. A., and Wilson, I. A. (1990). *Science* **248,** 712–719.
Stellwagen, E., Rysavy, R., and Babul, G. (1972). *J. Biol. Chem.* **247,** 8074–8077.
Suckau, D., Köhl, J., Karwath, G., Schneider, K., Casaretto, M., Bitter-Suermann, D., and Przybylski, M. (1990). *Proc. Natl. Acad. Sci. U.S.A.* **87,** 9848–9852.
Suckau, D., Mak, M., and Przybylski, M. (1992). *Proc. Natl. Acad. Sci. U.S.A.* **89,** 5630–5634.
Torensma, R. (1993). *Immunol. Today* **14,** 370–371.
Underwood, P. A. (1985). *J. Immunol. Methods* **85,** 295–307.
Vajda, S., Kataoka, R., DeLisi, C., Margalit, H., Berzofsky, J. A., and Cornette, J. L. (1990). *Annu. Rev. Biophys. Biophys. Chem.* **19,** 69–82.
Van Regenmortel, M. H. V. (1986). *Trends Biochem. Sci.* **11,** 36–39.
Van Regenmortel, M. H. V. (1989). *Philos Trans. R. Soc. London B* **323,** 451–466.
Waltho, J. P., Feher, V. A., Merutka, G., Dyson, H. J., and Wright, P. E. (1993). *Biochemistry* **32,** 6337–6347.
Wells, J. A. (1991). *In* "Methods in Enzymology" (J. L. Langone, ed.), Vol. 202, pp. 390–410. Academic Press, San Diego.
White-Scharf, M. E., Pott, B. J., Smith, L. M., Sokolowski, K. A., Rusche, J. R., and Silver, S. (1993). *Virology* **192,** 197–206.
Wilson, I. A., and Stanfield, R. L. (1994). *Curr. Opinion Struct. Biol.* **4,** 857–867.
Wilson, I. A., Haft, D. H., Getzoff, E. D., Tainer, J. A., Lerner, R. A., and Brenner, S. (1985). *Proc. Natl. Acad. Sci. U.S.A.* **82,** 5255–5259.
Zinn-Justin, S., Roumestand, C., Drevet, P., Ménez, A., and Toma, F. (1993). *Biochemistry* **32,** 6884–6891.

12

Synthesis and Applications of Branched Peptides in Immunological Methods and Vaccines

James P. Tam
Vanderbilt University Medical Center
Department of Microbiology and Immunology
Nashville, Tennessee 37232-2363

Peptides: Synthesis, Structures, and Applications

I. Introduction

Branched peptides such as multiple antigen peptides (MAPs) are artificial proteins that hold promise in biochemical and biomedical applications. This chapter describes methods for preparing MAPs and applications such as synthetic vaccines, serodiagnostics, and peptide inhibitors. Furthermore, we describe a new design of MAPs containing lipidated built-in adjuvant which can be delivered by oral administration to elicit systemic and mucosal immunoglobulins as well as cytotoxic T-lymphocytes. To provide the precision and chemical unambiguity of this class of artificial proteins, we also describe orthogonal approaches by thiol and carbonyl chemistries to facilitate synthesis using unprotected peptide segments as building blocks and ligating them to the core matrix.

Proteins as a class of biopolymers have the unusual ability of displaying a diverse variety of folds and forms predetermined by the primary sequences and correlated to protein functions. We have also known for some time that many of the functional sites are located on the surfaces and that some are continuous sequences consisting of 4–21 residues. This holds for the antigenic sites of proteins, which are referred to as epitopes. Other sites include those for attachment, recognition, location, and cellular compartment translocations and are known as motifs. We have taken advantage of these known functional peptides, attaching them to a template to form branched peptides.

II. Types of Branched Peptides

All branched peptides share a similar design of branching from a core matrix or template that gives them different architectures and dendrimeric character (Tomalia *et al.*, 1990; Hodge, 1993). Several variations of branched peptides have been developed which differ only in the design of the core matrix to give a cascade, pennant, or radial type of branching arrangement (Fig. 1). We (Posnett *et al.*, 1988; Tam, 1988; Tam and Lu, 1989) have developed a cascade type of multiple antigenic peptide (MAP) whose core contains two or three levels of geometrically branched lysines (Fig. 1). MAPs have found ap-

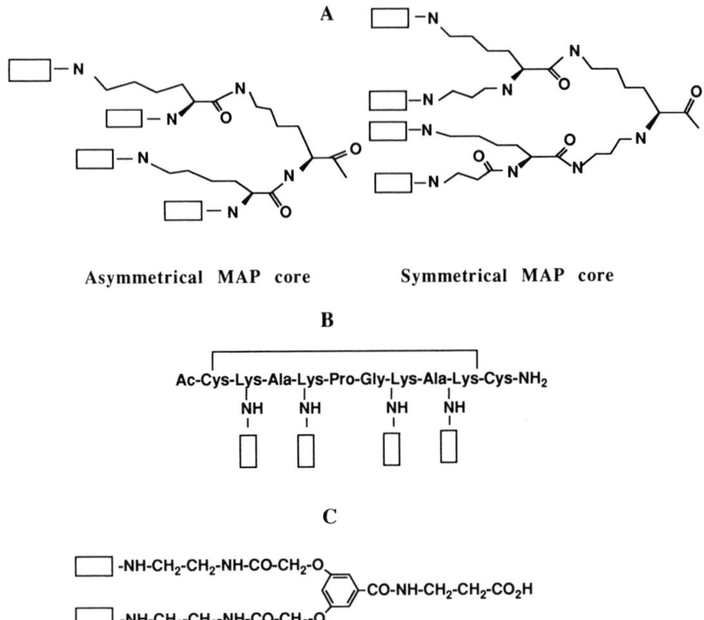

Figure 1 Design of core matrices or templates for different types of branching arrangements. (A) cascade type of multiple antigen peptides, (B) pennant type MAP such as template-assembled synthetic protein, and (C) radial. Open boxes represent antigens.

plications as immunogens (Tam, 1988; Tam and Lu, 1989), antigens for diagnostics (Tam and Zavala, 1989), intracellular delivery vehicles (Sheldon *et al.*, 1995), artificial enzymes (Hahn *et al.*, 1990), and inhibitors for cell attachment and entry (Sinnis *et al.*, 1994a,b; Ingham *et al.*, 1994). Others have also designed branched peptides with a pennant type of arrangement and emphasis on the topological aspects of template to study self-assembly of peptides and to form ordered structures. Mutter and Vuilleumier (1989) have designed a template consisting of a β-turn-forming peptide with lysyl side chains for peptide attachment for template assembled synthetic proteins (TASPs). Unson *et al.* (1984) and Sasaki and Kaiser (1989) used rigid organic templates to give a radial type of arrangement. Because of the similarity of these branched peptides to dendrimers (Tomalia *et al.*, 1990), we have also referred to them as peptide dendrimers in this paper.

This review focuses primarily on the design, preparation, and immunological applications of the cascade type MAPs. Because the preparative methods for MAPs are applicable to all branched peptides, and to our knowledge, have not been reviewed by others, we have reviewed this aspect in greater depth and placed strong emphasis on the orthogonal approach of ligating unprotected peptide segments to the core matrix or scaffolding.

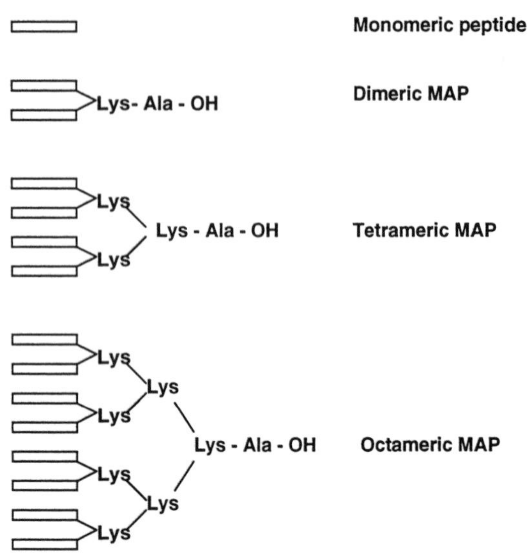

Figure 2 Schematic representation of MAP constructs.

III. Design and Physical Characteristics of Multiple Antigen Peptides

A key feature of the MAP system is the manyfold amplification of a peptide in a chemically defined manner. Unlike random polymerization, which usually leads to linear arrays of a wide range of polymers, the MAP system produces branched peptides in a controlled manner with an unambiguous structure. In the cascade type of MAP, this is achieved by utilizing a core matrix as a scaffolding consisting of several sequential levels of a trifunctional amino acid as a building unit (Tam, 1988). Lysine is most commonly used because it has two ends, the α- and ε-amino groups, available for the branching reactions. Other diamino acids, such as ornithine, have also been used with similar effects. However, when a Lys is used, the core matrix is asymmetrical, with a long arm consisting of the side chain and a short arm consisting of the α-amino group. In the case of an octameric lysyl core matrix containing three levels of branching, amino groups vary in distance from 7 to 18 carbon atoms from the first branched C^{α} atom. We have also designed a symmetrical core matrix consisting of Lys(β-Ala) as a building unit (Huang *et al.*, 1994). Branching occurs when a lysine or Lys(β-Ala) is coupled at the first level with one another to produce two reactive amino ends to give a bivalent MAP (Fig. 2). Further sequential propagation of lysines produces MAPs containing tetravalent or octavalent reactive amino ends, which are found to be sufficient for the purpose of eliciting high-titered antibody responses.

Table I Comparison of Multiple Antigen Peptides with Conventional Peptide–Protein Conjugates as Immunogens

Property	MAPs	Peptide–protein conjugates
Physical		
Structure	Branched	Branched
Composition	Defined	Ambiguous
Sites of branching	Known	Ambiguous
Amplification of antigens	Yes	Yes
Stoichiometric ratio of antigen to core matrix	Known	Variable
Aqueous solubility	High	Variable
Lyophilizable and stable	Yes	Unknown
Design and immunogenicity		
Need for conjugation	No	Yes
Flexibility to incorporate more than single epitope (e.g., T and B epitopes)	Yes	Maybe
Ability to incorporate built-in adjuvant	Yes	No
Probability of creating undesirable B epitopes	Low	High
Ability to be lipidated and of cytotoxic epitope to elicit CTL responses	Yes	Maybe
Need for processing	Maybe	Yes
Source of T-helper epitope	Not known	Yes
Persistent long-term memory response	Yes	Maybe

Multiple antigen peptides containing an octameric 15-mer antigenic peptide typically have a high-density cluster of peptides accounting for over 90% of the total molecular weight surrounding a small branching lysine core (Table I). Furthermore, MAPs are different from the conventional polylysyl conjugates because they contain oligomeric, branching peptide chains and are devoid of cationic lysine backbones. The backbone of MAP is made up of amide bonds, and most MAPs are remarkably stable in solution between pH 2 and 9. Thus, storage of MAPs should not constitute a significant problem; they can be stored or shipped as a lyophilized powder. These advantages are significant when compared with recombinant proteins or whole pathogen vaccine preparations that require a cold chain of storage to retain activity.

Because of the presence of the lysine core at the carboxyl end, which creates a polarity preference, the orientation of a peptide immunogen when attached to a core matrix is from a C to N when it is synthesized in a continuous sequence by the stepwise solid-phase method using Boc-benzyl-*tert*-butyloxycarbonyl chemistry or Fmoc-*tert*-butyl fluorenylmethoxycarbonyl chemistry, similar to that of a linear peptide. If the selected epitope is a carboxyl fragment of a protein, an indirect

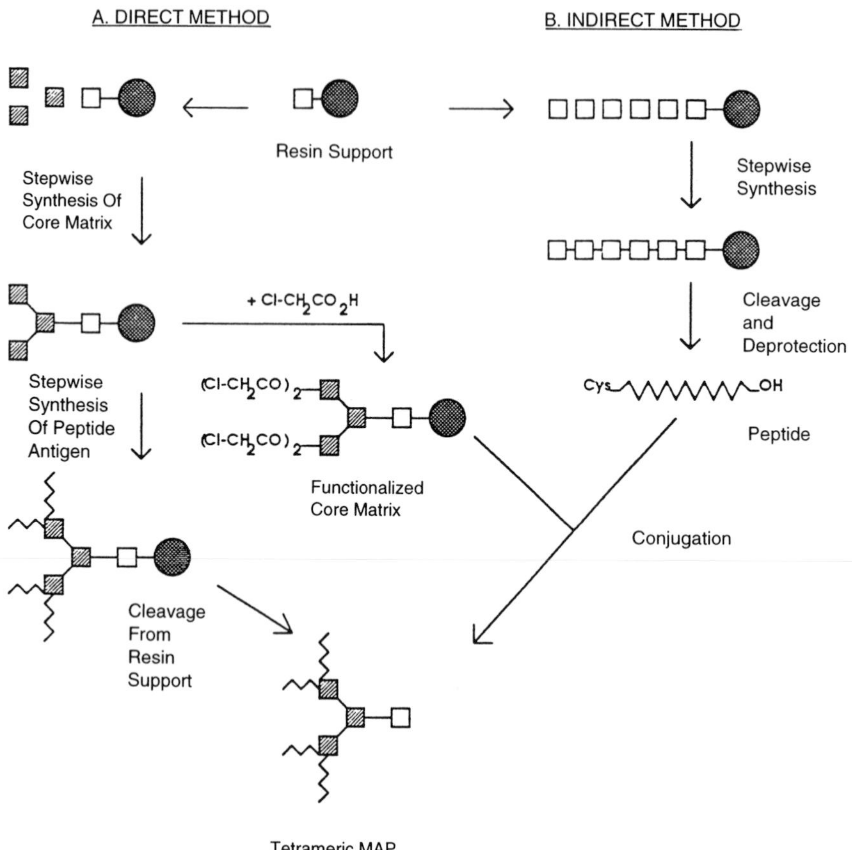

Figure 3 Direct and indirect synthesis of MAP constructs. Hatched boxes represent core matrix lysines, open boxes denote any amino acid, and zigzag lines represent peptide antigen.

modular method (Fig. 3) can be used instead of the direct stepwise approach to mimic the native protein conformation. The indirect approach overcomes the lack of flexibility in the orientation of a peptide and consists of the synthesis of a functionalized core matrix and unprotected peptide antigens separately, followed by chemoselective ligation of the two components (Lu *et al.*, 1991). Various aspects of chemoselective ligation are reviewed in Section VII.

IV. Multiple Antigen Peptide Immunogenicity

One of the popular uses of branched peptides is in immunology, because of simplicity of design and synthesis, versatility for investigating immunogenicity, reliability of generating site-specific antibodies, and generation of site-specific antibodies in the laboratory. Since the introduction of MAPs in 1988 by Tam, over 150 applications of MAPs have been documented in the literature (Table II), and it has become a popular method. Commercial companies specializing in peptide reagents now offer either reagents for synthesizing MAPs in laboratory or custom synthesis services. Before we review the application of MAPs in immunology, several points regarding the antigenic properties of the peptides are discussed.

A. Definition and Classification of Peptide Antigens

Common to all of the peptide-based approaches is the selection of a peptide sequence as an immunogen. Two criteria are considered in the peptide selection, immunogenicity and antigenicity. Immunogenicity provides the ability of the peptide to elicit high-titered antibody. Antigenicity provides the ability of the antibody to recognize the protein from which it is derived and is governed by the conformation or the shape of the peptide. In simple terms, one must choose a peptide sequence for the immune system to respond to and provide a correct shape so that the ensuing antibody will recognize the cognate protein. The portion or portions of the peptide or protein responsible to provide the "shape" of the antigen is operationally defined as a B-cell epitope while the portion responsible for inducing the antibody production is operationally defined as a T-cell epitope. Identification of T-cell epitopes can be accomplished by a systematic approach of synthesizing a large number of overlapping sequences that cover a stretch of protein, followed by sequential testing of each peptide for biologic activity. For example, T-helper peptides can induce proliferation, and T-cytotoxic peptides can induce CTL-directed lysis. The alternative approach to identifying T-cell epitopes is based on computer modeling programs with predictive algorithms, such as T-cell epitopes that have been selected on the basis of an amphipathic alpha helix.

B. Production of High-Titered Antibody Response

A linear peptide, ineffective in raising antibodies when administered alone in Freund's adjuvant (FA), can become immunogenic when given in a MAP format (Del Giudice *et al.*, 1990; Kamo *et al.*, 1992). Moreover, antibody titers induced by MAP peptides were found to be superior to those obtained by the corresponding linear peptide conjugated to a carrier protein. An 11-residue MAP construct of an internal sequence from the tyrosine kinase protein $p60^{5rc}$ elicited stronger primary and secondary responses than the same linear peptide conjugated to keyhole limpet hemocyanin (Tam, 1988). A plausible explanation for these observed results is that a MAP attains a polymeric and macromolecular structure when compared to a linear peptide, which in turn may be cleared from the body at a faster rate than the MAP.

Table II Applications of Multiple Antigen Peptides in Antibody Production and Vaccine Development

Antigen	Refs.
CS protein	
Plasmodium bergei	Romero *et al.* (1988); Migliorini (1990, 1993); Zavala and Chai (1990); Chai *et al.* (1992); Widmann *et al.* (1992); Valmori *et al.* (1994).
P. malariae	Del Guidice *et al.* (1990)
P. falciparum	Munesinghe *et al.* (1991); Pessi *et al.* (1991a,b); Valmori *et al.* (1992); Nardin and Nussenzweig (1993); Calvo-Calle *et al.* (1992, 1993); de Oliveira *et al.* (1994); Ahlborg (1995)
P. yoelii	Wang *et al.* (1995)
Surface antigen, hepatitis B	Tam and Lu (1989); Manivel *et al.* (1993)
Reductase, herpes simplex virus	Lankinen *et al.* (1989)
Trp protein, *Drosophila*	Wong *et al.* (1989)
UL47, UL8, Vmw63, and UL42	McLean *et al.* (1990); Parry *et al.* (1993); Sinclair *et al.* (1994); Marsden *et al.* (1994)
Lutropin, human	Troalen *et al.* (1990)
Ribosomal protein, plant	Szymkowski and Deering (1990)
Tandem repetitive motifs, *Leishmania*	Liew *et al.* (1990)
VP2 and VP7, bluetongue virus	Li and Yang (1990)
VP1 protein, foot-and-mouth disease virus	Francis *et al.* (1991); Lugovskoi *et al.* (1992); Brown (1992)
gp120, HIV-1	Wang *et al.* (1991); Defoort *et al.* (1992a,b); Nardelli *et al.* (1992a,b, 1994); Nardelli and Tam (1993); Levi *et al.* (1993); Kelker *et al.* (1994); De Santis *et al.* (1994); Huang *et al.* (1994); Fraisier *et al.* (1994); Vogel *et al.* (1994)
Fatty acid-binding protein, human	St. John *et al.* (1991)
Transforming growth factor α, human	Lu *et al.* (1991)
Alkyltransferase, human	Pegg *et al.* (1991)
P28 antigen, *Schistosoma mansoni*	Auriault *et al.* (1991); Khan *et al.* (1994); Pancre *et al.* (1994)
Cytochrome *P*-450IA, rat	Edwards *et al.* (1991)
SLT-1 protein, *Escherichia coli*	Boyd *et al.* (1991)
P30 antigen, *Toxoplasma gondii*	Darcy *et al.* (1992); Godard *et al.* (1994)
Mast cell protease-5′, mouse	McNeil *et al.* (1992)
ATPase, plant	Suzuki *et al.* (1992)
Calcium channel, rabbit	Malouf *et al.* (1992)
Sperm myoglobin, whale	McLean *et al.* (1992)
Sperm protein, human	Vanage *et al.* (1992, 1994)

(continues)

(continued)

Protein kinase p34^{cdc2}, eukaryote	Kamo *et al.* (1992)
Angiotensin II type-1 receptor, rat	Zelezna *et al.* (1992, 1994)
Nef, HIV-1	Estaquier *et al.* (1992, 1993)
Polymerase, influénza virus	Nieto *et al.* (1992)
Hemagglutinin, influenza virus	G. K. Toth *et al.* (1993); Naruse *et al.* (1994)
NSm protein, bunyamwera virus	Nakitare and Elliot (1993)
B19, parvovirus	Saikawa *et al.* (1993); Anderson *et al.* (1995)
Glucosyltransferase, *Streptococcus*	D. J. Smith *et al.* (1993, 1994)
Toxin T, *Bordetella pertussis*	Felici *et al.* (1993)
Toxin B, *Vibrio cholera*	Halimi and Rivaille (1993)
Membrane protein 1, *C. trachomatis*	Zhong *et al.* (1993)
Actin–fragmin kinase, plasmodia	Gettemans *et al.* (1993)
Seed protein, plant	Monsalve *et al.* (1993)
IGF binding protein, bovine	Arnold *et al.* (1993)
Glutamate receptor, rat	Molnar *et al.* (1993)
Glucose transporter, rat	Nagamatsu *et al.* (1993)
RNA-binding protein, rodent	Henderson *et al.* (1993)
Insulin receptor, human	Itoh *et al.* (1993)
Myotrophin, human	Sil *et al.* (1993)
Iron regulatory factor, human	Emery-Goodman *et al.* (1993); Gray *et al.* (1993)
Protein G., human	Raymond *et al.* (1993)
Tyrosine kinase, sponge	Schacke *et al.* (1994)
Core proteins, vaccinia virus	Vanslyke and Hruby (1994)
Membrane protein, *Neisseria meningitidis*	Christodoulides and Heckels (1994)
Cyclase, *Saccharomyces cerevisiae*	Shi *et al.* (1994)
Triose-phosphate isomerase, *S. mansoni*	Reynolds *et al.* (1994)
Chaperonin polypeptide, eukaryote	Lingappa *et al.* (1994)
Sperm protein, *Xenopus*	Bauer *et al.* (1994)
pP344 retinal cell, chicken	Iio *et al.* (1994)
Myosin phosphatase, chicken	Shimizu *et al.* (1994)
Prion protein, mouse	O'Rourke *et al.* (1994)
Amelogenin, mouse	Simmer *et al.* (1994)
Histones, mouse	Meziere *et al.* (1994)
GTPase, rat	Wilson *et al.* (1994)
Neurexins, rat	Perin (1994)
T-tubule, rabbit	Stout *et al.* (1994)
Membrane domain of band 3, human	Kang *et al.* (1992)
Ca^{2+} release channel, rabbit	Callaway *et al.* (1994)

(continues)

(continued)

Anion transport protein band 3, human	Crandall and Sherman (1994)
Guanylin, human	Kuhn *et al.* (1994); Cetin *et al.* (1994)
Profilin, human	Finkel *et al.* (1994)
Type V collagen, human	Moradi-Arneli *et al.* (1994)
Phosphatase 2A, human	Zolnierowicz (1994)
DR molecules, human	Demotz *et al.* (1994)
Elafin, human	Nara *et al.* (1994)
Gangliosides, human	Helling *et al.* (1994)
5α-Reductase 2, human	Eicheler *et al.* (1994)
Heme oxygenase-1, human	Kutty *et al.* (1994); M. A. Smith (1994)
gp46, HTLV-1	Baba *et al.* (1995)
Spliceosome protein, human	James *et al.* (1995)
Cyclin, human	Digweed *et al.* (1995)
Phospholipase A_2, human	Sa *et al.* (1995)
Kainate receptor, goldfish	Wo and Oewald (1995)

It should be pointed out that, in comparison to a carrier-conjugated peptide, the MAP macromolecule has the molecular weight of a small protein in which the antigen peptide represents more than 80% of the total structure. No immune response to the lysine residues has been detected (Posnett *et al.*, 1988; Del Giudice *et al.*, 1990).

C. Reactivity of Elicited Antisera against Cognate Proteins

Since the early experiments with MAP constructs we reported in conferences in 1987 (Tam, 1993), it was evident that immunization with the oligomeric MAP resulted in the production of antisera capable of recognizing the cognate sequence of the native protein (Tam, 1988). Six octameric MAP models were synthesized from sequences of unrelated proteins and injected in mice and rabbits. All MAPs elicited a strong antibody response against the immunizing peptide, and five of the generated antisera reacted with the native proteins. Among other reports, we mention the results obtained by Wolowczuk *et al.* (1991) with a MAP construct consisting of eight copies of a sequence from the Sm-28-GST antigen of the *Schistosoma mansoni*. Immunization with this construct elicited protein-specific antibodies which were able to mediate antibody-dependent cytotoxicity toward the parasite *in vitro*. Moreover, immunized rats were partially protected against challenge with infectious *Schistosoma mansoni* cercariae.

There are many comparative studies of MAPs with conventional peptide–conjugate systems. In some cases, there is a significant difference and a generally better quality of antisera in using MAPs versus peptide–protein conjugates (McLean *et al.*, 1991; Wang *et al.*, 1991; Estaquier *et al.*, Molinar *et al.*, 1993; Vogel *et al.*, 1994). The comparative study by McLean *et al.* (1991) with six different peptides appears to reflect on the general trend in the properties of antisera obtained by different presentations: sera obtained by immunization with MAPs of higher titers are obtained faster than those obtained with protein-conjugated peptides which in turn are higher than those obtained with resin-linked peptides. In several instances, immune sera generated by the administration of MAPs have been found to react with the cognate protein, whereas antisera generated by the corresponding linear peptide conjugated to a protein carrier had high antibody titers only against the immunizing peptide (Troalem *et al.*, 1990; Kamo *et al.*, 1992). The MAP format, furthermore, allowed the investigators to overcome the immunodominance of the linker and the carrier used in making the conjugated peptide, and to produce a monoclonal antibody directed against a highly conserved and very poorly immunogenic sequence (Kamo *et al.*, 1992). Comparison of MAPs with linear peptides has also been made (Francis *et al.*, 1991). In addition, the branched structure of MAPs may offer more resistance to proteolytic degradation and may be more suitable for oral administration than the monomeric peptide. Such a view is consistent with findings that oral immunization of mice with branched peptides such as MAPs elicited significantly higher IgA responses than those immunized with the monomer.

D. Correlating Multiple Antigen Peptide Structure and Immunogenicity

A frequently asked question concerns the optimal number of branches for MAPs. This issue was partially addressed in a comparative study on the major immunogenic epitope of foot-and-mouth disease virus (FMDV) serotype O, defined by amino acids 141–160 of the protein VP1. The peptide was found to be immunogenic in the absence of protein carrier when polymerized, thus indicating the presence, in the same sequence, of residues recognized by antibody-producing and T-helper cells (Francis *et al.*, 1987). The peptide was therefore used as a model to evaluate the effect of varying the number of the branching chains on MAP immunogenicity (Francis *et al.*, 1991). Guinea pigs were immunized with monomeric, dimeric, tetrameric, and octameric FMDV VP1 emulsified in incomplete Freund's adjuvant. Primary and secondary responses elicited by the tetrameric construct were similar to those obtained with the octamer, and they were significantly superior to that obtained with the dimer. The administration of the monomeric peptide did not generate an antibody response. The antisera were then tested against sets of peptides within the 141–160 linear sequence to map reactivity sites. It was found that the antibodies elicited by the octameric construct, in comparison to the others, were less specific for the N-terminal residues of the sequence, which are believed to be involved in neutralization of

Table III Promiscuous T-Helper Epitopes from Pathogens and Parasites

Source/protein	Sequence	Amino acids	Refs.
Malaria/circumsporozoite of *P. falciparum*	EKKIAKMEKASSVFNVVNS	380–398	Sinigaglia *et al.* (1988)
	IEQYLKKIKNSISTEWSPCS	331–350	Fern and Good (1992)
Clostridium tetani/tetanus toxin	QYIKANSKFIGITEL	830–844	Panina-Bordignon *et al.* (1989)
	FNNFTVSFWLRVPKVSASHLE	947–967	Panina-Bordignon *et al.* (1989)
	NSVDDALINSTKIYSYFPSV	580–599	Ho *et al.* (1990)
	PGINGKAIHLVNNESSE	916–932	Ho *et al.* (1990)
	QYIKANSKFIGITE	830–843	Demotz *et al.* (1989,a,b)
Influenza virus/hemagglutinin	SGPLKAEIAQRLE	17–29	Ceppellini *et al.* (1989)
	PKYVKQNTLKLAT	307–319	Roche and Cresswell (1990)
Measles virus/fusion protein	LSEIKGVIVHRLEGV	288–302	Partidos and Steward (1990)
Hepatitis B virus/HBsAg	FFLLTRILTIPQSLD	19–33	Celis *et al.* (1988)
Hepatitis A virus/VP1	NVPDPQVGITTMRDLKG	17–33	Ivanov *et al.* (1994)
	MSRIAAGDLESSVDDPRSEEDRR	276–298	Ivanov *et al.* (1994)
Papillomavirus/E7	DRAHYNI	48–54	Tindle *et al.* (1991)

the virus. It appears, therefore, that in the FMDV system the presentation of an antigenic peptide in a tetrameric structure is sufficient for an optimal response.

E. Enhanced Immunogenicity of Multiple Antigen Peptides Containing Chimeric B and T Epitopes

An important characteristic of the MAP system is the flexibility to incorporate multiple epitopes in the same construct. The generation of an optimal antibody response requires the recognition of the antigenic peptide fragments by T-helper cells that in turn promote the engagement of antigen-specific B cells. Selected samples of known potent T-helper epitopes from microbial sources are listed in Table III. Viral or microbial sources promote immunogenicity when covalently linked to the B-cell epitope. An effective synthetic immunogen should therefore contain sequences known to activate both T and B lymphocytes. The incorporation in the immunogen of T and B epitopes from the same pathogen might be particularly useful to enhance a specific immune response when infection occurs.

Chimeric MAP constructs have been extensively studied in the malaria synthetic vaccine model. An immunodominant B-cell epitope in the circumsporozoite (CS) protein of the rodent malaria parasite, *Plasmodium berghei*, consists of a repeat of a 16-amino acid sequence (residues 93–108). A T-helper epitope, recognized by several mouse strains, has been mapped in the same protein (residues 265–276). Ten MAP constructs were synthesized on the basis of the described T and B epitopes. The different constructs were designed to evaluate the relevance of number of copies, stoichiometry, and orientation of T and B sequences in a diepitope model (Tam *et al.*, 1990). They were tetrameric and octameric MAPs containing only the T and the B epitopes, or containing both epitopes linked in tandem. High antibody titers were elicited by the MAPs containing equimolar amounts of the T and B epitopes, whereas monoepitope MAPs and B–T monomers were not immunogenic in the A/J mice used in the experiments. In general, the results indicated that there was no real advantage in using an octameric over a tetrameric diepitope. In addition, the B–T configuration, in which the T epitope was next to the polylysine core, appeared to be the most efficient in eliciting antibody response, as 10-fold higher titers were produced compared to those achieved by immunization with irradiated sporozoites. Administration of B–T tetramer in alum, an adjuvant approved in humans, did raise a strong humoral response. However, the antibody titers were lower than those obtained by the same antigen in Freund's adjuvant emulsion (Chai *et al.*, 1992). Most importantly, the antibody response elicited by the MAP peptides was protective against intravenous challenge of immunized mice with 2000 *P. berghei* sporozoites. The degree of protection correlated with the antibody levels obtained by the immunization protocol.

Diepitope MAPs were synthesized in the design of a synthetic vaccine model for hepatitis B using a different chemical approach (Tam and Lu, 1989). A B-cell antigenic determinant (residues 139–147 of the S protein) and a T-cell antigen (residues 12–26 of the pre-S protein) were used in mono- and diepitope configurations. In the latter, the two peptides were connected in alternating forms on the lysine core, instead of being connected in tandem as in the malaria model. The B-cell determinant was immunogenic in rabbits only when presented in a MAP construct containing the pre-S peptide. Thus, these results confirmed that a diepitope MAP format can overcome the poor immunogenicity of a linear peptide.

The effectiveness of mono- and diepitope MAP configurations was further analyzed in a synthetic vaccine model for the human immunodeficiency virus type 1 (HIV-1) (Nardelli *et al.*, 1992a). In this paradigm, the monoepitope MAPs consisted of four copies of the major neutralizing determinant of HIV-1, which is localized in the third hypervariable region (V3) of the envelope protein gp120. Different MAPs were synthesized for three HIV-1 isolates (IIIB, RF, and MN). The diepitope MAPs contained a known T-helper sequence at the carboxyl terminus of the various B epitopes (residues 429–443 from gp120). The monoepitopes were found to elicit a species-dependent antibody response. Mice were nonresponders to the monoepitope MAP of RF and MN isolates and only poor responders to the IIIB sequences. The diepitopes were immunogenic in all species tested. The antisera generated in rabbits by the diepitopic MAPs had higher titers and neutralizing activity against homologous virus than those obtained by immunization with only the B epitope. Universal or promiscuous T-helper epitopes which are recognized by three or more strains of mice are particularly useful for the B–T chimeric construction (Table III). These epitopes are often derived from viral, bacterial, or parasitic origins and have been applied by others to enhance immunogenicity of B-cell–alone MAPs construct (McLean *et al.*, 1992; Munesinghe *et al.*, 1991; Marguerite *et al.*, 1992). Finally, Levi *et al.* (1993) have demonstrated that boosting with B and T MAP constructs induced a higher neutralization titer for HIV-1 and improved cellular responses than boosting with the recombinant gp160 protein.

F. Generation of Long-Term Antibody Responses

An example of induction of a long-lasting humoral response by MAPs can be found in the study of HIV-1 by Wang *et al.* (1991). Guinea pigs were immunized with a long linear peptide overlapping the V3 loop of gp120 (amino acids 297–329) conjugated with bovine serum albumin (BSA) or synthesized in an octameric format. Antibody titers and neutralizing activity of the antisera produced by the BSA-conjugated peptide peaked 4 months after the beginning of the immunization protocol. At the same time point, the antibody titers induced by the MAP were approximately in the same range. However, the neutralizing activity of the antisera induced by the MAP construct improved strikingly over time, reaching 30 to 50 times that of the BSA-conjugated peptide after 3 years. Moreover, with time the octamer-induced antisera started to show cross-neutralizing activity toward unrelated HIV-1 isolates. This study also illustrates the usefulness of a MAP that contains nearly pure immunogenic peptides without the burden of a protein carrier that may lead to immunological "switching" and in turn produce a heterologous response that does not give rise to long-term immunity.

G. Bypassing H-2 Restriction

Pessi *et al.* (1991a,b) have shown that the immunogenicity of an antigen not only can be improved but also significantly changed when presented in a MAP format. It is known that the immunogenicity of the repetitive sequence Asn-Ala-Asn-Pro in the malaria circumsporozoite protein of *Plasmodium flaciparum* is under the control of the murine major histocompatibility complex (MHC), H-2, since synthetic peptides encompassing this repeat induce responses only in *H-2b* mice. In contrast, administration of a MAP-Asn-Ala-Asn-Pro construct resulted in antibody production by five other *H-2* haplotypes. This lack of *H-2* restriction of an antigen presented as a MAP has not been reported for other circumsporozoite peptides (Munesinghe *et al.,* 1991). Nevertheless, the report by Pessi *et al.* (1991a) is consistent with our experience that short peptides in MAP formats elicit antibody response but are ineffective as monomeric forms. A plausible explanation may be that new helper epitopes are being generated in the MAP format, which are able to overcome the MHC restriction. This intriguing property of MAPs is currently under study in our laboratory.

V. Application of Multiple Antigen Peptides in Vaccines

Nearly all vaccines approved for human use consist of either killed or attenuated infectious agents. Despite the undisputed successes of such vaccines, development of vaccines by conventional methods is limited by several factors, namely, hazardous production, difficult storage that requires a cold chain, presence of contaminating materials, risk of infection in immunocompromised people, and unwanted side effects in the case of incomplete attenuation of the pathogen. To overcome the limitations, subunit vaccines consisting of either whole recombinant proteins or synthetic peptides are appealing alternatives because they are selective, chemically defined, and safe (Brown, 1990). Peptide-based vaccines have the additional advantage that relatively short stretches of amino acid sequence or peptide epitopes are capable of eliciting a protective immune response and can be selected in a protein, thus eliminating other epitopes potentially responsible for pathological effects due to nonspecific or undesirable stimulation (Adorini *et al.,* 1979; Robinson *et al.,* 1990; Robinson and Kehoe, 1992). Peptides can be synthesized to encompass antigenic determinants that are nonimmunogenic when the whole protein is used (Green *et al.,* 1982). Large quantities of chemically purified peptide vaccines can be prepared with automated methods. Furthermore, peptide-based immunogens are more likely to be resistant to denaturation, and they can be easily stored and transported without refrigeration.

A. New Design of Multiple Antigen Peptides for Vaccines

Our laboratory has focused on the development of MAPs that contain a built-in adjuvant and have the ability to form macromolecular complexes in a lipid matrix similar to the assembly of viral proteins on the coated surface of virions. To achieve this purpose, we developed a lipophilic membrane-anchoring moiety at the carboxyl terminus of MAPs to further their noncovalent amplification by a liposome or micellar form. Such noncovalent amplification of MAPs in a lipid matrix to many thousandfold has been called the macromolecular assemblage approach of MAPs (Huang *et al.*, 1994; Nardelli *et al.*, 1994). Furthermore, lipidated MAPs will have an added advantage for priming cytotoxic T lymphocytes (CTLs). The CTL component of the cellular immune responses mediates the activity of killer cells and macrophages, and CTLs provide effective protection against viral infection.

B. Multiple Antigen Peptides with Lipidated Built-In Adjuvant

Adjuvants are essential in any immunization protocol involving subunit vaccines in order to enhance a more directed immune response. Although Freund's adjuvant (FA), a water-in-oil emulsion of killed mycobacteria, elicits a strong antibody response in experimental animals in the laboratory, it often induces severe side effects including pyrogenicity and is not suitable for human use. The only group of substances currently licensed for vaccine use in humans are aluminum salts, which are less effective than FA in stimulating antibody response and produce only weak activation of the cellular immune response. This effect is probably due to the differential stimulatory activity of the two adjuvants: in mice FA stimulates preferentially the Th2 type of helper T cells, whereas alum induces the Th1 type. In searches for a compound retaining the stimulant activity of FA but devoid of its toxicity, lipopeptides were isolated from the outer cell membrane of *Escherichia coli* and synthetic analogs prepared (Wiesmuller *et al.*, 1983). The tripalmitoyl-*S*-glyceryl cysteine (P3C) was found to be particularly promising in potentiating the antibody response (Reitermann *et al.*, 1989) and was capable of inducing a strong cellular response *in vivo* when coupled to CTL epitopes (Deres *et al.*, 1989). Because of these properties, we have covalently linked P3C to MAPs to evaluate the combination in generating a completely synthetic peptide-based vaccine prototype without the need for carrier and adjuvant (Defoort *et al.*, 1992a,b).

We have also designed another type of adjuvant containing long alkyl fatty acid side chains (Huang *et al.*, 1994) such as palmitic acid conjugated to the side chains of Lys and Ser [Lys(Pal) and Ser(Pal)]. We found that the backbone of alternating D- and L-amino acids for the palmitoyl side chains is in a near parallel orientation for insertion or attachment to the lipid bilayer (Huang *et al.*, 1994). Dipeptides or tripeptides with fatty acid side chains such as Lys(Pal)-DLys(Pal)-Lys(Pal) mimic some of the immunological and adjuvant properties of P3C, are simple to prepare, and can be incorporated directly into the synthetic solid-phase scheme.

The value of lipidated peptide in eliciting cellular responses has been found by others. Pancre *et al.* (1994) found that immunization with a lipidated peptide derived from *Schistasoma mansoni* glutathione *S*-transerase provided a durable protective response in mice due to the induction of Th cells and CTLs.

To illustrate the utility of a MAP with a built-in adjuvant such as P3C, we used a model containing a sequence from the gp120 envelope protein of HIV-1,

IIIB isolate (amino acids 308–331), which overlaps a neutralizing B-cell epitope, a T-helper epitope and residues recognized by murine CTL of the H-2d phenotype (Takahashi *et al.*, 1988; Palker *et al.*, 1988, 1989; Rusche *et al.*, 1988). The model compound was synthesized in a tetrabranching MAP configuration (called B1M) and was coupled to the P3C moiety through a Ser-Ser spacer (Fig. 4). Mice and guinea pigs were injected with the immunogen, B1M–P3C, free or further amplified by the presentation on liposomes. Antibody titers of sera collected from immunized animals were lower than those obtained in previous experiments following immunization with B1M in FA; however, neutralization titers of the guinea pig antisera generated by the administration of B1M in FA or B1M-P3C were similar (Nardelli *et al.*, 1992a; Defoort *et al.*, 1992a,b).

The B1M–P3C construct induced a strong T-cell response, as measured by interleukin 2 (IL-2) production and cytolytic activity of splenocytes from the immunized mice. In particular, the cytotoxic response was induced by the lipidated MAP after just one immunization and was superior to the response induced by a full cycle of immunizations (four injections) of B1M in FA. We have demonstrated that the response was mediated by $CD8^+$ T lymphocytes and was MHC class I restricted (Nardelli and Tam, 1993). The induced lytic activity was specific. P815 cells expressing the HIV-1 glycoprotein gp160, following infection with a recombinant vaccinia virus or pulsed with the relevant peptide encompassing amino acids 308–331, were efficiently killed by the syngeneic effector cells. No cytotoxic activity was generated against P815 cells infected with wild-type vaccinia virus or pulsed with peptides encompassing other sequences of gp120. The reactivity was long-lasting and was still detectable 7 months after a single antigen injection (Table IV). Cytolytic activity against target cells presenting sequences from unrelated HIV-1 isolates was achieved by priming with a mixture of HIV-1 peptides (Table V). This result further illustrates the versatility of the macromolecular assemblage approach, where a mixture of lipidated MAPs can be used to overcome the deficiencies associated with a single peptide antigen.

The distinct advantages of lipidated MAPs in evoking a full range of responses without any extraneous adjuvant is also demonstrated with another MAP containing a sequence from the V3 loop and Lys(Pal)-DLys(Pal). The lipidated MAP was used to immunize mice and guinea pigs without any adjuvant. This form of lipidated MAPs also elicited strong antibody titers and lasting CTL responses.

Lipidated MAPs such as B1M–P3C were found to induce systemic antibody and cellular responses irrespective of the routes used for immunization (Nardelli *et al.*, 1992b, 1994). Intraperitoneal, intravenous, subcutaneous, and intragastric immunizations were able to generate cytotoxic activity in mouse spleens and immunoglobulin G (IgG) production in the sera. Moreover, after oral immunization specific IgA were detected in the mouse saliva. The ability of MAPs containing P3C to induce mucosal antibody response via oral administration adds a new dimension to applications of the MAP constructs, and it may be particularly useful in preventing transmission of pathogens, such as HIV, through mucosal surfaces.

```
Lys \
      Lys - Ser - Ser - NH - CH - CO - Ala - OH
Lys /                          |
                               CH2
                               |
                               CH2
                               |
                               CH2
                               |
                               CH2
                               |
                               NH - CO - CH - NH - CO - Palmitoyl
                                         |
                                         CH2
                                         |
                                         S
                                         |
                                         CH2
                                         |
                                         CH - O - CO - Palmitoyl
                                         |
                                         CH2 - O - CO - Palmitoyl
```

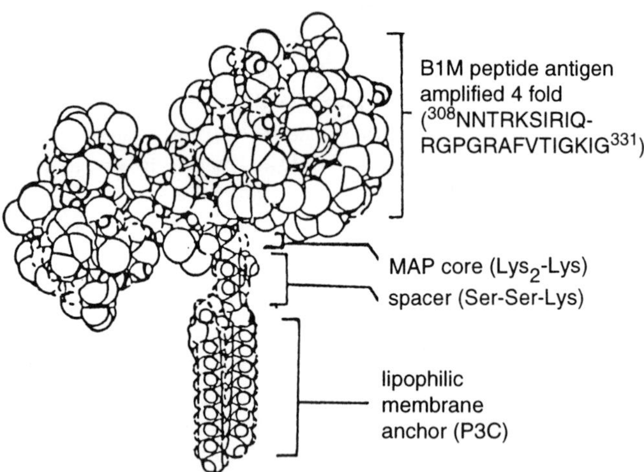

Figure 4 Schematic representation and computer-simulated model of B1M–P3C.

C. Macromolecular Assemblage as Mimicry of a Whole Organism

An impetus for the development of lipidated MAPs is the implementation of the macromolecular assemblage approach for vaccine design (Defoort *et al.*, 1992a). The idea is to present several similar or different closely packed peptide antigens as MAPs anchored on lipid vesicles that mimic the surface proteins of a whole organism. This approach would also allow a noncovalent mixture of lipi-

Table IV Induction of Peptide-Specific Cytotoxic T-Lymphocyte Memory by B2M–P3C Intraperitoneal or Intravenous Injection[a]

	Lysis (%) after priming period of			
Conditions	3 days	7 days	30 days	200 days
Intraperitoneal immunization				
20:1 E:T	56 (8)	94 (6)	95 (2)	58 (2)
10:1 E:T	41 (2)	92 (5)	81 (1)	45 (0)
5:1 E:T	27 (1)	85 (3)	62 (1)	30 (0)
Intravenous immunization				
20:1 E:T	94 (9)	92 (4)	75 (3)	ND[b]
10:1 E:T	86 (6)	85 (2)	52 (2)	
5:1 E:T	71 (3)	75 (1)	33 (2)	

[a]BALB/c mice were injected once with 100 μg B2M-P3C i.p. or i.v. Spleen cells were restimulated *in vitro* with B2M peptide (0.4 μ*M*) at different times after the priming. The CTL activity was determined on P815 target cells preincubated with B2M peptide using the effector to target cell ratios (E:T) indicated. In parenthesis are reported the values obtained with untreated P815 cells. From Nardelli and Tam (1993).
[b]ND, Not determined.

dated B- and T-cell epitopes assembled on a lipid matrix to enhance immunogenicity. At the same time, the combination of adjuvant effects of liposome and the built-in lipid anchor may replace the need for an extraneous adjuvant such as FA, which is toxic and unacceptable in humans. An advantage of lipidated MAPs is their efficiency in becoming entrapped in liposomes; nearly 80% of the lipidated MAPs were incorporated into liposomes compared to 2–5% of MAPs without the lipid anchor.

Table V Cytotoxic T-Lymphocyte Activity Induced by Priming with Single Peptide or Mixture[a]

		% Specific lysis of peptide-pulsed targets (E:T)			
		P815 + IIIB peptide		P815 + MN peptide	
In vivo priming	HIV-1 isolates	20:1	10:1	20:1	10:1
B2M–P3C	IIIB	84	60	30	18
B8M–P3C	MN	10	5	62	38
B2M–P3C + B8M–P3C	IIIB + MN	90	84	57	38

[a]BALB/c mice were immunized once i.p. with B2M–P3C (20 μg) and B8M–P3C (80 μg) alone or B2M–P3C plus B8M–P3C (20 + 80 μg). After 5 days of culture, the cytotoxic activity of IIIB- and MN-specific CTLs was tested against ^{51}Cr-labeled P815 targets pulsed with either 1 m*M* B2M or 4 μ*M* B8M at the effector:target cell ratios (E:T) indicated. From Nardelli and Tam (1993).

D. Oral Immunization Using Lipidated Multiple Antigen Peptides to Elicit Mucosal Immunity

In addition to demonstrating that parenteral immunization with a lipidated MAP (B1M–P3C) produces neutralizing antibody and cellular responses in mice and guinea pigs, we further showed the feasibility of using a lipidated MAP construct as an oral immunogen. Intragastric administration of B1M–P3C stimulated a secretory mucosal IgA response and systemic plasma IgG production. It also induced cell-mediated immunity as shown by lymphokine production and generation of a specific cytotoxic response in mice (Defoort *et al.*, 1992a,b). Moreover, intragastric delivery of B1M–P3C generated systemic T-lymphocyte stimulation and specific cytotoxic activity. The CTL response was eliminated by treatment with CD8-specific antibody (Defoort *et al.*, 1992a). The detection of cytotoxic activity in the spleen indicates that priming of gut-associated lymphoid tissues by lipopeptide feeding results in the stimulation of cellular systemic immunity. An improved method for B1M–P3C delivery to the IgA inductive sites will likely enhance the effectiveness of B1M–P3C.

VI. Other Applications

Multimerization of peptides by branching as shown in MAPs has found applications in areas other than as immunogens and vaccines (Table VI). These applications may be grouped into five areas: immunoassays and serodiagnosis, epitope mapping, inhibitors, artificial proteins, and various biochemical studies and purification methods. Several of these are elaborated in this section.

A. Immunoassays, Epitope Mapping, and Serodiagnosis

As a group, short synthetic peptides are usually ineffective antigens for solid-phase immunoassays owing to their poor ability to attach to solid surfaces. This is particularly true for short linear peptides lacking hydrophobic side chains (Tam and Zavala, 1989). Furthermore, synthetic peptides sometimes lose their antigenicity, presumably because the essential antigenic side chains are not exposed on binding to the solid surface. The multimeric nature of MAP constructs has been found to overcome these deficiencies and provide consistently reproducible results in increased surface-binding properties and sensitivity of detection (Tam and Zavala, 1989; Marguerite *et al.*, 1992; Briand *et al.*, 1992).

Comparative studies of MAPs and linear peptides have shown that MAPs display improved binding to plastic surfaces relative to the corresponding monomers. It was found that the branch peptide MAP constructs were able to bind a significantly greater amount of antibody than the same concentration of monomeric peptides. Using four MAP models, the sensitivity of detection was related to the number of branching chains. The dimer was the least reactive. The octameric MAP was the optimal configuration, showing a slightly better reactivity than the hexadecameric MAP (Tam and Zavala, 1989).

Table VI Applications of Multiple Antigen Peptides in Diagnosis and Biochemical Uses

Applications	Refs.
Immunoassays and serodiagnosis	
Malaria	Tam and Zavala (1989); Habluetzel *et al.* (1991)
Cirrhosis	Briand *et al.* (1992)
HIV-1	Marsden *et al.* (1992); Robertson *et al.* (1992); Estaquier *et al.* (1993); Vogel *et al.* (1994)
Schistosoma mansoni	Marguerite *et al.* (1992)
Systemic lupus erythematosus	Sabbatini *et al.* (1993a); Caponi *et al.* (1995)
Epstein-Barr virus	Marchini *et al.* (1994)
Epitope mapping and ligand binding	
Bluetongue virus	Yang *et al.* (1992)
Systemic lupus erythematosus	Sabbatini *et al.* (1993b)
Hepatitis C virus	Simmonds *et al.* (1993)
Inhibitors	
Macroautophagia and proteolysis	Miotto *et al.* (1994); Mortimore *et al.* (1994)
Tumor growth and metastasis	Nomizu *et al.* (1993); Kim *et al.* (1994)
Enzyme inhibitors	Fassina and Cassani (1993)
HIV-1 fusion and infection	Yahi *et al.* (1994a,b, 1995); Weeks *et al.* (1994)
Interleukin 6	Wallace *et al.* (1994)
Sporozoite, malaria	Sinnis *et al.* (1994a,b)
Fibronectin	Ingham *et al.* (1994)
Artificial proteins	
Minicollagen	Fields *et al.* (1993)
Synthetic enzyme	Hahn *et al.* (1990)
Biochemical studies	
Affinity purification of antibodies	Butz *et al.* (1994)
Presentation of T-cell epitopes	Grillot *et al.* (1993)
Affinity purifications	Fassina (1992); Fassina *et al.* (1992a,b); Yao *et al.* (1994)
Intracellular delivery	Sheldon *et al.* (1995)

The multimeric arrangement of MAPs may also improve the early detection of antibodies of low affinity, such as those of IgM isotype, during the early phase of infections. This ability may be particularly important in the screening process for contaminations in blood-derived products. By increasing the avidity and coating efficiency of linear peptides, MAPs allow detection of very low concentrations of antibodies in sera (Habluetzel *et al.*, 1991; Marsden *et al.*, 1992). In a dramatic example, a MAP shows a sensitivity increase of over 10^8-fold when compared to a peptide antigen in an enzyme-linked immunosorbent assay

(ELISA). The combined results indicate that the use of MAPs may be the method of choice to produce antigens for solid-phase immunoassays and is a promising tool for serodiagnosis in naturally immunized or infected individuals.

B. Inhibitors

A promising application of branched peptides using the MAP format has been found to be the general design of inhibitors. For example, branched peptides with clustered positive charges can lead to stronger binding than their monomer by allowing multiple points of contact. Thus, a weakly binding peptide bearing cationic charges has been found to increase binding. Clustering could be achieved by adsorption on a surface or by coupling to a carrier protein or Sepharose bead. However, clustering by using multermization of a branched peptide such as those developed by MAP is more controllable and provides an unambiguous structure that can be refined by analogs. The design of inhibitors includes those that may be useful for inhibiting entry of malaria to hepatocytes or HIV to $CD4^-$ epithelial cells.

Ingham *et al.* (1994) have used a branched tetrameric peptide with the sequence of NVSPPRRARVTDATETTITISW derived from the amino terminus of the type III module in the hep-2 domain of fibronectin to study its interaction with heparin. This tetrameric peptide causes a nearly hyperbolic increase in aniostropy of fibronectin–heparin with an apparent dissociation constant 10-fold lower than that of the monomer or of the protein domain from which the peptide is derived. Because heparin sulfate proteoglycans are used by malaria sporozoites to gain entry to hepatocytes, a conserved, but cationic sequence called region II-plus found on the surface protein of malaria, circumsporozoite (CS) protein, has been shown to require aggregation for the binding to the receptor. Sinnis *et al.* (1994a) have made use of this observation and prepared a branched region II-plus tetramer and found that it inhibits both CS protein binding to hepatocytes and CS clearance in mice.

The observation of increased binding of branched peptides to proteins or cell surfaces compared to that of the native protein has been exploited for the application as inhibitors. Nomizu *et al.* (1993) prepared a branched 16-mer of MAP containing a pentapeptide YIGSR, an active peptide derived from the laminin B1 chain, to enhance its activity in inhibiting tumor growth and metastasis. The inhibitory activity increases with increased branching; i.e., 16>8>4. Furthermore, the mechanism of this inhibition by the multimeric branched YIGSR has been elucidated by Kim *et al.* (1994) and has been shown to involve apoptosis of the tumor cells. Apoptosis, the programmed cell death, is a regulated self-destruction in normal cell growth. It will be interesting to determine how clustering of this simple peptide leads to a cascade of signals of gene transcription and protein synthesis leading to cell death.

VII. Preparation of Multiple Antigen Peptides

A. Stepwise Method for Preparing Multiple Antigen Peptides and Peptide Dendrimers

A common approach to the preparation of peptide dendrimers such as MAPs is by stepwise solid-phase synthesis, starting with the C-terminus core matrix using a diprotected Lys such as Boc-Lys(Boc) in Boc chemistry or Fmoc-Lys(Fmoc) in Fmoc chemistry to reach the desired branching level (Tam, 1988; Tam and Lu, 1989). The selected peptide antigen is then added stepwise to the resin-bound lysyl core matrix to create MAPs. This stepwise method of preparation can produce all antigens attached to the core matrix in a single operation with C → N orientation. Chimeric epitopes can be produced this way by tandem synthesis of B + T or T + B epitopes. Alternatively, T and B epitopes can be synthesized on different arms of the corer matrix using a core bearing two different amine-protecting groups such as Boc-Lys(Fmoc) or Boc-Lys(Npys) in Boc chemistry or Fmoc-Lys(Dde) or Fmoc-Lys(Npys) in Fmoc chemistry [Dde, 1-(4,4-dimethyl-2,6-dioxocyclohexyllidene)ethyl; and Npys, 3-nitro-2-sulfenyl pyridine] (Ahlborg, 1995). The Dde protecting group was developed by Bycroft and co-workers (1993) and is removed by 2% hydrazine in the presence of Fmoc groups. This is a significant addition to the base-driven Fmoc chemistry. Fmoc-Lys(Dde) and Dde-Lys(Fmoc) are commercially available and have become increasingly popular for the selective incorporation of two different peptides or functional groups in branched peptide chemistry. Npys is versatile because it can be removed under neutral conditions by a trivalent phosphine or a thiol. With these combinations of protecting groups, the peptide can be elongated differentially at either arm to obtain the diepitope MAP.

Multiple antigen peptides are high molecular weight macromolecules, often exceeding 15,000 Da, making their synthesis by stepwise solid-phase methods (Merrifield, 1963) and subsequent purification to high homogeneity challenging (Mutter *et al.,* 1992). Convergent or modular synthesis using protected peptide segments (Kaiser *et al.,* 1989) is a logical alternative but is limited by solubility of the hydrophobic protected segments and by low efficiency of coupling. To overcome these limitations to the modular approach, we (Liu and Tam, 1994) and others (Gaertner *et al.,* 1992; Schnolzer and Kent, 1992) have developed chemoselective, which we refer to as orthogonal, approaches to preparation of polypeptides by ligating unprotected peptide segments to the core matrix. Furthermore, the resulting peptide dendrimers do not contain bulky protecting groups and do not require harsh deprotecting treatments, which renders these aqueous soluble dendrimers amenable to the application of conventional protein purification methods. Furthermore, the modular approach has the flexibility of not only incorporating different types of epitopes but also of choosing the orientation of either C → N or N → C.

B. Modular, Orthogonal Approach Using Unprotected Peptides

At present, there are three general methods (Table VII) suitable for the preparation of peptide dendrimers using unprotected peptide segments. The first centers on thiol chemistry and includes thioalkylation, thiol addition, and thiol–disulfide exchange. The second is based on carbonyl chemistry and involves an addition–elimination reaction between a weak base and an aldehyde. The third uses enzymes in reverse proteolysis (Gaertner *et al.*, 1992). However, the application of reverse proteolysis in the synthesis of peptide dendrimers has not been fully developed and is not discussed further.

In both the thiol and carbonyl chemistries, a reactive pair consisting of a nucleophile and an electrophile is usually chosen and placed on the purified synthetic peptide monomer and the core matrix, respectively. These reactive groups are generally added to the peptide or core matrix during solid-phase synthesis. A common principle is the selection of a weak base as the nucleophile (Tables VIII and IX). Weak bases are useful because they have pK_a values significantly lower than those of the α- or ε-amines of lysine, and their reactivity can be distinguished from the strong amine bases below pH 7. Applicable weak bases include alkyl thiol, acyl thiol, 1,2-aminothiol (N-terminal cysteine), 1,2-aminoethanol (threonine), hydroxylamine, acylhydrazine, and arylhydrazine compounds. The other reactive component is usually an activated electrophile such as haloacetyl, activated unsymmetrical disulfide, maleimide, or aldehyde compounds. The chemo-

Table VII Orthogonal Methods for Ligating Unprotected Peptides to Form Branched or Linear Peptides

Methods	Reaction	Remarks
Thiol chemistry		
Thioalkylation	$R^1–SH + X–CH_2–CO–R^2 \rightarrow R^1–S–CH_2–CO–R^2$	X = Cl or Br
Thiol–disulfide exchange	$R^1–SH + R'–S–S–R^2 \rightarrow R^1–S–S–R^2$	R' = aromatic
Weak base–aldehyde	a) $R^1–C(=O)–H + NH_2–X–R^2 \rightarrow R^1–CH=N–X–R^2$	X = O or N
	b) $R^1–C(=O)–H + H_2N(HX)–R^2 \rightarrow R^1–\text{cyclic}(X–N)–R_2$	X = O or S
Reverse proteolysis	$R^1–CO–OH + NH_2NH–CO–R^2 \rightarrow R^1\text{-CO-NHNH-}R^2$	R_2 = peptide
	$R^1–CO–NHNH–CO–NHNH_2 + H–C(=O)–R^2 \rightarrow$	$R^1 = NHNH_2$ and R^3 = peptide
	$R^1–CO–NHNH–CO–NHN=CH–R^3$	

Table VIII Examples of Orthogonal Ligation by Thiol Chemistry via Thioalkylation and Thiol–Disulfide Exchange

Thiol nucleophile	Electrophile	Reaction pH	Product	Remarks
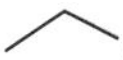 ⁄⁀SH	X–CH_2–C(=O)–	6–8	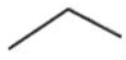 ⁄⁀S–CH_2–C(=O)–	X = Cl or Br
–C(=O)–SH	Br–CH_2–C(=O)–	4–5	–C(=O)–S–CH_2–C(=O)–	
 ⁄⁀SH	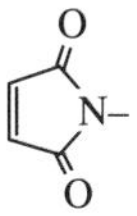 maleimide (N–)	7	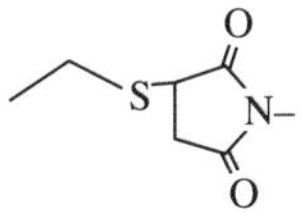 ⁄⁀S–succinimide (N–)	
 ⁄⁀SH	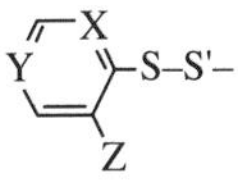 pyridyl(X, Y, Z)–S–S′–	8	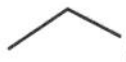 ⁄⁀S–S′–	X = N, Y = C, and Z = NO_2 or X = C, Y = N, and Z = H

selectivity is achieved when the mutually reactive groups are brought together under aqueous conditions with the weak base as the sole nucleophile to react with the electrophile in the presence of other functional groups on the peptides. Most of these methods produce nonpeptide bonds, but there are new amide-bond forming methods being developed that are applicable to the synthesis of peptide dendrimers (Liu and Tam, 1994; Kemp and Carey, 1991; Dawson *et al.,* 1994).

C. Thiol Chemistry

Thiol chemistry exploits the reactivity of sulfhydryls in the alkylation with α-halocarbonyls, addition to conjugated olefins, and sulfur–sulfur exchange with disulfides. In practice, only the alkylation and exchange reactions have been used for the formation of peptide dendrimers. However, all three reactions should be equally applicable.

1. Thioalkylation

Thioalkylation (Table VIII) is popular in protein chemistry as a means to attach ligands, peptides, or reporter groups (Means and Feeney, 1990) and as cross-linking reagents because of the easy access of thiol groups in proteins (Wilbur, 1992). For this reason, thioalkylation is also a convenient method for the synthesis of peptide dendrimers because the reactive thiol and haloacetyl (also referred to as α-halocarbonyl) groups can be added to the synthetic peptides during stepwise solid-phase synthesis. Both alkyl (RCH_2SH) and acyl (RCOSH) types of

Table IX Examples of Orthogonal Ligation by Weak Base–Aldehyde Chemistry

Weak base	Reaction pH	Product	Remarks
HX–CH(NH$_2$)–C(=O)– (N-terminal Cys/Ser-type 1,2-aminothiol/diamine)	3–5	5-membered ring (X, N–H) –C(=O)–	X = S or N
NH_2–OCH_2–C(=O)–	5	–CH=N–OCH_2–C(=O)–	
NH_2–NH–C(=O)–	5	–CH=N–NH–C(=O)–	
NH_2–NH–C_6H_4–C(=O)–	5	–CH_2–N=N–C_6H_4–C(=O)–	

thiols have been used. Alkyl thiols, generally derived from cysteine, could be placed at any position (Table X), whereas acyl thiol in the form of thiocarboxylic acid is limited at present to the C terminus. The haloacetyl moiety, whose C–X bond is activated by the acetyl group, can be attached to the N terminus or the side chain of lysine positioned anywhere in the sequence (Linder and Robey, 1987). It is stable to HF cleavage conditions in the absence of thiol scavengers. Bromoacetyl appears to be the reagent of choice because it is more reactive than the corresponding chloro analog, particularly in aqueous media, whereas the iodo analog is seldom used because of instability to cleavage conditions in solid-phase synthesis. Synthetic peptides with N^{α}-bromoacetyl and C-terminal cysteine have been polymerized at pH 7–8 to give "peptomers" connected by thioether bonds as shown by Robey and Fields (1989). The peptomers may be potentially useful as immunogens and vaccines; however, side reaction caused by intramolecular cyclization has been significant in some amino acid sequences (Robey and Fields, 1989).

Application of thioalkylation on peptide dendrimers was first demonstrated by Lu *et al.* (1991) and subsequently by Defoort *et al.* (1992a). In both cases, chloroacetyl groups were incorporated on the lysine core matrix and coupled to purified, synthetic N-terminal cysteinyl peptides to yield MAPs with unambiguous structures. MAPs with masses exceeding 10,000 Da could be obtained in high purity as determined by mass spectrometric analysis. The reverse placement with thiol on the core matrix could be achieved by using the *S*-acetyl group attached to the lysyl core matrix (Drijfhout and Bloemhoff, 1991) and haloacetyl groups on the peptide.

Table X Sites of Attachment or Transformation of Reactive Groups for Orthogonal Ligation of Unprotected Synthetic Peptides for Peptide Dendrimers

Reactive group	Site of attachment or transformation			Remarks
	NH_2	CO_2H	Side chain	
Alkyl SH	Cys	Cys	Cys	
$HS-CH_2-C(=O)-$	+	–	Lys	
$HS-C(=O)-$	–	+	Lys	
$X-CH_2-C(=O)-$	+		Lys	X = Br or Cl
R–S–S–	Cys	Cys	Cys	R = aromatic
$NH_2-O-CH_2-C(=O)-$	+	+	Lys	
$NH_2NH-C(=O)-$	+[a]	+	Lys	
$NH_2NH-C_6H_4-C(=O)-$	+	–	Lys	
HS–, H_2N– (Cys-type structure)	Cys	–	Lys	
HO–, CH_3, H_2N– (Thr-type structure)	Thr	–	Lys	
$H-C(=O)-$	Ser, Thr, Cys[b]	–	Lys, Cys[b]	
$(CH_3O)_2-CH-CH_2-$	+	+[c]	Lys	

[a] Shao and Tam (1995).

[b] Clamp and Hough (1965).

[c] Liu and Tam (1994).

The rate of thioalkylation to give thioethers is pH dependent and increases as the pH becomes more basic. Because side reactions, such as oxidation of thiol to disulfide (which consumes much of the starting materials) as well as hydrolysis of haloacetyl group, occur significantly at more alkaline pH, thioalkylation is commonly performed at pH 7.5–8. To minimize disulfide formation, Defoort *et al.* (1992b) utilize phosphine as a reducing agent, with significant success. We have found that S–S oxidation can be minimized by using a small amount of dithiothreitol prior to the reaction (Lu *et al.*, 1991) and including EDTA during the reaction (Shao and Tam, 1995; Spetzler and Tam, 1994).

Peptides containing thiocarboxylic acid are versatile intermediates because of the dual roles played by the acyl thiol as both electrophile and nucleophile. Blake (1981) and Yamashiro and Li (1988) have described the preparation of thiocarboxylic acid peptide derivatives by solid-phase methods and their application in a convergent scheme known as "aqueous strategy" in which partially protected peptide segments are used. Peptide thiocarboxylic acids have been successfully used by Schnolzer and Kent (1992) for thioalkylation in backbone engineering, and they were subsequently applied by Dawson and Kent (1993) in a template-assembled synthetic protein (TASP) molecule. The reaction is performed around pH 5 and is often assisted by guanidine as denaturant. Unlike the thioalkylation with alkyl thiols, which results in stable thioethers, thiocarboxylic acids yield thioesters that are susceptible to cleavage, particularly under neutral and basic pH. Using a peptide model, we have found that the thioester exchanges at pH 5.6 with nucleophiles at a slow rate of 10% in 24 hr. Decomposition becomes significant at pH 7.0 and 7.6 in aqueous solution, particularly in the presence of a small amount of trace heavy metals with cleavage arising from hydrolysis or aminolysis of unprotected peptides (>50% in 24 hr). As most immunogens are administered at pH 7, the stability of dendrimers containing thioesters may be problematic in such applications.

2. *Thiol–Disulfide Exchange*

Thiol–disulfide exchanges to form unsymmetrical disulfide bonds have been exploited extensively both in protein and peptide chemistry (Moore and Ward, 1956). King *et al.* (1978) have utilized an aromatic thiol, the 4-dithiopyridyl group, as a cysteinyl-activating group to cross-link proteins via intermolecular disulfide bonds. Similarly, other aromatic thiols such as 2-thiopyridyl and nitropyridyl sulfenyl (Npys) have been used to form unsymmetrical disulfides in peptide synthesis (Drijfhout and Bloemhoff, 1991). These and other similar principles could be applied to the ligation of unprotected peptides to a core matrix. A general scheme (Table VIII) is to activate one of the thiols as a mixed disulfide with an aromatic thiol such as 2-thiolpyridine and nitrothiopyridine analogs. Drijhout and Bloemhoff (1991) have successfully shown that this chemistry is effective for the synthesis of MAPs in which a one-pot reaction is performed with a protected *S*-acetyl MAP core matrix and an activated *S*-(Npys)-cysteinyl peptide in hydroxylamine at pH 8.

3. *Thiol Addition*

Thioethers can also be formed by adding a cysteine thiol to an activated double bond of a maleimido group (Moore and Ward, 1956) (Table VIII). This method is particularly popular for cross-linking proteins with reporter groups, and Kitagawa and Aikawa (1976) have shown that insulin-containing maleimido groups could couple to glycoproteins. The reaction is specific and is usually carried out in aqueous solution with an optimum pH around 7. At pH 7, lysine and histidine also react slowly with maleimido groups (Smyth *et al.*, 1964). Thus, long reaction times should be avoided. Furthermore, hydrolysis of maleimides to nonreactive maleic acids occurs above pH 8 (Ishi and Lehrer, 1966). However, the method is convenient because *N*-alkyl or *N*-aryl maleimide groups are available either as free carboxylic acids or as active esters such as *N*-hydroxysuccinimides (Keller and Rudinger, 1975; Carlsson *et al.*, 1978; King *et al.*, 1978; Yoshitake *et al.*, 1979; Wunsch *et al.*, 1985), so that they can be incorporated as a premade unit in solid-phase synthesis. The maleimido group on lysine and phenylalanine has been shown to be stable to 100% trifluoroacetic acid (TFA) for 3 hr (Keller and Rudinger, 1975), and it is fully compatible with the Fmoc chemistry in peptide synthesis when the maleimido group is added last to the peptide sequence.

D. Carbonyl Chemistry

Thus far, ligation methods based on thiol chemistry have exploited the versatility and unusual reactivity of the thiol group as a weak base to distinguish itself among other functional groups in the displacement or addition reaction with an electrophile. A broader approach along the same principles is to exploit the selectivity of other weak bases, particularly their strong tendency for condensation reactions with aldehydes (Table IX). In general, the desirable weak bases, except cysteine, do not occur naturally in amino acid sequences, and several types of weak bases for ligation to aldehydes have been developed for this purpose. A determining criterion for their suitability is the ability to react with aldehydes under acidic conditions so that other side-chain nucleophiles are protected by protonation. The first type consists of conjugated amines whose basicities are lowered by neighboring electron-withdrawing groups. These include hydroxylamine (Rose, 1994) to give oxime, and substituted hydrazines such as acylhydrazines (King *et al.*, 1986) and phenylhydrazines (Drijfhout and Bloemhoff, 1991; Tam *et al.*, 1994) to give hydrazone. The second type contains the 1,2-disubstituted pattern which in an addition reaction forms a cyclic compound. This includes derivatives of 1,2-aminoethanethiol and 1,2-aminoethanol such as those found in N-terminal cysteine and threonine. These 1,2-disubstituted weak bases react with aldehydes to form a proline-like ring such as thiazolidine from cysteine and oxazolidine from threonine (Tam *et al.*, 1994).

1. *Preparation of Aldehyde Groups*

Although aldehydes can be obtained by many organic transformations, three methods have been developed for this purpose. The first method exploits the popular $NaIO_4$ oxidation of N-terminal Ser, Thr, or Cys to give an α-oxoacyl group (Clamp and Hough, 1965; Dixon and Fields, 1972). The N-terminal amino acids serve as precursors for aldehydes and can be transformed before the ligation. Furthermore, it has been shown that periodate oxidation of a 2-amino alcohol such as Ser is about 1000 times faster than oxidation of a diol (Gaertner *et al.*, 1992). We have found that oxidation of Ser-containing peptides could be accomplished within a few minutes at neutral pH using imidazole buffer. Oxidation of disulfide bonds of cysteine, the thioether of Met, the indole of Trp, or other sensitive groups in the peptides could be minimized by using a large excess of Met as scavenger. Oxidation of C-terminal Cys to give the corresponding aldehyde is also utilized but requires a longer time for completion (Tam *et al.*, 1994). In practice, transformation of terminal 1,2-aminoalcohol (Ser, Thr) or 1,2-aminothiol (Cys) to an aldehyde is convenient and highly compatible with the overall scheme of peptide synthesis because both Boc and Fmoc chemistry can be used to generate the aldehyde-precursor moieties on either the peptide or core matrix.

The second method uses a protected aldehyde capped on the N terminus or the lysyl side chains. Because of the general lability of aldehydes to acids such as HF and to a lesser extent to TFA, the method is so far limited to synthesis using Fmoc chemistry. Suitable protected aldehyde derivatives include those that contain acetal-alkanoic acids. Chiang *et al.* (1994) use 5,5-dimethoxy-1-oxopentanoic acid to cap N-terminal amine and deblock the acetal protecting group with HCl/TFE/DCM (TFE, trifluoroethanol; DCM, dichloromethane) for 10 min to yield the desired aldehyde. Shao and Tam (1995) use 2,2-dimethoxyacetic acid in coupling to lysyl MAP core matrix. Because of the electron-withdrawing effect of the amide group adjacent to the acetal, it is stable to TFA and requires concentrated HCl for removal. Cyclic acetals are more resistant to TFA, but we have found that they are insufficiently stable to the usual TFA cleavage conditions in peptide synthesis. We have used the β-chloro-cyclic acetal to increase the TFA stability. Treatment with catalytic amount of vitamin B_{12} in Zn^{2+}-containing buffer at pH 7 releases the protecting group (Scheffold and Amble, 1980). These examples illustrate different strategies for protecting aldehydes in which masking or unmasking could be performed prior to, during, or after the TFA cleavage steps.

The third method developed by Liu and Tam (1994) and Liu and coworkers (1995) is a $n + 1$ method in which an amino acid containing a masked aldehyde is added to the C terminus of a purified peptide segment. This method bypasses problems associated with the instability of aldehydes during TFA cleavage or $NaIO_4$ oxidation of large peptide segments. Furthermore, this method will allow the synthesis of peptides by Boc chemistry. The method as shown in Fig. 5 involves the synthesis of a peptide containing either alkylester or thioester at the C terminus by solid-phase method with alkyl ester or alkyl thioester resin. A large excess of an amino acid containing a masked glycoaldehyde is then added to the peptide ester segment enzymatically or to peptide thioester chemically via the Ag^+–hydroxylsuccinimide method. Both types of addition are efficient and are favored by the large excess of amino acid derivatives, and they are usually complete in less than 1 hr even with large peptide segments. The protected aldehydes can then be unmasked by mild acidic conditions in the presence of the other component bearing the weak base to complete the ligation reaction.

2. Oxime

Although hydroxylamine–aldehyde chemistry to form oximes was used by Erlanger *et al.* (1957) and later by Pochon *et al.* (1989) for site-specific conjugation to proteins, this weak base was first utilized for the synthesis of peptide dendrimers by Rose (1994), Tuchscherer (1993), and Shao and Tam (1995). It is a convenient method because aminooxyacetic acid, $NH_2OCH_2CO_2H$, is commercially available and can be used as a protected unit to be added to any position in the peptide sequence during the solid-phase synthesis. NH_2OCH_2-containing peptides are stable to HF or TFA cleavage conditions and are ligated to the aldehyde-containing template to give oxime peptide dendrimers. As shown by Rose (1994), when performed at pH 4.6 with a 5-fold molar excess of hydroxylamine per aldehyde reactive group, this reaction gives a hexavalent oxime peptide dendrimer in 90% yield after 16–18 hr (Smyth *et al.*, 1964) and an even bigger octavalent oxime peptide dendrimer with a mass of 19,916 Da. Rose further showed the advantage of oxime derived from its stability in water at room temperature at pH 2–7; this is in contrast to the necessity of $NaCNH_3$ reduction to stabilize the imine bond when alkyl or aryl amines are used (King *et al.*, 1986).

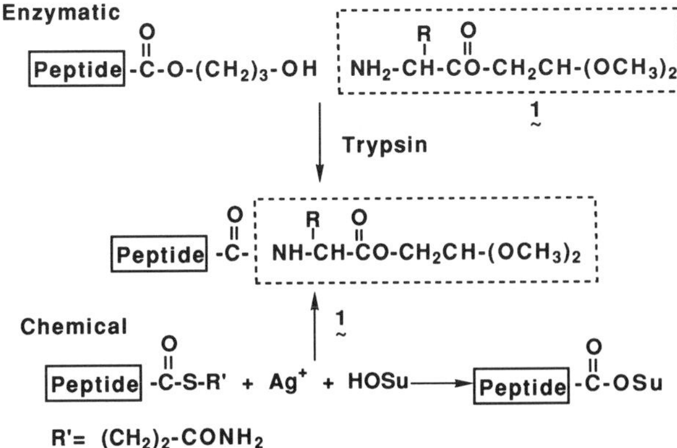

Figure 5 Schematic of $n + 1$ methods for introducing masked aldehydes at the C terminus of unprotected peptides.

3. Hydrazone

Ligation by hydrazone formation is perhaps the oldest method for synthesis of proteins and carbohydrates, and there is a wealth of information available (Bergbreiter and Momongan, 1991). In general, acyl- and aryl-substituted hydrazines have been used successfully. Offord, Rose, and colleagues (Gaertner *et al.*, 1992, 1994; Fisch *et al.*, 1992) have developed and applied acylhydrazine–aldehyde chemistry for site-specific conjugation to proteins, semisynthesis of proteins, and backbone engineering of proteins. Acylhydrazines are most conveniently generated at the C terminus via hydrazinolysis of peptide esters and reverse proteolysis with monoprotected hydrazine.

The formation of hydrazones by acylhydrazines and aldehydes is generally carried out at pH 4.5–5.0. King *et al.* (1978) have found that the conjugation of acetyl hydrazide with *p*-carbobenzaldehyde is best accomplished at pH 4.73. Higher or lower pH leads to lower yield and slower kinetics. Thus, hydrazone formation in aqueous medium usually proceeds through a bell-shaped curve, and it may be necessary to determine the optimal pH for a given pair of acylhydrazine and aldehyde. Hydrazones, particularly small molecules, are susceptible to hydrolysis at acidic pH. King *et al.* (1986) show that 38 and 41% of the acetylhydrazone of *p*-carboxybenzaldehyde decomposed at pH 4.2 and 5.0, respectively, after 18 hr. On the other hand, Geoghegan and Stroh (1992) have shown that hydrazone resulting from the ligation of acylhydrazine and α-oxoacyl peptide is stable at pH 6–8 for at least 12 hr at 22°C but relatively labile at acidic pH. At pH 2, the half-life for decomposition of the hydrazone bond is about 100 min. To achieve aqueous stability, the hydrazone bond is often reduced by $NaCNH_3$ to the corresponding hydrazine. In general, we have found that hydrazone bonds in peptide dendrimers are relatively stable at neutral pH, probably because of their macromolecular nature (Geoghegan and Stroh, 1992).

To utilize hydrazide–aldehyde chemistry for ligation of peptides on the N terminus, Shao and Tam (1995) have developed Boc-monohydrazide succinic acid for use in coupling to the amino groups. The reverse polarity of this chemistry allows facile incorporation of acylhydrazine into the α-amine and the side chain of lysine at any position. Thus, unprotected peptides containing the hydrazide succinyl group are used to ligate to the aldehyde-containing MAP core matrix with great efficiency. Similarly, we have developed 4-Boc-monohydrazinobenzoic (Hob) acid as a derivative to modify α- and side-chain amines in peptides during peptide synthesis (Spetzler and Tam, 1994). Like the hydroxylamine derivative, Hob-peptides are stable to usual cleavage conditions involving TFA or HF. With a 1.5-fold excess of Hob-peptide, the reaction proceeds rapidly and completely with a tetravalent α-oxoacyl-MAP core at pH 4.5 within 1 hr and in 10 min when a 2-fold excess of Hob-peptide is used. Furthermore, the progress of the phenylhydrazone reaction is conveniently monitored at 340 nm (Spetzler and Tam, 1994).

4. Thiazolidine and Oxazolidine Ring Formation

The reaction of aldehydes with N-terminal cysteine to give thiazolidine has long been known (Schubert, 1936), but it is only relatively recently that this reaction has been exploited (Liu and Tam, 1994; Rao and Tam, 1994; Shao and Tam, 1995; Spetzler and Tam, 1994; Tam *et al.*, 1994) for chemoselective ligation of unprotected peptides. Thiazolidine and oxazolidine are thia- and oxa-proline analogs. Thus, among all the methods, this method is unusual in providing a heterocyclic ring at the ligation site and may be useful in imparting conformational constraints to the peptides. Unlike thiol chemistry, thiazolidine ring formation requires both a thiol and an amine in a 1,2-substituted relationship, making the reaction highly specific.

We have developed a new method (Tam *et al.*, 1994) to show the specificity of this reaction using a dye-labeled alanyl ester aldehyde to react with libraries of 400 dipeptides that contained all dipeptide combinations of the 20 genetically coded amino acids. The libraries are spot-synthesized on a paper support (Frank and Doring, 1988). The thiazolidine or oxazolidine ring formation is visualized when the dye-labeled glyoxylyl ester aldehyde reacts with dipeptides in the library. The glyoxylyl ester aldehyde could react with dipeptides containing the following N-terminal amino acids: Cys, Thr, Trp, Ser, His, and Asn. However, the order and extent of reactivity are highly dependent on pH, solvent, and neighboring participation by the adjacent amino acid. In general, the amino acids could be divided into three categories. (1) N-terminal Cys and Thr are the most reactive. Cys reacts rapidly and completely within 30 min to form thiazolidine in both aqueous solutions and solutions containing a high content of water-miscible organic solvents. Thr reacts to form oxazolidine slowly in aqueous buffer ($t_{1/2} > 300$ hr) but rapidly and completely within 20 hr in organic–water solvent mixtures. (2) N-terminal Trp, His, and Ser are comparatively much less reactive than Cys or Thr. Trp reacts slowly and completely in aqueous buffers but significantly more slowly and incompletely in water–organic solvent mixtures. Both His and Ser react very slowly and incompletely in both solvent systems. (3) Finally, Asn reacts nearly insignificantly in both solvent systems. Other groups including the N termini and amino side chains of Lys and Arg do not show any reactivity. The oxazolidine formation of Thr is greatly accelerated by performing the reaction in 90% DMSO or DMF. The significant rate enhancement by the water-miscible organic solvent on Thr is particularly important since it allows the synthesis of disulfide-rich protein domains. Furthermore, ring formation with N-terminal Trp, His, and Asn provides a convenient route to prepare bicyclic and unusual heterocyclic derivatives for structure–activity studies.

Thiazolidine ring formation can be performed at pH 2–8, but generally pH 4.5–5.4 provides efficient rates without side reactions. For the synthesis of peptide dendrimers, we have found that, in the case of tetravalent species, the reaction is facile and complete in 1 hr for peptides shorter than 15 amino acid residues but requires longer for octavalent peptide dendrimers. Oxazolidine ring formation is comparatively slow and requires nearly anhydrous conditions at neutral to slightly basic pH to be effective.

To illustrate the utility of the thiaxolidine ligation (Fig. 6), Rao and Tam (1994) have synthesized an octabranched thiazolidinyl peptide dendrimer with a molecular weight of 24,205 by ligating the N-terminal cysteine of an unprotected 24-residue peptide, CI-24 (CNYNKRKRIHIGPGRAFYTTKNII), to a glyoxylyl octavalent lysyl scaffolding. The CI-24 peptide contains the principal neutralizing determinant of the surface coat protein gp120 of HIV-1, MN strain, which is a target for the development of HIV-1 vaccines. The alkyl aldehyde in the scaffolding is generated by oxidizing the 1,2-aminoethanol moiety of the N-terminal Ser on the scaffolding [$Ser_8Lys_4Lys_2$-Lys-β-Ala (Ser_8-MAP] with sodium periodate at pH 7 to yield a glyoxylyl derivative of $(HCO)_8$-MAP in nearly quantitative yield.

The thiazolidine ligation is adequately performed at pH 4.2 in water. However, we have found that the use of an organic cosolvent and elevated temperature (37°C) provides consistently better results than those in water alone because these conditions enhance the rate of formation and prevent various intermediate dendrimers from aggregating or precipitating during the course of the reaction.

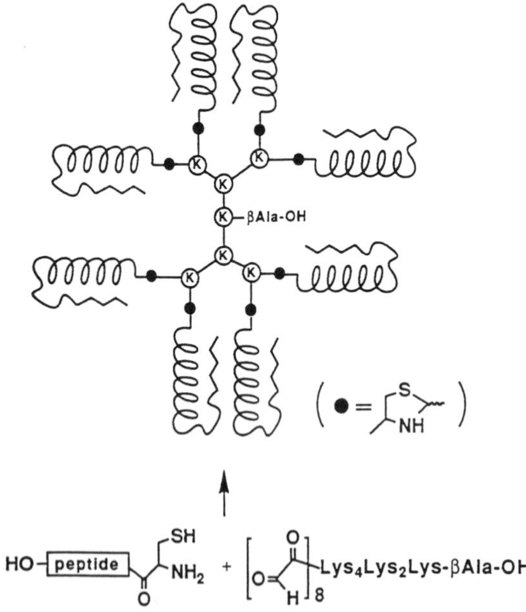

Figure 6 Schematic representation of a peptide dendrimer containing eight peptidyl branches anchored on a scaffolding of oligolysine (indicated by circled K's) via a thiazolidine linkage (filled circles) which is obtained by reacting the N-terminal cysteine with a glyoxylyl scaffolding.

The best combination is *N*-methylpyrrolidinone (NMP)–water (1:1, v/v). Other organic cosolvents such as dimethyl sulfoxide (DMSO) or dimethylformamide (DMF) are not suitable. DMSO is shown to be a mild oxidant, leading to disulfide formation of cysteinyl-containing peptides (Tam *et al.*, 1991), and the use of DMF results in formylation (M + 28) of the unprotected peptide as shown by mass spectrometric analysis of products containing M + 28 peaks. Using the optimized NMP–water mixture, the less-hindered tetra- and pentameric MAPs are obtained in less than 2 hr, whereas synthesis of the hexa- and heptameric MAPs requires 7 and 30 hr, respectively. The more hindered, fully substituted octameric MAP is found to give 82% yield in 67 hr.

The advantages of the thiazolidine ligation approach are apparent in the purification and characterization of the end products (Fig. 7). Each form of dendrimer can be clearly identified by reversed-phase high-performance liquid chromatography (HPLC) despite the large molecular weights. More importantly, elution occurs in the order of increased molecular weight, which allows easy monitoring and optimization. Furthermore, the molecular weights [octamer (calculated/found): 24,205 D/24,211 ± 24 u; heptamer: 21,358.8/21,361 ± 21.4u; hexamer: 18,512/18517 ± 18.5u; pentamer: 15,665/15,658 ± 15.7 u) have been unambiguously established by mass spectral analysis in addition to other conventional characterization. The octameric peptide dendrimer is believed to be the largest artificial protein obtained by controlled synthesis, with a molecular weight of 24,205, surpassing in size the octaoxime peptide dendrimer prepared by Rose (1994) having a molecular weight of 19,916.

E. Optimization of Carbonyl Chemistry

We have also studied optimal conditions for the weak base–aldehyde ligation (Shao and Tam, 1995). A common mechanism in this chemistry is the condensation reaction that eliminates a mole of water. Thus, we found that the reactions are greatly accelerated in mixtures of water with water-miscible organic solvents such as DMF, NMP, or DMSO (1:1, v/v) and at elevated temperature (37°C). A comparative study by Shao and Tam (1995) using a branched MAP core matrix with aldehydes and unprotected 20-amino acid peptides containing aminooxy, hydrazide, or cysteine at the N termini shows that conventional methods in aqueous solution at ambient temperature require 24–60 hr for completion. However, with optimized conditions of including water-miscible cosolvents and elevating the temperature, the reaction rates increase 12- to 27-fold, and completed reactions are obtained in 2–8 hr. A major concern in the use of thiols in thiazolidine formation is the side reaction caused by oxidation to disulfides. The side reaction is minimized by keeping the reaction mixture at acidic pH and by including a metal-chelating agent such as EDTA to achieve free radical-mediated oxidation. With these precautions, Shao and Tam (1995) have found that only 2% of the thiol is oxidized to disulfide at the completion of the reaction, compared to the usual greater than 50% oxidation at neutral conditions.

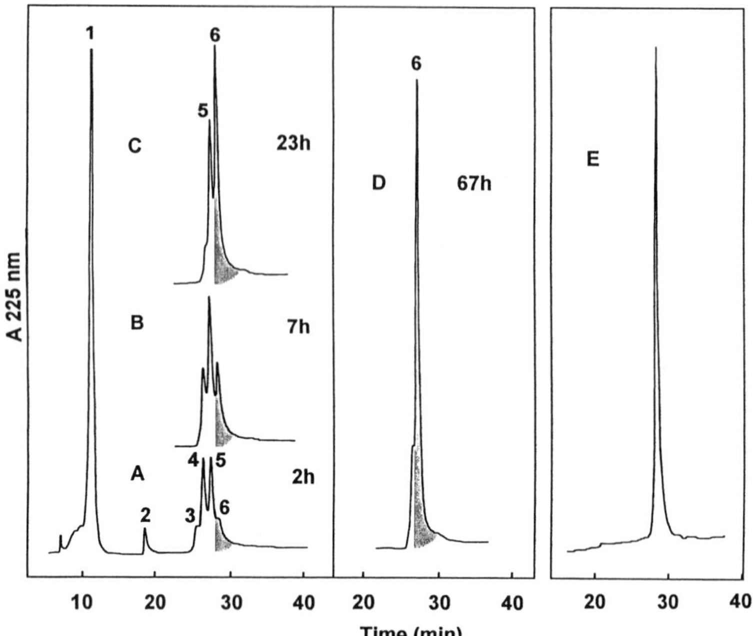

Figure 7 Time course of thiazolidine ligation of Cys-peptide (CNYNKRKRIHIGPGRAFYT-TKNII) and the glyoxylyl scaffolding [$(HCOCO)_8Lys_4Lys_2Lys$-βAla] analyzed by reversed-phase HPLC. (A) At 2 hr, peak 1 is the excess Cys-peptide; peak 2 is a side product of peak 1 in the disulfide form; peaks 3, 4, 5, and 6 correspond to penta-, hexa-, hepta-, and octameric forms. The molecular weights of these forms were confirmed by mass spectrometry (LD-MS). (B–D) Progression and completion of the reaction. (E) Purified octameric peptide dendrimer. HPLC conditions were as follows: a linear gradient of 0.67% B/min from 30% buffer B was applied, with buffer A being 100% water with 0.05% TFA and buffer B being 60% CH_3CN with 0.039% TFA.

However, because both thioalkylation and thiol–disulfide exchange reactions are performed at neutral–basic pH, thiol oxidation cannot be totally prevented, and a significant excess of thiols is therefore necessary.

F. Stability of Oxime, Hydrazone, and Thiazolidine Bonds in Peptide Dendrimers

The stability of oxime, hydrazone, and thiazolidine in peptide dendrimers was also studied by Shao and Tam (1995). Over 24 hr, the oxime bond is stable at acidic and neutral pH, but 21% of the peptide dendrimer decomposes at pH 9 to the peptide and lysyl MAP core. The hydrazone bond is stable at pH 5 and 7, whereas 32 and 26% of the peptide dendrimer are decomposed at pH 3 and 9, respectively. The thiazolidine linkage shows the highest stability, being stable between pH 3 and 9. These results are consistent with previous knowledge that thiazolidine is a stable masked aldehyde derivative (Ratner and Clarke, 1937) and would be suitable for use in immunization, in which stability of the peptide immunogen is a requirement.

In summary, carbonyl chemistry using weak base–aldehyde ligation of unprotected peptide segments to a core matrix to form peptide dendrimers is specific, efficient, and versatile. The weak base and aldehyde can be obtained from readily accessible starting materials, can be incorporated into synthetic peptides or the core matrix in many masked forms, and are compatible with synthetic and purification schemes of peptide synthesis.

G. Synthesis of Malaria Vaccine Containing Multiple Disulfides on Multiple Antigen Peptides

The development of different methods for the preparation of peptide dendrimers is necessary for providing the flexibility of N or C attachment to the core matrix. For immunization, the polarity of attachment often influences the desirable recognition of the cognate protein from which the peptide antigen is derived (Lu *et al.*, 1991). Furthermore, synthetic flexibility becomes necessary when one considers the ligation of peptide antigens containing multiple disulfide bonds to the MAP core matrix. This is the case in the synthesis of a potential malaria vaccine based on MAPs containing an epidermal growth factor (EGF)-like domain which has been identified as a protective antigen derived from the merozoite surface protein (MSP-1) in the asexual blood stage (Mackay *et al.*, 1985). The EGF-like domain contains 50 amino acids and three pairs of disulfide bonds. The presence of the disulfides precludes the methods of thiol chemistry (i.e., thioalkylation or thiol-disulfide exchange) for incorporating the preformed EGF domain into MAPs, and thus weak base–aldehyde chemistry was used (Fig. 8). The EGF domain was obtained by a two-step disulfide-forming strategy (Spetzler *et al.*, 1994). The molecule contains a synthetic weak base as a *p*-hydrazinobenzoic acid (Hob) group which does not interfere with the peptide synthesis. The MAP core matrix contains Ser on all the amine termini and is oxidized to an aldehyde of the α-oxoacyl derivative. The ligation of Hob-peptide and the α-oxoacyl MAP core matrix is performed at pH 5 to yield the desired peptide dendrimer in 24 hr.

H. Other Applications of Orthogonal Ligation

The orthogonal approach of ligating unprotected peptide segments has a distinct origin in protein chemistry, from which most of the methods mentioned in this chapter are derived. Although our concerns center on the synthesis of peptide dendrimers, the methods could be applied to proteins with a linear contiguous sequence in "backbone engineering." In this regard, thioester and hydrazone

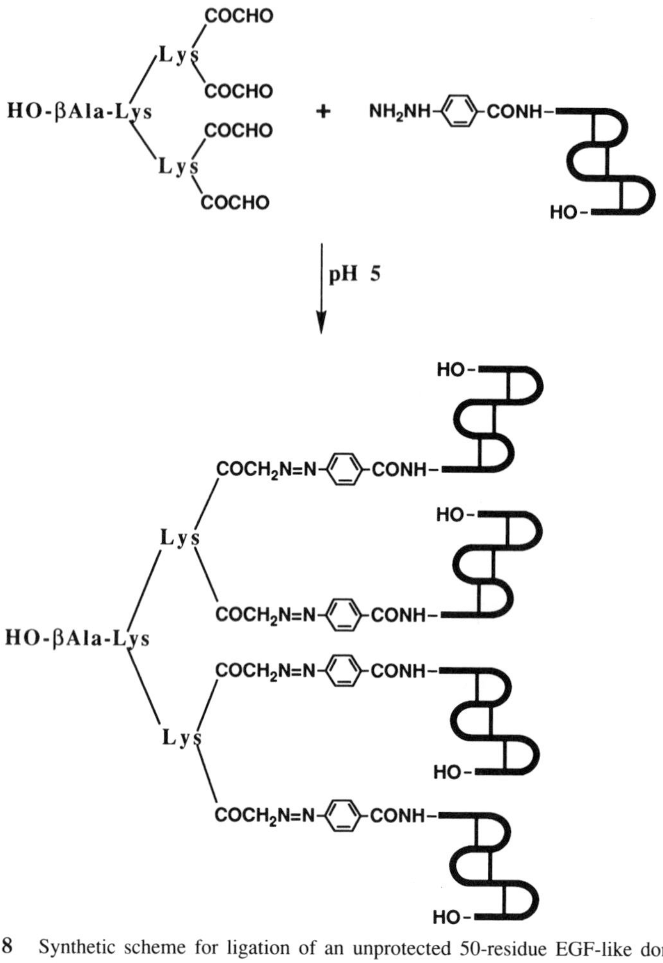

Figure 8 Synthetic scheme for ligation of an unprotected 50-residue EGF-like domain to the MAP core matrix.

bonds have been used as surrogate peptide bonds in the synthesis of proteins (Gaertner *et al.*, 1992; Schnolzer and Kent, 1992). New methods have been developed for the amide ligation of unprotected peptides (Kemp and Carey, 1991; Dawson *et al.*, 1994; Liu and Tam, 1994). All these methods make use of Cys as the nucleophile at the ligation site to form a covalent bond between the thiol side chain and the acyl segment followed by a proximity-driven intramolecular acyl transfer to form the amide bond as enunciated by Kemp and colleagues in their "thiol capture" scheme (Fotouhi *et al.*, 1989).

However, the schemes developed by Kemp (Fotouhi *et al.*, 1989; Kemp and Carey, 1991) and those by Dawson *et al.* (1994) produce X–Cys bonds with many free thiols that need to be addressed in the case of peptide dendrimers. On the other hand, the scheme developed by Liu and Tam (1994) produced an X–thioproline bond which may be more suitable for the synthesis of peptide dendrimers.

Another application of the orthogonal approach is the preparation of conformationally constrained peptide antigens (Conley *et al.*, 1993; Chiang *et al.*, 1994; Errhenius and Satterthwait, 1994) to mimic the native structure of the proteins from which they are derived. Thus, instead of intermolecular reaction to form the peptide dendrimer, intramolecular reaction of unprotected peptides will give cyclic peptides with end-to-end, end-to-side chain, side chain-to-side chain, end-to-backbone, and many other linkages (Botti *et al.*, 1995; Pallin *et al.*, 1995). This application is just developing as we realize the importance of the shape of the peptide antigens in eliciting high-affinity antibodies necessary for vaccine development. There are other applications of the chemoselective methods in the modification of proteins with peptides bearing reporter groups, lipids, carbohydrates, and receptor-specific ligands. Such modified proteins would have wide applications for therapeutic and diagnostic purposes.

The chemoselective approach is also important for the synthesis of newly designed MAPs to accommodate the increased sophistication of incorporating lipids and several types of epitopes. The newer design of MAPs containing lipidated built-in adjuvants eliminates the need for extraneous adjuvant and evokes a complete profile of immunological responses include B, T-helper, and T-cytotoxic immunities. More importantly, new MAPs can be administered orally to elicit mucosal responses. These developments in both methodology and design will further enhance applications of branched peptides for biomedical uses.

Acknowledgments

This work was supported by U.S. Public Health Service Grants CA 36544, AI 35577, and AI 28701.

References

Adorini, L., Harvey, M. A., Miller, A., and Sercarz, E. E. (1979). *J. Exp. Med.* **150,** 293–306.

Ahlborg, N. (1995). *J. Immunol. Methods* **179,** 269–275.

Anderson, S., Momoeda, M., Kawase, M., Kajigaya, S., and Young, N. (1995). *Virology* **206,** 626–632.

Arnold, D. R., Moshayedi, P., Schoen, T. J., Jones, B. E., Chader, G. J., and Waldbillig, R. J. (1993). *Exp. Eye Res.* **56,** 555–565.

Auriault, C., Wolowczuk, I., Gras-Masse, H., Marguerite, M., Boulanger, D., Capron, A., and Tartar, A. (1991). *Pept. Res.* **4,** 6–11.

Baba, E., Nakamura, M., Ohkuma, K., Kira, J., Tanaka, Y., Nakano, S., and Niho, Y. (1995). *J. Immunol.* **154,** 399–412.

Bauer, D., Murphy, C., Wu, Z., Wu, C., and Gall, J. (1994). *Mol. Biol. Cell* **5,** 633–644.

Bergbreiter, D. E., and Momongan, M. (1991). *In* "Comprehensive Organic Synthesis and Efficiency in Modern Organic Chemistry" (B. M. Trost and J. Flemming, eds.), Vol. 2, pp. 503–526. Pergamon, New York.

Blake, J. (1981). *Int. J. Pept. Protein Res.* **17,** 273–274.

Botti, P., Eom, K. D., and Tam, J. P. *In* "Proceedings of the Fourteenth American Peptide Symposium," in press. ESCOM, Leiden.

Boyd, B., Richardson, S., and Gariepy, J. (1991). *Infect. Immun.* **59,** 750–757.

Briand, J.-P., Muller, S., and Van Regenmortel, M. H. V. (1985). *J. Immunol. Methods* **78,** 59–69.

Briand, J.-P., Andre, C., Tuaillon, N., Herve, L., Neimark, J., and Muller, S. (1992). *Hepatology (Baltimore, MD)* **16,** 1395–1403.

Brown, F. J. (1990). *Semin. Virol.* **1,** 67–74.

Brown, F. J. (1992). *Vaccine* **10,** 1022–1026.

Butz, S., Rawer, S., Rapp, W., and Birsner, U. (1994). *Pept. Res.* **7,** 20–23.

Bycroft, B. W., Chane, W. C., Chhabra, S. R., and Hone, N. D. (1993). *J. Chem. Soc., Chem. Commun.,* 778.

Callaway, C., Seryshev, A., Wang, J. P., Slavik, K. J., Needleman, D., Cantu, C., Wu, Y., Jayaraman, T., Marks, A., and Hamilton, S. (1994). *J. Biol. Chem.* **269,** 15876–15884.

Calvo-Calle, M., Nardin, E., Clavijo, P., Boudin, C., Stuber, D., Takacs, B., Nussenzweig, R., and Cochrane, A. (1992). *J. Immunol.* **149,** 2695–2701.

Calvo-Calle, M., de Oliveira, G. A., Clavijo, P., Maracic, M., Tam, J. P., Lu, Y. A., Nardin, E. H., Nussenzweig, R. S., and Cochrane, A. H. (1993). *J. Immunol.* **150,** 1403–1412.

Caponi, L., Pegorano, S., Di Bartolo, V., Rovero, P., Revoltella, R., and Bombardieri, S. (1995). *J. Immunol. Methods* **179,** 193–202.

Carlsson, J., Drevin, H., and Axen, R. (1978). *Biochem. J.* **173,** 723–737.

Celis, E., Ou, D., and Otvos, L. (1988). *J. Immunol.* **140,** 1808–1815.

Ceppellini, R., Frumento, G., Ferrara, G., Tosi, R., Chersi, A., and Pernis, B. (1989). *Nature (London)* **339,** 392–394.

Cetin, Y., Kuhn, M., Kulalsiz, H., Adermann, K., Bargsten, G., Grube, D., and Forssmann, W. (1994). *Proc. Natl. Acad. Sci. U.S.A.* **91,** 2935–2939.

Chai, S. K., Clavijo, P., Tam, J. P., and Zavala, F. (1992). *J. Immunol.* **149,** 2385–2390.

Chiang, L.-C., Cabezas, E., Calvo, J. C., and Satterthwait, A. C. (1994). *In* "Proceedings of the Thirteenth American Peptide Symposium, Peptides: Chemistry, Structure and Biology" (R. S. Hodges and J. A. Smith, eds.), pp. 278–280. ESCOM, Leiden.

Christodoulides, M., and Heckels, J. E. (1994). *J. Gen. Microbiol.* **140,** 2951–2960.

Clamp, R., and Hough, L. (1965). *Biochem. J.* **94,** 17–24.

Conley, A. J., Tolman, R. L., Bednarek, M. A., Leanza, W. J., Marburg, S., Underwood, D. J., Emini, E. A., and Conley, A. J. (1993). *Int. J. Pept. Protein Res.* **41,** 455–466.

Crandall, I., and Sherman, I. W. (1994). *Parasitology* **108,** 389–396.

Darcy, F., Maes, P., Gras-Masse, H., Auriault, C., Bossus, M., Deslee, D., Godard, I., Cesbrom, M. F., Tartar, A., and Capron, A. (1992). *J. Immunol.* **149,** 3636–3641.

Dawson, P. E., and Kent, S. B. H. (1993). *J. Am. Chem. Soc.* **115,** 7263–7266.

Dawson, P. E., Muir, T. W., Clark-Lewis, I., and Kent, S. B. H. (1994). *Science* **266,** 776–779.

Defoort, J.-P, Nardelli, B., Huang, W., Ho, D. D., and Tam, J. P. (1992a). *Proc. Natl. Acad. Sci. U.S.A.* **89,** 3879–3883.

Defoort, J. P., Nardelli, B., Huang, W., and Tam, J. P. (1992b). *Int. J. Pept. Protein Res.* **40,** 214–221.

Del Giudice, G., Tougne, C., Louis, J. A., Lambert, P.-H., Bianchi, E., Bonelli, F., Chiappinelli, L., and Pessi, A. (1990). *Eur. J. Immunol.* **20,** 1619–1622.

Demotz, S., Lanzavecchia, A., Eisel, U., Niemann, H., Widmann, C., and Corradin, G. (1989a). *J. Immunol.* **142,** 394–402.

Demotz, S., Matricardi, P., Lanzavecchia, A., and Corragin, G. (1989b). *J. Immunol. Methods* **122,** 67–72.

Demotz, S., Danieli, C., Wallny, H. J., and Majdic, O. (1994). *Mol. Immunol.* **31,** 885–893.

de Oliveira, G. A., Clavijo, P., Nussenzweig, R., and Nardin, E. H. (1994). *Vaccine* **12,** 1012–1017.

Deres, K., Schild, H., Wiesmuller, K.-H., Jung, G., and Rammensee, H. G. (1989). *Nature (London)* **342,** 561–564.

De Santis, C., Lopalco, L., Robbioni, P., Longhi, R., Rapocciolo, G., Siccardi, A., and Beretta, A. (1994). *AIDS Res. Hum. Retroviruses* **10,** 157–162.

Digweed, M., Gunthert, U., Schneider, R., Seyschab, H., Friedl, R., and Sperling, K. (1995). *Mol. Cell Biol.* **15,** 305–314.

Dixon, H. B. F. and Fields, R. (1972). *In* "Methods in Enzymology" (C. H. W. Hirs and S. N. Timasheff, eds.), Vol. 25, pp. 409–419. Academic Press, New York.

Drijfhout, J. W., and Bloemhoff, W. (1991). *Int. J. Pept. Protein Res.* **37,** 27–32.

Dyrberg, T., and Oldstone, M. B. (1986). *J. Exp. Med.* **164,** 1344–1349.

Edwards, R., Singleton, A., Murray, B., Murray, S., Boobis, A., and Davies, D. (1991). *Biochem. J.* **278,** 749–757.

Eicheler, W., Tuohimaa, P., Vilja, P., Adermann, K., Forssmann, W. G., and Aumuller, G. (1994). *J. Histochem. Cytochem.* **42,** 667–675.

Emery-Goodman, A., Hirling, H., Scarpellino, L., Henderson, B., and Kuhu, L. (1993). *Nucleic Acids Res.* **21,** 1457–1461.

Erlanger, B. F., Borek, F., Beiser, S. M., and Lieberman, S. J. (1957). *Biol. Chem.* **228,** 713–727.

Errhenius, T., and Satterthwait, A. C. (1994). *In* "Proceedings of the Thirteenth American Peptide Symposium, Peptides: Chemistry, Structure and Biology" (J. E. Rivier and G. R. Marshall, eds.), pp. 870–872. ESCOM, Leiden.

Estaquier, J., Boutillon, C., Arneisen, J., Gras-Masse, H., Delanoye, A., Lecocq, J., Dixson, A., Tartar, A., Capron, A., and Ariault, C. (1992). *Mol. Immunol.* **29,** 1337–1345.

Estaquier, J., Boutillon, C., Gras-Masse, H., Ameisen, J., Capron, A., Tartar, A., and Auriault, C. (1993). *Vaccine* **11,** 1083–1091.

Fassina, G. (1992). *J. Chromatogr.* **591,** 99–106.

Fassina, G., and Cassani, G. (1993). *Pept. Res.* **6,** 73–78.

Fassina, G., Corti, A., and Cassani, G. (1992a). *Int. J. Pept. Protein Res.* **39,** 549–556.

Fassina, G., Cassani, G., and Corti, A. (1992b) *Arch. Biochem. Biophys.* **296,** 137–143.

Felici, F., Luzzago, A., Folgori, A., and Cortese, R. (1993). *Gene* **128,** 21–27.

Fern, J., and Good, M. F. (1992). *J. Immunol.* **148,** 907–913.

Fields, C. G., Mickelson, D. J., Drake, S., McCarthy, J. B., and Fields, G. B. (1993). *J. Biol. Chem.* **268,** 14153–14160.

Finkel, T., Theriot, J., Dise, K., Tomaselli, G., and Goldschmidt-Clermont, P. (1994). *Proc. Natl. Acad. Sci. U.S.A.* **91,** 1510–1514.

Fisch, I., Kunzi, G., Rose, K., and Offord, R. E. (1992). *Bioconjugate Chem.* **3,** 147–153.

Fotouhi, N., Galakatos, N. G., and Kemp, D. S. (1989). *J. Org. Chem.* **54,** 2803–2817.

Fraisier, C., Ebersold, A., Blomber, J., and Desgranges, C. (1994). *J. Immunol. Methods* **176,** 9–12.

Francis, M. J., Hasting, G. Z., Brown, F., McDermed, J., Lu, Y. A., and Tam, J. P. (1991). *Immunology* **73,** 249–254.

Francis, M. J., Fry, C. M., Rowlands, D. J., Bittle, J. L., Houghten, R. A., Lerner, R. A., and Brown, F. (1987). *Immunology* **61,** 1–6.

Francis, M. J., Hastings, G. Z., Brown, F., McDermed, J., Lu, Y.-A., and Tam, J. P. (1991). *Immunology* **73,** 249–254.

Frank, R., and Doring, R. (1988). *Tetrahedron* **44,** 6031–6040.

Gaertner, H. F., Rose, K., Cotton, R., Timms, D., Camble, R., and Offord, R. E. (1992). *Bioconjugate Chem.* **3,** 262–268.

Gaertner, H. F., Offord, R. E., Cotton, R., Timms, D., Camble, R., and Rose, K. (1994). *J. Biol. Chem.* **269,** 7224–7230.

Geoghegan, K. F., and Stroh, J. G. (1992). *Bioconjugate Chem.* **3,** 136–146.

Gettemans, J., De Ville, Y., Vandekerckhove, J., and Waelkens, E. (1993). *Eur. J. Biochem.* **214,** 111–119.

Godard, I., Estaquier, J., Zenner, L., Bossus, M., Auriault, C., Darcy, F., Gras-Masse, H., and Capron, A. (1994). *Mol. Immunol.* **31,** 1353–1363.

Gray, N., Quick, S., Goossen, B., Constable, A., Hirling, H., Kuhn, L. C., and Hentze, M. (1993). *Eur. J. Biochem.* **218,** 657–667.

Green, N., Alexander, H., Olson, A., Alexander, S., Shinnick, T. M., Sutcliffe, J. G., and Lerner, R. A. (1982). *Cell (Cambridge, Mass.)* **28,** 477–487.

Grillot, D., Valmori, D., Lambert, P. H., Corradin, G., and Del Giudice, G. (1993). *Infect. Immun.* **61,** 3064–3067.

Habluetzel, A., Pessi, A., Bianchi, E., Rotigliano, G., and Esposito, F. (1991). *Immunol. Lett.* **30,** 75–80.
Hahn, K. W., Klis, W. A., and Stewart, J. M. (1990). *Science* **248,** 1544–1547.
Halimi, H., and Rivaile, P. (1993). *Vaccine* **11,** 1233–1239.
Harding, C. V., and Unanue, E. R. (1990). *Cell. Regul.* **1,** 499–509.
Helling, F., Shang, A., Calves, M., Zhang, S., Ren, S., Yu, R. K., Oettgen, H. F., and Livingston, P. O. (1994). *Cancer Res.* **54,** 197–203.
Henderson, B. R., Seiser, C., and Kühns, L. C. (1993). *J. Biol. Chem.* **268,** 27327–27334.
Ho, P., Mutch, D., Winkel, K., Saul, A. J., Jones, G. I., Doran, T. J., and Rzepczyk, C. M. (1990). *Eur. J. Immunol.* **20,** 477–483.
Hodge, P., (1993). *Nature (London)* **362,** 18–19.
Huang, W., Nardelli, B., and Tam, J. P. (1994). *Mol. Immunol.* **31,** 1191–1199.
Ingham, K., Brew, S., and Migliorini, M. (1994). *Arch. Biochem. Biophys.* **314,** 242–246.
Ishi, S. S., and Lehrer, J. (1966). *Biophys. J.* **50,** 75.
Itoh, N., Jobo, K., Tsujimoto, K., Ohta, M., and Kawasaki, T. (1993). *J. Biol. Chem.* **268,** 17983–17986.
Ivanov, V. S., Kulik, L. N., Gabrielian A. E., Tchikin, L. D., Kozhich, A. T., and Ivanov, V. T. (1994). *FEBS Lett.* **345,** 159–161.
James, J. A., Gross, T., Scofield, H., and Harley, J. (1995). *J. Exp. Med.* **181,** 453-461.
Kaiser, E. T., Mihara, H., Laforet, G. A., Kelly, J. W., Walters, L., Findeis, M. A., and Sasaki, T. (1989). *Science* **243,** 187–192.
Kamo, K., Jordan, R., Hsu, H.-T., and Hudson, D. (1992). *J. Immunol. Methods* **156,** 163–170.
Kang, D., Okubo, K., Hamasaki, N., Kuroda, N., and Shiraki, H. (1992). *J. Biol. Chem.* **267,** 19211–19217.
Kelker, H. C., Schlesinger, D., and Valentine, F. T. (1994). *J. Immunol.* **152,** 4139–4148.
Keller, O., and Rudinger, J. (1975). *Helv. Chim. Acta.* **85,** 531–541.
Kemp, D. S., and Carey, R. I. (19910. *Tetrahedron Lett.* **32,** 2845–2848.
Khan, C. M., Villareal-Ramos, B., Pierce, R., Demarco de Hormaeche, R., McNeill, H., Ali, T., Chatfield, S., Capron, A., Dougan, G., and Hormaeche, C. (1994). *J. Immunol.* **153,** 5634–5642.
Kim, W., Schnaper, H. W., Nomizu, M., Yamada, Y., and Kleinman, H. K. (1994). *Cancer Res.* **54,** 5005–5010.
King, T. P., Li, Y., and Kochoumian, L. (1978). *Biochemistry* **171,** 499–506.
King, T. P., Zhao, S. W., and Lam, T. (1986). *Biochemistry* **25,** 5774–5779.
Kitagawa, T., and Aikawa, T. (1976). *J. Biochem. (Tokyo).* **79,** 233–236.
Kuhn, M., Kulaksiz, H., Adermann, K., Rechkemmer, G., and Forssmann, W. G. (1994). *FEBS Lett.* **341,** 218–222.
Kutty, R. K., Chandrasekharam, N., Nagineni, C. N., Kutty, G., Hooks, J. J., Chader, G. J., and Wiggert, B. J. (1994). *J. Cell. Physiol.* **159,** 371–378.
Lankinen, H., Telford, E., MacDonald, D., and Marsden, H. (1989). *J. Gen. Virol.* **70,** 3159–3169.
Levi, M., Ruden, U., Birx, D., Loomis, L., Redfield, R., Lovgren, K., Akerblom, L., Sandstrom, E., and Wahren, B. (1993). *J. AIDS* **6,** 855–864.
Li, J. K., and Yang, Y. Y. (1990). *Virology* **178,** 552–559.
Liew, F. Y., Millot, S. M., and Schmidt, J. A. (1990). *J. Exp. Med.* **172,** 1359–1365.
Linder, W., and Robey, F. A. (1987). *Int. J. Pept. Protein Res.* **30,** 794–800.
Lingappa, J. R., Martin, R., Wong, M. E., Ganem, D., Welch, W. J., and Lingappa, V. (1994). *J. Cell Biol.* **125,** 99–111.
Lio, A., Mochii, M., Agata, K., Kodoma, R., and Eguchi, G. (1994). *Dev. Growth Differ.* **36,** 155–164.
Liu, C.-F., and Tam, J. P. (1994). *Proc. Natl. Acad. Sci. U.S.A.* **91,** 6584–6588.
Liu, C. F., Rao, R., and Tam, J. P. (1995). *Proc. Natl. Acad. Sci. U.S.A.* in press.
Lu, Y.-A., Clavijo, P., Galantino, M., Shen, Z.-Y., and Tam, J. P. (1991a). *Mol. Immunol.* **28,** 623–630.
Lugovskoi, A. A., Rybakov, S. S., Ivaniushchenkov, V. N., Chepurkin, A. V., Petrov, V. N., Driagalin, N. N., and Burdov, A. N. (1992). *Bioorg. Khim.* **18,** 942–950.

Mackay, M., Goman, M., Bone, N., Hyde, J. E., Scaife, J., Certa, U., Stummenberg, H., and Bujard, H. (1985). *EMBO J.* **4,** 3823–3829.

McLean, G., Rixon, F., Langeland, N., Haarr, L., and Marsden, H. (1990). *J. Gen. Virol.* **71,** 2953–2960.

McLean, G., Owsianka, A., Subak-Sharpe, J., and Marsden, H. (1991). *J. Immunol. Methods* **137,** 149–157.

McLean, G., Cross, A., Munns, M., and Marsden, H. (1992). *J. Immunol. Methods* **155,** 113–120.

McNeil, H. P., Frenkel, D. P., Austen, K. F., Friend, D. S., and Stevens, R. I. (1992). *J. Immunol.* **149,** 2466–2472.

Malouf, N. N., McMahon, D. K., Hainsworth, C. N., and Kay, B. K. (1992). *Neuron* **8,** 899–906.

Manivel, V., Tripathy, A., Durgapal, H., Kumar, A., Panda, S., and Rao, K. (1993). *Vaccine* **11,** 366–371.

Marchini, B., Dolcher, M. P., Sabbatini, A., Klein, G., and Mogliorini, P. (1994). *J. Autoimmun.* **7,** 179–191.

Marguerite, M., Bossus, M., Mazingue, C., Wolowczuk, I., Grass-Masse, H., Tartar, A., Capron, A., and Auriault, C. (1992). *Mol. Immunol.* **29,** 793–800.

Marsden, H. S., Owsianka, A. M., Graham, S., McLean, G. W., Robertson, C. A., and Subak-Sharpe, J. H. (1992). *J. Immunol. Methods* **147,** 65–72.

Marsden, H. S., Murphy, M., Mevey, G., MacEachran, A., Owsianka, M., and Stown, N. (1994). *J. Gen. Virol.* **75,** 3127–3135.

Merrifield, R. B. (1963). *J. Am. Chem. Soc.* **85,** 2149–2154.

Means, G. E., and Feeney, R. E. (1990). *Bioconjugate Chem.* **1,** 2–12.

Meziere, C., Stockl, F., Batsford, S., Vogt, A., and Muller, S. (1994). *Clin. Exp. Immunol.* **98,** 287–294.

Migliorini, P., Boulanger, N., Betschart, B., and Corradin, G. (1990). *Scand. J. Immunol.* **31,** 237–242.

Migliorini, P., Betschart, B., and Corradin, G. (1993). *Eur. J. Immunol.* **23,** 582–585.

Miotto, G., Venerando, R., Marin, O., Siliprandi, N., and Mortimore, G. E. (1994). *J. Biol. Chem.* **269,** 25348–25353.

Molinar, E., Baude, A., Richmond, S. A., Patel, P. B., Somogyi, P., and McIlhinney, R. A. J. (1993). *Neuroscience (Oxford)* **53,** 307–326.

Monsalve, R., Gonzalez de la Pena, M. A., Menendez-Arias, L., Lopez-Otin, C., Villalba, M., and Rodriguez, R. (1993). *Biochem. J.* **293,** 625–632.

Moore, J. E., and Ward, W. H. (1956). *J. Am. Chem. Soc.* **78,** 2414–2418.

Moradi-Ameli, M., Rousseau, J., Kleman, J. P., Champliaud, M. F., Boutillon, M. M., Bernillon, J., Wallach, J., and Van Der Rest, M. (1994). *Eur. J. Biochem.* **221,** 987–995.

Mortimore, G. E., Wert, J. J., Miotto, G., Venerando, R., and Kadowaki, M. (1994). *Biochem. Biophys. Res. Commun.* **203,** 200–208.

Munesinghe, D. Y., Clavijo, P., Calvo Calle, M., Nussenzweig, R. S., and Nardin, E. (1991). *Eur. J. Immunol.* **21,** 3015–3020.

Mutter, M., and Vuilleumier, S. (1989). *Angew. Chem., Int. Ed. Engl.* **28,** 535–54.

Mutter, M., Tuschscherer, G. G., Miller, C., Altmann, K., Carey, R. I., Wyss, D. F., Labhardt, A. M., and Rivier, J. E. (1992). *J. Am. Chem. Soc.* **114,** 1463–1470.

Nagamatsu, S., Sawa, H., Kamada, K., Nakamichi, Y., Yoshimoto, K., and Hoshino, T. (1993). *FEBS Lett.* **334,** 289–295.

Nakitare, G., and Elliot, R. (1993). *Virology* **195,** 511–520.

Nara, K., Ito, S., Ito, T., Suzuki, Y., Ghoneim, M. A., Tachibana, S., and Hirose, S. (1994). *J. Biochem. (Tokyo)* **115,** 441–448.

Nardelli, B., and Tam, J. P. (1993). *Immunology* **79,** 355–361.

Nardelli, B., Defoort, J.-P., Huang, W., and Tam, J. P. (1992a). *AIDS Res. Hum. Retroviruses* **8,** 1405–1407.

Nardelli, B., Lu, Y.-A., Shiu, D. R., Delpierre-Defoort, C., Profy, A. T., and Tam, J. P. (1992b). *J. Immunol.* **148,** 914–920.

Nardelli, B., Haser, P. B., and Tam, J. (1994). *Vaccine* **12,** 1335–1339.

Nardin, E. H., and Nussensweig, R. S. (1993). *Annu. Rev. Immunol.* **11,** 687–727.

Naruse, H., Ogasawara, K., Kaneda, R., Hatakeyama, S., Itoh, T., Kida, H., Miyazaki, T., Good, R., and Onoe, K. (1994). *Proc. Natl. Acad. Sci. U.S.A.* **91,** 9588–9592.

Nieto, A., de la Luna, S., Barcena, J., Portela, A., Valcarcel, J., Melero, J. A., and Ortin, J. (1992). *Virus Res.* **24,** 65–75.

Nomizu, M., Yamamura, K., Kleinman, H. K., and Yamada, Y. (1993). *Cancer Res.* **53,** 3459–3461.

O'Rourke, K., Huff, T., Leathers, C., Robinson, M., and Gorham, J. (1994). *J. Gen. Virol.* **75,** 1511–1514.

Palker, T. J., Clark, M. E., Langlois, A. J., Matthews, T. J., Weinhold, K. J., Randall, R. R., Bolognesi, D. P., and Haynes, B. F. (1988). *Proc. Natl. Acad. Sci. U.S.A.* **85,** 1932–1936.

Pallin, T. D., Spetzler, J. C., and Tam, J. P. (1995). *In* "Proceedings of the Fourteenth American Peptide Symposium," in press. ESCOM, Leiden.

Pancre, V., Wolowczuk, I., Bossus, M., Gras-Masse, H., Guerret, S., Delanoye, A., Capron, A., and Auriault, C. (1994). *Mol. Immunol.* **31,** 1247–1256.

Panina-Bordignon, P., Tan, A., Termijtelen, A., Demotz, S., Corradin, G., and Lanzavecchia, A. (1989). *Eur. J. Immunol.* **19,** 2237–2242.

Parry, M. E., Stow, N. D., and Marsden, H. (1993). *J. Gen. Virol.* **74,** 607–612.

Partidos, C. D., and Steward, M. W. (1990). *J. Gen. Virol.* **71,** 2099–2105.

Pegg, A. E., Wiest, L., Mummert, C., and Dolan, M. E. (1991). *Carcinogenesis* **12,** 1671–1677.

Perin, M. S. (1994). *J. Biol. Chem.* **269,** 8576–8581.

Pessi, A., Valmori, D., Migliorini, P, Tougne, C., Bianchi, E., Lambert, P.-H., Corradin, G., and Del Giudice, G. (1991a). *Eur. J. Immunol.* **21,** 2273–2276.

Pessi, A., Bianchi, E., Chiappinelli, L., Bonelli, F., Tougne, C., Lambert, P. H., and Del Giudice, G. (1991b). *Parassitologia* **33,** 79–84.

Pochon, S., Buchegger, F., Pelegrin, A., Mach, J.-P., Offord, R. E., Ryser, J. E., and Rose, K. (1989). *Int. J. Cancer* **43,** 1188–1194.

Posnett, D. N., McGrath, H., and Tam, J. P. (1988). *J. Biol. Chem.* **263,** 1719–1725.

Rao, C., and Tam, J. P. (1994). *J. Am. Chem. Soc.* **116,** 6975–6976.

Ratner, S., and Clarke, H. T. (1937). *J. Am. Chem. Soc.* **59,** 200–206.

Raymond, J., Olsen, C., and Gettys, T. (1993). *Biochemistry* **32,** 11064–11075.

Reitermann, A., Metzger, J., Wiesmuller, K.-H., Jung, G., and Bessler, W. G. (1989). *Biol. Chem. Hoppe-Seyler* **370,** 343–352.

Reynolds, S., Dahal, C., and Harn, D. (1994). *J. Immunol.* **152,** 193–200.

Robertson, C., Mok, J. Y., Froebel, K., Simmonds, P., Burns, S., Marsden, H., and Graham, S. (1992). *J. Infect. Dis.* **166,** 704–709.

Robey, F. A., and Fields, R. L. (1989). *Anal. Biochem.* **177,** 373–377.

Robinson, J. H., and Kehoe, M. A. (1992). *Immunol. Today* **13,** 362–366.

Robinson, W. E., Jr., Kawamura, T., Gorny, M. K., Lake, D., Xu, J.-Y., Matsumoto, Y., Sugano, T., Masuho, Y., Mitchell, W. M., Hersh, E., and Zolla-Pazner, S. (1990). *Proc. Natl. Acad. Sci. U.S.A.* **87,** 3185–3189.

Roche, P. A., and Cresswell, P. (1990). *J. Immunol.* **144,** 1849–1856.

Romero, P., Tam, J. P., Schlesinger, D., Clavijo, P., Gibson, H., Barr, P., Nussenzweig, R., Nussenzweig, V., and Zavala, F. (1988). *Eur. J. Immunol.* **18,** 1951–1957.

Rose, K. (1994). *J. Am. Chem. Soc.* **116,** 30–33.

Rusche, J. R., Javaherian, K., McDanal, C., Petro, J., Lynn, D. L., Grimaila, R., Langlois, A., Gallo, R. C., Arthur, L. O., Fischinger, P. J., Bolognesi, D. P., Putney, S. D., and Matthews, T. J. (1988). *Proc. Natl. Acad. Sci. U.S.A.* **85,** 3198–3202.

Sa, G., Murugesan, G., Jaye, M., Ivashchenko, Y., and Fox, P. (1995). *J. Biol. Chem.* **270,** 2360–2366.

Sabbatini, A., Bombardieri, S., and Migliorini P. (1993a). *Eur. J. Immunol.* **23,** 1146–1152.

Sabbatini, A., Dolcher, M. P., Marchini, B., Bombardieri, S., and Migliorini, P. (1993b). *J. Rheumatol.* **20,** 1679–1683.

Saikawa, T., Anderson, S., Momoeda, M., Kajigaya, S., and Young, N. (1993). *J. Virol.* **67,** 3004–3009.

Sasaki, T., and Kaiser, E. T. (1989). *J. Am. Chem. Soc.* **111,** 380–381.
Schacke, H., Schroder, H., Gamulin, V., Rinkevich, B., Muller, I., and Muller, W. E. G. (1994). *Mol. Membr. Biol.* **11,** 101–107.
Scheffold, R., and Amble, E. (1980). *Angew. Chem.* **92,** 643.
Schnolzer, M., and Kent, S. B. H. (1992). *Science* **256,** 221–225.
Schubert, M. P. (1936). *J. Biol. Chem.* **114,** 341–350.
Schwartz, R. H. (1985). *Annu. Rev. Immunol.* **3,** 237–261.
Sheldon, K., Liu, D., Ferguson, J., and Gariepy, J. (1995). *Proc. Natl. Acad. Sci. U.S.A.* **92,** 2056–2060.
Shao, J., and Tam, J. P. (1995). *J. Am. Chem. Soc.* in press.
Shi, Z., Buntel, C., and Griffin, J. (1994). *Proc. Natl. Acad. Sci. U.S.A.* **91,** 7370–7374.
Shimizu, H., Ito, M., Miyahara, M., Ichikawa, K., Okubo, S., Konish, T., Naka, M., Tanaka, T., Hirano, K., Hartshorne, D., and Nakano, T. (1994). *J. Biol. Chem.* **269,** 30407–30411.
Sil, P., Misono, K., and Sen, S. (1993). *Circ. Res.* **73,** 98–108.
Simmer, J. P., Hu, C. C., Lau, E., Sarte, P., Slavkin, H., and Fincham, A. (1994). *Calcif. Tissue Int.* **55,** 302–310.
Simmonds, P., Rose, K. A., Graham, S., Chan, S. W., McOrnish, F., Dow, B., Follette, E. A., Yap, P., and Marsden, H. (1993). *J. Clin. Microbiol.* **31,** 1493–1503.
Sinclair, M., McLauchlan, J., Marsden, H., and Brown, M. (1994). *J. Gen. Virol.* **75,** 1083–1089.
Sinigaglia, F., Guttinger, M., Kilgus, J., Doran, D. M., Matilde, H., Etlinger, H., Trzeciak, A., Gillesen, D., and Pink, J. R. I. (1988). *Nature (London)* **336,** 778–780.
Sinnis, P., Clavijo, P., Fenyo, D., Chait, B. T., Cerami, C., and Nussenzweig. V. (1994a). *J. Exp. Med.* **180,** 297–306.
Sinnis, P., Rose, K. A., Graham, S., Chan, S. W., McOrnish, F., Dow, B., Follett, E. A., Yap, P., and Marsden, H. (1994b). *J. Exp. Med.* **180,** 297–306.
Smith, D. J., Taubman, M. A., Holmberg, C. F., Eastcort, J., King, W., and Ali-Salaam, P. (1993). *Infect. Immun.* **61,** 2899–2905.
Smith, D. J., Taubman, M., King, W., Eida, S., Powell, J., and Eastcott, J. (1994). *Infect. Immun.* **62.**
Smith, M. A., Kutty, K., Richey, P., Yan, S. D., Sterm, D., Chader, G., Wiggert, B., Petersen, R., and Perry, G. (1994). *Am. J. Pathol.* **145,** 42–47.
Smyth, D. G., Blumenfeld, O. O., and Konigsberg, W. (1964). *Biochem. J.* **91,** 589–595.
Spetzler, J. C., and Tam. J. P. (1994). *Int. J. Pept. Protein Res.* **45,** 78–85.
Spetzler, J. C., Rao, C., and Tam, J. P. (1994). *Int. J. Pept. Protein Res.* **43,** 351–358.
St. John, L. C., Bell, F. P., Kezdy, F. J., Vosters, A., Sharma, S. K., Kinner, J. H., and Smith, C. W. (1991). *Comp. Biochem. Physiol.* **99,** 431–435.
Stout, J. G., Brittsan, A., and Kirley, T. (1994). *Biochem. Mol. Biol. Int.* **33,** 1091–1098.
Suzuki, Y. S., Wang, Y., and Takemoto, J. Y. (1992). *Plant Physiol.* **99,** 1314–1320.
Szymkowski, D. E., and Deering, R. (1990). *Nucleic Acids Res.* **18,** 4695–4701.
Takahashi, H., Cohen, J., Hosmalin, A., Cease, K. B., Houghten, R., Cornette, J. L., DeLisi, C., Moss, B., Germain, R. N., and Berzofsky, J. A. (1988). *Proc. Natl. Acad. Sci. U.S.A.* **85,** 3105–3109.
Tam, J. P. (1988). *Proc. Natl. Acad. Sci. U.S.A.* **85,** 5409–5413.
Tam, J. P. (1993). U.S. Patent 5,229,490.
Tam, J. P., and Lu, Y.-A. (1989). *Proc. Natl. Acad. Sci. U.S.A.* **86,** 9084–9088.
Tam, J. P., and Zavala, F. (1989). *J. Immunol. Methods* **124,** 53–61.
Tam, J. P., Rao, C., Shao, J., and Liu, C.-F. (1994). *Int. J. Pept. Protein Res.* in press.
Tam, J. P., Wu, C.-R., Lui, W., and Zhang, J.-W. (1991). *J. Am. Chem. Soc,* **113,** 6659–6662.
Tindle, R. W., Fernando, G. J. P., Sterling, J. C., and Frazer, I. H. (1991). *Proc. Natl. Acad. Sci. U.S.A.* **88,** 5887–5891.
Tomalia, D. A., Naylor, A. M., and Goddard III, W. A. (1990). *Angew. Chem., Int. Ed. Engl.* **29,** 138–175.
Toth, G. K., Varadi, G., Nagy, Z., Monostori, E., Penke, B., Hegedus, Z., Ando, I., Fazekas, G., Kurucz, I., Mak, M., and Rajnavolgyi, E. (1993). *Pept. Res.* **6,** 272–280.

Toth, I., Danton, M., Flinn, N., and Gibbons, W. (1993). *Tetrahedron Lett.* **34,** 3925–3928.

Troalen, F., Razafindratsita, A., Puisieux, A., Voltzel, T., Bohuon, C., Bellet, D., and Bidart, J.-M. (1990). *Mol. Immunol.* **27,** 363–368.

Tuchscherer, G. (1993). *Tetrahedron Lett.* **34,** 8419–8422.

Unson, C. G., Erickson, B. W., Richardson, D. G., and Richardson, J. S. (1984). *Fed. Proc.* **43,** 1837.

Valmori, D., Pessi, A., Bianchi, E., and Corradin, G. (1992). *J. Immunol.* **149,** 717–721.

Valmori, D., Romero, J., Men, Y,. Maryanski, J., Romero, P., and Corradin, G. (1994). *Eur. J. Immunol.* **24,** 1458–1462.

Vanage, G., Lu, Y. A., Tam, J. P., and Koide, S. S. (1992). *Biochem. Biophys. Res. Commun.* **183,** 538–543.

Vanage, G., Jaiswal, Y. K., Lu, Y. A., Tam, J. P., Wang, L. F., and Koide, S. S. (1994). *Res. Commun. Chem. Pathol. Pharmacol.* **84,** 3–15.

Vanslyke, J. K., and Hruby, D. E. (1994). *Virology* **198,** 624–635.

Vogel, T., Kurth, R., and Norley, S. (1994). *J. Immunol.* **153,** 1895–1904.

Wallace, A., Altamura, S., Toniatti, C., Vitelli, A., Bianchi, E., Delmastro, P., Ciliberto, G., and Pessi, A. (1994). *Pept. Res.* **7,** 27–31.

Wang, C. Y., Looney, D. J., Li, M. L., Walfield, A. M., Ye, J., Hosein, B., Tam, J. P., and Wong-Staal, F. (1991). *Science* **254,** 285–288.

Wang, R., Charoenvit, Y., Corradin, G., Porrozzi, R., Hunter, R., Glenn, G., Alving, C., Church, P., and Hoffman, S. (1995). *J. Immunol.* **154,** 2784–2793.

Weeks, B., Nomizu, M., Otaka, A., Weston, C., Okusu, A., Tamamura, H., Matsumoto, A., Yamamoto, N., and Fujii, N. (1994). *Biochem. Biophys. Res. Commun.* **202,** 470–475.

Widmann, C., Romero, P., Maryanski, J. L., Corradin, G., and Valmori, D. (1992). *J. Immunol. Methods* **155,** 95–99.

Wiesmuller, K.-H., Bessler, W. G., and Jung, G. (1983). *Hoppe-Seyler Z. Physiol. Chem.* **364,** 593–606.

Wilbur, D. S. (1992). *Bioconjugate Chem.* **3,** 433–462.

Wilson, B. S., Nuoffer, C., Meinkoth, J. L., McCaffery, M., Feramisco, J. R., Balch, W. E., and Farquhar, M. G. (1994). *J. Cell Biol.* **125,** 557–571.

Wo, Z. G., and Oewald, R. (1995). *J. Biol. Chem.* **270,** 2000–2009.

Wolowczuk, I., Auriault, C., Bossus, M., Boulanger, D., Gras-Masse, H., Mazingue, C., Pierce, R. J., Grezel, D., Reid, G. D., Tartar, A., and Capron, A. (1991). *J. Immunol.* **146,** 1987–1995.

Wong, F., Schaefer, E. L., Roop, B. C., LaMendola, J. N., Johonson-Seaton, D., and Shao, D. (1989). *Neutron* **3,** 81–94.

Wunsch, E., Moroder, L., Nyfeler, R., Kalbacher, H., and Gemeiner, M. (1985). *Biol. Chem. Hoppe-Seyler* **366,** 35–61.

Yahi, N., Fantini, J., Mabrouk, K., Tamalet, C., De Micco, P., Van Rietschoten, J., Rochat, H., and Sabatier, J. M. (1994a). *J. Virol.* **68,** 5714–5720.

Yahi, N., Sabatier, J., Nickel, P., Mabrouk, K., Gonzalez-Scarano, F., and Fantini, J. (1994b). *J. Biol. Chem.* **269,** 24349–24353.

Yahi, N., Sabatier, J., Baghdiguian, S., Gonzalez-Scarano, F., and Fantini, J. (1995). *J. Virol.* **69,** 320–325.

Yamashiro, D., and Li, C. H. (1988). *Int. J. Pept. Protein Res.* **31,** 322–334.

Yang, Y., Johnson, T., Mecham, J., Tam, J., and Li, J. (1992). *Virology* **188,** 530–536.

Yao, Z. J., Kao, M. C., Loh, K. C., and Chung, M. C. (1994). *J. Chromatogr. A* **679,** 190–194.

Yoshitake, S., Yamada, Y., Ishikawa, E., and Masseyeff, R. (1979). *Eur. J. Biochem.* **101,** 396–399.

Zavala, F., and Chai, S. (1990). *Immunol. Lett.* **25,** 271–274.

Zelezna, B., Richards, E. M., Tang, W., Lu, D., Sumner, C., and Raizada, M. K. (1992). *Biochem. Biophys. Res. Commun.* **183,** 781–788.

Zelezna, B., Veselsky, L., Velek, J., Zicha, J., and Kunes, J. (1994). *Eur. J. Pharmacol.* **260,** 95–98.

Zhong, G., Toth, I., Reid, R., and Burnham, R. C. (1993). *J. Immunol.* **151,** 3728–3736.

Zolnierowickz, S., Csortos, C., Bondor, J., Verin, A., Mumby, M., and DePaoli-Roach, A. (1994). *Biochemistry* **33,** 11858–11867.

Index

B

C

D

I

R

S